Virtual Power Plants and Electricity Markets

Luis Baringo • Morteza Rahimiyan

Virtual Power Plants and Electricity Markets

Decision Making Under Uncertainty

Springer

Luis Baringo
Escuela Técnica Superior de Ingeniería Industrial
Universidad de Castilla-La Mancha
Ciudad Real, Spain

Morteza Rahimiyan
Department of Electrical Engineering
Shahrood University of Technology
Shahrood, Iran

ISBN 978-3-030-47601-4 ISBN 978-3-030-47602-1 (eBook)
https://doi.org/10.1007/978-3-030-47602-1

This Springer imprint is published by the registered company Springer Nature Switzerland AG.
The registered company address is: Gewerbestrasse 11, 6330 Cham, Switzerland

To Carmen and María
To my parents, my wife, Ali, and Hossein

Preface

This book "**Virtual Power Plants and Electricity Markets: Decision Making Under Uncertainty**" provides a number of relevant models to make optimal operation and planning strategies for virtual power plants (VPPs) participating in electricity markets.

There has been a significant rise of interest in renewable units based on wind, solar, and photovoltaic energy in the last decades. These renewable units started to be developed with the aim of reducing greenhouse gas emissions. However, despite their benefits, the uncertainty associated with most of the renewable generating units, whose production varies depending on the meteorological phenomena, limits the participation of the owners of these units in different electricity markets. Under this condition, the owners are exposed to a significant level of market risks that may reduce their profitability and discourage them from active participation in electricity markets.

A possible solution to mitigate the negative effects of the uncertainty in the production of renewable generating units is to combine them with other generating units, storage facilities, and flexible demands that can shift part of their energy consumption. This gives rise to the concept of VPP, which can be defined as a set of generating units, storage facilities, and demands that operate as a single entity to optimize the use of the energy resources. For example, conventional generating units may be used when the renewable production is low, while the surplus renewable energy, e.g., during the night, may be stored in the storage units to meet the demands during the following day. This book describes and illustrates different problems faced by VPPs participating in electricity markets. These problems are solved through advanced optimization techniques.

This book consists of seven chapters and two appendices.

Chapter 1 provides an introduction to both VPPs and electricity markets. Then, the chapter describes the main characteristics of the decision-making problems under uncertainty used in this book. Next, it includes a brief literature review of operation and planning problems faced by VPPs participating in electricity markets. The chapter concludes by clarifying the scope of the book.

Chapter 2 describes the main components of VPPs, including demands, conventional power plants, stochastic renewable production facilities, and storage units. The chapter provides models to simulate the working of these elements. Additionally, this chapter provides some insights about the role of VPPs in the development of smart grids.

Chapter 3 is devoted to the scheduling problem of a risk-neutral VPP participating in energy electricity markets, including the day-ahead, real-time, and futures markets. A simple deterministic model is firstly introduced to clarify this decision-making problem. This model is then extended using a stochastic programming problem that accounts for uncertainties in market prices and power production levels of stochastic renewable generating units.

Chapter 4 extends the models developed in Chap. 3 to account for a risk-averse VPP, i.e., a VPP that takes into account the profit risk associated with its scheduling decisions. With this aim, different models based on stochastic programming, static robust optimization, hybrid stochastic-robust optimization, and adaptive robust optimization are presented.

Chapter 5 considers the participation of VPPs not only in energy markets but also in reserve electricity markets. It is illustrated using a number of increasingly complex models how a VPP can use the flexibility of the energy resources to provide both up and down reserves requested by the system operator. Uncertainties in market prices, stochastic renewable production levels, and requests of reserve deployment are addressed in the decision-making problems using a stochastic programming problem and an adaptive robust optimization approach.

Chapter 6 considers price-maker VPPs, i.e., VPPs that have the capability to alter market prices. This requires the use of bi-level models comprising an upper-level problem, which represents the operation and offering decisions of the VPP, and a lower-level problem, which represents the market-clearing problem. This way, the formation of market prices is endogenously represented in the decision-making problem.

Chapter 7 describes the expansion planning problems faced by VPPs. The main aim of a VPP is maximizing its profit by participating in different electricity markets, as illustrated in Chaps. 3–6; however, this objective may be also attached by building new generating assets. Uncertainties in the expansion planning problem are modeled using a set of scenarios.

Finally, Appendix A provides an overview of the optimization problems used in this book, while Appendix B describes the main characteristics of the decision-making problems under uncertainty presented in the book.

The book is written in a tutorial style and modular format, including many illustrative examples to facilitate comprehension. This feature makes the book of interest to a diverse audience, including students and practitioners in the electric energy sector with a limited background on VPPs and decision making under uncertainty.

The benefits of reading this book include the comprehension of all the details of the operation and planning problems faced by VPPs, an insightful understanding of the formulation of decision-making problems under uncertainty, and the familiarization with solution algorithms to solve these problems.

To conclude, the authors would like to thank their colleagues and students at the Universidad de Castilla–La Mancha and the Shahrood University of Technology for their insightful observations, pertinent corrections, and helpful comments.

Ciudad Real, Spain
Shahrood, Iran
March 2020

Luis Baringo
Morteza Rahimiyan

Contents

Chapter 1
Virtual Power Plants

This chapter provides an overview of decision-making problems related to virtual power plants (VPPs). Section 1.1 defines the concept of VPP. Section 1.2 describes the basic organization of current electricity markets. Section 1.3 explains the strong relationship between VPPs and smart grids. Section 1.4 provides an overview of decision-making problems under uncertainty, including stochastic and robust optimization problems. Section 1.5 reviews the different approaches used in the technical literature to deal with operation and planning problems for VPPs. Finally, Sect. 1.6 clarifies the scope of this book.

1.1 Background

The energy consumption all over the world has rapidly increased in the last decades. This has motivated the building of new generating units to increase the available generation capacity. Due to some international agreements to reduce greenhouse gas emissions, such as the Kyoto Protocol in 1997 [38] or the Paris Agreement in 2015 [13], this new capacity is mainly based on renewable energy, such as wind, solar, and photovoltaic energy. Therefore, the penetration of renewable generating units in power systems has considerably increased over the last years.

Renewable generating units have many advantages in comparison with conventional power plants based on fossil fuels and nuclear energy, e.g., they are emission free, they require neither fuel nor water consumption, and they can be installed at many locations worldwide. However, the production of stochastic renewable generating units such as solar- or wind-power units is variable and uncertain since it depends on meteorological phenomena. Thus, the participation of these units in electricity markets is limited since they are penalized in case of deviating from the scheduled production. A possible solution to avoid these penalties is the combination of these stochastic renewable generating units with other technologies such as conventional power plants, energy storage facilities, and demands. In such a case, on one hand, the surplus power production in hours of high renewable production can be stored or used to supply the demands; on the other hand, a deficit in the power production in case of low renewable production levels can be compensated with the conventional power plants or the energy stored in the storage facilities.

Within this context, the concept of VPP appears in 1997 [3] as the aggregation of different energy resources that operate as a single entity with the aim of optimizing their use [22].

According to [43], the components of a VPP can be classified into two groups:

1. Renewable energy resources. This group includes wind-power, photovoltaic solar, and hydro-power units. Note that both large-scale and low-capacity generating units can be integrated within a VPP.
2. Nonrenewable energy resources. This group comprises conventional power plants, energy storage systems, electric vehicles, and demands.

Without loss of generality, this book considers VPPs that integrate stochastic renewable generating units, conventional power plants, energy storage facilities, and demands. A detailed model of the different components in the VPP is provided in Chap. 2.

As previously explained, the main aim of VPPs is the optimization of the energy resources. This can be achieved by maximizing the profit achieved by VPPs due to its participation in different electricity markets. This goal can be reached by optimizing both the operation and expansion planning strategies of VPPs. The operation strategies of VPPs are analyzed in Chaps. 3–6, while the expansion planning strategies of VPPs are reviewed in Chap. 7.

L. Baringo, M. Rahimiyan, *Virtual Power Plants and Electricity Markets*, https://doi.org/10.1007/978-3-030-47602-1_1

1.2 Overview of Electricity Markets

The production, transmission, distribution, and supply of electric energy generally turn around an electricity market. In most parts of the world, electricity markets include different trading floors, being the day-ahead (DA), the real-time (RT), and futures markets the most important.

On one hand, the DA market allows trading electricity commodities once a day, one day in advance, and on an hourly basis. On the other hand, the RT market, also known as the balancing market in some places, closes some minutes before the actual power delivery. In addition, the futures market allows transactions from one week to several years in advance. The optimal strategies of VPPs in the DA and RT energy markets are analyzed in Chaps. 3–6, respectively, while some insights about the participation of VPPs in futures markets are provided in Chap. 3.

In addition to energy trading, VPPs can also provide services to support transmission system management [29], e.g., the capacity, storage, and flexibility of VPPs can be used to enable the system balance. In particular, Chaps. 5 and 6 analyze the participation of VPPs in the DA reserve electricity market, also known as regulation market.

Participants in electricity markets can be classified into two groups, namely producers and consumers. Producers submit offers, consisting in a set of energy blocks [MWh] at increasing prices [$/MWh], while consumers submit bids, comprising a set of consumption blocks [MWh] at decreasing prices [$/MWh]. Then, the market operator, or the independent system operator depending on the market jurisdiction, uses the data of offers and bids to match supply and demand in the system. This is done using appropriate market-clearing tools. The outputs of these tools are the accepted production and consumption levels for producers and consumers, as well as the market-clearing prices paid to producers for their production and paid by consumers for their consumption.

At the time of participating in electricity markets, participants can behave as price takers or price makers. A price-taker agent does not influence market-clearing prices, while price-maker participants have the capability of altering market prices by strategically deciding their participation strategy in the market. Chapters 2–5 and 7 consider price-taker VPPs, while Chap. 6 describes the operation strategies of a price-maker VPP.

1.3 Virtual Power Plants and Smart Grids

VPPs are thought to provide power services to costumers through virtually sharing information:

1. There is an energy management system (EMS) in charge of the VPP that collects technical and economic information from the components in the VPP.
2. The EMS uses this information to decide the operation and expansion planning strategies in electricity markets of the VPP as a whole.
3. The EMS communicates the market operator the decisions of the VPP.
4. The market operator uses the information from the VPP and other market participants to clear the markets. Then, the market operator communicates the EMS the market outputs.
5. The EMS uses the market outputs and the information from the different components in the VPP to decide the actual dispatch of the different units.

Figure 1.1 clarifies this information flow.

The decision-making process of VPPs is dynamic due to the continuous sharing and processing of real-time information from its components. Therefore, VPPs are strongly related to smart grids [2] since they depend on the availability of smart meters and information technologies. For this reason, an efficient communication bandwidth and high-speed processing algorithms are needed to optimize the full potential of VPPs. Chapters 2 and 3 give some insights about the role of smart grids with VPPs.

VPPs are flexible at the time of geographically locating the energy resources, which enables VPPs to integrate components beyond the constraints of the power grid. Thus, the aggregation of assets provided by VPPs paves the way towards smarter power systems by reaping the benefits of economies of scale, thereby boosting efficiencies and minimizing costs.

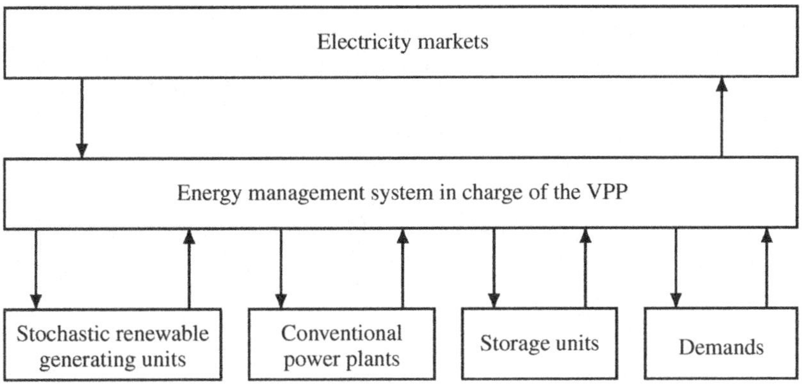

Fig. 1.1 Information flow in the VPP

1.4 Decision Making Under Uncertainty

This book describes different decision-making problems under uncertainty. Two different approaches are generally considered in the technical literature to deal with this type of problems, namely stochastic programming [7] and robust optimization [5].

1.4.1 Stochastic Programming

Scenario-based stochastic programming is used to solve decision-making problems under uncertainty in which unknown parameters are represented using a finite set of scenarios [7]. Each scenario realization comprises a plausible realization of the uncertain parameters and has an associated probability of occurrence.

Stochastic programming problems aim at minimizing (maximizing) an objective function representing the expected cost (profit) over all scenario realizations. This cost (profit) minimization (maximization) is carried out considering the constraints of the problem for all scenario realizations.

Such stochastic programming problems are interesting for a risk-neutral agent, who is indifferent with the risk affecting the objective function. A risk-averse agent can solve risk-constrained stochastic programming problems which aim at optimizing a multi-objective function to make a trade-off between the expected cost (profit) and a risk metric.

One of the main disadvantages of stochastic programming is that its performance depends on the knowledge of the probability distribution function of the uncertain data. Moreover, the feasibility of the problem is only guaranteed for the input scenarios considered. Therefore, a sufficient number of scenarios must be considered. However, a large number of scenarios may yield a computationally intractable problem.

Scenarios are usually generated using historical data and forecasting tools such as time series models [12], ARIMA models [10], and Monte Carlo simulation [16]. Then, the number of scenarios may be reduced using appropriate scenario reduction techniques [21].

Scenarios are used to build the so-called scenario trees that comprise nodes, which represent the decision-making stages, and branches, which represent the different scenario realizations.

Depending on the number of decision-making stages, stochastic programming problems are classified into two-stage and multi-stage problems:

1. Two-stage stochastic programming problems include two stages. First-stage decisions, generally known as here-and-now decisions, are made before knowing the actual uncertainty realization. Then, second-stage decisions, also known as wait-and-see decisions, are made after knowing the uncertainty realization. Figure 1.2 shows an example of scenario tree for a two-stage stochastic programming problem.
2. Multi-stage stochastic programming problems are used when decisions must be done in more than two stages. In this type of problems, it is important to establish the non-anticipativity of the decisions [7], i.e., given a decision stage, if the realizations of the uncertain parameters are identical up to this stage, then decisions made at this stage must be also identical. In other words, decisions at each stage depend on the previous uncertainty realizations but are unique for the future uncertainty realizations. Figure 1.3 shows an example of scenario tree for a multi-stage stochastic programming problem.

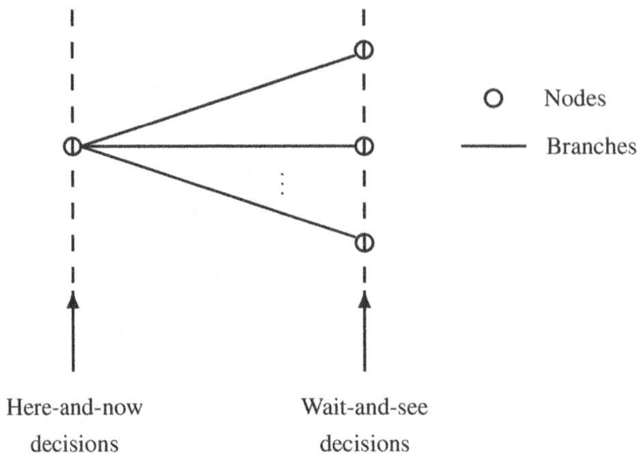

Fig. 1.2 Scenario tree for a two-stage stochastic programming problem

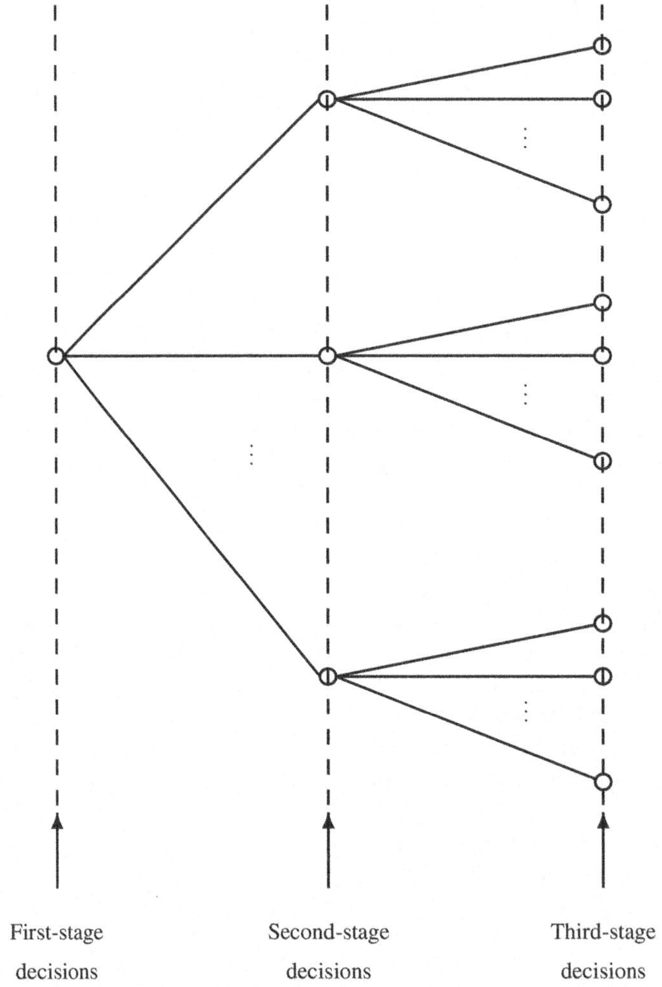

Fig. 1.3 Scenario tree for a multi-stage stochastic programming problem

Scenario-based stochastic programming problems can be transformed into equivalent deterministic problems that can be solved using standard mathematical programming solution methods.

Stochastic programming is used in this book to model both operation (in Chaps. 3–6) and expansion planning problems (in Chap. 7) for VPPs.

1.4.2 Robust Optimization

Robust optimization is another tool to solve decision-making problems under uncertainty. In this case, uncertainty is modeled using decision variables within a prespecified uncertainty set. The solution of a robust optimization problem considers all the plausible realizations of the uncertain parameters within the uncertainty set. Moreover, this solution is optimal for the worst-case realization of the uncertain parameters.

Accordingly, robust optimization allows taking a conservative strategy. However, a decision maker can also benefit from the flexibility of robust optimization by adapting the so-called uncertainty budgets, which allow controlling the conservatism of the solution.

Robust problems can be classified into two groups depending on the decisions made in the decision-making process:

1. Robust optimization without recourse, also known as static robust optimization, assumes that all decisions are made simultaneously and before knowing the actual uncertainty realization [5]. Thus, solutions are generally too conservative.
2. Robust optimization with recourse, also known as two-stage robust optimization, allows adjusting some decisions in response to the uncertainty realization [6]. Therefore, solutions are not so conservative as in the case of static robust optimization.

Two-stage robust optimization problems can be formulated as tri-level problems [39]. The upper-level problem determines the first-stage decisions, taking into account every realization of the uncertain parameters within the uncertainty set. Then, for a given upper level, the middle-level problem identifies the worst-case uncertainty realization within the uncertainty set. Finally, the lower-level problem determines the best reaction against the worst-case uncertainty realization identified in the middle-level problem.

The resulting tri-level problem can be solved using two decomposition-based solution procedures, namely Benders' decomposition [4] and the column-and-constraint generation algorithm [42].

Robust optimization is used in this book to model operation problems for VPPs in Chaps. 4 and 5.

1.5 Operation and Expansion Strategies for Virtual Power Plants

VPPs are expected to play a relevant role in future electric energy systems, which has motivated the development of different tools to manage them. This section reviews different approaches used in the technical literature for the operation and expansion planning strategies for VPPs.

1.5.1 Operation Strategies

There is an extensive technical literature on operation problems for VPPs [23, 24, 40, 43]. Most of the works consider a central entity that acts as an EMS that decides the optimal operation of each component within the VPP. These operating decisions are made with the aim of optimizing the energy resources in the VPP, e.g., with the aim of maximizing the profit of the VPP by participating in different electricity markets.

Operation models can be classified depending on the markets in which the VPP participates:

1. The participation of VPPs in the DA energy market is analyzed in [25, 28, 32, 37], as well as in Chaps. 3 and 4 of this book.
2. The participation of VPPs in the DA and RT energy markets is addressed in [11, 17, 26, 30, 31, 33, 35, 44], as well as in Chaps. 3 and 4 of this book.
3. The optimal trading of VPPs in both DA energy and reserve markets is examined in [14, 18, 19, 27, 34, 41], as well as in Chaps. 5 and 6 of this book.
4. The optimal trading of a VPP in the DA and RT energy and reserve markets is formulated in [11, 36].

1.5.2 *Expansion Planning Strategies*

Expansion planning can be tackled in two different ways, namely considering a market framework and adopting a centralized approach [9]. In a market framework, a profit-oriented agent makes the investment decisions with the aim of maximizing profit. On the other hand, a central planner makes the expansion decisions that are most efficient for the system when a centralized approach is considered. The VPP planning belongs to the first approach.

Moreover, two types of decisions can be distinguished in the expansion planning of VPPs. On one hand, a VPP can be expanded by attracting new participants to join it. In this vein, [20] uses a genetic algorithm to encourage the participation of distributed generation in the VPP, while [8] designs a mechanism for sustainable VPP formation using a combinatorial auction problem and considering strategic producers that use renewable energy sources. On the other hand, the VPP can be expanded by building new facilities, as done in [1], which proposes a stochastic approach to solve this problem.

Chapter 7 is devoted to the expansion planning strategies for VPPs.

1.6 Scope of the Book

Making operation and expansion planning decisions for VPPs requires the use of suitable computational models, considering all the technical and economical details that are relevant. This book provides detailed computational models to identify the optimal operation and expansion plans for VPPs with respect to the objectives and limitations of the models. Specifically, this book provides:

- A detailed model for the most common components in VPPs (Chap. 2).
- Different decision-making models to derive the optimal scheduling for VPPs participating in energy markets, considering both a risk-neutral (Chap. 3) and a risk-adverse (Chap. 4) behavior.
- Different decision-making models to determine the optimal participation of price-taker VPPs in both energy and reserve markets (Chap. 5).
- Different decision-making models to determine the optimal participation of price-maker VPPs in both energy and reserve markets (Chap. 6).
- Different decision-making models to select the optimal expansion plans for VPPs (Chap. 7).

The above models are formulated in a rigorous way. Moreover, this book includes algorithms to solve these models, GAMS [15] codes to solve them, as well as many numerical examples.

References

1. Akbari, T., Taheri, H., Abroshan, M., Kamali, M.: Scenario-based static DG expansion planning for a virtual power plant. Int. Rev. Model. Simul. **5**(1), 263–269 (2012)
2. Asmus, P.: Microgrids, virtual power plants and our distributed energy future. Electr. J. **23**(10), 72–82 (2010)
3. Awerbuch, S., Preston, A.: The Virtual Utility: Accounting, Technology and Competitive Aspects of the Emerging Industry. Springer, New York, NY (1997)
4. Benders, J.F.: Partitioning procedures for solving mixed-variables programming problems. Numer. Math. **4**(1), 238–252 (1962)
5. Bertsimas, D., Brown, D.B., Caramanis, C.: Theory and applications of robust optimization. SIAM Rev. **53**(3), 464–501 (2011)
6. Ben-Tal, A., Nemirovski, A.: Robust convex optimization. Math. Oper. Res. **23**(4), 769–805 (1998)
7. Birge, J.R., Louveaux, F.: Introduction to Stochastic Programming, 2nd edn. Springer, New York, NY (2011)
8. Biswas, S., Bagchi, D., Narahari, Y.: Mechanism design for sustainable virtual power plant formation. In: Presented at 2014 IEEE International Conference on Automation Science and Engineering (CASE), Taipei, pp. 67–72 (2014)
9. Conejo, A.J., Baringo, L., Kazempour, S.J., Siddiqui, A.S.: Investment in Electricity Generation and Transmission: Decision Making under Uncertainty. Springer, Switzerland (2016)
10. Conejo, A.J., Contreras, J., Espínola, R., Plazas, M.A.: Forecasting electricity prices for a day-ahead pool-based electric energy market. Int. J. Forecast. **21**(3), 435–462 (2005)
11. Dabbagh, S.R., Sheikh-El-Eslami, M.K.: Risk-based profit allocation to DERs integrated with a virtual power plant using cooperative game theory. Electr. Pow. Sys. Res. **121**, 368–378 (2015)
12. De Gooijer, J.G., Hyndman, R.J.: 25 years of time series forecasting. Int. J. Forecast. **22**(3), 443–473 (2006)
13. European Commission.: Available at https://ec.europa.eu/clima/policies/international/negotiations/paris_en (2020)
14. Fan, S., Ai, Q., Piao, L.: Fuzzy day-ahead scheduling of virtual power plant with optimal confidence level. IET Gener. Transm. Dis. **10**(1), 205–212 (2016)

15. GAMS.: Available at www.gams.com/ (2020)
16. Hajipour, E., Bozorg, M., Fotuhi-Firuzabad, M.: Stochastic capacity expansion planning of remote microgrids with wind farms and energy storage. IEEE Trans. Sustain. Energy. **6**(2), 491–498 (2015)
17. Kardakos, E.G., Simoglou, C.K., Bakirtzis, A.G.: Optimal offering strategy of a virtual power plant: a stochastic bi-level approach. IEEE Trans. Power Syst. **7**(2), 794–806 (2016).
18. Mahdavi, S.S., Javidi, M.H.: VPP decision making in power markets using Benders decomposition. Int. Trans. Electr. Energy Syst. **24**(7), 960–975 (2014)
19. Mashhour, E., Moghaddas-Tafreshi, S.M.: Bidding strategy of virtual power plant for participating in energy and spinning reserve markets – Part I: Problem formulation. IEEE Trans. Power Syst. **26**(2), 949–956 (2011)
20. Monyei, C.G., Fakolujo, O.A., Bolanle, M.K.: Virtual power plants: Stochastic techniques for effective costing and participation in a decentralized electricity network in Nigeria. In: Presented at 2013 IEEE International Conference on Emerging & Sustainable Technologies for Power & ICT in a Developing Society (NIGERCON), Owerri, pp. 374–377 (2013)
21. Morales, J.M., Pineda, S., Conejo, A.J., Carrión, M.: Scenario reduction for futures market trading in electricity markets. IEEE Trans. Power Syst. **24**(2), 878–888 (2009)
22. Morales, J.M., Conejo, A.J., Madsen, H., Pinson, P., Zugno, M.: Integrating Renewables in Electricity Markets – Operational Problems. Springer, New York, US (2014)
23. Nosratabadi, S.M., Hooshmand, R.-A., Gholipour, E.: A comprehensive review on microgrid and virtual power plant concepts employed for distributed energy resources scheduling in power systems. Renew. Sust. Energ. Rev. **67**, 341–363 (2016)
24. Othman, M.M., Hegazy, Y.G., Abdelaziz, A.Y.: A review of virtual power plant definitions, components, framework, and optimization. Int. Electr. Eng. J. **6**(9), 2010–2024 (2015)
25. Pandzic, H., Kuzle, I., Capuder, T.: Virtual power plant mid-term dispatch optimization. Appl. Energy. **101**, 134–141 (2013)
26. Pandzic, H., Morales, J.M., Conejo, A.J., Kuzle, I.: Offering model for a virtual power plant based on stochastic programming. Appl. Energy. **105**, 282–292 (2013)
27. Peik-Herfeh, M., Seifi, H., Sheikh-El-Eslami, M.K.: Decision making of a virtual power plant under uncertainties for bidding in a day-ahead market using point estimate method. Int. J. Electr. Power Energy Syst. **44**(1), 88–98 (2011)
28. Peik-Herfeh, M., Seifi, H., Sheikh-El-Eslami, M.K.: Two-stage approach for optimal dispatch of distributed energy resources in distribution networks considering virtual power plant concept. Int. J. Electr. Energy. **24**(1), 43–63 (2014)
29. Pudjianto, D., Ramsay, C., Strbac, G.: Virtual power plant and system integration of distributed energy resources. IET Renew. Power Gener. **1**(1), 10–16 (2007)
30. Rahimiyan, M., Baringo, L: Strategic bidding for a virtual power plant in the day-ahead and real-time markets: A price-taker robust optimization approach. IEEE Trans. Power Syst. **31**(4), 2676–2687 (2016)
31. Riveros, J.Z., Bruninx, K., Poncelet, K., D'haeseleer, W.: Bidding strategies for virtual power plants considering CHPs and intermittent renewables. Energy Convers. Manag. **103**, 408–418 (2015)
32. Shabanzadeh, M., Sheikh-El-Eslami, M.-K., Haghifam, M.-R.: The design of a risk-hedging tool for virtual power plants via robust optimization approach. Appl. Energy. **155**, 766–777 (2015)
33. Shabanzadeh, M., Sheikh-El-Eslami, M.-K., Haghifam, M.-R.: Risk-based medium-term trading strategy for a virtual power plant with first-order stochastic dominance constraints. IET Gener. Transm. Dis. **11**(2), 520–529 (2017)
34. Sowa, T., Krengel, S., Koopmann, S., Nowak, J.: Multi-criteria operation strategies of power-to-heat-systems in virtual power plants with a high penetration of renewable energies. Energy Procedia **46**, 237–245 (2014)
35. Tajeddinia, M.A., Rahimi-Kiana, A., Soroudi, A.: Risk averse optimal operation of a virtual power plant using two stage stochastic programming. Energy **73**, 958–967 (2014)
36. Tang, W., Yang, H. T.: Optimal operation and bidding strategy of a virtual power plant integrated with energy storage systems and elasticity demand response. IEEE Access **7**, 79798–79809 (2019)
37. Tascikaraoglu, A., Erdinc, O., Uzunoglu, M., Karakas, A.: An adaptive load dispatching and forecasting strategy for a virtual power plant including renewable energy conversion units. Appl. Energy **119**, 445–453 (2014)
38. United Nations, Climate Change.: Available at https://unfccc.int/kyoto_protocol (2020)
39. Vicente, L.N., P. H. Calamai, P.H.: Bilevel and multilevel programming: A bibliography review. J. Global Optim. **5**(3), 291–306 (1994)
40. Yavuz, L., Önen, A., Muyeen, S.M., Kamwa, I.: Transformation of microgrid to virtual power plant - A comprehensive review. IET Gener. Transm. Dis. **13**(11), 1994–2005 (2019)
41. Zamani, A.G., Zakariazadeh, A., Jadid, S.: Day-ahead resource scheduling of a renewable energy based virtual power plant. Appl. Energy **169**, 324–340 (2016)
42. Zeng, B., Zhao, L.: Solving two-stage robust optimization problems using a column-and-constraint generation method. Oper. Res. Lett. **41**(5), 457–461 (2013)
43. Zhang, G., Jiang, C., Wang, X.: Comprehensive review on structure and operation of virtual power plant in electrical system. IET Gener. Transm. Dis. **13**(2), 145–156 (2019)
44. Zhao, Q., Shen, Y., Li, M.: Control and bidding strategy for virtual power plants with renewable generation and inelastic demand in electricity markets. IEEE Trans. Sustain. Energy **7**(2), 562–575 (2016)

Chapter 2
Virtual Power Plant Model

This chapter provides models for the most common components of virtual power plants (VPPs). Components described below include demands, conventional power plants, stochastic renewable generating units, and energy storage facilities. These components can be interconnected in an electric network, which is also modeled in this chapter. The joint modeling of all these components constitutes the basic model for a VPP, which will be used in the following chapters of this book.

2.1 Introduction

A virtual power plant (VPP) can be defined as a set of electricity assets, including both generating and demand units, that are operated as a single entity in order to optimize the use of the energy resources [7]. In this chapter, the main characteristics and working of the most common components of VPPs are described.

Section 2.2 provides the main notation used in this chapter. Section 2.3 describes demands in VPPs. These demands are generally flexible so that they can shift part of their energy consumption due to both technical and economical reasons. Section 2.4 provides a model for conventional power plants such as combined cycle gas turbine (CCGT) generating units. Section 2.5 describes stochastic renewable generating units such as solar- and wind-power units. The main difference with conventional power plants described in Sect. 2.4 is that the available production level of stochastic renewable generating units depends on the availability of a weather resource such as the wind speed or the solar irradiation, which are generally subject to uncertainty. Section 2.6 describes energy storage units, i.e., facilities such as batteries that can store energy to be used in a latter period. All these components can be interconnected in an electric network as described in Sect. 2.7. Section 2.8 provides an overview on VPPs equipped with the smart grid technology. Sections 2.3–2.8 provide a number of clarifying examples. Section 2.9 summarizes the chapter and suggests some references for further reading. Finally, Sect. 2.10 provides the GAMS codes used for solving some of the examples of Sects. 2.3–2.8.

2.2 Notation

The main notation used in this chapter is provided below for quick reference. Other symbols are defined as needed throughout the chapter.

2.2.1 Indexes

The indexes of the VPP model are:

c Conventional power plants.
d Demands.

The original version of this chapter was revised: Wrong placement of section headings 2.10.2 and 2.10.4 and an incomplete GAMS code in section 2.10.4 have been updated. A correction to this chapter is available at https://doi.org/10.1007/978-3-030-47602-1_8

L. Baringo, M. Rahimiyan, *Virtual Power Plants and Electricity Markets*, https://doi.org/10.1007/978-3-030-47602-1_2

ℓ Network lines.
n Network nodes.
r Stochastic renewable generating units.
$r(\ell)$ Receiving-end node of network line ℓ.
s Storage units.
$s(\ell)$ Sending-end node of network line ℓ.
t Time periods.

2.2.2 Sets

The sets of the VPP model are:

Ω^{C} Conventional power plants.
Ω^{D} Demands.
Ω^{D}_n Demands located at node n.
Ω^{G} Generating units.
Ω^{G}_n Generating units located at node n.
Ω^{L} Network lines.
Ω^{N} Network nodes.
Ω^{R} Stochastic renewable generating units.
Ω^{S} Storage units.
Ω^{S}_n Storage units located at node n.
Ω^{T} Time periods.
Ψ^{D}_{dt} Feasibility set of demand d in time period t.
Ψ^{G}_{gt} Feasibility set of generating unit g in time period t.

2.2.3 Parameters

The parameters of the VPP model are:

$\underline{E}^{\mathrm{D}}_d$ Minimum energy consumption of demand d throughout the planning horizon [MWh].
$\underline{E}^{\mathrm{S}}_{st}$ Lower bound of the energy stored in storage unit s in time period t [MWh].
$\overline{E}^{\mathrm{S}}_{st}$ Upper bound of the energy stored in storage unit s in time period t [MWh].
$\underline{P}^{\mathrm{C}}_c$ Minimum power production of conventional power plant c [MW].
$\overline{P}^{\mathrm{C}}_c$ Maximum power production of conventional power plant c [MW].
$\underline{P}^{\mathrm{D}}_{dt}$ Lower bound of the power consumption of demand d in time period t [MW].
$\overline{P}^{\mathrm{D}}_{dt}$ Upper bound of the power consumption of demand d in time period t [MW].
$\overline{P}^{\mathrm{L}}_\ell$ Capacity of network line ℓ [MW].
$\overline{P}^{\mathrm{R}}_{rt}$ Available production level of stochastic renewable generating unit r in time period t [MW].
$\overline{P}^{\mathrm{S,C}}_s$ Charging capacity of storage unit s [MW].
$\overline{P}^{\mathrm{S,D}}_s$ Discharging capacity of storage unit s [MW].
$R^{\mathrm{C,D}}_c$ Down ramping limit of conventional power plant c [MW/h].
$R^{\mathrm{C,SD}}_c$ Shut-down ramping limit of conventional power plant c [MW/h],
$R^{\mathrm{C,SU}}_c$ Start-up ramping limit of conventional power plant c [MW/h],
$R^{\mathrm{C,U}}_c$ Up ramping limit of conventional power plant c [MW/h].
$R^{\mathrm{D,D}}_d$ Down ramping limit of demand d [MW/h].
$R^{\mathrm{D,U}}_d$ Up ramping limit of demand d [MW/h].
X_ℓ Reactance of network line ℓ [Ω].
Δt Duration of time periods [h].

$\eta_s^{\text{S,C}}$ Charging efficiency of storage unit s [%].
$\eta_s^{\text{S,D}}$ Discharging efficiency of storage unit s [%].

2.2.4 Variables

The variables of the VPP model are:

e_{st}^{S} Energy stored in storage unit s in time period t [MWh].
p_{ct}^{C} Power generation of conventional power plant c in time period t [MW].
p_{dt}^{D} Power consumption of demand d in time period t [MW].
p_{gt}^{G} Power generation of generating unit g in time period t [MW].
$p_{\ell t}^{\text{L}}$ Power flow through line ℓ in time period t [MW].
p_{rt}^{R} Production of stochastic renewable generating unit r in time period t [MW].
$p_{st}^{\text{S,C}}$ Charging power level of storage unit s in time period t [MW].
$p_{st}^{\text{S,D}}$ Discharging power level of storage unit s in time period t [MW].
u_{ct}^{C} Binary variable that is equal to 1 if conventional power plant c is generating electricity in time period t, being 0 otherwise.
u_{st}^{S} Binary variable used to prevent the simultaneous charging and discharging of storage unit s in time period t.
$v_{ct}^{\text{C,SD}}$ Binary variable that is equal to 1 if conventional power plant c is shut down in time period t, being 0 otherwise.
$v_{ct}^{\text{C,SU}}$ Binary variable that is equal to 1 if conventional power plant c is started up in time period t, being 0 otherwise.
δ_{nt} Voltage angle at bus n in time period t [rad].

2.3 Demands

Consumption assets constitute a basic element in VPPs. In general, demands in VPPs are flexible, i.e., they can shift part of their energy consumption according to the needs in the VPP. These needs can be motivated by both technical and economical issues. For example, it may be needed to shift part of the energy consumption of a demand because the renewable production in a given time period is low. However, it may be advisable to shift part of the energy consumption because the price in the market is very high.

A basic model for flexible demands comprises the following equations:

$$\underline{P}_{dt}^{\text{D}} \le p_{dt}^{\text{D}} \le \overline{P}_{dt}^{\text{D}}, \quad \forall d \in \Omega^{\text{D}}, \forall t \in \Omega^{\text{T}}, \tag{2.1a}$$

$$p_{dt}^{\text{D}} - p_{d(t-1)}^{\text{D}} \le R_d^{\text{D,U}} \Delta t, \quad \forall d \in \Omega^{\text{D}}, \forall t \in \Omega^{\text{T}}, \tag{2.1b}$$

$$p_{d(t-1)}^{\text{D}} - p_{dt}^{\text{D}} \le R_d^{\text{D,D}} \Delta t, \quad \forall d \in \Omega^{\text{D}}, \forall t \in \Omega^{\text{T}}, \tag{2.1c}$$

$$\sum_{t \in \Omega^{\text{T}}} p_{dt}^{\text{D}} \Delta t \ge \underline{E}_d^{\text{D}}, \quad \forall d \in \Omega^{\text{D}}. \tag{2.1d}$$

Equations (2.1a) limit the demand power consumption in each time period. Equations (2.1b) and (2.1c) impose up and down ramping limits on the demand power consumption of two consecutive time periods, respectively. Finally, Eqs. (2.1d) establish that the energy consumption of each demand during the planning horizon has to be at least $\underline{E}_d^{\text{D}}$.

Note that Eqs. (2.1) allow also considering inflexible demands. In such a case, the lower and upper bounds of a demand d in a time period t used in Eqs. (2.1a) would be the same, i.e., $\underline{P}_{dt}^{\text{D}} = \overline{P}_{dt}^{\text{D}}$.

Illustrative Example 2.1 Flexible Demand

A 6-h planning horizon is considered. There is a flexible demand whose lower and upper consumption bounds are provided in Fig. 2.1. The power demand consumption in the time period previous to the beginning of the planning horizon was 10 MW, the minimum energy consumption throughout the planning horizon is 120 MWh, while both the up and down ramping limits of the demand are 10 MW/h.

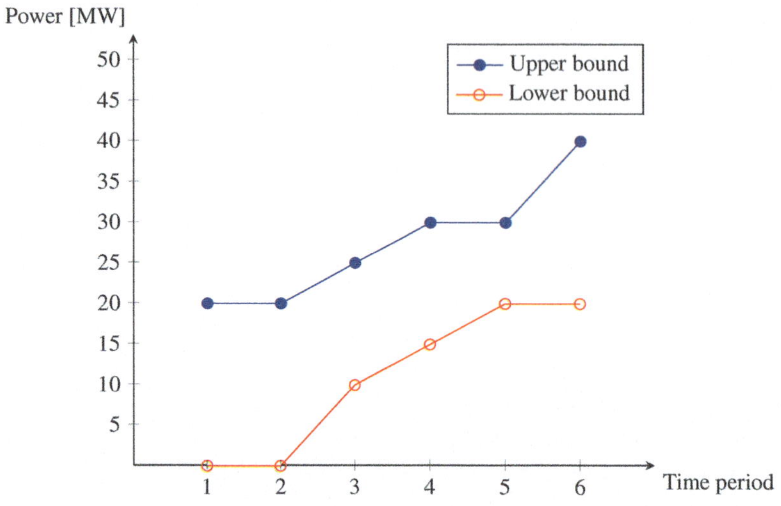

Fig. 2.1 Illustrative Example 2.1: lower and upper bounds of the power consumption

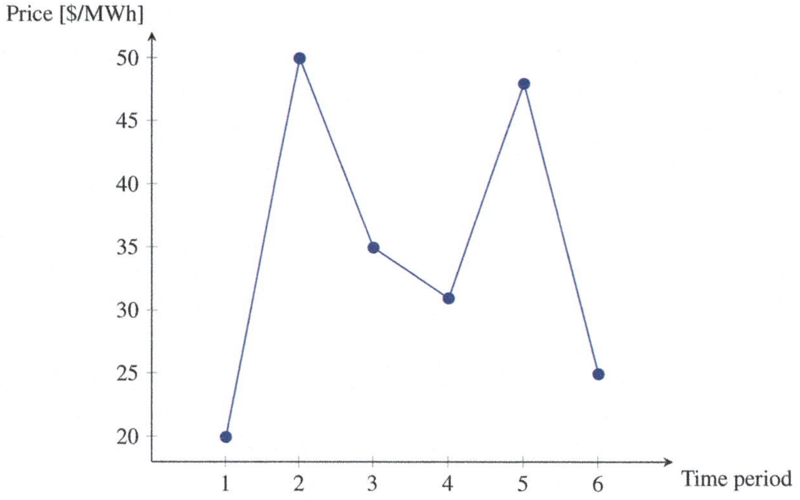

Fig. 2.2 Illustrative Example 2.1: market prices

This demand is supplied by purchasing energy in an electricity market whose market prices throughout the planning horizon are provided in Fig. 2.2.

The objective is to supply the flexible demand at minimum cost while complying with its technical constraints.

First, the objective function to be minimized is:

$$z = 20p_1^{\mathrm{D}} + 50p_2^{\mathrm{D}} + 35p_3^{\mathrm{D}} + 31p_4^{\mathrm{D}} + 48p_5^{\mathrm{D}} + 25p_6^{\mathrm{D}}.$$

This objective function represents the cost of the energy purchased in the market and used to supply the demand. Next, the constraints of the problem are formulated:

- Power consumption limits:

$$0 \leq p_1^{\mathrm{D}} \leq 20,$$
$$0 \leq p_2^{\mathrm{D}} \leq 20,$$
$$10 \leq p_3^{\mathrm{D}} \leq 25,$$
$$15 \leq p_4^{\mathrm{D}} \leq 30,$$

$$20 \leq p_5^{\mathrm{D}} \leq 30,$$
$$20 \leq p_6^{\mathrm{D}} \leq 40.$$

- Up ramping limits:

$$p_1^{\mathrm{D}} - 10 \leq 10,$$
$$p_2^{\mathrm{D}} - p_1^{\mathrm{D}} \leq 10,$$
$$p_3^{\mathrm{D}} - p_2^{\mathrm{D}} \leq 10,$$
$$p_4^{\mathrm{D}} - p_3^{\mathrm{D}} \leq 10,$$
$$p_5^{\mathrm{D}} - p_4^{\mathrm{D}} \leq 10,$$
$$p_6^{\mathrm{D}} - p_5^{\mathrm{D}} \leq 10.$$

- Down ramping limits:

$$10 - p_1^{\mathrm{D}} \leq 10,$$
$$p_1^{\mathrm{D}} - p_2^{\mathrm{D}} \leq 10,$$
$$p_2^{\mathrm{D}} - p_3^{\mathrm{D}} \leq 10,$$
$$p_3^{\mathrm{D}} - p_4^{\mathrm{D}} \leq 10,$$
$$p_4^{\mathrm{D}} - p_5^{\mathrm{D}} \leq 10,$$
$$p_5^{\mathrm{D}} - p_6^{\mathrm{D}} \leq 10.$$

- Minimum energy consumption throughout the planning horizon:

$$p_1^{\mathrm{D}} + p_2^{\mathrm{D}} + p_3^{\mathrm{D}} + p_4^{\mathrm{D}} + p_5^{\mathrm{D}} + p_6^{\mathrm{D}} \geq 120.$$

Finally, the optimization variables of the above problem are those included in set $\Phi^{\mathrm{IE2.1}} = \left\{p_1^{\mathrm{D}}, p_2^{\mathrm{D}}, p_3^{\mathrm{D}}, p_4^{\mathrm{D}}, p_5^{\mathrm{D}}, p_6^{\mathrm{D}}\right\}$. Note that these optimization variables include a subscript that indicates the time period.

The above problem is a linear programming (LP) problem [10] that can be solved, for instance, using CPLEX [6] under GAMS [5]. Results are provided in Fig. 2.3 that shows the optimal power consumption throughout the planning horizon.

As observed, the power consumption is within the bounds in every time period. Demand is preferably supplied in time periods with the lowest market prices, namely time periods 1, 4, and 6, in which the power consumption is close to the upper bound. However, the upper bound is never reached due to the down ramping limits. For example, consuming 20 MW in time period 1 would imply consuming at least 10 MW in time period 2, which has the highest market price.

The optimal value of the objective function is \$3900, which represents the total cost of supplying the flexible demand throughout the planning horizon.

□

A GAMS [5] code for solving Illustrative Example 2.1 is provided in Sect. 2.10.

Illustrative Example 2.2 Impact of Demand Flexibility

The data of Illustrative Example 2.1 are considered again. Next, the impact of demand flexibility on the total costs of supplying the demand is analyzed. To do so, three cases are considered. First, the demand is assumed to be inflexible, i.e., the power consumption levels are fixed in every time period without considering the market prices. Second, the up and down ramping limits are considered to be equal to 50 MW/h so that the associated constraints are not active. These two cases are denoted as inflexible and very flexible demand, respectively, while the third case is the base case analyzed in Illustrative Example 2.1, which is denoted as flexible demand.

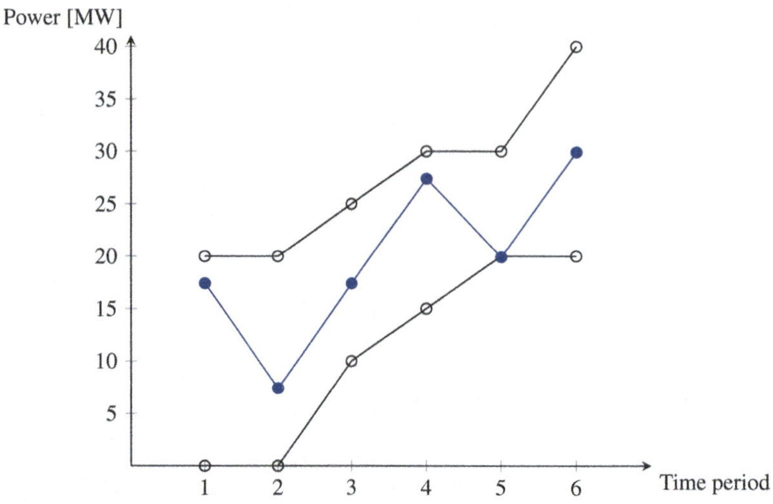

Fig. 2.3 Illustrative Example 2.1: optimal power consumption

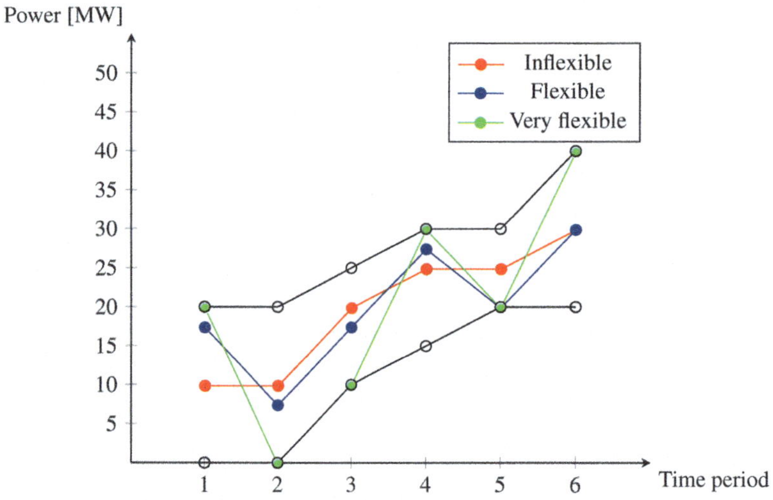

Fig. 2.4 Illustrative Example 2.2: optimal power consumption profiles

Table 2.1 Illustrative Example 2.2: total costs

Case	Cost [$]
Inflexible demand	4125
Flexible demand	3900
Very flexible demand	3640

The power consumption profiles for these three cases are shown in Fig. 2.4, while the total costs are provided in Table 2.1.

The total cost is increased by 5.76% in the case of an inflexible demand in comparison with the case of a flexible demand, while the total cost is reduced by 6.67% in the case of a very flexible demand. This highlights the importance of flexibility in the demands to adapt to fluctuations in market prices.

□

2.4 Conventional Power Plants

Conventional power plants such as CCGT, nuclear, or carbon units can also participate in VPPs. A basic model for these conventional power plants is provided below:

$$u_{ct}^{\mathrm{C}}, v_{ct}^{\mathrm{C,SD}}, v_{ct}^{\mathrm{C,SU}} \in \{0,1\}, \quad \forall c \in \Omega^{\mathrm{C}}, \forall t \in \Omega^{\mathrm{T}}, \tag{2.2a}$$

$$u_{ct}^{\mathrm{C}} - u_{c(t-1)}^{\mathrm{C}} = v_{ct}^{\mathrm{C,SU}} - v_{ct}^{\mathrm{C,SD}}, \quad \forall c \in \Omega^{\mathrm{C}}, \forall t \in \Omega^{\mathrm{T}}, \tag{2.2b}$$

$$v_{ct}^{\mathrm{C,SU}} + v_{ct}^{\mathrm{C,SD}} \leq 1, \quad \forall c \in \Omega^{\mathrm{C}}, \forall t \in \Omega^{\mathrm{T}}, \tag{2.2c}$$

$$\underline{P}_{c}^{\mathrm{C}} u_{ct}^{\mathrm{C}} \leq p_{ct}^{\mathrm{C}} \leq \overline{P}_{c}^{\mathrm{C}} u_{ct}^{\mathrm{C}}, \quad \forall c \in \Omega^{\mathrm{C}}, \forall t \in \Omega^{\mathrm{T}}, \tag{2.2d}$$

$$p_{ct}^{\mathrm{C}} - p_{c(t-1)}^{\mathrm{C}} \leq \left(R_{c}^{\mathrm{C,U}} u_{c(t-1)}^{\mathrm{C}} + R_{c}^{\mathrm{C,SU}} v_{ct}^{\mathrm{C,SU}}\right) \Delta t, \quad \forall c \in \Omega^{\mathrm{C}}, \forall t \in \Omega^{\mathrm{T}}, \tag{2.2e}$$

$$p_{c(t-1)}^{\mathrm{C}} - p_{ct}^{\mathrm{C}} \leq \left(R_{c}^{\mathrm{C,D}} u_{ct}^{\mathrm{C}} + R_{c}^{\mathrm{C,SD}} v_{ct}^{\mathrm{C,SD}}\right) \Delta t, \quad \forall c \in \Omega^{\mathrm{C}}, \forall t \in \Omega^{\mathrm{T}}. \tag{2.2f}$$

Equations (2.2a) define binary variables u_{ct}^{C}, $v_{ct}^{\mathrm{C,SD}}$, and $v_{ct}^{\mathrm{C,SU}}$ used to represent the on/off statuses, the shut-downs, and the start-ups of conventional power plants in a given planning horizon, respectively. The relationship between these variables is established in Eqs. (2.2b) and (2.2c) . Equations (2.2d) impose bounds on the power production of conventional power plants. Finally, Eqs. (2.2e) and (2.2f) limit the variation on the power production of conventional power plants in two consecutive time periods based on their up- and down-ramping capabilities, respectively.

Illustrative Example 2.3 Conventional Power Plant

A power producer that owns a conventional power plant with the technical and economic data provided in Table 2.2 is considered.

The power producer solves the self-scheduling problem of its conventional power plant for a 6-h planning horizon. The market prices for these 6 h are the same than those considered in Illustrative Example 2.1 and provided in Fig. 2.2.

Considering these data, the problem is formulated next.

First, the objective function to be maximized, which represents the total profit achieved by the power producer throughout the planning horizon using the conventional power plant:

$$\begin{aligned} z = {} & 20p_1^{\mathrm{C}} + 50p_2^{\mathrm{C}} + 35p_3^{\mathrm{C}} + 31p_4^{\mathrm{C}} + 48p_5^{\mathrm{C}} + 25p_6^{\mathrm{C}} \\ & - 33\left(p_1^{\mathrm{C}} + p_2^{\mathrm{C}} + p_3^{\mathrm{C}} + p_4^{\mathrm{C}} + p_5^{\mathrm{C}} + p_6^{\mathrm{C}}\right) \\ & - 6\left(u_1^{\mathrm{C}} + u_2^{\mathrm{C}} + u_3^{\mathrm{C}} + u_4^{\mathrm{C}} + u_5^{\mathrm{C}} + u_6^{\mathrm{C}}\right) \\ & - 120\left(v_1^{\mathrm{C,SU}} + v_2^{\mathrm{C,SU}} + v_3^{\mathrm{C,SU}} + v_4^{\mathrm{C,SU}} + v_5^{\mathrm{C,SU}} + v_6^{\mathrm{C,SU}}\right) \\ & - 60\left(v_1^{\mathrm{C,SD}} + v_2^{\mathrm{C,SD}} + v_3^{\mathrm{C,SD}} + v_4^{\mathrm{C,SD}} + v_5^{\mathrm{C,SD}} + v_6^{\mathrm{C,SD}}\right). \end{aligned}$$

Table 2.2 Illustrative Example 2.3: data of the conventional power plant

Capacity	60 MW
Minimum power output	10 MW
Ramping-up limit	20 MW/h
Ramping-down limit	15 MW/h
Start-up ramping limit	30 MW/h
Shut-down ramping limit	15 MW/h
Initial status	OFF
Initial power output	0
Online cost	$6/h
Variable cost	$33/MWh
Start-up cost	$120
Shut-down cost	$60

The first line represents the revenues achieved by the power producer, while the second, third, fourth, and fifth lines are the variable, online, start-up, and shut-down costs, respectively.

Next, the constraints of the problem are formulated:

- Declaration of binary variables:

$$u_1^C, u_2^C, u_3^C, u_4^C, u_5^C, u_1^C \in \{0, 1\},$$
$$v_1^{C,SD}, v_2^{C,SD}, v_3^{C,SD}, v_4^{C,SD}, v_5^{C,SD}, v_6^{C,SD} \in \{0, 1\},$$
$$v_1^{C,SU}, v_2^{C,SU}, v_3^{C,SU}, v_4^{C,SU}, v_5^{C,SU}, v_6^{C,SU} \in \{0, 1\}.$$

- Logic expressions:

$$u_1^C - 0 = v_1^{C,SU} - v_1^{C,SD},$$
$$u_2^C - u_1^C = v_2^{C,SU} - v_2^{C,SD},$$
$$u_3^C - u_2^C = v_3^{C,SU} - v_3^{C,SD},$$
$$u_4^C - u_3^C = v_4^{C,SU} - v_4^{C,SD},$$
$$u_5^C - u_4^C = v_5^{C,SU} - v_5^{C,SD},$$
$$u_6^C - u_5^C = v_6^{C,SU} - v_6^{C,SD},$$
$$v_1^{C,SU} + v_1^{C,SD} \leq 1,$$
$$v_2^{C,SU} + v_2^{C,SD} \leq 1,$$
$$v_3^{C,SU} + v_3^{C,SD} \leq 1,$$
$$v_4^{C,SU} + v_4^{C,SD} \leq 1,$$
$$v_5^{C,SU} + v_5^{C,SD} \leq 1,$$
$$v_6^{C,SU} + v_6^{C,SD} \leq 1.$$

- Power output limits:

$$10u_1^C \leq p_1^C \leq 60u_1^C,$$
$$10u_2^C \leq p_2^C \leq 60u_2^C,$$
$$10u_3^C \leq p_3^C \leq 60u_3^C,$$
$$10u_4^C \leq p_4^C \leq 60u_4^C,$$
$$10u_5^C \leq p_5^C \leq 60u_5^C,$$
$$10u_6^C \leq p_6^C \leq 60u_6^C.$$

- Ramping-up limits:

$$p_1^C - 0 \leq \left(0 + 30v_1^{C,SU}\right),$$
$$p_2^C - p_1^C \leq \left(20u_1^C + 30v_2^{C,SU}\right),$$
$$p_3^C - p_2^C \leq \left(20u_2^C + 30v_3^{C,SU}\right),$$

$$p_4^C - p_3^C \le \left(20u_3^C + 30v_4^{C,SU}\right),$$

$$p_5^C - p_4^C \le \left(20u_4^C + 30v_5^{C,SU}\right),$$

$$p_6^C - p_5^C \le \left(20u_5^C + 30v_6^{C,SU}\right).$$

- Ramping-down limits:

$$0 - p_1^C \le \left(15u_1^C + 15v_1^{C,SD}\right),$$

$$p_1^C - p_2^C \le \left(15u_2^C + 15v_2^{C,SD}\right),$$

$$p_2^C - p_3^C \le \left(15u_3^C + 15v_3^{C,SD}\right),$$

$$p_3^C - p_4^C \le \left(15u_4^C + 15v_4^{C,SD}\right),$$

$$p_4^C - p_5^C \le \left(15u_5^C + 15v_5^{C,SD}\right),$$

$$p_5^C - p_6^C \le \left(15u_6^C + 15v_6^{C,SD}\right).$$

The optimization variables of the above problem are those included in set $\Phi^{\mathrm{IE2.3}} = \left\{p_1^C, p_2^C, p_3^C, p_4^C, p_5^C, p_6^C, u_1^C, u_2^C, u_3^C, u_4^C, u_5^C, u_6^C, v_1^{C,SD}, v_2^{C,SD}, v_3^{C,SD}, v_4^{C,SD}, v_5^{C,SD}, v_6^{C,SD}, v_1^{C,SU}, v_2^{C,SU}, v_3^{C,SU}, v_4^{C,SU}, v_5^{C,SU}, v_6^{C,SU}\right\}$. Note that these variables include a subscript that denotes the time period.

The above problem is generally known as the self-scheduling problem of a power producer [2]. It is a mixed-integer linear programming (MILP) problem [10] that can be solved using appropriate software, e.g., CPLEX [6] under GAMS [5].

The optimal solution is provided in Table 2.3. The first column identifies the time period. The second, third, and fourth columns respectively indicate the optimal on/off, start-up, and shut-down statuses of the conventional power plant, while the fifth column gives the optimal power output.

It is optimal to start up the conventional power plant in time period 2 due to the relatively high market price in this time period. Then, the power plant remains on and producing electricity until the end of the planning horizon.

The optimal value of the objective function, i.e., the total profit achieved by the power producer in charge of the conventional power plant, is \$920. Table 2.4 shows how this profit splits among time periods. It should be noted that the profit is negative in time periods 4 and 6 due to the low market prices in comparison with the variable cost of the conventional

Table 2.3 Illustrative Example 2.3: results

Time period	On/off status	Start-up status	Shut-down status	Power output [MW]
1	0	0	0	0
2	1	1	0	30
3	1	0	0	50
4	1	0	0	40
5	1	0	0	60
6	1	0	0	45

Table 2.4 Illustrative Example 2.3: profit of the conventional power plant in different time periods

Time period	Profit [\$]
1	0
2	384
3	94
4	−86
5	894
6	−366

power plant. However, it is not optimal to turn off the power plant in these time periods due to the start-up and shut-down costs, as well as due to the ramping limits.

□

A GAMS [5] code for solving Illustrative Example 2.3 is provided in Sect. 2.10.

2.5 Stochastic Renewable Production Facilities

Besides conventional power plants, stochastic renewable production facilities such as solar- and wind-power units can also participate in VPPs. The working of these generating units is modeled using the following equations:

$$0 \leq p^{\mathrm{R}}_{rt} \leq \overline{P}^{\mathrm{R}}_{rt}, \quad \forall r \in \Omega^{\mathrm{R}}, \forall t \in \Omega^{\mathrm{T}}. \tag{2.3}$$

Equations (2.3) impose limits on the production of stochastic renewable generating units. The main difference with Eqs. (2.2) defined for conventional power plants is that the production of stochastic renewable production facilities is limited by the available production level, which varies over time periods since it depends on meteorological phenomena such as the solar irradiation or the wind speed.

2.6 Energy Storage Units

Another essential components of VPPs are energy storage facilities. These units are used to store energy, e.g., in the case of a surplus in the production of stochastic renewable generating units. This stored energy can be used in a later time period if needed, e.g., in the case of low renewable production or in the case of a peak in the demand consumption. A basic model for storage facilities comprises the following equations:

$$u^{\mathrm{S}}_{st} \in \{0, 1\}, \quad \forall s \in \Omega^{\mathrm{S}}, \forall t \in \Omega^{\mathrm{T}}, \tag{2.4a}$$

$$0 \leq p^{\mathrm{S,C}}_{st} \leq \overline{P}^{\mathrm{S,C}}_{st} u^{\mathrm{S}}_{st}, \quad \forall s \in \Omega^{\mathrm{S}}, \forall t \in \Omega^{\mathrm{T}}, \tag{2.4b}$$

$$0 \leq p^{\mathrm{S,D}}_{st} \leq \overline{P}^{\mathrm{S,D}}_{st} \left(1 - u^{\mathrm{S}}_{st}\right), \quad \forall s \in \Omega^{\mathrm{S}}, \forall t \in \Omega^{\mathrm{T}}, \tag{2.4c}$$

$$e^{\mathrm{S}}_{st} = e^{\mathrm{S}}_{s(t-1)} + p^{\mathrm{S,C}}_{st} \Delta t \eta^{\mathrm{S,C}}_{s} - \frac{p^{\mathrm{S,D}}_{st} \Delta t}{\eta^{\mathrm{S,D}}_{s}}, \quad \forall s \in \Omega^{\mathrm{S}}, \forall t \in \Omega^{\mathrm{T}}, \tag{2.4d}$$

$$\underline{E}^{\mathrm{S}}_{st} \leq e^{\mathrm{S}}_{st} \leq \overline{E}^{\mathrm{S}}_{st}, \quad \forall s \in \Omega^{\mathrm{S}}, \forall t \in \Omega^{\mathrm{T}}. \tag{2.4e}$$

Equations (2.4a) define binary variables u^{S}_{st}, which are used to prevent the simultaneous charging and discharging of storage units. Equations (2.4b) and (2.4c) impose bounds on the charging and discharging power levels, respectively. Equations (2.4d) define the energy evolution in the storage unit. The energy stored in a given time period is equal to the energy stored in the previous time period, plus the charging energy minus the discharging energy. Note that charging and discharging efficiencies are considered in both the charging and discharging power levels, respectively. Finally, Eqs. (2.4e) impose bounds on the energy stored in every time period.

Illustrative Example 2.4 Stochastic Renewable and Storage Units

A wind-power unit is considered, whose available production levels during a 6-h planning horizon are provided in Fig. 2.5. The power producer in charge of this wind-power unit participates in an electricity market whose prices during the planning horizon are the same than those considered in Illustrative Example 2.1 and provided in Fig. 2.2. For technical reasons, the maximum amount of power that the power producer can trade (sell or buy) in the market is 40 MW. Thus, the wind-power unit is combined with a storage facility that can be used to store the surplus energy if needed. The technical data of this storage unit are provided in Table 2.5. Finally, the minimum energy level in the storage unit in time period 6 is 20 MWh to prevent this storage unit to be depleted at the end of the planning horizon.

Considering these data, the power producer determines the optimal power to be traded in the market in each time period.

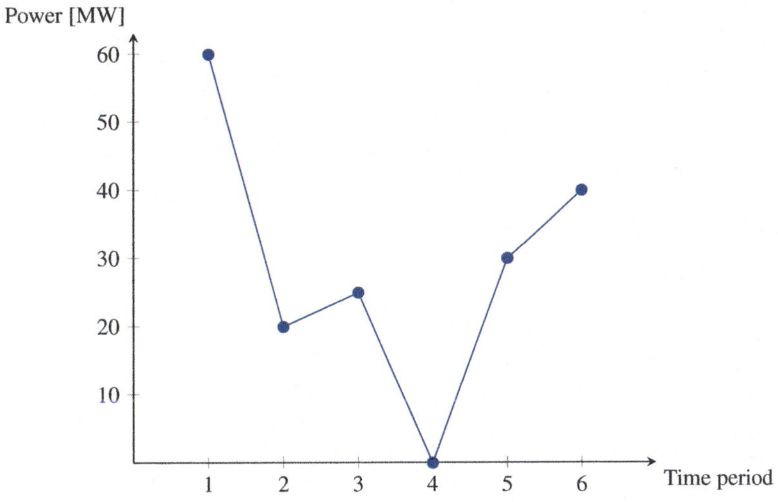

Fig. 2.5 Illustrative Example 2.4: available wind-power production levels

Table 2.5 Illustrative Example 2.4: data of the storage unit

Maximum charging power	20 MW
Maximum discharging power	20 MW
Maximum energy level	40 MWh
Minimum energy level	0
Initial energy level	20 MWh
Charging efficiency	0.9
Discharging efficiency	0.9

First, the objective function to be maximized is as follows:

$$z = 20p_1^{\mathrm{M}} + 50p_2^{\mathrm{M}} + 35p_3^{\mathrm{M}} + 31p_4^{\mathrm{M}} + 48p_5^{\mathrm{M}} + 25p_6^{\mathrm{M}}.$$

This objective function represents the profit achieved by the power producer in charge of the wind-power and storage units. Note that variables p_1^{M}, p_2^{M}, p_3^{M}, p_4^{M}, p_5^{M}, and p_6^{M} can take both positive and negative values. Positive values indicate that the power producer is selling energy to the electricity market, while negative values indicate that the power producer buys energy from the market.

Next, the constraints of the problem are formulated:

- Power balance:

$$p_1^{\mathrm{M}} = p_1^{\mathrm{R}} + p_1^{\mathrm{S,D}} - p_1^{\mathrm{S,C}},$$
$$p_2^{\mathrm{M}} = p_2^{\mathrm{R}} + p_2^{\mathrm{S,D}} - p_2^{\mathrm{S,C}},$$
$$p_3^{\mathrm{M}} = p_3^{\mathrm{R}} + p_3^{\mathrm{S,D}} - p_3^{\mathrm{S,C}},$$
$$p_4^{\mathrm{M}} = p_4^{\mathrm{R}} + p_4^{\mathrm{S,D}} - p_4^{\mathrm{S,C}},$$
$$p_5^{\mathrm{M}} = p_5^{\mathrm{R}} + p_5^{\mathrm{S,D}} - p_5^{\mathrm{S,C}},$$
$$p_6^{\mathrm{M}} = p_6^{\mathrm{R}} + p_6^{\mathrm{S,D}} - p_6^{\mathrm{S,C}}.$$

- Market limits:

$$-40 \leq p_1^{\mathrm{M}} \leq 40,$$
$$-40 \leq p_2^{\mathrm{M}} \leq 40,$$

$$-40 \leq p_3^{\mathrm{M}} \leq 40,$$
$$-40 \leq p_4^{\mathrm{M}} \leq 40,$$
$$-40 \leq p_5^{\mathrm{M}} \leq 40,$$
$$-40 \leq p_6^{\mathrm{M}} \leq 40.$$

- Wind-power production limits:

$$0 \leq p_1^{\mathrm{R}} \leq 60,$$
$$0 \leq p_2^{\mathrm{R}} \leq 20,$$
$$0 \leq p_3^{\mathrm{R}} \leq 25,$$
$$0 \leq p_4^{\mathrm{R}} \leq 0,$$
$$0 \leq p_5^{\mathrm{R}} \leq 30,$$
$$0 \leq p_6^{\mathrm{R}} \leq 40.$$

- Declaration of binary variables:

$$u_1^{\mathrm{S}}, u_2^{\mathrm{S}}, u_3^{\mathrm{S}}, u_4^{\mathrm{S}}, u_5^{\mathrm{S}}, u_6^{\mathrm{S}} \in \{0, 1\}.$$

- Charging power limits:

$$0 \leq p_1^{\mathrm{S,C}} \leq 20u_1^{\mathrm{S}},$$
$$0 \leq p_2^{\mathrm{S,C}} \leq 20u_2^{\mathrm{S}},$$
$$0 \leq p_3^{\mathrm{S,C}} \leq 20u_3^{\mathrm{S}},$$
$$0 \leq p_4^{\mathrm{S,C}} \leq 20u_4^{\mathrm{S}},$$
$$0 \leq p_5^{\mathrm{S,C}} \leq 20u_5^{\mathrm{S}},$$
$$0 \leq p_6^{\mathrm{S,C}} \leq 20u_6^{\mathrm{S}}.$$

- Discharging power limits:

$$0 \leq p_1^{\mathrm{S,D}} \leq 20\left(1 - u_1^{\mathrm{S}}\right),$$
$$0 \leq p_2^{\mathrm{S,D}} \leq 20\left(1 - u_2^{\mathrm{S}}\right),$$
$$0 \leq p_3^{\mathrm{S,D}} \leq 20\left(1 - u_3^{\mathrm{S}}\right),$$
$$0 \leq p_4^{\mathrm{S,D}} \leq 20\left(1 - u_4^{\mathrm{S}}\right),$$
$$0 \leq p_5^{\mathrm{S,D}} \leq 20\left(1 - u_5^{\mathrm{S}}\right),$$
$$0 \leq p_6^{\mathrm{S,D}} \leq 20\left(1 - u_6^{\mathrm{S}}\right).$$

- Energy balance in the storage unit:

$$e_1^{\mathrm{S}} = 20 + 0.9p_1^{\mathrm{S,C}} - \frac{p_1^{\mathrm{S,D}}}{0.9},$$

$$e_2^{\mathrm{S}} = e_1^{\mathrm{S}} + 0.9p_2^{\mathrm{S,C}} - \frac{p_2^{\mathrm{S,D}}}{0.9},$$

$$e_3^{\mathrm{S}} = e_2^{\mathrm{S}} + 0.9p_3^{\mathrm{S,C}} - \frac{p_3^{\mathrm{S,D}}}{0.9},$$

$$e_4^{\mathrm{S}} = e_3^{\mathrm{S}} + 0.9p_4^{\mathrm{S,C}} - \frac{p_4^{\mathrm{S,D}}}{0.9},$$

$$e_5^{\mathrm{S}} = e_4^{\mathrm{S}} + 0.9p_5^{\mathrm{S,C}} - \frac{p_5^{\mathrm{S,D}}}{0.9},$$

$$e_6^{\mathrm{S}} = e_5^{\mathrm{S}} + 0.9p_6^{\mathrm{S,C}} - \frac{p_6^{\mathrm{S,D}}}{0.9}.$$

- Energy stored limits:

$$0 \le e_1^{\mathrm{S}} \le 40,$$

$$0 \le e_2^{\mathrm{S}} \le 40,$$

$$0 \le e_3^{\mathrm{S}} \le 40,$$

$$0 \le e_4^{\mathrm{S}} \le 40,$$

$$0 \le e_5^{\mathrm{S}} \le 40,$$

$$20 \le e_6^{\mathrm{S}} \le 40.$$

The optimization variables of the problem are those included in set $\Phi^{\mathrm{IE2.4}} = \{p_1^{\mathrm{M}}, p_2^{\mathrm{M}}, p_3^{\mathrm{M}}, p_4^{\mathrm{M}}, p_5^{\mathrm{M}}, p_6^{\mathrm{M}}, p_1^{\mathrm{R}}, p_2^{\mathrm{R}}, p_3^{\mathrm{R}}, p_4^{\mathrm{R}}, p_5^{\mathrm{R}}, p_6^{\mathrm{R}}, p_1^{\mathrm{S,C}}, p_2^{\mathrm{S,C}}, p_3^{\mathrm{S,C}}, p_4^{\mathrm{S,C}}, p_5^{\mathrm{S,C}}, p_6^{\mathrm{S,C}}, p_1^{\mathrm{S,D}}, p_2^{\mathrm{S,D}}, p_3^{\mathrm{S,D}}, p_4^{\mathrm{S,D}}, p_5^{\mathrm{S,D}}, p_6^{\mathrm{S,D}}, u_1^{\mathrm{S,C}}, u_2^{\mathrm{S,C}}, u_3^{\mathrm{S,C}}, u_4^{\mathrm{S,C}}, u_5^{\mathrm{S,C}}, u_6^{\mathrm{S,C}}, e_1^{\mathrm{S}}, e_2^{\mathrm{S}}, e_3^{\mathrm{S}}, e_4^{\mathrm{S}}, e_5^{\mathrm{S}}, e_6^{\mathrm{S}}\}$. Note that the subscript in these variables denotes the time period.

The optimal power traded in the market is shown in Fig. 2.6, while the rest of the results are provided in Table 2.6. The first column identifies the time period. The second column provides the power production of the wind-power unit. The third

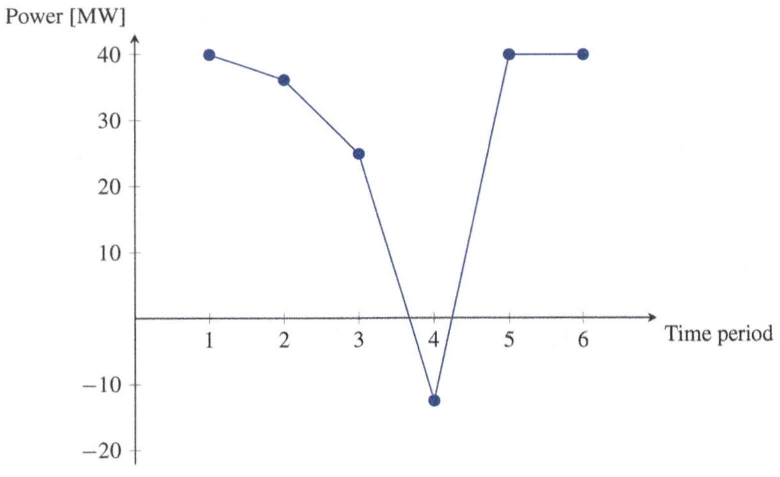

Fig. 2.6 Illustrative Example 2.4: optimal power traded in the market

Table 2.6 Illustrative Example 2.4: results

t	p_t^{R} [MW]	$p_t^{\mathrm{S,C}}$ [MW]	$p_t^{\mathrm{S,D}}$ [MW]	e_t^{S} [MWh]
1	60	20	0	38
2	20	0	16.2	20
3	25	0	0	20
4	0	12.3	0	31.1
5	30	0	10	20
6	40	0	0	20

and fourth columns give the charging and discharging power levels of the energy storage facility, while the fifth column provides the energy stored in this unit.

Note that the available wind-power production is used in all time periods. Since in time period 1 the available production level is 60 MW and the maximum power that can be traded is 40 MW, the remaining 20 MW are used to charge the storage unit. The storage facility is also charged in time period 4 in which the power producer buys 12.3 MW due to the comparatively low market prices. On the other hand, the storage unit is discharged in time periods 2 and 5 due to the comparatively high market prices and low wind-power production levels.

Finally, the optimal value on the objective function is \$6022.28, which represents the total profit achieved by the power producer in charge of the wind-power unit and the storage facility.

□

A GAMS [5] code for solving Illustrative Example 2.4 is provided in Sect. 2.10.

2.7 Network

Sections 2.3–2.6 provide models for consumption and generating assets in VPPs. These elements of a VPP are located within an electric network. In some cases, it is necessary to represent the network constraints in the model. For the sake of simplicity, a dc power flow model without losses [2] is considered:

$$\sum_{g \in \Omega_n^{\mathrm{G}}} p_{gt}^{\mathrm{G}} - \sum_{\ell | s(\ell)=n} p_{\ell t}^{\mathrm{L}} + \sum_{\ell | r(\ell)=n} p_{\ell t}^{\mathrm{L}} = \sum_{d \in \Omega_n^{\mathrm{D}}} p_{dt}^{\mathrm{D}}, \quad \forall n \in \Omega^{\mathrm{N}}, \forall t \in \Omega^{\mathrm{T}}, \tag{2.5a}$$

$$p_{\ell t}^{\mathrm{L}} = \frac{1}{X_\ell} \left(\delta_{s(\ell)t} - \delta_{r(\ell)t} \right), \quad \forall \ell \in \Omega^{\mathrm{L}}, \forall t \in \Omega^{\mathrm{T}}, \tag{2.5b}$$

$$-\overline{P}_\ell^{\mathrm{L}} \leq p_{\ell t}^{\mathrm{L}} \leq \overline{P}_\ell^{\mathrm{L}}, \quad \forall \ell \in \Omega^{\mathrm{L}}, \forall t \in \Omega^{\mathrm{T}}, \tag{2.5c}$$

$$\delta_{nt} = 0, \quad n : \text{ref.}, \forall t \in \Omega^{\mathrm{T}}, \tag{2.5d}$$

$$p_{gt}^{\mathrm{G}} \in \Psi_{gt}^{\mathrm{G}}, \quad \forall g \in \Omega^{\mathrm{G}}, \forall t \in \Omega^{\mathrm{T}}, \tag{2.5e}$$

$$p_{dt}^{\mathrm{D}} \in \Psi_{dt}^{\mathrm{D}}, \quad \forall d \in \Omega^{\mathrm{D}}, \forall t \in \Omega^{\mathrm{T}}. \tag{2.5f}$$

Equations (2.5a) establish the power balance at each node of the network. The power generated plus the net power flow through network lines connected to each node must be equal to the power demand. Equations (2.5b) define the power flows in network lines, which are limited to the corresponding capacities using Eqs. (2.5c). Equations (2.5d) fix the voltage angle at the reference node. Finally, Eqs. (2.5e) and (2.5f) impose that the power generated and the power demand must be within their corresponding feasibility sets, respectively. Note that set Ω^{G} includes conventional power plants, stochastic renewable generating units, and storage facilities working as generating units (i.e., when they are discharging), while set Ω^{D} refers to flexible demands and storage facilities working as demands (i.e., when they are charging).

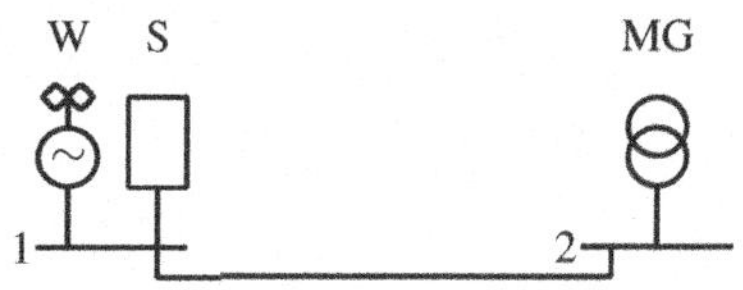

Fig. 2.7 Illustrative Example 2.5: two-node system

Illustrative Example 2.5 Network Constraints

The data of Illustrative Example 2.4 are considered again. The wind-power (W) and storage (S) units are located within the two-node network depicted in Fig. 2.7. These two units are located at node 1, while this network is connected to the main grid (MG) through node 2.

The network line connecting nodes 1 and 2 has a reactance of 0.2 Ω, being its capacity limit equal to 30 MW. Base impedance and base power are 1 Ω and 1 MW, respectively.

The objective function of this problem is the same as that considered in Illustrative Example 2.4. Moreover, the constraints provided in Illustrative Example 2.4 are also needed here. The difference is that it is necessary to consider the following additional constraints to represent the network:

- Nodal power balance:

$$p_1^{\mathrm{R}} + p_1^{\mathrm{S,D}} - p_1^{\mathrm{L}} = p_1^{\mathrm{S,C}},$$
$$p_1^{\mathrm{L}} = p_1^{\mathrm{M}},$$
$$p_2^{\mathrm{R}} + p_2^{\mathrm{S,D}} - p_2^{\mathrm{L}} = p_2^{\mathrm{S,C}},$$
$$p_2^{\mathrm{L}} = p_2^{\mathrm{M}},$$
$$p_3^{\mathrm{R}} + p_3^{\mathrm{S,D}} - p_3^{\mathrm{L}} = p_3^{\mathrm{S,C}},$$
$$p_3^{\mathrm{L}} = p_3^{\mathrm{M}},$$
$$p_4^{\mathrm{R}} + p_4^{\mathrm{S,D}} - p_4^{\mathrm{L}} = p_4^{\mathrm{S,C}},$$
$$p_4^{\mathrm{L}} = p_4^{\mathrm{M}},$$
$$p_5^{\mathrm{R}} + p_5^{\mathrm{S,D}} - p_5^{\mathrm{L}} = p_5^{\mathrm{S,C}},$$
$$p_5^{\mathrm{L}} = p_5^{\mathrm{M}},$$
$$p_6^{\mathrm{R}} + p_6^{\mathrm{S,D}} - p_6^{\mathrm{L}} = p_6^{\mathrm{S,C}},$$
$$p_6^{\mathrm{L}} = p_6^{\mathrm{M}}.$$

- Power flows:

$$p_1^{\mathrm{L}} = \frac{1}{0.2}\left(\delta_{11} - \delta_{21}\right),$$
$$p_2^{\mathrm{L}} = \frac{1}{0.2}\left(\delta_{12} - \delta_{22}\right),$$
$$p_3^{\mathrm{L}} = \frac{1}{0.2}\left(\delta_{13} - \delta_{23}\right),$$
$$p_4^{\mathrm{L}} = \frac{1}{0.2}\left(\delta_{14} - \delta_{24}\right),$$
$$p_5^{\mathrm{L}} = \frac{1}{0.2}\left(\delta_{15} - \delta_{25}\right),$$
$$p_6^{\mathrm{L}} = \frac{1}{0.2}\left(\delta_{16} - \delta_{26}\right).$$

- Power flow limits:

$$-30 \le p_1^{\mathrm{L}} \le 30,$$
$$-30 \le p_2^{\mathrm{L}} \le 30,$$
$$-30 \le p_3^{\mathrm{L}} \le 30,$$
$$-30 \le p_4^{\mathrm{L}} \le 30,$$
$$-30 \le p_5^{\mathrm{L}} \le 30,$$
$$-30 \le p_6^{\mathrm{L}} \le 30.$$

- Voltage angle at the reference node:

$$\delta_{11}, \delta_{12}, \delta_{13}, \delta_{14}, \delta_{15}, \delta_{16} = 0.$$

The optimization variables of the problem are those included in set $\Phi^{\mathrm{IE2.5}} = \{p_1^{\mathrm{M}}, p_2^{\mathrm{M}}, p_3^{\mathrm{M}}, p_4^{\mathrm{M}}, p_5^{\mathrm{M}}, p_6^{\mathrm{M}}, p_1^{\mathrm{R}}, p_2^{\mathrm{R}}, p_3^{\mathrm{R}}, p_4^{\mathrm{R}}, p_5^{\mathrm{R}}, p_6^{\mathrm{R}}, p_1^{\mathrm{S,C}}, p_2^{\mathrm{S,C}}, p_3^{\mathrm{S,C}}, p_4^{\mathrm{S,C}}, p_5^{\mathrm{S,C}}, p_6^{\mathrm{SC}}, p_1^{\mathrm{S,D}}, p_2^{\mathrm{S,D}}, p_3^{\mathrm{S,D}}, p_4^{\mathrm{SD}}, p_5^{\mathrm{S,D}}, p_6^{\mathrm{S,D}}, u_1^{\mathrm{S}}, u_2^{\mathrm{S}}, u_3^{\mathrm{S}}, u_4^{\mathrm{S}}, u_5^{\mathrm{S}}, u_6^{\mathrm{S}}, e_1^{\mathrm{S}}, e_2^{\mathrm{S}}, e_3^{\mathrm{S}}, e_4^{\mathrm{S}}, e_5^{\mathrm{S}}, e_6^{\mathrm{S}}, p_1^{\mathrm{L}}, p_2^{\mathrm{L}}, p_3^{\mathrm{L}}, p_4^{\mathrm{L}}, p_5^{\mathrm{L}}, p_6^{\mathrm{L}}, \delta_{11}, \delta_{12}, \delta_{13}, \delta_{14}, \delta_{15}, \delta_{16}, \delta_{21}, \delta_{22}, \delta_{23}, \delta_{24}, \delta_{25}, \delta_{26}\}$. These variables include a single subscript that denotes the time period or two subscripts in the case of voltage angles, the first one denoting the node and the second one the time period.

The optimal power traded in the market is shown in Fig. 2.8, while the rest of the results are provided in Table 2.7. The first column identifies the time period. The second column provides the power production of the wind-power unit. The third and fourth columns give the charging and discharging power levels of the storage facility, while the fifth column provides the energy stored in this unit.

As observed, the power traded in the market follows a pattern similar to that obtained in Illustrative Example 2.4. However, in this case the power traded in the market is bounded by the 30-MW capacity of the network line.

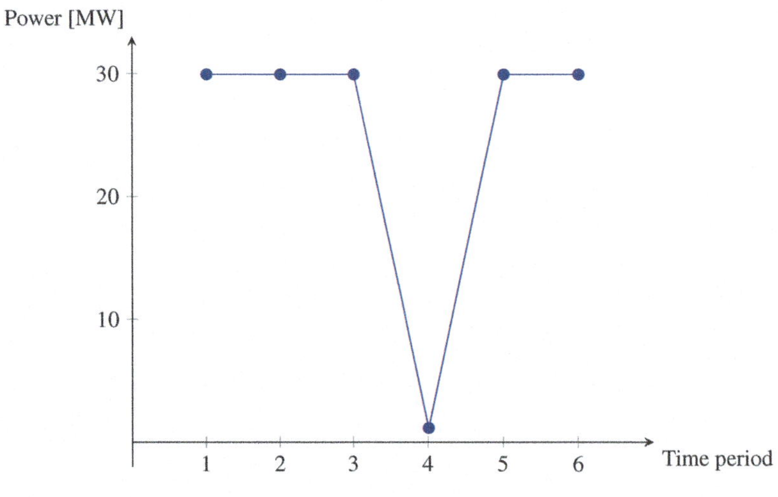

Fig. 2.8 Illustrative Example 2.5: optimal power traded in the market

Table 2.7 Illustrative Example 2.5: results

t	p_t^{R} [MW]	$p_t^{\mathrm{S,C}}$ [MW]	$p_t^{\mathrm{S,D}}$ [MW]	e_t^{S} [MWh]
1	50	20	0	38
2	20	0	10	26.9
3	25	0	5	21.33
4	0	0	1.2	20
5	30	0	0	20
6	30	0	0	20

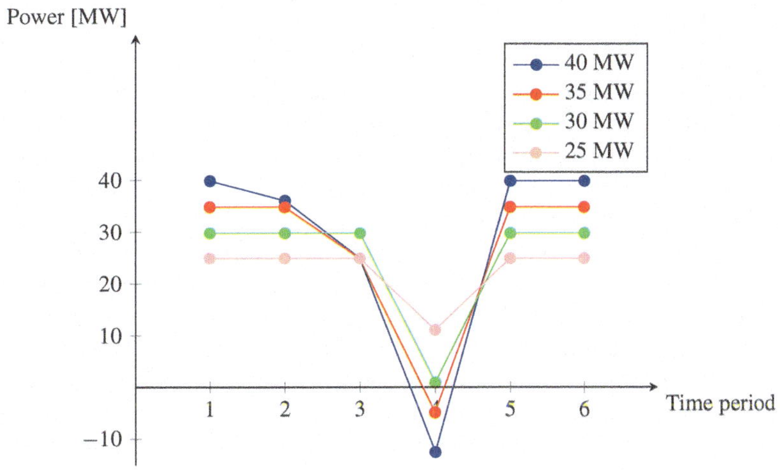

Fig. 2.9 Illustrative Example 2.5: optimal power traded in the market for different capacity limits

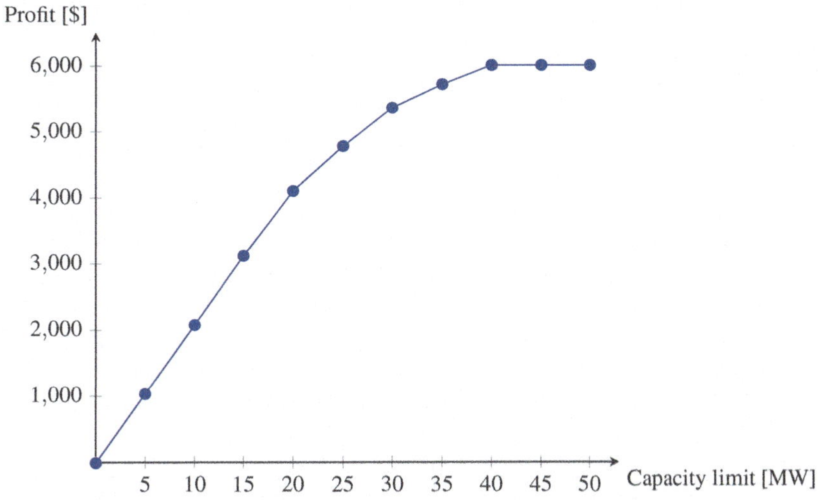

Fig. 2.10 Illustrative Example 2.5: optimal profit for different capacity limits

Finally, the optimal value on the objective function is \$5377.2, which is the total profit achieved by the power producer in charge of the wind-power unit and the storage facility.

□

A GAMS [5] code for solving Illustrative Example 2.5 is provided in Sect. 2.10.

Illustrative Example 2.6 Impact of Network Limits

Illustrative Example 2.5 is solved again for different values of the capacity limit of the network line connecting nodes 1 and 2.

Results are provided in Fig. 2.9, which shows the power traded in the market for different values of the network capacity limit. As observed, the pattern is similar in all cases; however, if the capacity limit is 25 or 30 MW the power producer only sells energy to the market. Moreover, if the capacity limit is 40 MW, which is the maximum power that can be traded with the main grid, the results are identical to those obtained in Illustrative Example 2.4.

On the other hand, Fig. 2.10 provides the profit achieved by the power producer in charge of the wind-power and storage units for different network capacity limits. As observed, the profit increases as the capacity limit is increased until the 40-MW capacity limit is reached. Values larger than 40 MW do not have any impact on the results since the maximum power that can be traded in the market through the main grid connected at node 2 is 40 MW and all the assets managed by the power producer are located at node 1.

□

2.8 Smart Grid Technology

Sections 2.3–2.7 introduce models for the operation of the consumption and generating assets in a VPP. The output of these models is the scheduling of the different units for a given planning horizon considering both their technical and economic constraints. These models are generally solved at the beginning of the planning horizon. For example, if the VPP participates in the day-ahead energy market, decisions are made one day in advance for all the hours of the following day. Thus, the values of the parameters used in these models are determined according to the available information at this point in time, including the economic and technical data of the generating units and demands, as well as the forecasts of uncertain data such as the prices in the market and the production levels of stochastic renewable generating units in the VPP.

With the evolution of the smart grid technology, VPPs can also benefit from a bidirectional communication with their units, as well as with the market. Under this condition, the VPP receives the required information from its units and the market near real-time, and then the VPP can update the values of the parameters used in the decision-making models. Additionally, the VPP can make scheduling decisions of the units time by time, communicating them to the units and the market in real-time. Thus, taking the advantages of the smart grid technology, the decision-making models differ from the models explained in Sects. 2.3–2.7 in two main aspects:

1. Models can be run time by time throughout a planning horizon.
2. Parameters used in the models can be updated based on the new information available in each time step.

The following example clarifies the role of smart grid technology in VPPs.

Illustrative Example 2.7 Scheduling with Smart Grid Technology

The data of Illustrative Example 2.4, which analyzes a power producer in charge of a wind-power plant and an energy storage unit, are considered again. The power producer decides the optimal scheduling of these two units for a 6-h planning horizon. To do so, it solves an optimization problem at the beginning of the planning horizon that considers as input parameters the forecasts of market prices and the production levels of the wind-power unit. However, what happens if the actual market prices and/or the actual available production levels of the wind-power plant are different to the forecasts considered at the time the problem is solved? This may have a great impact on the profit achieved by the power producer.

Next it is assumed that the power producer is equipped with the smart grid technology, which allows bidirectional communication between the power producer and the units, as well as between the power producer and the market in real time. Under these circumstances, the power producer solves the optimization problem described in Illustrative Example 2.4; however, only the decisions for the first time period are actually implemented. Then, just before the second time period, the power producer receives updated information about the market prices and the available production levels of the wind-power unit. With this updated information, the power producer solves a new optimization problem, which is very similar to the one formulated in Illustrative Example 2.4 but with two main differences:

1. Decisions in the first time period are fixed since they have been already implemented.
2. Model parameters are updated based on the new information available.

Then, this process is repeated for each time period of the planning horizon. This way, the power producer can update the scheduling decisions and adapt to the new conditions in both the units it manages and the market.

Tables 2.8 and 2.9 provide the updated data of market prices and available wind-power production levels at each time step, respectively. The first column identifies the time period. The second column shows the forecast values just before the first time period, which match the values considered in Illustrative Example 2.4, while the third, fourth, fifth, sixth, and seventh columns provide the forecast values just before the second, third, fourth, fifth, and sixth time periods, respectively. Note that bold values in Tables 2.8 and 2.9 refer to actual data of market prices and available wind-power production levels. For

Table 2.8 Illustrative Example 2.7: updated market prices

Time period	1	2	3	4	5	6
1	**20**	**20**	**20**	**20**	**20**	**20**
2	50	**55**	**55**	**55**	**55**	**55**
3	35	32	**30**	**30**	**30**	**30**
4	31	34	28	**30**	**30**	**30**
5	48	46	42	44	**45**	**45**
6	25	30	30	38	36	**35**

Table 2.9 Illustrative Example 2.7: updated available wind-power production levels

Time period	1	2	3	4	5	6
1	**60**	**60**	**60**	**60**	**60**	**60**
2	20	**25**	**25**	**25**	**25**	**25**
3	25	35	**30**	**30**	**30**	**30**
4	0	0	0	**5**	**5**	**5**
5	30	30	25	18	**20**	**20**
6	40	32	28	22	23	**25**

example, at the beginning of time period 1, the power producer knows with certainty the available wind-power production level at this time period, which is denoted by a bold value.

Using the above data, the power producer solves six problems, one for each time period. The first one, just before time period 1, is exactly the same than that solved in Illustrative Example 2.4. The only difference is that now the power producer only implements decisions obtained for time period 1, i.e., at this time period the power producer uses the available 60 MW of wind-power production to sell 40 MW to the market and to charge 20 MW the storage unit, so that the energy level is 38 MWh. Then, just before time period 2, the power producer solves a new optimization problem using the updated data it gets from the units and the market (third columns of Tables 2.8 and 2.9) and with fixed values for the decisions made for the first time period. This problem is formulated below:

$$\max_{\Phi_2^{\text{IE2.7}}} \; 55p_2^{\text{M}} + 32p_3^{\text{M}} + 34p_4^{\text{M}} + 46p_5^{\text{M}} + 30p_6^{\text{M}}$$

subject to

$$p_2^{\text{M}} = p_2^{\text{R}} + p_2^{\text{S,D}} - p_2^{\text{S,C}},$$
$$p_3^{\text{M}} = p_3^{\text{R}} + p_3^{\text{S,D}} - p_3^{\text{S,C}},$$
$$p_4^{\text{M}} = p_4^{\text{R}} + p_4^{\text{S,D}} - p_4^{\text{S,C}},$$
$$p_5^{\text{M}} = p_5^{\text{R}} + p_5^{\text{S,D}} - p_5^{\text{S,C}},$$
$$p_6^{\text{M}} = p_6^{\text{R}} + p_6^{\text{S,D}} - p_6^{\text{S,C}},$$
$$-40 \le p_2^{\text{M}} \le 40,$$
$$-40 \le p_3^{\text{M}} \le 40,$$
$$-40 \le p_4^{\text{M}} \le 40,$$
$$-40 \le p_5^{\text{M}} \le 40,$$
$$-40 \le p_6^{\text{M}} \le 40,$$
$$0 \le p_2^{\text{R}} \le 25,$$
$$0 \le p_3^{\text{R}} \le 35,$$
$$0 \le p_4^{\text{R}} \le 0,$$
$$0 \le p_5^{\text{R}} \le 30,$$
$$0 \le p_6^{\text{R}} \le 32.$$
$$u_2^{\text{S}}, u_3^{\text{S}}, u_4^{\text{S}}, u_5^{\text{S}}, u_6^{\text{S}} \in \{0, 1\},$$
$$0 \le p_2^{\text{S,C}} \le 20u_2^{\text{S}},$$
$$0 \le p_3^{\text{S,C}} \le 20u_3^{\text{S}},$$
$$0 \le p_4^{\text{S,C}} \le 20u_4^{\text{S}},$$

$$0 \le p_5^{S,C} \le 20u_5^S,$$
$$0 \le p_6^{S,C} \le 20u_6^S,$$
$$0 \le p_2^{S,D} \le 20\left(1 - u_2^S\right),$$
$$0 \le p_3^{S,D} \le 20\left(1 - u_3^S\right),$$
$$0 \le p_4^{S,D} \le 20\left(1 - u_4^S\right),$$
$$0 \le p_5^{S,D} \le 20\left(1 - u_5^S\right),$$
$$0 \le p_6^{S,D} \le 20\left(1 - u_6^S\right),$$
$$e_2^S = 38 + 0.9p_2^{S,C} - \frac{p_2^{S,D}}{0.9},$$
$$e_3^S = e_2^S + 0.9p_3^{S,C} - \frac{p_3^{S,D}}{0.9},$$
$$e_4^S = e_3^S + 0.9p_4^{S,C} - \frac{p_4^{S,D}}{0.9},$$
$$e_5^S = e_4^S + 0.9p_5^{S,C} - \frac{p_5^{S,D}}{0.9},$$
$$e_6^S = e_5^S + 0.9p_6^{S,C} - \frac{p_6^{S,D}}{0.9},$$
$$0 \le e_2^S \le 40,$$
$$0 \le e_3^S \le 40,$$
$$0 \le e_4^S \le 40,$$
$$0 \le e_5^S \le 40,$$
$$20 \le e_6^S \le 40.$$

The optimization variables of the problem are those included in set $\Phi_2^{\mathrm{IE2.7}} = \{p_2^M, p_3^M, p_4^M, p_5^M, p_6^M, p_2^R, p_3^R, p_4^R, p_5^R, p_6^R, p_2^{S,C}, p_3^{S,C}, p_4^{S,C}, p_5^{S,C}, p_6^{S,C}, p_2^{S,D}, p_3^{S,D}, p_4^{S,D}, p_5^{S,D}, p_6^{S,D}, u_2^{S,C}, u_3^{S,C}, u_4^{S,C}, u_5^{S,C}, u_6^{S,C}, e_2^S, e_3^S, e_4^S, e_5^S, e_6^S\}$. Note that the above problem uses the updated information of input parameters, as well as the energy level of the storage unit in time period 1.

The power producer solves the problem and obtain the decisions for time periods 2–6; however, only decisions for time period 2 are implemented. In particular, the power producer uses all the available wind-power production and discharges 15 MW the storage unit, which allows it to sell 40 MW in the market, being 21.33 MWh the energy level of the storage facility. Note that these decisions for time period 2 are different from those obtained in Illustrative Example 2.4.

Next, the power producer receives updated information for the third time period and solves the following problem:

$$\max_{\Phi_3^{\mathrm{IE2.7}}} \; 30p_3^M + 28p_4^M + 42p_5^M + 30p_6^M$$

subject to

$$p_3^M = p_3^R + p_3^{S,D} - p_3^{S,C},$$
$$p_4^M = p_4^R + p_4^{S,D} - p_4^{S,C},$$
$$p_5^M = p_5^R + p_5^{S,D} - p_5^{S,C},$$

$$
\begin{aligned}
& p_6^{\mathrm{M}} = p_6^{\mathrm{R}} + p_6^{\mathrm{S,D}} - p_6^{\mathrm{S,C}}, \\
& -40 \le p_3^{\mathrm{M}} \le 40, \\
& -40 \le p_4^{\mathrm{M}} \le 40, \\
& -40 \le p_5^{\mathrm{M}} \le 40, \\
& -40 \le p_6^{\mathrm{M}} \le 40, \\
& 0 \le p_3^{\mathrm{R}} \le 30, \\
& 0 \le p_4^{\mathrm{R}} \le 0, \\
& 0 \le p_5^{\mathrm{R}} \le 25, \\
& 0 \le p_6^{\mathrm{R}} \le 28. \\
& u_3^{\mathrm{S}}, u_4^{\mathrm{S}}, u_5^{\mathrm{S}}, u_6^{\mathrm{S}} \in \{0, 1\}, \\
& 0 \le p_3^{\mathrm{S,C}} \le 20u_3^{\mathrm{S}}, \\
& 0 \le p_4^{\mathrm{S,C}} \le 20u_4^{\mathrm{S}}, \\
& 0 \le p_5^{\mathrm{S,C}} \le 20u_5^{\mathrm{S}}, \\
& 0 \le p_6^{\mathrm{S,C}} \le 20u_6^{\mathrm{S}}, \\
& 0 \le p_3^{\mathrm{S,D}} \le 20\left(1 - u_3^{\mathrm{S}}\right), \\
& 0 \le p_4^{\mathrm{S,D}} \le 20\left(1 - u_4^{\mathrm{S}}\right), \\
& 0 \le p_5^{\mathrm{S,D}} \le 20\left(1 - u_5^{\mathrm{S}}\right), \\
& 0 \le p_6^{\mathrm{S,D}} \le 20\left(1 - u_6^{\mathrm{S}}\right), \\
& e_3^{\mathrm{S}} = 21.33 + 0.9p_3^{\mathrm{S,C}} - \frac{p_3^{\mathrm{S,D}}}{0.9}, \\
& e_4^{\mathrm{S}} = e_3^{\mathrm{S}} + 0.9p_4^{\mathrm{S,C}} - \frac{p_4^{\mathrm{S,D}}}{0.9}, \\
& e_5^{\mathrm{S}} = e_4^{\mathrm{S}} + 0.9p_5^{\mathrm{S,C}} - \frac{p_5^{\mathrm{S,D}}}{0.9}, \\
& e_6^{\mathrm{S}} = e_5^{\mathrm{S}} + 0.9p_6^{\mathrm{S,C}} - \frac{p_6^{\mathrm{S,D}}}{0.9}, \\
& 0 \le e_3^{\mathrm{S}} \le 40, \\
& 0 \le e_4^{\mathrm{S}} \le 40, \\
& 0 \le e_5^{\mathrm{S}} \le 40, \\
& 20 \le e_6^{\mathrm{S}} \le 40.
\end{aligned}
$$

The optimization variables of the problem are those included in set $\Phi_3^{\mathrm{IE2.7}} = \{p_3^{\mathrm{M}}, p_4^{\mathrm{M}}, p_5^{\mathrm{M}}, p_6^{\mathrm{M}}, p_3^{\mathrm{R}}, p_4^{\mathrm{R}}, p_5^{\mathrm{R}}, p_6^{\mathrm{R}}, p_3^{\mathrm{S,C}}, p_4^{\mathrm{S,C}}, p_5^{\mathrm{S,C}}, p_6^{\mathrm{S,C}}, p_3^{\mathrm{S,D}}, p_4^{\mathrm{S,D}}, p_5^{\mathrm{S,D}}, p_6^{\mathrm{S,D}}, u_3^{\mathrm{S}}, u_4^{\mathrm{S}}, u_5^{\mathrm{S}}, u_6^{\mathrm{S}}, e_3^{\mathrm{S}}, e_4^{\mathrm{S}}, e_5^{\mathrm{S}}, e_6^{\mathrm{S}}\}$.

The problem is solved and the power producer decides to sell 30 MW in the third time period, while the storage unit is not operated in this case.

Next, the problem is formulated for the fourth time period:

$$\max_{\Phi_4^{\mathrm{IE2.7}}} 30p_4^{\mathrm{M}} + 44p_5^{\mathrm{M}} + 38p_6^{\mathrm{M}}$$

subject to

$$p_4^{\mathrm{M}} = p_4^{\mathrm{R}} + p_4^{\mathrm{S,D}} - p_4^{\mathrm{S,C}},$$
$$p_5^{\mathrm{M}} = p_5^{\mathrm{R}} + p_5^{\mathrm{S,D}} - p_5^{\mathrm{S,C}},$$
$$p_6^{\mathrm{M}} = p_6^{\mathrm{R}} + p_6^{\mathrm{S,D}} - p_6^{\mathrm{S,C}},$$
$$-40 \le p_4^{\mathrm{M}} \le 40,$$
$$-40 \le p_5^{\mathrm{M}} \le 40,$$
$$-40 \le p_6^{\mathrm{M}} \le 40,$$
$$0 \le p_4^{\mathrm{R}} \le 5,$$
$$0 \le p_5^{\mathrm{R}} \le 18,$$
$$0 \le p_6^{\mathrm{R}} \le 22.$$
$$u_4^{\mathrm{S}}, u_5^{\mathrm{S}}, u_6^{\mathrm{S}} \in \{0, 1\},$$
$$0 \le p_4^{\mathrm{S,C}} \le 20u_4^{\mathrm{S}},$$
$$0 \le p_5^{\mathrm{S,C}} \le 20u_5^{\mathrm{S}},$$
$$0 \le p_6^{\mathrm{S,C}} \le 20u_6^{\mathrm{S}},$$
$$0 \le p_4^{\mathrm{S,D}} \le 20\left(1 - u_4^{\mathrm{S}}\right),$$
$$0 \le p_5^{\mathrm{S,D}} \le 20\left(1 - u_5^{\mathrm{S}}\right),$$
$$0 \le p_6^{\mathrm{S,D}} \le 20\left(1 - u_6^{\mathrm{S}}\right),$$
$$e_4^{\mathrm{S}} = 21.33 + 0.9p_4^{\mathrm{S,C}} - \frac{p_4^{\mathrm{S,D}}}{0.9},$$
$$e_5^{\mathrm{S}} = e_4^{\mathrm{S}} + 0.9p_5^{\mathrm{S,C}} - \frac{p_5^{\mathrm{S,D}}}{0.9},$$
$$e_6^{\mathrm{S}} = e_5^{\mathrm{S}} + 0.9p_6^{\mathrm{S,C}} - \frac{p_6^{\mathrm{S,D}}}{0.9},$$
$$0 \le e_4^{\mathrm{S}} \le 40,$$
$$0 \le e_5^{\mathrm{S}} \le 40,$$
$$20 \le e_6^{\mathrm{S}} \le 40.$$

The optimization variables of the problem are those included in set $\Phi_4^{\mathrm{IE2.7}} = \{p_4^{\mathrm{M}}, p_5^{\mathrm{M}}, p_6^{\mathrm{M}}, p_4^{\mathrm{R}}, p_5^{\mathrm{R}}, p_6^{\mathrm{R}}, p_4^{\mathrm{S,C}}, p_5^{\mathrm{S,C}}, p_6^{\mathrm{S,C}}, p_4^{\mathrm{S,D}}, p_5^{\mathrm{S,D}}, p_6^{\mathrm{S,D}}, u_4^{\mathrm{S}}, u_5^{\mathrm{S}}, u_6^{\mathrm{S}}, e_4^{\mathrm{S}}, e_5^{\mathrm{S}}, e_6^{\mathrm{S}}\}$.

In time period 4, the available wind-power production level is only 5 MW, and the power producer decides to buy 15 MW and charge 20 MW the storage unit, reaching an energy level of 39.33 MWh.

Then, the problem is formulated for the fifth time period using the updated data:

$$\max_{\Phi_5^{\mathrm{IE2.7}}} 45p_5^{\mathrm{M}} + 36p_6^{\mathrm{M}}$$

subject to

$$
\begin{aligned}
& p_5^{\mathrm{M}} = p_5^{\mathrm{R}} + p_5^{\mathrm{S,D}} - p_5^{\mathrm{S,C}}, \\
& p_6^{\mathrm{M}} = p_6^{\mathrm{R}} + p_6^{\mathrm{S,D}} - p_6^{\mathrm{S,C}}, \\
& -40 \le p_5^{\mathrm{M}} \le 40, \\
& -40 \le p_6^{\mathrm{M}} \le 40, \\
& 0 \le p_5^{\mathrm{R}} \le 20, \\
& 0 \le p_6^{\mathrm{R}} \le 23. \\
& u_5^{\mathrm{S}}, u_6^{\mathrm{S}} \in \{0, 1\}, \\
& 0 \le p_5^{\mathrm{S,C}} \le 20 u_5^{\mathrm{S}}, \\
& 0 \le p_6^{\mathrm{S,C}} \le 20 u_6^{\mathrm{S}}, \\
& 0 \le p_5^{\mathrm{S,D}} \le 20\left(1 - u_5^{\mathrm{S}}\right), \\
& 0 \le p_6^{\mathrm{S,D}} \le 20\left(1 - u_6^{\mathrm{S}}\right), \\
& e_5^{\mathrm{S}} = 39.33 + 0.9 p_5^{\mathrm{S,C}} - \frac{p_5^{\mathrm{S,D}}}{0.9}, \\
& e_6^{\mathrm{S}} = e_5^{\mathrm{S}} + 0.9 p_6^{\mathrm{S,C}} - \frac{p_6^{\mathrm{S,D}}}{0.9}, \\
& 0 \le e_5^{\mathrm{S}} \le 40, \\
& 20 \le e_6^{\mathrm{S}} \le 40.
\end{aligned}
$$

The optimization variables of the problem are those included in set $\Phi_5^{\mathrm{IE2.7}} = \{p_5^{\mathrm{M}}, p_6^{\mathrm{M}}, p_5^{\mathrm{R}}, p_6^{\mathrm{R}}, p_5^{\mathrm{S,C}}, p_6^{\mathrm{S,C}}, p_5^{\mathrm{S,D}}, p_6^{\mathrm{S,D}}, u_5^{\mathrm{S}}, u_6^{\mathrm{S}}, e_5^{\mathrm{S}}, e_6^{\mathrm{S}}\}$.

In the fifth time period, the power producer uses 20 MW of available wind-power production and discharges 17.4 MW the storage unit, being 20 MWh the resulting energy level. In this case, the power producer sells 37.4 MW in the market.

Finally, the problem is formulated for the last time period:

$$
\max_{\Phi_6^{\mathrm{IE2.7}}} \; 35 p_6^{\mathrm{M}}
$$

subject to

$$
\begin{aligned}
& p_6^{\mathrm{M}} = p_6^{\mathrm{R}} + p_6^{\mathrm{S,D}} - p_6^{\mathrm{S,C}}, \\
& -40 \le p_6^{\mathrm{M}} \le 40, \\
& 0 \le p_6^{\mathrm{R}} \le 25. \\
& u_6^{\mathrm{S}} \in \{0, 1\}, \\
& 0 \le p_6^{\mathrm{S,C}} \le 20 u_6^{\mathrm{S}}, \\
& 0 \le p_6^{\mathrm{S,D}} \le 20\left(1 - u_6^{\mathrm{S}}\right), \\
& e_6^{\mathrm{S}} = 20 + 0.9 p_6^{\mathrm{S,C}} - \frac{p_6^{\mathrm{S,D}}}{0.9}, \\
& 20 \le e_6^{\mathrm{S}} \le 40.
\end{aligned}
$$

The optimization variables of the problem are those included in set $\Phi_6^{\mathrm{IE2.7}} = \{p_6^{\mathrm{M}}, p_6^{\mathrm{R}}, p_6^{\mathrm{S,C}}, p_6^{\mathrm{S,D}}, u_6^{\mathrm{S}}, e_6^{\mathrm{S}}\}$.

In this case, the power producer sells the available 25 MW of wind-power production to the market.

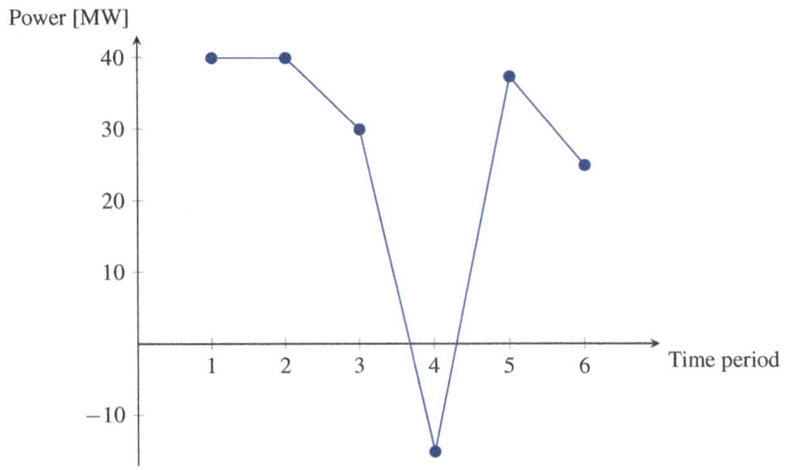

Fig. 2.11 Illustrative Example 2.7: optimal power traded in the market

Table 2.10 Illustrative Example 2.7: results

t	p_t^{R} [MW]	$p_t^{\mathrm{S,C}}$ [MW]	$p_t^{\mathrm{S,D}}$ [MW]	e_t^{S} [MWh]
1	60	20	0	38
2	25	0	15	21.33
3	30	0	0	21.33
4	5	0	20	39.33
5	20	17.4	0	20
6	25	0	0	20

Results are summarized in Fig. 2.11 and Table 2.10, which provide the optimal power traded in the market and the rest of results, respectively. Note that the actual schedules and power traded in the market are different to those obtained in Illustrative Example 2.4.

□

2.9 Summary and Further Reading

This chapter describes basic models for the most common components in a VPP, namely flexible demands, conventional power plants, stochastic renewable generating units, and storage facilities. To do so, the operation of components is described, which is illustrated using a number of examples.

The models described in this chapter will be used in the remaining chapters of this book devoted to operation and expansion planning problems for VPPs.

Some of the models described in this chapter are basic and may be extended by incorporating additional details of the components. In this sense, it may be possible to consider more sophisticated demand response programs [1] or the specific characteristics of demand assets such as a fleet of electric vehicles [9]. Regarding generating units, it is possible to include additional details such as minimum up and down times [8]. In the case of energy storage facilities, it is possible to include battery cycle degradation [4].

The network is modeled in this chapter using a simple dc power flow model that may be extended using a complete ac power flow model [2] or a linearized ac power flow model [3].

2.10 GAMS Codes

This section provides the GAMS codes used to solve some of the illustrative examples of this chapter.

2.10.1 Flexible Demand

An input GAMS [5] code for solving Illustrative Example 2.1 is provided below:

```
SETS
d          'demands'                    /d1*d1/
t          'periods'                    /t1*t6/;
PARAMETER price(t)        'price'
/t1        20
t2         50
t3         35
t4         31
t5         48
t6         25/;
TABLE DEMDATA(d,*)        'demand data'
           RU       RD       Emin     PD0
d1         10       10       120      10;
TABLE PDmin(d,t)   'demand lower bound'
           t1       t2       t3       t4       t5       t6
d1         0        0        10       15       20       20;
TABLE PDmax(d,t)   'demand upper bound'
           t1       t2       t3       t4       t5       t6
d1         20       20       25       30       30       40;
VARIABLES
Z          'objective function'
PD(d,t)    'power consumption of demand d in time period t';
EQUATIONS
cost       'objective function'
pdlim1     'demand limits I'
pdlim2     'demand limits II'
rampup1    'up ramping limit I'
rampup2    'up ramping limit II'
rampdw1    'down ramping limit I'
rampdw2    'down ramping limit II'
minE       'minimum energy consumption';
*
cost..                 z=e=sum((d,t),price(t)*PD(d,t));
pdlim1(d,t)..          PDmin(d,t)=l=PD(d,t);
pdlim2(d,t)..          PD(d,t)=l=PDmax(d,t);
rampup1(d)..           PD(d,'t1')-DEMDATA(d,'PD0')=l=DEMDATA(d,'RU');
rampup2(d,t)$(ORD(t) GT 1).. PD(d,t)-PD(d,t-1)=l=DEMDATA(d,'RU');
rampdw1(d)..           DEMDATA(d,'PD0')-PD(d,'t1')=l=DEMDATA(d,'RD');
rampdw2(d,t)$(ORD(t) GT 1).. PD(d,t-1)-PD(d,t)=l=DEMDATA(d,'RD');
minE(d)..              sum(t,PD(d,t))=g=DEMDATA(d,'Emin');
MODEL IE1 /all/;
OPTION optcr=0;
SOLVE IE1 using LP minimizing z;
DISPLAY z.l, PD.l;
```

2.10.2 Conventional Power Plant

An input GAMS [5] code for solving Illustrative Example 2.3 is provided below:

```
SETS
c          'conventional power plants'       /c1*c1/
t          'periods'                         /t1*t6/;
PARAMETER price(t)      'price'
/t1        20
t2         50
t3         35
t4         31
t5         48
t6         25/;
TABLE GENDATA(c,*)                'generating unit data'
       Pmin     Pmax     RU       RD       RSU      RSD      UC0      P0
c1     10       60       20       15       30       15       0        0;
TABLE GENCOSTDATA(c,*)   'cost data'
         VC       FC       SUC      SD
c1       33       6        120      60;
VARIABLES
Z             'objective function'
PCPP(c,t)     'power production of cpp c in time period t';
BINARY VARIABLES
UC(c,t)       'status of cpp c in time period t'
VSD(c,t)      'shut down status of cpp c in time period t'
VSU(c,t)      'start up status of cpp c in time period t';
EQUATIONS
prof          'objective function'
log1          'logic expressions I'
log2          'logic expressions II'
log3          'logic expressions III'
pclim1        'production limits I'
pclim2        'production limits II'
rampGup1      'up ramping limit I'
rampGup2      'up ramping limit II'
rampGdw1      'down ramping limit I'
rampGdw2      'down ramping limit II';
*
prof..      z=e=sum((c,t),price(t)*PCPP(c,t)-GENCOSTDATA(c,'VC')*
    PCPP(c,t)-GENCOSTDATA(c,'FC')*UC(c,t)-GENCOSTDATA(c,'SUC')*
    VSU(c,t)-GENCOSTDATA(c,'SDC')*VSD(c,t));
log1(c)..  UC(c,'t1')-GENDATA(c,'UC0')=e=VSU(c,'t1')-VSD(c,'t1');
log2(c,t)$(ORD(t) GT 1)..  UC(c,t)-UC(c,t-1)=e=VSU(c,t)-VSD(c,t);
log3(c,t).. VSU(c,t)+VSD(c,t)=l=1;
pclim1(c,t).. GENDATA(c,'Pmin')*UC(c,t)=l=PCPP(c,t);
pclim2(c,t).. PCPP(c,t)=l=GENDATA(c,'Pmax')*UC(c,t);
rampGup1(c).. PCPP(c,'t1')-GENDATA(c,'P0')=l=GENDATA(c,'RU')*
    GENDATA(c,'UC0')+GENDATA(c,'RSU')*VSU(c,'t1');
rampGup2(c,t)$(ORD(t) GT 1).. PCPP(c,t)-PCPP(c,t-1)=l=GENDATA(c,'
    RU')*UC(c,t-1)+GENDATA(c,'RSU')*VSU(c,t);
rampGdw1(c).. GENDATA(c,'P0')-PCPP(c,'t1')=l=GENDATA(c,'RD')*UC(c
    ,'t1')+GENDATA(c,'RSD')*VSD(c,'t1');
rampGdw2(c,t)$(ORD(t) GT 1).. PCPP(c,t-1)-PCPP(c,t)=l=GENDATA(c,'
    RD')*UC(c,t)+GENDATA(c,'RSD')*VSD(c,t);
MODEL IE3 /all/;
OPTION optcr=0;
SOLVE IE3 using MIP maximizing z;
PARAMETER Profit(c,t)    profit per period;
Profit(c,t)=price(t)*PCPP.l(c,t)-GENCOSTDATA(c,'VC')*PCPP.l(c,t)-
    GENCOSTDATA(c,'FC')*UC.l(c,t)-GENCOSTDATA(c,'SUC')*VSU.l(c,t)
    -GENCOSTDATA(c,'SDC')*VSD.l(c,t)
DISPLAY z.l, PCPP.l, UC.l, VSU.l, VSD.l, Profit;
```

2.10.3 Stochastic Renewable and Storage Units

An input GAMS [5] code for solving Illustrative Example 2.4 is provided below:

```
SETS
r        'stochastic renewable units'      /r1*r1/
s        'storage units'                   /s1*s1/
t        'time periods'                    /t1*t6/;
SCALAR PmarketMAX        'maximum power traded in the market'
    /40/;
PARAMETER price(t)      'price'
/t1      20
t2       50
t3       35
t4       31
t5       48
t6       25/;
TABLE PRAV(r,t)   'available renewable production level'
         t1      t2      t3      t4      t5      t6
r1       60      20      25      0       30      40;
TABLE STORAGEDATA(s,*)  'storage data'
         PCmax   PDmax   EMAX    E0      EFFC    EFFD
s1       20      20      40      20      0.9     0.9;
VARIABLES
Z        'objective function'
PM(t)    'power traded in the market';
POSITIVE VARIABLES
ES(s,t)  'energy stored in storage c in time period t'
PR(r,t)  'power production of renewable unit r in time period t'
PSC(s,t) 'charging power of storage c in time period t'
PSD(s,t) 'discharging power of storage c in time period t'
BINARY VARIABLES
US(s,t)  'avoid the simultaneous charging and discharging of
    storage s in time period t';
EQUATIONS
prof     'objective function'
balance  'power balance'
market1  'market limits I'
market2  'market limits II'
RPlim    'renewable production limits'
CPlim    'charging power limits'
DPlim    'discharging power limits'
ESbal1   'energy balance in the storage units I'
ESbal2   'energy balance in the storage units II'
ESlim1   'energy limits in the storage unit I'
ESlim2   'energy limits in the storage unit II';
*
prof..                      z=e=sum(t,price(t)*PM(t));
balance(t)..                PM(t)=e=sum(r,PR(r,t))+sum(s,PSD(s,t)
    -PSC(s,t));
market1(t)..                -PmarketMAX=l=PM(t);
market2(t)..                PM(t)=l=PmarketMAX;
RPlim(r,t)..                PR(r,t)=l=PRAV(r,t);
CPlim(s,t)..                PSC(s,t)=l=STORAGEDATA(s,'PCmax')*US(
    s,t);
DPlim(s,t)..                PSD(s,t)=l=STORAGEDATA(s,'PDmax')*(1-
    US(s,t));
ESbal1(s)..                 ES(s,'t1')=e=STORAGEDATA(s,'E0')+
    STORAGEDATA(s,'EFFC')*PSC(s,'t1')-PSD(s,'t1')/STORAGEDATA(s,'
    EFFD');
```

```
ESbal2(s,t)$(ord(t) GT 1).. ES(s,t)=e=ES(s,t-1)+STORAGEDATA(s,'
    EFFC')*PSC(s,t)-PSD(s,t)/STORAGEDATA(s,'EFFD');
ESlim1(s,t)..              ES(s,t)=l=STORAGEDATA(s,'EMAX');
ESlim2(s,t)..              ES(s,t)=g=STORAGEDATA(s,'E0');
MODEL IE4 /all/;
OPTION optcr=0;
SOLVE IE4 using MIP maximizing z;
DISPLAY z.l, PM.l, PR.l, PSC.l, PSD.l, ES.l;
```

2.10.4 Network Constraints

An input GAMS [5] file for solving Illustrative Example 2.5 is provided below:

```
SETS
l               'lines'                           /l1*l1/
n               'nodes'                           /n1*n2/
r               'stochastic renewable units'      /r1*r1/
s               'storage units'                   /s1*s1/
t               'time periods'                    /t1*t6/
re(l,n)         'receiving-end node of line'      /l1.n2/
se(l,n)         'sending-end node of line'        /l1.n1/
mapR(r,n)       'location of renewable units'     /r1.n1/
mapS(s,n)       'location of storage units'       /s1.n1/
mapMG(n)        'location of main grid'           /n2/
ref(n)          'reference node'                  /n1/;
SCALAR PmarketMAX     'maximum power traded in the market' /40/;
PARAMETER price(t)    'price'
/t1       20
t2        50
t3        35
t4        31
t5        48
t6        25/;
TABLE PRAV(r,t)    'available renewable production level'
          t1      t2      t3      t4      t5      t6
r1        60      20      25      0       30      40;
TABLE STORAGEDATA(s,*)  'storage data'
          PCmax   PDmax   EMAX    E0      EFFC    EFFD
s1        20      20      40      20      0.9     0.9;
TABLE LINEDATA(l,*)     'line data'
          X       PLmax
l1        0.2     30;
VARIABLES
Z         'objective function'
PM(t)     'power traded in the market'
PL(l,t)   'power flow through line l in time period t'
ANG(n,t)  'voltage angle at node n in time period t';
POSITIVE VARIABLES
ES(s,t)   'energy stored in storage c in time period t'
PR(r,t)   'power production of renewable unit r in time period t'
PSC(s,t)  'charging power of storage c in time period t'
PSD(s,t)  'discharging power of storage c in time period t'
BINARY VARIABLES
US(s,t)   'avoid the simultaneous charging and discharging of
    storage s in time period t';
EQUATIONS
prof        'objective function'
balanceN1   'power balance at nodes (main grid connection)'
balanceN2   'power balance at nodes (remaining nodes)'
pflow       'power flows through lines'
pflowlim1   'power flow limits I'
pflowlim2   'power flow limits II'
refN        'reference node'
market1     'market limits I'
market2     'market limits II'
RPlim       'renewable production limits'
```

```
CPlim      'charging power limits'
DPlim      'discharging power limits'
ESbal1     'energy balance in the storage units I'
ESbal2     'energy balance in the storage units II'
ESlim1     'energy limits in the storage unit I'
ESlim2     'energy limits in the storage unit II';
*
prof..                           z=e=sum(t,price(t)*PM(t));
balanceN1(n,t)$mapMG(n)..    SUM(r$mapR(r,n),PR(r,t))+sum(s$mapS(s
    ,n),PSD(s,t))-SUM(l$se(l,n),PL(l,t))+SUM(l$re(l,n),PL(l,t))=e
    =PM(t)+sum(s$mapS(s,n),PSC(s,t));
balanceN2(n,t)$(not mapMG(n))..  SUM(r$mapR(r,n),PR(r,t))+sum(
    s$mapS(s,n),PSD(s,t))-SUM(l$se(l,n),PL(l,t))+SUM(l$re(l,n),PL
    (l,t))=e=sum(s$mapS(s,n),PSC(s,t));
pflow(l,t)..                 PL(l,t)=e=LINEDATA(l,'X')*(sum(n$se(l
    ,n),ANG(n,t))-sum(n$re(l,n),ANG(n,t)));
pflowlim1(l,t)..             -LINEDATA(l,'PLmax')=l=PL(l,t);
pflowlim2(l,t)..             PL(l,t)=l=LINEDATA(l,'PLmax');
refN(n,t)$ref(n)..           ANG(n,t)=e=0;
market1(t)..                 -PmarketMAX=l=PM(t);
market2(t)..                 PM(t)=l=PmarketMAX;
RPlim(r,t)..                 PR(r,t)=l=PRAV(r,t);
CPlim(s,t)..                 PSC(s,t)=l=STORAGEDATA(s,'PCmax')*US(
    s,t);
DPlim(s,t)..                 PSD(s,t)=l=STORAGEDATA(s,'PDmax')*(1-
    US(s,t));
ESbal1(s)..                  ES(s,'t1')=e=STORAGEDATA(s,'E0')+
    STORAGEDATA(s,'EFFC')*PSC(s,'t1')-PSD(s,'t1')/STORAGEDATA(s,'
    EFFD');
ESbal2(s,t)$(ord(t) GT 1).. ES(s,t)=e=ES(s,t-1)+STORAGEDATA(s,'
    EFFC')*PSC(s,t)-PSD(s,t)/STORAGEDATA(s,'EFFD');
ESlim1(s,t)..                ES(s,t)=l=STORAGEDATA(s,'EMAX');
ESlim2(s,t)..                ES(s,t)=g=STORAGEDATA(s,'E0');
MODEL IE5 /all/;
OPTION optcr=0;
SOLVE IE5 using MIP maximizing z;
DISPLAY z.l, PM.l, PR.l, PSC.l, PSD.l, ES.l, PL.l, ANG.l;
```

References

1. Chen, Z., Wu, L., Fu, Y.: Real-time price-based demand response management for residential appliances via stochastic optimization and robust optimization. IEEE Trans. Smart Grid. **3**(4), 1822–1831 (2012)
2. Conejo, A.J., Baringo, L.: Power System Operations. Springer, Switzerland (2018)
3. Farivar, M., Low, S.J.: Branch flow model: relaxations and convexification: Part I. IEEE Trans. Power Syst. **28**(3), 2554–2564 (2012)
4. Foggo, B., Yu, N.: Improved battery storage valuation through degradation reduction. IEEE Trans. Smart Grid. **9**(6), 5721–5732 (2018)
5. GAMS.: Available at www.gams.com/ (2020)
6. ILOG CPLEX.: Available at www.ilog.com/products/cplex/ (2020)
7. Morales, J.M., Conejo, A.J., Madsen, H., Pinson, P., Zugno, M.: Integrating Renewables in Electricity Markets – Operational Problems. Springer, New York, US (2014)
8. Ostrowski, J., Anjos, M.F., Vannelli, A.: Tight mixed integer linear programming formulations for the unit commitment problem. IEEE Trans. Power Syst. **27**(1), 39–46 (2012)
9. Rajakaruna, S., Shahnia, F., Ghosh, A.: Plug In Electric Vehicles in Smart Grids. Springer, Singapore (2015)
10. Sioshansi, R., Conejo, A.J.: Optimization in Engineering. Models and Algorithms. Springer, New York, US (2017)

Chapter 3
Optimal Scheduling of a Risk-Neutral Virtual Power Plant in Energy Markets

This chapter addresses the scheduling problem of a virtual power plant (VPP) in different energy markets. This problem is solved to determine the optimal energy traded by the VPP in these markets, as well as the production and consumption of the energy resources that comprise the VPP with the aim of maximizing the profit of the VPP while complying with the technical constraints of these generation and consumption assets. Therefore, the scheduling problem is essential to achieve an efficient and secure operation of the VPP. Due to uncertainties in the market prices and the available production levels of stochastic renewable generating units, this is a decision-making problem under uncertainty. To accomplish these tasks, the scheduling problem is formulated through two practical options based on deterministic and stochastic approaches. The deterministic approach models the uncertainties involved in the scheduling problem by single-point forecasts, which are used as a single scenario. On the other hand, the stochastic approach models the uncertainties through a number of scenarios. A risk-neutral VPP that is indifferent to financial risks resulting from uncertainties is considered.

3.1 Introduction

The self-scheduling of a virtual power plant (VPP) in energy markets refers to the problem of determining the optimal energy traded in these markets, e.g., the day-ahead (DA) and the real-time (RT) energy markets, as well as the optimal dispatch of the generation and consumption assets that comprise the VPP. As the VPP includes both generation and consumption units, the VPP can both buy from and sell to the energy markets at suitable times.

The self-scheduling decisions are made with the aim of maximizing the profit of the VPP. To do so, the VPP usually participates in a sequence of energy markets, including the DA energy market for which decisions are made one day in advance and the RT energy market that is cleared some minutes before power delivery [2, 14, 17].

Generally, a significant part of the energy is traded by the VPP in the DA energy market, while the VPP utilizes the opportunity of participating in the RT energy market to compensate the possible deviations from the schedules in the DA energy market. However, a small-size VPP may prefer to trade energy only in the RT energy market [16, 18]. Additionally, as the VPP share increases, VPP owners may also participate in long-term markets such as futures markets, which are generally characterized with more stable prices compared to pool markets.

In practice, the VPP seeks to optimize the integrated operation of the energy resources including demands, conventional power plants, storage units, and stochastic renewable production facilities. These units are typically characterized by diverse techno-economic features. Generally, each unit seeks to achieve its own economic objective subject to its own technical constraints. Moreover, the units in the VPP are interconnected within a network that imposes limitations on power balance, power flows through lines, voltage angles, and power injections at nodes connected to the main grid. These limitations, generally known as externalities, couple the constraints of the units, and thus influence the schedules of the VPP [20]. Thus, the integrated scheduling of the VPP as a single entity is a complex decision-making problem due to the diverse techno-economic characteristics of the units and the externalities imposed by the network. Additionally, note that this problem can be further complicated in the case of a VPP that includes units working with multiple energy carriers such as electricity, natural gas, and heat [10].

At the time of solving the self-scheduling problem of a VPP, there are a number of uncertainties, e.g., the market prices and the available production levels of stochastic renewable generating units. Therefore, the scheduling problem of a VPP is a decision-making problem under uncertainty. These uncertainties may have a great impact on the profit achieved by the VPP, as well as on the technical operation of the different units. Thus, to obtain efficient and practical schedules, it is essential to

L. Baringo, M. Rahimiyan, *Virtual Power Plants and Electricity Markets*, https://doi.org/10.1007/978-3-030-47602-1_3

accurately handle the uncertainties. To achieve this goal, there are alternative options that can be generally categorized into deterministic, stochastic programming, and robust optimization approaches.

The deterministic approach determines the optimal solution under a single future realization of the uncertainties. This uncertainty modeling is quite simple and can be easily implemented in a short computation time. However, results are generally inaccurate.

Alternatively, the scenario-based stochastic programming approach models the uncertainty in the decision-making process using a finite number of future realizations of the uncertain parameters. This modeling framework generally results in more accurate results at the expense of a higher computation cost [7]. Using a stochastic programming approach, the VPP aims at maximizing its expected profit over all considered scenario realizations while complying with the technical constraints for all scenarios. In most cases, the stochastic programming approach outperforms the deterministic approach if the decision maker is equipped with advanced forecasting tools. This can be validated using an out-of-sample analysis [15].

In those cases with highly intensive variability of the uncertainties, the scenario-based uncertainty model may be not accurate enough, and even the deterministic approach may yield higher economic performance than the stochastic programming approach. As a trade-off between the accuracy of the uncertainty model and the computation time, it is possible to implement a robust optimization approach, which uses prespecified uncertainty sets to represent the uncertainties [3]. The robust optimization approach seeks to optimize the economic objective under the worst realization of the uncertainties. This approach guarantees also that the solution is feasible provided that the uncertain parameters take values within the considered uncertainty sets. This is not the case of the solution achieved using stochastic programming, whose feasibility is only guaranteed for the scenarios considered.

The optimal scheduling of the VPP in the energy markets has been analyzed in the technical literature using stochastic programming [1, 8, 14] and robust optimization approaches [12, 16, 17]. To increase the performance of the stochastic programming and robust optimization approaches, these two approaches can be combined as the stochastic robust optimization approach proposed in [18] and extended to develop an adaptive robust optimization approach in [2].

In this chapter, a risk-neutral VPP that is indifferent with the financial risk associated with its scheduling decisions is considered. Thus, this chapter focuses on the deterministic and stochastic programming approaches without any risk control strategy. Risk-averse VPPs and the use of robust optimization approaches are analyzed in Chap. 4.

With the evolution of the smart grid technology, the information flow and the decision-making process have been experiencing rapid changes. On one hand, the decision-making process is conditioned by the available information that can be updated real-time. On the other hand, the decisions can be implemented near real-time. Under this framework, the scheduling decisions can be made through two different centralized and decentralized structures. In a centralized structure, an energy management system (EMS) in charge of the VPP receives the required information from the units and centrally decides on the schedules of each unit, as well as on the schedule of the VPP as a whole [16]. On the contrary, a decentralized structure constitutes a hierarchical framework [19]. In a *local* management level, each unit is equipped with a *local* EMS deciding on its schedule and sending these decisions to the *central* EMS. Then, in a *central* management level, the *central* EMS coordinates the *local* decisions to prevent any constraint violation and to increase the total gain by making a trade-off among economic objectives of the units in the VPP. In this chapter, the first structure is chosen for the scheduling of the VPP.

The remaining parts of this chapter are organized as follows. Section 3.2 describes the scheduling problem of the VPP, as well as the information flow and decision-making process used for making decisions. Section 3.3 formulates the scheduling problem of the risk-neutral VPP in the DA and RT energy markets using a deterministic model. This model is extended in Sect. 3.4 using a stochastic programming model that considers the potential impact of a finite number of scenario realizations on the scheduling problem of the risk-neutral VPP in the DA and RT energy markets. In these two sections, the working of the deterministic and stochastic programming models is clarified through a number of illustrative examples. Section 3.5 extends the scheduling problem of a risk-neutral VPP that participates also in futures markets. Section 3.6 summarizes the models and the outcomes drawn from this chapter, and suggests further reading on the subject of this chapter. Finally, Sect. 3.7 provides the GAMS codes used for solving the illustrative examples.

3.2 Problem Description

This section describes the main characteristics of the scheduling problem of a risk-neutral VPP in energy markets.

3.2.1 Notation

The main notation used in this chapter is provided below for quick reference. Other symbols are defined as needed throughout the chapter. If symbol $\sim$ appears over any parameter/variable below, it denotes the actual value of this parameter/variable.

3.2.1.1 Indexes

The indexes of the scheduling problem are:

c	Conventional power plants.
d	Demands.
ℓ	Network lines.
n	Network nodes.
r	Stochastic renewable generating units.
$r(\ell)$	Receiving-end node of network line ℓ.
s	Storage units.
$s(\ell)$	Sending-end node of network line ℓ.
t	Time periods.

3.2.1.2 Sets

The sets of the scheduling problem are:

Ω^{C}	Conventional power plants.
Ω^{C}_n	Conventional power plants located at node n.
Ω^{D}	Demands.
Ω^{D}_n	Demands located at node n.
Ω^{L}	Network lines.
Ω^{M}	Network nodes connected to the main grid.
Ω^{N}	Network nodes.
Ω^{R}	Stochastic renewable generating units.
Ω^{R}_n	Stochastic renewable generating units located at node n.
Ω^{S}	Storage units.
Ω^{S}_n	Storage units located at node n.
Ω^{T}	Time periods.
Π^{DA}	Scenario realizations of uncertain parameters between the closure of the DA energy market and the dispatch of the units in the VPP.
Π^{RT}	Scenario realizations of uncertain parameters between the closure of the RT energy market and the dispatch of the units in the VPP.

3.2.1.3 Parameters

The parameters of the scheduling problem are:

B_ℓ	Susceptance of line ℓ [S].
$\underline{C}^{\mathrm{C}}_c$	Shut-down cost of conventional power plant c [\$].
$\overline{C}^{\mathrm{C}}_c$	Start-up cost of conventional power plant c [\$].
$C^{\mathrm{C,V}}_c$	Variable production cost of conventional power plant c [\$/MWh].
$\underline{E}^{\mathrm{D}}_d$	Minimum energy consumption of demand d throughout the planning horizon [MWh].
$\underline{E}^{\mathrm{S}}_{st}$	Lower bound of the energy stored in storage unit s in time period t [MWh].
$\overline{E}^{\mathrm{S}}_{st}$	Upper bound of the energy stored in storage unit s in time period t [MWh].
$\overline{P}^{\mathrm{C}}_c$	Maximum power production of conventional power plant c [MW].

$\underline{P}^{\mathrm{C}}_{c}$ Minimum power production of conventional power plant c [MW].
$\underline{P}^{\mathrm{D}}_{dt}$ Lower bound of the power consumption of demand d in time period t [MW].
$\overline{P}^{\mathrm{D}}_{dt}$ Upper bound of the power consumption of demand d in time period t [MW].
$\overline{P}^{\mathrm{L}}_{\ell}$ Capacity of line ℓ [MW].
$\overline{P}^{\mathrm{M}}_{n}$ Maximum power that can be traded with the grid at node n [MW].
$P^{\mathrm{R,A}}_{rt}$ Available power production level of stochastic renewable generating unit r in time period t [MW].
$\overline{P}^{\mathrm{S,C}}_{s}$ Charging capacity of storage unit s [MW].
$\overline{P}^{\mathrm{S,D}}_{s}$ Discharging capacity of storage unit s [MW].
$\overline{P}^{\mathrm{V}}$ Power capacity of generation assets (positive) in the VPP [MW].
$\underline{P}^{\mathrm{V}}$ Power capacity of consumption assets (negative) in the VPP [MW].
$R^{\mathrm{C,D}}_{c}$ Down ramping limit of conventional power plant c [MW/h].
$R^{\mathrm{C,U}}_{c}$ Up ramping limit of conventional power plant c [MW/h].
$R^{\mathrm{D,D}}_{d}$ Down ramping limit of demand d [MW/h].
$R^{\mathrm{D,U}}_{d}$ Up ramping limit of demand d [MW/h].
U^{D}_{dt} Utility of demand d from its energy consumption in time period t [\$/MWh].
Δt Duration of time periods [h].
$\eta^{\mathrm{S,C}}_{s}$ Charging efficiency of storage unit s [%].
$\eta^{\mathrm{S,D}}_{s}$ Discharging efficiency of storage unit s [%].
$\lambda^{\mathrm{DA}}_{t}$ DA energy market price in time period t [\$/MWh].
$\lambda^{\mathrm{RT}}_{t}$ RT energy market price in time period t [\$/MWh].

3.2.1.4 Variables

The variables of the scheduling problem are:

$\overline{c}^{\mathrm{C}}_{ct}$ Auxiliary variable used to linearize the start-up cost of conventional power plant c in time period t [\$].
$\underline{c}^{\mathrm{C}}_{ct}$ Auxiliary variable used to linearize the shut-down cost of conventional power plant c in time period t [\$].
$e^{\mathrm{S,C}}_{st}$ Energy stored in storage unit s in time period t [MWh].
p^{C}_{ct} Power generation of conventional power plant c in time period t [MW].
p^{D}_{dt} Power consumption of demand d in time period t [MW].
p^{DA}_{t} Power scheduled to be bought from/sold to (if negative/positive) the DA energy market in time period t [MW].
$p^{\mathrm{L}}_{\ell t}$ Power flow through line ℓ in time period t [MW].
p^{M}_{nt} Power scheduled to be bought from/sold to (if negative/positive) the DA and RT energy markets at node n in time period t [MW].
p^{R}_{rt} Production of stochastic renewable generating unit r in time period t [MW].
p^{RT}_{t} Power scheduled to be bought from/sold to (if negative/positive) the RT energy market in time period t [MW].
p^{S}_{st} Net power level of storage unit s in time period t [MW].
$p^{\mathrm{S,C}}_{st}$ Charging power level of storage unit s in time period t [MW].
$p^{\mathrm{S,D}}_{st}$ Discharging power level of storage unit s in time period t [MW].
u^{C}_{ct} Binary variable that is equal to 1 if conventional power plant c is generating electricity in time period t, being 0 otherwise.
δ_{nt} Voltage angle at node n in time period t [rad].

3.2.2 Scheduling Problem

The scheduling of the VPP is a decision-making problem that determines the optimal dispatch of the units in the VPP, as well as the power traded in the energy markets. The VPP under consideration includes a number of units with different technical constraints and economic objectives as described below:

1. Demands generally have a minimum energy consumption throughout a planning horizon, while their power consumption is limited by boundary values and ramping constraints. Additionally, each demand is willing to achieve maximum utility from energy consumption. These features form their operational flexibility that should be appropriately utilized in the scheduling of the VPP.
2. The scheduling of conventional power plants can be seen as a unit-commitment problem in which both the commitment status and the power production are determined. The power bounds and ramping constraints are important technical limitations of conventional power plants. Additionally, the fuel cost of these plants influences the financial objective of the scheduling problem.
3. The storage units are scheduled to produce and store energy at suitable times. The level of energy stored, as well as the charging and discharging power levels are bounded by the corresponding capacities. As the VPP is the owner of the storage units, their operation cost can be considered negligible.
4. The power produced by the stochastic renewable production facilities is limited by the available power production level. The operation cost of the stochastic renewable production facilities can be also considered negligible.

All the units that comprise the VPP may be interconnected within an electricity network, which is connected to the main grid at different nodes. Thus, the power productions and consumptions of the units in the VPP are also influenced by power flow constraints. Therefore, the scheduling of the VPP is a complex operation problem.

On the other hand, the scheduling of the VPP is a decision-making problem under uncertainty. The uncertainties involved in this problem are the available production levels of stochastic renewable generating units, as well as the prices of the energy markets. The available stochastic production level is imposed as an upper bound on the power production. Note that if this uncertainty is ignored, the schedules may not be applicable in practice. Additionally, the VPP buys/sells energy from/to the energy markets and pays/is paid for this at the corresponding market prices. Therefore, the uncertainty in the prices influences the profit achieved by the VPP. Ignoring this uncertainty may result in an additional cost in the scheduling of the VPP or in a reduction in its profit.

The VPP can participate strategically in different energy markets. For example, the VPP can decide the offering and bidding strategies in the DA energy market, one day in advance. Then, it can participate in the RT energy market time by time, if needed, deciding corrective actions on the energy scheduled in the DA market. Therefore, it is necessary to consider the potential ability of the units in the VPP for deriving the best offering and bidding strategies in the DA and RT energy markets.

The EMS in charge of the VPP is responsible of submitting the offering and bidding decisions of the VPP to the market operator, as well as of scheduling the units in the VPP. As VPPs can be equipped with smart grid technologies, it is important to take advantage of this technology to establish a bidirectional communication between the EMS and the units in the VPP, as well as between the EMS and the market operator near real-time:

1. The information related to the technical and economic parameters of the units in the VPP is collected and sent to the EMS through the communication infrastructure.
2. The EMS also informs the units about their schedules.
3. Additionally, the EMS submits the offering and bidding decisions of the VPP in the energy markets to the market operator, and receives the outcomes of the energy markets such as the prices and the schedules of the VPP.

3.2.3 Decision and Information Flow

The EMS in charge of the VPP makes decisions for participating in energy markets in two stages, namely the scheduling decisions in the DA and the RT energy markets.

In the first stage, the EMS decides the offering and bidding strategies of the VPP in the DA energy market. These decisions are made one day in advance and for all time periods of the following day. Then, once the DA energy market is cleared, the EMS is informed about the schedules of the VPP and the prices in this energy market. The decision and information flow for the scheduling of the VPP in the DA energy market are summarized below:

1. The EMS knows the schedules of the VPP and the prices in the DA energy market for the current day D. Additionally, the schedules of the VPP and the prices in the RT energy market are known for the time periods prior to the current one of day D.
2. The EMS receives the required information from the units in the VPP for day $D + 1$.

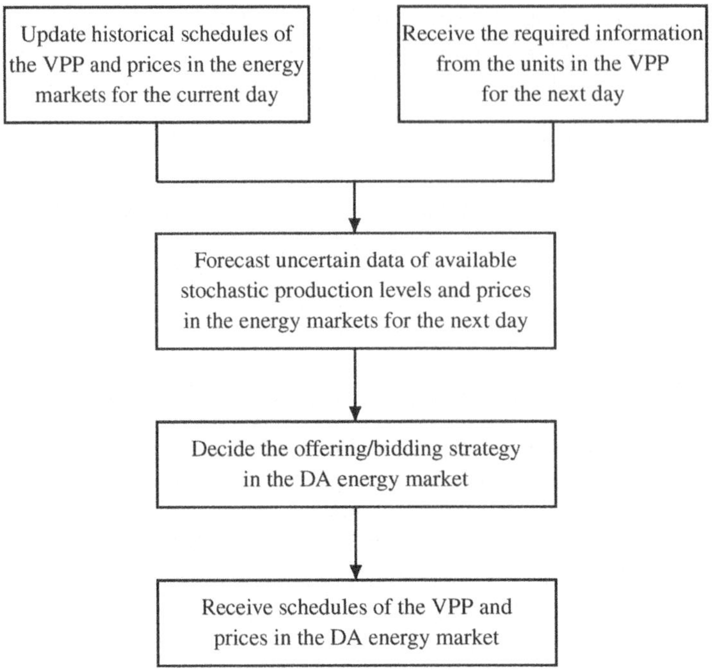

Fig. 3.1 Decision and information flow for the scheduling of the VPP in the DA energy market

3. The EMS forecasts the uncertain data related to the available stochastic renewable production levels and the prices in the energy markets for all time periods of the following day $D + 1$.
4. The EMS decides the offering and bidding strategies of the VPP in the DA energy market for day $D + 1$ and sends these decisions to the market operator.
5. The EMS receives the schedules of the VPP and the prices in the DA energy market for day $D + 1$.

The above procedure is clarified through the flowchart depicted in Fig. 3.1.

In the second stage, the EMS submits the offering and bidding decisions in the RT energy market time by time. When the RT energy market is cleared in each time period, the EMS receives the scheduled power and the price. At this time, the EMS dispatches the units in the VPP. These dispatch decisions include the level of power loads, the commitment status and the power production of the conventional power plants, the power produced by/transferred to the storage units, and the power produced by the stochastic renewable production facilities. The decision and information flow for the scheduling of the VPP in the RT energy market is summarized below:

1. The EMS knows the schedules of the VPP and the prices in the energy markets cleared in the previous time periods.
2. Set the time period $\tau = 1$.
3. The EMS receives updated information of the units in the VPP some minutes prior to the time period τ.
4. The EMS forecasts the uncertain data related to the available stochastic renewable production levels and the prices in the RT energy market for the remaining time periods of the current day. It is possible to assume that the EMS knows the available power production level of the stochastic renewable generating units in time period τ with almost certainty.
5. The EMS submits the offering and bidding decisions in the RT energy market for time period τ.
6. The EMS is informed about the schedules of the VPP and the price in the RT energy market for time period τ.
7. The EMS dispatches the units in the VPP for time period τ.
8. Update $\tau = \tau + 1$. If $\tau \leq |\Omega^{\mathrm{T}}|$, go to step 3. Otherwise, current day is finished.

The above procedure is presented through a flowchart in Fig. 3.2.

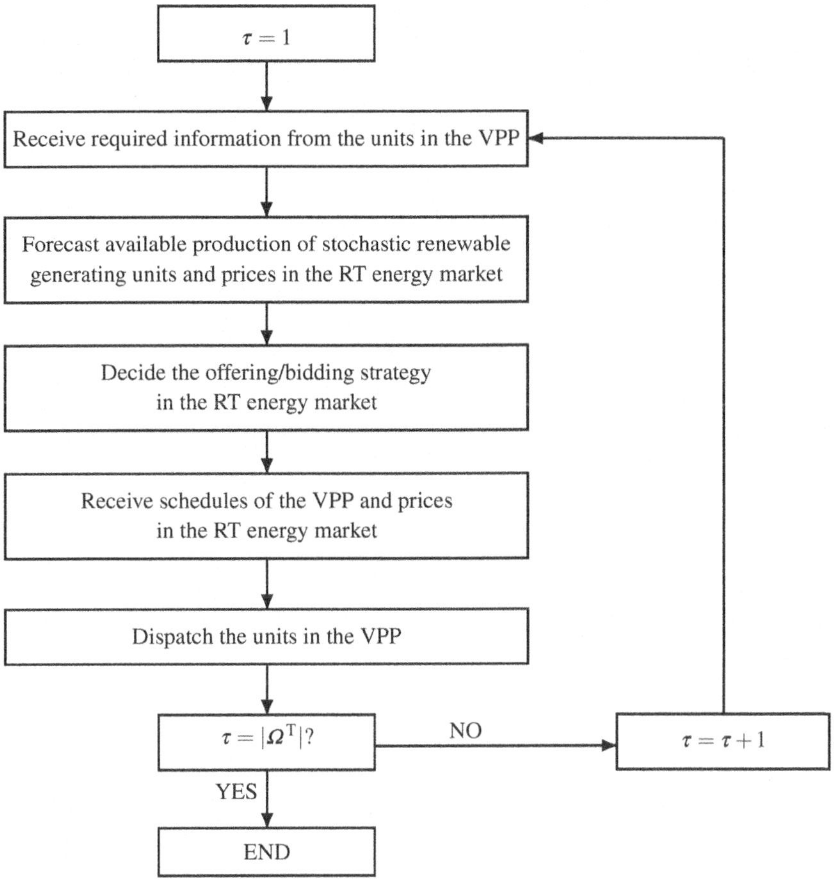

Fig. 3.2 Decision and information flow for the scheduling of the VPP in the RT energy market

3.3 Deterministic Approach

This section provides the formulation of the scheduling problem of a VPP in energy markets using a deterministic approach.

3.3.1 Deterministic Scheduling in the Day-Ahead Energy Market

The deterministic scheduling problem of a risk-neutral VPP in the DA energy market can be formulated as the following mixed-integer linear programming (MILP) problem:

$$\max_{\Phi^{\mathrm{DA}}} \sum_{t \in \Omega^{\mathrm{T}}} \Bigg[\lambda_t^{\mathrm{DA}} p_t^{\mathrm{DA}} \Delta t + \lambda_t^{\mathrm{RT}} p_t^{\mathrm{RT}} \Delta t + \sum_{d \in \Omega^{\mathrm{D}}} U_{dt}^{\mathrm{D}} p_{dt}^{\mathrm{D}} \Delta t - \sum_{c \in \Omega^{\mathrm{C}}} \left(C_c^{\mathrm{C,V}} p_{ct}^{\mathrm{C}} \Delta t + \overline{c}_{ct}^{\mathrm{C}} + \underline{c}_{ct}^{\mathrm{C}} \right) \Bigg] \tag{3.1a}$$

subject to

$$\sum_{r \in \Omega_n^{\mathrm{R}}} p_{rt}^{\mathrm{R}} + \sum_{s \in \Omega_n^{\mathrm{S}}} p_{st}^{\mathrm{S}} - \sum_{\ell | s(\ell)=n} p_{\ell t}^{\mathrm{L}} + \sum_{\ell | r(\ell)=n} p_{\ell t}^{\mathrm{L}} = p_{nt}^{\mathrm{M}} + \sum_{d \in \Omega_n^{\mathrm{D}}} p_{dt}^{\mathrm{D}}, \quad \forall n \in \Omega^{\mathrm{M}}, \forall t \in \Omega^{\mathrm{T}}, \tag{3.1b}$$

$$\sum_{r\in\Omega_n^{\mathrm{R}}} p_{rt}^{\mathrm{R}} + \sum_{s\in\Omega_n^{\mathrm{S}}} p_{st}^{\mathrm{S}} - \sum_{\ell|s(\ell)=n} p_{\ell t}^{\mathrm{L}} + \sum_{\ell|r(\ell)=n} p_{\ell t}^{\mathrm{L}}$$

$$= \sum_{d\in\Omega_n^{\mathrm{D}}} p_{dt}^{\mathrm{D}}, \quad \forall n \in \Omega^{\mathrm{N}} \setminus n \in \Omega^{\mathrm{M}}, \forall t \in \Omega^{\mathrm{T}}, \tag{3.1c}$$

$$p_t^{\mathrm{DA}} + p_t^{\mathrm{RT}} = \sum_{n\in\Omega^{\mathrm{M}}} p_{nt}^{\mathrm{M}}, \quad \forall t \in \Omega^{\mathrm{T}}, \tag{3.1d}$$

$$p_{\ell t}^{\mathrm{L}} = B_\ell(\delta_{s(\ell)t} - \delta_{r(\ell)t}), \quad \forall \ell \in \Omega^{\mathrm{L}}, \forall t \in \Omega^{\mathrm{T}}, \tag{3.1e}$$

$$-\overline{P}_\ell^{\mathrm{L}} \le p_{\ell t}^{\mathrm{L}} \le \overline{P}_\ell^{\mathrm{L}}, \quad \forall \ell \in \Omega^{\mathrm{L}}, \forall t \in \Omega^{\mathrm{T}} \tag{3.1f}$$

$$\delta_{nt} = 0, \quad n : \text{ref.}, \forall t \in \Omega^{\mathrm{T}}, \tag{3.1g}$$

$$-\pi \le \delta_{nt} \le \pi, \quad \forall n \in \Omega^{\mathrm{N}} \setminus n : \text{ref.}, \forall t \in \Omega^{\mathrm{T}}, \tag{3.1h}$$

$$\underline{P}^{\mathrm{V}} \le p_t^{\mathrm{DA}} \le \overline{P}^{\mathrm{V}}, \quad \forall t \in \Omega^{\mathrm{T}}, \tag{3.1i}$$

$$\underline{P}^{\mathrm{V}} \le p_t^{\mathrm{DA}} + p_t^{\mathrm{RT}} \le \overline{P}^{\mathrm{V}}, \quad \forall t \in \Omega^{\mathrm{T}}, \tag{3.1j}$$

$$-\overline{P}_n^{\mathrm{M}} \le p_{nt}^{\mathrm{M}} \le \overline{P}_n^{\mathrm{M}}, \quad \forall n \in \Omega^{\mathrm{M}}, \forall t \in \Omega^{\mathrm{T}}, \tag{3.1k}$$

$$\sum_{t\in\Omega^{\mathrm{T}}} p_{dt}^{\mathrm{D}} \Delta t \ge \underline{E}_d^{\mathrm{D}}, \quad \forall d \in \Omega^{\mathrm{D}}, \tag{3.1l}$$

$$\underline{P}_{dt}^{\mathrm{D}} \le p_{dt}^{\mathrm{D}} \le \overline{P}_{dt}^{\mathrm{D}}, \quad \forall d \in \Omega^{\mathrm{D}}, \forall t \in \Omega^{\mathrm{T}}, \tag{3.1m}$$

$$-R_d^{\mathrm{D,D}} \Delta t \le p_{dt}^{\mathrm{D}} - p_{d(t-1)}^{\mathrm{D}} \le R_d^{\mathrm{D,U}} \Delta t, \quad \forall d \in \Omega^{\mathrm{D}}, \forall t \in \Omega^{\mathrm{T}}, \tag{3.1n}$$

$$\underline{P}_c^{\mathrm{C}} u_{ct}^{\mathrm{C}} \le p_{ct}^{\mathrm{C}} \le \overline{P}_c^{\mathrm{C}} u_{ct}^{\mathrm{C}}, \quad \forall c \in \Omega^{\mathrm{C}}, \forall t \in \Omega^{\mathrm{T}}, \tag{3.1o}$$

$$-R_c^{\mathrm{C,D}} \Delta t \le p_{ct}^{\mathrm{C}} - p_{c(t-1)}^{\mathrm{C}} \le R_c^{\mathrm{C,U}} \Delta t, \quad \forall c \in \Omega^{\mathrm{C}}, \forall t \in \Omega^{\mathrm{T}}, \tag{3.1p}$$

$$\overline{C}_c^{\mathrm{C}} (u_{ct}^{\mathrm{C}} - u_{c(t-1)}^{\mathrm{C}}) \le \overline{c}_{ct}^{\mathrm{C}}, \quad \forall c \in \Omega^{\mathrm{C}}, \forall t \in \Omega^{\mathrm{T}}, \tag{3.1q}$$

$$0 \le \overline{c}_{ct}^{\mathrm{C}}, \quad \forall c \in \Omega^{\mathrm{C}}, \forall t \in \Omega^{\mathrm{T}}, \tag{3.1r}$$

$$\underline{C}_c^{\mathrm{C}} (u_{c(t-1)}^{\mathrm{C}} - u_{ct}^{\mathrm{C}}) \le \underline{c}_{ct}^{\mathrm{C}}, \quad \forall c \in \Omega^{\mathrm{C}}, \forall t \in \Omega^{\mathrm{T}}, \tag{3.1s}$$

$$0 \le \underline{c}_{ct}^{\mathrm{C}}, \quad \forall c \in \Omega^{\mathrm{C}}, \forall t \in \Omega^{\mathrm{T}}, \tag{3.1t}$$

$$e_{st}^{\mathrm{S}} = e_{s(t-1)}^{\mathrm{S}} + \eta_s^{\mathrm{S,C}} p_{st}^{\mathrm{S,C}} \Delta t - \frac{p_{st}^{\mathrm{S,D}}}{\eta_s^{\mathrm{S,D}}} \Delta t, \quad \forall s \in \Omega^{\mathrm{S}}, \forall t \in \Omega^{\mathrm{T}}, \tag{3.1u}$$

$$\underline{E}_s^{\mathrm{S}} \le e_{st}^{\mathrm{S}} \le \overline{E}_s^{\mathrm{S}}, \quad \forall s \in \Omega^{\mathrm{S}}, \forall t \in \Omega^{\mathrm{T}}, \tag{3.1v}$$

$$0 \le p_{st}^{\mathrm{S,D}} \le \overline{P}_s^{\mathrm{S,D}}, \quad \forall s \in \Omega^{\mathrm{S}}, \forall t \in \Omega^{\mathrm{T}}, \tag{3.1w}$$

$$0 \le p_{st}^{\mathrm{S,C}} \le \overline{P}_s^{\mathrm{S,C}}, \quad \forall s \in \Omega^{\mathrm{S}}, \forall t \in \Omega^{\mathrm{T}}, \tag{3.1x}$$

$$p_{st}^{\mathrm{S}} = p_{st}^{\mathrm{S,D}} - p_{st}^{\mathrm{S,C}}, \quad \forall s \in \Omega^{\mathrm{S}}, \forall t \in \Omega^{\mathrm{T}}, \tag{3.1y}$$

$$0 \le p_{rt}^{\mathrm{R}} \le P_{rt}^{\mathrm{R,A}}, \quad \forall r \in \Omega^{\mathrm{R}}, \forall t \in \Omega^{\mathrm{T}}, \tag{3.1z}$$

where variables in set $\Phi^{\mathrm{DA}} = \Big\{ p_t^{\mathrm{DA}}, p_t^{\mathrm{RT}}, \forall t \in \Omega^{\mathrm{T}}; \overline{c}_{ct}^{\mathrm{C}}, \underline{c}_{ct}^{\mathrm{C}}, p_{ct}^{\mathrm{C}}, u_{ct}^{\mathrm{C}}, \forall c \in \Omega^{\mathrm{C}}, \forall t \in \Omega^{\mathrm{T}}; p_{dt}^{\mathrm{D}}, \forall d \in \Omega^{\mathrm{D}}, \forall t \in \Omega^{\mathrm{T}}; p_{rt}^{\mathrm{R}}, \forall r \in \Omega^{\mathrm{R}}, \forall t \in \Omega^{\mathrm{T}}; e_{st}^{\mathrm{S}}, p_{st}^{\mathrm{S}}, p_{st}^{\mathrm{S,C}}, p_{st}^{\mathrm{S,D}}, \forall s \in \Omega^{\mathrm{S}}, \forall t \in \Omega^{\mathrm{T}}; p_{\ell t}^{\mathrm{L}}, \forall \ell \in \Omega^{\mathrm{L}}, \forall t \in \Omega^{\mathrm{T}}; p_{nt}^{\mathrm{M}}, \forall n \in \Omega^{\mathrm{M}}, \forall t \in \Omega^{\mathrm{T}}; \delta_{nt}, \forall n \in \Omega^{\mathrm{N}}, \forall t \in \Omega^{\mathrm{T}} \Big\}$ are the optimization variables of problem (3.1).

The EMS schedules the VPP to achieve maximum utility. On one hand, the EMS aims at maximizing the utility of demands as well as the revenue gained by selling energy in the DA and RT markets. On the other hand, the EMS minimizes the energy procurement cost. To reach these goals, the objective function (3.1a) comprises the following terms:

1. Term $\sum_{t \in \Omega^{\mathrm{T}}} \left(\lambda_t^{\mathrm{DA}} p_t^{\mathrm{DA}} \Delta t + \lambda_t^{\mathrm{RT}} p_t^{\mathrm{RT}} \Delta t \right)$ is the revenue (or cost, if negative) due to selling (or buying) energy to (from) the DA and RT energy markets.
2. Term $\sum_{t \in \Omega^{\mathrm{T}}} \sum_{d \in \Omega^{\mathrm{D}}} U_{dt}^{\mathrm{D}} p_{dt}^{\mathrm{D}} \Delta t$ is the utility of the demands obtained by consuming energy throughout the day.
3. Term $\sum_{t \in \Omega^{\mathrm{T}}} \sum_{c \in \Omega^{\mathrm{C}}} \left(C_c^{\mathrm{C,V}} p_{ct}^{\mathrm{C}} \Delta t + \overline{c}_{ct}^{\mathrm{C}} + \underline{c}_{ct}^{\mathrm{C}} \right)$ is the production cost of the conventional power plants located in the VPP.

Constraints (3.1b)–(3.1k) and (3.1l)–(3.1z) define the feasible operating region of the network and the units in the VPP, respectively.

Constraints (3.1b) and (3.1c) constitute the power balance of production and consumption at nodes connected to the main grid and the remaining nodes, respectively. Equations (3.1d) define the total power traded in the DA and RT energy markets as the summation of the power injection at nodes connected to the main grid. Equations (3.1e) determine the power flows through the network, which are limited by the line capacities using Eqs. (3.1f). A dc power flow without losses is considered for the sake of simplicity [5]. Equations (3.1g) set to zero the voltage angle at the reference node. Constraints (3.1h) impose bounds on the voltage angles of other nodes. Constraints (3.1i) state that the total power traded in the DA energy market is limited by the available energy production and consumption in the VPP. Similarly, constraints (3.1j) impose limits on the total power that can be traded in both the DA and the RT energy markets. Constraints (3.1k) limit the power injection from the main grid at each node. Constraints (3.1l) impose a minimum energy consumption for the demands during the planning horizon. Constraints (3.1m) limit the power consumption of the demands within their minimum and maximum bounds. Constraints (3.1n) impose bounds on the variation of power consumption in successive time periods. Constraints (3.1o) limit the power production of the conventional power plants. Constraints (3.1p) consider the ramping limitation of the conventional power plants. Constraints (3.1q) and (3.1r) define the start-up cost of the conventional power plants. Similarly, constraints (3.1s) and (3.1t) compute the shut-down cost of the conventional power plants. Equations (3.1u) establish the energy balance of the storage units. The energy levels of these units are bounded using constraints (3.1v). Constraints (3.1w) and (3.1x) impose bounds on the discharging and charging power levels of storage units, respectively. Constraints (3.1y) define the actual dispatch of the storage units as the difference between the discharging and charging power levels. Constraints (3.1z) state that the power produced by the stochastic renewable production facilities is lower than or equal to their available values, i.e., power spillage is possible if needed.

In the deterministic model (3.1), uncertain parameters modeling the available production levels of stochastic renewable generating units and the prices in the energy markets are replaced by the deterministic values obtained from a single-point forecast. Solving the deterministic model (3.1) provides the power quantity $p_t^{\mathrm{DA}}, \forall t \in \Omega^{\mathrm{T}}$, to be submitted to the DA energy market. One option is to submit the power quantity with the single-point forecast of the price in the DA energy market for the scheduling horizon. Considering this strategy, the power quantity may not be accepted in the DA energy market for some time periods As another option, a price-taker VPP may submit the power quantity with high/zero price to be sure for buying/selling energy in the DA energy market. This second option is considered in this deterministic approach.

Illustrative Example 3.1 Deterministic Scheduling in the DA Energy Market

The deterministic model is derived for a VPP that includes 5 units interconnected within the three-node network depicted in Fig. 3.3. Two demands 1 and 2 are located at nodes 2 and 3, respectively. One conventional power plant is also installed at node 2. One photovoltaic (PV) production facility and one storage unit are located at node 3. The VPP can trade energy with the main grid through node 1. The maximum power injection at node 1 connected to the main grid is 100 MW. The network data are provided in Table 3.1. The first column identifies the line number. The second and third columns determine the sending and receiving nodes of each line, respectively. The fourth and fifth columns show the values of susceptance and capacity of each line, respectively. The base values of apparent power and line-to-line voltage are considered 100 MVA and 10 kV, respectively. Therefore, the base value of impedance is 1 Ω.

The techno-economic data of the demands, the conventional power plant, and the storage unit are presented in Tables 3.2, 3.3, and 3.4, respectively. Based on generation and consumption assets in the VPP, the maximum power that can be sold to/bought from the DA and RT energy markets is 40/60 MW in each time period.

The single-point forecasts and actual values of the uncertain parameters are given in Table 3.5.

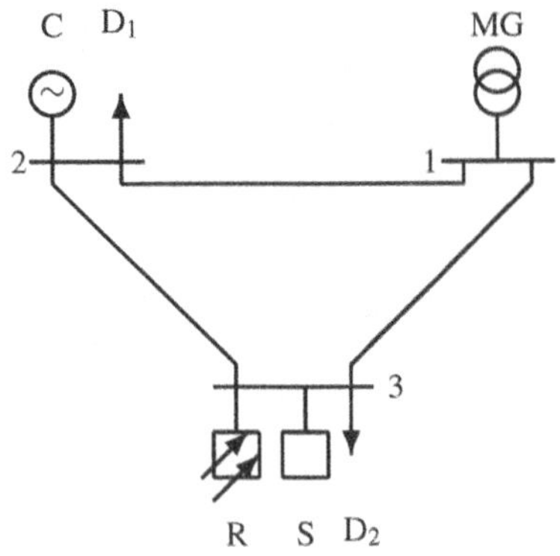

Fig. 3.3 Illustrative Example 3.1: three-node system

Table 3.1 Illustrative Example 3.1: data of lines

Line	From	To	Susceptance [S]	Capacity [MW]
1	1	2	200	50
2	1	3	200	50
3	2	3	200	50

Table 3.2 Illustrative Example 3.1: data of the demands

Parameter	Demand 1	Demand 2
Minimum energy consumption	30 MWh	50 MWh
Maximum power load in time periods 1, 2, 3	20, 20, 10 MW	30, 30, 15 MW
Minimum power load in time periods 1, 2, 3	0, 0, 0 MW	5, 5, 5 MW
Utility in time periods 1, 2, 3	40, 50, 40 \$/MWh	50, 60, 50 \$/MWh
Ramping up limit	15 MW/h	25 MW/h
Ramping down limit	15 MW/h	25 MW/h
Initial power load	10 MW	20 MW

Table 3.3 Illustrative Example 3.1: data of the conventional power plant

Variable production cost	\$45/MWh
Start-up/shut-down cost	\$5
Maximum power production limit	20 MW
Minimum power production limit	0 MW
Ramping up limit	20 MW/h
Ramping down limit	20 MW/h
Initial commitment status	off
Initial power production	0 MW

Table 3.4 Illustrative Example 3.1: data of the storage unit

Maximum charging power	10 MW
Maximum discharging power	10 MW
Maximum energy level	10 MWh
Minimum energy level	0 MWh
Initial energy level	5 MWh
Charging efficiency	0.9
Discharging efficiency	0.9

Using the above data, the deterministic model for the scheduling of this VPP in the DA energy market for the 3-h planning horizon is formulated. The length of each time period is considered to be equal to 1 h. First, the objective function to be maximized is:

$$
\begin{aligned}
z = {} & 27p_1^{\mathrm{DA}} + 47p_2^{\mathrm{DA}} + 37p_3^{\mathrm{DA}} + 27.6p_1^{\mathrm{RT}} + 51.1p_2^{\mathrm{RT}} + 33.6p_3^{\mathrm{RT}} \\
& + 40p_{11}^{\mathrm{D}} + 50p_{12}^{\mathrm{D}} + 40p_{13}^{\mathrm{D}} + 50p_{21}^{\mathrm{D}} + 60p_{22}^{\mathrm{D}} + 50p_{23}^{\mathrm{D}} \\
& - (45p_1^{\mathrm{C}} + \overline{c}_1^{\mathrm{C}} + \underline{c}_1^{\mathrm{C}}) - (45p_2^{\mathrm{C}} + \overline{c}_2^{\mathrm{C}} + \underline{c}_2^{\mathrm{C}}) - (45p_3^{\mathrm{C}} + \overline{c}_3^{\mathrm{C}} + \underline{c}_3^{\mathrm{C}}).
\end{aligned}
$$

Table 3.5 Illustrative Example 3.1: single-point forecasts and actual data

Parameter	Time period 1	Time period 2	Time period 3
Price forecasts in the DA energy market	\$27/MWh	\$47/MWh	\$37/MWh
Actual prices in the DA energy market	\$25/MWh	\$50/MWh	\$35/MWh
Price forecasts in the RT energy market	\$27.6/MWh	\$51.1/MWh	\$33.6/MWh
Actual prices in the RT energy market	\$35/MWh	\$60/MWh	\$28/MWh
Available PV production forecasts	6.9 MW	5.8 MW	1.1 MW
Actual values of available PV production	6 MW	8.5 MW	2 MW

The first line is the revenue/cost obtained from trading energy in the DA and RT energy markets in the planning horizon. The second and third lines introduce the utility of demands and the production cost of the conventional power plant in the planning horizon, respectively.

Then, the constraints of the problem are formulated:

- Power balance constraints:

$$
\begin{aligned}
& -p^{\mathrm{L}}_{11} - p^{\mathrm{L}}_{21} = p^{\mathrm{M}}_{1}, \\
& -p^{\mathrm{L}}_{12} - p^{\mathrm{L}}_{22} = p^{\mathrm{M}}_{2}, \\
& -p^{\mathrm{L}}_{13} - p^{\mathrm{L}}_{23} = p^{\mathrm{M}}_{3}, \\
& p^{\mathrm{C}}_{1} - p^{\mathrm{L}}_{31} + p^{\mathrm{L}}_{11} = p^{\mathrm{D}}_{11}, \\
& p^{\mathrm{C}}_{2} - p^{\mathrm{L}}_{32} + p^{\mathrm{L}}_{12} = p^{\mathrm{D}}_{12}, \\
& p^{\mathrm{C}}_{3} - p^{\mathrm{L}}_{33} + p^{\mathrm{L}}_{13} = p^{\mathrm{D}}_{13}, \\
& p^{\mathrm{R}}_{1} + p^{\mathrm{S}}_{1} + p^{\mathrm{L}}_{21} + p^{\mathrm{L}}_{31} = p^{\mathrm{D}}_{21}, \\
& p^{\mathrm{R}}_{2} + p^{\mathrm{S}}_{2} + p^{\mathrm{L}}_{22} + p^{\mathrm{L}}_{32} = p^{\mathrm{D}}_{22}, \\
& p^{\mathrm{R}}_{3} + p^{\mathrm{S}}_{3} + p^{\mathrm{L}}_{23} + p^{\mathrm{L}}_{33} = p^{\mathrm{D}}_{23}.
\end{aligned}
$$

- Power traded in the DA and RT energy markets:

$$
\begin{aligned}
& p^{\mathrm{DA}}_{1} + p^{\mathrm{RT}}_{1} = p^{\mathrm{M}}_{1}, \\
& p^{\mathrm{DA}}_{2} + p^{\mathrm{RT}}_{2} = p^{\mathrm{M}}_{2}, \\
& p^{\mathrm{DA}}_{3} + p^{\mathrm{RT}}_{3} = p^{\mathrm{M}}_{3}.
\end{aligned}
$$

- Definition of power flows through lines:

$$
\begin{aligned}
& \frac{p^{\mathrm{L}}_{11}}{100} = 200(\delta_{11} - \delta_{21}), \\
& \frac{p^{\mathrm{L}}_{12}}{100} = 200(\delta_{12} - \delta_{22}), \\
& \frac{p^{\mathrm{L}}_{13}}{100} = 200(\delta_{13} - \delta_{23}), \\
& \frac{p^{\mathrm{L}}_{21}}{100} = 200(\delta_{11} - \delta_{31}), \\
& \frac{p^{\mathrm{L}}_{22}}{100} = 200(\delta_{12} - \delta_{32}),
\end{aligned}
$$

$$\frac{p^{\mathrm{L}}_{23}}{100} = 200(\delta_{13} - \delta_{33}),$$
$$\frac{p^{\mathrm{L}}_{31}}{100} = 200(\delta_{21} - \delta_{31}),$$
$$\frac{p^{\mathrm{L}}_{32}}{100} = 200(\delta_{22} - \delta_{32}),$$
$$\frac{p^{\mathrm{L}}_{33}}{100} = 200(\delta_{23} - \delta_{33}).$$

- Limits on power flows through lines:

$$-50 \leq p^{\mathrm{L}}_{11} \leq 50,$$
$$-50 \leq p^{\mathrm{L}}_{12} \leq 50,$$
$$-50 \leq p^{\mathrm{L}}_{13} \leq 50,$$
$$-50 \leq p^{\mathrm{L}}_{21} \leq 50,$$
$$-50 \leq p^{\mathrm{L}}_{22} \leq 50,$$
$$-50 \leq p^{\mathrm{L}}_{23} \leq 50,$$
$$-50 \leq p^{\mathrm{L}}_{31} \leq 50,$$
$$-50 \leq p^{\mathrm{L}}_{32} \leq 50,$$
$$-50 \leq p^{\mathrm{L}}_{33} \leq 50.$$

- Definition of voltage angle at reference node:

$$\delta_{11} = 0,$$
$$\delta_{12} = 0,$$
$$\delta_{13} = 0.$$

- Limits on voltage angles:

$$-\pi \leq \delta_{21} \leq \pi,$$
$$-\pi \leq \delta_{22} \leq \pi,$$
$$-\pi \leq \delta_{23} \leq \pi,$$
$$-\pi \leq \delta_{31} \leq \pi,$$
$$-\pi \leq \delta_{32} \leq \pi,$$
$$-\pi \leq \delta_{33} \leq \pi.$$

- Limits on the power traded in DA and RT energy markets:

$$-60 \leq p^{\mathrm{DA}}_{1} \leq 40,$$
$$-60 \leq p^{\mathrm{DA}}_{2} \leq 40,$$
$$-60 \leq p^{\mathrm{DA}}_{3} \leq 40,$$
$$-60 \leq p^{\mathrm{DA}}_{1} + p^{\mathrm{RT}}_{1} \leq 40,$$

$$-60 \leq p_2^{\text{DA}} + p_2^{\text{RT}} \leq 40,$$
$$-60 \leq p_3^{\text{DA}} + p_3^{\text{RT}} \leq 40.$$

- Limits on the power traded in the market:

$$-100 \leq p_1^{\text{M}} \leq 100,$$
$$-100 \leq p_2^{\text{M}} \leq 100,$$
$$-100 \leq p_3^{\text{M}} \leq 100.$$

- Minimum energy consumption of demands:

$$p_{11}^{\text{D}} + p_{12}^{\text{D}} + p_{13}^{\text{D}} \geq 30,$$
$$p_{21}^{\text{D}} + p_{22}^{\text{D}} + p_{23}^{\text{D}} \geq 50.$$

- Limits on power consumption of demands:

$$0 \leq p_{11}^{\text{D}} \leq 20,$$
$$0 \leq p_{12}^{\text{D}} \leq 20,$$
$$0 \leq p_{13}^{\text{D}} \leq 10,$$
$$5 \leq p_{21}^{\text{D}} \leq 30,$$
$$5 \leq p_{22}^{\text{D}} \leq 30,$$
$$5 \leq p_{23}^{\text{D}} \leq 15.$$

- Ramping limits of demands:

$$-15 \leq p_{11}^{\text{D}} - 10 \leq 15,$$
$$-15 \leq p_{12}^{\text{D}} - p_{11}^{\text{D}} \leq 15,$$
$$-15 \leq p_{13}^{\text{D}} - p_{12}^{\text{D}} \leq 15,$$
$$-25 \leq p_{21}^{\text{D}} - 20 \leq 25,$$
$$-25 \leq p_{22}^{\text{D}} - p_{21}^{\text{D}} \leq 25,$$
$$-25 \leq p_{23}^{\text{D}} - p_{22}^{\text{D}} \leq 25.$$

- Limits on the production of conventional power plants:

$$0 \leq p_1^{\text{C}} \leq 20u_1^{\text{C}},$$
$$0 \leq p_2^{\text{C}} \leq 20u_2^{\text{C}},$$
$$0 \leq p_3^{\text{C}} \leq 20u_3^{\text{C}}.$$

- Ramping limits of conventional power plants:

$$-20 \leq p_1^{\text{C}} - 0 \leq 20,$$
$$-20 \leq p_2^{\text{C}} - p_1^{\text{C}} \leq 20,$$
$$-20 \leq p_3^{\text{C}} - p_2^{\text{C}} \leq 20.$$

- Definition of start-up and shot-down costs of conventional power plants:

$$\begin{aligned}
&5(u_1^{\mathrm{C}} - 0) \le \overline{c}_1^{\mathrm{C}},\\
&5(u_2^{\mathrm{C}} - u_1^{\mathrm{C}}) \le \overline{c}_2^{\mathrm{C}},\\
&5(u_3^{\mathrm{C}} - u_2^{\mathrm{C}}) \le \overline{c}_3^{\mathrm{C}},\\
&0 \le \overline{c}_1^{\mathrm{C}},\\
&0 \le \overline{c}_2^{\mathrm{C}},\\
&0 \le \overline{c}_3^{\mathrm{C}},\\
&5(0 - u_1^{\mathrm{C}}) \le \underline{c}_1^{\mathrm{C}},\\
&5(u_1^{\mathrm{C}} - u_2^{\mathrm{C}}) \le \underline{c}_2^{\mathrm{C}},\\
&5(u_2^{\mathrm{C}} - u_3^{\mathrm{C}}) \le \underline{c}_3^{\mathrm{C}},\\
&0 \le \underline{c}_1^{\mathrm{C}},\\
&0 \le \underline{c}_2^{\mathrm{C}},\\
&0 \le \underline{c}_3^{\mathrm{C}}.
\end{aligned}$$

- Energy evolution in the storage unit:

$$\begin{aligned}
&e_1^{\mathrm{S}} = 5 + 0.9p_1^{\mathrm{S,C}} - \frac{p_1^{\mathrm{S,D}}}{0.9},\\
&e_2^{\mathrm{S}} = e_1^{\mathrm{S}} + 0.9p_2^{\mathrm{S,C}} - \frac{p_2^{\mathrm{S,D}}}{0.9},\\
&e_3^{\mathrm{S}} = e_2^{\mathrm{S}} + 0.9p_3^{\mathrm{S,C}} - \frac{p_3^{\mathrm{S,D}}}{0.9}.
\end{aligned}$$

- Energy limits of the storage unit:

$$\begin{aligned}
&0 \le e_1^{\mathrm{S}} \le 10,\\
&0 \le e_2^{\mathrm{S}} \le 10,\\
&0 \le e_3^{\mathrm{S}} \le 10.
\end{aligned}$$

- Charging and discharging power limits of the storage unit:

$$\begin{aligned}
&0 \le p_1^{\mathrm{S,C}} \le 10,\\
&0 \le p_2^{\mathrm{S,C}} \le 10,\\
&0 \le p_3^{\mathrm{S,C}} \le 10,\\
&0 \le p_1^{\mathrm{S,D}} \le 10,\\
&0 \le p_2^{\mathrm{S,D}} \le 10,\\
&0 \le p_3^{\mathrm{S,D}} \le 10.
\end{aligned}$$

- Net power of the storage unit:

$$p_1^{\mathrm{S}} = p_1^{\mathrm{S,D}} - p_1^{\mathrm{S,C}},$$
$$p_2^{\mathrm{S}} = p_2^{\mathrm{S,D}} - p_2^{\mathrm{S,C}},$$
$$p_3^{\mathrm{S}} = p_3^{\mathrm{S,D}} - p_3^{\mathrm{S,C}}.$$

- Limits on the production of the stochastic renewable generating unit:

$$0 \leq p_1^{\mathrm{R}} \leq 6.9,$$
$$0 \leq p_2^{\mathrm{R}} \leq 5.8,$$
$$0 \leq p_3^{\mathrm{R}} \leq 1.1.$$

Finally, the optimization variables of the problem are those included in set $\Phi^{3.1}=\{\overline{c}_1^{\mathrm{C}}, \overline{c}_2^{\mathrm{C}}, \overline{c}_3^{\mathrm{C}}, \underline{c}_1^{\mathrm{C}}, \underline{c}_2^{\mathrm{C}}, \underline{c}_3^{\mathrm{C}}, e_1^{\mathrm{S}}, e_2^{\mathrm{S}}, e_3^{\mathrm{S}}, p_{11}^{\mathrm{D}}, p_{12}^{\mathrm{D}}, p_{13}^{\mathrm{D}}, p_{21}^{\mathrm{D}}, p_{22}^{\mathrm{D}}, p_{23}^{\mathrm{D}}, p_1^{\mathrm{C}}, p_2^{\mathrm{C}}, p_3^{\mathrm{C}}, p_1^{\mathrm{R}}, p_2^{\mathrm{R}}, p_3^{\mathrm{R}}, p_1^{\mathrm{S}}, p_2^{\mathrm{S}}, p_3^{\mathrm{S}}, p_1^{\mathrm{S,D}}, p_2^{\mathrm{S,D}}, p_3^{\mathrm{S,D}}, p_1^{\mathrm{S,C}}, p_2^{\mathrm{S,C}}, p_3^{\mathrm{S,C}}, p_1^{\mathrm{DA}}, p_2^{\mathrm{DA}}, p_3^{\mathrm{DA}}, p_1^{\mathrm{M}}, p_2^{\mathrm{M}}, p_3^{\mathrm{M}}, p_1^{\mathrm{RT}}, p_2^{\mathrm{RT}}, p_3^{\mathrm{RT}}, u_1^{\mathrm{C}}, u_2^{\mathrm{C}}, u_3^{\mathrm{C}}, p_{11}^{\mathrm{L}}, p_{12}^{\mathrm{L}}, p_{13}^{\mathrm{L}}, p_{21}^{\mathrm{L}}, p_{22}^{\mathrm{L}}, p_{23}^{\mathrm{L}}, p_{31}^{\mathrm{L}}, p_{32}^{\mathrm{L}}, p_{33}^{\mathrm{L}}, \delta_{11}, \delta_{12}, \delta_{13}, \delta_{21}, \delta_{22}, \delta_{23}, \delta_{31}, \delta_{32}, \delta_{33}\}$.

The above model is an MILP problem that can be solved using the CPLEX solver [11] under GAMS [9]. The optimal value of the objective function is equal to \$2856.8. The power quantity-price submitted in the DA energy market for the first, second, and third time periods are (−60 MW,\$100/MWh), (−60 MW,\$100/MWh), and (40 MW,\$0/MWh), respectively. The bidding decisions in time periods 1 and 2 include high prices, while the offering decision in time period 3 is made at zero price to guarantee that the bids and offers are accepted. The VPP is willing to buy energy in the DA energy market in time periods 1 and 2, i.e., the VPP behaves as a demand in these time periods. However, the VPP behaves as a generating unit in time period 3 and sells energy in the DA energy market. This behavior is due to the difference between the single-point forecasts of the prices in the DA and RT energy markets, as well as to the needs in the VPP. In time periods 1 and 2, the prices in the DA energy market are lower than those values in the RT energy market. Under such condition, it is beneficial to buy energy in the DA energy market, and then sell some energy in the RT energy market. That is, the VPP uses arbitrage opportunities to increase its profit. Results show also that the VPP sells 11.3 and 59.8 MW in the RT energy market in time periods 1 and 2, respectively. On the contrary, the price in the DA energy market in time period 3 is higher than the price in the RT energy market, and the VPP seeks to sell energy in the former market and then buy energy in the latter. The inspection of the results verifies that the VPP is willing to buy 63.9 MW in the RT energy market in this time period.

Note that the solution of the deterministic model provides the scheduling decisions in both the DA and RT energy markets, as well as the dispatch of the different units in the VPP; however, the EMS in charge of the VPP implements only the offering and bidding strategies in the DA market, and additional results on the units and scheduling in the RT energy market are not utilized at this stage.

Once the DA energy market is cleared, the actual prices in Table 3.5 are known and communicated to the VPP. Note that due to the offer and bid prices of the VPP, all the offer and bids are accepted in the DA energy market. The VPP buys 60 MW in both time periods 1 and 2, while it sells 40 MW in time period 3. According to the scheduled power and market prices, the VPP pays a net value of \$3100.

□

A GAMS [9] code for solving Illustrative Example 3.1 is provided in Sect. 3.7.

3.3.2 Deterministic Scheduling in the Real-Time Energy Market

Once the DA energy market is cleared, the EMS in charge of the VPP knows the scheduled power and prices in this market. Then, the VPP has the opportunity to participate in the RT energy market time by time to compensate any possible deviation from the scheduled power in the DA energy market. The EMS submits the offering and bidding decisions in the RT energy market for each time period $\tau \in \Omega^{\mathrm{T}}$ by solving the following deterministic scheduling model, which is formulated as an

MILP problem:

$$\max_{\Phi^{\mathrm{RT}}} \quad \sum_{t=\tau}^{|\Omega^{\mathrm{T}}|} \left[\lambda_t^{\mathrm{RT}} p_t^{\mathrm{RT}} \Delta t + \sum_{d \in \Omega^{\mathrm{D}}} U_{dt}^{\mathrm{D}} p_{dt}^{\mathrm{D}} \Delta t - \sum_{c \in \Omega^{\mathrm{C}}} (C_c^{\mathrm{C,V}} p_{ct}^{\mathrm{C}} \Delta t + \overline{c}_{ct}^{\mathrm{C}} + \underline{c}_{ct}^{\mathrm{C}}) \right] \tag{3.2a}$$

subject to

$$\sum_{r \in \Omega_n^{\mathrm{R}}} p_{rt}^{\mathrm{R}} + \sum_{s \in \Omega_n^{\mathrm{S}}} p_{st}^{\mathrm{S}} - \sum_{\ell | s(\ell)=n} p_{\ell t}^{\mathrm{L}} + \sum_{\ell | r(\ell)=n} p_{\ell t}^{\mathrm{L}} = p_{nt}^{\mathrm{M}} + \sum_{d \in \Omega_n^{\mathrm{D}}} p_{dt}^{\mathrm{D}}, \quad \forall n \in \Omega^{\mathrm{M}}, \forall t \geq \tau, \tag{3.2b}$$

$$\sum_{r \in \Omega_n^{\mathrm{R}}} p_{rt}^{\mathrm{R}} + \sum_{s \in \Omega_n^{\mathrm{S}}} p_{st}^{\mathrm{S}} - \sum_{\ell | s(\ell)=n} p_{\ell t}^{\mathrm{L}} + \sum_{\ell | r(\ell)=n} p_{\ell t}^{\mathrm{L}} = \sum_{d \in \Omega_n^{\mathrm{D}}} p_{dt}^{\mathrm{D}}, \quad \forall n \in \Omega^{\mathrm{N}} \setminus n \in \Omega^{\mathrm{M}}, \forall t \geq \tau, \tag{3.2c}$$

$$\tilde{p}_t^{\mathrm{DA}} + p_t^{\mathrm{RT}} = \sum_{n \in \Omega^{\mathrm{M}}} p_{nt}^{\mathrm{M}}, \quad \forall t \geq \tau, \tag{3.2d}$$

$$p_{\ell t}^{\mathrm{L}} = B_\ell (\delta_{s(\ell)t} - \delta_{r(\ell)t}), \quad \forall \ell \in \Omega^{\mathrm{L}}, \forall t \geq \tau, \tag{3.2e}$$

$$-\overline{P}_\ell^{\mathrm{L}} \leq p_{\ell t}^{\mathrm{L}} \leq \overline{P}_\ell^{\mathrm{L}}, \quad \forall \ell \in \Omega^{\mathrm{L}}, \forall t \geq \tau, \tag{3.2f}$$

$$\delta_{nt} = 0, \quad n : \text{ref.}, \forall t \geq \tau, \tag{3.2g}$$

$$-\pi \leq \delta_{nt} \leq \pi, \quad \forall n \in \Omega^{\mathrm{N}} \setminus n : \text{ref.}, \forall t \geq \tau, \tag{3.2h}$$

$$\underline{P}^{\mathrm{V}} \leq \tilde{p}_t^{\mathrm{DA}} + p_t^{\mathrm{RT}} \leq \overline{P}^{\mathrm{V}}, \quad \forall t \geq \tau, \tag{3.2i}$$

$$-\overline{P}_n^{\mathrm{M}} \leq p_{nt}^{\mathrm{M}} \leq \overline{P}_n^{\mathrm{M}}, \quad \forall n \in \Omega^{\mathrm{M}}, \forall t \geq \tau \tag{3.2j}$$

$$\sum_{t=1}^{\tau-1} \tilde{p}_{dt}^{\mathrm{D}} \Delta t + \sum_{t=\tau}^{|\Omega^{\mathrm{T}}|} p_{dt}^{\mathrm{D}} \Delta t \geq \underline{E}_d^{\mathrm{D}}, \quad \forall d \in \Omega^{\mathrm{D}}, \tag{3.2k}$$

$$\underline{P}_{dt}^{\mathrm{D}} \leq p_{dt}^{\mathrm{D}} \leq \overline{P}_{dt}^{\mathrm{D}}, \quad \forall d \in \Omega^{\mathrm{D}}, \forall t \geq \tau, \tag{3.2l}$$

$$-R_d^{\mathrm{D,D}} \Delta t \leq p_{dt}^{\mathrm{D}} - p_{d(t-1)}^{\mathrm{D}} \leq R_d^{\mathrm{D,U}} \Delta t, \quad \forall d \in \Omega^{\mathrm{D}}, \forall t \geq \tau, \tag{3.2m}$$

$$\underline{P}_c^{\mathrm{C}} u_{ct}^{\mathrm{C}} \leq p_{ct}^{\mathrm{C}} \leq \overline{P}_c^{\mathrm{C}} u_{ct}^{\mathrm{C}}, \quad \forall c \in \Omega^{\mathrm{C}}, \forall t \geq \tau, \tag{3.2n}$$

$$-R_c^{\mathrm{C,D}} \Delta t \leq p_{ct}^{\mathrm{C}} - p_{c(t-1)}^{\mathrm{C}} \leq R_c^{\mathrm{C,U}} \Delta t, \quad \forall c \in \Omega^{\mathrm{C}}, \forall t \geq \tau, \tag{3.2o}$$

$$\overline{C}_c^{\mathrm{C}} (u_{ct}^{\mathrm{C}} - u_{c(t-1)}^{\mathrm{C}}) \leq \overline{c}_{ct}^{\mathrm{C}}, \quad \forall c \in \Omega^{\mathrm{C}}, \forall t \geq \tau, \tag{3.2p}$$

$$0 \leq \overline{c}_{ct}^{\mathrm{C}}, \quad \forall c \in \Omega^{\mathrm{C}}, \forall t \geq \tau, \tag{3.2q}$$

$$\underline{C}_c^{\mathrm{C}} (u_{c(t-1)}^{\mathrm{C}} - u_{ct}^{\mathrm{C}}) \leq \underline{c}_{ct}^{\mathrm{C}}, \quad \forall c \in \Omega^{\mathrm{C}}, \forall t \geq \tau, \tag{3.2r}$$

$$0 \leq \underline{c}_{ct}^{\mathrm{C}}, \quad \forall c \in \Omega^{\mathrm{C}}, \forall t \geq \tau, \tag{3.2s}$$

$$e_{st}^{\mathrm{S}} = e_{s(t-1)}^{\mathrm{S}} + \eta_s^{\mathrm{S,C}} p_{st}^{\mathrm{S,C}} \Delta t - \frac{p_{st}^{\mathrm{S,D}}}{\eta_s^{\mathrm{S,D}}} \Delta t, \quad \forall s \in \Omega^{\mathrm{S}}, \forall t \geq \tau, \tag{3.2t}$$

$$\underline{E}_s^{\mathrm{S}} \leq e_{st}^{\mathrm{S}} \leq \overline{E}_s^{\mathrm{S}}, \quad \forall s \in \Omega^{\mathrm{S}}, \forall t \geq \tau, \tag{3.2u}$$

$$0 \leq p_{st}^{\mathrm{S,D}} \leq \overline{P}_{s}^{\mathrm{S,D}}, \quad \forall s \in \Omega^{\mathrm{S}}, \forall t \geq \tau, \tag{3.2v}$$

$$0 \leq p_{st}^{\mathrm{S,C}} \leq \overline{P}_{s}^{\mathrm{S,C}}, \quad \forall s \in \Omega^{\mathrm{S}}, \forall t \geq \tau, \tag{3.2w}$$

$$p_{st}^{\mathrm{S}} = p_{st}^{\mathrm{S,D}} - p_{st}^{\mathrm{S,C}}, \quad \forall s \in \Omega^{\mathrm{S}}, \forall t \geq \tau, \tag{3.2x}$$

$$0 \leq p_{rt}^{\mathrm{R}} \leq \tilde{P}_{rt}^{\mathrm{R,A}}, \quad \forall r \in \Omega^{\mathrm{R}}, t = \tau, \tag{3.2y}$$

$$0 \leq p_{rt}^{\mathrm{R}} \leq P_{rt}^{\mathrm{R,A}}, \quad \forall r \in \Omega^{\mathrm{R}}, \forall t > \tau, \tag{3.2z}$$

where set $\Phi^{\mathrm{RT}} = \Big\{ p_t^{\mathrm{RT}}; \overline{c}_{ct}^{\mathrm{C}}, \underline{c}_{ct}^{\mathrm{C}}, p_{ct}^{\mathrm{C}}, u_{ct}^{\mathrm{C}}, \forall c \in \Omega^{\mathrm{C}}; p_{dt}^{\mathrm{D}}, \forall d \in \Omega^{\mathrm{D}}; p_{rt}^{\mathrm{R}}, \forall r \in \Omega^{\mathrm{R}}; e_{st}^{\mathrm{S}}, p_{st}^{\mathrm{S}}, p_{st}^{\mathrm{S,C}}, p_{st}^{\mathrm{S,D}}, \forall s \in \Omega^{\mathrm{S}}; p_{\ell t}^{\mathrm{L}}, \forall \ell \in \Omega^{\mathrm{L}}; p_{nt}^{\mathrm{M}}, \delta_{nt}, \forall t \in \Omega^{\mathrm{T}}; \forall n \in \Omega^{\mathrm{N}}, \forall n \in \Omega^{\mathrm{M}}, \Big\}, \forall t \geq \tau$, includes the optimization variables of problem (3.2).

The objective function (3.2a) to be maximized is the utility obtained by participating in the RT energy market during time period τ and the remaining time periods of the day. Constraints (3.2b)–(3.2j) constitute the power flow through the electric network considering its limitations during the scheduling horizon $[\tau, |\Omega^{\mathrm{T}}|]$. Constraints (3.2k)–(3.2m), (3.2n)–(3.2s), and (3.2t)–(3.2x) impose the technical limitations of the demands, the conventional power plants, and the storage units for the scheduling horizon $[\tau, |\Omega^{\mathrm{T}}|]$. Constraints (3.2y) and (3.2z) limit the power produced by the stochastic renewable production facilities by the available power production level for the actual time period τ and the remaining scheduling horizon $[\tau + 1, |\Omega^{\mathrm{T}}|]$, respectively. Note that the available production level is known with most certainty some minutes before the actual time period τ. However, the available stochastic production level is unknown for the remaining time periods.

Model (3.2) is similar to the deterministic model (3.1) used to derive the scheduling decisions in the DA energy market. The main differences are due to the new information available to the EMS between the closure of the DA energy market and the closure of the RT energy market before actual time period τ. This new information includes:

1. The schedules of the VPP and the prices in the DA energy market.
2. The available power production level of the stochastic renewable production facilities before time period τ. The available stochastic production level in time period τ is also known with almost certainty.
3. The prices in the RT energy market cleared before the actual time period τ.
4. The power dispatch of the units in the VPP before the actual time period τ.

This new information is added to the historical data to improve the forecast of the uncertain parameters during the scheduling horizon. In the deterministic model (3.2), note that the uncertain parameters representing the available stochastic production levels and the prices in the RT energy market during the time intervals $[\tau + 1, |\Omega^{\mathrm{T}}|]$ and $[\tau, |\Omega^{\mathrm{T}}|]$, respectively, are modeled as single-point forecasts.

It is important to note that the deterministic model (3.2) is solved and the decision variables are obtained for the scheduling horizon $[\tau, |\Omega^{\mathrm{T}}|]$. However, only decision variables in time period τ are actually implemented to determine the offering and bidding strategy of the VPP in the RT energy market for time period τ. That is, the EMS in charge of the VPP decides a power quantity p_{τ}^{RT} for time period τ in the RT energy market some minutes before the upcoming time period. This way, the VPP compensates any possible deviation from the scheduled power in the DA energy market.

Illustrative Example 3.2 Deterministic Scheduling in the RT Energy Market

The VPP described in Illustrative Example 3.1 is considered again. This VPP seeks to identify its offering/bidding strategy in the RT energy market during the scheduling horizon. To accomplish this task, the VPP participates three times in this RT market:

1. In the first round, the VPP uses the single-point forecasts of the uncertain parameters for the scheduling horizon some minutes before the power delivery in the first time period. Note that the VPP knows the available PV production level in this first time period with most certainty. Once the RT energy market is cleared, the VPP knows its schedule and the market price in this time period.
2. In the second round, the VPP takes the advantage of the new available information for participating in the RT energy market in the second time period.
3. In third round, the VPP uses the single-point forecast of the uncertain parameters in the third and last time period.

Considering the above explanation, the single-point forecasts are updated time by time as shown in Table 3.6. Note that bold numbers indicate actual data known by the VPP.

Considering the procedure explained in Sect. 3.3.2 and the scheduled energy in the DA energy market obtained in Illustrative Example 3.1, the deterministic model (3.2) is implemented and run three times to obtain the offering and bidding

Table 3.6 Illustrative Example 3.2: single-point forecasts and actual data

Round	1		2		3	
Time period	λ_t^{RT} [$/MWh]	$P_t^{R,A}$ [MW]	λ_t^{RT} [$/MWh]	$P_t^{R,A}$ [MW]	λ_t^{RT} [$/MWh]	$P_t^{R,A}$ [MW]
1	32.6	**6**	**35**	**6**	**35**	**6**
2	56.8	9.6	60.5	**8.5**	**60**	**8.5**
3	31.2	3.6	30	2.6	27.8	**2**

decisions for the scheduling horizon. In the first round, some minutes before time period 1, the VPP solves the scheduling problem in the RT market, which is very similar to the problem solved in Illustrative Example 3.1. The only differences are due to the information known between the closure of the DA and RT energy markets, namely:

1. the scheduled power in the DA energy market for the scheduling horizon,
2. the RT market prices prior to time period 1, and
3. the available PV power production level prior to time period 1.

The VPP also knows the available production level of the PV generating unit in time period 1 with almost certainty. Given the data in Table 3.6 for the first round, the problem is formulated below:

$$
\begin{aligned}
\max_{\Phi_1^{3.2}} \quad & 32.6p_1^{RT} + 56.8p_2^{RT} + 31.2p_3^{RT} \\
& + 40p_{11}^{D} + 50p_{12}^{D} + 40p_{13}^{D} + 50p_{21}^{D} + 60p_{22}^{D} + 50p_{23}^{D} \\
& - (45p_1^{C} + \overline{c}_1^{C} + \underline{c}_1^{C}) - (45p_2^{C} + \overline{c}_2^{C} + \underline{c}_2^{C}) - (45p_3^{C} + \overline{c}_3^{C} + \underline{c}_3^{C})
\end{aligned}
$$

subject to

$$
\begin{aligned}
& -p_{11}^{L} - p_{21}^{L} = p_1^{M}, \\
& -p_{12}^{L} - p_{22}^{L} = p_2^{M}, \\
& -p_{13}^{L} - p_{23}^{L} = p_3^{M}, \\
& p_1^{C} - p_{31}^{L} + p_{11}^{L} = p_{11}^{D}, \\
& p_2^{C} - p_{32}^{L} + p_{12}^{L} = p_{12}^{D}, \\
& p_3^{C} - p_{33}^{L} + p_{13}^{L} = p_{13}^{D}, \\
& p_1^{R} + p_1^{S} + p_{21}^{L} + p_{31}^{L} = p_{21}^{D}, \\
& p_2^{R} + p_2^{S} + p_{22}^{L} + p_{32}^{L} = p_{22}^{D}, \\
& p_3^{R} + p_3^{S} + p_{23}^{L} + p_{33}^{L} = p_{23}^{D}, \\
& -60 + p_1^{RT} = p_1^{M}, \\
& -60 + p_2^{RT} = p_2^{M}, \\
& 40 + p_3^{RT} = p_3^{M}, \\
& \frac{p_{11}^{L}}{100} = 200(\delta_{11} - \delta_{21}), \\
& \frac{p_{12}^{L}}{100} = 200(\delta_{12} - \delta_{22}), \\
& \frac{p_{13}^{L}}{100} = 200(\delta_{13} - \delta_{23}), \\
& \frac{p_{21}^{L}}{100} = 200(\delta_{11} - \delta_{31}),
\end{aligned}
$$

$$
\begin{aligned}
&\frac{p^{\mathrm{L}}_{22}}{100} = 200(\delta_{12} - \delta_{32}),\\
&\frac{p^{\mathrm{L}}_{23}}{100} = 200(\delta_{13} - \delta_{33}),\\
&\frac{p^{\mathrm{L}}_{31}}{100} = 200(\delta_{21} - \delta_{31}),\\
&\frac{p^{\mathrm{L}}_{32}}{100} = 200(\delta_{22} - \delta_{32}),\\
&\frac{p^{\mathrm{L}}_{33}}{100} = 200(\delta_{23} - \delta_{33}),\\
&-50 \le p^{\mathrm{L}}_{11} \le 50,\\
&-50 \le p^{\mathrm{L}}_{12} \le 50,\\
&-50 \le p^{\mathrm{L}}_{13} \le 50,\\
&-50 \le p^{\mathrm{L}}_{21} \le 50,\\
&-50 \le p^{\mathrm{L}}_{22} \le 50,\\
&-50 \le p^{\mathrm{L}}_{23} \le 50,\\
&-50 \le p^{\mathrm{L}}_{31} \le 50,\\
&-50 \le p^{\mathrm{L}}_{32} \le 50,\\
&-50 \le p^{\mathrm{L}}_{33} \le 50,\\
&\delta_{11} = 0,\\
&\delta_{12} = 0,\\
&\delta_{13} = 0,\\
&-\pi \le \delta_{21} \le \pi,\\
&-\pi \le \delta_{22} \le \pi,\\
&-\pi \le \delta_{23} \le \pi,\\
&-\pi \le \delta_{31} \le \pi,\\
&-\pi \le \delta_{32} \le \pi,\\
&-\pi \le \delta_{33} \le \pi,\\
&-60 \le -60 + p^{\mathrm{RT}}_{1} \le 40,\\
&-60 \le -60 + p^{\mathrm{RT}}_{2} \le 40,\\
&-60 \le 40 + p^{\mathrm{RT}}_{3} \le 40,\\
&-100 \le p^{\mathrm{M}}_{1} \le 100,\\
&-100 \le p^{\mathrm{M}}_{2} \le 100,\\
&-100 \le p^{\mathrm{M}}_{3} \le 100,\\
&p^{\mathrm{D}}_{11} + p^{\mathrm{D}}_{12} + p^{\mathrm{D}}_{13} \ge 30,\\
&p^{\mathrm{D}}_{21} + p^{\mathrm{D}}_{22} + p^{\mathrm{D}}_{23} \ge 50,
\end{aligned}
$$

$$0 \le p_{11}^{\mathrm{D}} \le 20,$$
$$0 \le p_{12}^{\mathrm{D}} \le 20,$$
$$0 \le p_{13}^{\mathrm{D}} \le 10,$$
$$5 \le p_{21}^{\mathrm{D}} \le 30,$$
$$5 \le p_{22}^{\mathrm{D}} \le 30,$$
$$5 \le p_{23}^{\mathrm{D}} \le 15,$$
$$-15 \le p_{11}^{\mathrm{D}} - 10 \le 15,$$
$$-15 \le p_{12}^{\mathrm{D}} - p_{11}^{\mathrm{D}} \le 15,$$
$$-15 \le p_{13}^{\mathrm{D}} - p_{12}^{\mathrm{D}} \le 15,$$
$$-25 \le p_{21}^{\mathrm{D}} - 20 \le 25,$$
$$-25 \le p_{22}^{\mathrm{D}} - p_{21}^{\mathrm{D}} \le 25,$$
$$-25 \le p_{23}^{\mathrm{D}} - p_{22}^{\mathrm{D}} \le 25,$$
$$0 \le p_{1}^{\mathrm{C}} \le 20u_{1}^{\mathrm{C}},$$
$$0 \le p_{2}^{\mathrm{C}} \le 20u_{2}^{\mathrm{C}},$$
$$0 \le p_{3}^{\mathrm{C}} \le 20u_{3}^{\mathrm{C}},$$
$$-20 \le p_{1}^{\mathrm{C}} - 0 \le 20,$$
$$-20 \le p_{2}^{\mathrm{C}} - p_{1}^{\mathrm{C}} \le 20,$$
$$-20 \le p_{3}^{\mathrm{C}} - p_{2}^{\mathrm{C}} \le 20,$$
$$5(u_{1}^{\mathrm{C}} - 0) \le \overline{c}_{1}^{\mathrm{C}},$$
$$5(u_{2}^{\mathrm{C}} - u_{1}^{\mathrm{C}}) \le \overline{c}_{2}^{\mathrm{C}},$$
$$5(u_{3}^{\mathrm{C}} - u_{2}^{\mathrm{C}}) \le \overline{c}_{3}^{\mathrm{C}},$$
$$0 \le \overline{c}_{1}^{\mathrm{C}},$$
$$0 \le \overline{c}_{2}^{\mathrm{C}},$$
$$0 \le \overline{c}_{3}^{\mathrm{C}},$$
$$5(0 - u_{1}^{\mathrm{C}}) \le \underline{c}_{1}^{\mathrm{C}},$$
$$5(u_{1}^{\mathrm{C}} - u_{2}^{\mathrm{C}}) \le \underline{c}_{2}^{\mathrm{C}},$$
$$5(u_{2}^{\mathrm{C}} - u_{3}^{\mathrm{C}}) \le \underline{c}_{3}^{\mathrm{C}},$$
$$0 \le \underline{c}_{1}^{\mathrm{C}},$$
$$0 \le \underline{c}_{2}^{\mathrm{C}},$$
$$0 \le \underline{c}_{3}^{\mathrm{C}},$$
$$e_{1}^{\mathrm{S}} = 5 + 0.9p_{1}^{\mathrm{S,C}} - \frac{p_{1}^{\mathrm{S,D}}}{0.9},$$
$$e_{2}^{\mathrm{S}} = e_{1}^{\mathrm{S}} + 0.9p_{2}^{\mathrm{S,C}} - \frac{p_{2}^{\mathrm{S,D}}}{0.9},$$

$$
\begin{aligned}
& e_3^{\mathrm{S}} = e_2^{\mathrm{S}} + 0.9 p_3^{\mathrm{S,C}} - \frac{p_3^{\mathrm{S,D}}}{0.9}, \\
& 0 \leq e_1^{\mathrm{S}} \leq 10, \\
& 0 \leq e_2^{\mathrm{S}} \leq 10, \\
& 0 \leq e_3^{\mathrm{S}} \leq 10, \\
& 0 \leq p_1^{\mathrm{S,C}} \leq 10, \\
& 0 \leq p_2^{\mathrm{S,C}} \leq 10, \\
& 0 \leq p_3^{\mathrm{S,C}} \leq 10, \\
& 0 \leq p_1^{\mathrm{S,D}} \leq 10, \\
& 0 \leq p_2^{\mathrm{S,D}} \leq 10, \\
& 0 \leq p_3^{\mathrm{S,D}} \leq 10, \\
& p_1^{\mathrm{S}} = p_1^{\mathrm{S,D}} - p_1^{\mathrm{S,C}}, \\
& p_2^{\mathrm{S}} = p_2^{\mathrm{S,D}} - p_2^{\mathrm{S,C}}, \\
& p_3^{\mathrm{S}} = p_3^{\mathrm{S,D}} - p_3^{\mathrm{S,C}}, \\
& 0 \leq p_1^{\mathrm{R}} \leq 6, \\
& 0 \leq p_2^{\mathrm{R}} \leq 9.6, \\
& 0 \leq p_3^{\mathrm{R}} \leq 3.6,
\end{aligned}
$$

where the optimization variables of the problem are included in set $\Phi_1^{3.2}=\{\overline{c}_1^{\mathrm{C}}, \overline{c}_2^{\mathrm{C}}, \overline{c}_3^{\mathrm{C}}, \underline{c}_1^{\mathrm{C}}, \underline{c}_2^{\mathrm{C}}, \underline{c}_3^{\mathrm{C}}, e_1^{\mathrm{S}}, e_2^{\mathrm{S}}, e_3^{\mathrm{S}}, p_{11}^{\mathrm{D}}, p_{12}^{\mathrm{D}}, p_{13}^{\mathrm{D}}, p_{21}^{\mathrm{D}}, p_{22}^{\mathrm{D}}, p_{23}^{\mathrm{D}}, p_1^{\mathrm{C}}, p_2^{\mathrm{C}}, p_3^{\mathrm{C}}, p_1^{\mathrm{R}}, p_2^{\mathrm{R}}, p_3^{\mathrm{R}}, p_1^{\mathrm{S}}, p_2^{\mathrm{S}}, p_3^{\mathrm{S}}, p_1^{\mathrm{S,D}}, p_2^{\mathrm{S,D}}, p_3^{\mathrm{S,D}}, p_1^{\mathrm{S,C}}, p_2^{\mathrm{S,C}}, p_3^{\mathrm{S,C}}, p_1^{\mathrm{M}}, p_2^{\mathrm{M}}, p_3^{\mathrm{M}}, p_1^{\mathrm{RT}}, p_2^{\mathrm{RT}}, p_3^{\mathrm{RT}}, u_1^{\mathrm{C}}, u_2^{\mathrm{C}}, u_3^{\mathrm{C}}, p_{11}^{\mathrm{L}}, p_{12}^{\mathrm{L}}, p_{13}^{\mathrm{L}}, p_{21}^{\mathrm{L}}, p_{22}^{\mathrm{L}}, p_{23}^{\mathrm{L}}, p_{31}^{\mathrm{L}}, p_{32}^{\mathrm{L}}, p_{33}^{\mathrm{L}}, \delta_{11}, \delta_{12}, \delta_{13}, \delta_{21}, \delta_{22}, \delta_{23}, \delta_{31}, \delta_{32}, \delta_{33}\}$.

The above model is an MILP problem that can be solved using CPLEX [11] under GAMS [9], obtaining the power quantities traded by the VPP in the RT market in the planning horizon. However, only the power quantity in time period 1 is submitted. The VPP submits a power quantity of 10.4 MW with zero price for time period 1 to compensate the deviation from the scheduled power (−60 MW) in the DA energy market. Once the RT energy market is cleared, the market price is known as \$35/MWh, and the VPP gains \$364 for trading in the RT energy market. Thus, the net injection from node 1 connected to the main grid is 49.6 MW. The VPP dispatches its own energy assets in time period 1 as provided in Table 3.7. Therefore, the VPP obtains \$2300 as the utility of the energy consumption minus the production cost of the conventional power plant.

In the second round, the information on the RT market price and the available production level of the PV generating unit in time period 1 is used to update the single-point forecast as given in Table 3.6. Then, the EMS in charge of the VPP solves the following scheduling problem for time period 2:

$$
\begin{aligned}
\max_{\Phi_2^{3.2}} \quad & 60.5 p_2^{\mathrm{RT}} + 30 p_3^{\mathrm{RT}} \\
& + 50 p_{12}^{\mathrm{D}} + 40 p_{13}^{\mathrm{D}} + 60 p_{22}^{\mathrm{D}} + 50 p_{23}^{\mathrm{D}} \\
& - (45 p_2^{\mathrm{C}} + \overline{c}_2^{\mathrm{C}} + \underline{c}_2^{\mathrm{C}}) - (45 p_3^{\mathrm{C}} + \overline{c}_3^{\mathrm{C}} + \underline{c}_3^{\mathrm{C}})
\end{aligned}
$$

subject to

$$
\begin{aligned}
& -p_{12}^{\mathrm{L}} - p_{22}^{\mathrm{L}} = p_2^{\mathrm{M}}, \\
& -p_{13}^{\mathrm{L}} - p_{23}^{\mathrm{L}} = p_3^{\mathrm{M}}, \\
& p_2^{\mathrm{C}} - p_{32}^{\mathrm{L}} + p_{12}^{\mathrm{L}} = p_{12}^{\mathrm{D}}, \\
& p_3^{\mathrm{C}} - p_{33}^{\mathrm{L}} + p_{13}^{\mathrm{L}} = p_{13}^{\mathrm{D}}, \\
& p_2^{\mathrm{R}} + p_2^{\mathrm{S}} + p_{22}^{\mathrm{L}} + p_{32}^{\mathrm{L}} = p_{22}^{\mathrm{D}}, \\
& p_3^{\mathrm{R}} + p_3^{\mathrm{S}} + p_{23}^{\mathrm{L}} + p_{33}^{\mathrm{L}} = p_{23}^{\mathrm{D}}, \\
& -60 + p_2^{\mathrm{RT}} = p_2^{\mathrm{M}}, \\
& 40 + p_3^{\mathrm{RT}} = p_3^{\mathrm{M}}, \\
& \frac{p_{12}^{\mathrm{L}}}{100} = 200(\delta_{12} - \delta_{22}), \\
& \frac{p_{13}^{\mathrm{L}}}{100} = 200(\delta_{13} - \delta_{23}), \\
& \frac{p_{22}^{\mathrm{L}}}{100} = 200(\delta_{12} - \delta_{32}), \\
& \frac{p_{23}^{\mathrm{L}}}{100} = 200(\delta_{13} - \delta_{33}), \\
& \frac{p_{32}^{\mathrm{L}}}{100} = 200(\delta_{22} - \delta_{32}), \\
& \frac{p_{33}^{\mathrm{L}}}{100} = 200(\delta_{23} - \delta_{33}), \\
& -50 \le p_{12}^{\mathrm{L}} \le 50, \\
& -50 \le p_{13}^{\mathrm{L}} \le 50, \\
& -50 \le p_{22}^{\mathrm{L}} \le 50, \\
& -50 \le p_{23}^{\mathrm{L}} \le 50, \\
& -50 \le p_{32}^{\mathrm{L}} \le 50, \\
& -50 \le p_{33}^{\mathrm{L}} \le 50, \\
& \delta_{12} = 0, \\
& \delta_{13} = 0, \\
& -\pi \le \delta_{22} \le \pi, \\
& -\pi \le \delta_{23} \le \pi, \\
& -\pi \le \delta_{32} \le \pi, \\
& -\pi \le \delta_{33} \le \pi, \\
& -60 \le -60 + p_2^{\mathrm{RT}} \le 40, \\
& -60 \le 40 + p_3^{\mathrm{RT}} \le 40, \\
& -100 \le p_2^{\mathrm{M}} \le 100, \\
& -100 \le p_3^{\mathrm{M}} \le 100,
\end{aligned}
$$

$$
\begin{aligned}
&20 + p_{12}^{\mathrm{D}} + p_{13}^{\mathrm{D}} \geq 30,\\
&30 + p_{22}^{\mathrm{D}} + p_{23}^{\mathrm{D}} \geq 50,\\
&0 \leq p_{12}^{\mathrm{D}} \leq 20,\\
&0 \leq p_{13}^{\mathrm{D}} \leq 10,\\
&5 \leq p_{22}^{\mathrm{D}} \leq 30,\\
&5 \leq p_{23}^{\mathrm{D}} \leq 15,\\
&-15 \leq p_{12}^{\mathrm{D}} - 20 \leq 15,\\
&-15 \leq p_{13}^{\mathrm{D}} - p_{12}^{\mathrm{D}} \leq 15,\\
&-25 \leq p_{22}^{\mathrm{D}} - 30 \leq 25,\\
&-25 \leq p_{23}^{\mathrm{D}} - p_{22}^{\mathrm{D}} \leq 25,\\
&0 \leq p_2^{\mathrm{C}} \leq 20u_2^{\mathrm{C}},\\
&0 \leq p_3^{\mathrm{C}} \leq 20u_3^{\mathrm{C}},\\
&-20 \leq p_2^{\mathrm{C}} - 0 \leq 20,\\
&-20 \leq p_3^{\mathrm{C}} - p_2^{\mathrm{C}} \leq 20,\\
&5(u_2^{\mathrm{C}} - 0) \leq \overline{c}_2^{\mathrm{C}},\\
&5(u_3^{\mathrm{C}} - u_2^{\mathrm{C}}) \leq \overline{c}_3^{\mathrm{C}},\\
&0 \leq \overline{c}_2^{\mathrm{C}},\\
&0 \leq \overline{c}_3^{\mathrm{C}},\\
&5(0 - u_2^{\mathrm{C}}) \leq \underline{c}_2^{\mathrm{C}},\\
&5(u_2^{\mathrm{C}} - u_3^{\mathrm{C}}) \leq \underline{c}_3^{\mathrm{C}},\\
&0 \leq \underline{c}_2^{\mathrm{C}},\\
&0 \leq \underline{c}_3^{\mathrm{C}},\\
&e_2^{\mathrm{S}} = 10 + 0.9p_2^{\mathrm{S,C}} - \frac{p_2^{\mathrm{S,D}}}{0.9},\\
&e_3^{\mathrm{S}} = e_2^{\mathrm{S}} + 0.9p_3^{\mathrm{S,C}} - \frac{p_3^{\mathrm{S,D}}}{0.9},\\
&0 \leq e_2^{\mathrm{S}} \leq 10,\\
&0 \leq e_3^{\mathrm{S}} \leq 10,\\
&0 \leq p_2^{\mathrm{S,C}} \leq 10,\\
&0 \leq p_3^{\mathrm{S,C}} \leq 10,\\
&0 \leq p_2^{\mathrm{S,D}} \leq 10,\\
&0 \leq p_3^{\mathrm{S,D}} \leq 10,\\
&p_2^{\mathrm{S}} = p_2^{\mathrm{S,D}} - p_2^{\mathrm{S,C}},\\
&p_3^{\mathrm{S}} = p_3^{\mathrm{S,D}} - p_3^{\mathrm{S,C}},
\end{aligned}
$$

Table 3.7 Illustrative Example 3.2: optimal schedules of the units in the VPP

Time period	p_t^{DA} [MW]	p_t^{RT} [MW]	p_{1t}^{D} [MW]	p_{2t}^{D} [MW]	p_t^{C} [MW]	p_t^{R} [MW]	p_t^{S} [MW]	$p_t^{\mathrm{S,D}}$ [MW]	$p_t^{\mathrm{S,C}}$ [MW]	e_t^{S} [MWh]	Profit [$]
1	−60	10.4	20	30	0	6	−5.6	0	5.6	10	2300
2	−60	87.5	5	5	20	8.5	9	9	0	0	−355
3	40	−63	10	15	0	2	0	0	0	0	1150

$$0 \le p_2^{\mathrm{R}} \le 8.5,$$
$$0 \le p_3^{\mathrm{R}} \le 2.6,$$

where the optimization variables of the problem are those included in set $\Phi_2^{3.2}=\{\overline{c}_2^{\mathrm{C}}, \overline{c}_3^{\mathrm{C}}, \underline{c}_2^{\mathrm{C}}, \underline{c}_3^{\mathrm{C}}, e_2^{\mathrm{S}}, e_3^{\mathrm{S}}, p_{12}^{\mathrm{D}}, p_{13}^{\mathrm{D}}, p_{22}^{\mathrm{D}}, p_{23}^{\mathrm{D}}, p_2^{\mathrm{C}}, p_3^{\mathrm{C}}, p_2^{\mathrm{R}}, p_3^{\mathrm{R}}, p_2^{\mathrm{S}}, p_3^{\mathrm{S}}, p_2^{\mathrm{S,D}}, p_3^{\mathrm{S,D}}, p_2^{\mathrm{S,C}}, p_3^{\mathrm{S,C}}, p_2^{\mathrm{M}}, p_3^{\mathrm{M}}, p_2^{\mathrm{RT}}, p_3^{\mathrm{RT}}, u_2^{\mathrm{C}}, u_3^{\mathrm{C}}, p_{12}^{\mathrm{L}}, p_{13}^{\mathrm{L}}, p_{22}^{\mathrm{L}}, p_{23}^{\mathrm{L}}, p_{32}^{\mathrm{L}}, p_{33}^{\mathrm{L}}, \delta_{12}, \delta_{13}, \delta_{22}, \delta_{23}, \delta_{32}, \delta_{33}\}$.

Solving the problem, the VPP finds a power quantity of 87.5 MW offered at zero price to be submitted in the RT energy market. Then, it is obtained that the RT market price in time period 2 is $60/MWh and thus the VPP sells 87.5 MW. As the scheduled power in this time period in the DA energy market is −60 MW, the net injection to node 1 connected to the main grid is 27.5 MW. Then, the dispatch decisions on the energy assets of the VPP are obtained as shown in Table 3.7, and the utility resulting from the energy consumption minus the production cost is −$355.

In the third round, the VPP uses the updated forecasts provided in Table 3.6 and solves the problem below:

$$\begin{aligned} \max_{\Phi_3^{3.2}} \quad & 27.8 p_3^{\mathrm{RT}} \\ & + 40 p_{13}^{\mathrm{D}} + 50 p_{23}^{\mathrm{D}} \\ & - (45 p_3^{\mathrm{C}} + \overline{c}_3^{\mathrm{C}} + \underline{c}_3^{\mathrm{C}}) \end{aligned}$$

subject to

$$\begin{aligned}
& - p_{13}^{\mathrm{L}} - p_{23}^{\mathrm{L}} = p_3^{\mathrm{M}}, \\
& p_3^{\mathrm{C}} - p_{33}^{\mathrm{L}} + p_{13}^{\mathrm{L}} = p_{13}^{\mathrm{D}}, \\
& p_3^{\mathrm{R}} + p_3^{\mathrm{S}} + p_{23}^{\mathrm{L}} + p_{33}^{\mathrm{L}} = p_{23}^{\mathrm{D}}, \\
& 40 + p_3^{\mathrm{RT}} = p_3^{\mathrm{M}}, \\
& \frac{p_{13}^{\mathrm{L}}}{100} = 200(\delta_{13} - \delta_{23}), \\
& \frac{p_{23}^{\mathrm{L}}}{100} = 200(\delta_{13} - \delta_{33}), \\
& \frac{p_{33}^{\mathrm{L}}}{100} = 200(\delta_{23} - \delta_{33}), \\
& - 50 \le p_{13}^{\mathrm{L}} \le 50, \\
& - 50 \le p_{23}^{\mathrm{L}} \le 50, \\
& - 50 \le p_{33}^{\mathrm{L}} \le 50, \\
& \delta_{13} = 0, \\
& - \pi \le \delta_{23} \le \pi, \\
& - \pi \le \delta_{33} \le \pi, \\
& - 60 \le 40 + p_3^{\mathrm{RT}} \le 40,
\end{aligned}$$

$$
\begin{aligned}
& -100 \le p_3^{\mathrm{M}} \le 100, \\
& 20 + 5 + p_{13}^{\mathrm{D}} \ge 30, \\
& 30 + 5 + p_{23}^{\mathrm{D}} \ge 50, \\
& 0 \le p_{13}^{\mathrm{D}} \le 10, \\
& 5 \le p_{23}^{\mathrm{D}} \le 15, \\
& -15 \le p_{13}^{\mathrm{D}} - 5 \le 15, \\
& -25 \le p_{23}^{\mathrm{D}} - 5 \le 25, \\
& 0 \le p_3^{\mathrm{C}} \le 20 u_3^{\mathrm{C}}, \\
& -20 \le p_3^{\mathrm{C}} - 20 \le 20, \\
& 5(u_3^{\mathrm{C}} - 1) \le \overline{c}_3^{\mathrm{C}}, \\
& 0 \le \overline{c}_3^{\mathrm{C}}, \\
& 5(1 - u_3^{\mathrm{C}}) \le \underline{c}_3^{\mathrm{C}}, \\
& 0 \le \underline{c}_3^{\mathrm{C}}, \\
& e_3^{\mathrm{S}} = 0 + 0.9 p_3^{\mathrm{S,C}} - \frac{p_3^{\mathrm{S,D}}}{0.9}, \\
& 0 \le e_3^{\mathrm{S}} \le 10, \\
& 0 \le p_3^{\mathrm{S,C}} \le 10, \\
& 0 \le p_3^{\mathrm{S,D}} \le 10, \\
& p_3^{\mathrm{S}} = p_3^{\mathrm{S,D}} - p_3^{\mathrm{S,C}}, \\
& 0 \le p_3^{\mathrm{R}} \le 2,
\end{aligned}
$$

where the optimization variables of the problem are those included in set $\Phi_3^{3.2}=\{\overline{c}_3^{\mathrm{C}}, \underline{c}_3^{\mathrm{C}}, e_3^{\mathrm{S}}, p_{13}^{\mathrm{D}}, p_{23}^{\mathrm{D}}, p_3^{\mathrm{C}}, p_3^{\mathrm{R}}, p_3^{\mathrm{S}}, p_3^{\mathrm{S,D}}, p_3^{\mathrm{S,C}}, p_3^{\mathrm{M}}, p_3^{\mathrm{RT}}, u_3^{\mathrm{C}}, p_{13}^{\mathrm{L}}, p_{23}^{\mathrm{L}}, p_{33}^{\mathrm{L}}, \delta_{13}, \delta_{23}, \delta_{33}\}$.

The VPP submits a power quantity of −63 MW in round 3 to compensate the deviation from the scheduled power of 40 MW in the DA energy market. The VPP considers a very high bid price, e.g., 100/MWh, to guarantee that its bid is accepted. Then, the RT energy market is cleared and the resulting market price is \$28/MWh in time period 3. Therefore, the scheduled power of −63 MW in the RT market is communicated to the VPP. Finally, the VPP dispatches its own energy assets provided in the last row of Table 3.7, and the resulting utility is equal to \$1150. Note that the profit provided in the last column of Table 3.7 is the utility of demands minus the production cost of the conventional power plant in the VPP.

□

A GAMS [9] code for solving Illustrative Example 3.2 is provided in Sect. 3.7.

3.4 Stochastic Programming Approach

In Sect. 3.3, the scheduling problem of a risk-neutral VPP is formulated using a deterministic approach in which the uncertain parameters in future time periods are replaced by deterministic values based on single-point forecasts. These deterministic models are simple and easy to implement. However, the uncertainty modeling is generally inaccurate. On one hand, the solutions of the deterministic models (3.1) and (3.2) are optimal provided that the actual realizations of the uncertain parameters are equal to the deterministic values. On the other hand, the deviation of the actual realization of the uncertain parameters from the deterministic values may lead to the infeasibility of the solution in real application.

To overcome these two issues, this section describes a stochastic programming approach that uses a more accurate uncertainty representation by incorporating a finite number of scenarios with their occurrence probabilities. This way, the risk-neutral VPP seeks to achieve the maximum expected utility over all considered scenario realizations. Additionally, the obtained solution is feasible for all considered scenario realizations. In the next subsection, the uncertainty model used by this stochastic programming approach is explained.

3.4.1 Uncertainty Model

In the stochastic programming approach, uncertain parameters, including the prices in the DA and RT energy markets and the available stochastic production levels of renewable generating units, are represented through a finite number of scenarios. The empirical analysis carried out in [17] indicates that the prices in the DA and RT energy markets are highly correlated. Moreover, this study verifies that the correlation between the available power levels of small-size *local* stochastic renewable production facilities and the prices in the energy markets is negligible. Thus, scenarios modeling uncertain parameters need to represent the correlation of energy market prices. Note, however, that in those cases with a high penetration of stochastic renewable production, a meaningful negative correlation between the available stochastic production levels and the prices in the energy markets may be observed. Under such a condition, scenarios should also address this correlation.

An appropriate uncertainty model should include the information obtained between the closure of the DA and the RT energy markets, including the market prices in the DA market and in the previous RT markets, as well as the available stochastic production levels. This means that the scenario realizations of the RT market prices are conditioned by DA energy market prices. Additionally, the available stochastic production level depends on the scenario realization of the available stochastic production levels between the closure of the DA and the related RT markets.

Given the above explanation, each scenario can be built through each plausible combination of uncertain parameter realizations between the closure of the DA and RT energy markets, indexed by ω, and realizations between the closure of the RT energy market and the actual dispatch decisions, indexed by γ. Accordingly, the scenario set Π^{DA} includes the plausible combinations of ω and γ, i.e., $\Pi^{\mathrm{DA}} = \{(\omega, \gamma), \forall \omega, \forall \gamma\}$. Each pair (ω, γ) has a probability $\pi_{\omega\gamma}$, while $\sum_{(\omega,\gamma)\in\Pi^{\mathrm{DA}}} \pi_{\omega\gamma} = 1$. The corresponding scenario tree is shown in Fig. 3.4.

Considering the scenario tree, the decision sequence for making decisions in the energy markets is described below:

1. The EMS makes the offering and bidding decisions in the DA energy market. The decisions are independent of the scenario realizations as they are made before the actual realization of the uncertain parameters is known. Thus, the offering and bidding decisions in the DA energy market are *here-and-now* decisions.
2. When a scenario ω occurs, the EMS receives information about the schedule of the VPP and the prices in the DA energy market, as well as the prices in the RT energy market and the available stochastic production levels between the closure of the DA and the RT energy markets.
3. The EMS makes the offering and bidding decisions in the RT energy market. These decisions depend on each scenario realization ω. Thus, these decisions are *wait-and-see* decisions with respect to the scenario realization ω.
4. Once a scenario γ is realized, the EMS knows the schedule of the VPP and the price in the RT energy market, as well as the available stochastic production levels.
5. The EMS dispatches the units in the VPP. Thus, the dispatch decisions are *wait-and-see* decisions with respect to the scenario realization γ.
6. The EMS updates scenarios and decides in the RT energy market time by time through steps 3 to 5 till the current day is finished.

Illustrative Example 3.3 Uncertainty Model for Stochastic Scheduling of VPP in the DA Energy Market

A VPP including only a PV production facility with a capacity of 10 MW is considered. The EMS decides on the schedules within a 3-h scheduling horizon for day $D + 1$. Time period -1 between the closure of the DA and RT energy markets at current day D is considered. Thus, the EMS knows the available PV power production level in time period -1 with most certainty.

Two scenarios of the prices in the DA energy market for the scheduling horizon are considered. Note that the actual price in the DA energy market in time period -1 is known at day D.

On the other hand, four scenarios for the difference between the prices in the RT and DA energy markets during the time horizon [1,3] are generated as explained below:

1. Two scenarios of the price difference between the closure of the DA and RT energy markets, i.e., in time period -1, are considered.

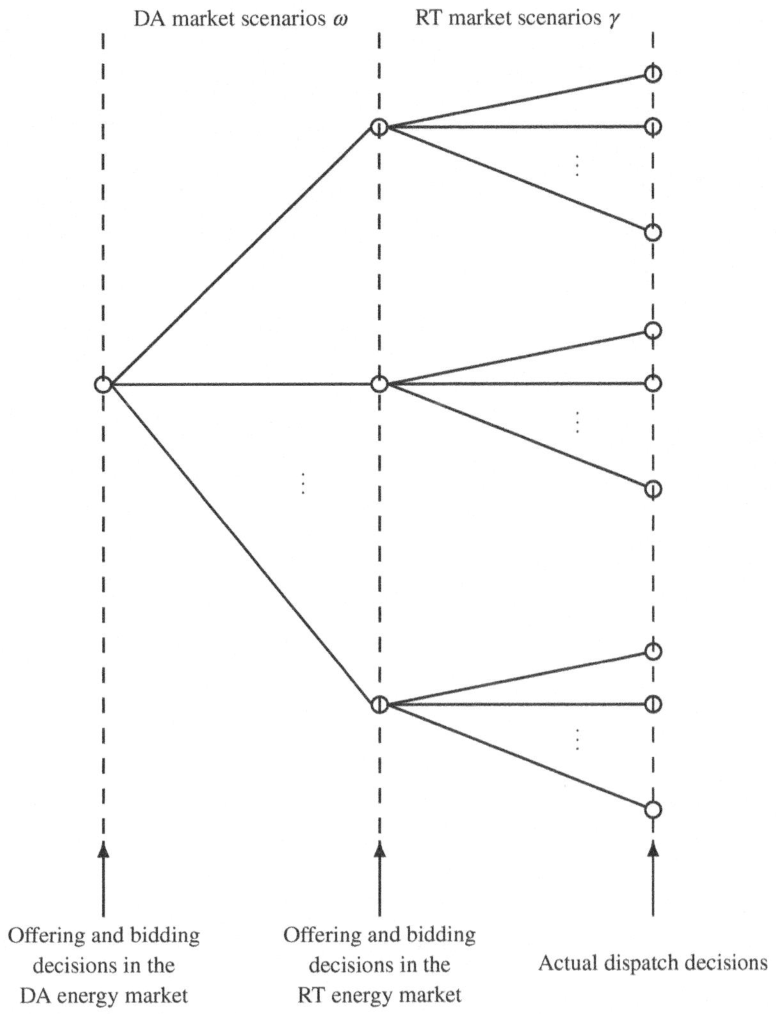

Fig. 3.4 Scenario tree

Table 3.8 Illustrative Example 3.3: generated scenarios for each uncertain parameter

Time period	λ_t^{DA} [\$/MWh]		$(\lambda_t^{\mathrm{RT}} - \lambda_t^{\mathrm{DA}})$ [\$/MWh]				$P_t^{\mathrm{R,A}}$ [MW]			
−1	**20**	**20**	+2	+2	0	0	**5**	**5**	**5**	**5**
1	20	30	+5	0	+2	−2	9	8	4	5
2	40	50	+10	+5	+4	−1	7	5	6	5
3	30	40	−5	−2	−10	−2	1	2	0	1
Probability	0.3	0.7	0.2	0.4	0.1	0.3	0.3	0.3	0.2	0.2

2. For each scenario realization in step 1, two scenarios of the price difference during the scheduling horizon, i.e., in time periods 1, 2, and 3, are generated.

Additionally, four scenarios for the available PV production levels during the time horizon [1,3] are generated. These scenarios with their occurrence probability are presented in Table 3.8. Bold values in Table 3.8 indicate known actual data.

Given the scenarios for uncertain parameters, the scenario tree for making decisions in the DA energy market is built. The scenario set includes a total of 32 scenarios. Table 3.9 indicates how each scenario realization ω is paired with each scenario realization γ to build the scenario set $\Pi^{\mathrm{DA}} = \{(\omega, \gamma), \ \forall \omega, \forall \gamma\}$. Each scenario ω presents a combination of 2 scenario realizations of the prices in the DA market during time periods 1, 2, and 3, and 2 scenario realizations of the price difference in time period −1. This means that 4 scenario realizations represented by ω are considered. For each scenario realization ω, each scenario γ is obtained as the combination of 2 scenario realizations of the price difference and 4 scenario realizations of the available PV production during time periods 1, 2, and 3. Thus, 8 scenario realizations presented by γ are obtained.

Table 3.9 Illustrative Example 3.3: scenario tree

	λ_t^{DA} [$/MWh]			λ_t^{RT} [$/MWh]				$P_t^{R,A}$ [MW]			
Scenario	1	2	3	−1	1	2	3	1	2	3	Probability
$\omega_1\gamma_1$	20	40	30	22	25	50	25	9	7	1	0.018
$\omega_1\gamma_2$	20	40	30	22	25	50	25	8	5	2	0.018
$\omega_1\gamma_3$	20	40	30	22	25	50	25	4	6	0	0.012
$\omega_1\gamma_4$	20	40	30	22	25	50	25	5	5	1	0.012
$\omega_1\gamma_5$	20	40	30	22	20	45	28	9	7	1	0.036
$\omega_1\gamma_6$	20	40	30	22	20	45	28	8	5	2	0.036
$\omega_1\gamma_7$	20	40	30	22	20	45	28	4	6	0	0.024
$\omega_1\gamma_8$	20	40	30	22	20	45	28	5	5	1	0.024
$\omega_2\gamma_1$	20	40	30	20	22	44	20	9	7	1	0.009
$\omega_2\gamma_2$	20	40	30	20	22	44	20	8	5	2	0.009
$\omega_2\gamma_3$	20	40	30	20	22	44	20	4	6	0	0.006
$\omega_2\gamma_4$	20	40	30	20	22	44	20	5	5	1	0.006
$\omega_2\gamma_5$	20	40	30	20	18	39	28	9	7	1	0.027
$\omega_2\gamma_6$	20	40	30	20	18	39	28	8	5	2	0.027
$\omega_2\gamma_7$	20	40	30	20	18	39	28	4	6	0	0.018
$\omega_2\gamma_8$	20	40	30	20	18	39	28	5	5	1	0.018
$\omega_3\gamma_1$	30	50	40	22	35	60	35	9	7	1	0.042
$\omega_3\gamma_2$	30	50	40	22	35	60	35	8	5	2	0.042
$\omega_3\gamma_3$	30	50	40	22	35	60	35	4	6	0	0.028
$\omega_3\gamma_4$	30	50	40	22	35	60	35	5	5	1	0.028
$\omega_3\gamma_5$	30	50	40	22	30	55	38	9	7	1	0.084
$\omega_3\gamma_6$	30	50	40	22	30	55	38	8	5	2	0.084
$\omega_3\gamma_7$	30	50	40	22	30	55	38	4	6	0	0.056
$\omega_3\gamma_8$	30	50	40	22	30	55	38	5	5	1	0.056
$\omega_4\gamma_1$	30	50	40	20	32	54	30	9	7	1	0.021
$\omega_4\gamma_2$	30	50	40	20	32	54	30	8	5	2	0.021
$\omega_4\gamma_3$	30	50	40	20	32	54	30	4	6	0	0.014
$\omega_4\gamma_4$	30	50	40	20	32	54	30	5	5	1	0.014
$\omega_4\gamma_5$	30	50	40	20	28	49	38	9	7	1	0.063
$\omega_4\gamma_6$	30	50	40	20	28	49	38	8	5	2	0.063
$\omega_4\gamma_7$	30	50	40	20	28	49	38	4	6	0	0.042
$\omega_4\gamma_8$	30	50	40	20	28	49	38	5	5	1	0.042

□

Illustrative Example 3.4 Uncertainty Model for Stochastic Scheduling of VPP in the RT Energy Market

The VPP makes its decisions in the RT energy market for the scheduling horizon. In first round, some minutes before the power delivery in time period 1, the prices in the DA energy market during the scheduling horizon, the price in the RT energy market in time period −1, and the available PV production level in time period −1 are known. Additionally, the EMS knows the available PV production level in time period 1 with most certainty. Using the new information that is available between the closure of the DA and RT energy markets, a number of scenarios are generated for each uncertain parameter as provided in Table 3.10. Using these scenarios, the scenario tree is built as shown in Table 3.11. In fact, set Π^{RT} contains all the scenario realizations required for making decision in the RT energy market in time period 1.

In the second round, the VPP makes its decisions in the RT energy market for time period 2 some minutes before the power delivery. As a previous step, the VPP updates the uncertainty model considering the new available information in time period 1. The price in the RT energy market and the available PV production level in time period 1 are known. Additionally, the EMS knows the available PV production level in time period 2 with most certainty. Accordingly, a number of scenarios are generated for each uncertain parameter as shown in Table 3.12. Using these scenarios, the scenario tree shown in Table 3.13 is obtained. Note that set Π^{RT} contains all the scenario realization required for making decision in the RT energy market in time period 2.

Table 3.10 Illustrative Example 3.4: generated scenarios for each uncertain parameter at first round

Time period	λ_t^{DA} [$/MWh]	$(\lambda_t^{\mathrm{RT}} - \lambda_t^{\mathrm{DA}})$ [$/MWh]		$P_t^{\mathrm{R,A}}$[MW]	
−1	**20**	**+1**	**+1**	**5**	**5**
1	**25**	+10	+4	**6**	**6**
2	**50**	+8	+5	10	8
3	**35**	−5	−2	4	2
Probability	1	0.6	0.4	0.8	0.2

Table 3.11 Illustrative Example 3.4: scenario tree at first round

	λ_t^{RT} [$/MWh]			$P_t^{\mathrm{R,A}}$[MW]		
Scenario	1	2	3	2	3	Probability
γ_1	35	58	30	10	4	0.48
γ_2	35	58	30	8	2	0.12
γ_3	29	55	33	10	4	0.32
γ_4	29	55	33	8	2	0.08

Table 3.12 Illustrative Example 3.4: generated scenarios for each uncertain parameter at second round

Time period	λ_t^{DA} [$/MWh]	$(\lambda_t^{\mathrm{RT}} - \lambda_t^{\mathrm{DA}})$		[$/MWh] $P_t^{\mathrm{R,A}}$ [MW]	
−1	**20**	**+1**	**+1**	**5**	**5**
1	**25**	**+10**	**+10**	**6**	**6**
2	**50**	+9	+12	**8.5**	**8.5**
3	**35**	−6	−4	3	2
Probability	1	0.5	0.5	0.6	0.4

Table 3.13 Illustrative Example 3.4: scenario tree at second round

	λ_t^{RT} [$/MWh]		$P_t^{\mathrm{R,A}}$ [MW]	
Scenario	2	3	3	Probability
γ_1	59	29	3	0.3
γ_2	59	29	2	0.2
γ_3	62	31	3	0.3
γ_4	62	31	2	0.2

In third round, for making decisions in time period 3, the available PV production level is known. Only the market price is unknown. The uncertainty in the price difference is represented using 4 scenarios, namely +5, −13, −10, and −6$/MWh with probabilities 0.1, 0.1, 0.4, and 0.4, respectively, and consequently the RT market price may take the values 40, 22, 25, and 29 $/MWh, respectively.

□

3.4.2 *Stochastic Scheduling in the Day-Ahead Energy Market*

The stochastic scheduling problem of a risk-neutral VPP for making decisions in the DA energy market is formulated as the following MILP problem:

$$\max_{\Phi^{\mathrm{DA,S}}} \quad \sum_{\omega}\sum_{\gamma} \pi_{\omega\gamma} \sum_{t\in\Omega^{\mathrm{T}}} \Bigg[\lambda_{t\omega}^{\mathrm{DA}} p_t^{\mathrm{DA}} \Delta t + \lambda_{t\omega\gamma}^{\mathrm{RT}} p_{t\omega}^{\mathrm{RT}} \Delta t$$

$$+ \sum_{d\in\Omega^{\mathrm{D}}} U_{dt}^{\mathrm{D}} p_{dt\omega\gamma}^{\mathrm{D}} \Delta t - \sum_{c\in\Omega^{\mathrm{C}}} \left(C_c^{\mathrm{C,V}} p_{ct\omega\gamma}^{\mathrm{C}} \Delta t + \overline{c}_{ct\omega\gamma}^{\mathrm{C}} + \underline{c}_{ct\omega\gamma}^{\mathrm{C}} \right) \Bigg] \tag{3.3a}$$

subject to

$$\sum_{c\in\Omega_n^{\mathrm{C}}} p_{ct\omega\gamma}^{\mathrm{C}} + \sum_{r\in\Omega_n^{\mathrm{R}}} p_{rt\omega\gamma}^{\mathrm{R}} + \sum_{s\in\Omega_n^{\mathrm{S}}} p_{st\omega\gamma}^{\mathrm{S}} - \sum_{\ell|s(\ell)=n} p_{\ell t\omega\gamma}^{\mathrm{L}} + \sum_{\ell|r(\ell)=n} p_{\ell t\omega\gamma}^{\mathrm{L}}$$

$$= p_{nt\omega\gamma}^{\mathrm{M}} + \sum_{d\in\Omega_n^{\mathrm{D}}} p_{dt\omega\gamma}^{\mathrm{D}}, \quad \forall n \in \Omega^{\mathrm{M}}, \forall t \in \Omega^{\mathrm{T}}, \forall\omega, \forall\gamma, \tag{3.3b}$$

$$\sum_{c\in\Omega_n^{\mathrm{C}}} p_{ct\omega\gamma}^{\mathrm{C}} + \sum_{r\in\Omega_n^{\mathrm{R}}} p_{rt\omega\gamma}^{\mathrm{R}} + \sum_{s\in\Omega_n^{\mathrm{S}}} p_{st\omega\gamma}^{\mathrm{S}} - \sum_{\ell|s(\ell)=n} p_{\ell t\omega\gamma}^{\mathrm{L}} + \sum_{\ell|r(\ell)=n} p_{\ell t\omega\gamma}^{\mathrm{L}}$$

$$= \sum_{d\in\Omega_n^{\mathrm{D}}} p_{dt\omega\gamma}^{\mathrm{D}}, \quad \forall n \in \Omega^{\mathrm{N}} \setminus n \in \Omega^{\mathrm{M}}, \forall t \in \Omega^{\mathrm{T}}, \forall\omega, \forall\gamma, \tag{3.3c}$$

$$p_t^{\mathrm{DA}} + p_{t\omega}^{\mathrm{RT}} = \sum_{n\in\Omega^{\mathrm{M}}} p_{nt\omega\gamma}^{\mathrm{M}}, \quad \forall t \in \Omega^{\mathrm{T}}, \forall\omega, \forall\gamma, \tag{3.3d}$$

$$p_{\ell t\omega\gamma}^{\mathrm{L}} = B_\ell(\delta_{s(\ell)t\omega\gamma} - \delta_{r(\ell)t\omega\gamma}), \quad \forall \ell \in \Omega^{\mathrm{L}}, \forall t \in \Omega^{\mathrm{T}}, \forall\omega, \forall\gamma, \tag{3.3e}$$

$$-\overline{P}_\ell^{\mathrm{L}} \le p_{\ell t\omega\gamma}^{\mathrm{L}} \le \overline{P}_\ell^{\mathrm{L}} \quad \forall \ell \in \Omega^{\mathrm{L}}, \forall t \in \Omega^{\mathrm{T}}, \forall\omega, \forall\gamma, \tag{3.3f}$$

$$\delta_{nt\omega\gamma} = 0, \quad n : \text{ref.}, \forall t \in \Omega^{\mathrm{T}}, \forall\omega, \forall\gamma, \tag{3.3g}$$

$$-\pi \le \delta_{nt\omega\gamma} \le \pi, \quad \forall n \in \Omega^{\mathrm{N}} \setminus n : \text{ref.}, \forall t \in \Omega^{\mathrm{T}}, \forall\omega, \forall\gamma, \tag{3.3h}$$

$$\underline{P}^{\mathrm{V}} \le p_t^{\mathrm{DA}} \le \overline{P}^{\mathrm{V}}, \quad \forall t \in \Omega^{\mathrm{T}}, \tag{3.3i}$$

$$\underline{P}^{\mathrm{V}} \le p_t^{\mathrm{DA}} + p_{t\omega}^{\mathrm{RT}} \le \overline{P}^{\mathrm{V}}, \quad \forall t \in \Omega^{\mathrm{T}}, \forall\omega, \tag{3.3j}$$

$$-\overline{P}_n^{\mathrm{M}} \le p_{nt\omega\gamma}^{\mathrm{M}} \le \overline{P}_n^{\mathrm{M}}, \quad \forall n \in \Omega^{\mathrm{M}}, \forall t \in \Omega^{\mathrm{T}}, \forall\omega, \forall\gamma, \tag{3.3k}$$

$$\sum_{t\in\Omega^{\mathrm{T}}} p_{dt\omega\gamma}^{\mathrm{D}} \Delta t \ge \underline{E}_d^{\mathrm{D}}, \quad \forall d \in \Omega^{\mathrm{D}}, \forall\omega, \forall\gamma, \tag{3.3l}$$

$$\underline{P}_{dt}^{\mathrm{D}} \le p_{dt\omega\gamma}^{\mathrm{D}} \le \overline{P}_{dt}^{\mathrm{D}}, \quad \forall d \in \Omega^{\mathrm{D}}, \forall t \in \Omega^{\mathrm{T}}, \forall\omega, \forall\gamma, \tag{3.3m}$$

$$-R_d^{\mathrm{D,D}} \Delta t \le p_{dt\omega\gamma}^{\mathrm{D}} - p_{d(t-1)\omega\gamma}^{\mathrm{D}} \le R_d^{\mathrm{D,U}} \Delta t, \quad \forall d \in \Omega^{\mathrm{D}}, \forall t \in \Omega^{\mathrm{T}}, \forall\omega, \forall\gamma, \tag{3.3n}$$

$$\underline{P}_c^{\mathrm{C}} u_{ct\omega\gamma}^{\mathrm{C}} \le p_{ct\omega\gamma}^{\mathrm{C}} \le \overline{P}_c^{\mathrm{C}} u_{ct\omega\gamma}^{\mathrm{C}}, \quad \forall c \in \Omega^{\mathrm{C}}, \forall t \in \Omega^{\mathrm{T}}, \forall\omega, \forall\gamma, \tag{3.3o}$$

$$-R_c^{\mathrm{C,D}} \Delta t \le p_{ct\omega\gamma}^{\mathrm{C}} - p_{c(t-1)\omega\gamma}^{\mathrm{C}} \le R_c^{\mathrm{C}} \Delta t, \quad \forall c \in \Omega^{\mathrm{C}}, \forall t \in \Omega^{\mathrm{T}}, \forall\omega, \forall\gamma, \tag{3.3p}$$

$$\overline{C}_c^{\mathrm{C}}(u_{ct\omega\gamma}^{\mathrm{C}} - u_{c(t-1)\omega\gamma}^{\mathrm{C}}) \le \overline{c}_{ct\omega\gamma}^{\mathrm{C}}, \quad \forall c \in \Omega^{\mathrm{C}}, \forall t \in \Omega^{\mathrm{T}}, \forall\omega, \forall\gamma, \tag{3.3q}$$

$$0 \le \overline{c}_{ct\omega\gamma}^{\mathrm{C}}, \quad \forall c \in \Omega^{\mathrm{C}}, \forall t \in \Omega^{\mathrm{T}}, \forall\omega, \forall\gamma, \tag{3.3r}$$

$$\underline{C}_c^{\mathrm{C}}(u_{c(t-1)\omega\gamma}^{\mathrm{C}} - u_{ct\omega\gamma}^{\mathrm{C}}) \le \underline{c}_{ct\omega\gamma}^{\mathrm{C}}, \quad \forall c \in \Omega^{\mathrm{C}}, \forall t \in \Omega^{\mathrm{T}}, \forall\omega, \forall\gamma, \tag{3.3s}$$

$$0 \le \underline{c}_{ct\omega\gamma}^{\mathrm{C}} \quad \forall c \in \Omega^{\mathrm{C}}, \forall t \in \Omega^{\mathrm{T}}, \forall\omega, \forall\gamma, \tag{3.3t}$$

$$e_{st\omega\gamma}^{\mathrm{S}} = e_{s(t-1)\omega\gamma}^{\mathrm{S}} + \eta_s^{\mathrm{S,C}} p_{st\omega\gamma}^{\mathrm{S,C}} \Delta t - \frac{p_{st\omega\gamma}^{\mathrm{S,D}}}{\eta_s^{\mathrm{S,D}}} \Delta t, \quad \forall s \in \Omega^{\mathrm{S}}, \forall t \in \Omega^{\mathrm{T}}, \forall\omega, \forall\gamma, \tag{3.3u}$$

$$\underline{E}_{st}^{\mathrm{S}} \le e_{st\omega\gamma}^{\mathrm{S}} \le \overline{E}_{st}^{\mathrm{S}}, \quad \forall s \in \Omega^{\mathrm{S}}, \forall t \in \Omega^{\mathrm{T}}, \forall\omega, \forall\gamma, \tag{3.3v}$$

$$0 \le p_{st\omega\gamma}^{\mathrm{S,C}} \le \overline{P}_s^{\mathrm{S,C}}, \quad \forall s \in \Omega^{\mathrm{S}}, \forall t \in \Omega^{\mathrm{T}}, \forall\omega, \forall\gamma, \tag{3.3w}$$

$$0 \le p_{st\omega\gamma}^{\mathrm{S,D}} \le \overline{P}_s^{\mathrm{S,D}}, \quad \forall s \in \Omega^{\mathrm{S}}, \forall t \in \Omega^{\mathrm{T}}, \forall\omega, \forall\gamma, \tag{3.3x}$$

$$p^{\mathrm{S}}_{st\omega\gamma} = p^{\mathrm{S,D}}_{st\omega\gamma} - p^{\mathrm{S,C}}_{st\omega\gamma}, \quad \forall s \in \Omega^{\mathrm{S}}, \forall t \in \Omega^{\mathrm{T}}, \forall \omega, \forall \gamma, \tag{3.3y}$$

$$0 \le p^{\mathrm{R}}_{rt\omega\gamma} \le P^{\mathrm{R,A}}_{rt\omega\gamma}, \quad \forall r \in \Omega^{\mathrm{R}}, \forall t \in \Omega^{\mathrm{T}}, \forall \omega, \forall \gamma, \tag{3.3z}$$

where variables in set $\Phi^{\mathrm{DA,S}}=\Big\{e^{\mathrm{S}}_{st\omega\gamma}, p^{\mathrm{S}}_{st\omega\gamma}, p^{\mathrm{S,C}}_{st\omega\gamma}, p^{\mathrm{S,D}}_{st\omega\gamma}, \forall s \in \Omega^{\mathrm{S}}, \forall t \in \Omega^{\mathrm{T}}, \forall \omega, \forall \gamma; \overline{c}^{\mathrm{C}}_{ct\omega\gamma}, \underline{c}^{\mathrm{C}}_{ct\omega\gamma}, p^{\mathrm{C}}_{ct\omega\gamma}, u^{\mathrm{C}}_{ct\omega\gamma}, \forall c \in \Omega^{\mathrm{C}}, \forall t \in \Omega^{\mathrm{T}}, \forall \omega, \forall \gamma; p^{\mathrm{D}}_{dt\omega\gamma}, \forall d \in \Omega^{\mathrm{D}}, \forall t \in \Omega^{\mathrm{T}}, \forall \omega, \forall \gamma; p^{\mathrm{R}}_{rt\omega\gamma}, \forall r \in \Omega^{\mathrm{R}}, \forall t \in \Omega^{\mathrm{T}}, \forall \omega, \forall \gamma; p^{\mathrm{DA}}_{t}, \forall t \in \Omega^{\mathrm{T}}; p^{\mathrm{RT}}_{t\omega}, \forall t, \forall \omega; p^{\mathrm{L}}_{\ell t\omega\gamma}, \forall \ell \in \Omega^{\mathrm{L}}, \forall t \in \Omega^{\mathrm{T}}, \forall \omega, \forall \gamma; p^{\mathrm{M}}_{nt\omega\gamma}, \delta_{nt\omega\gamma}, \forall n \in \Omega^{\mathrm{M}}, \forall t \in \Omega^{\mathrm{T}}, \forall \omega, \forall \gamma; \forall n \in \Omega^{\mathrm{N}}, \forall t \in \Omega^{\mathrm{T}}, \forall \omega, \forall \gamma\Big\}$, are the optimization variables of problem (3.3).

The structure of the stochastic model (3.3) is very similar to the structure of the deterministic model (3.1), while considering the potential impact of the scenario realizations on the objective function and the constraints. The main differences between these two models are explained below:

1. The utility of the VPP is uncertain and depends on the scenario realization. In this case, the objective function (3.3a) is the expected utility over all considered scenario realizations. Thus, the utility for each scenario realization multiplied by the corresponding occurrence probability is calculated.
2. The feasible operating region of the VPP depends on the scenario realization. Constraints (3.3b)–(3.3z) provide the feasible operating region for all scenario realizations.

Solving (3.3) provides the optimal power quantity during the scheduling horizon $[1, |\Omega^{\mathrm{T}}|]$ to be submitted in the DA energy market. However, it may be more practical to submit offering and bidding curves (pairs of power-price) instead of a single power quantity. With the purpose of building offering and bidding curves [13], variable p^{DA}_{t} is made dependent with the scenario realization ω, and thus it is replaced by $p^{\mathrm{DA}}_{t\omega}$ in the stochastic model (3.3). Additionally, the following constraints are added to the stochastic model (3.3):

$$p^{\mathrm{DA}}_{t\omega} \le p^{\mathrm{DA}}_{t\omega'}, \quad \text{if } \lambda^{\mathrm{DA}}_{t\omega} \le \lambda^{\mathrm{DA}}_{t\omega'}, \quad \forall t \in \Omega^{\mathrm{T}}, \forall \omega, \forall \omega', \tag{3.4a}$$

$$p^{\mathrm{DA}}_{t\omega} = p^{\mathrm{DA}}_{t\omega'}, \quad \text{if } \lambda^{\mathrm{DA}}_{t\omega} = \lambda^{\mathrm{DA}}_{t\omega'}, \quad \forall t \in \Omega^{\mathrm{T}}, \forall \omega, \forall \omega'. \tag{3.4b}$$

In practice, a nondecreasing curve is required to be submitted in the DA energy market. As both negative and positive values for buying and selling energy are considered, constraints (3.4a) constitute the nondecreasing curves, while constraints (3.4b) consider the nonanticipativity constraints that impose that a single power quantity must be submitted for two scenario realizations with the same price in the DA market.

Illustrative Example 3.5 Stochastic Scheduling of the VPP in the DA Energy Market: Offering and Bidding Strategies

The VPP characterized by the techno-economic data in Illustrative Example 3.1 is considered. The VPP decides the offering and bidding strategies in the DA energy market for the 3-h scheduling horizon. As explained in Illustrative Example 3.3, the uncertainties involved are the DA market prices during the scheduling horizon, the RT market prices, and the available PV production levels spanning from the closure of the DA energy market till the end of the scheduling horizon. The uncertainty model in Illustrative Example 3.3 including 32 scenarios is considered.

The stochastic model (3.3) is implemented and run to obtain the offering and bidding decisions in the DA energy market. The pairs of quantity-price to be submitted for time periods 1, 2, and 3 are obtained as:

- Time period 1: (−60 MW, \$20/MWh) and (−60 MW, \$30/MWh).
- Time period 2: (−60 MW, \$40/MWh) and (−60 MW, \$50/MWh).
- Time period 3: (40 MW, \$30/MWh) and (40 MW, \$40/MWh).

Figure 3.5 depicts the quantity-price curves submitted in the DA energy market. As the clearing prices in the DA energy market are \$25/MWh, \$50/MWh, and \$35/MWh in time periods 1, 2, and 3, respectively, the scheduled power is −60 MW, −60 MW, and 40 MW, respectively.

As indicated in Table 3.8 of Illustrative Example 3.3, the difference between the prices in RT and DA energy markets during the first and second time periods is positive in most scenarios. In this condition, the VPP prefers to buy energy in the DA energy market, and sell it to the RT energy market. However, the price difference in the third time period is negative for all scenarios, which motivates the VPP to sell in the DA market and buy in the RT market. The VPP pays \$3100 for trading energy in the DA energy market.

□

A GAMS [9] code for solving Illustrative Example 3.5 is provided in Sect. 3.7.

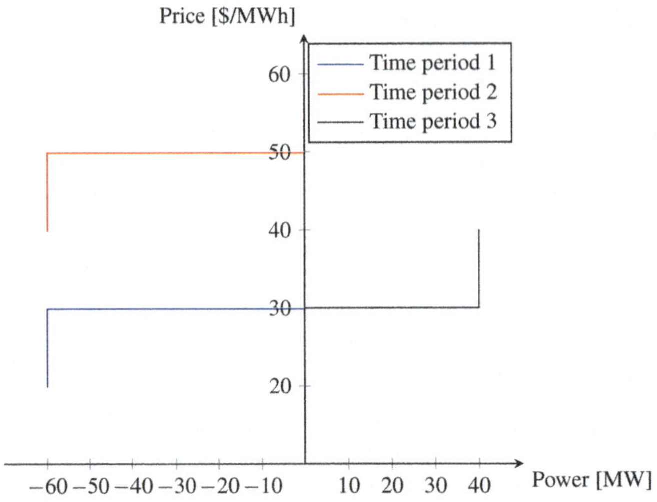

Fig. 3.5 Illustrative Example 3.5: quantity-price curves submitted in the DA energy market

3.4.3 Stochastic Scheduling in the Real-Time Energy Market

The stochastic scheduling problem of a risk-neutral VPP to determine its optimal offering and bidding strategy in the RT energy market in each time period $\tau \in \Omega^{\mathrm{T}}$ can be formulated as the MILP problem below:

$$\max_{\Phi^{\mathrm{RT,S}}} \sum_{\gamma \in \Pi^{\mathrm{RT}}} \pi_\gamma \sum_{t=\tau}^{|\Omega^{\mathrm{T}}|} \Big[\lambda^{\mathrm{RT}}_{t\gamma} p^{\mathrm{RT}}_t \Delta t + \sum_{d \in \Omega^{\mathrm{D}}} U^{\mathrm{D}}_{dt} p^{\mathrm{D}}_{dt\gamma} \Delta t$$

$$- \sum_{c \in \Omega^{\mathrm{C}}} (C^{\mathrm{C,V}}_c p^{\mathrm{C}}_{ct\gamma} \Delta t + \overline{c}^{\mathrm{C}}_{ct\gamma} + \underline{c}^{\mathrm{C}}_{ct\gamma}) \Big] \tag{3.5a}$$

subject to

$$\sum_{c \in \Omega^{\mathrm{C}}_n} p^{\mathrm{C}}_{ct\gamma} + \sum_{r \in \Omega^{\mathrm{R}}_n} p^{\mathrm{R}}_{rt\gamma} + \sum_{s \in \Omega^{\mathrm{S}}_n} p^{\mathrm{S}}_{st\gamma} - \sum_{\ell | s(\ell)=n} p^{\mathrm{L}}_{\ell t\gamma} + \sum_{\ell | r(\ell)=n} p^{\mathrm{L}}_{\ell t\gamma}$$

$$= p^{\mathrm{M}}_{nt\gamma} + \sum_{d \in \Omega^{\mathrm{D}}_n} p^{\mathrm{D}}_{dt\gamma}, \quad \forall n \in \Omega^{\mathrm{M}}, \forall t \geq \tau, \forall \gamma, \tag{3.5b}$$

$$\sum_{c \in \Omega^{\mathrm{C}}_n} p^{\mathrm{C}}_{ct\gamma} + \sum_{r \in \Omega^{\mathrm{R}}_n} p^{\mathrm{R}}_{rt\gamma} + \sum_{s \in \Omega^{\mathrm{S}}_n} p^{\mathrm{S}}_{st\gamma} - \sum_{\ell | s(\ell)=n} p^{\mathrm{L}}_{\ell t\gamma} + \sum_{\ell | r(\ell)=n} p^{\mathrm{L}}_{\ell t\gamma}$$

$$= \sum_{d \in \Omega^{\mathrm{D}}_n} p^{\mathrm{D}}_{dt\gamma}, \quad \forall n \in \Omega^{\mathrm{N}} \setminus n \in \Omega^{\mathrm{M}}, \forall t \geq \tau, \forall \gamma, \tag{3.5c}$$

$$\tilde{p}^{\mathrm{DA}}_t + p^{\mathrm{RT}}_t = \sum_{n \in \Omega^{\mathrm{M}}} p^{\mathrm{M}}_{nt\gamma}, \quad \forall t \geq \tau, \forall \gamma, \tag{3.5d}$$

$$p^{\mathrm{L}}_{\ell t\gamma} = B_\ell (\delta_{s(\ell)t\gamma} - \delta_{r(\ell)t\gamma}), \quad \forall \ell \in \Omega^{\mathrm{L}}, \forall t \geq \tau, \forall \gamma, \tag{3.5e}$$

$$-\overline{P}^{\mathrm{L}}_\ell \leq p^{\mathrm{L}}_{\ell t\gamma} \leq \overline{P}^{\mathrm{L}}_\ell, \quad \forall \ell \in \Omega^{\mathrm{L}}, \forall t \geq \tau, \forall \gamma, \tag{3.5f}$$

$$\delta_{nt\gamma} = 0, \quad n : \text{ref.}, \forall t \geq \tau, \forall \gamma, \tag{3.5g}$$

$$-\pi \leq \delta_{nt\gamma} \leq \pi, \quad \forall n \in \Omega^{\mathrm{N}} \setminus n : \text{ref.}, \forall t \geq \tau, \forall \gamma, \tag{3.5h}$$

$$\underline{P}^{\mathrm{V}} \leq \tilde{p}_t^{\mathrm{DA}} + p_t^{\mathrm{RT}} \leq \overline{P}^{\mathrm{V}} \quad \forall t \geq \tau, \tag{3.5i}$$

$$-\overline{P}_n^{\mathrm{M}} \leq p_{nt\gamma}^{\mathrm{M}} \leq \overline{P}_n^{\mathrm{M}}, \quad \forall n \in \Omega^{\mathrm{M}}, \forall t \geq \tau, \forall \gamma, \tag{3.5j}$$

$$\sum_{t=1}^{\tau-1} \tilde{p}_{dt}^{\mathrm{D}} + \sum_{t=\tau}^{|\Omega^{\mathrm{T}}|} p_{dt\gamma}^{\mathrm{D}} \Delta t \geq \underline{E}_d^{\mathrm{D}}, \quad \forall d \in \Omega^{\mathrm{D}}, \forall \gamma, \tag{3.5k}$$

$$\underline{P}_{dt}^{\mathrm{D}} \leq p_{dt\gamma}^{\mathrm{D}} \leq \overline{P}_{dt}^{\mathrm{D}}, \quad \forall d \in \Omega^{\mathrm{D}}, \forall t \geq \tau, \forall \gamma, \tag{3.5l}$$

$$-R_d^{\mathrm{D,D}} \Delta t \leq p_{dt\gamma}^{\mathrm{D}} - p_{d(t-1)\gamma}^{\mathrm{D}} \leq R_d^{\mathrm{D,U}} \Delta t, \quad \forall d \in \Omega^{\mathrm{D}}, \forall t \geq \tau, \forall \gamma, \tag{3.5m}$$

$$\underline{P}_c^{\mathrm{C}} u_{ct\gamma}^{\mathrm{C}} \leq p_{ct\gamma}^{\mathrm{C}} \leq \overline{P}_c^{\mathrm{C}} u_{ct\gamma}^{\mathrm{C}}, \quad \forall c \in \Omega^{\mathrm{C}}, \forall t \geq \tau, \forall \gamma, \tag{3.5n}$$

$$-R_c^{\mathrm{C,D}} \Delta t \leq p_{ct\gamma}^{\mathrm{C}} - p_{c(t-1)\gamma}^{\mathrm{C}} \leq R_c^{\mathrm{C,U}} \Delta t, \quad \forall c \in \Omega^{\mathrm{C}}, \forall t \geq \tau, \forall \gamma, \tag{3.5o}$$

$$\overline{C}_c^{\mathrm{C}} (u_{ct\gamma}^{\mathrm{C}} - u_{c(t-1)\gamma}^{\mathrm{C}}) \leq \overline{c}_{ct\gamma}^{\mathrm{C}}, \quad \forall c \in \Omega^{\mathrm{C}}, \forall t \geq \tau, \forall \gamma, \tag{3.5p}$$

$$0 \leq \overline{c}_{ct\gamma}^{\mathrm{C}}, \quad \forall c \in \Omega^{\mathrm{C}}, \forall t \geq \tau, \forall \gamma, \tag{3.5q}$$

$$\underline{C}_c^{\mathrm{C}} (u_{c(t-1)\gamma}^{\mathrm{C}} - u_{ct\gamma}^{\mathrm{C}}) \leq \underline{c}_{ct\gamma}^{\mathrm{C}}, \quad \forall c \in \Omega^{\mathrm{C}}, \forall t \geq \tau, \forall \gamma, \tag{3.5r}$$

$$0 \leq \underline{c}_{ct\gamma}^{\mathrm{C}}, \quad \forall c \in \Omega^{\mathrm{C}}, \forall t \geq \tau, \forall \gamma, \tag{3.5s}$$

$$e_{st\gamma}^{\mathrm{S}} = e_{s(t-1)\gamma}^{\mathrm{S}} + \eta_s^{\mathrm{S,C}} p_{st\gamma}^{\mathrm{S,C}} \Delta t - \frac{p_{st\gamma}^{\mathrm{S,D}}}{\eta_s^{\mathrm{S,D}}} \Delta t, \quad \forall s \in \Omega^{\mathrm{S}}, \forall t \geq \tau, \forall \gamma, \tag{3.5t}$$

$$\underline{E}_{st}^{\mathrm{S}} \leq e_{st\gamma}^{\mathrm{S}} \leq \overline{E}_{st}^{\mathrm{S}}, \quad \forall s \in \Omega^{\mathrm{S}}, \forall t \geq \tau, \forall \gamma, \tag{3.5u}$$

$$0 \leq p_{st\gamma}^{\mathrm{S,C}} \leq \overline{P}_s^{\mathrm{S,C}}, \quad \forall s \in \Omega^{\mathrm{S}}, \forall t \geq \tau, \forall \gamma, \tag{3.5v}$$

$$0 \leq p_{st\gamma}^{\mathrm{S,D}} \leq \overline{P}_s^{\mathrm{S,D}}, \quad \forall s \in \Omega^{\mathrm{S}}, \forall t \geq \tau, \forall \gamma, \tag{3.5w}$$

$$p_{st\gamma}^{\mathrm{S}} = p_{st\gamma}^{\mathrm{S,D}} - p_{st\gamma}^{\mathrm{S,C}}, \quad \forall s \in \Omega^{\mathrm{S}}, \forall t \geq \tau, \forall \gamma, \tag{3.5x}$$

$$0 \leq p_{rt\gamma}^{\mathrm{R}} \leq \tilde{P}_{rt}^{\mathrm{R,A}}, \quad \forall r \in \Omega^{\mathrm{R}}, t = \tau, \forall \gamma, \tag{3.5y}$$

$$0 \leq p_{rt\gamma}^{\mathrm{R}} \leq P_{rt\gamma}^{\mathrm{R,A}}, \quad \forall r \in \Omega^{\mathrm{R}}, \forall t > \tau, \forall \gamma, \tag{3.5z}$$

where variables in set $\Phi^{\mathrm{RT,S}} = \Big\{ e_{st\gamma}^{\mathrm{S}}, p_{st\gamma}^{\mathrm{S}}, p_{st\gamma}^{\mathrm{S,C}}, p_{st\gamma}^{\mathrm{S,D}}, \forall s \in \Omega^{\mathrm{S}}, \forall \gamma; p_{ct\gamma}^{\mathrm{C}}, u_{ct\gamma}^{\mathrm{C}}, \forall c \in \Omega^{\mathrm{C}}, \forall \gamma; p_{dt\gamma}^{\mathrm{D}}, \forall d \in \Omega^{\mathrm{D}}, \forall \gamma; p_{rt\gamma}^{\mathrm{R}}, \forall r \in \Omega^{\mathrm{R}}, \forall \gamma; p_t^{\mathrm{RT}}, \forall t \in \Omega^{\mathrm{T}}; p_{\ell t\gamma}^{\mathrm{L}}, \forall \ell \in \Omega^{\mathrm{L}}, \forall \gamma; p_{nt\gamma}^{\mathrm{M}}, \forall n \in \Omega^{\mathrm{M}}, \forall \omega, \forall \gamma; \delta_{nt\gamma}, \forall n \in \Omega^{\mathrm{N}}, \forall \gamma \Big\}$, $\forall t \geq \tau$, are the optimization variables of problem (3.5).

The main differences between the stochastic programming model (3.5) and the corresponding deterministic model (3.2) are summarized below:

1. The objective function (3.5a) is the expected utility over different scenario realizations of the uncertain parameters during the scheduling horizon $[\tau, |\Omega^{\mathrm{T}}|]$.
2. The feasible operating region of the units in the VPP and the traded power with the main grid are made dependent with the scenario realizations of the uncertain parameters during the scheduling horizon $[\tau, |\Omega^{\mathrm{T}}|]$.

Note that the optimal power quantities p_t^{RT} are obtained for the scheduling horizon $[\tau, |\Omega^{\mathrm{T}}|]$ by solving the stochastic programming model (3.5). However, only the single power quantity p_t^{RT} in time period τ is submitted in the RT energy market.

As in the case of the DA market, in order to submit a quantity-price curve, decision variables p_t^{RT} are made dependent with the scenario realization γ. In other words, decision variables p_t^{RT} in the stochastic programming model (3.5) are replaced by $p_{t\gamma}^{\text{RT}}$. Additionally, the following constraints are also added to the stochastic model (3.5):

$$p_{t\gamma}^{\text{RT}} \leq p_{t\gamma'}^{\text{RT}}, \quad \text{if } \lambda_{t\gamma}^{\text{RT}} \leq \lambda_{t\gamma'}^{\text{RT}}, \quad \forall t \in \Omega^{\text{T}}, \forall \gamma, \forall \gamma', \tag{3.6a}$$

$$p_{t\gamma}^{\text{RT}} = p_{t\gamma'}^{\text{RT}}, \quad \text{if } \lambda_{t\gamma}^{\text{RT}} = \lambda_{t\gamma'}^{\text{RT}}, \quad \forall t \in \Omega^{\text{T}}, \forall \gamma, \forall \gamma'. \tag{3.6b}$$

Constraints (3.6a) are imposed to build a nondecreasing curve as practically submitted in energy markets. Constraints (3.6b) guarantee a single power quantity corresponding to the same price scenario realizations in each time period.

Illustrative Example 3.6 Stochastic Scheduling in the RT Energy Market: Offering and Bidding Strategies

The VPP described in Illustrative Example 3.1 is considered. As obtained in Illustrative Example 3.5, the VPP trades −60, −60, and 40 MW in time periods 1, 2, and 3, respectively, in the DA energy market. Next, the VPP takes the opportunity of participating in the RT energy market during the scheduling horizon. In the first round (i.e., $\tau = 1$), the EMS makes decision some minutes before power delivery in this time period. Considering the new information of the DA market price during the scheduling horizon, the RT market price in time period −1, and the available PV production level in the time period −1, the uncertainty model is built similar to Table 3.11 in Illustrative Example 3.4. In addition, the VPP knows the available PV production level in time period 1 with most certainty.

The stochastic model (3.5) is implemented and run to determine the offering and bidding decisions in the first time period. The quantity-price curve submitted in this time period is built by the pairs (12.9 MW, \$35/MWh) and (10.4 MW, \$29/MWh) shown in Fig. 3.6. As the market price in this time period is \$35/MWh, the scheduled power is 12.9 MW, which means that the VPP behaves as a seller in the RT energy market, achieving \$453.1 from selling energy in the RT energy market. Note that the VPP has already bought 60 MW from the DA energy market in the first time period. Thus, the EMS dispatches the units in the VPP considering a net injection of 47.1 MW from node 1 connected to the main grid and the initial values of the power dispatch of the units.

Before the second time period, new information about the RT market price in the first time period and the available PV production level in the second time period is available, and the uncertainty model shown in Table 3.13 of Illustrative Example 3.4 is updated. In the second round (i.e., $\tau = 2$), the quantity-price curve is obtained as (65 MW, \$59/MWh) and (90 MW, \$62/MWh) for the second time period. Figure 3.6 shows the quantity-price curve submitted to the RT energy market. Once the market price is declared as \$60/MWh, the scheduled power is 65 MW and the VPP gains \$3900 by selling in the RT energy market. Again, the VPP behaves as a seller in the RT energy market in the second time period, while it has already bought 60 MW from the DA energy market in this time period. Considering a net injection of 5 MW to the main grid and the power dispatch of the units in the first time period, the EMS dispatches the units.

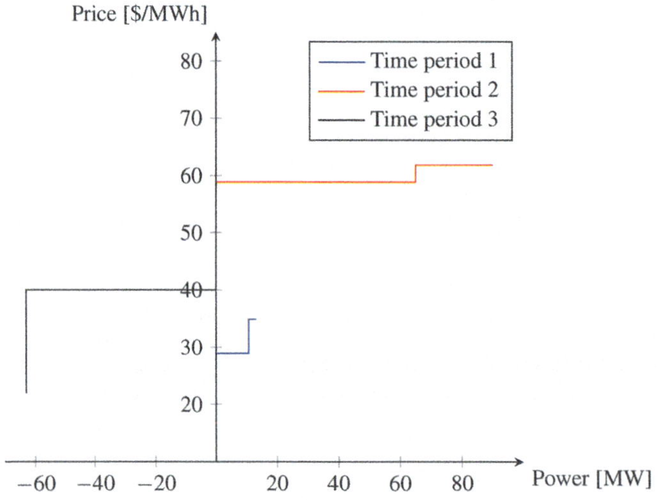

Fig. 3.6 Illustrative Example 3.6: quantity-price curves submitted in the RT energy market

Table 3.14 Illustrative Example 3.6: optimal schedules of the units in the VPP

Time period	p_t^{DA} [MW]	p_t^{RT} [MW]	p_{1t}^{D} [MW]	p_{2t}^{D} [MW]	p_t^{C} [MW]	p_t^{R} [MW]	p_t^{S} [MW]	$p_t^{\mathrm{S,D}}$ [MW]	$p_t^{\mathrm{S,C}}$ [MW]	e_t^{S} [MWh]	Profit [$]
1	−60	12.9	17.5	30	0	6	−5.6	0	5.6	10	2200
2	−60	65	2.5	30	20	8.5	9	9	0	0	1020
3	40	−63	10	15	0	2	0	0	0	0	1150

Before time period 3, the uncertainty model is updated similar to Illustrative Example 3.4. In the third round (i.e., $\tau = 3$), the quantity-price curve is built using the power-price pairs (−63 MW, \$22/MWh), (−63 MW, \$25/MWh), (−63 MW, \$29/MWh), and (−63 MW, \$40/MWh) depicted in Fig. 3.6. The scheduled power is −63 MW due to a market price of \$28/MWh, and then the VPP pays \$1764 in the RT energy market. Thus, the VPP buys 63 MW from the RT energy market, while it sells 40 MW to the DA market in the third time period. This means that the VPP is dispatched so that the net injection from the main grid is 23 MW. The results of the power dispatch of the units in each time period are presented in Table 3.14. Note that the profit shown in the last column is obtained as the utility of energy consumption minus the production cost of the conventional power plant.

□

A GAMS [9] code for solving Illustrative Example 3.6 is provided in Sect. 3.7.

3.5 Participation of Virtual Power Plants in Futures Markets

In the previous sections of this chapter, the scheduling problem of a risk-neutral VPP in pool-based energy markets (e.g., the DA and RT energy markets) is modeled and solved. However, in reality, market participants can also sign forward contracts in the futures market for a medium-term horizon (e.g., a year on the basis of monthly contracts) to hedge against the potential risk of the volatile signals in the pool markets, mainly in market prices. For example, stochastic programming models are used by a power retailer [4] and a power producer [6] to choose among a set of available forward contracts in the futures market.

By the expansion of VPPs in the near future, it is expected that they also participate in futures markets. The VPP can take the opportunity of participating in these futures markets to hedge against the financial risk that results from the volatility of pool-based market prices. In this sense, the VPP can decide to sell energy in the futures market at lower prices than the expected prices in the pool markets. However, reducing the risk level is achieved at the expense of lower expected utility. Additionally, the VPP can act as buyer in the futures market, while buying energy with higher prices than the expected prices in the pool markets. Afterward, the VPP can sell or buy energy in the pool market to achieve a higher utility or procure requirements near the time of power delivery.

The scheduling problem of a risk-neutral VPP in the futures market is the problem of selecting among a set of forward contracts. For example, under a yearly framework including a number of contract time periods (e.g., 12 months), the VPP can select among a set of available forward contracts to sign in the futures market for each contract time period. The prices of forward contracts are generally known and the VPP decides the amount of energy to be bought/sold in each forward contract.

On the other hand, this scheduling problem is under the uncertainty that stems from the prices in the pool-based markets and the available stochastic production levels in a medium-term horizon. Thus, the VPP decides on the power bought/sold in each forward contract for the future time periods, being these decisions made before the actual realizations of the uncertainties, and thus *here-and-now* decisions with respect to the uncertainties. Once a scenario ω is realized, the prices in the pool markets (e.g., the DA energy market) and the available stochastic production levels throughout the year are known.

The stochastic scheduling of a risk-neutral VPP in the futures market seeks to maximize the expected utility that can be obtained by the VPP over future realizations of the uncertain parameters in the pool-based markets plus/minus the revenue/cost due to forward selling/buying contracts in the futures market during a medium-term horizon (e.g., one year). This problem is formulated as the following MILP problem:

$$\max_{\Phi^{\mathrm{FT}}} \quad \sum_{f \in \Omega^{\mathrm{FT}}} (\lambda_f^{\mathrm{FT,S}} p_f^{\mathrm{FT,S}} \Delta f - \lambda_f^{\mathrm{FT,B}} p_f^{\mathrm{FT,B}} \Delta f)$$
$$+ \sum_{\omega \in \Pi^{\mathrm{FT}}} \pi_\omega \sum_{t=1}^{365|\Omega^{\mathrm{T}}|} \left[\lambda_{t\omega}^{\mathrm{DA}} p_{t\omega}^{\mathrm{DA}} \Delta t + \sum_{d \in \Omega^{\mathrm{D}}} U_{dt}^{\mathrm{D}} p_{dt\omega}^{\mathrm{D}} \Delta t - \sum_{c \in \Omega^{\mathrm{C}}} C_{ct}^{\mathrm{C,V}} p_{ct\omega}^{\mathrm{C}} \Delta t \right]. \tag{3.7}$$

The above objective function is maximized over the feasible operating region of the VPP during the medium-term horizon. The feasible operating region is formed by three groups of constraints as explained below:

1. Technical constraints of the units in the VPP depending on each scenario realization ω. These constraints include the minimum daily energy consumption, as well as lower and upper power consumption limits for demands, lower and upper generation limits for conventional power plants, available production levels as upper bound of power production by stochastic renewable generating units, and energy balance, energy capacity, and charging/discharging power limits of storage units.
2. Constraints to guarantee the power balance in the VPP considering the power traded in the futures and pool-based markets as:

$$\sum_{c \in \Omega^{\mathrm{C}}} p_{ct\omega}^{\mathrm{C}} + \sum_{r \in \Omega^{\mathrm{R}}} p_{rt\omega}^{\mathrm{R}} + \sum_{s \in \Omega^{\mathrm{S}}} p_{st\omega}^{\mathrm{S}} - \sum_{d \in \Omega^{\mathrm{D}}} p_{dt\omega}^{\mathrm{D}}$$
$$= \sum_{f \in \Omega_t^{\mathrm{FT}}} (p_f^{\mathrm{FT,S}} - p_f^{\mathrm{FT,B}}) + p_{t\omega}^{\mathrm{DA}}, \quad \forall t \in \Omega^{\mathrm{T}}, \forall \omega. \tag{3.8}$$

3. Constraints that impose bounds on the power bought/sold through each forward buying/selling contract:

$$0 \le p_f^{\mathrm{FT,B}} \le \overline{P}_f^{\mathrm{FT,B}} y_f, \quad \forall f \in \Omega^{\mathrm{FT}}, \tag{3.9a}$$
$$0 \le p_f^{\mathrm{FT,S}} \le \overline{P}_f^{\mathrm{FT,S}} (1 - y_f), \quad \forall f \in \Omega^{\mathrm{FT}}, \tag{3.9b}$$
$$y_f \in \{0, 1\}, \quad \forall f \in \Omega^{\mathrm{FT}}, \tag{3.9c}$$

where:

- Π^{FT} is a set that includes all scenario realizations of uncertain parameters during one year,
- Ω^{FT} is the set of available forward contracts throughout the year,
- Ω_t^{FT} is the set of available forward contracts during time period t,
- Δf is the power-to-energy conversion factor for forward contract f,
- $\overline{P}_f^{\mathrm{FT,B}}$ is the maximum power that can be bought by forward contract f,
- $\overline{P}_f^{\mathrm{FT,S}}$ is the maximum power that can be sold by forward contract f,
- $p_f^{\mathrm{FT,B}}/p_f^{\mathrm{FT,S}}$ is the power bought/sold by forward contract f, and
- $\Phi^{\mathrm{FT}} = \Big\{ e_{st\omega}^{\mathrm{S}}, p_{st\omega}^{\mathrm{S}}, p_{st\omega}^{\mathrm{S,C}}, p_{st\omega}^{\mathrm{S,D}}, \forall s \in \Omega^{\mathrm{S}}, \forall T \in \Omega^{\mathrm{T}}, \forall \omega; p_{ct\omega}^{\mathrm{C}}, \forall c \in \Omega^{\mathrm{C}}, \forall t \in \Omega^{\mathrm{T}}, \forall \omega; p_{rt\omega}^{\mathrm{R}}, \forall r \in \Omega^{\mathrm{R}}, \forall t \in \Omega^{\mathrm{T}}, \forall \omega; p_{dt\omega}^{\mathrm{D}}, \forall d \in \Omega^{\mathrm{C,D}}, \forall t \in \Omega^{\mathrm{T}}, \forall \omega; p_{t\omega}^{\mathrm{DA}}, \forall t \in \Omega^{\mathrm{T}}, \forall \omega; y_f, p_f^{\mathrm{FT,S}}, p_f^{\mathrm{FT,B}}, \forall f \in \Omega^{\mathrm{FT}} \Big\}$ is the set of optimization variables.

Solving the above stochastic programming model, the amount of power that is bought from/sold to the futures market through each forward buying/selling contract, i.e., variables $p_f^{\text{FT,B}}/p_f^{\text{FT,S}}$, $\forall f \in \Omega^{\text{FT}}$, is determined. Note that the power traded in the futures market needs to be taken into account when the VPP participates in the DA and RT energy markets.

3.6 Summary and Further Reading

This chapter provides practical mathematical models used for scheduling a risk-neutral VPP in energy markets. The scheduling problem of the VPP is the problem of determining the power to be traded by the VPP in different energy markets, e.g., the DA and RT energy markets, and then dispatching the units in the VPP. On one hand, a VPP includes a variety of units characterized with different techno-economic properties. On the other hand, uncertainties in the market prices and the available stochastic production levels impose risk on the scheduling problem. This means that the scheduling of the VPP is not only a complex operation problem, but also a decision-making problem under uncertainty.

In order to schedule the VPP, two classes of optimization models are used in this chapter:

1. A deterministic model that allows to decide offering and bidding decisions under a single scenario realization of uncertain parameters, provided by single-point forecasts.
2. A stochastic programming model in which the VPP seeks to maximize the expected utility over a predefined set of scenario realizations of uncertain parameters, while operating the units under each scenario realization.

A number of illustrative examples are presented to clarify the structure and applicability of the deterministic and stochastic programming models. The main conclusions that can be drawn from the illustrative examples and the mathematical models provided in this chapter are as follows:

1. The single scenario realization of uncertain parameters used in the deterministic model makes this model easy to implement and run in a highly efficient computation. However, the stochastic programming model uses a number of scenario realizations, which allows more accurate representation of uncertainties at a higher computation cost.
2. Due to the low accurate representation of uncertain market prices in the deterministic model, the power traded by the VPP in the energy markets may be zero in some time periods. Under such a condition, the VPP accounts only on its own available energy resources that may lead to an infeasible operation.
3. The willingness of the VPP for buying/selling energy in the DA and RT energy markets depends on the difference between the prices in these two markets. The behavior of the VPP in the energy markets depends on the accuracy of the uncertainty representation of the price difference between these two markets.
4. The stochastic programming model generally needs a large number of scenarios that exponentially intensifies the computational cost of the problem. This is a crucial issue for scheduling the VPP in the RT energy market. In order to solve this issue, robust optimization is an alternative option that will be analyzed in Chap. 4.

Readers who are not familiar enough with the mathematical optimization models used throughout this chapter such as linear programming and mixed-integer linear programming are referred to Appendix A of this book. Detailed explanations on how to build scenario trees and use stochastic programming is presented in [7]. Additionally, readers can refer to Appendix B of this book for the mathematics of the stochastic programming.

3.7 GAMS Codes

This section provides the GAMS codes used to solve some of the illustrative examples of this chapter.

3.7.1 Deterministic Scheduling in the Day-Ahead Energy Market

An input GAMS [9] code for solving Illustrative Example 3.1 is provided below:

```
* Deterministic scheduling of the risk-neutral VPP in the day-ahead energy market
SET b 'Bus' /b1*b3/;
SET l 'Line' /l1*l3/;
SET t 'Time period' /t1*t3/;
SET u 'Unit' /u1*u5/;
SET incDES(l,b) 'destination bus' /l1.b2,l2.b3,l3.b3/;
SET incORI(l,b) 'origin bus' /l1.b1,l2.b1,l3.b2/;
SET incD(u) 'demand' /u1*u2/;
SET incG(u) 'conventional power plant' /u3/;
SET incR(u) 'stochastic production facility' /u4/;
SET incS(u) 'storage unit' /u5/;
SET incMB(b) 'buses connected to the main grid' /b1/;
SET incREF(b) 'reference bus' /b1/;
SET incDB(u,b) 'demand at bus' /u1.b2,u2.b3/;
SET incGB(u,b) 'conventional power plant at bus' /u3.b2/;
SET incRB(u,b) 'stochastic production facility at bus' /u4.b3/;
SET incSB(u,b) 'storage unit at bus' /u5.b3/;
SCALAR TT 'Number of trading hours in a day' /3/;
SCALAR delta 'Power-energy conversion factor' /1/;
SCALAR mup 'Output conversion efficiency of the storage unit' /0.9/;
SCALAR mun 'Input conversion efficiency of the storage unit' /0.9/;
SCALAR Sbase 'base apparent power' /100/;
SCALAR Vbase 'base voltage' /10/;
SCALAR PV_max 'Maximum power that can be produced by the VPP' /40/;
SCALAR PV_min 'Maximum power that can be consumed by the VPP' /-60/;
***** Demands data
TABLE PD_max(u,t) 'Maximum load level for unit u in time period t'
         t1      t2      t3
u1       20      20      10
u2       30      30      15;
TABLE PD_min(u,t) 'Minimum load level for unit u in time period t'
         t1      t2      t3
u1       0       0       0
u2       5       5       5;
TABLE Utility(u,t) 'Utility of unit u for energy consumption in time period t'
         t1       t2       t3
u1       40       50       40
u2       50       60       50;
PARAMETER ED_min(u) 'Minimum daily energy consumption requested by unit i'
/u1       30
u2        50/;
```

```
PARAMETER RD_up(u) 'Load drop ramping limits for unit u'
/u1        15
u2         25/;
PARAMETER RD_do(u) 'Pick-up ramping limit for unit u'
/u1        15
u2         25/;
PARAMETER PD0(u) 'Initial load level for each unit'
/u1       10
u2        20/;
***** Conventional power plants data
PARAMETER PG_max(u) 'Maximum power production limit for unit u in time period t'
/u3        20/;
PARAMETER PG_min(u) 'Minimum power production limit for unit u in time period t'
/u3        0/;
PARAMETER RG_up(u) 'Ramping up limits for unit u'
/u3         20/;
PARAMETER RG_do(u) 'Ramping down limit for unit u'
/u3         20/;
TABLE CG(u,t) 'Production cost for unit u'
         t1     t2      t3
u3         45     45      45;
PARAMETER CG_start(u) 'Start up cost for unit u'
/u3         5/;
PARAMETER CG_shot(u) 'Shot down cost for unit u'
/u3         5/;
PARAMETER PG0(u) 'Initial power production of unit u'
/u3         0/;
PARAMETER V0(u) 'Initial status of unit u'
/u3         0/;
***** Storage units data
PARAMETER ES_max(u) 'Energy capacity of the storage unit u'
/u5       10/;
PARAMETER PS_p(u) 'Maximum power that can be produced by unit u'
/u5       10/;
PARAMETER PS_n(u) 'Maximum power that can be transferred to unit u'
/u5       10/;
PARAMETER ES0(u) 'Initial energy stored in the storage unit'
/u5       5/;
***** Network data
PARAMETER PL_max(l) 'Capacity of line l'
/l1         50
l2          50
l3          50/;
PARAMETER BL(l) 'Susceptance of line l'
/l1        200
l2         200
l3         200/;
BL(l)=BL(l)/(Sbase/(Vbase**2));
PARAMETER PM_max(b) 'Maximum power that can be bought from the main grid'
/b1       100/;
```

```
***** Single-point forecasts of uncertain parameters
TABLE PR_max(u,t) 'Point forecasts of available stochastic
    production in time period t'
          t1     t2      t3
u4        6.9    5.8     1.1;
PARAMETER lambda_DA(t) 'Point forecasts of prices in the day-
    ahead energy market in time period t'
/t1        27
t2         47
t3         37/;
PARAMETER lambda_RT(t) 'Point forecasts of prices in the real-
    time energy market in time period t'
/t1        27.6
t2         51.1
t3         33.6/;
***** Actual data of the day-ahead market prices
PARAMETER lambda_DA_act(t) 'Actual day-ahead market price in time
     period t'
/t1        25
t2         50
t3         35/;
VARIABLES
pl(l,t) 'Power flow through line l in time period t'
pda(t) 'Power traded by the VPP in the day-ahead energy market in
     time period t'
prt(t) 'Power traded by the VPP in the real-time energy market in
     time period t'
pm(b,t) 'Power injection at bus b connected to the main grid in
    time period t'
teta(b,t) 'Voltage angle at bus b in time period t'
z 'Objective'
p(u,t)  'Power dispatch of unit u in time period t';
POSITIVE VARIABLES
es(u,t)  'Energy stored at the end of time period t'
psp(u,t) 'Power produced by the storage unit in time period t'
psn(u,t) 'Power transferred to the storage unit in time period t'
cg_st(u,t) 'Auxiliary variable to linearize the start-up cost of
    unit u in time period t'
cg_sh(u,t) 'Auxiliary variable to linearize the shot-down cost of
     unit u in time period t';
BINARY VARIABLES
v(u,t) 'Binary variable to state the commitment status of unit u
    in time period t';
*****************************************************************
**************** Equations definition ***********************
EQUATIONS
obj 'The objective function'
nodal_balance_mg 'power balance at the main grid buses'
nodal_balance 'power balance at the remaining buses'
trade 'power traded in the day-ahead and real-time markets'
line_power 'power flow through lines'
line_max 'upper bound of lines flows'
line_min 'lower bound of lines flows'
voltage_angle_ref 'voltage angle at reference bus'
```

```
voltage_angle_max 'upper bound of voltage angles'
voltage_angle_min 'lower bound of voltage angles'
tradeda_min 'lower bound of power traded in the day-ahead market'
tradeda_max 'upper bound of power traded in the day-ahead market'
tradedart_min 'lower bound of power traded in the energy markets'
tradedart_max 'upper bound of power traded in the energy markets'
trade_min 'lower bound of power injection at each bus'
trade_max 'upper bound of power injection at each bus'
dailyenergy_min 'daily minimum energy consumption for each demand
    '
demand_max 'upper bound of load level'
demand_min 'lower bound of load level'
demand_ramp_down_initial 'load ramp down limitation'
demand_ramp_pick_initial 'load ramp up limitation'
demand_ramp_down 'load ramp down limitation'
demand_ramp_pick 'load ramp up limitation'
CPP_max 'maximum power production limitation'
CPP_min 'minimum power production limitation'
CPP_ramp_down_initial 'ramp down production limitation'
CPP_ramp_up_initial 'ramp up production limitation'
CCP_ramp_down 'ramp down production limitation'
CCP_ramp_up 'ramp up production limitation'
CCP_startcost_initial 'start up cost of unit'
CCP_shotcost_initial  'shot down cost of unit'
CCP_startcost 'start up cost of unit'
CCP_shotcost 'shot down cost of unit'
storage_balance_initial 'power balance equation in the storage
    unit'
storage_balance 'power balance equation in the storage unit'
storage_capacity 'upper bound of energy stored in the storage
    unit'
storage_psp_max 'maxmimum power that can be produced by storage
    unit'
storage_psn_max 'maxmimum power that can be transferred to
    storage unit'
storage_injection 'power injection of storage unit'
stochastic_max 'available stochastic production limitation';
*
obj..       z=E=SUM(t,lambda_DA(t)*pda(t)*delta+lambda_RT(t)*prt(
    t)*delta+SUM(u$incD(u),Utility(u,t)*p(u,t)*delta)-SUM(u$incG(
    u),CG(u,t)*p(u,t)*delta+cg_st(u,t)+cg_sh(u,t)));
nodal_balance_mg(b,t)$incMB(b)..    SUM(u$incGB(u,b),p(u,t))+SUM(
    u$incRB(u,b),p(u,t))+SUM(u$incSB(u,b),p(u,t))-SUM(l$incORI(l,
    b),pl(l,t))+SUM(l$incDES(l,b),pl(l,t))=E=pm(b,t)+SUM(u$incDB(
    u,b),p(u,t));
nodal_balance(b,t)$(NOT incMB(b)).. SUM(u$incGB(u,b),p(u,t))+SUM(
    u$incRB(u,b),p(u,t))+SUM(u$incSB(u,b),p(u,t))-SUM(l$incORI(l,
    b),pl(l,t))+SUM(l$incDES(l,b),pl(l,t))=E=SUM(u$incDB(u,b),p(u
    ,t));
trade(t)..          pda(t)+prt(t)=E=SUM(b$incMB(b),pm(b,t));
line_power(l,t)..   pl(l,t)=E=Sbase*BL(l)*(SUM(b$incORI(l,b),teta(
    b,t))-SUM(b$incDES(l,b),teta(b,t)));
line_max(l,t)..     pl(l,t)=L=PL_max(l);
line_min(l,t)..     -PL_max(l)=L=pl(l,t);
```

```
voltage_angle_ref(b,t)$incREF(b)..    teta(b,t)=E=0;
voltage_angle_max(b,t)..              teta(b,t)=L=Pi;
voltage_angle_min(b,t)..              -Pi=L=teta(b,t);
tradeda_min(t)..                      PV_min=L=pda(t);
tradeda_max(t)..                      pda(t)=L=PV_max;
tradedart_min(t)..                    PV_min=L=pda(t)+prt(t);
tradedart_max(t)..                    pda(t)+prt(t)=L=PV_max;
trade_min(b,t)$incMB(b)..             -PM_max(b)=L=pm(b,t);
trade_max(b,t)$incMB(b)..             pm(b,t)=L=PM_max(b);
dailyenergy_min(u)$incD(u)..    SUM(t,p(u,t)*delta)=G=ED_min(u);
demand_max(u,t)$incD(u)..             p(u,t)=L=PD_max(u,t);
demand_min(u,t)$incD(u)..             PD_min(u,t)=L=p(u,t);
demand_ramp_down_initial(u,t)$((incD(u)) AND (ORD(t) EQ 1)).. -
    RD_do(u)=L=p(u,t)-PD0(u);
demand_ramp_pick_initial(u,t)$((incD(u)) AND (ORD(t) EQ 1))..
    p(u,t)-PD0(u)=L=RD_up(u);
demand_ramp_down(u,t)$((incD(u)) AND (ORD(t) GE 2)).. -RD_do(u)=
    L=p(u,t)-p(u,t-1);
demand_ramp_pick(u,t)$((incD(u)) AND (ORD(t) GE 2))..   p(u,t)-p(
    u,t-1)=L=RD_up(u);
CPP_max(u,t)$incG(u)..    p(u,t)=L=PG_max(u)*v(u,t);
CPP_min(u,t)$incG(u)..    v(u,t)*PG_min(u)=L=p(u,t);
CPP_ramp_down_initial(u,t)$((incG(u)) AND (ORD(t) EQ 1)).. -RG_do
    (u)=L=p(u,t)-PG0(u);
CPP_ramp_up_initial(u,t)$((incG(u)) AND (ORD(t) EQ 1)).. p(u,t)-
    PG0(u)=L=RG_up(u);
CCP_ramp_down(u,t)$((incG(u)) AND (ORD(t) GE 2)).. -RG_do(u)=L=p
    (u,t)-p(u,t-1);
CCP_ramp_up(u,t)$((incG(u)) AND (ORD(t) GE 2)).. p(u,t)-p(u,t
    -1)=L=RG_up(u);
CCP_startcost_initial(u,t)$((incG(u)) AND (ORD(t) EQ 1))..
    CG_start(u)*(v(u,t)-V0(u))=L=cg_st(u,t);
CCP_shotcost_initial(u,t)$((incG(u)) AND (ORD(t) EQ 1))..
    CG_shot(u)*(V0(u)-v(u,t))=L=cg_sh(u,t);
CCP_startcost(u,t)$((incG(u)) AND (ORD(t) GE 2)).. CG_start(u)*(
    v(u,t)-v(u,t-1))=L=cg_st(u,t);
CCP_shotcost(u,t)$((incG(u)) AND (ORD(t) GE 2)).. CG_shot(u)*(v(
    u,t-1)-v(u,t))=L=cg_sh(u,t);
storage_balance_initial(u,t)$((incS(u)) AND (ORD(t) EQ 1)).. es(u
    ,t)=E=ES0(u)+mun*psn(u,t)*delta-psp(u,t)*delta/mup;
storage_balance(u,t)$((incS(u)) AND (ORD(t) GE 2)).. es(u,t)=E=es
    (u,t-1)+mun*psn(u,t)*delta-psp(u,t)*delta/mup;
storage_capacity(u,t)$incS(u)..    es(u,t)=L=ES_max(u);
storage_psp_max(u,t)$incS(u)..     psp(u,t)=L=PS_p(u);
storage_psn_max(u,t)$incS(u)..     psn(u,t)=L=PS_n(u);
storage_injection(u,t)$incS(u)..   p(u,t)=E=psp(u,t)-psn(u,t);
stochastic_max(u,t)$incR(u)..      p(u,t)=L=PR_max(u,t);
*********************************************************
**************** Model definition ***************************
MODEL OSDA       /all/;
OPTION OPTCR=0;
OPTION OPTCA=0;
OPTION iterlim=1e8;
****************************************************
```

```
**Parameter definition for building quantity-price curve ****
PARAMETER state 'convergence status';
PARAMETER pda_act 'actual power traded in the day-ahead market';
PARAMETER revenueda_act 'revenue in the day-ahead market';
*******************************************************
**************** Solution ****************************
**************** Solving the scheduling problem**********
SOLVE OSDA USING MIP MAXIMIZING z;
state = OSDA.modelstat;
**************** Day-ahead market clearing*****************
pda_act(t)=pda.l(t);
revenueda_act=SUM(t,pda_act(t)*lambda_DA_act(t));
****************************************
display state, pda_act, revenueda_act;
```

3.7.2 Deterministic Scheduling in the Real-Time Energy Market

An input GAMS [9] code for solving Illustrative Example 3.2 is provided below:

```
* Deterministic scheduling of the risk-neutral VPP in the real-
    time energy market
SET b 'Bus' /b1*b3/;
SET l 'Line' /l1*l3/;
SET t 'Time period' /t1*t3/;
SET h 'Time period' /h1*h3/;
SET u 'Unit' /u1*u5/;
SET incDES(l,b) 'destination bus' /l1.b2,l2.b3,l3.b3/;
SET incORI(l,b) 'origin bus' /l1.b1,l2.b1,l3.b2/;
SET incD(u) 'demand' /u1*u2/;
SET incG(u) 'conventional power plant' /u3/;
SET incR(u) 'stochastic production facility' /u4/;
SET incS(u) 'storage unit' /u5/;
SET incMB(b) 'buses connected to the main grid' /b1/;
SET incREF(b) 'reference bus' /b1/;
SET incDB(u,b) 'demand at bus' /u1.b2,u2.b3/;
SET incGB(u,b) 'conventional power plant at bus' /u3.b2/;
SET incRB(u,b) 'stochastic production facility at bus' /u4.b3/;
SET incSB(u,b) 'storage unit at bus' /u5.b3/;
SCALAR TT 'Number of trading hours in a day' /3/;
SCALAR delta 'Power-energy conversion factor' /1/;
SCALAR mup 'Output conversion efficiency of the storage unit'
    /0.9/;
SCALAR mun 'Input conversion efficiency of the storage unit'
    /0.9/;
SCALAR Sbase 'base apparent power' /100/;
SCALAR Vbase 'base voltage' /10/;
SCALAR PV_max 'Maximum power that can be produced by the VPP'
    /40/;
SCALAR PV_min 'Maximum power that can be consumed by the VPP'
    /-60/;
```

```
SCALAR tau 'iteration' /1/;
***** Demands data
TABLE PD_max(u,t) 'Maximum load level for unit u in time period t
    '
          t1      t2      t3
u1        20      20      10
u2        30      30      15;
TABLE PD_min(u,t) 'Minimum load level for unit u in time period t
    '
          t1      t2      t3
u1        0       0       0
u2        5       5       5;
TABLE Utility(u,t) 'Utility of unit u for energy consumption in
    time period t'
          t1       t2      t3
u1        40       50      40
u2        50       60      50;
PARAMETER ED_min(u) 'Minimum daily energy consumption requested
    by unit i'
/u1        30
u2         50/;
PARAMETER RD_up(u) 'Load drop ramping limits for unit u'
/u1        15
u2         25/;
PARAMETER RD_do(u) 'Pick-up ramping limit for unit u'
/u1        15
u2         25/;
PARAMETER PD0(u) 'Initial load level for each unit'
/u1       10
u2        20/;
***** Conventional power plants data
PARAMETER PG_max(u) 'Maximum power production limit for unit u in
     time period t'
/u3        20/;
PARAMETER PG_min(u) 'Minimum power production limit for unit u in
     time period t'
/u3        0/;
PARAMETER RG_up(u) 'Ramping up limits for unit u'
/u3         20/;
PARAMETER RG_do(u) 'Ramping down limit for unit u'
/u3         20/;
TABLE CG(u,t) 'Production cost for unit u'
           t1      t2      t3
u3         45      45      45;
PARAMETER CG_start(u) 'Start up cost for unit u'
/u3         5/;
PARAMETER CG_shot(u) 'Shot down cost for unit u'
/u3         5/;
PARAMETER PG0(u) 'Initial power production of unit u'
/u3         0/;
PARAMETER V0(u) 'Initial status of unit u'
/u3         0/;
***** Storage units data
PARAMETER ES_max(u) 'Energy capacity of the storage unit u'
```

```
/u5        10/;
PARAMETER PS_p(u) 'Maximum power that can be produced by unit u'
/u5        10/;
PARAMETER PS_n(u) 'Maximum power that can be transferred to unit u'
/u5        10/;
PARAMETER ES0(u) 'Initial energy stored in the storage unit'
/u5        5/;
***** Network data
PARAMETER PL_max(l) 'Capacity of line l'
/l1          50
l2           50
l3           50/;
PARAMETER BL(l) 'Susceptance of line l'
/l1          200
l2           200
l3           200/;
BL(l)=BL(l)/(Sbase/(Vbase**2));
PARAMETER PM_max(b) 'Maximum power that can be bought from the main grid'
/b1        100/;
***** Single-point forecasts of uncertain parameters
TABLE DATA_PR_max(h,t) 'Point forecasts of available stochastic production in time period t'
           t1     t2      t3
h1         6.0    9.6     3.6
h2         6.0    8.5     2.6
h3         6      8.5     2;
PARAMETER PR_max(u,t) 'Point forecasts of available stochastic production in time period t';
PARAMETER lambda_RT(t) 'Point forecasts of prices in the real-time energy market in time period t';
TABLE DATA_lambda_RT(h,t) 'Point forecasts of prices in the real-time energy market in time period t'
           t1     t2      t3
h1         32.6   56.8    31.2
h2         35     60.5    30
h3         35     60    27.8;
***** Actual data of schedules and uncertain parameters
PARAMETER PDA_act(t) 'Power scheduled in the day-ahead energy market'
/t1        -60
t2         -60
t3         40/;
TABLE PR_max_act(u,t) 'Actual available stochastic production in time period t'
           t1      t2      t3
u4         6       8.5     2;
PARAMETER lambda_RT_act(t) 'Actual real-time market price in time period t'
/t1          35
t2         60
t3         28/;
VARIABLES
```

```
pl(l,t) 'Power flow through line l in time period t'
pda(t) 'Power traded by the VPP in the day-ahead energy market in
     time period t'
prt(t) 'Power traded by the VPP in the real-time energy market in
     time period t'
pm(b,t) 'Power injection at bus b connected to the main grid in
    time period t'
teta(b,t) 'Voltage angle at bus b in time period t'
z 'Objective'
p(u,t)  'Power dispatch of unit u in time period t';
POSITIVE VARIABLES
es(u,t)  'Energy stored at the end of time period t'
psp(u,t) 'Power produced by the storage unit in time period t'
psn(u,t) 'Power transferred to the storage unit in time period t'
cg_st(u,t) 'Auxiliary variable to linearize the start-up cost of
    unit u in time period t'
cg_sh(u,t) 'Auxiliary variable to linearize the shot-down cost of
     unit u in time period t';
BINARY VARIABLES
v(u,t) 'Binary variable to state the commitment status of unit u
    in time period t';
*********************************************************
******Equations definition *************
EQUATIONS
obj 'The objective function'
nodal_balance_mg 'power balance at the main grid connection'
nodal_balance 'power balance at the remaining buses'
trade 'power traded in the day-ahead and real-time energy markets
     '
line_power 'power flow through lines'
line_max 'upper bound of lines flows'
line_min 'lower bound of lines flows'
voltage_angle_ref 'voltage angle at reference bus'
voltage_angle_max 'upper bound of voltage angles'
voltage_angle_min 'lower bound of voltage angles'
tradedart_min 'lower bound of power traded in the energy markets'
tradedart_max 'upper bound of power traded in the energy markets'
trade_min 'lower bound of power injection at each bus'
trade_max 'upper bound of power injection at each bus'
dailyenergy_min 'daily minimum energy consumption for each demand
     '
demand_max 'upper bound of load level'
demand_min 'lower bound of load level'
demand_ramp_down_initial 'load ramp down limitation'
demand_ramp_pick_initial 'load ramp up limitation'
demand_ramp_down 'load ramp down limitation'
demand_ramp_pick 'load ramp up limitation'
CPP_max 'maximum power production limitation'
CPP_min 'minimum power production limitation'
CPP_ramp_down_initial 'ramp down production limitation'
CPP_ramp_up_initial 'ramp up production limitation'
CCP_ramp_down 'ramp down production limitation'
CCP_ramp_up 'ramp up production limitation'
CCP_startcost_initial 'start up cost of unit'
```

```
CCP_shotcost_initial  'shot down cost of unit'
CCP_startcost 'start up cost of unit'
CCP_shotcost 'shot down cost of unit'
storage_balance_initial 'power balance equation in the storage
    unit'
storage_balance 'power balance equation in the storage unit'
storage_capacity 'upper bound of energy stored in the storage
    unit'
storage_psp_max 'maxmimum power that can be produced by storage
    unit'
storage_psn_max 'maxmimum power that can be transferred to
    storage unit'
storage_injection 'power injection of storage unit'
stochastic_max1 'maximum power production limitation'
stochastic_max2 'maximum power production limitation';
*
obj..   z=E=SUM(t$(ORD(t) GE tau),lambda_RT(t)*prt(t)*delta+SUM(
    u$incD(u),Utility(u,t)*p(u,t)*delta)-SUM(u$incG(u),CG(u,t)*p(
    u,t)*delta+cg_st(u,t)+cg_sh(u,t)));
nodal_balance_mg(b,t)$((incMB(b)) AND (ORD(t) GE tau))..  SUM(
    u$incGB(u,b),p(u,t))+SUM(u$incRB(u,b),p(u,t))+SUM(u$incSB(u,b
    ),p(u,t))-SUM(l$incORI(l,b),pl(l,t))+SUM(l$incDES(l,b),pl(l,t
    ))=E=pm(b,t)+SUM(u$incDB(u,b),p(u,t));
nodal_balance(b,t)$((NOT incMB(b)) AND (ORD(t) GE tau))..   SUM(
    u$incGB(u,b),p(u,t))+SUM(u$incRB(u,b),p(u,t))+SUM(u$incSB(u,b
    ),p(u,t))-SUM(l$incORI(l,b),pl(l,t))+SUM(l$incDES(l,b),pl(l,t
    ))=E=SUM(u$incDB(u,b),p(u,t));
trade(t)$(ORD(t) GE tau)..    pda(t)+prt(t)=E=SUM(b$incMB(b),pm(b
    ,t));
line_power(l,t)$(ORD(t) GE tau)..  pl(l,t)=E=Sbase*BL(l)*(SUM(
    b$incORI(l,b),teta(b,t))-SUM(b$incDES(l,b),teta(b,t)));
line_max(l,t)$(ORD(t) GE tau)..  pl(l,t)=L=PL_max(l);
line_min(l,t)$(ORD(t) GE tau)..   -PL_max(l)=L=pl(l,t);
voltage_angle_ref(b,t)$((incREF(b)) AND (ORD(t) GE tau)).. teta(b
    ,t)=E=0;
voltage_angle_max(b,t)$(ORD(t) GE tau)..  teta(b,t)=L=Pi;
voltage_angle_min(b,t)$(ORD(t) GE tau)..  -Pi=L=teta(b,t);
tradedart_min(t)$(ORD(t) GE tau).. PV_min=L=pda(t)+prt(t);
tradedart_max(t)$(ORD(t) GE tau).. pda(t)+prt(t)=L=PV_max;
trade_min(b,t)$((incMB(b)) AND (ORD(t) GE tau)).. -PM_max(b)=L=pm
    (b,t);
trade_max(b,t)$((incMB(b)) AND (ORD(t) GE tau)).. pm(b,t)=L=
    PM_max(b);
dailyenergy_min(u)$incD(u)..    SUM(t,p(u,t)*delta)=G=ED_min(u);
demand_max(u,t)$((incD(u)) AND (ORD(t) GE tau))..  p(u,t)=L=
    PD_max(u,t);
demand_min(u,t)$((incD(u)) AND (ORD(t) GE tau))..  PD_min(u,t)=L=
    p(u,t);
demand_ramp_down_initial(u,t)$((incD(u)) AND (ORD(t) EQ 1) AND (
    ORD(t) GE tau))..   -RD_do(u)=L=p(u,t)-PD0(u);
demand_ramp_pick_initial(u,t)$((incD(u)) AND (ORD(t) EQ 1) AND (
    ORD(t) GE tau))..  p(u,t)-PD0(u)=L=RD_up(u);
demand_ramp_down(u,t)$((incD(u)) AND (ORD(t) GE 2) AND (ORD(t) GE
    tau))..   -RD_do(u)=L=p(u,t)-p(u,t-1);
```

```
demand_ramp_pick(u,t)$((incD(u)) AND (ORD(t) GE 2) AND (ORD(t) GE tau)).. p(u,t)-p(u,t-1)=L=RD_up(u);
CPP_max(u,t)$((incG(u)) AND (ORD(t) GE tau)).. p(u,t)=L=PG_max(u)*v(u,t);
CPP_min(u,t)$((incG(u)) AND (ORD(t) GE tau)).. v(u,t)*PG_min(u)=L=p(u,t);
CPP_ramp_down_initial(u,t)$((incG(u)) AND (ORD(t) EQ 1) AND (ORD(t) GE tau)).. -RG_do(u)=L=p(u,t)-PG0(u);
CPP_ramp_up_initial(u,t)$((incG(u)) AND (ORD(t) EQ 1) AND (ORD(t) GE tau)).. p(u,t)-PG0(u)=L=RG_up(u);
CCP_ramp_down(u,t)$((incG(u)) AND (ORD(t) GE 2) AND (ORD(t) GE tau)).. -RG_do(u)=L=p(u,t)-p(u,t-1);
CCP_ramp_up(u,t)$((incG(u)) AND (ORD(t) GE 2) AND (ORD(t) GE tau)).. p(u,t)-p(u,t-1)=L=RG_up(u);
CCP_startcost_initial(u,t)$((incG(u)) AND (ORD(t) EQ 1) AND (ORD(t) GE tau)).. CG_start(u)*(v(u,t)-V0(u))=L=cg_st(u,t);
CCP_shotcost_initial(u,t)$((incG(u)) AND (ORD(t) EQ 1) AND (ORD(t) GE tau)).. CG_shot(u)*(V0(u)-v(u,t))=L=cg_sh(u,t);
CCP_startcost(u,t)$((incG(u)) AND (ORD(t) GE 2) AND (ORD(t) GE tau)).. CG_start(u)*(v(u,t)-v(u,t-1))=L=cg_st(u,t);
CCP_shotcost(u,t)$((incG(u)) AND (ORD(t) GE 2) AND (ORD(t) GE tau)).. CG_shot(u)*(v(u,t-1)-v(u,t))=L=cg_sh(u,t);
storage_balance_initial(u,t)$((incS(u)) AND (ORD(t) EQ 1) AND (ORD(t) GE tau)).. es(u,t)=E=ES0(u)+mun*psn(u,t)*delta-psp(u,t)*delta/mup;
storage_balance(u,t)$((incS(u)) AND (ORD(t) GE 2) AND (ORD(t) GE tau)).. es(u,t)=E=es(u,t-1)+mun*psn(u,t)*delta-psp(u,t)*delta/mup;
storage_capacity(u,t)$((incS(u)) AND (ORD(t) GE tau)).. es(u,t)=L=ES_max(u);
storage_psp_max(u,t)$((incS(u)) AND (ORD(t) GE tau)).. psp(u,t)=L=PS_p(u);
storage_psn_max(u,t)$((incS(u)) AND (ORD(t) GE tau)).. psn(u,t)=L=PS_n(u);
storage_injection(u,t)$((incS(u)) AND (ORD(t) GE tau)).. p(u,t)=E=psp(u,t)-psn(u,t);
stochastic_max1(u,t)$((incR(u)) AND (ORD(t) EQ tau)).. p(u,t)=L=PR_max_act(u,t);
stochastic_max2(u,t)$((incR(u)) AND (ORD(t) GT tau)).. p(u,t)=L=PR_max(u,t);
***********************************************
******** Model definition *****************
MODEL DSRT /all/;
OPTION OPTCR=0;
OPTION OPTCA=0;
OPTION iterlim=1e8;
******************************************
** Parameter definition for building quantity-price curve ****
PARAMETER state 'convergence status';
PARAMETER prt_act 'actual power traded in the real-time market';
PARAMETER revenuert_act 'revenue in the real-time market';
PARAMETER utilityrt_act 'utilty gained after dispatch';
PARAMETER data_p 'power dispatch of units';
PARAMETER data_es 'energy stored';
```

```
PARAMETERS data_v 'status of conventional power plant';
**********************************************
pda.fx(t)=PDA_act(t);
**********************************************
*****VPP Scheduling *****************
LOOP(h$(ORD(h) LE 3),
tau=ORD(h);
PR_max(u,t)=DATA_PR_max(h,t);
lambda_RT(t)=DATA_lambda_RT(h,t);
SOLVE DSRT USING MIP MAXIMIZING z;
state(h) = DSRT.modelstat;
LOOP(t$(ORD(t) EQ tau),
**** Real-time market clearing**************
prt_act(t)=prt.l(t);
revenuert_act(t)=prt_act(t)*lambda_RT_act(t);
);
**** VPP Dispatching*********************
SOLVE DSRT USING MIP MAXIMIZING z;
state(h) = DSRT.modelstat;
LOOP(t$(ORD(t) EQ tau),
data_p(u,t)=p.l(u,t);
p.fx(u,t)=data_p(u,t);
data_es(u,t)=es.l(u,t);
es.fx(u,t)=data_es(u,t);
data_v(u,t)=v.l(u,t);
v.fx(u,t)=data_v(u,t);
);
utilityrt_act(h)=SUM(t$(ORD(t) EQ tau),SUM(u$incD(u),Utility(u,t)
    *p.l(u,t)*delta)-SUM(u$incG(u),CG(u,t)*p.l(u,t)*delta+cg_st.l
    (u,t)+cg_sh.l(u,t)));
);
*********************************
display state, prt_act, revenuert_act, utilityrt_act, data_p,
    data_es, data_v;
```

3.7.3 Stochastic Scheduling in the Day-Ahead Energy Market

An input GAMS [9] code for solving Illustrative Example 3.5 is provided below:

```
* Stochastic scheduling of the risk-neutral VPP in the day-ahead
    energy market
SET b 'Bus' /b1*b3/;
SET l 'Line' /l1*l3/;
SET t 'Time period' /t1*t3/;
SET u 'Unit' /u1*u5/;
SET o 'scenarios' /o1*o2/;
SET f 'scenarios' /f1*f2/;
SET g 'scenarios' /g1*g2/;
SET i 'scenarios' /i1*i1/;
SET j 'scenarios' /j1*j4/;
```

```
SET incDES(l,b) 'destination bus' /l1.b2,l2.b3,l3.b3/;
SET incORI(l,b) 'origin bus' /l1.b1,l2.b1,l3.b2/;
SET incD(u) 'demand' /u1*u2/;
SET incG(u) 'conventional power plant' /u3/;
SET incR(u) 'stochastic production facility' /u4/;
SET incS(u) 'storage unit' /u5/;
SET incMB(b) 'buses connected to the main grid' /b1/;
SET incREF(b) 'reference bus' /b1/;
SET incDB(u,b) 'demand at bus' /u1.b2,u2.b3/;
SET incGB(u,b) 'conventional power plant at bus' /u3.b2/;
SET incRB(u,b) 'stochastic production facility at bus' /u4.b3/;
SET incSB(u,b) 'storage unit at bus' /u5.b3/;
ALIAS(o,op);
SCALAR TT 'Number of trading hours in a day' /3/;
SCALAR delta 'Power-energy conversion factor' /1/;
SCALAR mup 'Output conversion efficiency of the storage unit'
    /0.9/;
SCALAR mun 'Input conversion efficiency of the storage unit'
    /0.9/;
SCALAR Sbase 'base apparent power' /100/;
SCALAR Vbase 'base voltage' /10/;
SCALAR PV_max 'Maximum power that can be produced by the VPP'
    /40/;
SCALAR PV_min 'Maximum power that can be consumed by the VPP'
    /-60/;
***** Demands data
TABLE PD_max(u,t) 'Maximum load level for unit u in time period t
    '
         t1     t2     t3
u1       20     20     10
u2       30     30     15;
TABLE PD_min(u,t) 'Minimum load level for unit u in time period t
    '
         t1     t2     t3
u1       0      0      0
u2       5      5      5;
TABLE Utility(u,t) 'Utility of unit u for energy consumption in
    time period t'
         t1      t2     t3
u1       40      50     40
u2       50      60     50;
PARAMETER ED_min(u) 'Minimum daily energy consumption requested
    by unit i'
/u1       30
u2        50/;
PARAMETER RD_up(u) 'Load drop ramping limits for unit u'
/u1       15
u2        25/;
PARAMETER RD_do(u) 'Pick-up ramping limit for unit u'
/u1       15
u2        25/;
PARAMETER PD0(u) 'Initial load level for each unit'
/u1      10
u2       20/;
```

```
***** Conventional power plants data
PARAMETER PG_max(u) 'Maximum power production limit for unit u in
    time period t'
/u3        20/;
PARAMETER PG_min(u) 'Minimum power production limit for unit u in
    time period t'
/u3        0/;
PARAMETER RG_up(u) 'Ramping up limits for unit u'
/u3         20/;
PARAMETER RG_do(u) 'Ramping down limit for unit u'
/u3         20/;
TABLE CG(u,t) 'Production cost for unit u'
         t1     t2      t3
u3        45     45      45;
PARAMETER CG_start(u) 'Start up cost for unit u'
/u3         5/;
PARAMETER CG_shot(u) 'Shot down cost for unit u'
/u3         5/;
PARAMETER PG0(u) 'Initial power production of unit u'
/u3         0/;
PARAMETER V0(u) 'Initial status of unit u'
/u3         0/;
***** Storage units data
PARAMETER ES_max(u) 'Energy capacity of the storage unit u'
/u5       10/;
PARAMETER PS_p(u) 'Maximum power that can be produced by unit u'
/u5       10/;
PARAMETER PS_n(u) 'Maximum power that can be transferred to unit
    u'
/u5       10/;
PARAMETER ES0(u) 'Initial energy stored in the storage unit'
/u5       5/;
***** Network data
PARAMETER PL_max(l) 'Capacity of line l'
/l1         50
l2          50
l3          50/;
PARAMETER BL(l) 'Susceptance of line l'
/l1        200
l2         200
l3         200/;
BL(l)=BL(l)/(Sbase/(Vbase**2));
PARAMETER PM_max(b) 'Maximum power that can be bought from the
    main grid'
/b1       100/;
***** Scenarios of uncertain parameters
TABLE PR_max 'Scenarios of available stochastic production in
    time period t'
              u4.t1    u4.t2    u4.t3
i1.j1         9        7        1
i1.j2         8        5        2
i1.j3         4        6        0
i1.j4         5        5        1;
```

```
TABLE lambda_DA 'Scenarios of prices in the day-ahead energy
    market in time period t'
        t1        t2        t3
o1      20        40        30
o2      30        50        40;
TABLE lambda_RT 'Scenarios of prices in the real-time energy
    market in time period t'
            t1        t2        t3
o1.f1.g1    25        50        25
o1.f1.g2    20        45        28
o1.f2.g1    22        44        20
o1.f2.g2    18        39        28
o2.f1.g1    35        60        35
o2.f1.g2    30        55        38
o2.f2.g1    32        54        30
o2.f2.g2    28        49        38;
PARAMETER prob(o,f,g,i,j) 'probability of scenarios'
/o1.f1.g1.i1.j1        0.018
o1.f1.g1.i1.j2        0.018
o1.f1.g1.i1.j3        0.012
o1.f1.g1.i1.j4        0.012
o1.f1.g2.i1.j1        0.036
o1.f1.g2.i1.j2        0.036
o1.f1.g2.i1.j3        0.024
o1.f1.g2.i1.j4        0.024
o1.f2.g1.i1.j1        0.009
o1.f2.g1.i1.j2        0.009
o1.f2.g1.i1.j3        0.006
o1.f2.g1.i1.j4        0.006
o1.f2.g2.i1.j1        0.027
o1.f2.g2.i1.j2        0.027
o1.f2.g2.i1.j3        0.018
o1.f2.g2.i1.j4        0.018
o2.f1.g1.i1.j1        0.042
o2.f1.g1.i1.j2        0.042
o2.f1.g1.i1.j3        0.028
o2.f1.g1.i1.j4        0.028
o2.f1.g2.i1.j1        0.084
o2.f1.g2.i1.j2        0.084
o2.f1.g2.i1.j3        0.056
o2.f1.g2.i1.j4        0.056
o2.f2.g1.i1.j1        0.021
o2.f2.g1.i1.j2        0.021
o2.f2.g1.i1.j3        0.014
o2.f2.g1.i1.j4        0.014
o2.f2.g2.i1.j1        0.063
o2.f2.g2.i1.j2        0.063
o2.f2.g2.i1.j3        0.042
o2.f2.g2.i1.j4        0.042/;
***** Actual data of the day-ahead market prices
PARAMETER lambda_DA_act(t) 'Actual day-ahead market price in time
    period t'
/t1        25
t2        50
```

```
t3        35/;
VARIABLES
pl(l,t,o,f,g,i,j) 'Power flow through line l in time period t'
pda(t,o) 'Power traded by the VPP in the day-ahead energy market in time period t'
prt(t,o,f,i) 'Power traded by the VPP in the real-time energy market in time period t'
pm(b,t,o,f,g,i,j) 'Power injection at bus b connected to the main grid in time period t'
teta(b,t,o,f,g,i,j) 'Voltage angle at bus b in time period t'
z 'Objective'
zsc(o,f,g,i,j) 'Objective in each scenario'
p(u,t,o,f,g,i,j)  'Power dispatch of unit u in time period t';
POSITIVE VARIABLES
es(u,t,o,f,g,i,j)  'Energy stored at the end of time period t'
psp(u,t,o,f,g,i,j) 'Power produced by the storage unit in time period t'
psn(u,t,o,f,g,i,j) 'Power transferred to the storage unit in time period t'
cg_st(u,t,o,f,g,i,j) 'Auxiliary variable to linearize the start-up cost of unit u in time period t'
cg_sh(u,t,o,f,g,i,j) 'Auxiliary variable to linearize the shot-down cost of unit u in time period t';
BINARY VARIABLES
v(u,t,o,f,g,i,j) 'Binary variable to state the commitment status of unit u in time period t';
*****************************************************
********Equations definition *****************
EQUATIONS
obj 'The objective function'
objscen 'the objective in each scenario'
nodal_balance_mg 'power balance at the main grid buses'
nodal_balance 'power balance at the remaining buses'
trade 'power traded in the day-ahead and real-time markets'
line_power 'power flow through lines'
line_max 'upper bound of lines flows'
line_min 'lower bound of lines flows'
voltage_angle_ref 'voltage angle at reference bus'
voltage_angle_max 'upper bound of voltage angles'
voltage_angle_min 'lower bound of voltage angles'
tradeda_min 'lower bound of power traded in the day-ahead market'
tradeda_max 'upper bound of power traded in the day-ahead market'
tradedart_min 'lower bound of power traded in the energy markets'
tradedart_max 'upper bound of power traded in the energy markets'
trade_min 'lower bound of power injection at each bus'
trade_max 'upper bound of power injection at each bus'
dailyenergy_min 'daily minimum energy consumption for each demand'
demand_max 'upper bound of load level'
demand_min 'lower bound of load level'
demand_ramp_down_initial 'load ramp down limitation'
demand_ramp_pick_initial 'load ramp up limitation'
demand_ramp_down 'load ramp down limitation'
demand_ramp_pick 'load ramp up limitation'
```

```
CPP_max 'maximum power production limitation'
CPP_min 'minimum power production limitation'
CPP_ramp_down_initial 'ramp down production limitation'
CPP_ramp_up_initial 'ramp up production limitation'
CCP_ramp_down 'ramp down production limitation'
CCP_ramp_up 'ramp up production limitation'
CCP_startcost_initial 'start up cost of unit'
CCP_shotcost_initial  'shot down cost of unit'
CCP_startcost 'start up cost of unit'
CCP_shotcost 'shot down cost of unit'
storage_balance_initial 'power balance equation in the storage
    unit'
storage_balance 'power balance equation in the storage unit'
storage_capacity 'upper bound of energy stored in the storage
    unit'
storage_psp_max 'maxmimum power that can be produced by storage
    unit'
storage_psn_max 'maxmimum power that can be transferred to
    storage unit'
storage_injection 'power injection of storage unit'
stochastic_max 'maximum power production limitation'
curve_nondec 'building nondecreasing offering and bidding curves'
curve_nonant 'nonanticipativity constraints';
*
obj..     z=E=SUM((o,f,g,i,j),prob(o,f,g,i,j)*zsc(o,f,g,i,j));
objscen(o,f,g,i,j)..  zsc(o,f,g,i,j)=E=SUM(t,lambda_DA(o,t)*pda(t
    ,o)*delta+lambda_RT(o,f,g,t)*prt(t,o,f,i)*delta+SUM(u$incD(u)
    ,Utility(u,t)*p(u,t,o,f,g,i,j)*delta)-SUM(u$incG(u),CG(u,t)*p
    (u,t,o,f,g,i,j)*delta+cg_st(u,t,o,f,g,i,j)+cg_sh(u,t,o,f,g,i,
    j)));
nodal_balance_mg(b,t,o,f,g,i,j)$(incMB(b))..            SUM(
    u$incGB(u,b),p(u,t,o,f,g,i,j))+SUM(u$incRB(u,b),p(u,t,o,f,g,i
    ,j))+SUM(u$incSB(u,b),p(u,t,o,f,g,i,j))-SUM(l$incORI(l,b),pl(
    l,t,o,f,g,i,j))+SUM(l$incDES(l,b),pl(l,t,o,f,g,i,j))=E=pm(b,t
    ,o,f,g,i,j)+SUM(u$incDB(u,b),p(u,t,o,f,g,i,j));
nodal_balance(b,t,o,f,g,i,j)$(NOT incMB(b))..      SUM(u$incGB(u,
    b),p(u,t,o,f,g,i,j))+SUM(u$incRB(u,b),p(u,t,o,f,g,i,j))+SUM(
    u$incSB(u,b),p(u,t,o,f,g,i,j))-SUM(l$incORI(l,b),pl(l,t,o,f,g
    ,i,j))+SUM(l$incDES(l,b),pl(l,t,o,f,g,i,j))=E=SUM(u$incDB(u,b
    ),p(u,t,o,f,g,i,j));
trade(t,o,f,g,i,j)..    pda(t,o)+prt(t,o,f,i)=E=SUM(b$incMB(b),pm(
    b,t,o,f,g,i,j));
line_power(l,t,o,f,g,i,j).. pl(l,t,o,f,g,i,j)=E=Sbase*BL(l)*(SUM(
    b$incORI(l,b),teta(b,t,o,f,g,i,j))-SUM(b$incDES(l,b),teta(b,t
    ,o,f,g,i,j)));
line_max(l,t,o,f,g,i,j)..    pl(l,t,o,f,g,i,j)=L=PL_max(l);
line_min(l,t,o,f,g,i,j)..    -PL_max(l)=L=pl(l,t,o,f,g,i,j);
voltage_angle_ref(b,t,o,f,g,i,j)$(incREF(b)).. teta(b,t,o,f,g,i,j
    )=E=0;
voltage_angle_max(b,t,o,f,g,i,j)..  teta(b,t,o,f,g,i,j)=L=Pi;
voltage_angle_min(b,t,o,f,g,i,j)..  -Pi=L=teta(b,t,o,f,g,i,j);
tradeda_min(t,o).. PV_min=L=pda(t,o);
tradeda_max(t,o).. pda(t,o)=L=PV_max;
tradedart_min(t,o,f,i).. PV_min=L=pda(t,o)+prt(t,o,f,i);
```

```
tradedart_max(t,o,f,i).. pda(t,o)+prt(t,o,f,i)=L=PV_max;
trade_min(b,t,o,f,g,i,j)$(incMB(b)).. -PM_max(b)=L=pm(b,t,o,f,g,i,j);
trade_max(b,t,o,f,g,i,j)$(incMB(b)).. pm(b,t,o,f,g,i,j)=L=PM_max(b);
dailyenergy_min(u,o,f,g,i,j)$(incD(u)).. SUM(t,p(u,t,o,f,g,i,j)*delta)=G=ED_min(u);
demand_max(u,t,o,f,g,i,j)$(incD(u)).. p(u,t,o,f,g,i,j)=L=PD_max(u,t);
demand_min(u,t,o,f,g,i,j)$(incD(u)).. PD_min(u,t)=L=p(u,t,o,f,g,i,j);
demand_ramp_down_initial(u,t,o,f,g,i,j)$((incD(u)) AND (ORD(t) EQ 1)).. -RD_do(u)=L=p(u,t,o,f,g,i,j)-PD0(u);
demand_ramp_pick_initial(u,t,o,f,g,i,j)$((incD(u)) AND (ORD(t) EQ 1)).. p(u,t,o,f,g,i,j)-PD0(u)=L=RD_up(u);
demand_ramp_down(u,t,o,f,g,i,j)$((incD(u)) AND (ORD(t) GE 2)).. -RD_do(u)=L=p(u,t,o,f,g,i,j)-p(u,t-1,o,f,g,i,j);
demand_ramp_pick(u,t,o,f,g,i,j)$((incD(u)) AND (ORD(t) GE 2)).. p(u,t,o,f,g,i,j)-p(u,t-1,o,f,g,i,j)=L=RD_up(u);
CPP_max(u,t,o,f,g,i,j)$(incG(u)).. p(u,t,o,f,g,i,j)=L=PG_max(u)*v(u,t,o,f,g,i,j);
CPP_min(u,t,o,f,g,i,j)$(incG(u)).. v(u,t,o,f,g,i,j)*PG_min(u)=L=p(u,t,o,f,g,i,j);
CPP_ramp_down_initial(u,t,o,f,g,i,j)$((incG(u)) AND (ORD(t) EQ 1)).. -RG_do(u)=L=p(u,t,o,f,g,i,j)-PG0(u);
CPP_ramp_up_initial(u,t,o,f,g,i,j)$((incG(u)) AND (ORD(t) EQ 1)).. p(u,t,o,f,g,i,j)-PG0(u)=L=RG_up(u);
CCP_ramp_down(u,t,o,f,g,i,j)$((incG(u)) AND (ORD(t) GE 2)).. -RG_do(u)=L=p(u,t,o,f,g,i,j)-p(u,t-1,o,f,g,i,j);
CCP_ramp_up(u,t,o,f,g,i,j)$((incG(u)) AND (ORD(t) GE 2)).. p(u,t,o,f,g,i,j)-p(u,t-1,o,f,g,i,j)=L=RG_up(u);
CCP_startcost_initial(u,t,o,f,g,i,j)$((incG(u)) AND (ORD(t) EQ 1)).. CG_start(u)*(v(u,t,o,f,g,i,j)-V0(u))=L=cg_st(u,t,o,f,g,i,j);
CCP_shotcost_initial(u,t,o,f,g,i,j)$((incG(u)) AND (ORD(t) EQ 1)).. CG_shot(u)*(V0(u)-v(u,t,o,f,g,i,j))=L=cg_sh(u,t,o,f,g,i,j);
CCP_startcost(u,t,o,f,g,i,j)$((incG(u)) AND (ORD(t) GE 2)).. CG_start(u)*(v(u,t,o,f,g,i,j)-v(u,t-1,o,f,g,i,j))=L=cg_st(u,t,o,f,g,i,j);
CCP_shotcost(u,t,o,f,g,i,j)$((incG(u)) AND (ORD(t) GE 2)).. CG_shot(u)*(v(u,t-1,o,f,g,i,j)-v(u,t,o,f,g,i,j))=L=cg_sh(u,t,o,f,g,i,j);
storage_balance_initial(u,t,o,f,g,i,j)$((incS(u)) AND (ORD(t) EQ 1)).. es(u,t,o,f,g,i,j)=E=ES0(u)+mun*psn(u,t,o,f,g,i,j)*delta-psp(u,t,o,f,g,i,j)*delta/mup;
storage_balance(u,t,o,f,g,i,j)$((incS(u)) AND (ORD(t) GE 2)).. es(u,t,o,f,g,i,j)=E=es(u,t-1,o,f,g,i,j)+mun*psn(u,t,o,f,g,i,j)*delta-psp(u,t,o,f,g,i,j)*delta/mup;
storage_capacity(u,t,o,f,g,i,j)$(incS(u)).. es(u,t,o,f,g,i,j)=L=ES_max(u);
storage_psp_max(u,t,o,f,g,i,j)$(incS(u)).. psp(u,t,o,f,g,i,j)=L=PS_p(u);
```

```
storage_psn_max(u,t,o,f,g,i,j)$(incS(u))..  psn(u,t,o,f,g,i,j)=L=
    PS_n(u);
storage_injection(u,t,o,f,g,i,j)$(incS(u))..  p(u,t,o,f,g,i,j)=E=
    psp(u,t,o,f,g,i,j)-psn(u,t,o,f,g,i,j);
stochastic_max(u,t,o,f,g,i,j)$(incR(u))..   p(u,t,o,f,g,i,j)=L=
    PR_max(i,j,u,t);
curve_nondec(t,o,op)$(lambda_DA(o,t) LT lambda_DA(op,t)).. pda(t,
    o) =L= pda(t,op);
curve_nonant(t,o,op)$(lambda_DA(o,t) EQ lambda_DA(op,t))..  pda(t
    ,o)=E=pda(t,op);
*********************************************************
****** Model definition **************
MODEL SSDA       /all/;
OPTION OPTCR=0;
OPTION OPTCA=0;
OPTION iterlim=1e8;
*******************************************
**Parameter definition for building quantity-price curve ****
PARAMETER state 'convergence status';
PARAMETER pda_act 'actual power traded in the day-ahead market';
PARAMETER curve_pda 'offering-bidding power in the day-ahead
    market';
PARAMETER curve_priceda 'offering-bidding price in the day-ahead
    market';
PARAMETER revenueda_act 'revenue in the day-ahead market';
*********************************************
******* Solution ****************
******* Building bid curve********
SOLVE SSDA USING MIP MAXIMIZING z;
state = SSDA.modelstat;
curve_pda(t,o)=pda.l(t,o);
curve_priceda(t,o)=lambda_DA(o,t);
******* Day-ahead market clearing****************
pda_act(t)=0;
LOOP(t$(ORD(t) LE TT),
LOOP(o,
if (curve_priceda(t,o)< lambda_DA_act(t),
if (curve_pda(t,o)>0,
pda_act(t)=max(curve_pda(t,o),pda_act(t));););
);
if (curve_priceda(t,o)> lambda_DA_act(t),
if (curve_pda(t,o)<0,
pda_act(t)=min(curve_pda(t,o),pda_act(t));););
);
if (curve_priceda(t,o)=lambda_DA_act(t),
pda_act(t)=curve_pda(t,o);
);
);
);
revenueda_act=SUM(t,pda_act(t)*lambda_DA_act(t));
**************************************
display state, curve_pda, curve_priceda, lambda_DA_act, pda_act,
    revenueda_act;
```

3.7.4 Stochastic Scheduling in the Real-Time Energy Market

An input GAMS [9] code for solving Illustrative Example 3.6 is provided below:

```
* Stochastic scheduling of the risk-neutral VPP in the real-time
   energy market
SET b 'Bus' /b1*b3/;
SET l 'Line' /l1*l3/;
SET t 'Time period' /t1*t3/;
SET h 'Time period' /h1*h3/;
SET g 'scenarios' /g1*g4/;
SET u 'Unit' /u1*u5/;
SET incDES(l,b) 'destination bus' /l1.b2,l2.b3,l3.b3/;
SET incORI(l,b) 'origin bus' /l1.b1,l2.b1,l3.b2/;
SET incD(u) 'demand' /u1*u2/;
SET incG(u) 'conventional power plant' /u3/;
SET incR(u) 'stochastic production facility' /u4/;
SET incS(u) 'storage unit' /u5/;
SET incMB(b) 'buses connected to the main grid' /b1/;
SET incREF(b) 'reference bus' /b1/;
SET incDB(u,b) 'demand at bus' /u1.b2,u2.b3/;
SET incGB(u,b) 'conventional power plant at bus' /u3.b2/;
SET incRB(u,b) 'stochastic production facility at bus' /u4.b3/;
SET incSB(u,b) 'storage unit at bus' /u5.b3/;
ALIAS(g,gp);
SCALAR TT 'Number of trading hours in a day' /3/;
SCALAR delta 'Power-energy conversion factor' /1/;
SCALAR mup 'Output conversion efficiency of the storage unit'
   /0.9/;
SCALAR mun 'Input conversion efficiency of the storage unit'
   /0.9/;
SCALAR Sbase 'base apparent power' /100/;
SCALAR Vbase 'base voltage' /10/;
SCALAR PV_max 'Maximum power that can be produced by the VPP'
   /40/;
SCALAR PV_min 'Maximum power that can be consumed by the VPP'
   /-60/;
SCALAR tau 'iteration' /1/;
***** Demands data
TABLE PD_max(u,t) 'Maximum load level for unit u in time period t
   '
         t1      t2      t3
u1       20      20      10
u2       30      30      15;
TABLE PD_min(u,t) 'Minimum load level for unit u in time period t
   '
         t1      t2      t3
u1       0       0       0
u2       5       5       5;
TABLE Utility(u,t) 'Utility of unit u for energy consumption in
   time period t'
         t1       t2       t3
u1       40       50       40
u2       50       60       50;
```

```
PARAMETER ED_min(u) 'Minimum daily energy consumption requested
    by unit i'
/u1       30
u2        50/;
PARAMETER RD_up(u) 'Load drop ramping limits for unit u'
/u1       15
u2        25/;
PARAMETER RD_do(u) 'Pick-up ramping limit for unit u'
/u1       15
u2        25/;
PARAMETER PD0(u) 'Initial load level for each unit'
/u1      10
u2       20/;
***** Conventional power plants data
PARAMETER PG_max(u) 'Maximum power production limit for unit u in
     time period t'
/u3       20/;
PARAMETER PG_min(u) 'Minimum power production limit for unit u in
     time period t'
/u3       0/;
PARAMETER RG_up(u) 'Ramping up limits for unit u'
/u3        20/;
PARAMETER RG_do(u) 'Ramping down limit for unit u'
/u3        20/;
TABLE CG(u,t) 'Production cost for unit u'
         t1      t2       t3
u3        45      45       45;
PARAMETER CG_start(u) 'Start up cost for unit u'
/u3        5/;
PARAMETER CG_shot(u) 'Shot down cost for unit u'
/u3        5/;
PARAMETER PG0(u) 'Initial power production of unit u'
/u3        0/;
PARAMETER V0(u) 'Initial status of unit u'
/u3        0/;
***** Storage units data
PARAMETER ES_max(u) 'Energy capacity of the storage unit u'
/u5      10/;
PARAMETER PS_p(u) 'Maximum power that can be produced by unit u'
/u5      10/;
PARAMETER PS_n(u) 'Maximum power that can be transferred to unit
    u'
/u5      10/;
PARAMETER ES0(u) 'Initial energy stored in the storage unit'
/u5      5/;
***** Network data
PARAMETER PL_max(l) 'Capacity of line l'
/l1        50
l2         50
l3         50/;
PARAMETER BL(l) 'Susceptance of line l'
/l1       200
l2        200
l3        200/;
```

```
BL(l)=BL(l)/(Sbase/(Vbase**2));
PARAMETER PM_max(b) 'Maximum power that can be bought from the
    main grid'
/b1      100/;
***** Scenarios of uncertain parameters
PARAMETER PR_max(u,t,g) 'Point forecasts of available stochastic
    production in time period t';
TABLE DATA_PR_max(h,g,u,t) 'Point forecasts of available
    stochastic production in time period t'
         u4.t1    u4.t2    u4.t3
h1.g1    6        10       4
h1.g2    6        8        2
h1.g3    6        10       4
h1.g4    6        8        2
h2.g1    6        8.5      3
h2.g2    6        8.5      2
h2.g3    6        8.5      3
h2.g4    6        8.5      2
h3.g1    6        8.5      2
h3.g2    6        8.5      2
h3.g3    6        8.5      2
h3.g4    6        8.5      2;
PARAMETER lambda_RT(t,g) 'Point forecasts of prices in the real-
    time energy market in time period t';
TABLE DATA_lambda_RT(h,g,t) 'Point forecasts of prices in the
    real-time energy market in time period t'
         t1        t2        t3
h1.g1    35        58        30
h1.g2    35        58        30
h1.g3    29        55        33
h1.g4    29        55        33
h2.g1    35        59        29
h2.g2    35        59        29
h2.g3    35        62        31
h2.g4    35        62        31
h3.g1    35        60        40
h3.g2    35        60        22
h3.g3    35        60        25
h3.g4    35        60        29;
PARAMETER prob(g) 'probability of scenarios';
PARAMETER DATA_prob(h,g) 'probability of scenarios'
/h1.g1       0.48
h1.g2        0.12
h1.g3        0.32
h1.g4        0.08
h2.g1        0.3
h2.g2        0.2
h2.g3        0.3
h2.g4        0.2
h3.g1        0.1
h3.g2        0.1
h3.g3        0.4
h3.g4        0.4/;
***** Actual data of schedules and uncertain parameters
```

```
PARAMETER PDA_act(t) 'Power scheduled in the day-ahead energy
    market'
/t1      -60
t2       -60
t3       40/;
TABLE PR_max_act(u,t) 'Actual available stochastic production in
    time period t'
        t1      t2      t3
u4      6       8.5     2;
PARAMETER lambda_RT_act(t) 'Actual real-time market price in time
    period t'
/t1       35
t2        60
t3        28/;
VARIABLES
pl(l,t,g) 'Power flow through line l in time period t'
pda(t) 'Power traded by the VPP in the day-ahead energy market in
    time period t'
prt(t,g) 'Power traded by the VPP in the real-time energy market
    in time period t'
pm(b,t,g) 'Power injection at bus b connected to the main grid in
    time period t'
teta(b,t,g) 'Voltage angle at bus b in time period t'
z 'Objective'
zsc(g) 'Objective in each scenario'
p(u,t,g)  'Power dispatch of unit u in time period t';
POSITIVE VARIABLES
es(u,t,g)  'Energy stored at the end of time period t'
psp(u,t,g) 'Power produced by the storage unit in time period t'
psn(u,t,g) 'Power transferred to the storage unit in time period
    t'
cg_st(u,t,g) 'Auxiliary variable to linearize the start-up cost
    of unit u in time period t'
cg_sh(u,t,g) 'Auxiliary variable to linearize the shot-down cost
    of unit u in time period t';
BINARY VARIABLES
v(u,t,g) 'Binary variable to state the commitment status of unit
    u in time period t';
****************************************************
******Equations definition *********
EQUATIONS
obj 'The objective function'
objscen 'the objective in each scenario'
nodal_balance_mg 'power balance at the main grid connection'
nodal_balance 'power balance at the remaining buses'
trade 'power traded in the day-ahead and real-time energy markets
    '
line_power 'power flow through lines'
line_max 'upper bound of lines flows'
line_min 'lower bound of lines flows'
voltage_angle_ref 'voltage angle at reference bus'
voltage_angle_max 'upper bound of voltage angles'
voltage_angle_min 'lower bound of voltage angles'
tradedart_min 'lower bound of power traded in the energy markets'
```

```
tradedart_max 'upper bound of power traded in the energy markets'
trade_min 'lower bound of power injection at each bus'
trade_max 'upper bound of power injection at each bus'
dailyenergy_min 'daily minimum energy consumption for each demand'
demand_max 'upper bound of load level'
demand_min 'lower bound of load level'
demand_ramp_down_initial 'load ramp down limitation'
demand_ramp_pick_initial 'load ramp up limitation'
demand_ramp_down 'load ramp down limitation'
demand_ramp_pick 'load ramp up limitation'
CPP_max 'maximum power production limitation'
CPP_min 'minimum power production limitation'
CPP_ramp_down_initial 'ramp down production limitation'
CPP_ramp_up_initial 'ramp up production limitation'
CCP_ramp_down 'ramp down production limitation'
CCP_ramp_up 'ramp up production limitation'
CCP_startcost_initial 'start up cost of unit'
CCP_shotcost_initial  'shot down cost of unit'
CCP_startcost 'start up cost of unit'
CCP_shotcost 'shot down cost of unit'
storage_balance_initial 'power balance equation in the storage unit'
storage_balance 'power balance equation in the storage unit'
storage_capacity 'upper bound of energy stored in the storage unit'
storage_psp_max 'maxmimum power that can be produced by storage unit'
storage_psn_max 'maxmimum power that can be transferred to storage unit'
storage_injection 'power injection of storage unit'
stochastic_max1 'maximum power production limitation'
stochastic_max2 'maximum power production limitation'
curve_nondec 'building nondecreasing offering and bidding curves'
curve_nonant 'nonanticipativity constraints'
curve_nonant_p 'nonanticipativity constraints'
curve_nonant_v 'nonanticipativity constraints'
curve_nonant_cgst 'nonanticipativity constraints'
curve_nonant_cgsh 'nonanticipativity constraints'
curve_nonant_psp 'nonanticipativity constraints'
curve_nonant_psn 'nonanticipativity constraints';
*
obj..      z=E=SUM(g,prob(g)*zsc(g));
objscen(g)..    zsc(g)=E=SUM(t$(ORD(t) GE tau),lambda_RT(t,g)*prt(t,g)*delta+SUM(u$incD(u),Utility(u,t)*p(u,t,g)*delta)-SUM(u$incG(u),CG(u,t)*p(u,t,g)*delta+cg_st(u,t,g)+cg_sh(u,t,g)));
nodal_balance_mg(b,t,g)$((incMB(b)) AND (ORD(t) GE tau))..  SUM(u$incGB(u,b),p(u,t,g))+SUM(u$incRB(u,b),p(u,t,g))+SUM(u$incSB(u,b),p(u,t,g))-SUM(l$incORI(l,b),pl(l,t,g))+SUM(l$incDES(l,b),pl(l,t,g))=E=pm(b,t,g)+SUM(u$incDB(u,b),p(u,t,g));
nodal_balance(b,t,g)$((NOT incMB(b)) AND (ORD(t) GE tau))..  SUM(u$incGB(u,b),p(u,t,g))+SUM(u$incRB(u,b),p(u,t,g))+SUM(u$incSB(u,b),p(u,t,g))-SUM(l$incORI(l,b),pl(l,t,g))+SUM(l$incDES(l,b),pl(l,t,g))=E=SUM(u$incDB(u,b),p(u,t,g));
```

```
trade(t,g)$(ORD(t) GE tau).. pda(t)+prt(t,g)=E=SUM(b$incMB(b),pm(b,t,g));
line_power(l,t,g)$(ORD(t) GE tau).. pl(l,t,g)=E=Sbase*BL(l)*(SUM(b$incORI(l,b),teta(b,t,g))-SUM(b$incDES(l,b),teta(b,t,g)));
line_max(l,t,g)$(ORD(t) GE tau).. pl(l,t,g)=L=PL_max(l);
line_min(l,t,g)$(ORD(t) GE tau).. -PL_max(l)=L=pl(l,t,g);
voltage_angle_ref(b,t,g)$((incREF(b)) AND (ORD(t) GE tau)).. teta(b,t,g)=E=0;
voltage_angle_max(b,t,g)$(ORD(t) GE tau).. teta(b,t,g)=L=Pi;
voltage_angle_min(b,t,g)$(ORD(t) GE tau).. -Pi=L=teta(b,t,g);
tradedart_min(t,g)$(ORD(t) GE tau).. PV_min=L=pda(t)+prt(t,g);
tradedart_max(t,g)$(ORD(t) GE tau).. pda(t)+prt(t,g)=L=PV_max;
trade_min(b,t,g)$((incMB(b)) AND (ORD(t) GE tau)).. -PM_max(b)=L=pm(b,t,g);
trade_max(b,t,g)$((incMB(b)) AND (ORD(t) GE tau)).. pm(b,t,g)=L=PM_max(b);
dailyenergy_min(u,g)$incD(u).. SUM(t,p(u,t,g)*delta)=G=ED_min(u);
demand_max(u,t,g)$((incD(u)) AND (ORD(t) GE tau)).. p(u,t,g)=L=PD_max(u,t);
demand_min(u,t,g)$((incD(u)) AND (ORD(t) GE tau)).. PD_min(u,t)=L=p(u,t,g);
demand_ramp_down_initial(u,t,g)$((incD(u)) AND (ORD(t) EQ 1) AND (ORD(t) GE tau)).. -RD_do(u)=L=p(u,t,g)-PD0(u);
demand_ramp_pick_initial(u,t,g)$((incD(u)) AND (ORD(t) EQ 1) AND (ORD(t) GE tau)).. p(u,t,g)-PD0(u)=L=RD_up(u);
demand_ramp_down(u,t,g)$((incD(u)) AND (ORD(t) GE 2) AND (ORD(t) GE tau)).. -RD_do(u)=L=p(u,t,g)-p(u,t-1,g);
demand_ramp_pick(u,t,g)$((incD(u)) AND (ORD(t) GE 2) AND (ORD(t) GE tau)).. p(u,t,g)-p(u,t-1,g)=L=RD_up(u);
CPP_max(u,t,g)$((incG(u)) AND (ORD(t) GE tau)).. p(u,t,g)=L=PG_max(u)*v(u,t,g);
CPP_min(u,t,g)$((incG(u)) AND (ORD(t) GE tau)).. v(u,t,g)*PG_min(u)=L=p(u,t,g);
CPP_ramp_down_initial(u,t,g)$((incG(u)) AND (ORD(t) EQ 1) AND (ORD(t) GE tau)).. -RG_do(u)=L=p(u,t,g)-PG0(u);
CPP_ramp_up_initial(u,t,g)$((incG(u)) AND (ORD(t) EQ 1) AND (ORD(t) GE tau)).. p(u,t,g)-PG0(u)=L=RG_up(u);
CCP_ramp_down(u,t,g)$((incG(u)) AND (ORD(t) GE 2) AND (ORD(t) GE tau)).. -RG_do(u)=L=p(u,t,g)-p(u,t-1,g);
CCP_ramp_up(u,t,g)$((incG(u)) AND (ORD(t) GE 2) AND (ORD(t) GE tau)).. p(u,t,g)-p(u,t-1,g)=L=RG_up(u);
CCP_startcost_initial(u,t,g)$((incG(u)) AND (ORD(t) EQ 1) AND (ORD(t) GE tau)).. CG_start(u)*(v(u,t,g)-V0(u))=L=cg_st(u,t,g);
CCP_shotcost_initial(u,t,g)$((incG(u)) AND (ORD(t) EQ 1) AND (ORD(t) GE tau)).. CG_shot(u)*(V0(u)-v(u,t,g))=L=cg_sh(u,t,g);
CCP_startcost(u,t,g)$((incG(u)) AND (ORD(t) GE 2) AND (ORD(t) GE tau)).. CG_start(u)*(v(u,t,g)-v(u,t-1,g))=L=cg_st(u,t,g);
CCP_shotcost(u,t,g)$((incG(u)) AND (ORD(t) GE 2) AND (ORD(t) GE tau)).. CG_shot(u)*(v(u,t-1,g)-v(u,t,g))=L=cg_sh(u,t,g);
storage_balance_initial(u,t,g)$((incS(u)) AND (ORD(t) EQ 1) AND (ORD(t) GE tau)).. es(u,t,g)=E=ES0(u)+mun*psn(u,t,g)*delta-psp(u,t,g)*delta/mup;
```

```
storage_balance(u,t,g)$((incS(u)) AND (ORD(t) GE 2) AND (ORD(t) GE tau)).. es(u,t,g)=E=es(u,t-1,g)+mun*psn(u,t,g)*delta-psp(u,t,g)*delta/mup;
storage_capacity(u,t,g)$((incS(u)) AND (ORD(t) GE tau)).. es(u,t,g)=L=ES_max(u);
storage_psp_max(u,t,g)$((incS(u)) AND (ORD(t) GE tau)).. psp(u,t,g)=L=PS_p(u);
storage_psn_max(u,t,g)$((incS(u)) AND (ORD(t) GE tau)).. psn(u,t,g)=L=PS_n(u);
storage_injection(u,t,g)$((incS(u)) AND (ORD(t) GE tau)).. p(u,t,g)=E=psp(u,t,g)-psn(u,t,g);
stochastic_max1(u,t,g)$((incR(u)) AND (ORD(t) EQ tau)).. p(u,t,g)=L=PR_max_act(u,t);
stochastic_max2(u,t,g)$((incR(u)) AND (ORD(t) GE tau)).. p(u,t,g)=L=PR_max(u,t,g);
curve_nondec(t,g,gp)$((lambda_RT(t,g) LT lambda_RT(t,gp)) AND (ORD(t) GE tau)).. prt(t,g)=L=prt(t,gp);
curve_nonant(t,g,gp)$((lambda_RT(t,g) EQ lambda_RT(t,gp)) AND (ORD(t) GE tau)).. prt(t,g)=E=prt(t,gp);
curve_nonant_p(u,t,g,gp)$(ORD(t) EQ tau).. p(u,t,g)=E=p(u,t,gp);
curve_nonant_v(u,t,g,gp)$((incG(u)) AND (ORD(t) EQ tau)).. v(u,t,g)=E=v(u,t,gp);
curve_nonant_cgst(u,t,g,gp)$((incG(u)) AND (ORD(t) EQ tau)).. cg_st(u,t,g)=E=cg_st(u,t,gp);
curve_nonant_cgsh(u,t,g,gp)$((incG(u)) AND (ORD(t) EQ tau)).. cg_sh(u,t,g)=E=cg_sh(u,t,gp);
curve_nonant_psp(u,t,g,gp)$((incS(u)) AND (ORD(t) EQ tau)).. psp(u,t,g)=E=psp(u,t,gp);
curve_nonant_psn(u,t,g,gp)$((incS(u)) AND (ORD(t) EQ tau)).. psn(u,t,g)=E=psn(u,t,gp);
*******************************************************
********* Model definition *************
MODEL SSRT /obj,objscen,nodal_balance_mg,nodal_balance,trade,line_max,line_min,voltage_angle_ref,voltage_angle_max,voltage_angle_min,
tradedart_min,tradedart_max,trade_min,trade_max,dailyenergy_min,demand_max,demand_min,demand_ramp_down_initial,demand_ramp_pick_initial,
demand_ramp_down,demand_ramp_pick,CPP_max,CPP_min,CPP_ramp_down_initial,CPP_ramp_up_initial,CCP_ramp_down,CCP_ramp_up,CCP_startcost_initial,
CCP_shotcost_initial,CCP_startcost,CCP_shotcost,storage_balance_initial,storage_balance,storage_capacity,storage_psp_max,storage_psn_max,
storage_injection,stochastic_max1,stochastic_max2,curve_nondec,curve_nonant/;
MODEL SDRT /obj,objscen,nodal_balance_mg,nodal_balance,trade,line_max,line_min,voltage_angle_ref,voltage_angle_max,voltage_angle_min,
tradedart_min,tradedart_max,trade_min,trade_max,dailyenergy_min,demand_max,demand_min,demand_ramp_down_initial,demand_ramp_pick_initial,
```

```
demand_ramp_down,demand_ramp_pick,CPP_max,CPP_min,
    CPP_ramp_down_initial,CPP_ramp_up_initial,CCP_ramp_down,
    CCP_ramp_up,CCP_startcost_initial,
CCP_shotcost_initial,CCP_startcost,CCP_shotcost,
    storage_balance_initial,storage_balance,storage_capacity,
    storage_psp_max,storage_psn_max,
storage_injection,stochastic_max1,stochastic_max2,curve_nonant_p,
    curve_nonant_v,curve_nonant_cgst,curve_nonant_cgsh,
    curve_nonant_psp,curve_nonant_psn/;
OPTION OPTCR=0;
OPTION OPTCA=0;
OPTION iterlim=1e8;
*******************************************
** Parameter definition for building quantity-price curve *****
PARAMETER state 'convergence status';
PARAMETER state_dis 'convergence status at dispatch';
PARAMETER prt_act 'actual power traded in the real-time market';
PARAMETER curve_prt 'offering-bidding power in the real-time
    market';
PARAMETER curve_pricert 'offering-bidding price in the real-time
    market';
PARAMETER revenuert_act 'revenue in the real-time market';
PARAMETER utilityrt_act 'utilty gained after dispatch';
PARAMETER data_p 'power dispatch of units';
PARAMETER data_es 'energy stored';
PARAMETER data_v 'commitment status';
**************************
pda.fx(t)=PDA_act(t);
**********************************
** VPP Scheduling ******************
LOOP(h$(ORD(h) LE 3),
tau=ORD(h);
PR_max(u,t,g)=DATA_PR_max(h,g,u,t);
lambda_RT(t,g)=DATA_lambda_RT(h,g,t);
prob(g)=DATA_prob(h,g);
SOLVE SSRT USING MIP MAXIMIZING z;
state(h) = SSRT.modelstat;
LOOP(t$(ORD(t) EQ tau),
curve_prt(t,g)=prt.l(t,g);
curve_pricert(t,g)=lambda_RT(t,g);
**Real-time market clearing*******************
prt_act(t)=0;
LOOP(g,
if (curve_pricert(t,g)< lambda_RT_act(t),
if (curve_prt(t,g)>=0, prt_act(t)=max(curve_prt(t,g),prt_act(t))
    ;);
);
if (curve_pricert(t,g)> lambda_RT_act(t),
if (curve_prt(t,g)<=0, prt_act(t)=min(curve_prt(t,g),prt_act(t))
    ;);
);
if (curve_pricert(t,g)= lambda_RT_act(t),
prt_act(t)=curve_prt(t,g);
);
```

```
);
revenuert_act(t)=prt_act(t)*lambda_RT_act(t);
prt.fx(t,g)=prt_act(t);
);
** VPP Dispatching****************
SOLVE SDRT USING MIP MAXIMIZING z;
state_dis(h) = SDRT.modelstat;
LOOP(t$(ORD(t) EQ tau),
data_p(u,h)=p.l(u,t,'g1');
p.fx(u,t,g)=data_p(u,h);
data_es(u,h)=es.l(u,t,'g1');
es.fx(u,t,g)=data_es(u,h);
data_v(u,h)=v.l(u,t,'g1');
v.fx(u,t,g)=data_v(u,h);
);
utilityrt_act(h)=SUM(t$(ORD(t) EQ tau),SUM(u$incD(u),Utility(u,t)
    *p.l(u,t,'g1')*delta)-SUM(u$incG(u),CG(u,t)*p.l(u,t,'g1')*
    delta+cg_st.l(u,t,'g1')+cg_sh.l(u,t,'g1')));
);
****************************
display state, state_dis, curve_prt, curve_pricert, lambda_RT_act
    , prt_act, revenuert_act, data_p,data_es,utilityrt_act;
```

References

1. Alharbi, W., Raahemifar, K.: Probabilistic coordination of microgrid energy resources operation considering uncertainties. Electr. Pow. Syst. Res. **128** 1–10 (2015)
2. Baringo, A., Baringo, L.: A stochastic adaptive robust optimization approach for the offering strategy of a virtual power plant. IEEE Trans. Power Syst. **32**(5), 3492–3504 (2017)
3. Bertsimas, D., Sim, M.: The price of robustness. Oper. Res. **52**, 35–53 (2004)
4. Carrión, M. Arroyo, J.M., Conejo, A.J.: A bilevel stochastic programming approach for retailer futures market trading. IEEE Trans. Power Syst. **24**(3), 1446–1456 (2009)
5. Conejo, A.J., Baringo, L.: Power System Operations. Springer, Switzerland (2018)
6. Conejo, A.J., Garcáa-Bertrand, R., Carrión, M., Caballero, Á., Andrés, A.D.: Optimal involvement in futures markets of a power producer. IEEE Trans. Power Syst. **23**(2), 703–711 (2008)
7. Conejo, A.J., Carrión, M., Morales, J.M.: Decision Making Under Uncertainty in Electricity Markets. Springer, New York (2010)
8. Dabbagh, S.R., Sheikh-El-Eslami, M.K.: Risk assessment of virtual power plants offering in energy and reserve markets. IEEE Trans. Power Syst. **31**, 3572–3582 (2016)
9. GAMS.: Available at www.gams.com/ (2020)
10. Ghasemi, A., Banejad, M., Rahimiyan, M.: Integrated energy scheduling under uncertainty in a micro energy grid. IET Gener. Transm. Distrib. **12**(12), 2887–2896 (2018)
11. ILOG CPLEX.: Available at www.ilog.com/products/cplex/ (2020)
12. Malysz, P., Sirouspour, S., Emadi, A.: An optimal energy storage control strategy for grid-connected microgrids. IEEE Trans. Smart Grid **5**, 1785—1796 (2014)
13. Morales, J.M., Conejo, A.J., Pérez-Ruiz, J.: Short-term trading for a wind power producer. IEEE Trans. Power Syst. **25**(1), 554–564 (2010)
14. Pandžić, H., Morales, J.M., Conejo, A.J., Kuzle, I.: Offering model for a virtual power plant based on stochastic programming. Appl. Energy **105**, 282–292 (2013)
15. Rahimiyan, M., Morales, J.M., Conejo, A.J.: Evaluating alternative offering strategies for wind producers in a pool. Appl. Energy **88**, 4918–4926 (2011)
16. Rahimiyan, M., Baringo, L., Conejo, A.J.: Energy management of a cluster of interconnected price-responsive demands. IEEE Trans. Power Syst. **29**(2), 645–655 (2014)
17. Rahimiyan, M., Baringo, L.: Strategic bidding for a virtual power plant in the day-ahead and real-time markets: a price-taker robust optimization approach. IEEE Trans. Power Syst. **31**(4), 2676–2687 (2016)
18. Rahimiyan, M., Baringo, L.: Real-time energy management of a smart virtual power plant. IET Gener. Transm. Distrib. **13**(11), 2015–2023 (2018)
19. Shi, W., Xie, X., Chu, C.C., Gadh, R.: Distributed optimal energy management in microgrids. IEEE Trans. Smart Grid **6**(3), 1137–1146 (2015)
20. Vahabi, A.R., Latify, M.A., Rahimiyan, M., Yousefi, G.R.: An equitable and efficient energy management approach for a cluster of interconnected price responsive demands. Appl. Energy **219**, 276–289 (2018)

Chapter 4
Optimal Scheduling of a Risk-Averse Virtual Power Plant in Energy Markets

This chapter describes the optimal scheduling problem of a virtual power plant (VPP) participating in energy markets. A risk-averse VPP is considered, i.e., some metrics are introduced on the problem to control the risk associated with the scheduling decisions. With this purpose, four different models are provided and described, namely a stochastic programming problem, a static robust model, a hybrid stochastic-robust problem, and an adaptive robust optimization approach.

4.1 Introduction

As described in Chap. 3, the scheduling problem of a virtual power plant (VPP) consists in determining the optimal power traded in the market as well as the power output of the energy assets that comprise the VPP. As in Chap. 3, the VPP under consideration comprises both production and consumption energy resources, including conventional power plants, stochastic renewable generating units, storage facilities, and flexible demands.

In this chapter, the VPP under consideration participates only in energy markets, including the day-ahead (DA) and real-time (RT) energy markets. At the time of making the scheduling decisions in these markets, the VPP faces a number of uncertainties, e.g., market prices and stochastic renewable production levels. These uncertainties must be appropriately accounted for in order to make informed decisions. To do so, uncertainties are modeled using two different approaches:

1. Using scenarios: each uncertain parameter takes values within a predefined set of discrete realizations known as scenarios.
2. Using confidence bounds: each uncertain parameter takes values within an interval for a given confidence level.

The scheduling problem described in this chapter is a decision-making problem under uncertainty similar to the one described in Chap. 3. The main difference is that here a risk-averse VPP is considered, i.e., a VPP that determines its scheduling decisions with the aim of maximizing its profit but also minimizing the profit risk associated with these decisions. To do so, four different models are described, which include risk metrics that allow the VPP to control its profit risk.

The first model is a stochastic programming problem in which uncertainties are modeled using scenarios. The conditional value-at-risk (CVaR) metric is introduced in this model to control the profit risk.

The second option is a static robust optimization approach in which uncertainties are modeled using confidence bounds. In this case, the so-called uncertainty budgets are introduced to control the profit risk in the scheduling decisions.

The third model consists in the combination of the stochastic programming and robust optimization approaches, resulting in a hybrid stochastic-robust approach in which some uncertainties are modeled using scenarios and others using confidence bounds. This way, each source of uncertainty is modeled using the approach that better adapts to it.

Finally, the problem is formulated using an adaptive robust optimization approach. In this case, uncertainties are also modeled using confidence bounds. The main difference with the static robust optimization model is that this case assumes that the VPP can make corrective actions after the actual uncertainty realization is known.

The remainder of this chapter is organized as follows. Section 4.2 defines the notation used in the chapter. Section 4.3 provides a stochastic programming model for the self-scheduling problem of a risk-averse VPP participating in the DA and RT energy markets. Section 4.4 extends the formulation described in Sect. 4.3 considering a static robust approach. Next, the two models described in Sects. 4.3 and 4.4 are combined in Sect. 4.5 using a hybrid stochastic-robust model. Then, the static robust model described in Sect. 4.4 is extended in Sect. 4.6 considering an adaptive robust optimization model. Sections 4.3–4.6 include a number of clarifying examples. Section 4.7 summarizes the chapter and suggests some references for further reading. Finally, Sect. 4.8 provides the GAMS codes used to solve some of the examples described in Sects. 4.3–4.6.

L. Baringo, M. Rahimiyan, *Virtual Power Plants and Electricity Markets*, https://doi.org/10.1007/978-3-030-47602-1_4

4.2 Notation

The main notation used in this chapter is provided below for quick reference. Other symbols are defined as needed throughout the chapter. Note that if the symbols below include a subscript ω, it denotes their values in scenario ω.

4.2.1 *Indexes*

The indexes of the scheduling problem of a risk-averse VPP are:

c	Conventional power plants.
d	Demands.
ℓ	Network lines.
n	Network nodes.
r	Stochastic renewable generating units.
$r(\ell)$	Receiving-end node of network line ℓ.
s	Storage units.
$s(\ell)$	Sending-end node of network line ℓ.
t	Time periods.
ν	Iterations of the column-and-constraint generation algorithm.
ω	Scenarios

4.2.2 *Sets*

The sets of the scheduling problem of a risk-averse VPP are:

Ω^{C}	Conventional power plants.
Ω^{C}_n	Conventional power plants located at node n.
Ω^{D}	Demands.
Ω^{D}_n	Demands located at node n.
Ω^{L}	Network lines.
Ω^{M}	Network nodes connected to the main grid.
Ω^{N}	Network nodes.
Ω^{R}	Stochastic renewable generating units.
Ω^{R}_n	Stochastic renewable generating units located at node n.
Ω^{S}	Storage units.
Ω^{S}_n	Storage units located at node n.
Ω^{T}	Time periods.
Π^{DA}	Scenario realizations of uncertain parameters faced by the VPP at the time of determining the scheduling decisions in the DA energy market.

4.2.3 *Parameters*

The parameters of the scheduling problem of a risk-averse VPP are:

B_ℓ	Susceptance of line ℓ [S].
$C^{\mathrm{C,F}}_c$	Online cost of conventional power plant c [$/h].
$C^{\mathrm{C,V}}_c$	Variable cost of conventional power plant c [$/MWh].
$\underline{E}^{\mathrm{D}}_d$	Minimum energy consumption of demand d throughout the planning horizon [MWh].
$\underline{E}^{\mathrm{S}}_{st}$	Lower bound of the energy stored in storage unit s in time period t [MWh].

$\overline{E}^{\mathrm{S}}_{st}$ Upper bound of the energy stored in storage unit s in time period t [MWh].
$\underline{P}^{\mathrm{C}}_{c}$ Minimum power production of conventional power plant c [MW].
$\overline{P}^{\mathrm{C}}_{c}$ Maximum power production of conventional power plant c [MW].
$\underline{P}^{\mathrm{D}}_{dt}$ Lower bound of the power consumption of demand d in time period t [MW].
$\overline{P}^{\mathrm{D}}_{dt}$ Upper bound of the power consumption of demand d in time period t [MW].
$\underline{P}^{\mathrm{DA}}$ Lower bound of the power traded in the DA energy market [MW].
$\overline{P}^{\mathrm{DA}}$ Upper bound of the power traded in the DA energy market [MW].
$P^{\mathrm{R,A}}_{rt}$ Available generation level of the stochastic renewable generating unit r in time period t [MW].
$\tilde{P}^{\mathrm{R,A}}_{rt}$ Average level of the available generation of the stochastic renewable generating unit r in time period t [MW].
$\hat{P}^{\mathrm{R,A}}_{rt}$ Fluctuation level of the available generation of the stochastic renewable generating unit r in time period t [MW].
$\overline{P}^{\mathrm{S,C}}_{s}$ Charging capacity of storage unit s [MW].
$\overline{P}^{\mathrm{S,D}}_{s}$ Discharging capacity of storage unit s [MW].
$\underline{P}^{\mathrm{V}}$ Power capacity of consumption assets in the VPP [MW].
$\overline{P}^{\mathrm{V}}$ Power capacity of generation assets in the VPP [MW].
$R^{\mathrm{C,D}}_{c}$ Down ramping limit of conventional power plant c [MW/h].
$R^{\mathrm{C,U}}_{c}$ Up ramping limit of conventional power plant c [MW/h].
$R^{\mathrm{D,D}}_{d}$ Down ramping limit of demand d [MW/h].
$R^{\mathrm{D,U}}_{d}$ Up ramping limit of demand d [MW/h].
U^{D}_{dt} Utility of demand d acquired from the energy consumption in time period t [$/MWh].
α Confidence level used to compute the CVaR.
β Weighting parameter to model the trade-off between expected profit and CVaR.
Δt Duration of time periods [h].
$\eta^{\mathrm{S,C}}_{s}$ Charging efficiency of storage unit s [%].
$\eta^{\mathrm{S,D}}_{s}$ Discharging efficiency of storage unit s [%].
Γ^{R}_{r} Uncertainty budget of the stochastic renewable generating unit r.
$\lambda^{\mathrm{DA}}_{t}$ DA energy market price in time period t [$/MWh].
$\lambda^{\mathrm{RT}}_{t}$ RT energy market price in time period t [$/MWh].

4.2.4 *Variables*

The variables of the scheduling problem of a risk-averse VPP are:

e^{S}_{st} Energy stored in storage unit s in time period t [MWh].
p^{C}_{ct} Power generation of conventional power plant c in time period t [MW].
p^{D}_{dt} Power consumption of demand d in time period t [MW].
p^{DA}_{t} Power traded in the DA energy market in time period t [MW].
$p^{\mathrm{L}}_{\ell t}$ Power flow through line ℓ in time period t [MW].
p^{M}_{nt} Power scheduled to be bought from/sold to (if negative/positive) the DA and RT energy markets at node n in time period t [MW].
p^{R}_{rt} Production of stochastic renewable generating unit r in time period t [MW].
$p^{\mathrm{R,A}}_{rt}$ Available production level of stochastic renewable generating unit r in time period t [MW].
p^{RT}_{t} Power traded in the RT energy market in time period t [MW].
$p^{\mathrm{S,C}}_{st}$ Charging power level of storage unit s in time period t [MW].
$p^{\mathrm{S,D}}_{st}$ Discharging power level of storage unit s in time period t [MW].
u^{C}_{ct} Binary variable that is equal to 1 if conventional power plant c is generating electricity in time period t, being 0 otherwise.
δ_{nt} Voltage angle at bus n in time period t [rad].
μ Auxiliary variable used to linearize the CVaR metric [$].
ρ Value-at-risk measure [$].
ζ Total profit obtained by the VPP throughout the planning horizon [$].

4.3 Stochastic Programming Approach

This section provides a stochastic programming model for the scheduling problem of a risk-averse VPP participating in energy markets.

4.3.1 Problem Description

The VPP under consideration participates in pool markets, including the DA and RT energy markets. First, the VPP determines its scheduling decisions in the DA energy market, one day in advance. Second, the VPP decides its scheduling decisions in the RT energy market for each time period. Third, the VPP dispatches its units considering their techno-economic properties and the scheduled energy in the DA and RT energy markets. The VPP is considered to be equipped with the smart grid technology, which enables the VPP to repeat the second and third decision stages time by time.

As explained in Chap. 3, the VPP faces uncertainties at the time of solving the scheduling problem. These uncertainties generally have a great impact on the profit achieved by the VPP; moreover, uncertainties may threaten the operation feasibility of the VPP. In this section, uncertainties are handled using a stochastic programming approach in which each uncertain parameter is represented using a number of scenarios. This way, the VPP aims at maximizing its expected profit over the predefined set of scenario realizations and considering the feasibility of decisions taken for each scenario realization.

Note also that these uncertainties impose financial risks on the scheduling problem. Thus, this chapter considers a risk-averse VPP that is conservative against the profit risk associated with its scheduling decisions. In this section, the CVaR metric is considered to account for the profit risk [13]. In a profit maximization problem, the CVaR for a given α confidence level is defined as the expected profit of the $(1-\alpha)100\%$ scenarios yielding the lowest profits.

4.3.2 Uncertainty Model

As explained in the previous section, an accurate modeling of the uncertainties in the scheduling problem of a risk-averse VPP in the DA and RT energy markets is necessary. In particular, the uncertain parameters considered are the market prices and the available production levels of stochastic renewable generating units. The stochastic programming model described in this section uses a scenario-based uncertainty model that represents uncertain parameters through a number of scenarios. Moreover, these scenarios should appropriately address the correlation between uncertain parameters since a meaningful correlation is observed between the prices in the DA and RT energy markets [11]. However, the correlation between the prices and the available production levels of *small-size local* stochastic renewable generating units can be considered negligible.

A scenario set Π^{DA} containing a finite number of scenarios is considered. Each scenario ω has a probability of occurrence π_ω, while the summation of probabilities is equal to one. Each scenario ω describes the information related to the prices in the DA and RT energy markets, as well as the available production levels of stochastic renewable generating units.

Accordingly, the sequential decision-making procedure is carried out as explained below:

1. The VPP makes the scheduling decisions in the DA energy market and determines the status of its own conventional power plants, one day in advance. The decisions on the power traded in the DA energy market and the status of its own conventional power plants are *here-and-now* decisions with respect to the scenario realizations.
2. Once a scenario ω is realized, the VPP knows the prices in the DA and RT energy markets, as well as the available production levels of stochastic renewable generating units for the planning horizon.
3. The VPP decides the schedules in the RT energy market, and dispatches the units in the VPP, some minutes before the power delivery in each time period. The scheduling decisions in the RT energy market and the dispatch decisions are *wait-and-see* decisions with respect to the scenario realization ω.

Figure 4.1 shows the scenario tree used to represent uncertainties in the scheduling problem of a risk-averse VPP in the DA energy market.

Considering the above procedure, the scheduling problem of a risk-averse VPP in the energy markets is formulated in the next section.

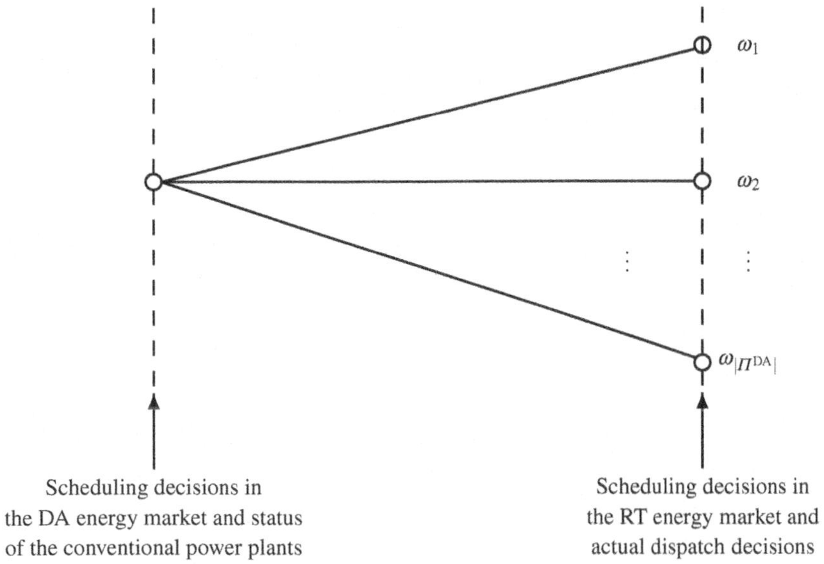

Fig. 4.1 Scenario tree used in the scheduling problem of a risk-averse VPP in the DA energy market

4.3.3 Formulation

Considering the uncertainty model described in Sect. 4.3.2, the scheduling problem of the risk-averse VPP in the DA energy market using a stochastic programming approach is formulated as the following mixed-integer linear programming (MILP) model:

$$\max_{\Phi^{DA,S}} \quad \sum_{\omega \in \Pi^{DA}} \pi_\omega \zeta_\omega + \beta \left[\rho - \frac{1}{1-\alpha} \sum_{\omega \in \Pi^{DA}} \pi_\omega \mu_\omega \right] \tag{4.1a}$$

subject to

$$\zeta_\omega = \sum_{t \in \Omega^T} \Bigg[\lambda^{DA}_{t\omega} p^{DA}_t \Delta t + \lambda^{RT}_{t\omega} p^{RT}_{t\omega} \Delta t + \sum_{d \in \Omega^D} U^D_{dt} p^D_{dt\omega} \Delta t - \sum_{c \in \Omega^C} (C^{C,F}_c u^C_{ct} + C^{C,V}_c p^C_{ct\omega} \Delta t) \Bigg], \quad \forall \omega \in \Pi^{DA}, \tag{4.1b}$$

$$-\zeta_\omega + \rho - \mu_\omega \leq 0, \quad \forall \omega \in \Pi^{DA}, \tag{4.1c}$$

$$0 \leq \mu_\omega, \quad \forall \omega \in \Pi^{DA}, \tag{4.1d}$$

$$\sum_{c \in \Omega^C_n} p^C_{ct\omega} + \sum_{r \in \Omega^R_n} p^R_{rt\omega} + \sum_{s \in \Omega^S_n} (p^{S,D}_{st\omega} - p^{S,C}_{st\omega}) - \sum_{\ell | s(\ell) = n} p^L_{\ell t\omega} + \sum_{\ell | r(\ell) = n} p^L_{\ell t\omega} = p^M_{nt\omega} + \sum_{d \in \Omega^D_n} p^D_{dt\omega}, \quad \forall n \in \Omega^M, \forall t \in \Omega^T, \forall \omega \in \Pi^{DA}, \tag{4.1e}$$

$$\sum_{c \in \Omega^C_n} p^C_{ct\omega} + \sum_{r \in \Omega^R_n} p^R_{rt\omega} + \sum_{s \in \Omega^S_n} (p^{S,D}_{st\omega} - p^{S,C}_{st\omega}) - \sum_{\ell | s(\ell) = n} p^L_{\ell t\omega} + \sum_{\ell | r(\ell) = n} p^L_{\ell t\omega} = \sum_{d \in \Omega^D_n} p^D_{dt\omega}, \quad \forall n \in \Omega^N \setminus n \in \Omega^M, \forall t \in \Omega^T, \forall \omega \in \Pi^{DA}, \tag{4.1f}$$

$$p_t^{\text{DA}} + p_{t\omega}^{\text{RT}} = \sum_{n \in \Omega^{\text{M}}} p_{nt\omega}^{\text{M}}, \quad \forall t \in \Omega^{\text{T}}, \forall \omega \in \Pi^{\text{DA}}, \tag{4.1g}$$

$$p_{\ell t\omega}^{\text{L}} = B_\ell(\delta_{s(\ell)t\omega} - \delta_{r(\ell)t\omega}), \quad \forall \ell \in \Omega^{\text{L}}, \forall t \in \Omega^{\text{T}}, \forall \omega \in \Pi^{\text{DA}}, \tag{4.1h}$$

$$-\overline{P}_\ell^{\text{L}} \le p_{\ell t\omega}^{\text{L}} \le \overline{P}_\ell^{\text{L}}, \quad \forall \ell \in \Omega^{\text{L}}, \forall t \in \Omega^{\text{T}}, \forall \omega \in \Pi^{\text{DA}}, \tag{4.1i}$$

$$\delta_{nt\omega} = 0, \quad n : \text{ref.}, \forall t \in \Omega^{\text{T}}, \forall \omega \in \Pi^{\text{DA}}, \tag{4.1j}$$

$$-\pi \le \delta_{nt\omega} \le \pi, \quad \forall n \in \Omega^{\text{N}} \setminus n : \text{ref.}, \forall t \in \Omega^{\text{T}}, \forall \omega \in \Pi^{\text{DA}}, \tag{4.1k}$$

$$\underline{P}^{\text{V}} \le p_t^{\text{DA}} \le \overline{P}^{\text{V}}, \quad \forall t \in \Omega^{\text{T}}, \tag{4.1l}$$

$$\underline{P}^{\text{V}} \le p_t^{\text{DA}} + p_{t\omega}^{\text{RT}} \le \overline{P}^{\text{V}}, \quad \forall t \in \Omega^{\text{T}}, \forall \omega, \tag{4.1m}$$

$$-\overline{P}_n^{\text{M}} \le p_{nt\omega}^{\text{M}} \le \overline{P}_n^{\text{M}}, \quad \forall n \in \Omega^{\text{M}}, \forall t \in \Omega^{\text{T}}, \forall \omega \in \Pi^{\text{DA}}, \tag{4.1n}$$

$$\sum_{t \in \Omega^{\text{T}}} p_{dt\omega}^{\text{D}} \Delta t \ge \underline{E}_d^{\text{D}}, \quad \forall d \in \Omega^{\text{D}}, \forall \omega \in \Pi^{\text{DA}}, \tag{4.1o}$$

$$\underline{P}_{dt}^{\text{D}} \le p_{dt\omega}^{\text{D}} \le \overline{P}_{dt}^{\text{D}}, \quad \forall d \in \Omega^{\text{D}}, \forall t \in \Omega^{\text{T}}, \forall \omega \in \Pi^{\text{DA}}, \tag{4.1p}$$

$$-R_d^{\text{D,D}} \Delta t \le p_{dt\omega}^{\text{D}} - p_{d(t-1)\omega}^{\text{D}} \le R_d^{\text{D,U}} \Delta t, \quad \forall d \in \Omega^{\text{D}}, \forall t \in \Omega^{\text{T}}, \forall \omega \in \Pi^{\text{DA}}, \tag{4.1q}$$

$$\underline{P}_c^{\text{C}} u_{ct}^{\text{C}} \le p_{ct\omega}^{\text{C}} \le \overline{P}_c^{\text{C}} u_{ct}^{\text{C}}, \quad \forall c \in \Omega^{\text{C}}, \forall t \in \Omega^{\text{T}}, \forall \omega \in \Pi^{\text{DA}}, \tag{4.1r}$$

$$-R_c^{\text{C,D}} \Delta t \le p_{ct\omega}^{\text{C}} - p_{c(t-1)\omega}^{\text{C}} \le R_c^{\text{C,U}} \Delta t, \quad \forall c \in \Omega^{\text{C}}, \forall t \in \Omega^{\text{T}}, \forall \omega \in \Pi^{\text{DA}}, \tag{4.1s}$$

$$e_{st\omega}^{\text{S}} = e_{s(t-1)\omega}^{\text{S}} + \eta_s^{\text{S,C}} p_{st\omega}^{\text{S,C}} \Delta t - \frac{p_{st\omega}^{\text{S,D}}}{\eta_s^{\text{S,D}}} \Delta t, \quad \forall s \in \Omega^{\text{S}}, \forall t \in \Omega^{\text{T}}, \forall \omega \in \Pi^{\text{DA}}, \tag{4.1t}$$

$$\underline{E}_{st}^{\text{S}} \le e_{st\omega}^{\text{S}} \le \overline{E}_{st}^{\text{S}}, \quad \forall s \in \Omega^{\text{S}}, \forall t \in \Omega^{\text{T}}, \forall \omega \in \Pi^{\text{DA}}, \tag{4.1u}$$

$$0 \le p_{st\omega}^{\text{S,C}} \le \overline{P}_s^{\text{S,C}}, \quad \forall s \in \Omega^{\text{S}}, \forall t \in \Omega^{\text{T}}, \forall \omega \in \Pi^{\text{DA}}, \tag{4.1v}$$

$$0 \le p_{st\omega}^{\text{S,D}} \le \overline{P}_s^{\text{S,D}}, \quad \forall s \in \Omega^{\text{S}}, \forall t \in \Omega^{\text{T}}, \forall \omega \in \Pi^{\text{DA}}, \tag{4.1w}$$

$$0 \le p_{rt\omega}^{\text{R}} \le P_{rt\omega}^{\text{R,A}}, \quad \forall r \in \Omega^{\text{R}}, \forall t \in \Omega^{\text{T}}, \forall \omega \in \Pi^{\text{DA}}, \tag{4.1x}$$

where variables in set $\Phi^{\text{DA,S}} = \Big\{ e_{st\omega}^{\text{S}}, p_{st\omega}^{\text{S,C}}, p_{st\omega}^{\text{S,D}}, \forall s \in \Omega^{\text{S}}, \forall t \in \Omega^{\text{T}}, \forall \omega \in \Pi^{\text{DA}}; p_{ct\omega}^{\text{C}}, u_{ct}^{\text{C}}, \forall c \in \Omega^{\text{C}}, \forall t \in \Omega^{\text{T}}, \forall \omega \in \Pi^{\text{DA}};$ $p_{dt\omega}^{\text{D}}, \forall d \in \Omega^{\text{D}}, \forall t \in \Omega^{\text{T}}, \forall \omega \in \Pi^{\text{DA}}; p_{rt\omega}^{\text{R}}, \forall r \in \Omega^{\text{R}}, \forall t \in \Omega^{\text{T}}, \forall \omega \in \Pi^{\text{DA}}; p_t^{\text{DA}}, \forall t \in \Omega^{\text{T}}; p_{nt\omega}^{\text{M}}, \forall n \in \Omega^{\text{M}}, \forall t \in \Omega^{\text{T}};$ $\delta_{nt\omega}, \forall n \in \Omega^{\text{N}}, \forall t \in \Omega^{\text{T}}, \forall \omega \in \Pi^{\text{DA}}; p_t^{\text{DA}}, \forall t \in \Omega^{\text{T}}; p_{t\omega}^{\text{RT}}, \forall t \in \Omega^{\text{T}}, \forall \omega \in \Pi^{\text{DA}}; p_{\ell t\omega}^{\text{L}}, \forall \ell \in \Omega^{\text{L}}, \forall t \in \Omega^{\text{T}}, \forall \omega \in \Pi^{\text{DA}}; \rho;$ $\zeta_\omega, \mu_\omega, \forall \omega \in \Pi^{\text{DA}} \Big\}$ are the optimization variables of the stochastic programming model (4.1).

The structure of the stochastic model (4.1) is similar to the stochastic model for the risk-neutral VPP described in Chap. 3. The main difference is the multi-objective function (4.1a) that includes two terms, namely:

1. Term $\sum_{\omega \in \Pi^{\text{DA}}} \pi_\omega \zeta_\omega$ that represents the expected profit achieved by the VPP.
2. Term $\rho - \frac{1}{1-\alpha} \sum_{\omega \in \Pi^{\text{DA}}} \pi_\omega \mu_\omega$ that is the CVaR.

Note that the CVaR is multiplied by nonnegative parameter β, which allows the VPP to model the trade-off between the expected profit and the CVaR. For $\beta = 0$, the VPP behaves as a risk-neutral player that seeks to achieve the maximum expected profit and neglects the profit risk associated with its scheduling decisions. By increasing the value of parameter β, a risk-averse VPP that reduces the risk at the cost of a lower expected profit is considered. Equalities (4.1b) define the profit for each scenario realization. Constraints (4.1c)–(4.1d) allow incorporating the CVaR metric into the model as a set of linear expressions.

Constraints (4.1e)–(4.1x) define the feasible operating region of the VPP for each scenario realization. Constraints (4.1e) impose the balance of power generation and consumption at the nodes connected to the main grid. Similarly, constraints (4.1f) establish the power balance at the remaining network nodes. Equations (4.1g) define the total power traded in the DA and RT energy markets as the summation of power injection to/from the nodes connected to the main grid. Constraints (4.1h) consider the power flow through network lines bounded by their capacity (4.1i). Constraints (4.1j) fix to zero the voltage angle at the reference node, while the voltage angles at other nodes are bounded by (4.1k). Constraints (4.1l) and (4.1m) limit the power traded in the DA energy market, as well as the total power traded in the DA and RT energy markets by the available production and consumption capacity in the VPP. Constraints (4.1n) impose bounds on the power injection at the nodes connected to the main grid. Constraints (4.1o), (4.1p), and (4.1q) state the technical limitations of the demands including the minimum energy consumption during the planning horizon, power load, and ramping limitations, respectively. Constraints (4.1r) and (4.1s) consider the power production and ramping limitations of the conventional power plants, respectively. Constraints (4.1t) establish the energy balance in the storage units bounded by the corresponding energy capacities in constraints (4.1u). Constraints (4.1v) and (4.1w) limit the charging and discharging power levels of storage units, respectively. Constraints (4.1x) state that the power production of stochastic renewable generating units must be lower than or equal to the available one.

The stochastic model (4.1) provides the traded power quantity p_t^{DA} in the DA energy market and the status of the conventional power plants in the VPP for all time periods of the day.

Illustrative Example 4.1 Stochastic Scheduling Problem of a Risk-Averse VPP

A VPP comprises a conventional power plant, two flexible demands, a storage units, and a renewable photovoltaic (PV) generating unit. The technical and economic data of the conventional power plant, the demands, and the storage unit are provided in Tables 4.1, 4.2, and 4.3, respectively, while the installed capacity of the PV generating unit is 10 MW. Based on generation and consumption assets in the VPP, the maximum power that can be sold to/bought from the DA and RT energy markets is 40/60 MW in each time period.

The units in the VPP are interconnected within the three-node network depicted in Fig. 4.2. Node 1 is connected with the main grid and is bounded by an injection capacity of 100 MW. The conventional power plant and demand 1 are located at node 2. Demand 2, the storage unit, and the PV generating unit are located at node 3. The network data are given in Table 4.4.

Table 4.1 Illustrative Example 4.1: data of the conventional power plant

Variable production cost	\$45/MWh
Online production cost	\$10/h
Maximum power production limit	20 MW
Minimum power production limit	0 MW
Ramping up limit	20 MW/h
Ramping down limit	20 MW/h
Initial commitment status	off
Initial power production	0 MW

Table 4.2 Illustrative Example 4.1: data of the demands

Parameter	Demand 1	Demand 2
Minimum energy consumption	30 MWh	50 MWh
Maximum power load in time periods 1, 2, 3	20, 20, 10 MW	30, 30, 15 MW
Minimum power load in time periods 1, 2, 3	0, 0, 0 MW	5, 5, 5 MW
Utility in time periods 1, 2, 3	40, 50, 40 \$/MWh	50, 60, 50 \$/MWh
Ramping up limit	15 MW/h	25 MW/h
Ramping down limit	15 MW/h	25 MW/h
Initial power load	10 MW	20 MW

Table 4.3 Illustrative Example 4.1: data of the storage unit

Maximum charging power	10 MW
Maximum discharging power	10 MW
Maximum energy level	10 MWh
Minimum energy level	5 MWh
Initial energy level	5 MWh
Charging efficiency	0.9
Discharging efficiency	0.9

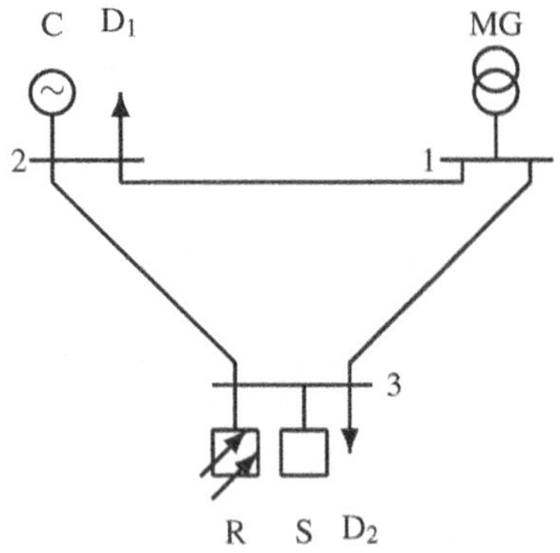

Fig. 4.2 Illustrative Example 4.1: three-node system

Table 4.4 Illustrative Example 4.1: network data

Line	From	To	Susceptance [S]	Capacity [MW]
1	1	2	200	50
2	1	3	200	50
3	2	3	200	50

Table 4.5 Illustrative Example 4.1: data for scenarios

	λ_t^{DA} [$/MWh]			λ_t^{RT} [$/MWh]			
Scenario	1	2	3	1	2	3	Probability
1	20	40	30	25	50	25	$0.4 \times 0.4 = 0.16$
2	20	40	30	10	45	28	$0.4 \times 0.6 = 0.24$
3	30	50	40	35	60	35	$0.6 \times 0.4 = 0.24$
4	30	50	40	20	55	38	$0.6 \times 0.6 = 0.36$

Table 4.6 Illustrative Example 4.1: data of the PV generating unit

Time period	Forecast level [MW]
1	9
2	7
3	1

First column identifies each line number. Second and third columns indicate the sending- and receiving-end nodes of each line, respectively. Fourth and fifth columns present susceptance and capacity of each line, respectively. The base values of apparent power and line-to-line voltage are considered 100 MVA and 10 kV, respectively. Consequently, the base value of impedance is equal to 1 Ω.

The scenarios are presented in Table 4.5. First column identifies the scenarios. Second, third, and fourth columns indicate the DA market prices in time periods 1, 2, and 3, respectively. Fifth, sixth, and seventh columns provide the RT market prices during the time periods 1, 2 and 3, respectively. The last column determines the probability of scenarios. The scenario tree is built considering 4 scenarios; each one presenting information on the prices in the DA and RT energy markets. First, 2 scenarios of the prices in the DA energy market are considered, with a probability of occurrence of 0.4 and 0.6. Second, for each scenario of the DA market price, 2 scenarios of the RT market price are considered; each one with a probability of occurrence of 0.4 and 0.6.

For the sake of simplicity, the available production levels of the PV generating unit are modeled through the single-point forecasts provided in Table 4.6.

Given the techno-economic data of the units and the uncertainty model, the stochastic scheduling problem of the risk-averse VPP in the DA energy market is formulated below. The value of β is set equal to 1 to consider a risk-averse strategy. In addition, the CVaR is calculated for a confidence level α=0.8.

The objective function to be maximized is formulated below:

$$z = 0.16\zeta_1 + 0.24\zeta_2 + 0.24\zeta_3 + 0.36\zeta_4$$
$$+ \left[\rho - \frac{1}{1-0.8}(0.16\mu_1 + 0.24\mu_2 + 0.24\mu_3 + 0.36\mu_4)\right].$$

The first line is the expected profit over the considered scenarios. The second line is the weighted CVaR.

The constraints of the problem are formulated below:

- Profit calculation:

$$\zeta_1 = \Big[20p_1^{\text{DA}} + 25p_{11}^{\text{RT}} + 40p_{111}^{\text{D}} + 50p_{211}^{\text{D}} - (10u_1^{\text{C}} + 45p_{11}^{\text{C}})$$
$$40p_2^{\text{DA}} + 50p_{21}^{\text{RT}} + 50p_{121}^{\text{D}} + 60p_{221}^{\text{D}} - (10u_2^{\text{C}} + 45p_{21}^{\text{C}})$$
$$30p_3^{\text{DA}} + 25p_{31}^{\text{RT}} + 40p_{131}^{\text{D}} + 50p_{231}^{\text{D}} - (10u_3^{\text{C}} + 45p_{31}^{\text{C}})\Big],$$

$$\zeta_2 = \Big[20p_1^{\text{DA}} + 10p_{12}^{\text{RT}} + 40p_{112}^{\text{D}} + 50p_{212}^{\text{D}} - (10u_1^{\text{C}} + 45p_{12}^{\text{C}})$$
$$40p_2^{\text{DA}} + 45p_{22}^{\text{RT}} + 50p_{122}^{\text{D}} + 60p_{222}^{\text{D}} - (10u_2^{\text{C}} + 45p_{22}^{\text{C}})$$
$$30p_3^{\text{DA}} + 28p_{32}^{\text{RT}} + 40p_{132}^{\text{D}} + 50p_{232}^{\text{D}} - (10u_3^{\text{C}} + 45p_{32}^{\text{C}})\Big],$$

$$\zeta_3 = \Big[30p_1^{\text{DA}} + 35p_{13}^{\text{RT}} + 40p_{113}^{\text{D}} + 50p_{213}^{\text{D}} - (10u_1^{\text{C}} + 45p_{13}^{\text{C}})$$
$$50p_2^{\text{DA}} + 60p_{23}^{\text{RT}} + 50p_{123}^{\text{D}} + 60p_{223}^{\text{D}} - (10u_2^{\text{C}} + 45p_{23}^{\text{C}})$$
$$40p_3^{\text{DA}} + 35p_{33}^{\text{RT}} + 40p_{133}^{\text{D}} + 50p_{233}^{\text{D}} - (10u_3^{\text{C}} + 45p_{33}^{\text{C}})\Big],$$

$$\zeta_4 = \Big[30p_1^{\text{DA}} + 20p_{14}^{\text{RT}} + 40p_{114}^{\text{D}} + 50p_{214}^{\text{D}} - (10u_1^{\text{C}} + 45p_{14}^{\text{C}})$$
$$50p_2^{\text{DA}} + 55p_{24}^{\text{RT}} + 50p_{124}^{\text{D}} + 60p_{224}^{\text{D}} - (10u_2^{\text{C}} + 45p_{24}^{\text{C}})$$
$$40p_3^{\text{DA}} + 38p_{34}^{\text{RT}} + 40p_{134}^{\text{D}} + 50p_{234}^{\text{D}} - (10u_3^{\text{C}} + 45p_{34}^{\text{C}})\Big].$$

- Constraints for calculating the CVaR:

$$-\zeta_1 + \rho - \mu_1 \leq 0,$$
$$-\zeta_2 + \rho - \mu_2 \leq 0,$$
$$-\zeta_3 + \rho - \mu_3 \leq 0,$$
$$-\zeta_4 + \rho - \mu_4 \leq 0,$$
$$\mu_1 \geq 0,$$
$$\mu_2 \geq 0,$$
$$\mu_3 \geq 0,$$
$$\mu_4 \geq 0.$$

- Power balance constraints:

$$-p_{111}^{\text{L}} - p_{211}^{\text{L}} = p_{11}^{\text{M}},$$
$$-p_{112}^{\text{L}} - p_{212}^{\text{L}} = p_{12}^{\text{M}},$$
$$-p_{113}^{\text{L}} - p_{213}^{\text{L}} = p_{13}^{\text{M}},$$
$$-p_{114}^{\text{L}} - p_{214}^{\text{L}} = p_{14}^{\text{M}},$$

$$
\begin{aligned}
&-p^{\mathrm{L}}_{121}-p^{\mathrm{L}}_{221}=p^{\mathrm{M}}_{21},\\
&-p^{\mathrm{L}}_{122}-p^{\mathrm{L}}_{222}=p^{\mathrm{M}}_{22},\\
&-p^{\mathrm{L}}_{123}-p^{\mathrm{L}}_{223}=p^{\mathrm{M}}_{23},\\
&-p^{\mathrm{L}}_{124}-p^{\mathrm{L}}_{224}=p^{\mathrm{M}}_{24},\\
&-p^{\mathrm{L}}_{131}-p^{\mathrm{L}}_{231}=p^{\mathrm{M}}_{31},\\
&-p^{\mathrm{L}}_{132}-p^{\mathrm{L}}_{232}=p^{\mathrm{M}}_{32},\\
&-p^{\mathrm{L}}_{133}-p^{\mathrm{L}}_{233}=p^{\mathrm{M}}_{33},\\
&-p^{\mathrm{L}}_{134}-p^{\mathrm{L}}_{234}=p^{\mathrm{M}}_{34},\\
&p^{\mathrm{C}}_{11}-p^{\mathrm{L}}_{311}+p^{\mathrm{L}}_{111}=p^{\mathrm{D}}_{111},\\
&p^{\mathrm{C}}_{12}-p^{\mathrm{L}}_{312}+p^{\mathrm{L}}_{112}=p^{\mathrm{D}}_{112},\\
&p^{\mathrm{C}}_{13}-p^{\mathrm{L}}_{313}+p^{\mathrm{L}}_{113}=p^{\mathrm{D}}_{113},\\
&p^{\mathrm{C}}_{14}-p^{\mathrm{L}}_{314}+p^{\mathrm{L}}_{114}=p^{\mathrm{D}}_{114},\\
&p^{\mathrm{C}}_{21}-p^{\mathrm{L}}_{321}+p^{\mathrm{L}}_{121}=p^{\mathrm{D}}_{121},\\
&p^{\mathrm{C}}_{22}-p^{\mathrm{L}}_{322}+p^{\mathrm{L}}_{122}=p^{\mathrm{D}}_{122},\\
&p^{\mathrm{C}}_{23}-p^{\mathrm{L}}_{323}+p^{\mathrm{L}}_{123}=p^{\mathrm{D}}_{123},\\
&p^{\mathrm{C}}_{24}-p^{\mathrm{L}}_{324}+p^{\mathrm{L}}_{124}=p^{\mathrm{D}}_{124},\\
&p^{\mathrm{C}}_{31}-p^{\mathrm{L}}_{331}+p^{\mathrm{L}}_{131}=p^{\mathrm{D}}_{131},\\
&p^{\mathrm{C}}_{32}-p^{\mathrm{L}}_{332}+p^{\mathrm{L}}_{132}=p^{\mathrm{D}}_{132},\\
&p^{\mathrm{C}}_{33}-p^{\mathrm{L}}_{333}+p^{\mathrm{L}}_{133}=p^{\mathrm{D}}_{133},\\
&p^{\mathrm{C}}_{34}-p^{\mathrm{L}}_{334}+p^{\mathrm{L}}_{134}=p^{\mathrm{D}}_{134},\\
&p^{\mathrm{R}}_{11}+p^{\mathrm{S,D}}_{11}-p^{\mathrm{S,C}}_{11}+p^{\mathrm{L}}_{211}+p^{\mathrm{L}}_{311}=p^{\mathrm{D}}_{211},\\
&p^{\mathrm{R}}_{12}+p^{\mathrm{S,D}}_{12}-p^{\mathrm{S,C}}_{12}+p^{\mathrm{L}}_{212}+p^{\mathrm{L}}_{312}=p^{\mathrm{D}}_{212},\\
&p^{\mathrm{R}}_{13}+p^{\mathrm{S,D}}_{13}-p^{\mathrm{S,C}}_{13}+p^{\mathrm{L}}_{213}+p^{\mathrm{L}}_{313}=p^{\mathrm{D}}_{213},\\
&p^{\mathrm{R}}_{14}+p^{\mathrm{S,D}}_{14}-p^{\mathrm{S,C}}_{14}+p^{\mathrm{L}}_{214}+p^{\mathrm{L}}_{314}=p^{\mathrm{D}}_{214},\\
&p^{\mathrm{R}}_{21}+p^{\mathrm{S,D}}_{21}-p^{\mathrm{S,C}}_{21}+p^{\mathrm{L}}_{221}+p^{\mathrm{L}}_{321}=p^{\mathrm{D}}_{221},\\
&p^{\mathrm{R}}_{22}+p^{\mathrm{S,D}}_{22}-p^{\mathrm{S,C}}_{22}+p^{\mathrm{L}}_{222}+p^{\mathrm{L}}_{322}=p^{\mathrm{D}}_{222},\\
&p^{\mathrm{R}}_{23}+p^{\mathrm{S,D}}_{23}-p^{\mathrm{S,C}}_{23}+p^{\mathrm{L}}_{223}+p^{\mathrm{L}}_{323}=p^{\mathrm{D}}_{223},\\
&p^{\mathrm{R}}_{24}+p^{\mathrm{S,D}}_{24}-p^{\mathrm{S,C}}_{24}+p^{\mathrm{L}}_{224}+p^{\mathrm{L}}_{324}=p^{\mathrm{D}}_{224},\\
&p^{\mathrm{R}}_{31}+p^{\mathrm{S,D}}_{31}-p^{\mathrm{S,C}}_{31}+p^{\mathrm{L}}_{231}+p^{\mathrm{L}}_{331}=p^{\mathrm{D}}_{231},\\
&p^{\mathrm{R}}_{32}+p^{\mathrm{S,D}}_{32}-p^{\mathrm{S,C}}_{32}+p^{\mathrm{L}}_{232}+p^{\mathrm{L}}_{332}=p^{\mathrm{D}}_{232},\\
&p^{\mathrm{R}}_{33}+p^{\mathrm{S,D}}_{33}-p^{\mathrm{S,C}}_{33}+p^{\mathrm{L}}_{233}+p^{\mathrm{L}}_{333}=p^{\mathrm{D}}_{233},\\
&p^{\mathrm{R}}_{34}+p^{\mathrm{S,D}}_{34}-p^{\mathrm{S,C}}_{34}+p^{\mathrm{L}}_{234}+p^{\mathrm{L}}_{334}=p^{\mathrm{D}}_{234}.
\end{aligned}
$$

- Power traded in the DA and RT energy markets:

$$p_1^{DA} + p_{11}^{RT} = p_{11}^{M},$$
$$p_1^{DA} + p_{12}^{RT} = p_{12}^{M},$$
$$p_1^{DA} + p_{13}^{RT} = p_{13}^{M},$$
$$p_1^{DA} + p_{14}^{RT} = p_{14}^{M},$$
$$p_2^{DA} + p_{21}^{RT} = p_{21}^{M},$$
$$p_2^{DA} + p_{22}^{RT} = p_{22}^{M},$$
$$p_2^{DA} + p_{23}^{RT} = p_{23}^{M},$$
$$p_2^{DA} + p_{24}^{RT} = p_{24}^{M},$$
$$p_3^{DA} + p_{31}^{RT} = p_{31}^{M},$$
$$p_3^{DA} + p_{32}^{RT} = p_{32}^{M},$$
$$p_3^{DA} + p_{33}^{RT} = p_{33}^{M},$$
$$p_3^{DA} + p_{34}^{RT} = p_{34}^{M}.$$

- Definition of power flows through lines:

$$\frac{p_{111}^{L}}{100} = 200(\delta_{111} - \delta_{211}),$$
$$\frac{p_{112}^{L}}{100} = 200(\delta_{112} - \delta_{212}),$$
$$\frac{p_{113}^{L}}{100} = 200(\delta_{113} - \delta_{213}),$$
$$\frac{p_{114}^{L}}{100} = 200(\delta_{114} - \delta_{214}),$$
$$\frac{p_{121}^{L}}{100} = 200(\delta_{121} - \delta_{221}),$$
$$\frac{p_{122}^{L}}{100} = 200(\delta_{122} - \delta_{222}),$$
$$\frac{p_{123}^{L}}{100} = 200(\delta_{123} - \delta_{223}),$$
$$\frac{p_{124}^{L}}{100} = 200(\delta_{124} - \delta_{224}),$$
$$\frac{p_{131}^{L}}{100} = 200(\delta_{131} - \delta_{231}),$$
$$\frac{p_{132}^{L}}{100} = 200(\delta_{132} - \delta_{232}),$$
$$\frac{p_{133}^{L}}{100} = 200(\delta_{133} - \delta_{233}),$$
$$\frac{p_{134}^{L}}{100} = 200(\delta_{134} - \delta_{234}),$$

$$\frac{p^{\mathrm{L}}_{211}}{100} = 200(\delta_{111} - \delta_{311}),$$
$$\frac{p^{\mathrm{L}}_{212}}{100} = 200(\delta_{112} - \delta_{312}),$$
$$\frac{p^{\mathrm{L}}_{213}}{100} = 200(\delta_{113} - \delta_{313}),$$
$$\frac{p^{\mathrm{L}}_{214}}{100} = 200(\delta_{114} - \delta_{314}),$$
$$\frac{p^{\mathrm{L}}_{221}}{100} = 200(\delta_{121} - \delta_{321}),$$
$$\frac{p^{\mathrm{L}}_{222}}{100} = 200(\delta_{122} - \delta_{322}),$$
$$\frac{p^{\mathrm{L}}_{223}}{100} = 200(\delta_{123} - \delta_{323}),$$
$$\frac{p^{\mathrm{L}}_{224}}{100} = 200(\delta_{124} - \delta_{324}),$$
$$\frac{p^{\mathrm{L}}_{231}}{100} = 200(\delta_{131} - \delta_{331}),$$
$$\frac{p^{\mathrm{L}}_{232}}{100} = 200(\delta_{132} - \delta_{332}),$$
$$\frac{p^{\mathrm{L}}_{233}}{100} = 200(\delta_{133} - \delta_{333}),$$
$$\frac{p^{\mathrm{L}}_{234}}{100} = 200(\delta_{134} - \delta_{334}),$$
$$\frac{p^{\mathrm{L}}_{311}}{100} = 200(\delta_{211} - \delta_{311}),$$
$$\frac{p^{\mathrm{L}}_{312}}{100} = 200(\delta_{212} - \delta_{312}),$$
$$\frac{p^{\mathrm{L}}_{313}}{100} = 200(\delta_{213} - \delta_{313}),$$
$$\frac{p^{\mathrm{L}}_{314}}{100} = 200(\delta_{214} - \delta_{314}),$$
$$\frac{p^{\mathrm{L}}_{321}}{100} = 200(\delta_{221} - \delta_{321}),$$
$$\frac{p^{\mathrm{L}}_{322}}{100} = 200(\delta_{222} - \delta_{322}),$$
$$\frac{p^{\mathrm{L}}_{323}}{100} = 200(\delta_{223} - \delta_{323}),$$
$$\frac{p^{\mathrm{L}}_{324}}{100} = 200(\delta_{224} - \delta_{324}),$$
$$\frac{p^{\mathrm{L}}_{331}}{100} = 200(\delta_{231} - \delta_{331}),$$
$$\frac{p^{\mathrm{L}}_{332}}{100} = 200(\delta_{232} - \delta_{332}),$$

$$\frac{p_{333}^{\mathrm{L}}}{100} = 200(\delta_{233} - \delta_{333}),$$
$$\frac{p_{334}^{\mathrm{L}}}{100} = 200(\delta_{234} - \delta_{334}).$$

- Limits on power flows through lines:

$$-50 \le p_{111}^{\mathrm{L}} \le 50,$$
$$-50 \le p_{112}^{\mathrm{L}} \le 50,$$
$$-50 \le p_{113}^{\mathrm{L}} \le 50,$$
$$-50 \le p_{114}^{\mathrm{L}} \le 50,$$
$$-50 \le p_{121}^{\mathrm{L}} \le 50,$$
$$-50 \le p_{122}^{\mathrm{L}} \le 50,$$
$$-50 \le p_{123}^{\mathrm{L}} \le 50,$$
$$-50 \le p_{124}^{\mathrm{L}} \le 50,$$
$$-50 \le p_{131}^{\mathrm{L}} \le 50,$$
$$-50 \le p_{132}^{\mathrm{L}} \le 50,$$
$$-50 \le p_{133}^{\mathrm{L}} \le 50,$$
$$-50 \le p_{134}^{\mathrm{L}} \le 50,$$
$$-50 \le p_{211}^{\mathrm{L}} \le 50,$$
$$-50 \le p_{212}^{\mathrm{L}} \le 50,$$
$$-50 \le p_{213}^{\mathrm{L}} \le 50,$$
$$-50 \le p_{214}^{\mathrm{L}} \le 50,$$
$$-50 \le p_{221}^{\mathrm{L}} \le 50,$$
$$-50 \le p_{222}^{\mathrm{L}} \le 50,$$
$$-50 \le p_{223}^{\mathrm{L}} \le 50,$$
$$-50 \le p_{224}^{\mathrm{L}} \le 50,$$
$$-50 \le p_{231}^{\mathrm{L}} \le 50,$$
$$-50 \le p_{232}^{\mathrm{L}} \le 50,$$
$$-50 \le p_{233}^{\mathrm{L}} \le 50,$$
$$-50 \le p_{234}^{\mathrm{L}} \le 50,$$
$$-50 \le p_{311}^{\mathrm{L}} \le 50,$$
$$-50 \le p_{312}^{\mathrm{L}} \le 50,$$
$$-50 \le p_{313}^{\mathrm{L}} \le 50,$$
$$-50 \le p_{314}^{\mathrm{L}} \le 50,$$
$$-50 \le p_{321}^{\mathrm{L}} \le 50,$$

$$
\begin{aligned}
&-50 \leq p^{\mathrm{L}}_{322} \leq 50,\\
&-50 \leq p^{\mathrm{L}}_{323} \leq 50,\\
&-50 \leq p^{\mathrm{L}}_{324} \leq 50,\\
&-50 \leq p^{\mathrm{L}}_{331} \leq 50,\\
&-50 \leq p^{\mathrm{L}}_{332} \leq 50,\\
&-50 \leq p^{\mathrm{L}}_{333} \leq 50,\\
&-50 \leq p^{\mathrm{L}}_{334} \leq 50.
\end{aligned}
$$

- Definition of voltage angle at reference node:

$$
\begin{aligned}
&\delta_{111} = 0,\\
&\delta_{112} = 0,\\
&\delta_{113} = 0,\\
&\delta_{114} = 0,\\
&\delta_{121} = 0,\\
&\delta_{122} = 0,\\
&\delta_{123} = 0,\\
&\delta_{124} = 0,\\
&\delta_{131} = 0,\\
&\delta_{132} = 0,\\
&\delta_{133} = 0,\\
&\delta_{134} = 0.
\end{aligned}
$$

- Limits on voltage angles:

$$
\begin{aligned}
&-\pi \leq \delta_{211} \leq \pi,\\
&-\pi \leq \delta_{212} \leq \pi,\\
&-\pi \leq \delta_{213} \leq \pi,\\
&-\pi \leq \delta_{214} \leq \pi,\\
&-\pi \leq \delta_{221} \leq \pi,\\
&-\pi \leq \delta_{222} \leq \pi,\\
&-\pi \leq \delta_{223} \leq \pi,\\
&-\pi \leq \delta_{224} \leq \pi,\\
&-\pi \leq \delta_{231} \leq \pi,\\
&-\pi \leq \delta_{232} \leq \pi,\\
&-\pi \leq \delta_{233} \leq \pi,\\
&-\pi \leq \delta_{234} \leq \pi,\\
&-\pi \leq \delta_{311} \leq \pi,\\
&-\pi \leq \delta_{312} \leq \pi,
\end{aligned}
$$

$$
\begin{aligned}
&-\pi \le \delta_{313} \le \pi,\\
&-\pi \le \delta_{314} \le \pi,\\
&-\pi \le \delta_{321} \le \pi,\\
&-\pi \le \delta_{322} \le \pi,\\
&-\pi \le \delta_{323} \le \pi,\\
&-\pi \le \delta_{324} \le \pi,\\
&-\pi \le \delta_{331} \le \pi,\\
&-\pi \le \delta_{332} \le \pi,\\
&-\pi \le \delta_{333} \le \pi,\\
&-\pi \le \delta_{334} \le \pi.
\end{aligned}
$$

- Limits on the power traded in DA and RT energy markets:

$$
\begin{aligned}
&-60 \le p_1^{\mathrm{DA}} \le 40,\\
&-60 \le p_2^{\mathrm{DA}} \le 40,\\
&-60 \le p_3^{\mathrm{DA}} \le 40,\\
&-60 \le p_1^{\mathrm{DA}} + p_{11}^{\mathrm{RT}} \le 40,\\
&-60 \le p_1^{\mathrm{DA}} + p_{12}^{\mathrm{RT}} \le 40,\\
&-60 \le p_1^{\mathrm{DA}} + p_{13}^{\mathrm{RT}} \le 40,\\
&-60 \le p_1^{\mathrm{DA}} + p_{14}^{\mathrm{RT}} \le 40,\\
&-60 \le p_2^{\mathrm{DA}} + p_{21}^{\mathrm{RT}} \le 40,\\
&-60 \le p_2^{\mathrm{DA}} + p_{22}^{\mathrm{RT}} \le 40,\\
&-60 \le p_2^{\mathrm{DA}} + p_{23}^{\mathrm{RT}} \le 40,\\
&-60 \le p_2^{\mathrm{DA}} + p_{24}^{\mathrm{RT}} \le 40,\\
&-60 \le p_3^{\mathrm{DA}} + p_{31}^{\mathrm{RT}} \le 40,\\
&-60 \le p_3^{\mathrm{DA}} + p_{32}^{\mathrm{RT}} \le 40,\\
&-60 \le p_3^{\mathrm{DA}} + p_{33}^{\mathrm{RT}} \le 40,\\
&-60 \le p_3^{\mathrm{DA}} + p_{34}^{\mathrm{RT}} \le 40.
\end{aligned}
$$

- Limits on the power traded in the market:

$$
\begin{aligned}
&-100 \le p_{11}^{\mathrm{M}} \le 100,\\
&-100 \le p_{12}^{\mathrm{M}} \le 100,\\
&-100 \le p_{13}^{\mathrm{M}} \le 100,\\
&-100 \le p_{14}^{\mathrm{M}} \le 100,\\
&-100 \le p_{21}^{\mathrm{M}} \le 100,\\
&-100 \le p_{22}^{\mathrm{M}} \le 100,
\end{aligned}
$$

$$-100 \leq p^{\mathrm{M}}_{23} \leq 100,$$
$$-100 \leq p^{\mathrm{M}}_{24} \leq 100,$$
$$-100 \leq p^{\mathrm{M}}_{31} \leq 100,$$
$$-100 \leq p^{\mathrm{M}}_{32} \leq 100,$$
$$-100 \leq p^{\mathrm{M}}_{33} \leq 100,$$
$$-100 \leq p^{\mathrm{M}}_{34} \leq 100.$$

- Minimum energy consumption of demands:

$$p^{\mathrm{D}}_{111} + p^{\mathrm{D}}_{121} + p^{\mathrm{D}}_{131} \geq 30,$$
$$p^{\mathrm{D}}_{112} + p^{\mathrm{D}}_{122} + p^{\mathrm{D}}_{132} \geq 30,$$
$$p^{\mathrm{D}}_{113} + p^{\mathrm{D}}_{123} + p^{\mathrm{D}}_{133} \geq 30,$$
$$p^{\mathrm{D}}_{114} + p^{\mathrm{D}}_{124} + p^{\mathrm{D}}_{134} \geq 30,$$
$$p^{\mathrm{D}}_{211} + p^{\mathrm{D}}_{221} + p^{\mathrm{D}}_{231} \geq 50,$$
$$p^{\mathrm{D}}_{212} + p^{\mathrm{D}}_{222} + p^{\mathrm{D}}_{232} \geq 50,$$
$$p^{\mathrm{D}}_{213} + p^{\mathrm{D}}_{223} + p^{\mathrm{D}}_{233} \geq 50,$$
$$p^{\mathrm{D}}_{214} + p^{\mathrm{D}}_{224} + p^{\mathrm{D}}_{234} \geq 50.$$

- Limits on power consumption of demands:

$$0 \leq p^{\mathrm{D}}_{111} \leq 20,$$
$$0 \leq p^{\mathrm{D}}_{112} \leq 20,$$
$$0 \leq p^{\mathrm{D}}_{113} \leq 20,$$
$$0 \leq p^{\mathrm{D}}_{114} \leq 20,$$
$$0 \leq p^{\mathrm{D}}_{121} \leq 20,$$
$$0 \leq p^{\mathrm{D}}_{122} \leq 20,$$
$$0 \leq p^{\mathrm{D}}_{123} \leq 20,$$
$$0 \leq p^{\mathrm{D}}_{124} \leq 20,$$
$$0 \leq p^{\mathrm{D}}_{131} \leq 10,$$
$$0 \leq p^{\mathrm{D}}_{132} \leq 10,$$
$$0 \leq p^{\mathrm{D}}_{133} \leq 10,$$
$$0 \leq p^{\mathrm{D}}_{134} \leq 10,$$
$$5 \leq p^{\mathrm{D}}_{211} \leq 30,$$
$$5 \leq p^{\mathrm{D}}_{212} \leq 30,$$
$$5 \leq p^{\mathrm{D}}_{213} \leq 30,$$
$$5 \leq p^{\mathrm{D}}_{214} \leq 30,$$

$$5 \le p^{\mathrm{D}}_{221} \le 30,$$
$$5 \le p^{\mathrm{D}}_{222} \le 30,$$
$$5 \le p^{\mathrm{D}}_{223} \le 30,$$
$$5 \le p^{\mathrm{D}}_{224} \le 30,$$
$$5 \le p^{\mathrm{D}}_{231} \le 15,$$
$$5 \le p^{\mathrm{D}}_{232} \le 15,$$
$$5 \le p^{\mathrm{D}}_{233} \le 15,$$
$$5 \le p^{\mathrm{D}}_{234} \le 15.$$

- Ramping limits of demands:

$$-15 \le p^{\mathrm{D}}_{111} - 10 \le 15,$$
$$-15 \le p^{\mathrm{D}}_{112} - 10 \le 15,$$
$$-15 \le p^{\mathrm{D}}_{113} - 10 \le 15,$$
$$-15 \le p^{\mathrm{D}}_{114} - 10 \le 15,$$
$$-15 \le p^{\mathrm{D}}_{121} - p^{\mathrm{D}}_{111} \le 15,$$
$$-15 \le p^{\mathrm{D}}_{122} - p^{\mathrm{D}}_{112} \le 15,$$
$$-15 \le p^{\mathrm{D}}_{123} - p^{\mathrm{D}}_{113} \le 15,$$
$$-15 \le p^{\mathrm{D}}_{124} - p^{\mathrm{D}}_{114} \le 15,$$
$$-15 \le p^{\mathrm{D}}_{131} - p^{\mathrm{D}}_{121} \le 15,$$
$$-15 \le p^{\mathrm{D}}_{132} - p^{\mathrm{D}}_{122} \le 15,$$
$$-15 \le p^{\mathrm{D}}_{133} - p^{\mathrm{D}}_{123} \le 15,$$
$$-15 \le p^{\mathrm{D}}_{134} - p^{\mathrm{D}}_{124} \le 15,$$
$$-25 \le p^{\mathrm{D}}_{211} - 20 \le 25,$$
$$-25 \le p^{\mathrm{D}}_{212} - 20 \le 25,$$
$$-25 \le p^{\mathrm{D}}_{213} - 20 \le 25,$$
$$-25 \le p^{\mathrm{D}}_{214} - 20 \le 25,$$
$$-25 \le p^{\mathrm{D}}_{221} - p^{\mathrm{D}}_{211} \le 25,$$
$$-25 \le p^{\mathrm{D}}_{222} - p^{\mathrm{D}}_{212} \le 25,$$
$$-25 \le p^{\mathrm{D}}_{223} - p^{\mathrm{D}}_{213} \le 25,$$
$$-25 \le p^{\mathrm{D}}_{224} - p^{\mathrm{D}}_{214} \le 25,$$
$$-25 \le p^{\mathrm{D}}_{231} - p^{\mathrm{D}}_{221} \le 25,$$
$$-25 \le p^{\mathrm{D}}_{232} - p^{\mathrm{D}}_{222} \le 25,$$
$$-25 \le p^{\mathrm{D}}_{233} - p^{\mathrm{D}}_{223} \le 25,$$
$$-25 \le p^{\mathrm{D}}_{234} - p^{\mathrm{D}}_{224} \le 25.$$

- Limits on the production of conventional power plant:

$$0 \leq p^{C}_{11} \leq 20u^{C}_{1},$$
$$0 \leq p^{C}_{12} \leq 20u^{C}_{1},$$
$$0 \leq p^{C}_{13} \leq 20u^{C}_{1},$$
$$0 \leq p^{C}_{14} \leq 20u^{C}_{1},$$
$$0 \leq p^{C}_{21} \leq 20u^{C}_{2},$$
$$0 \leq p^{C}_{22} \leq 20u^{C}_{2},$$
$$0 \leq p^{C}_{23} \leq 20u^{C}_{2},$$
$$0 \leq p^{C}_{24} \leq 20u^{C}_{2},$$
$$0 \leq p^{C}_{31} \leq 20u^{C}_{3},$$
$$0 \leq p^{C}_{32} \leq 20u^{C}_{3},$$
$$0 \leq p^{C}_{33} \leq 20u^{C}_{3},$$
$$0 \leq p^{C}_{34} \leq 20u^{C}_{3}.$$

- Ramping limits of conventional power plants:

$$-20 \leq p^{C}_{11} - 0 \leq 20,$$
$$-20 \leq p^{C}_{12} - 0 \leq 20,$$
$$-20 \leq p^{C}_{13} - 0 \leq 20,$$
$$-20 \leq p^{C}_{14} - 0 \leq 20,$$
$$-20 \leq p^{C}_{21} - p^{C}_{11} \leq 20,$$
$$-20 \leq p^{C}_{22} - p^{C}_{12} \leq 20,$$
$$-20 \leq p^{C}_{23} - p^{C}_{13} \leq 20,$$
$$-20 \leq p^{C}_{24} - p^{C}_{14} \leq 20,$$
$$-20 \leq p^{C}_{31} - p^{C}_{21} \leq 20,$$
$$-20 \leq p^{C}_{32} - p^{C}_{22} \leq 20,$$
$$-20 \leq p^{C}_{33} - p^{C}_{23} \leq 20,$$
$$-20 \leq p^{C}_{34} - p^{C}_{24} \leq 20.$$

- Energy evolution in the storage unit:

$$e^{S}_{11} = 5 + 0.9p^{S,C}_{11} - \frac{p^{S,D}_{11}}{0.9},$$
$$e^{S}_{12} = 5 + 0.9p^{S,C}_{12} - \frac{p^{S,D}_{12}}{0.9},$$
$$e^{S}_{13} = 5 + 0.9p^{S,C}_{13} - \frac{p^{S,D}_{13}}{0.9},$$

$$e_{14}^{\mathrm{S}} = 5 + 0.9p_{14}^{\mathrm{S,C}} - \frac{p_{14}^{\mathrm{S,D}}}{0.9},$$

$$e_{21}^{\mathrm{S}} = e_{11}^{\mathrm{S}} + 0.9p_{21}^{\mathrm{S,C}} - \frac{p_{21}^{\mathrm{S,D}}}{0.9},$$

$$e_{22}^{\mathrm{S}} = e_{12}^{\mathrm{S}} + 0.9p_{22}^{\mathrm{S,C}} - \frac{p_{22}^{\mathrm{S,D}}}{0.9},$$

$$e_{23}^{\mathrm{S}} = e_{13}^{\mathrm{S}} + 0.9p_{23}^{\mathrm{S,C}} - \frac{p_{23}^{\mathrm{S,D}}}{0.9},$$

$$e_{24}^{\mathrm{S}} = e_{14}^{\mathrm{S}} + 0.9p_{24}^{\mathrm{S,C}} - \frac{p_{24}^{\mathrm{S,D}}}{0.9},$$

$$e_{31}^{\mathrm{S}} = e_{21}^{\mathrm{S}} + 0.9p_{31}^{\mathrm{S,C}} - \frac{p_{31}^{\mathrm{S,D}}}{0.9},$$

$$e_{32}^{\mathrm{S}} = e_{22}^{\mathrm{S}} + 0.9p_{32}^{\mathrm{S,C}} - \frac{p_{32}^{\mathrm{S,D}}}{0.9},$$

$$e_{33}^{\mathrm{S}} = e_{23}^{\mathrm{S}} + 0.9p_{33}^{\mathrm{S,C}} - \frac{p_{33}^{\mathrm{S,D}}}{0.9},$$

$$e_{34}^{\mathrm{S}} = e_{24}^{\mathrm{S}} + 0.9p_{34}^{\mathrm{S,C}} - \frac{p_{34}^{\mathrm{S,D}}}{0.9}.$$

- Energy limits of the storage unit:

$$5 \le e_{11}^{\mathrm{S}} \le 10,$$
$$5 \le e_{12}^{\mathrm{S}} \le 10,$$
$$5 \le e_{13}^{\mathrm{S}} \le 10,$$
$$5 \le e_{14}^{\mathrm{S}} \le 10,$$
$$5 \le e_{21}^{\mathrm{S}} \le 10,$$
$$5 \le e_{22}^{\mathrm{S}} \le 10,$$
$$5 \le e_{23}^{\mathrm{S}} \le 10,$$
$$5 \le e_{24}^{\mathrm{S}} \le 10,$$
$$5 \le e_{31}^{\mathrm{S}} \le 10,$$
$$5 \le e_{32}^{\mathrm{S}} \le 10,$$
$$5 \le e_{33}^{\mathrm{S}} \le 10,$$
$$5 \le e_{34}^{\mathrm{S}} \le 10.$$

- Charging and discharging power limits of the storage unit:

$$0 \le p_{11}^{\mathrm{S,C}} \le 10,$$
$$0 \le p_{12}^{\mathrm{S,C}} \le 10,$$
$$0 \le p_{13}^{\mathrm{S,C}} \le 10,$$
$$0 \le p_{14}^{\mathrm{S,C}} \le 10,$$

$$0 \leq p_{21}^{\mathrm{S,C}} \leq 10,$$
$$0 \leq p_{22}^{\mathrm{S,C}} \leq 10,$$
$$0 \leq p_{23}^{\mathrm{S,C}} \leq 10,$$
$$0 \leq p_{24}^{\mathrm{S,C}} \leq 10,$$
$$0 \leq p_{31}^{\mathrm{S,C}} \leq 10,$$
$$0 \leq p_{32}^{\mathrm{S,C}} \leq 10,$$
$$0 \leq p_{33}^{\mathrm{S,C}} \leq 10,$$
$$0 \leq p_{34}^{\mathrm{S,C}} \leq 10,$$
$$0 \leq p_{11}^{\mathrm{S,D}} \leq 10,$$
$$0 \leq p_{12}^{\mathrm{S,D}} \leq 10,$$
$$0 \leq p_{13}^{\mathrm{S,D}} \leq 10,$$
$$0 \leq p_{14}^{\mathrm{S,D}} \leq 10,$$
$$0 \leq p_{21}^{\mathrm{S,D}} \leq 10,$$
$$0 \leq p_{22}^{\mathrm{S,D}} \leq 10,$$
$$0 \leq p_{23}^{\mathrm{S,D}} \leq 10,$$
$$0 \leq p_{24}^{\mathrm{S,D}} \leq 10,$$
$$0 \leq p_{31}^{\mathrm{S,D}} \leq 10,$$
$$0 \leq p_{32}^{\mathrm{S,D}} \leq 10,$$
$$0 \leq p_{33}^{\mathrm{S,D}} \leq 10,$$
$$0 \leq p_{34}^{\mathrm{S,D}} \leq 10.$$

- Limits on the production of the stochastic PV generating unit:

$$0 \leq p_{11}^{\mathrm{R}} \leq 9,$$
$$0 \leq p_{12}^{\mathrm{R}} \leq 9,$$
$$0 \leq p_{13}^{\mathrm{R}} \leq 9,$$
$$0 \leq p_{14}^{\mathrm{R}} \leq 9,$$
$$0 \leq p_{21}^{\mathrm{R}} \leq 7,$$
$$0 \leq p_{22}^{\mathrm{R}} \leq 7,$$
$$0 \leq p_{23}^{\mathrm{R}} \leq 7,$$
$$0 \leq p_{24}^{\mathrm{R}} \leq 7,$$
$$0 \leq p_{31}^{\mathrm{R}} \leq 1,$$
$$0 \leq p_{32}^{\mathrm{R}} \leq 1,$$
$$0 \leq p_{33}^{\mathrm{R}} \leq 1,$$
$$0 \leq p_{34}^{\mathrm{R}} \leq 1.$$

Finally, the optimization variables of the problem are those included in set $\Phi^{\mathrm{IE4.1}}=\{e^{\mathrm{S}}_{11}, e^{\mathrm{S}}_{12}, e^{\mathrm{S}}_{13}, e^{\mathrm{S}}_{14}, e^{\mathrm{S}}_{21}, e^{\mathrm{S}}_{22}, e^{\mathrm{S}}_{23}, e^{\mathrm{S}}_{24}, e^{\mathrm{S}}_{31}, e^{\mathrm{S}}_{32}, e^{\mathrm{S}}_{33}, e^{\mathrm{S}}_{34}, p^{\mathrm{D}}_{111}, p^{\mathrm{D}}_{112}, p^{\mathrm{D}}_{113}, p^{\mathrm{D}}_{114}, p^{\mathrm{D}}_{121}, p^{\mathrm{D}}_{122}, p^{\mathrm{D}}_{123}, p^{\mathrm{D}}_{124}, p^{\mathrm{D}}_{131}, p^{\mathrm{D}}_{132}, p^{\mathrm{D}}_{133}, p^{\mathrm{D}}_{134}, p^{\mathrm{D}}_{211}, p^{\mathrm{D}}_{212}, p^{\mathrm{D}}_{213}, p^{\mathrm{D}}_{214}, p^{\mathrm{D}}_{221}, p^{\mathrm{D}}_{222}, p^{\mathrm{D}}_{223}, p^{\mathrm{D}}_{224}, p^{\mathrm{D}}_{231}, p^{\mathrm{D}}_{232}, p^{\mathrm{D}}_{233}, p^{\mathrm{D}}_{234}, p^{\mathrm{C}}_{11}, p^{\mathrm{C}}_{12}, p^{\mathrm{C}}_{13}, p^{\mathrm{C}}_{14}, p^{\mathrm{C}}_{21}, p^{\mathrm{C}}_{22}, p^{\mathrm{C}}_{23}, p^{\mathrm{C}}_{24}, p^{\mathrm{C}}_{31}, p^{\mathrm{C}}_{32}, p^{\mathrm{C}}_{33}, p^{\mathrm{C}}_{34}, p^{\mathrm{R}}_{11}, p^{\mathrm{R}}_{12}, p^{\mathrm{R}}_{13}, p^{\mathrm{R}}_{14}, p^{\mathrm{R}}_{21}, p^{\mathrm{R}}_{22}, p^{\mathrm{R}}_{23}, p^{\mathrm{R}}_{24}, p^{\mathrm{R}}_{31}, p^{\mathrm{R}}_{32}, p^{\mathrm{R}}_{33}, p^{\mathrm{R}}_{34}, p^{\mathrm{S,C}}_{11}, p^{\mathrm{S,C}}_{12}, p^{\mathrm{S,C}}_{13}, p^{\mathrm{S,C}}_{14}, p^{\mathrm{S,C}}_{21}, p^{\mathrm{S,C}}_{22}, p^{\mathrm{S,C}}_{23}, p^{\mathrm{S,C}}_{24}, p^{\mathrm{S,C}}_{31}, p^{\mathrm{S,C}}_{32}, p^{\mathrm{S,C}}_{33}, p^{\mathrm{S,C}}_{34}, p^{\mathrm{S,D}}_{11}, p^{\mathrm{S,D}}_{12}, p^{\mathrm{S,D}}_{13}, p^{\mathrm{S,D}}_{14}, p^{\mathrm{S,D}}_{21}, p^{\mathrm{S,D}}_{22}, p^{\mathrm{S,D}}_{23}, p^{\mathrm{S,D}}_{24}, p^{\mathrm{S,D}}_{31}, p^{\mathrm{S,D}}_{32}, p^{\mathrm{S,D}}_{33}, p^{\mathrm{S,D}}_{34}, p^{\mathrm{DA}}_{1}, p^{\mathrm{DA}}_{2}, p^{\mathrm{DA}}_{3}, p^{\mathrm{M}}_{11}, p^{\mathrm{M}}_{12}, p^{\mathrm{M}}_{13}, p^{\mathrm{M}}_{14}, p^{\mathrm{M}}_{21}, p^{\mathrm{M}}_{22}, p^{\mathrm{M}}_{23}, p^{\mathrm{M}}_{24}, p^{\mathrm{M}}_{31}, p^{\mathrm{M}}_{32}, p^{\mathrm{M}}_{33}, p^{\mathrm{M}}_{34}, p^{\mathrm{RT}}_{11}, p^{\mathrm{RT}}_{12}, p^{\mathrm{RT}}_{13}, p^{\mathrm{RT}}_{14}, p^{\mathrm{RT}}_{21}, p^{\mathrm{RT}}_{22}, p^{\mathrm{RT}}_{23}, p^{\mathrm{RT}}_{24}, p^{\mathrm{RT}}_{31}, p^{\mathrm{RT}}_{32}, p^{\mathrm{RT}}_{33}, p^{\mathrm{RT}}_{34}, u^{\mathrm{C}}_{1}, u^{\mathrm{C}}_{2}, u^{\mathrm{C}}_{3}, p^{\mathrm{L}}_{111}, p^{\mathrm{L}}_{112}, p^{\mathrm{L}}_{113}, p^{\mathrm{L}}_{114}, p^{\mathrm{L}}_{121}, p^{\mathrm{L}}_{122}, p^{\mathrm{L}}_{123}, p^{\mathrm{L}}_{124}, p^{\mathrm{L}}_{131}, p^{\mathrm{L}}_{132}, p^{\mathrm{L}}_{133}, p^{\mathrm{L}}_{134}, p^{\mathrm{L}}_{211}, p^{\mathrm{L}}_{212}, p^{\mathrm{L}}_{213}, p^{\mathrm{L}}_{214}, p^{\mathrm{L}}_{221}, p^{\mathrm{L}}_{222}, p^{\mathrm{L}}_{223}, p^{\mathrm{L}}_{224}, p^{\mathrm{L}}_{231}, p^{\mathrm{L}}_{232}, p^{\mathrm{L}}_{233}, p^{\mathrm{L}}_{234}, p^{\mathrm{L}}_{311}, p^{\mathrm{L}}_{312}, p^{\mathrm{L}}_{313}, p^{\mathrm{L}}_{314}, p^{\mathrm{L}}_{321}, p^{\mathrm{L}}_{322}, p^{\mathrm{L}}_{323}, p^{\mathrm{L}}_{324}, p^{\mathrm{L}}_{331}, p^{\mathrm{L}}_{332}, p^{\mathrm{L}}_{333}, p^{\mathrm{L}}_{334}, \delta_{111}, \delta_{112}, \delta_{113}, \delta_{114}, \delta_{121}, \delta_{122}, \delta_{123}, \delta_{124}, \delta_{131}, \delta_{132}, \delta_{133}, \delta_{134}, \delta_{211}, \delta_{212}, \delta_{213}, \delta_{214}, \delta_{221}, \delta_{222}, \delta_{223}, \delta_{224}, \delta_{231}, \delta_{232}, \delta_{233}, \delta_{234}, \delta_{311}, \delta_{312}, \delta_{313}, \delta_{314}, \delta_{321}, \delta_{322}, \delta_{323}, \delta_{324}, \delta_{331}, \delta_{332}, \delta_{333}, \delta_{334}\}$.

The above problem is a two-stage stochastic programming model. In the first stage, the VPP decides the scheduled power and the status of the conventional power plant for the planning horizon in the DA energy market. As the scenarios of the prices in the DA and RT energy markets are known after making the scheduling decisions in the DA energy market, these decisions are *here-and-now* decisions with respect to the uncertainties. However, the scheduling decisions in the RT energy market and the dispatch decisions are made after knowing the actual scenario realizations, and thus are *wait-and-see* decisions with respect to the uncertainties.

The above MILP problem is solved using CPLEX [9] under GAMS [8]. The optimal value of the objective function is \$5899.2. That is the summation of the expected profit in the DA and RT energy markets, and the weighted CVaR, which are equal to \$3105.7 and \$2793.5, respectively.

The power schedules of the VPP in the DA energy market are provided in Table 4.7. Accordingly, the VPP behaves as a consumer in time periods 1 and 2. However, it behaves as a producer in time period 3. The conventional power plant is also turned on in time period 2.

The power traded in the RT energy market is also depicted in Fig. 4.3. Accordingly, the VPP buys energy in the RT energy market in all scenarios in time periods 1 and 3. Additionally, the VPP behaves as a producer in the RT energy market in all scenarios in time period 2.

The actual prices of the DA energy market are 25, 50, and 35 \$/MWh in time periods 1, 2, and 3, respectively, which are used to calculate the revenue obtained in the DA energy market. According to the scheduled power and the actual prices, the VPP pays \$1952.2 in the DA energy market. Table 4.8 shows the actual profit achieved by the VPP in the DA energy market in each time period.

□

Table 4.7 Illustrative Example 4.1: schedules in the DA energy market

Time period	Scheduled power [MW]	Status of conventional power plant
1	−14.1	0
2	−60	1
3	40	0

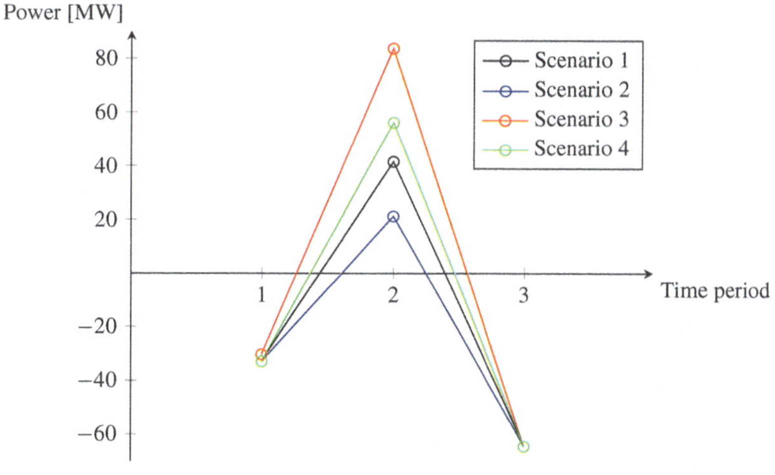

Fig. 4.3 Illustrative Example 4.9: power traded in the RT energy market

Table 4.8 Illustrative Example 4.1: profit in each time period

Time period	Profit [$]
1	−352.5
2	−3000
3	1400

Table 4.9 Illustrative Example 4.2: power scheduled in each time period [MW]

	β				
Time period	0	0.5	1	1.5	2
1	40	40	−14.1	−14.1	−14.1
2	−60	−60	−60	−60	−60
3	40	40	40	40	40

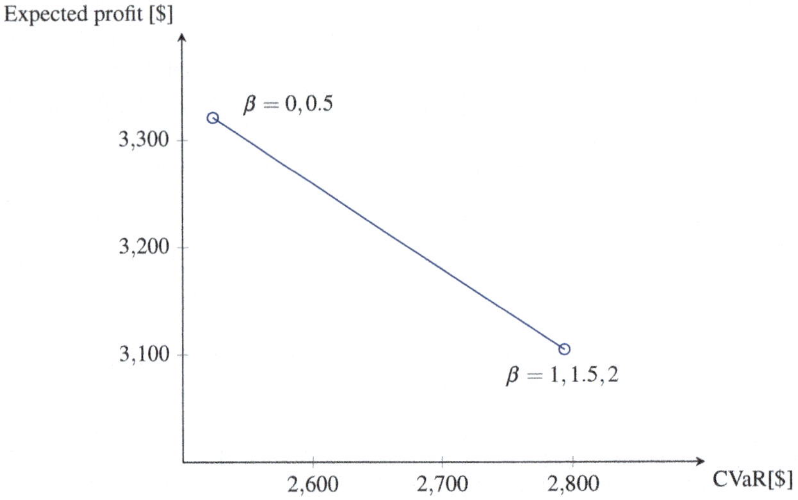

Fig. 4.4 Illustrative Example 4.9: expected profit with respect to CVaR for different values of β

A GAMS [8] code for solving Illustrative Example 4.1 is provided in Sect. 4.8.

Illustrative Example 4.2 Impact of the Risk Strategy on the Stochastic Scheduling of a VPP

The data of Illustrative Example 4.1 are considered again. Next, the impact of the parameter β used to model the trade-off between expected profit and CVaR on the results is analyzed. To do so, parameter β varies from 0 to 2, for the given confidence level α of 0.8.

Table 4.9 shows the power scheduled in the DA energy market for different values of β. Results reveal that the VPP behaves as a consumer/producer in time period 2/3 independently of the risk strategy. In time period 1, the VPP behaves as a producer for the values of β=0 and 0.5. In this time period, it behaves as a consumer for the values of β=1, 1.5, and 2. For all values of β, the status of the conventional power plant is on in time period 2, and off in time periods 1 and 3.

By the inspection of the scenario data provided in Table 4.5 in Illustrative Example 4.1, the RT market prices are greater/lower than the DA market prices in all scenarios during time period 2/3. Thus, in these time periods, a risk-averse VPP seeks a maximum profit by buying/selling energy in the DA energy market and selling/buying energy in the RT energy market with higher/lower prices. In time period 1, the RT market prices are greater than the DA market prices for scenarios 1 and 3. However, the RT market prices are lower than the DA market prices for scenarios 2 and 4. Under the given scenarios, considering a less risk-averse strategy, i.e., if β=0 and 0.5, the VPP prefers to sell energy in the DA energy market and to buy energy in the RT energy market. For a risk-averse strategy, i.e., for β=1, 1.5, and 2, the VPP acts different in time period 1.

Figure 4.4 shows the efficient frontier, which represents the expected profit versus the CVaR for different values of β. The efficient frontier clarifies that the expected profit decreases as β increases. In the contrary, the CVaR increases by the increase in parameter β. This means that by taking a more risk-averse strategy, the expected profit of the VPP in the energy markets is lower; however, the VPP benefits from a higher CVaR.

□

4.4 Robust Optimization Approach

This section describes a robust optimization approach to handle uncertainties in the scheduling problem of a risk-averse VPP participating in energy markets.

4.4.1 *Problem Description*

Section 4.3 describes a stochastic programming model for the scheduling problem of a VPP in which uncertainties are modeled using a set of scenarios. This has some disadvantages:

1. Generating accurate scenarios requires knowing the probability distribution functions of uncertain parameters. However, this information is not always available.
2. A large number of scenarios is generally needed to have a good representation of uncertainties, which may lead to intractable problems.
3. The feasibility of the operation of the VPP is only guaranteed for the considered scenarios. However, decisions made may not be feasible if the actual realizations of uncertain parameters do not belong to the considered scenario set.

To overcome these issues, this section describes a robust optimization approach, which provides a solution that optimizes the scheduling problem under the worst realization of uncertainties, i.e., the solution is optimal under the worst realization of uncertainties. In this case, uncertain parameters are represented using confidence bounds. This robust optimization approach has the following advantages in comparison with the stochastic programming model described in Sect. 4.3, namely:

1. Uncertainty sets based on confidence bounds are generally easier to obtain than generating accurate scenarios.
2. Robust optimization problems generally require moderate computation cost. This is specifically relevant in decision-making processes that need to be accomplished in a limited time, e.g., the RT scheduling problem.
3. Robust optimization provides solutions that are feasible provided that the actual realizations of uncertain parameters are within the considered confidence bounds. This means that the robust optimization approach is more reliable to implement practical decisions under uncertainty.

The decision-making process based on a robust optimization approach considers that the VPP makes the scheduling decisions in the DA and RT energy markets, as well as the dispatch decisions of its own units with the aim of maximizing the profit, anticipating that once these decisions are made, the worst-case realization of uncertainties occurs.

4.4.2 *Uncertainty Model*

Uncertainties in energy market prices and stochastic renewable power production levels are modeled through confidence bounds. As explained in [11], market prices and the stochastic power production levels of local renewable generating units are not meaningfully correlated; however, the prices in the DA and RT energy markets are correlated. Thus, the price in the RT energy market can be defined as the summation of the price in the DA energy market and an auxiliary parameter called residual price, i.e.:

$$\lambda_t^{\mathrm{RT}} = \lambda_t^{\mathrm{DA}} + r_t^{\mathrm{RT}}, \quad \forall t \in \Omega^{\mathrm{T}}. \tag{4.2}$$

As corroborated by the analysis carried out in [11], DA energy market prices, residual prices, and stochastic power production levels are not meaningfully correlated.

In the case that uncertain parameters are correlated, the uncertainty model needs to address the correlation. Accordingly, the robust optimization model should be extended to directly incorporate the correlation [5].

Uncertain parameters take values within known confidence bounds for a given confidence level, i.e.:

$$\lambda_t^{\mathrm{DA}} \in [\tilde{\lambda}_t^{\mathrm{DA}} - \hat{\lambda}_t^{\mathrm{DA}}, \tilde{\lambda}_t^{\mathrm{DA}} + \hat{\lambda}_t^{\mathrm{DA}}], \quad \forall t \in \Omega^{\mathrm{T}}, \tag{4.3a}$$

$$r_t^{\mathrm{RT}} \in [\tilde{r}_t^{\mathrm{RT}} - \hat{r}_t^{\mathrm{RT}}, \tilde{r}_t^{\mathrm{RT}} + \hat{r}_t^{\mathrm{RT}}], \quad \forall t \in \Omega^{\mathrm{T}}, \tag{4.3b}$$

$$P_{rt}^{\mathrm{R,A}} \in [\tilde{P}_{rt}^{\mathrm{R,A}} - \hat{P}_{rt}^{\mathrm{R,A}}, \tilde{P}_{rt}^{\mathrm{R,A}} + \hat{P}_{rt}^{\mathrm{R,A}}], \quad \forall t \in \Omega^{\mathrm{T}}, \forall r \in \Omega^{\mathrm{R}}, \tag{4.3c}$$

where parameters $\tilde{\lambda}_t^{\mathrm{DA}}$, $\tilde{r}_t^{\mathrm{RT}}$, and $\tilde{P}_{rt}^{\mathrm{R,A}}$ refer to the nominal values of uncertain parameters, while parameters $\hat{\lambda}_t^{\mathrm{DA}} \geq 0$, $\hat{r}_t^{\mathrm{RT}} \geq 0$, and $\hat{P}_{rt}^{\mathrm{R,A}} \geq 0$ are deviations around the nominal values.

Set J^{DA} is defined as $J^{\mathrm{DA}}=\{\lambda_t^{\mathrm{DA}}, r_t^{\mathrm{RT}}, \forall t \in \Omega^{\mathrm{T}}\}$, including uncertain prices throughout the day. Therefore, the cardinality of set J^{DA}, denoted by $|J^{\mathrm{DA}}|$, is equal to $2|\Omega^{\mathrm{T}}|$. Similarly, sets $J_{rt}^{\mathrm{R}}=\{P_{rt}^{\mathrm{R,A}}\}$, $\forall r \in \Omega^{\mathrm{R}}$, $\forall t \in \Omega^{\mathrm{T}}$, include the uncertain available production levels of each stochastic renewable generating unit in each time period, whose cardinalities are equal to 1, i.e., $|J_{rt}^{\mathrm{R}}|=1$, $\forall r \in \Omega^{\mathrm{R}}$, $\forall t \in \Omega^{\mathrm{T}}$.

It is unlikely that all uncertain parameters deviate from their nominal values simultaneously. Under this assumption, the number of uncertain parameters that deviate from their nominal values is limited using parameters Γ^{DA} and Γ_{rt}^{R}, $\forall r \in \Omega^{\mathrm{R}}$, $\forall t \in \Omega^{\mathrm{T}}$. Thus, parameters Γ^{DA} and Γ_{rt}^{R} can take values within the intervals $[0, 2|\Omega^{\mathrm{T}}|]$ and $[0, 1]$, respectively. These parameters allow adjusting the robustness of the solution with respect to the variation of uncertain data belonging to the uncertainty sets J^{DA} and J_{rt}^{R}, $\forall r \in \Omega^{\mathrm{R}}$, $\forall t \in \Omega^{\mathrm{T}}$, respectively. Considering a conservative strategy, a risk-averse VPP would tend to choose the maximum values of $2|\Omega^{\mathrm{T}}|$ and 1 for parameters Γ^{DA} and Γ_{rt}^{R}, respectively. This strategy results in a solution that is robust against the deviation of all uncertain prices and the uncertain available production levels of stochastic renewable generating units in each time period. The VPP can reduce the conservatism level of the solution by decreasing the values of these parameters within their admissible intervals. In a risky strategy, the VPP ignores the potential impact of the deviations around the nominal values by choosing parameters Γ^{DA} and Γ_{rt}^{R} equal to 0.

4.4.3 Formulation

Considering the decision-making sequence described in Sect. 4.4.1 and the uncertainty model described in Sect. 4.4.2, the scheduling problem of a risk-averse VPP in the DA energy market can be formulated as the max-min problem below:

$$\max_{\Phi^{\mathrm{DA}}} \sum_{t\in\Omega^{\mathrm{T}}} \left[\sum_{d\in\Omega^{\mathrm{D}}} U_{dt}^{\mathrm{D}} p_{dt}^{\mathrm{D}} \Delta t - \sum_{c\in\Omega^{\mathrm{C}}} (C_c^{\mathrm{C,F}} u_{ct}^{\mathrm{C}} + C_c^{\mathrm{C,V}} p_{ct}^{\mathrm{C}} \Delta t) \right]$$
$$+ \min_{\Phi^{\mathrm{DA}}, \mathrm{U} \subseteq J^{\mathrm{DA}}} \sum_{t\in\Omega^{\mathrm{T}}} \left[\lambda_t^{\mathrm{DA}} p_t^{\mathrm{DA}} \Delta t + \left(\lambda_t^{\mathrm{DA}} + r_t^{\mathrm{RT}}\right) p_t^{\mathrm{RT}} \Delta t \right] \tag{4.4a}$$

subject to

$$\sum_{c\in\Omega_n^{\mathrm{C}}} p_{ct}^{\mathrm{C}} + \sum_{r\in\Omega_n^{\mathrm{R}}} p_{rt}^{\mathrm{R}} + \sum_{s\in\Omega_n^{\mathrm{S}}} (p_{st}^{\mathrm{S,D}} - p_{st}^{\mathrm{S,C}}) - \sum_{\ell|s(\ell)=n} p_{\ell t}^{\mathrm{L}} + \sum_{\ell|r(\ell)=n} p_{\ell t}^{\mathrm{L}}$$
$$= p_{nt}^{\mathrm{M}} + \sum_{d\in\Omega_n^{\mathrm{D}}} p_{dt}^{\mathrm{D}}, \quad \forall n \in \Omega^{\mathrm{M}}, \forall t \in \Omega^{\mathrm{T}}, \tag{4.4b}$$

$$\sum_{c\in\Omega_n^{\mathrm{C}}} p_{ct}^{\mathrm{C}} + \sum_{r\in\Omega_n^{\mathrm{R}}} p_{rt}^{\mathrm{R}} + \sum_{s\in\Omega_n^{\mathrm{S}}} (p_{st}^{\mathrm{S,D}} - p_{st}^{\mathrm{S,C}}) - \sum_{\ell|s(\ell)=n} p_{\ell t}^{\mathrm{L}} + \sum_{\ell|r(\ell)=n} p_{\ell t}^{\mathrm{L}}$$
$$= \sum_{d\in\Omega_n^{\mathrm{D}}} p_{dt}^{\mathrm{D}}, \quad \forall n \in \Omega^{\mathrm{N}} \setminus n \in \Omega^{\mathrm{M}}, \forall t \in \Omega^{\mathrm{T}}, \tag{4.4c}$$

$$p_t^{\mathrm{DA}} + p_t^{\mathrm{RT}} = \sum_{n\in\Omega^{\mathrm{M}}} p_{nt}^{\mathrm{M}}, \quad \forall t \in \Omega^{\mathrm{T}}, \tag{4.4d}$$

$$p_{\ell t}^{\mathrm{L}} = B_\ell(\delta_{s(\ell)t} - \delta_{r(\ell)t}), \quad \forall \ell \in \Omega^{\mathrm{L}}, \forall \in \Omega^{\mathrm{T}}, \tag{4.4e}$$

$$-\overline{P}_\ell^{\mathrm{L}} \leq p_{\ell t}^{\mathrm{L}} \leq \overline{P}_\ell^{\mathrm{L}}, \quad \forall \ell \in \Omega^{\mathrm{L}}, \forall t \in \Omega^{\mathrm{T}}, \tag{4.4f}$$

$$\delta_{nt} = 0, \quad n : \text{ref.}, \forall t \in \Omega^{\mathrm{T}}, \tag{4.4g}$$

$$-\pi \leq \delta_{nt} \leq \pi, \quad \forall n \in \Omega^{\mathrm{N}} \setminus n : \text{ref.}, \forall t \in \Omega^{\mathrm{T}}, \tag{4.4h}$$

$$\underline{P}^{\mathrm{V}} \leq p_t^{\mathrm{DA}} \leq \overline{P}^{\mathrm{V}}, \quad \forall t \in \Omega^{\mathrm{T}}, \tag{4.4i}$$

$$\underline{P}^{\mathrm{V}} \leq p_t^{\mathrm{DA}} + p_t^{\mathrm{RT}} \leq \overline{P}^{\mathrm{V}}, \quad \forall t \in \Omega^{\mathrm{T}}, \tag{4.4j}$$

$$-\overline{P}_n^{\mathrm{M}} \leq p_{nt}^{\mathrm{M}} \leq \overline{P}_n^{\mathrm{M}}, \quad \forall n \in \Omega^{\mathrm{M}}, \forall t \in \Omega^{\mathrm{T}}, \tag{4.4k}$$

$$\sum_{t \in \Omega^{\mathrm{T}}} p_{dt}^{\mathrm{D}} \Delta t \geq \underline{E}_d^{\mathrm{D}}, \quad \forall d \in \Omega^{\mathrm{D}}, \tag{4.4l}$$

$$\underline{P}_{dt}^{\mathrm{D}} \leq p_{dt}^{\mathrm{D}} \leq \overline{P}_{dt}^{\mathrm{D}}, \quad \forall d \in \Omega^{\mathrm{D}}, \forall t \in \Omega^{\mathrm{T}}, \tag{4.4m}$$

$$-R_d^{\mathrm{D,D}} \Delta t \leq p_{dt}^{\mathrm{D}} - p_{d(t-1)}^{\mathrm{D}} \leq R_d^{\mathrm{D,U}} \Delta t, \quad \forall d \in \Omega^{\mathrm{D}}, \forall t \in \Omega^{\mathrm{T}}, \tag{4.4n}$$

$$\underline{P}_c^{\mathrm{C}} u_{ct}^{\mathrm{C}} \leq p_{ct}^{\mathrm{C}} \leq \overline{P}_c^{\mathrm{C}} u_{ct}^{\mathrm{C}}, \quad \forall c \in \Omega^{\mathrm{C}}, \forall t \in \Omega^{\mathrm{T}}, \tag{4.4o}$$

$$-R_c^{\mathrm{C,D}} \Delta t \leq p_{ct}^{\mathrm{C}} - p_{c(t-1)}^{\mathrm{C}} \leq R_c^{\mathrm{C,U}} \Delta t, \quad \forall c \in \Omega^{\mathrm{C}}, \forall t \in \Omega^{\mathrm{T}}, \tag{4.4p}$$

$$e_{st}^{\mathrm{S}} = e_{s(t-1)}^{\mathrm{S}} + \eta_s^{\mathrm{S,C}} p_{st}^{\mathrm{S,C}} \Delta t - \frac{p_{st}^{\mathrm{S,D}}}{\eta_s^{\mathrm{S,D}}} \Delta t, \quad \forall s \in \Omega^{\mathrm{S}}, \forall t \in \Omega^{\mathrm{T}}, \tag{4.4q}$$

$$\underline{E}_s^{\mathrm{S}} \leq e_{st}^{\mathrm{S}} \leq \overline{E}_s^{\mathrm{S}}, \quad \forall s \in \Omega^{\mathrm{S}}, \forall t \in \Omega^{\mathrm{T}}, \tag{4.4r}$$

$$0 \leq p_{st}^{\mathrm{S,C}} \leq \overline{P}_s^{\mathrm{S,C}}, \quad \forall s \in \Omega^{\mathrm{S}}, \forall t \in \Omega^{\mathrm{T}}, \tag{4.4s}$$

$$0 \leq p_{st}^{\mathrm{S,D}} \leq \overline{P}_s^{\mathrm{S,D}}, \quad \forall s \in \Omega^{\mathrm{S}}, \forall t \in \Omega^{\mathrm{T}}, \tag{4.4t}$$

$$0 \leq p_{rt}^{\mathrm{R}} \leq \min_{\Phi_{rt}^{\mathrm{R,U}} \subseteq J_{rt}^{\mathrm{R}}} \; P_{rt}^{\mathrm{R,A}}, \quad \forall r \in \Omega^{\mathrm{R}}, \forall t \in \Omega^{\mathrm{T}}, \tag{4.4u}$$

where the optimization variables in the upper-level problem are those in set $\Phi^{\mathrm{DA}} = \Big\{ e_{st}^{\mathrm{S}}, p_{st}^{\mathrm{S,C}}, p_{st}^{\mathrm{S,D}}, \forall s \in \Omega^{\mathrm{S}}, \forall t \in \Omega^{\mathrm{T}}; p_{ct}^{\mathrm{C}}, u_{ct}^{C}, \forall c \in \Omega^{\mathrm{C}}, \forall t \in \Omega^{\mathrm{T}}; p_{dt}^{\mathrm{D}}, \forall d \in \Omega^{\mathrm{D}}, \forall t \in \Omega^{\mathrm{T}}; p_{nt}^{\mathrm{M}}, \forall n \in \Omega^{\mathrm{M}}, \forall t \in \Omega^{\mathrm{T}}; \delta_{nt}, \forall n \in \Omega^{\mathrm{N}}, \forall t \in \Omega^{\mathrm{T}}; p_{\ell t}^{\mathrm{L}}, \forall \ell \in \Omega^{\mathrm{L}}, \forall t \in \Omega^{\mathrm{T}}; p_{rt}^{\mathrm{R}}, \forall r \in \Omega^{\mathrm{R}}, \forall t \in \Omega^{\mathrm{T}}; p_t^{\mathrm{DA}}, p_t^{\mathrm{RT}}, \forall t \in \Omega^{\mathrm{T}} \Big\}$, while the optimization variables in the lower-level problems are those belonging to sets $\Phi^{\mathrm{DA,U}} \subseteq J^{\mathrm{DA}}$ and $\Phi_{rt}^{\mathrm{R,U}} \subseteq J_{rt}^{\mathrm{R}}, \forall r \in \Omega^{\mathrm{R}}, \forall t \in \Omega^{\mathrm{T}}$.

Problem (4.4) is an optimization problem under the uncertainty of the market prices and the available stochastic renewable production levels, which is formulated as a bi-level model as explained below:

1. In the upper-level problem, the scheduling decisions in the energy markets and the dispatch decisions on the units of the VPP are taken to maximize the worst-case profit, anticipating the worst-case uncertainty realization.
2. Given the scheduling and dispatch decisions, the lower-level problems aim at minimizing the profit under the uncertainty of market prices and the upper bound of the available power production levels of stochastic renewable generating units. Note that the uncertain DA and residual prices appear in the objective function (4.4a) affect the profit achieved by the VPP, while the uncertain available stochastic renewable power production levels are involved in constraints (4.4u) and therefore influence the feasible operation of the VPP.

The remaining constraints (4.4b)–(4.4t) do not include any uncertain parameters.

Considering the uncertainty model presented in Sect. 4.4.2 and the robust optimization approach explained in the Appendix B, problem (4.4) is formulated as a robust optimization model. Additionally, this robust optimization model can be transformed into an MILP model through the linearization method explained in [4]. Therefore, the scheduling problem of the risk-averse VPP in the DA energy market is formulated as the following robust MILP problem:

$$\begin{aligned}
\max_{\Phi^{\mathrm{DA,R}}} \quad & \sum_{t \in \Omega^{\mathrm{T}}} \Bigg[\sum_{d \in \Omega^{\mathrm{D}}} U_{dt}^{\mathrm{D}} p_{dt}^{\mathrm{D}} \Delta t - \sum_{c \in \Omega^{\mathrm{C}}} (C_c^{\mathrm{C,F}} u_{ct}^{\mathrm{C}} + C_c^{\mathrm{C,V}} p_{ct}^{\mathrm{C}} \Delta t) \Bigg] \\
& + \sum_{t \in \Omega^{\mathrm{T}}} \Big[\tilde{\lambda}_t^{\mathrm{DA}} (p_t^{\mathrm{DA}} \Delta t + p_t^{\mathrm{RT}} \Delta t) + \tilde{r}_t^{\mathrm{RT}} p_t^{\mathrm{RT}} \Delta t \Big] \\
& - \Bigg[\Gamma^{\mathrm{DA}} \kappa^{\mathrm{DA}} + \sum_{t \in \Omega^{\mathrm{T}}} (\psi_t^{\mathrm{DA}} + \psi_t^{\mathrm{RT}}) \Bigg]
\end{aligned} \tag{4.5a}$$

subject to

$$\text{Constraints (4.4b)–(4.4t)}, \tag{4.5b}$$

$$\kappa^{\mathrm{DA}} + \psi_t^{\mathrm{DA}} \geq \hat{\lambda}_t^{\mathrm{DA}} y_t^{\mathrm{DA}}, \quad \forall t \in \Omega^{\mathrm{T}}, \tag{4.5c}$$

$$-y_t^{\mathrm{DA}} \leq p_t^{\mathrm{DA}} \Delta t + p_t^{\mathrm{RT}} \Delta t \leq y_t^{\mathrm{DA}}, \quad \forall t \in \Omega^{\mathrm{T}}, \tag{4.5d}$$

$$\kappa^{\mathrm{DA}} + \psi_t^{\mathrm{RT}} \geq \hat{r}_t^{\mathrm{RT}} y_t^{\mathrm{RT}}, \quad \forall t \in \Omega^{\mathrm{T}}, \tag{4.5e}$$

$$-y_t^{\mathrm{RT}} \leq p_t^{\mathrm{RT}} \Delta t \leq y_t^{\mathrm{RT}}, \quad \forall t \in \Omega^{\mathrm{T}}, \tag{4.5f}$$

$$p_{rt}^{\mathrm{R}} + \Gamma_{rt}^{\mathrm{R}} \kappa_{rt}^{\mathrm{R}} + \psi_{rt}^{\mathrm{R}} \leq \tilde{P}_{rt}^{\mathrm{R,A}}, \quad \forall r \in \Omega^{\mathrm{R}}, \forall t \in \Omega^{\mathrm{T}}, \tag{4.5g}$$

$$\kappa_{rt}^{\mathrm{R}} + \psi_{rt}^{\mathrm{R}} \geq \hat{P}_{rt}^{\mathrm{R,A}} y_{rt}^{\mathrm{R,A}}, \quad \forall r \in \Omega^{\mathrm{R}}, \forall t \in \Omega^{\mathrm{T}}, \tag{4.5h}$$

$$1 \leq y_{rt}^{\mathrm{R,A}}, \quad \forall r \in \Omega^{\mathrm{R}}, \forall t \in \Omega^{\mathrm{T}}, \tag{4.5i}$$

$$p_{rt}^{\mathrm{R}}, \kappa^{\mathrm{DA}}, \psi_t^{\mathrm{DA}}, y_t^{\mathrm{DA}}, \psi_t^{\mathrm{RT}}, y_t^{\mathrm{RT}}, \kappa_{rt}^{\mathrm{R}}, \psi_{rt}^{\mathrm{R}}, y_{rt}^{\mathrm{R}} \geq 0, \quad \forall r \in \Omega^{\mathrm{R}}, \forall t \in \Omega^{\mathrm{T}}, \tag{4.5j}$$

where variables in set $\Phi^{\mathrm{DA,R}} = \Big\{ e_{st}^{\mathrm{S}}, p_{st}^{\mathrm{S,C}}, p_{st}^{\mathrm{S,D}}, \forall s \in \Omega^{\mathrm{S}}, \forall t \in \Omega^{\mathrm{T}}; p_{ct}^{\mathrm{C}}, u_{ct}^{\mathrm{C}}, \forall c \in \Omega^{\mathrm{C}}, \forall t \in \Omega^{\mathrm{T}}; p_{dt}^{\mathrm{D}}, \forall d \in \Omega^{\mathrm{D}}, \forall t \in \Omega^{\mathrm{T}}; p_{nt}^{\mathrm{M}}, \forall n \in \Omega^{\mathrm{M}}, \forall t \in \Omega^{\mathrm{T}}; \delta_{nt}, \forall n \in \Omega^{\mathrm{N}}, \forall t \in \Omega^{\mathrm{T}}; p_{\ell t}^{\mathrm{L}}, \forall \ell \in \Omega^{\mathrm{L}}, \forall t \in \Omega^{\mathrm{T}}; p_{rt}^{\mathrm{R}}, \kappa_{rt}^{\mathrm{R}}, \psi_{rt}^{\mathrm{R}}, y_{rt}^{\mathrm{R}}, \forall r \in \Omega^{\mathrm{R}}, \forall t \in \Omega^{\mathrm{T}}; p_t^{\mathrm{DA}}, p_t^{\mathrm{RT}}, \psi_t^{\mathrm{DA}}, y_t^{\mathrm{DA}}, \psi_t^{\mathrm{RT}}, y_t^{\mathrm{RT}}, \forall t \in \Omega^{\mathrm{T}}; \kappa^{\mathrm{DA}} \Big\}$ are the optimization variables of the robust optimization model (4.5).

The objective function (4.5a) maximizes the worst-case profit achieved by the VPP in the DA and RT energy markets due to the deviation of Γ^{DA} uncertain prices from their nominal values. Constraints (4.5b) define the feasible operating region of the network, the conventional power plants, the demands, and the storage units. Constraints (4.5c)–(4.5d) handle the uncertainties in the DA market prices affecting the objective function. Similarly, constraints (4.5e)–(4.5f) handle the uncertainties in the residual prices that influence the objective function. Constraints (4.5g)–(4.5i) protect the power production of each stochastic renewable generating unit against the deviation of Γ_{rt}^{R} uncertain available production levels in each time period. Constraints (4.5j) define nonnegative variables used to recast problem (4.4) as an MILP model (4.5).

Solving the robust optimization model (4.5) provides the scheduled power quantity in the DA and RT energy markets, as well as the dispatch power of the units in the VPP during the planning horizon. However, only the scheduled power in the DA energy market and the status of the conventional power plants are implemented.

Illustrative Example 4.3 Robust Scheduling Problem of a Risk-Averse VPP

A VPP similar to the one described in Illustrative Example 4.1 is considered, including one conventional power plant, two flexible demands, one storage unit, and one renewable PV generating unit. The technical and economic data are similar to those provided in Illustrative Example 4.1. The only difference is that the DA and RT market prices, as well as the available production levels of the PV generating unit are represented through the confidence bounds provided in Tables 4.10 and 4.11, respectively. Thus, the uncertainty set $J^{\mathrm{DA}} = \{\lambda_1^{\mathrm{DA}}, \lambda_2^{\mathrm{DA}}, \lambda_3^{\mathrm{DA}}, r_1^{\mathrm{RT}}, r_2^{\mathrm{RT}}, r_3^{\mathrm{RT}}\}$ includes 6 uncertain prices affecting the objective function. Additionally, the uncertainty sets $J_1^{\mathrm{R}} = \{P_1^{\mathrm{R,A}}\}$, $J_2^{\mathrm{R}} = \{P_2^{\mathrm{R,A}}\}$, and $J_3^{\mathrm{R}} = \{P_3^{\mathrm{R,A}}\}$ include the uncertain available production levels of the PV generating unit in time periods 1, 2, and 3, respectively, which influence the constraints of the PV power limitation.

Table 4.10 Illustrative Example 4.3: nominal and deviation values of the DA market and residual prices in each time period

	DA market prices [$/MWh]		Residual prices [$/MWh]	
Time period	Nominal value	Deviation value	Nominal value	Deviation value
1	26	3	7	4
2	53	5	5	7
3	31	6	−4	8

Table 4.11 Illustrative Example 4.3: nominal and deviation values of the available PV production in each time period

Time period	Nominal value [MW]	Deviation value [MW]
1	6	2
2	7	3
3	5	4

Using the technical and economic data of the VPP, the scheduling problem of the VPP in the DA energy market is formulated. A risk-averse strategy is assumed by considering the maximum values of parameters $\Gamma^{\mathrm{DA}}=6$, $\Gamma_1^{\mathrm{R}}=1$, $\Gamma_2^{\mathrm{R}}=1$, and $\Gamma_3^{\mathrm{R}}=1$.

The worst-case of the objective function to be maximized under the uncertainty of the DA and RT market prices is formulated below:

$$\begin{aligned}
\max_{\Phi^{\mathrm{IE4.3}}} \quad & \Big[40p_{11}^{\mathrm{D}} + 50p_{21}^{\mathrm{D}} - (10u_1^{\mathrm{C}} + 45p_1^{\mathrm{C}}) + 50p_{12}^{\mathrm{D}} + 60p_{22}^{\mathrm{D}} - (10u_2^{\mathrm{C}} + 45p_2^{\mathrm{C}}) \\
& + 40p_{13}^{\mathrm{D}} + 50p_{23}^{\mathrm{D}} - (10u_3^{\mathrm{C}} + 45p_3^{\mathrm{C}})\Big] \\
& + \Big[26(p_1^{\mathrm{DA}} + p_1^{\mathrm{RT}}) + 7p_1^{\mathrm{RT}} + 53(p_2^{\mathrm{DA}} + p_2^{\mathrm{RT}}) + 5p_2^{\mathrm{RT}} + 31(p_3^{\mathrm{DA}} + p_3^{\mathrm{RT}}) - 4p_3^{\mathrm{RT}}\Big] \\
& - \Big[6\kappa^{\mathrm{DA}} + \psi_1^{\mathrm{DA}} + \psi_1^{\mathrm{RT}} + \psi_2^{\mathrm{DA}} + \psi_2^{\mathrm{RT}} + \psi_3^{\mathrm{DA}} + \psi_3^{\mathrm{RT}}\Big].
\end{aligned}$$

The above objective function comprises three terms:

1. The profit calculated as the utility of energy consumption minus the production cost of the conventional power plant (first and second lines),
2. The profit obtained from trading in the DA and RT energy markets under the nominal values of market prices (third line),
3. The minus worst cost of robustness against the deviations of all market prices (i.e., $\Gamma^{\mathrm{DA}}=6$) from their nominal values (fourth line).

Then, the constraints without uncertain data are incorporated as:

$$\begin{aligned}
& -p_{11}^{\mathrm{L}} - p_{21}^{\mathrm{L}} = p_1^{\mathrm{M}}, \\
& -p_{12}^{\mathrm{L}} - p_{22}^{\mathrm{L}} = p_2^{\mathrm{M}}, \\
& -p_{13}^{\mathrm{L}} - p_{23}^{\mathrm{L}} = p_3^{\mathrm{M}}, \\
& p_1^{\mathrm{C}} - p_{31}^{\mathrm{L}} + p_{11}^{\mathrm{L}} = p_{11}^{\mathrm{D}}, \\
& p_2^{\mathrm{C}} - p_{32}^{\mathrm{L}} + p_{12}^{\mathrm{L}} = p_{12}^{\mathrm{D}}, \\
& p_3^{\mathrm{C}} - p_{33}^{\mathrm{L}} + p_{13}^{\mathrm{L}} = p_{13}^{\mathrm{D}}, \\
& p_1^{\mathrm{R}} + p_1^{\mathrm{S,D}} - p_1^{\mathrm{S,C}} + p_{21}^{\mathrm{L}} + p_{31}^{\mathrm{L}} = p_{21}^{\mathrm{D}}, \\
& p_2^{\mathrm{R}} + p_2^{\mathrm{S,D}} - p_2^{\mathrm{S,C}} + p_{22}^{\mathrm{L}} + p_{32}^{\mathrm{L}} = p_{22}^{\mathrm{D}}, \\
& p_3^{\mathrm{R}} + p_3^{\mathrm{S,D}} - p_3^{\mathrm{S,C}} + p_{23}^{\mathrm{L}} + p_{33}^{\mathrm{L}} = p_{23}^{\mathrm{D}}, \\
& p_1^{\mathrm{DA}} + p_1^{\mathrm{RT}} = p_1^{\mathrm{M}}, \\
& p_2^{\mathrm{DA}} + p_2^{\mathrm{RT}} = p_2^{\mathrm{M}}, \\
& p_3^{\mathrm{DA}} + p_3^{\mathrm{RT}} = p_3^{\mathrm{M}}, \\
& \frac{p_{11}^{\mathrm{L}}}{100} = 200(\delta_{11} - \delta_{21}), \\
& \frac{p_{12}^{\mathrm{L}}}{100} = 200(\delta_{12} - \delta_{22}), \\
& \frac{p_{13}^{\mathrm{L}}}{100} = 200(\delta_{13} - \delta_{23}),
\end{aligned}$$

$$\frac{p^{\mathrm{L}}_{21}}{100} = 200(\delta_{11} - \delta_{31}),$$
$$\frac{p^{\mathrm{L}}_{22}}{100} = 200(\delta_{12} - \delta_{32}),$$
$$\frac{p^{\mathrm{L}}_{23}}{100} = 200(\delta_{13} - \delta_{33}),$$
$$\frac{p^{\mathrm{L}}_{31}}{100} = 200(\delta_{21} - \delta_{31}),$$
$$\frac{p^{\mathrm{L}}_{32}}{100} = 200(\delta_{22} - \delta_{32}),$$
$$\frac{p^{\mathrm{L}}_{33}}{100} = 200(\delta_{23} - \delta_{33}),$$
$$-50 \le p^{\mathrm{L}}_{11} \le 50,$$
$$-50 \le p^{\mathrm{L}}_{12} \le 50,$$
$$-50 \le p^{\mathrm{L}}_{13} \le 50,$$
$$-50 \le p^{\mathrm{L}}_{21} \le 50,$$
$$-50 \le p^{\mathrm{L}}_{22} \le 50,$$
$$-50 \le p^{\mathrm{L}}_{23} \le 50,$$
$$-50 \le p^{\mathrm{L}}_{31} \le 50,$$
$$-50 \le p^{\mathrm{L}}_{32} \le 50,$$
$$-50 \le p^{\mathrm{L}}_{33} \le 50,$$
$$\delta_{11} = 0,$$
$$\delta_{12} = 0,$$
$$\delta_{13} = 0,$$
$$-\pi \le \delta_{21} \le \pi,$$
$$-\pi \le \delta_{22} \le \pi,$$
$$-\pi \le \delta_{23} \le \pi,$$
$$-\pi \le \delta_{31} \le \pi,$$
$$-\pi \le \delta_{32} \le \pi,$$
$$-\pi \le \delta_{33} \le \pi,$$
$$-60 \le p^{\mathrm{DA}}_{1} \le 40,$$
$$-60 \le p^{\mathrm{DA}}_{2} \le 40,$$
$$-60 \le p^{\mathrm{DA}}_{3} \le 40,$$
$$-60 \le p^{\mathrm{DA}}_{1} + p^{\mathrm{RT}}_{1} \le 40,$$
$$-60 \le p^{\mathrm{DA}}_{2} + p^{\mathrm{RT}}_{2} \le 40,$$
$$-60 \le p^{\mathrm{DA}}_{3} + p^{\mathrm{RT}}_{3} \le 40,$$
$$-100 \le p^{\mathrm{M}}_{1} \le 100,$$

$$
\begin{aligned}
& -100 \le p_2^{\mathrm{M}} \le 100, \\
& -100 \le p_3^{\mathrm{M}} \le 100, \\
& p_{11}^{\mathrm{D}} + p_{12}^{\mathrm{D}} + p_{13}^{\mathrm{D}} \ge 30, \\
& p_{21}^{\mathrm{D}} + p_{22}^{\mathrm{D}} + p_{23}^{\mathrm{D}} \ge 50, \\
& 0 \le p_{11}^{\mathrm{D}} \le 20, \\
& 0 \le p_{12}^{\mathrm{D}} \le 20, \\
& 0 \le p_{13}^{\mathrm{D}} \le 10, \\
& 5 \le p_{21}^{\mathrm{D}} \le 30, \\
& 5 \le p_{22}^{\mathrm{D}} \le 30, \\
& 5 \le p_{23}^{\mathrm{D}} \le 15, \\
& -15 \le p_{11}^{\mathrm{D}} - 10 \le 15, \\
& -15 \le p_{12}^{\mathrm{D}} - p_{11}^{\mathrm{D}} \le 15, \\
& -15 \le p_{13}^{\mathrm{D}} - p_{12}^{\mathrm{D}} \le 15, \\
& -25 \le p_{21}^{\mathrm{D}} - 20 \le 25, \\
& -25 \le p_{22}^{\mathrm{D}} - p_{21}^{\mathrm{D}} \le 25, \\
& -25 \le p_{23}^{\mathrm{D}} - p_{22}^{\mathrm{D}} \le 25, \\
& 0 \le p_1^{\mathrm{C}} \le 20 u_1^{\mathrm{C}}, \\
& 0 \le p_2^{\mathrm{C}} \le 20 u_2^{\mathrm{C}}, \\
& 0 \le p_3^{\mathrm{C}} \le 20 u_3^{\mathrm{C}}, \\
& -20 \le p_1^{\mathrm{C}} - 0 \le 20, \\
& -20 \le p_2^{\mathrm{C}} - p_1^{\mathrm{C}} \le 20, \\
& -20 \le p_3^{\mathrm{C}} - p_2^{\mathrm{C}} \le 20, \\
& e_1^{\mathrm{S}} = 5 + 0.9 p_1^{\mathrm{S,C}} - \frac{p_1^{\mathrm{S,D}}}{0.9}, \\
& e_2^{\mathrm{S}} = e_1^{\mathrm{S}} + 0.9 p_2^{\mathrm{S,C}} - \frac{p_2^{\mathrm{S,D}}}{0.9}, \\
& e_3^{\mathrm{S}} = e_2^{\mathrm{S}} + 0.9 p_3^{\mathrm{S,C}} - \frac{p_3^{\mathrm{S,D}}}{0.9}, \\
& 5 \le e_1^{\mathrm{S}} \le 10, \\
& 5 \le e_2^{\mathrm{S}} \le 10, \\
& 5 \le e_3^{\mathrm{S}} \le 10, \\
& 0 \le p_1^{\mathrm{S,C}} \le 10, \\
& 0 \le p_2^{\mathrm{S,C}} \le 10, \\
& 0 \le p_3^{\mathrm{S,C}} \le 10,
\end{aligned}
$$

$$0 \leq p_1^{\mathrm{S,D}} \leq 10,$$
$$0 \leq p_2^{\mathrm{S,D}} \leq 10,$$
$$0 \leq p_3^{\mathrm{S,D}} \leq 10.$$

Additionally, the constraints to control the robustness of the solution with respect to the uncertain prices are presented below:

$$\kappa^{\mathrm{DA}} + \psi_1^{\mathrm{DA}} \geq 3y_1^{\mathrm{DA}},$$
$$\kappa^{\mathrm{DA}} + \psi_2^{\mathrm{DA}} \geq 5y_1^{\mathrm{DA}},$$
$$\kappa^{\mathrm{DA}} + \psi_3^{\mathrm{DA}} \geq 6y_1^{\mathrm{DA}},$$
$$-y_1^{\mathrm{DA}} \leq p_1^{\mathrm{DA}} + p_1^{\mathrm{RT}} \leq y_1^{\mathrm{DA}},$$
$$-y_2^{\mathrm{DA}} \leq p_2^{\mathrm{DA}} + p_2^{\mathrm{RT}} \leq y_2^{\mathrm{DA}},$$
$$-y_3^{\mathrm{DA}} \leq p_3^{\mathrm{DA}} + p_3^{\mathrm{RT}} \leq y_3^{\mathrm{DA}},$$
$$\kappa^{\mathrm{DA}} + \psi_1^{\mathrm{RT}} \geq 4y_1^{\mathrm{RT}},$$
$$\kappa^{\mathrm{DA}} + \psi_2^{\mathrm{RT}} \geq 7y_2^{\mathrm{RT}},$$
$$\kappa^{\mathrm{DA}} + \psi_3^{\mathrm{RT}} \geq 8y_3^{\mathrm{RT}},$$
$$-y_1^{\mathrm{RT}} \leq p_1^{\mathrm{RT}} \leq y_1^{\mathrm{RT}},$$
$$-y_2^{\mathrm{RT}} \leq p_2^{\mathrm{RT}} \leq y_2^{\mathrm{RT}},$$
$$-y_3^{\mathrm{RT}} \leq p_3^{\mathrm{RT}} \leq y_3^{\mathrm{RT}},$$
$$\kappa^{\mathrm{DA}}, \psi_1^{\mathrm{DA}}, \psi_2^{\mathrm{DA}}, \psi_3^{\mathrm{DA}}, y_1^{\mathrm{DA}}, y_2^{\mathrm{DA}}, y_3^{\mathrm{DA}}, \psi_1^{\mathrm{RT}}, \psi_2^{\mathrm{RT}}, \psi_3^{\mathrm{RT}}, y_1^{\mathrm{RT}}, y_2^{\mathrm{RT}}, y_3^{\mathrm{RT}} \geq 0.$$

Finally, the constraints to protect the technical limitation of the PV generating unit against the uncertainty in its available production level are included below:

$$p_1^{\mathrm{R}} + \kappa_1^{\mathrm{R}} + \psi_1^{\mathrm{R}} \leq 6,$$
$$p_2^{\mathrm{R}} + \kappa_2^{\mathrm{R}} + \psi_2^{\mathrm{R}} \leq 7,$$
$$p_3^{\mathrm{R}} + \kappa_3^{\mathrm{R}} + \psi_3^{\mathrm{R}} \leq 5,$$
$$\kappa_1^{\mathrm{R}} + \psi_1^{\mathrm{R}} \geq 2y_1^{\mathrm{R}},$$
$$\kappa_2^{\mathrm{R}} + \psi_2^{\mathrm{R}} \geq 3y_2^{\mathrm{R}},$$
$$\kappa_3^{\mathrm{R}} + \psi_3^{\mathrm{R}} \geq 4y_3^{\mathrm{R}},$$
$$y_1^{\mathrm{R}} \geq 1,$$
$$y_2^{\mathrm{R}} \geq 1,$$
$$y_3^{\mathrm{R}} \geq 1,$$
$$p_1^{\mathrm{R}}, p_2^{\mathrm{R}}, p_3^{\mathrm{R}}, \kappa_1^{\mathrm{R}}, \kappa_2^{\mathrm{R}}, \kappa_3^{\mathrm{R}}, \psi_1^{\mathrm{R}}, \psi_2^{\mathrm{R}}, \psi_3^{\mathrm{R}}, y_1^{\mathrm{R}}, y_2^{\mathrm{R}}, y_3^{\mathrm{R}} \geq 0.$$

Finally, note that variables in set $\Phi^{\mathrm{IE4.3}} = \Big\{ e_1^{\mathrm{S}}, e_2^{\mathrm{S}}, e_3^{\mathrm{S}}, p_1^{\mathrm{S,C}}, p_2^{\mathrm{S,C}}, p_3^{\mathrm{S,C}}, p_1^{\mathrm{S,D}}, p_2^{\mathrm{S,D}}, p_3^{\mathrm{S,D}}, p_1^{\mathrm{C}}, p_2^{\mathrm{C}}, p_3^{\mathrm{C}}, u_1^{\mathrm{C}}, u_2^{\mathrm{C}}, u_3^{\mathrm{C}}, p_{11}^{\mathrm{D}},$ $p_{12}^{\mathrm{D}}, p_{13}^{\mathrm{D}}, p_{21}^{\mathrm{D}}, p_{22}^{\mathrm{D}}, p_{23}^{\mathrm{D}}, p_1^{\mathrm{M}}, p_2^{\mathrm{M}}, p_3^{\mathrm{M}}, \delta_{11}, \delta_{12}, \delta_{13}, \delta_{21}, \delta_{22}, \delta_{23}, \delta_{31}, \delta_{32}, \delta_{33}, p_{11}^{\mathrm{L}}, p_{12}^{\mathrm{L}}, p_{13}^{\mathrm{L}}, p_{21}^{\mathrm{L}}, p_{22}^{\mathrm{L}}, p_{23}^{\mathrm{L}}, p_{31}^{\mathrm{L}}, p_{32}^{\mathrm{L}}, p_{33}^{\mathrm{L}}, p_1^{\mathrm{R}},$

$p_2^{\text{R}}, p_3^{\text{R}}, p_1^{\text{DA}}, p_2^{\text{DA}}, p_3^{\text{DA}}, p_1^{\text{RT}}, p_2^{\text{RT}}, p_3^{\text{RT}}, \kappa^{\text{DA}}, \psi_1^{\text{DA}}, \psi_2^{\text{DA}}, \psi_3^{\text{DA}}, \psi_1^{\text{RT}}, \psi_2^{\text{RT}}, \psi_3^{\text{RT}}, y_1^{\text{DA}}, y_2^{\text{DA}}, y_3^{\text{DA}}, y_1^{\text{RT}}, y_2^{\text{RT}}, y_3^{\text{RT}}, \kappa_1^{\text{R}}, \kappa_2^{\text{R}},$
$\kappa_3^{\text{R}}, \psi_1^{\text{R}}, \psi_2^{\text{R}}, \psi_3^{\text{R}}, y_1^{\text{R}}, y_2^{\text{R}}, y_3^{\text{R}} \}$ are the optimization variables of the scheduling model.

The above MILP problem is solved using CPLEX [9] under GAMS [8]. The optimal value of the objective function is equal to \$1855.2, which is calculated as the profit under nominal values of market prices \$2210.2 minus the worst cost of robustness \$355. The optimal power quantities are $p_1^{\text{DA}}=-60$, $p_2^{\text{DA}}=-4$, and $p_3^{\text{DA}}=-24$ MW. This means that the VPP behaves as a consumer in the DA energy market during all the planning horizon. Additionally, the status of the conventional power plant is on only in time period 2.

The scheduled power in the RT energy market is 11 MW in time period 1, and zero in the remaining time periods. These schedules in the DA and RT energy market result in the maximum profit under the worst-case realization of market prices during the planning horizon. The scheduling decisions strongly depend on the residual prices, defined as the difference between the RT and DA market prices. In time period 1, the residual prices are anticipated to vary in the interval [3,11] \$/MWh. Thus, the residual price is positive under any deviation from its nominal value, and the VPP tends to buy energy in the DA market and sell energy in the RT market. However, the residual prices in time periods 2 and 3 can vary in the prediction intervals [−2,12] \$/MWh and [−12,4] \$/MWh, respectively. Under such a condition, the residual price may take positive or negative values, and the VPP does not have a tendency to trade in the RT market under the worst-case realization of market prices. Figure 4.5 depicts the dispatch decisions on the units in the VPP. Note that the power level of the storage unit represents its net injection into the VPP, i.e., the discharging power level minus the charging power level.

Once the DA energy market is cleared, the prices in the DA energy market are known as 25, 50, and 35 \$/MWh. Thus, the VPP pays \$2540 for buying energy in the DA energy market. The profit in each time period is given in Table 4.12.

Considering the nominal and deviation values provided in Table 4.11, the available PV production levels may vary within the intervals [4,8], [4,10], and [1,9] MW, in time periods 1, 2, and 3, respectively. Therefore, the worst-case power production levels of the PV generating unit is equal to 4, 4, and 1 MW in time periods 1, 2, and 3, respectively. This is due to the fact that the maximum value of parameters $\Gamma_1^{\text{R}}=\Gamma_2^{\text{R}}=\Gamma_3^{\text{R}}=1$ is considered.

□

A GAMS [8] code for solving Illustrative Example 4.3 is provided in Sect. 4.8.

Illustrative Example 4.4 Impact of the Risk Strategy on the Robust Scheduling of a VPP

In Illustrative Example 4.3, the scheduling problem of a risk-averse VPP is solved. This example is intended for assessing the impact of the risk strategy characterized by parameters Γ^{DA} and $\Gamma_t^{\text{R}}, \forall t \in \Omega^{\text{T}}$. Figure 4.6 shows the scheduled power in the DA energy market for different values of Γ^{DA}, while parameters $\Gamma_t^{\text{R}}, \forall t \in \Omega^{\text{T}}$, are fixed to 1. Results indicate that by

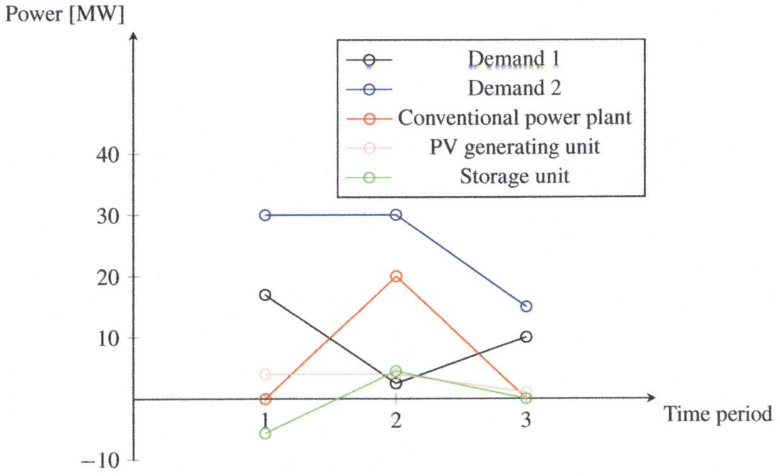

Fig. 4.5 Illustrative Example 4.9: dispatch decisions on the units in the VPP

Table 4.12 Illustrative Example 4.3: profit achieved by the VPP in each time period

Time period	Profit [\$]
1	−1500
2	−200
3	−840

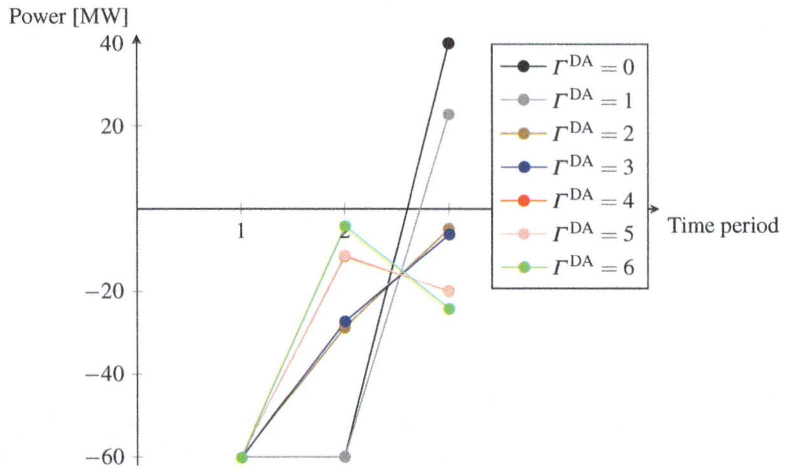

Fig. 4.6 Illustrative Example 4.4: power scheduled in the DA energy market for $\Gamma_t^R = 1$

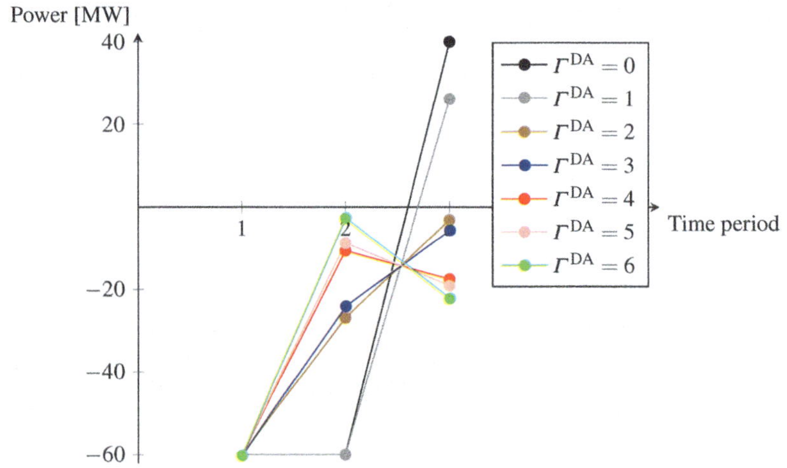

Fig. 4.7 Illustrative Example 4.4: power scheduled in the DA energy market for $\Gamma_t^R = 0.5$

decreasing parameter Γ^{DA} from 6 to 0, the VPP has more willingness to buy energy in time period 2. The opposite tendency is observed in time period 3, in which the VPP buys 24 MW for Γ^{DA}=6, while by decreasing Γ^{DA} to 0, the VPP prefers to sell 40 MW.

A similar trend is observed in Fig. 4.7 for $\Gamma_t^R = 0.5, \forall t \in \Omega^T$. In this case, the VPP has more willingness to sell energy, or in other words, has less tendency to buy energy. The worst-case available levels of PV power production are 4, 4, and 1 MW for $\Gamma_t^R = 1$ and $\Gamma^{DA} = 6$, while by decreasing Γ_t^R to 0.5, these worst-case levels increase to 5, 5.5, and 3 MW in time periods 1, 2, and 3, respectively.

The actual profit obtained in the DA energy market is shown in Fig. 4.8. The profit for low values of Γ^{DA} is lower than that for high values of Γ^{DA}. This means that taking a risk-averse strategy against the uncertain prices is more beneficial in the DA energy market. However, the profit for Γ_t^R=0.5 is higher than the one for Γ_t^R=1. In other words, the VPP can get a higher profit by choosing a less risk-averse strategy against the uncertain PV production in the DA energy market compared to a risk-averse strategy. However, by choosing a less risk-averse strategy, the VPP ignores the potential impact of a part of uncertainties on the constraints. This risky strategy may result in additional costs in the real-time operation, and even may threaten the operation feasibility.

□

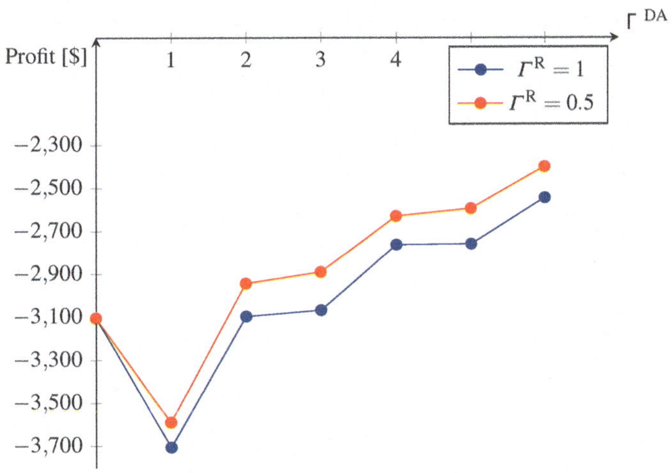

Fig. 4.8 Illustrative Example 4.4: actual profit with respect to parameter Γ^{DA}

4.5 Hybrid Stochastic-Robust Optimization Approach

This section models the scheduling problem of the risk-averse VPP in the energy markets using a hybrid stochastic-robust optimization approach.

4.5.1 Problem Description

Sections 4.3 and 4.4 present the scheduling problem of a risk-averse VPP in the energy markets using stochastic programming and robust optimization approaches, respectively. On one hand, as the stochastic programming approach represents uncertainties through a number of scenarios, the decisions may not be feasible if the actual realizations of uncertainties involved in technical constraints do not belong to the considered scenario set. This is a relevant issue when the VPP includes stochastic renewable generating units whose available production levels are highly variable and uncertain. Under this condition, the scenario-based uncertainty model cannot appropriately cover such uncertainty which may lead to solution infeasibility. On the other hand, the robust optimization approach provides solutions which are optimal only for the worst-case of the profit over the confidence bounds, which may result in very conservative results.

To remove the drawbacks of the stochastic programming and robust optimization approaches, a hybrid stochastic-robust optimization approach is used in this section. Here, the uncertainties in the objective function and the constraints are presented through a number of scenarios and confidence bounds, respectively. Thus, the hybrid stochastic-robust optimization approach maximizes the expected profit over the considered scenario set. Additionally, this hybrid approach considers that the available production levels of stochastic renewable generating units may vary within the confidence bounds.

The decision-making process is described below:

1. The VPP makes its decisions in the DA energy market, one day in advance. The VPP maximizes the expected profit over the scenarios of market prices, while forecasting the worst realization of the available stochastic renewable power production levels. When the DA energy market is cleared, the schedules of the VPP are known. Then, the scenario representing the prices in the DA and RT energy markets is realized. This means that the scheduling decisions in the DA energy market are independent of the scenarios and are *here-and-now* decisions with respect to the scenarios.
2. The VPP makes its decisions in the RT energy market and dispatches the units. Thus, these decisions are *wait-and-see* with respect to the scenarios.
3. The scheduling and dispatch decisions are made by anticipating that the worst-case realization of the available stochastic renewable power production levels occurs.

The scenario tree used in the above decision-making process is similar to the structure used in the stochastic programming approach as shown in Fig. 4.1. The main difference is that each scenario contains only information on the market prices, and the available stochastic renewable power production levels are represented through the confidence bounds. The next section explains how the uncertainty model combines scenarios and confidence bounds.

4.5.2 Uncertainty Model

The uncertainties in the DA and RT market prices influence the objective function of the scheduling problem, and thus are modeled by a number of scenarios. Each scenario realization presents information about the DA market prices, as well as about the RT market prices. The scenarios can be generated as explained below:

1. Generate a number of scenarios for the DA market prices.
2. For each scenario of the DA market price, generate a number of scenarios for the RT market prices.

Thus, set Π^{DA} includes the scenarios of the market prices generated through the above procedure. Each scenario ω provides information on the prices in the DA and RT energy markets, i.e., $\lambda_{t\omega}^{\mathrm{DA}}$ and $\lambda_{t\omega}^{\mathrm{RT}}$, and has a probability of occurrence π_{ω}. The summation of probabilities over the scenarios is equal to one.

The uncertainties in the available production levels of the stochastic renewable generating units are involved in the technical constraints of the scheduling problem, and thus are presented by the following confidence bounds:

$$P_{rt}^{\mathrm{R,A}} \in [\tilde{P}_{rt}^{\mathrm{R,A}} - \hat{P}_{rt}^{\mathrm{R,A}}, \tilde{P}_{rt}^{\mathrm{R,A}} + \hat{P}_{rt}^{\mathrm{R,A}}], \quad \forall r \in \Omega^{\mathrm{R}}, \forall t \in \Omega^{\mathrm{T}}, \tag{4.6}$$

where parameters $\tilde{P}_{rt}^{\mathrm{R,A}}$ and $\hat{P}_{rt}^{\mathrm{R,A}} \geq 0$ refer to the nominal value and the deviation around the nominal value of the uncertain parameters, respectively.

The uncertain data are included in the uncertainty sets defined as J_{rt}^{R}=$\{P_{rt}^{\mathrm{R,A}}\}$, $\forall r \in \Omega^{\mathrm{R}}$, $\forall t \in \Omega^{\mathrm{T}}$. Each one includes the available production level of each stochastic renewable generating unit in each time period. The cardinality of sets J_{rt}^{R}, $\forall r \in \Omega^{\mathrm{R}}$, $\forall t \in \Omega^{\mathrm{T}}$, is equal to 1, i.e., $|J_{rt}^{\mathrm{R}}|$=1, $\forall r \in \Omega^{\mathrm{R}}$, $\forall t \in \Omega^{\mathrm{T}}$.

As in the robust optimization approach, parameters Γ_{rt}^{R}, $\forall r \in \Omega^{\mathrm{R}}$, $\forall t \in \Omega^{\mathrm{T}}$, are introduced to control the protection level of the constraints against the uncertain data of the available stochastic renewable power production levels. These nonnegative parameters are bounded by the cardinality of sets J_{rt}^{R}, $\forall r \in \Omega^{\mathrm{R}}$, $\forall t \in \Omega^{\mathrm{T}}$, and thus can be chosen within the interval [0,1]. A value of Γ_{rt}^{R}=1 protects fully the constraints against the variation of the uncertain data within the confidence bounds. In other words, the solution is guaranteed to be feasible provided that the uncertain production levels take values within the considered confidence bounds. For Γ_{rt}^{R}=0, the VPP ignores the potential impact of the variation of the uncertain data on the constraints. Choosing a value of Γ_{rt}^{R} less than 1 guarantees the feasibility of the solution if the available production levels of the stochastic generating units deviate from their nominal values by at most $\Gamma_{rt}^{\mathrm{R}}\,\hat{P}_{rt}^{\mathrm{R,A}}$.

4.5.3 Formulation

Considering the decision-making process explained in Sect. 4.5.1 and the uncertainty model described in Sect. 4.5.2, the scheduling problem of a risk-averse VPP in the DA energy market is formulated as the following hybrid stochastic-robust problem:

$$\max_{\Phi^{\mathrm{DA,S}}} \quad \sum_{\omega \in \Pi^{\mathrm{DA}}} \pi_{\omega} \zeta_{\omega} + \beta \left[\rho - \frac{1}{1-\alpha} \sum_{\omega \in \Pi^{\mathrm{DA}}} \pi_{\omega} \mu_{\omega} \right] \tag{4.7a}$$

subject to

$$\zeta_{\omega} = \sum_{t \in \Omega^{\mathrm{T}}} \left[\lambda_{t\omega}^{\mathrm{DA}} p_{t}^{\mathrm{DA}} \Delta t + \lambda_{t\omega}^{\mathrm{RT}} p_{t\omega}^{\mathrm{RT}} \Delta t + \sum_{d \in \Omega^{\mathrm{D}}} U_{dt}^{\mathrm{D}} p_{dt\omega}^{\mathrm{D}} \Delta t, \right.$$

$$\left. - \sum_{c \in \Omega^{\mathrm{C}}} (C_{c}^{\mathrm{C,F}} u_{ct}^{\mathrm{C}} + C_{c}^{\mathrm{C,V}} p_{ct\omega}^{\mathrm{C}} \Delta t) \right], \quad \forall \omega \in \Pi^{\mathrm{DA}}, \tag{4.7b}$$

$$-\zeta_\omega + \rho - \mu_\omega \le 0, \quad \forall \omega \in \Pi^{\mathrm{DA}}, \tag{4.7c}$$

$$0 \le \mu_\omega, \quad \forall \omega \in \Pi^{\mathrm{DA}}, \tag{4.7d}$$

$$\sum_{c \in \Omega_n^{\mathrm{C}}} p_{ct\omega}^{\mathrm{C}} + \sum_{r \in \Omega_n^{\mathrm{R}}} p_{rt\omega}^{\mathrm{R}} + \sum_{s \in \Omega_n^{\mathrm{S}}} (p_{st\omega}^{\mathrm{S,D}} - p_{st\omega}^{\mathrm{S,C}}) - \sum_{\ell | s(\ell)=n} p_{\ell t\omega}^{\mathrm{L}} + \sum_{\ell | r(\ell)=n} p_{\ell t\omega}^{\mathrm{L}}$$
$$= p_{nt\omega}^{\mathrm{M}} + \sum_{d \in \Omega_n^{\mathrm{D}}} p_{dt\omega}^{\mathrm{D}}, \quad \forall n \in \Omega^{\mathrm{M}}, \forall t \in \Omega^{\mathrm{T}}, \forall \omega \in \Pi^{\mathrm{DA}}, \tag{4.7e}$$

$$\sum_{c \in \Omega_n^{\mathrm{C}}} p_{ct\omega}^{\mathrm{C}} + \sum_{r \in \Omega_n^{\mathrm{R}}} p_{rt\omega}^{\mathrm{R}} + \sum_{s \in \Omega_n^{\mathrm{S}}} (p_{st\omega}^{\mathrm{S,D}} - p_{st\omega}^{\mathrm{S,C}}) - \sum_{\ell | s(\ell)=n} p_{\ell t\omega}^{\mathrm{L}} + \sum_{\ell | r(\ell)=n} p_{\ell t\omega}^{\mathrm{L}}$$
$$= \sum_{d \in \Omega_n^{\mathrm{D}}} p_{dt\omega}^{\mathrm{D}}, \quad \forall n \in \Omega^{\mathrm{N}} \setminus n \in \Omega^{\mathrm{M}}, \forall t \in \Omega^{\mathrm{T}}, \forall \omega \in \Pi^{\mathrm{DA}}, \tag{4.7f}$$

$$p_t^{\mathrm{DA}} + p_{t\omega}^{\mathrm{RT}} = \sum_{n \in \Omega^{\mathrm{M}}} p_{nt\omega}^{\mathrm{M}}, \quad \forall t \in \Omega^{\mathrm{T}}, \forall \omega \in \Pi^{\mathrm{DA}}, \tag{4.7g}$$

$$p_{\ell t\omega}^{\mathrm{L}} = B_\ell (\delta_{s(\ell)t\omega} - \delta_{r(\ell)t\omega}), \quad \forall \ell \in \Omega^{\mathrm{L}}, \forall \in \Omega^{\mathrm{T}}, \forall \omega \in \Pi^{\mathrm{DA}}, \tag{4.7h}$$

$$-\overline{P}_\ell^{\mathrm{L}} \le p_{\ell t\omega}^{\mathrm{L}} \le \overline{P}_\ell^{\mathrm{L}}, \quad \forall \ell \in \Omega^{\mathrm{L}}, \forall t \in \Omega^{\mathrm{T}}, \forall \omega \in \Pi^{\mathrm{DA}}, \tag{4.7i}$$

$$\delta_{nt\omega} = 0, \quad n: \text{ref.}, \forall t \in \Omega^{\mathrm{T}}, \forall \omega \in \Pi^{\mathrm{DA}}, \tag{4.7j}$$

$$-\pi \le \delta_{nt\omega} \le \pi, \quad \forall n \in \Omega^{\mathrm{N}} \setminus n: \text{ref.}, \forall t \in \Omega^{\mathrm{T}}, \forall \omega \in \Pi^{\mathrm{DA}}, \tag{4.7k}$$

$$\underline{P}^{\mathrm{V}} \le p_t^{\mathrm{DA}} \le \overline{P}^{\mathrm{V}}, \quad \forall t \in \Omega^{\mathrm{T}}, \tag{4.7l}$$

$$\underline{P}^{\mathrm{V}} \le p_t^{\mathrm{DA}} + p_{t\omega}^{\mathrm{RT}} \le \overline{P}^{\mathrm{V}}, \quad \forall t \in \Omega^{\mathrm{T}}, \forall \omega, \in \Pi^{\mathrm{DA}} \tag{4.7m}$$

$$-\overline{P}_n^{\mathrm{M}} \le p_{nt\omega}^{\mathrm{M}} \le \overline{P}_n^{\mathrm{M}}, \quad \forall n \in \Omega^{\mathrm{M}}, \forall t \in \Omega^{\mathrm{T}}, \forall \omega \in \Pi^{\mathrm{DA}}, \tag{4.7n}$$

$$\sum_{t \in \Omega^{\mathrm{T}}} p_{dt\omega}^{\mathrm{D}} \Delta t \ge \underline{E}_d^{\mathrm{D}}, \quad \forall d \in \Omega^{\mathrm{D}}, \ \forall \omega \in \Pi^{\mathrm{DA}} \tag{4.7o}$$

$$\underline{P}_{dt}^{\mathrm{D}} \le p_{dt\omega}^{\mathrm{D}} \le \overline{P}_{dt}^{\mathrm{D}}, \quad \forall d \in \Omega^{\mathrm{D}}, \forall t \in \Omega^{\mathrm{T}}, \forall \omega \in \Pi^{\mathrm{DA}}, \tag{4.7p}$$

$$-R_d^{\mathrm{D,D}} \Delta t \le p_{dt\omega}^{\mathrm{D}} \quad p_{d(t-1)\omega}^{\mathrm{D}} \le R_d^{\mathrm{D,U}} \Delta t, \quad \forall d \in \Omega^{\mathrm{D}}, \forall t \in \Omega^{\mathrm{T}}, \forall \omega \in \Pi^{\mathrm{DA}}, \tag{4.7q}$$

$$\underline{P}_c^{\mathrm{C}} u_{ct}^{\mathrm{C}} \le p_{ct\omega}^{\mathrm{C}} \le \overline{P}_c^{\mathrm{C}} u_{ct}^{\mathrm{C}}, \quad \forall c \in \Omega^{\mathrm{C}}, \forall t \in \Omega^{\mathrm{T}}, \forall \omega \in \Pi^{\mathrm{DA}}, \tag{4.7r}$$

$$-R_c^{\mathrm{C,D}} \Delta t \le p_{ct\omega}^{\mathrm{C}} - p_{c(t-1)\omega}^{\mathrm{C}} \le R_c^{\mathrm{C,U}} \Delta t, \quad \forall c \in \Omega^{\mathrm{C}}, \forall t \in \Omega^{\mathrm{T}}, \forall \omega \in \Pi^{\mathrm{DA}}, \tag{4.7s}$$

$$e_{st\omega}^{\mathrm{S}} = e_{s(t-1)\omega}^{\mathrm{S}} + \eta_s^{\mathrm{S,C}} p_{st\omega}^{\mathrm{S,C}} \Delta t - \frac{p_{st\omega}^{\mathrm{S,D}}}{\eta_s^{\mathrm{S,D}}} \Delta t, \quad \forall s \in \Omega^{\mathrm{S}}, \forall t \in \Omega^{\mathrm{T}}, \forall \omega \in \Pi^{\mathrm{DA}}, \tag{4.7t}$$

$$\underline{E}_s^{\mathrm{S}} \le e_{st\omega}^{\mathrm{S}} \le \overline{E}_s^{\mathrm{S}}, \quad \forall s \in \Omega^{\mathrm{S}}, \forall t \in \Omega^{\mathrm{T}}, \forall \omega \in \Pi^{\mathrm{DA}}, \tag{4.7u}$$

$$0 \le p_{st\omega}^{\mathrm{S,C}} \le \overline{P}_s^{\mathrm{S,C}}, \quad \forall s \in \Omega^{\mathrm{S}}, \forall t \in \Omega^{\mathrm{T}}, \forall \omega \in \Pi^{\mathrm{DA}}, \tag{4.7v}$$

$$0 \le p_{st\omega}^{\mathrm{S,D}} \le \overline{P}_s^{\mathrm{S,D}}, \quad \forall s \in \Omega^{\mathrm{S}}, \forall t \in \Omega^{\mathrm{T}}, \forall \omega \in \Pi^{\mathrm{DA}}, \tag{4.7w}$$

$$0 \le p_{rt\omega}^{\mathrm{R}} \le \min_{\Phi_{rt}^{\mathrm{R,U}} \subseteq J_{rt}^{\mathrm{R}}} \ P_{rt}^{\mathrm{R,A}}, \quad \forall r \in \Omega^{\mathrm{R}}, \forall t \in \Omega^{\mathrm{T}}, \forall \omega \in \Pi^{\mathrm{DA}}, \tag{4.7x}$$

where the optimization variables in the upper-level problem are those in set $\Phi^{\mathrm{DA,S}} = \Big\{ e_{st\omega}^{\mathrm{S}}, p_{st\omega}^{\mathrm{S,C}}, p_{st\omega}^{\mathrm{S,D}}, \forall s \in \Omega^{\mathrm{S}}, \forall t \in \Omega^{\mathrm{T}}, \forall \omega \in \Pi^{\mathrm{DA}}; p_{ct\omega}^{\mathrm{C}}, u_{ct}^{C}, \forall c \in \Omega^{\mathrm{C}}, \forall t \in \Omega^{\mathrm{T}}, \forall \omega \in \Pi^{\mathrm{DA}}; p_{dt\omega}^{\mathrm{D}}, \forall d \in \Omega^{\mathrm{D}}, \forall t \in \Omega^{\mathrm{T}}, \forall \omega \in \Omega^{\mathrm{W}}; p_{rt\omega}^{\mathrm{R}}, \forall r \in \Omega^{\mathrm{R}}, \forall t \in \Omega^{\mathrm{T}}, \forall \omega \in \Omega^{\mathrm{W}}; p_t^{\mathrm{DA}}, \forall t \in \Omega^{\mathrm{T}}; p_{t\omega}^{\mathrm{RT}}, \forall t \in \Omega^{\mathrm{T}}, \forall \omega \in \Pi^{\mathrm{DA}}; p_{nt\omega}^{\mathrm{M}}, \forall n \in \Omega^{\mathrm{M}}, \forall t \in \Omega^{\mathrm{T}}; \delta_{nt\omega}, \forall n \in \Omega^{\mathrm{N}}, \forall t \in \Omega^{\mathrm{T}}, \forall \omega \Pi^{\mathrm{DA}}; p_{\ell t\omega}^{\mathrm{L}},$

$\forall \ell \in \Omega^{\mathrm{L}}, \forall t \in \Omega^{\mathrm{T}}, \forall \omega \in \Pi^{\mathrm{DA}}; \rho; \zeta_\omega, \mu_\omega, \forall \omega \in \Pi^{\mathrm{DA}}\}$, and the optimization variables in the lower-level problems are those in sets $\Phi_{rt}^{\mathrm{R,U}} \subseteq J_{rt}^{\mathrm{R}}, \forall r \in \Omega^{\mathrm{R}}, \forall t \in \Omega^{\mathrm{T}}$.

Problem (4.7) aims at maximizing the expected profit over the considered scenarios of market prices and is constrained by the operating region of the VPP under the worst realization of stochastic renewable power production levels:

1. The upper-level problem seeks to find the schedules in the DA and RT energy markets, as well as the dispatch of the units in the VPP. The VPP aims at maximizing the objective function (4.7a) including the expected profit plus the weighted CVaR over the considered scenarios. The scheduling and dispatch decisions are sought to be feasible under the worst-case of available stochastic renewable power production levels.
2. Given the scheduling and dispatch decisions, the lower-level problems minimize the upper bound of the power produced by the stochastic renewable generating units in (4.7x) under the uncertainty of the available power production levels.

The structure of the stochastic bi-level problem (4.7) is similar to the stochastic model (4.1). The objective function (4.7a) to be maximized is the expected profit plus the weighted CVaR calculated over the scenario realizations. Constraints (4.7b)–(4.7d) are used to define the profit depending on each scenario realization and to calculate the CVaR. Constraints (4.7e)–(4.7w) consider the feasible operating region of the network, the demands, the conventional power plants, and the storage units for each scenario realization.

The only difference is on the handling of the uncertainty in the available production levels of the stochastic renewable generating units through constraints (4.7x). These constraints seek to protect the power production of the stochastic renewable generating units against the worst-case realization of the available power production levels.

Considering the robust optimization theory explained in Appendix B, the robust counterpart of the scheduling problem (4.7) can be formulated as a hybrid stochastic-robust problem, which can be converted into the following equivalent MILP problem using some linearization techniques proposed in [4]:

$$\max_{\Phi^{\mathrm{DA,SR}}} \quad \sum_{\omega \in \Pi^{\mathrm{DA}}} \pi_\omega \zeta_\omega + \beta \left[\rho - \frac{1}{1-\alpha} \sum_{\omega \in \Pi^{\mathrm{DA}}} \pi_\omega \mu_\omega \right] \tag{4.8a}$$

subject to

$$\text{Constraints (4.7b)–(4.7w)}, \tag{4.8b}$$

$$p_{rt\omega}^{\mathrm{R}} + \Gamma_{rt}^{\mathrm{R}} \kappa_{rt\omega}^{\mathrm{R}} + \psi_{rt\omega}^{\mathrm{R}} \leq \tilde{P}_{rt}^{\mathrm{R,A}}, \quad \forall r \in \Omega^{\mathrm{R}}, \forall t \in \Omega^{\mathrm{T}}, \forall \omega \in \Pi^{\mathrm{DA}}, \tag{4.8c}$$

$$\kappa_{rt\omega}^{\mathrm{R}} + \psi_{rt\omega}^{\mathrm{R}} \geq \hat{P}_{rt}^{\mathrm{R,A}} y_{rt\omega}^{\mathrm{R,A}}, \quad \forall r \in \Omega^{\mathrm{R}}, \forall t \in \Omega^{\mathrm{T}}, \forall \omega \in \Pi^{\mathrm{DA}}, \tag{4.8d}$$

$$p_{rt\omega}^{\mathrm{R}}, \kappa_{rt\omega}^{\mathrm{R}}, \psi_{rt\omega}^{\mathrm{R}} \geq 0, \; y_{rt\omega}^{\mathrm{R}} \geq 1, \quad \forall r \in \Omega^{\mathrm{R}}, \forall t \in \Omega^{\mathrm{T}}, \forall \omega \in \Pi^{\mathrm{DA}}, \tag{4.8e}$$

where variables in set $\Phi^{\mathrm{DA,SR}} = \{ e_{st\omega}^{\mathrm{S}}, p_{st\omega}^{\mathrm{S,C}}, p_{st\omega}^{\mathrm{S,D}}, \forall s \in \Omega^{\mathrm{S}}, \forall t \in \Omega^{\mathrm{T}}, \forall \omega \in \Pi^{\mathrm{DA}}; p_{ct\omega}^{\mathrm{C}}, u_{ct}^{C}, \forall c \in \Omega^{\mathrm{C}}, \forall t \in \Omega^{\mathrm{T}}, \forall \omega \in \Pi^{\mathrm{DA}};$ $p_{dt\omega}^{\mathrm{D}}, \forall d \in \Omega^{\mathrm{D}}, \forall t \in \Omega^{\mathrm{T}}, \forall \omega \in \Pi^{\mathrm{DA}}; p_{rt\omega}^{\mathrm{R}}, \kappa_{rt\omega}^{\mathrm{R}}, \psi_{rt\omega}^{\mathrm{R}}, y_{rt\omega}^{\mathrm{R}}, \forall r \in \Omega^{\mathrm{R}}, \forall t \in \Omega^{\mathrm{T}}, \forall \omega \in \Pi^{\mathrm{DA}}; p_t^{\mathrm{DA}}, \forall t \in \Omega^{\mathrm{T}}; p_{t\omega}^{\mathrm{RT}},$ $\forall t \in \Omega^{\mathrm{T}}, \forall \omega \in \Pi^{\mathrm{DA}}; p_{nt\omega}^{\mathrm{M}}, \delta_{nt\omega}, \forall n \in \Omega^{\mathrm{M}}, \forall t \in \Omega^{\mathrm{T}}; \forall n \in \Pi^{\mathrm{DA}}, \forall t \in \Omega^{\mathrm{T}}, \forall \omega \in \Pi^{\mathrm{DA}}; p_{\ell t\omega}^{\mathrm{L}}, \forall \ell \in \Omega^{\mathrm{L}}, \forall t \in \Omega^{\mathrm{T}},$ $\forall \omega \in \Pi^{\mathrm{DA}}; \rho; \zeta_\omega, \mu_\omega, \forall \omega \in \Pi^{\mathrm{DA}} \}$ are the optimization variables of the hybrid stochastic-robust optimization model (4.8).

Problem (4.7) is converted into the MILP problem (4.8) by replacing constraints (4.7x) with constraints (4.8c)–(4.8e). These constraints protect the feasibility of the stochastic renewable power production against the uncertainty in the available production levels based on the robust optimization theory. Parameters $\Gamma_{rt}^{\mathrm{R}}, \forall r \in \Omega^{\mathrm{R}}, \forall t \in \Omega^{\mathrm{T}}$, are used to control the protection level of these constraints against the uncertainty in the available stochastic renewable power production levels.

The outputs of problem (4.8) are the scheduled power levels in the DA energy market, and the status of the conventional power plants in the VPP.

Illustrative Example 4.5 Stochastic-Robust Scheduling Problem of a Risk-Averse VPP

A VPP similar to the one described in Illustrative Example 4.1 is considered, including one conventional power plant, two flexible demands, one storage unit, and one renewable PV generating unit. The technical and economic data are similar to those provided in Illustrative Example 4.1. The only difference is the uncertainty model used. The uncertainties in the DA and RT market prices are modeled through a predefined set of scenarios as provided in Table 4.13, while the uncertainty in the available production levels of the PV generating unit is modeled using the confidence bounds given in Table 4.14.

Table 4.13 Illustrative Example 4.1: scenario data

	λ_t^{DA} [$/MWh]			λ_t^{RT} [$/MWh]			
Scenario	1	2	3	1	2	3	Probability
1	20	40	30	25	50	25	0.4 × 0.4 = 0.16
2	20	40	30	10	45	28	0.4 × 0.6 = 0.24
3	30	50	40	35	60	35	0.6 × 0.4 = 0.24
4	30	50	40	20	55	38	0.6 × 0.6 = 0.36

Table 4.14 Illustrative Example 4.5: nominal and deviation values of the available PV production in each time period

Time period	Nominal value [MW]	Deviation value [MW]
1	6	2
2	7	3
3	5	4

The VPP takes a risk-averse strategy considering high values of parameters β=1 and Γ_t^R=1. In addition, the CVaR is calculated for the confidence level α=0.8.

The formulation of the hybrid stochastic-robust problem is very similar to the stochastic programming problem formulated in Illustrative Example 4.1. The only difference is the handling of the uncertainty in the available production levels of the PV generating unit through the following constraints:

$$p_{11}^R + \kappa_{11}^R + \psi_{11}^R \leq 6,$$
$$p_{12}^R + \kappa_{12}^R + \psi_{12}^R \leq 6,$$
$$p_{13}^R + \kappa_{13}^R + \psi_{13}^R \leq 6,$$
$$p_{14}^R + \kappa_{14}^R + \psi_{14}^R \leq 6,$$
$$p_{21}^R + \kappa_{21}^R + \psi_{21}^R \leq 7,$$
$$p_{22}^R + \kappa_{22}^R + \psi_{22}^R \leq 7,$$
$$p_{23}^R + \kappa_{23}^R + \psi_{23}^R \leq 7,$$
$$p_{24}^R + \kappa_{24}^R + \psi_{24}^R \leq 7,$$
$$p_{31}^R + \kappa_{31}^R + \psi_{31}^R \leq 5,$$
$$p_{32}^R + \kappa_{32}^R + \psi_{32}^R \leq 5,$$
$$p_{33}^R + \kappa_{33}^R + \psi_{33}^R \leq 5,$$
$$p_{34}^R + \kappa_{34}^R + \psi_{34}^R \leq 5,$$
$$\kappa_{11}^R + \psi_{11}^R \geq 2y_{11}^R,$$
$$\kappa_{12}^R + \psi_{12}^R \geq 2y_{12}^R,$$
$$\kappa_{13}^R + \psi_{13}^R \geq 2y_{13}^R,$$
$$\kappa_{14}^R + \psi_{14}^R \geq 2y_{14}^R,$$
$$\kappa_{21}^R + \psi_{21}^R \geq 3y_{21}^R,$$
$$\kappa_{22}^R + \psi_{22}^R \geq 3y_{22}^R,$$
$$\kappa_{23}^R + \psi_{23}^R \geq 3y_{23}^R,$$
$$\kappa_{24}^R + \psi_{24}^R \geq 3y_{24}^R,$$
$$\kappa_{31}^R + \psi_{31}^R \geq 4y_{31}^R,$$
$$\kappa_{32}^R + \psi_{32}^R \geq 4y_{32}^R,$$

$$\kappa^{\mathrm{R}}_{33} + \psi^{\mathrm{R}}_{33} \geq 4y^{\mathrm{R}}_{33},$$

$$\kappa^{\mathrm{R}}_{34} + \psi^{\mathrm{R}}_{34} \geq 4y^{\mathrm{R}}_{34},$$

$$y^{\mathrm{R}}_{11} \geq 1,$$

$$y^{\mathrm{R}}_{12} \geq 1,$$

$$y^{\mathrm{R}}_{13} \geq 1,$$

$$y^{\mathrm{R}}_{14} \geq 1,$$

$$y^{\mathrm{R}}_{21} \geq 1,$$

$$y^{\mathrm{R}}_{22} \geq 1,$$

$$y^{\mathrm{R}}_{23} \geq 1,$$

$$y^{\mathrm{R}}_{24} \geq 1,$$

$$y^{\mathrm{R}}_{31} \geq 1,$$

$$y^{\mathrm{R}}_{32} \geq 1,$$

$$y^{\mathrm{R}}_{33} \geq 1,$$

$$y^{\mathrm{R}}_{34} \geq 1,$$

$$p^{\mathrm{R}}_{11}, p^{\mathrm{R}}_{21}, p^{\mathrm{R}}_{31}, \kappa^{\mathrm{R}}_{11}, \kappa^{\mathrm{R}}_{21}, \kappa^{\mathrm{R}}_{31}, \psi^{\mathrm{R}}_{11}, \psi^{\mathrm{R}}_{21}, \psi^{\mathrm{R}}_{31}, y^{\mathrm{R}}_{11}, y^{\mathrm{R}}_{21}, y^{\mathrm{R}}_{31} \geq 0,$$

$$p^{\mathrm{R}}_{12}, p^{\mathrm{R}}_{22}, p^{\mathrm{R}}_{32}, \kappa^{\mathrm{R}}_{12}, \kappa^{\mathrm{R}}_{22}, \kappa^{\mathrm{R}}_{32}, \psi^{\mathrm{R}}_{12}, \psi^{\mathrm{R}}_{22}, \psi^{\mathrm{R}}_{32}, y^{\mathrm{R}}_{12}, y^{\mathrm{R}}_{22}, y^{\mathrm{R}}_{32} \geq 0,$$

$$p^{\mathrm{R}}_{13}, p^{\mathrm{R}}_{23}, p^{\mathrm{R}}_{33}, \kappa^{\mathrm{R}}_{13}, \kappa^{\mathrm{R}}_{23}, \kappa^{\mathrm{R}}_{33}, \psi^{\mathrm{R}}_{13}, \psi^{\mathrm{R}}_{23}, \psi^{\mathrm{R}}_{33}, y^{\mathrm{R}}_{13}, y^{\mathrm{R}}_{23}, y^{\mathrm{R}}_{33} \geq 0,$$

$$p^{\mathrm{R}}_{14}, p^{\mathrm{R}}_{24}, p^{\mathrm{R}}_{34}, \kappa^{\mathrm{R}}_{14}, \kappa^{\mathrm{R}}_{24}, \kappa^{\mathrm{R}}_{34}, \psi^{\mathrm{R}}_{14}, \psi^{\mathrm{R}}_{24}, \psi^{\mathrm{R}}_{34}, y^{\mathrm{R}}_{14}, y^{\mathrm{R}}_{24}, y^{\mathrm{R}}_{34} \geq 0.$$

The hybrid stochastic-robust optimization model is solved to decide the schedules in the DA energy market. The optimal value of the objective function is \$5281.2. The objective function is obtained by the summation of the expected profit \$2812.7 and the weighted CVaR \$2468.5. The scheduled power in the DA energy market and the status of the conventional power plant are provided in Table 4.15. Thus, the VPP behaves as a consumer in time periods 1 and 2, but as a producer in time period 3.

The scheduled power in the RT energy market is shown in Fig. 4.9. The VPP buys energy in all scenarios in time periods 1 and 3. However, it sells energy in all scenarios in time period 2. Note that the scheduled power in the RT energy market is not implemented at this stage.

The PV power production level is equal to 4, 4, and 1 MW in time periods 1, 2, and 3, respectively, for all scenario realizations. Thus, the problem is feasible if the available production levels lie on the confidence bounds provided in Table 4.14.

Based on the scheduled power and the actual prices, the VPP pays \$2102.2 in the DA energy market. Table 4.16 shows the profit in each time period.

□

A GAMS [8] code for solving Illustrative Example 4.5 is provided in Sect. 4.8.

Illustrative Example 4.6 Impact of the Risk Strategy on the Stochastic-Robust Scheduling of a VPP

Table 4.15 Illustrative Example 4.5: schedules in the DA energy market

Time period	Scheduled power [MW]	Status of conventional power plant
1	−21.1	0
2	−60	1
3	40	0

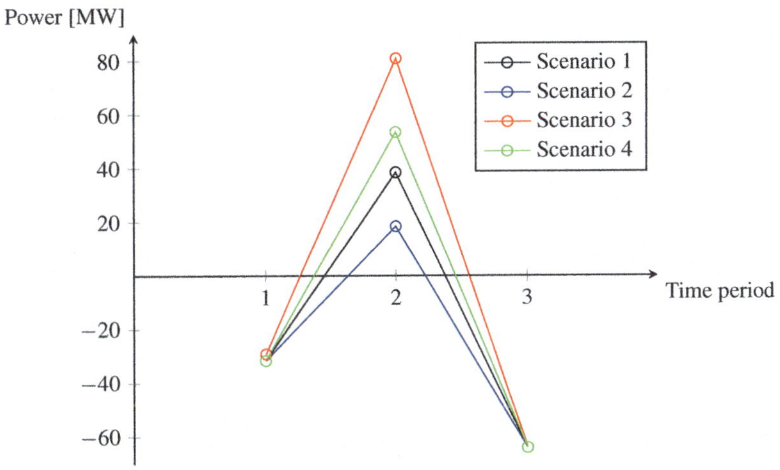

Fig. 4.9 Illustrative Example 4.5: power traded in the RT energy market

Table 4.16 Illustrative Example 4.5: profit in each time period

Time period	Profit [$]
1	−502.5
2	−3000
3	1400

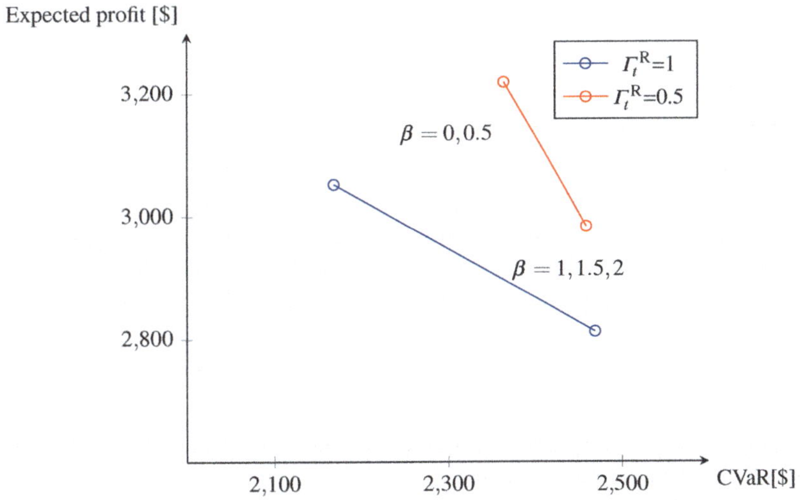

Fig. 4.10 Illustrative Example 4.6: expected profit with respect to CVaR for different values of β and Γ_t^R

In Illustrative Example 4.5, a risk-averse VPP is scheduled in the DA energy market, while parameters β and Γ_t^R are fixed to 1. This example analyzes how changing these parameters affects the schedules. First, parameter β is varied from 0 to 2, while parameter Γ^R is set equal to 1. The terms of the objective function are provided in Fig. 4.10 for different values of β. Observe that the expected profit decreases by the increase in β. However, the CVaR increases with respect to β.

The scheduled power is provided in Table 4.17 for different values of β. The VPP buys and sells energy in the DA energy market for different risk strategies in time periods 2 and 3, respectively. This can be explained by the fact that the DA market price in time period 2/3 is lower/higher than the RT market price for all considered scenarios as provided in Table 4.13. In time period 1, for a risk-averse strategy, i.e., for β=1, 1.5, and 2, the VPP decides to buy energy in the DA energy market. However, by taking a less risk-averse strategy, i.e., for β=0 and 0.5, the VPP behaves as a producer in the DA energy market.

Next, the model is solved for different values of β and parameter Γ_t^R=0.5. Similarly, the expected profit/the CVaR decreases/increases for higher values of β based on the results provided in Fig 4.10. Note that the expected profit and CVaR are greater than in the case with Γ_t^R=1; however, the protection level of the PV production limitation decreases. The

Table 4.17 Illustrative Example 4.6: power scheduled in each time period for Γ_t^{R}=1 and different values of β [MW]

	β				
Time period	0	0.5	1	1.5	2
1	40	40	−21.1	−21.1	−21.1
2	−60	−60	−60	−60	−60
3	40	40	40	40	40

Table 4.18 Illustrative Example 4.6: power scheduled in each time period for Γ_t^{R}=0.5 and different values of β [MW]

	β				
Time period	0	0.5	1	1.5	2
1	40	40	−19	−19	−19
2	−60	−60	−60	−60	−60
3	40	40	40	40	40

scheduled power in the DA energy market for Γ_t^{R}=0.5 is provided in Table 4.18, which is similar to the scheduled power in Table 4.17 for Γ_t^{R}=1.

□

4.6 Adaptive Robust Optimization Approach

This section describes the self-scheduling problem of a VPP considering an adaptive robust optimization approach. For the sake of simplicity, the VPP participates only in the DA energy market.

4.6.1 Problem Description

Robust models allow VPPs to make scheduling decisions by anticipating the worst uncertainty realizations within predefined uncertainty sets. Two types of robust models can be distinguished, namely the static robust optimization model described in Sect. 4.4 and the adaptive robust optimization model described in this section. The main difference between these two models is that the latter considers that it is possible to implement some corrective actions after knowing the actual realization of the uncertainty.

In particular, when making scheduling decisions in the DA energy market the decision-making process is generally as follows:

1. The optimal scheduling decisions are sought by maximizing the profit of the VPP. These scheduling decisions comprise the participation of the VPP in the DA energy market.
2. These optimal scheduling decisions are sought by anticipating that once they are made, the worst uncertainty realization will occur.
3. The worst-case uncertainty realization is considered by anticipating that once this worst-case is realized, the VPP adapts to it. That is, assuming that the scheduling decisions and uncertain parameters are fixed, the VPP decides the operation of its energy resources that maximizes its profit.

Note that the above decision sequence is consistent with reality. First, the VPP makes its scheduling decisions in the DA energy market. These decisions are made one day in advance and before knowing the actual uncertainty realization. Second, a worst-case occurs, e.g., very low renewable production. Finally, the VPP can use the flexibility of its energy resources to maximize its profit, e.g., if the available production level of the stochastic renewable generating units is low, the VPP can discharge the storage unit.

4.6.2 Formulation

The hierarchical structure described in the previous section can be represented using the following three-level optimization problem:

$$\max_{\Phi^{\mathrm{UL}}} \sum_{t\in\Omega^{\mathrm{T}}} \left(\lambda_t^{\mathrm{DA}} p_t^{\mathrm{DA}} \Delta t - \sum_{c\in\Omega^{\mathrm{C}}} C_c^{\mathrm{C,F}} u_{ct}^{\mathrm{C}} \right) - \min_{\Phi^{\mathrm{ML}}\in\Lambda} \max_{\Phi^{\mathrm{LL}}\in\Theta} \sum_{t\in\Omega^{\mathrm{T}}} \sum_{c\in\Omega^{\mathrm{C}}} C_c^{\mathrm{C,V}} p_{ct}^{\mathrm{C}} \Delta t \tag{4.9a}$$

subject to

$$\underline{P}^{\mathrm{DA}} \le p_t^{\mathrm{DA}} \le \overline{P}^{\mathrm{DA}}, \quad \forall t \in \Omega^{\mathrm{T}}, \tag{4.9b}$$

$$u_{ct}^{\mathrm{C}} \in \{0, 1\}, \quad \forall c \in \Omega^{\mathrm{C}}, \forall t \in \Omega^{\mathrm{T}}, \tag{4.9c}$$

where the optimization variables of the upper-, middle-, and lower-level problems are those included in sets $\Phi^{\mathrm{UL}} = \left\{p_t^{\mathrm{DA}}, \forall t \in \Omega^{\mathrm{T}}; u_{ct}^{\mathrm{C}}, \forall c \in \Omega^{\mathrm{C}}, \forall t \in \Omega^{\mathrm{T}}\right\}$, $\Phi^{\mathrm{ML}} = \left\{p_{rt}^{\mathrm{R,A}}, \forall r \in \Omega^{\mathrm{R}}, \forall t \in \Omega^{\mathrm{T}}\right\}$, and $\Phi^{\mathrm{LL}} = \left\{p_{dt}^{\mathrm{D}}, \forall d \in \Omega^{\mathrm{D}}, \forall t \in \Omega^{\mathrm{T}}; p_{ct}^{\mathrm{C}}, \forall c \in \Omega^{\mathrm{C}}, \forall t \in \Omega^{\mathrm{T}}; p_{rt}^{\mathrm{R}}, \forall r \in \Omega^{\mathrm{R}}, \forall t \in \Omega^{\mathrm{T}}; e_{st}^{\mathrm{S,C}}, p_{st}^{\mathrm{S,C}}, p_{st}^{\mathrm{S,D}}, \forall s \in \Omega^{\mathrm{S}}, \forall t \in \Omega^{\mathrm{T}}\right\}$, respectively.

Problem (4.9) is driven by the maximization of the worst-case profit of the VPP (4.9a) and involves three optimization problems:

1. The upper-level problem determines the scheduling decisions of the VPP in the DA energy market and the scheduling of the conventional power plant, maximizing the worst-case profit of the VPP and considering the limits on the power traded in the energy market (4.9b), as well as the definition of binary variables to determine the on/off status of the conventional power plants (4.9c).
2. Given the upper-level decisions, the middle-level problem identifies the worst-case realization of available stochastic renewable generation levels, i.e., those levels minimizing the profit of the VPP. Note that the middle-level decision variables in set Φ^{ML} must be within the uncertainty set Λ, which is defined in Sect. 4.6.3.
3. Given upper- and middle-level decisions, the lower-level problem models the operation of the VPP maximizing its profit. Note that the lower-level decision variables in set Φ^{LL} must be within the feasibility set Θ, which is defined in Sect. 4.6.4.

4.6.3 Uncertainty Set

The uncertainty in the available stochastic renewable production levels is modeled using variables $p_{rt}^{\mathrm{R,A}}, \forall r \in \Omega^{\mathrm{R}}, \forall t \in \Omega^{\mathrm{T}}$, which take values within known confidence bounds for a given confidence level, i.e.:

$$p_{rt}^{\mathrm{R,A}} \in \left[\tilde{P}_{rt}^{\mathrm{R,A}} - \hat{P}_{rt}^{\mathrm{R,A}}, \tilde{P}_{rt}^{\mathrm{R,A}} + \hat{P}_{rt}^{\mathrm{R,A}}\right], \quad \forall r \in \Omega^{\mathrm{R}}, \forall t \in \Omega^{\mathrm{T}}. \tag{4.10}$$

The so-called uncertainty budgets $\Gamma_r^{\mathrm{R}}, \forall r \in \Omega^{\mathrm{R}}$, are defined to control the conservativeness of the model. These parameters represent the maximum number of periods in which the available generation level of stochastic renewable generating unit r experiences fluctuations with respect to its average value. These uncertainty budgets take values between zero and $|\Omega^{\mathrm{T}}|$:

1. If $\Gamma_r^{\mathrm{R}} = 0$, it means that the available production level of stochastic renewable generating unit r is considered to be equal to its average level at all time periods, i.e., $p_{rt}^{\mathrm{R,A}} = \tilde{P}_{rt}^{\mathrm{R,A}}, \forall t \in \Omega^{\mathrm{T}}$. This can be seen as a risky strategy since it assumes that the available stochastic renewable production level is known.
2. If $\Gamma_r^{\mathrm{R}} = |\Omega^{\mathrm{T}}|$, it means that the available production level of stochastic renewable generating unit r can deviate from its average level at any time period of the planning horizon. This can be seen as the most conservative strategy.
3. Intermediate values indicate that the available production level of stochastic renewable generating unit r can deviate from its average level in a maximum of Γ_r^{R} periods.

Note that a polyhedral uncertainty set is considered. As shown in [10], in these cases the worst-case uncertainty realization corresponds to an extreme or vertex of the polyhedron representing the uncertainty sets. Therefore, the polyhedral uncertainty set can be equivalently represented by a binary-variable-based equivalent uncertainty set that only models the finite set of extremes or vertexes of the polyhedron and is formulated as follows:

$$u_{rt}^{\mathrm{R},-}, u_{rt}^{\mathrm{R},+} \in \{0,1\}, \quad \forall r \in \Omega^{\mathrm{R}}, \forall t \in \Omega^{\mathrm{T}}, \tag{4.11a}$$

$$p_{rt}^{\mathrm{R,A}} = \tilde{P}_{rt}^{\mathrm{R,A}} - u_{rt}^{\mathrm{R},-}\hat{P}_{rt}^{\mathrm{R,A}} + u_{rt}^{\mathrm{R},+}\hat{P}_{rt}^{\mathrm{R,A}}, \quad \forall r \in \Omega^{\mathrm{R}}, \forall t \in \Omega^{\mathrm{T}}, \tag{4.11b}$$

$$u_{rt}^{\mathrm{R},-} + u_{rt}^{\mathrm{R},+} \leq 1, \quad \forall r \in \Omega^{\mathrm{R}}, \forall t \in \Omega^{\mathrm{T}}, \tag{4.11c}$$

$$\sum_{t\in\Omega^{\mathrm{T}}} \left(u_{rt}^{\mathrm{R},-} + u_{rt}^{\mathrm{R},+}\right) \leq \Gamma_r^{\mathrm{R}}, \quad \forall r \in \Omega^{\mathrm{R}}. \tag{4.11d}$$

Binary variables $u_{rt}^{\mathrm{R},-}$ and $u_{rt}^{\mathrm{R},+}$ defined in (4.11a) are associated with the worst-case available renewable generation level. Equations (4.11b) express the available renewable generation levels in terms of the average and fluctuation levels associated with the corresponding confidence bounds. Equations (4.11c) guarantee that the available renewable generation level is not simultaneously equal to the minimum and the maximum values at every time period. Finally, constraints (4.11d) control the conservativeness of the model through the uncertainty budgets $\Gamma_r^{\mathrm{R}}, \forall r \in \Omega^{\mathrm{R}}$. Note that if for a given stochastic renewable generating unit r the uncertainty budget Γ_r^{R} is 0, then $p_{rt}^{\mathrm{R,A}} = \tilde{P}_{rt}^{\mathrm{R,A}}, \forall t \in \Omega^{\mathrm{T}}$. On the other hand, if $\Gamma_r^{\mathrm{R}} = |\Omega^{\mathrm{T}}|$, then $p_{rt}^{\mathrm{R,A}}$ can be at its lower ($p_{rt}^{\mathrm{R,A}} = \tilde{P}_{rt}^{\mathrm{R,A}} - \hat{P}_{rt}^{\mathrm{R,A}}$) or upper bound ($p_{rt}^{\mathrm{R,A}} = \tilde{P}_{rt}^{\mathrm{R,A}} + \hat{P}_{rt}^{\mathrm{R,A}}$) at any time period.

4.6.4 Feasibility of Operating Decision Variables

The feasibility of operating decision variables is modeled through set Θ as follows:

$$\Theta = \Big\{\Phi^{\mathrm{LL}}:$$

$$p_t^{\mathrm{DA}} = \sum_{c\in\Omega^{\mathrm{C}}} p_{ct}^{\mathrm{C}} + \sum_{r\in\Omega^{\mathrm{R}}} p_{rt}^{\mathrm{R}} + \sum_{s\in\Omega^{\mathrm{S}}} \left(p_{st}^{\mathrm{S,D}} - p_{st}^{\mathrm{S,C}}\right) - \sum_{d\in\Omega^{\mathrm{D}}} p_{dt}^{\mathrm{D}}, \quad \forall t \in \Omega^{\mathrm{T}}, \tag{4.12a}$$

$$\underline{P}_{dt}^{\mathrm{D}} \leq p_{dt}^{\mathrm{D}} \leq \overline{P}_{dt}^{\mathrm{D}}, \quad \forall d \in \Omega^{\mathrm{D}}, \forall t \in \Omega^{\mathrm{T}}, \tag{4.12b}$$

$$\sum_{t\in\Omega^{\mathrm{T}}} p_{dt}^{\mathrm{D}} \Delta t \geq \underline{E}_d^{\mathrm{D}}, \quad \forall d \in \Omega^{\mathrm{D}}, \tag{4.12c}$$

$$\overline{P}_{ct}^{\mathrm{C}} u_{ct}^{\mathrm{C}} \leq p_{ct}^{\mathrm{C}} \leq \overline{P}_{ct}^{\mathrm{C}} u_{ct}^{\mathrm{C}}, \quad \forall c \in \Omega^{\mathrm{C}}, \forall t \in \Omega^{\mathrm{T}}, \tag{4.12d}$$

$$0 \leq p_{rt}^{\mathrm{R}} \leq p_{rt}^{\mathrm{R,A}}, \quad \forall r \in \Omega^{\mathrm{R}}, \forall t \in \Omega^{\mathrm{T}}, \tag{4.12e}$$

$$0 \leq p_{st}^{\mathrm{S,C}} \leq \overline{P}_{st}^{\mathrm{S,C}}, \quad \forall s \in \Omega^{\mathrm{S}}, \forall t \in \Omega^{\mathrm{T}}, \tag{4.12f}$$

$$0 \leq p_{st}^{\mathrm{S,D}} \leq \overline{P}_{st}^{\mathrm{S,D}}, \quad \forall s \in \Omega^{\mathrm{S}}, \forall t \in \Omega^{\mathrm{T}}, \tag{4.12g}$$

$$e_{st}^{\mathrm{S}} = e_{s(t-1)}^{\mathrm{S}} + p_{st}^{\mathrm{S,C}} \Delta t \eta_s^{\mathrm{S,C}} - \frac{p_{st}^{\mathrm{S,D}} \Delta t}{\eta_s^{\mathrm{S,D}}}, \quad \forall s \in \Omega^{\mathrm{S}}, \forall t \in \Omega^{\mathrm{T}}, \tag{4.12h}$$

$$\underline{E}_{st}^{\mathrm{S}} \leq e_{st}^{\mathrm{S}} \leq \overline{E}_{st}^{\mathrm{S}}, \quad \forall s \in \Omega^{\mathrm{S}}, \forall t \in \Omega^{\mathrm{T}}\Big\}. \tag{4.12i}$$

Constraints (4.12a) define the power balance in the VPP. Constraints (4.12b) impose bounds on the power consumption of the demands, while constraints (4.12c) guarantee a minimum energy consumption throughout the planning horizon. Constraints (4.12d) and (4.12e) impose bounds on the production of the conventional power plants and the stochastic renewable generating units, respectively. Constraints (4.12f)–(4.12i) define the working of the storage facilities. Constraints (4.12f) and (4.12g) impose bounds on the charging and discharging power levels, respectively. Constraints (4.12h) define the energy level evolution of the storage units, while constraints (4.12i) impose bounds on the energy levels.

Note that the feasibility set Θ is parameterized in terms of upper-level decision variables in set Φ^{UL}ML and middle-level decision variables in set Φ^{ML}.

4.6.5 Detailed Formulation

For the sake of clarity, the detailed formulation of the adaptive robust optimization model for the scheduling problem of a VPP in the DA energy market is provided below:

$$\max_{\Phi^{\mathrm{UL}}} \min_{\Phi^{\mathrm{ML}}} \max_{\Phi^{\mathrm{LL}}} \sum_{t\in\Omega^{\mathrm{T}}} \left[\lambda_t^{\mathrm{DA}} p_t^{\mathrm{DA}} \Delta t - \sum_{c\in\Omega^{\mathrm{C}}} \left(C_c^{\mathrm{C,F}} u_{ct}^{\mathrm{C}} + C_c^{\mathrm{C,V}} p_{ct}^{\mathrm{C}} \Delta t \right) \right] \tag{4.13a}$$

subject to

$$\underline{P}^{\mathrm{DA}} \le p_t^{\mathrm{DA}} \le \overline{P}^{\mathrm{DA}}, \quad \forall t \in \Omega^{\mathrm{T}}, \tag{4.13b}$$

$$u_{ct}^{\mathrm{C}} \in \{0,1\}, \quad \forall c \in \Omega^{\mathrm{C}}, \forall t \in \Omega^{\mathrm{T}}, \tag{4.13c}$$

subject to

$$u_{rt}^{\mathrm{R},-}, u_{rt}^{\mathrm{R},+} \in \{0,1\}, \quad \forall r \in \Omega^{\mathrm{R}}, \forall t \in \Omega^{\mathrm{T}}, \tag{4.13d}$$

$$p_{rt}^{\mathrm{R,A}} = \tilde{P}_{rt}^{\mathrm{R,A}} - u_{rt}^{\mathrm{R},-} \hat{P}_{rt}^{\mathrm{R,A}} + u_{rt}^{\mathrm{R},+} \hat{P}_{rt}^{\mathrm{R,A}}, \quad \forall r \in \Omega^{\mathrm{R}}, \forall t \in \Omega^{\mathrm{T}}, \tag{4.13e}$$

$$u_{rt}^{\mathrm{R},-} + u_{rt}^{\mathrm{R},+} \le 1, \quad \forall r \in \Omega^{\mathrm{R}}, \forall t \in \Omega^{\mathrm{T}}, \tag{4.13f}$$

$$\sum_{t\in\Omega^{\mathrm{T}}} \left(u_{rt}^{\mathrm{R},-} + u_{rt}^{\mathrm{R},+} \right) \le \Gamma_r^{\mathrm{R}}, \quad \forall r \in \Omega^{\mathrm{R}}, \tag{4.13g}$$

subject to

$$p_t^{\mathrm{DA}} = \sum_{c\in\Omega^{\mathrm{C}}} p_{ct}^{\mathrm{C}} + \sum_{r\in\Omega^{\mathrm{R}}} p_{rt}^{\mathrm{R}} + \sum_{s\in\Omega^{\mathrm{S}}} \left(p_{st}^{\mathrm{S,D}} - p_{st}^{\mathrm{S,C}} \right) - \sum_{d\in\Omega^{\mathrm{D}}} p_{dt}^{\mathrm{D}}, \quad \forall t \in \Omega^{\mathrm{T}}, \tag{4.13h}$$

$$\underline{P}_{dt}^{\mathrm{D}} \le p_{dt}^{\mathrm{D}} \le \overline{P}_{dt}^{\mathrm{D}}, \quad \forall d \in \Omega^{\mathrm{D}}, \forall t \in \Omega^{\mathrm{T}}, \tag{4.13i}$$

$$\sum_{t\in\Omega^{\mathrm{T}}} p_{dt}^{\mathrm{D}} \Delta t \ge \underline{E}_d^{\mathrm{D}}, \quad \forall d \in \Omega^{\mathrm{D}}, \tag{4.13j}$$

$$\overline{P}_c^{\mathrm{C}} u_{ct}^{\mathrm{C}} \le p_{ct}^{\mathrm{C}} \le \overline{P}_c^{\mathrm{C}} u_{ct}^{\mathrm{C}}, \quad \forall c \in \Omega^{\mathrm{C}}, \forall t \in \Omega^{\mathrm{T}}, \tag{4.13k}$$

$$0 \le p_{rt}^{\mathrm{R}} \le p_{rt}^{\mathrm{R,A}}, \quad \forall r \in \Omega^{\mathrm{R}}, \forall t \in \Omega^{\mathrm{T}}, \tag{4.13l}$$

$$0 \le p_{st}^{\mathrm{S,C}} \le \overline{P}_{st}^{\mathrm{S,C}}, \quad \forall s \in \Omega^{\mathrm{S}}, \forall t \in \Omega^{\mathrm{T}}, \tag{4.13m}$$

$$0 \le p_{st}^{\mathrm{S,D}} \le \overline{P}_{st}^{\mathrm{S,D}}, \quad \forall s \in \Omega^{\mathrm{S}}, \forall t \in \Omega^{\mathrm{T}}, \tag{4.13n}$$

$$e_{st}^{\mathrm{S}} = e_{s(t-1)}^{\mathrm{S}} + p_{st}^{\mathrm{S,C}} \Delta t \eta_s^{\mathrm{S,C}} - \frac{p_{st}^{\mathrm{S,D}} \Delta t}{\eta_s^{\mathrm{S,D}}}, \quad \forall s \in \Omega^{\mathrm{S}}, \forall t \in \Omega^{\mathrm{T}}, \tag{4.13o}$$

$$\underline{E}_{st}^{\mathrm{S}} \le e_{st}^{\mathrm{S}} \le \overline{E}_{st}^{\mathrm{S}}, \quad \forall s \in \Omega^{\mathrm{S}}, \forall t \in \Omega^{\mathrm{T}}. \tag{4.13p}$$

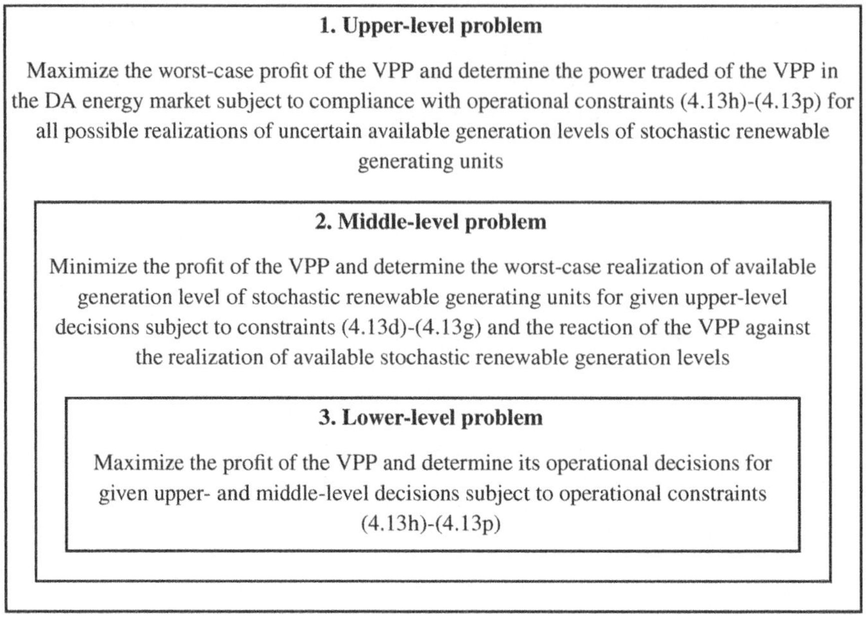

Fig. 4.11 Nested structure of the three-level optimization problem (4.13)

Problem (4.13) aims at determining the scheduling decisions of the VPP that maximize its worst-case profit (4.13a). The upper-level constraints (4.13b)–(4.13c) impose conditions on the power traded in the DA energy market and the on/off status of the conventional power plants. The middle-level constraints (4.13d)–(4.13g) define the uncertainty set. Finally, the lower-level constraints (4.13h)–(4.13p) describe the operation feasibility set.

Figure 4.11 clarifies the nested structure of the three-level optimization problem (4.13).

4.6.6 Solution Procedure

The three-level optimization problem (4.13) can be solved using a column-and-constraint generation algorithm [15]. This algorithm is based on the iterative solution of a master problem and a subproblem that exchange information on primal decision variables, and guarantees finite convergence to the global optimum. The formulation of the master problem and subproblem, as well as the algorithm are described in the following sections.

4.6.6.1 Master Problem

The master problem at iteration ν is formulated as follows:

$$\max_{\Phi^{\mathrm{M}}} \sum_{t \in \Omega^{\mathrm{T}}} \left(\lambda_t^{\mathrm{DA}} p_t^{\mathrm{DA}} \Delta t - \sum_{c \in \Omega^{\mathrm{C}}} C_c^{\mathrm{C,F}} u_{ct}^{\mathrm{C}} \right) + \theta \tag{4.14a}$$

subject to

$$\underline{P}^{\mathrm{DA}} \leq p_t^{\mathrm{DA}} \leq \overline{P}^{\mathrm{DA}}, \quad \forall t \in \Omega^{\mathrm{T}}, \tag{4.14b}$$

$$u_{ct}^{\mathrm{C}} \in \{0, 1\}, \quad \forall c \in \Omega^{\mathrm{C}}, \forall t \in \Omega^{\mathrm{T}}, \tag{4.14c}$$

$$\theta \leq - \sum_{t \in \Omega^{\mathrm{T}}} \sum_{c \in \Omega^{\mathrm{C}}} C_c^{\mathrm{C,V}} p_{ct\nu'}^{\mathrm{C}} \Delta t, \quad \nu' = 1, \dots, \nu, \tag{4.14d}$$

$$p_t^{\mathrm{DA}} = \sum_{c \in \Omega^{\mathrm{C}}} p_{ctv'}^{\mathrm{C}} + \sum_{r \in \Omega^{\mathrm{R}}} p_{rtv'}^{\mathrm{R}} + \sum_{s \in \Omega^{\mathrm{S}}} \left(p_{stv'}^{\mathrm{S,D}} - p_{stv'}^{\mathrm{S,C}} \right)$$

$$- \sum_{d \in \Omega^{\mathrm{D}}} p_{dtv'}^{\mathrm{D}}, \quad \forall t \in \Omega^{\mathrm{T}}, v' = 1, \ldots, v, \tag{4.14e}$$

$$\underline{P}_{dt}^{\mathrm{D}} \le p_{dtv'}^{\mathrm{D}} \le \overline{P}_{dt}^{\mathrm{D}}, \quad \forall d \in \Omega^{\mathrm{D}}, \forall t \in \Omega^{\mathrm{T}}, v' = 1, \ldots, v, \tag{4.14f}$$

$$\sum_{t \in \Omega^{\mathrm{T}}} p_{dtv'}^{\mathrm{D}} \Delta t \ge \underline{E}_d^{\mathrm{D}}, \quad \forall d \in \Omega^{\mathrm{D}}, v' = 1, \ldots, v, \tag{4.14g}$$

$$\underline{P}_c^{\mathrm{C}} u_{ct}^{\mathrm{C}} \le p_{ctv'}^{\mathrm{C}} \le \overline{P}_c^{\mathrm{C}} u_{ct}^{\mathrm{C}}, \quad \forall c \in \Omega^{\mathrm{C}}, \forall t \in \Omega^{\mathrm{T}}, v' = 1, \ldots, v, \tag{4.14h}$$

$$0 \le p_{rtv'}^{\mathrm{R}} \le p_{rt}^{\mathrm{R,A}(v')}, \quad \forall r \in \Omega^{\mathrm{R}}, \forall t \in \Omega^{\mathrm{T}}, v' = 1, \ldots, v, \tag{4.14i}$$

$$0 \le p_{stv'}^{\mathrm{S,C}} \le \overline{P}_{st}^{\mathrm{S,C}}, \quad \forall s \in \Omega^{\mathrm{S}}, \forall t \in \Omega^{\mathrm{T}}, v' = 1, \ldots, v, \tag{4.14j}$$

$$0 \le p_{stv'}^{\mathrm{S,D}} \le \overline{P}_{st}^{\mathrm{S,D}}, \quad \forall s \in \Omega^{\mathrm{S}}, \forall t \in \Omega^{\mathrm{T}}, v' = 1, \ldots, v, \tag{4.14k}$$

$$e_{stv'}^{\mathrm{S}} = e_{s(t-1)v'}^{\mathrm{S}} + p_{stv'}^{\mathrm{S,C}} \Delta t \eta_s^{\mathrm{S,C}} - \frac{p_{stv'}^{\mathrm{S,D}} \Delta t}{\eta_s^{\mathrm{S,D}}}, \quad \forall s \in \Omega^{\mathrm{S}}, \forall t \in \Omega^{\mathrm{T}}, v' = 1, \ldots, v, \tag{4.14l}$$

$$\underline{E}_{st}^{\mathrm{S}} \le e_{stv'}^{\mathrm{S}} \le \overline{E}_{st}^{\mathrm{S}}, \quad \forall s \in \Omega^{\mathrm{S}}, \forall t \in \Omega^{\mathrm{T}}, v' = 1, \ldots, v, \tag{4.14m}$$

where variables in set $\Phi^{\mathrm{M}} = \Big\{ \Phi^{\mathrm{UL}}; \theta; p_{dtv'}^{\mathrm{D}}, \forall d \in \Omega^{\mathrm{D}}, \forall t \in \Omega^{\mathrm{T}}, v' = 1, \ldots, v; p_{ctv'}^{\mathrm{C}}, \forall c \in \Omega^{\mathrm{C}}, \forall t \in \Omega^{\mathrm{T}}, v' = 1, \ldots, v;$ $p_{rtv'}^{\mathrm{R}}, \forall r \in \Omega^{\mathrm{R}}, \forall t \in \Omega^{\mathrm{T}}, v' = 1, \ldots, v; e_{stv'}^{\mathrm{S}}, p_{stv'}^{\mathrm{S,C}}, p_{st}^{\mathrm{S,D}}, \forall s \in \Omega^{\mathrm{RS}}, \forall t \in \Omega^{\mathrm{RT}}, v' = 1, \ldots, v \Big\}$ are the optimization variables of problem (4.14).

Master problem (4.14) is a relaxed version of the three-level optimization problem (4.13) wherein the auxiliary variable θ iteratively approximates the worst-case of the middle-level objective function. Thus, the size of master problem (4.14) increases with the iteration counter v since a new set of constraints (4.14d)–(4.14m) are included. Note that if $v = 0$, then constraints (4.14d)–(4.14m) are not included. Finally, observe that parameters $p_{rt}^{\mathrm{R,A}(v')}$, $\forall r \in \Omega^{\mathrm{R}}$, $\forall t \in \Omega^{\mathrm{T}}$, denote the optimal values for $p_{rt}^{\mathrm{R,A}}$, $\forall r \in \Omega^{\mathrm{R}}, \forall t \in \Omega^{\mathrm{T}}$, obtained from the solution of the subproblem at iteration v'.

4.6.6.2 Subproblem

The subproblem is a bi-level problem that corresponds to the two lowermost optimization levels of the original three-level problem (4.13) and is parameterized in terms of $p_t^{\mathrm{DA}(v)}$, $\forall t \in \Omega^{\mathrm{T}}$, and $u_{ct}^{\mathrm{C}(v)}$, $\forall c \in \Omega^{\mathrm{C}}, \forall t \in \Omega^{\mathrm{T}}$, which are obtained from the solution of the master problem (4.14) at iteration v. The formulation of the subproblem is provided below:

$$\min_{\Phi^{\mathrm{ML}}} \max_{\Phi^{\mathrm{LL}'}} - \sum_{t \in \Omega^{\mathrm{T}}} \left[\sum_{c \in \Omega^{\mathrm{C}}} C_c^{\mathrm{C,V}} p_{ct}^{\mathrm{C}} \Delta t + M \left(h_t^+ + h_t^- \right) \right] \tag{4.15a}$$

subject to

$$u_{rt}^{\mathrm{R},-}, u_{rt}^{\mathrm{R},+} \in \{0, 1\}, \quad \forall r \in \Omega^{\mathrm{R}}, \forall t \in \Omega^{\mathrm{T}}, \tag{4.15b}$$

$$p_{rt}^{\mathrm{R,A}} = \tilde{P}_{rt}^{\mathrm{R,A}} - u_{rt}^{\mathrm{R},-} \hat{P}_{rt}^{\mathrm{R,A}} + u_{rt}^{\mathrm{R},+} \hat{P}_{rt}^{\mathrm{R,A}}, \quad \forall r \in \Omega^{\mathrm{R}}, \forall t \in \Omega^{\mathrm{T}}, \tag{4.15c}$$

$$u_{rt}^{\mathrm{R},-} + u_{rt}^{\mathrm{R},+} \le 1, \quad \forall r \in \Omega^{\mathrm{R}}, \forall t \in \Omega^{\mathrm{T}}, \tag{4.15d}$$

$$\sum_{t \in \Omega^{\mathrm{T}}} \left(u_{rt}^{\mathrm{R},-} + u_{rt}^{\mathrm{R},+} \right) \le \Gamma_r^{\mathrm{R}}, \quad \forall r \in \Omega^{\mathrm{R}}, \tag{4.15e}$$

subject to

$$p_t^{\mathrm{DA}(\nu)} = \sum_{c\in\Omega^{\mathrm{C}}} p_{ct}^{\mathrm{C}} + \sum_{r\in\Omega^{\mathrm{R}}} p_{rt}^{\mathrm{R}} + \sum_{s\in\Omega^{\mathrm{S}}} \left(p_{st}^{\mathrm{S,D}} - p_{st}^{\mathrm{S,C}}\right) - \sum_{d\in\Omega^{\mathrm{D}}} p_{dt}^{\mathrm{D}} + h_t^+ - h_t^- \ : \lambda_t, \quad \forall t \in \Omega^{\mathrm{T}}, \tag{4.15f}$$

$$\underline{P}_{dt}^{\mathrm{D}} \le p_{dt}^{\mathrm{D}} \le \overline{P}_{dt}^{\mathrm{D}} \ : \underline{\mu}_{dt}^{\mathrm{D}}, \overline{\mu}_{dt}^{\mathrm{D}}, \quad \forall d \in \Omega^{\mathrm{D}}, \forall t \in \Omega^{\mathrm{T}}, \tag{4.15g}$$

$$\sum_{t\in\Omega^{\mathrm{T}}} p_{dt}^{\mathrm{D}} \Delta t \ge \underline{E}_d^{\mathrm{D}} \ : \xi_d, \quad \forall d \in \Omega^{\mathrm{D}}, \tag{4.15h}$$

$$\underline{P}_c^{\mathrm{C}} u_{ct}^{\mathrm{C}(\nu)} \le p_{ct}^{\mathrm{C}} \le \overline{P}_c^{\mathrm{C}} u_{ct}^{\mathrm{C}(\nu)} \ : \underline{\mu}_{ct}^{\mathrm{C}}, \overline{\mu}_{ct}^{\mathrm{C}}, \quad \forall c \in \Omega^{\mathrm{C}}, \forall t \in \Omega^{\mathrm{T}}, \tag{4.15i}$$

$$0 \le p_{rt}^{\mathrm{R}} \le p_{rt}^{\mathrm{R,A}} \ : \overline{\mu}_{rt}^{\mathrm{R}}, \quad \forall r \in \Omega^{\mathrm{R}}, \forall t \in \Omega^{\mathrm{T}}, \tag{4.15j}$$

$$0 \le p_{st}^{\mathrm{S,C}} \le \overline{P}_{st}^{\mathrm{S,C}} \ : \overline{\mu}_{st}^{\mathrm{S,C}}, \quad \forall s \in \Omega^{\mathrm{S}}, \forall t \in \Omega^{\mathrm{T}}, \tag{4.15k}$$

$$0 \le p_{st}^{\mathrm{S,D}} \le \overline{P}_{st}^{\mathrm{S,D}} \ : \overline{\mu}_{st}^{\mathrm{S,D}}, \quad \forall s \in \Omega^{\mathrm{S}}, \forall t \in \Omega^{\mathrm{T}}, \tag{4.15l}$$

$$e_{st}^{\mathrm{S}} = e_{s(t-1)}^{\mathrm{S}} + p_{st}^{\mathrm{S,C}} \Delta t \eta_s^{\mathrm{S,C}} - \frac{p_{st}^{\mathrm{S,D}} \Delta t}{\eta_s^{\mathrm{S,D}}} \ : \phi_{st}^{\mathrm{S}}, \quad \forall s \in \Omega^{\mathrm{S}}, \forall t \in \Omega^{\mathrm{T}}, \tag{4.15m}$$

$$\underline{E}_{st}^{\mathrm{S}} \le e_{st}^{\mathrm{S}} \le \overline{E}_{st}^{\mathrm{S}} \ : \underline{\mu}_{st}^{\mathrm{S}}, \overline{\mu}_{st}^{\mathrm{S}}, \quad \forall s \in \Omega^{\mathrm{S}}, \forall t \in \Omega^{\mathrm{T}}, \tag{4.15n}$$

$$h_t^+, h_t^- \ge 0, \quad \forall t \in \Omega^{\mathrm{T}}, \tag{4.15o}$$

where $\Phi^{\mathrm{LL}'} = \left\{\Phi^{\mathrm{LL}}; h_t^+, h_t^-, \forall t \in \Omega^{\mathrm{T}}\right\}$ whereas the lower-level dual variables are provided following a colon.

At each iteration ν, subproblem (4.15) determines the worst-case available generation levels of the stochastic renewable generating units, i.e., those levels that minimize the profit of the VPP for a given upper-level decision vector provided by the solution of the master problem (4.14). Constraints (4.15b)–(4.15e) represent the uncertainty set while constraints (4.15f)–(4.15o) represent the feasibility set. In order to guarantee the feasibility of the subproblem along the iterative process, power balance constraints (4.15f) are relaxed with nonnegative slack variables h_t^+ and h_t^-, $\forall t \in \Omega^{\mathrm{T}}$, that are penalized in the objective function (4.15a) using a sufficiently large positive constant M.

Subproblem (4.15) is a bi-level programming problem whose lower-level problem is continuous and linear on its decision variables. Such a property allows using the duality theory of linear programming to convert bi-level problem (4.15) into an equivalent single-level problem [4]. To do so:

1. The lower-level problem in (4.15) is replaced by its dual constraints.
2. The objective function (4.15a) is replaced by the dual-lower-level objective function.

Therefore, subproblem (4.15) is recast as the following single-level problem:

$$\begin{aligned}\min_{\Phi^{\mathrm{S}}} - \Bigg[\sum_{t\in\Omega^{\mathrm{T}}} \Bigg[\lambda_t p_t^{\mathrm{DA}(\nu)} &+ \sum_{d\in\Omega^{\mathrm{D}}} \left(\underline{\mu}_{dt}^{\mathrm{D}} \underline{P}_{dt}^{\mathrm{D}} - \overline{\mu}_{dt}^{\mathrm{D}} \overline{P}_{dt}^{\mathrm{D}}\right) + \sum_{c\in\Omega^{\mathrm{C}}} \Big(\underline{\mu}_{ct}^{\mathrm{C}} \underline{P}_c^{\mathrm{C}} u_{ct}^{\mathrm{C}(\nu)} \\ &- \overline{\mu}_{ct}^{\mathrm{C}} \overline{P}_c^{\mathrm{C}} u_{ct}^{\mathrm{C}(\nu)}\Big) - \sum_{r\in\Omega^{\mathrm{R}}} \overline{\mu}_{rt}^{\mathrm{R}} p_{rt}^{\mathrm{R,A}} - \sum_{s\in\Omega^{\mathrm{S}}} \Big(\overline{\mu}_{st}^{\mathrm{S,C}} \overline{P}_s^{\mathrm{S,C}} + \overline{\mu}_{st}^{\mathrm{S,D}} \overline{P}_s^{\mathrm{S,D}} \\ &- \underline{\mu}_{st}^{\mathrm{S}} \underline{E}_{st}^{\mathrm{S}} + \overline{\mu}_{st}^{\mathrm{S}} \overline{E}_{st}^{\mathrm{S}}\Big) \Bigg] + \sum_{d\in\Omega^{\mathrm{D}}} \xi_d \underline{E}_d^{\mathrm{D}} + \sum_{s\in\Omega^{\mathrm{S}}} \phi_{s1}^{\mathrm{S}} E_{s0}^{\mathrm{S}} \Bigg]\end{aligned} \tag{4.16a}$$

subject to

$$u_{rt}^{\mathrm{R},-}, u_{rt}^{\mathrm{R},+} \in \{0, 1\}, \quad \forall r \in \Omega^{\mathrm{R}}, \forall t \in \Omega^{\mathrm{T}}, \tag{4.16b}$$

$$p_{rt}^{\mathrm{R,A}} = \tilde{P}_{rt}^{\mathrm{R,A}} - u_{rt}^{\mathrm{R},-} \hat{P}_{rt}^{\mathrm{R,A}} + u_{rt}^{\mathrm{R},+} \hat{P}_{rt}^{\mathrm{R,A}}, \quad \forall r \in \Omega^{\mathrm{R}}, \forall t \in \Omega^{\mathrm{T}}, \tag{4.16c}$$

$$u_{rt}^{\mathrm{R},-} + u_{rt}^{\mathrm{R},+} \leq 1, \quad \forall r \in \Omega^{\mathrm{R}}, \forall t \in \Omega^{\mathrm{T}}, \tag{4.16d}$$

$$\sum_{t \in \Omega^{\mathrm{T}}} \left(u_{rt}^{\mathrm{R},-} + u_{rt}^{\mathrm{R},+} \right) \leq \Gamma_r^{\mathrm{R}}, \quad \forall r \in \Omega^{\mathrm{R}}, \tag{4.16e}$$

$$\lambda_t + \underline{\mu}_{ct}^{\mathrm{C}} - \overline{\mu}_{ct}^{\mathrm{C}} = C_c^{\mathrm{C,V}}, \quad \forall c \in \Omega^{\mathrm{C}}, \forall t \in \Omega^{\mathrm{T}}, \tag{4.16f}$$

$$\lambda_t - \overline{\mu}_{rt}^{\mathrm{R}} \leq 0, \quad \forall r \in \Omega^{\mathrm{R}}, \forall t \in \Omega^{\mathrm{T}}, \tag{4.16g}$$

$$\lambda_t - \overline{\mu}_{st}^{\mathrm{S,D}} + \frac{\Delta t}{\eta_s^{\mathrm{S,D}}} \phi_{st}^{\mathrm{S}} \leq 0, \quad \forall s \in \Omega^{\mathrm{S}}, \forall t \in \Omega^{\mathrm{T}}, \tag{4.16h}$$

$$-\lambda_t - \overline{\mu}_{st}^{\mathrm{S,C}} + \eta_s^{\mathrm{S,C}} \Delta t \phi_{st}^{\mathrm{S}} \leq 0, \quad \forall s \in \Omega^{\mathrm{S}}, \forall t \in \Omega^{\mathrm{T}}, \tag{4.16i}$$

$$-\lambda_t + \underline{\mu}_{dt}^{\mathrm{D}} - \overline{\mu}_{dt}^{\mathrm{D}} + \Delta t \xi_d = 0, \quad \forall d \in \Omega^{\mathrm{D}}, \forall t \in \Omega^{\mathrm{T}}, \tag{4.16j}$$

$$\lambda_t \leq M, \quad \forall t \in \Omega^{\mathrm{T}}, \tag{4.16k}$$

$$-\lambda_t \leq M, \quad \forall t \in \Omega^{\mathrm{T}}, \tag{4.16l}$$

$$\phi_{st}^{\mathrm{S}} - \phi_{st+1}^{\mathrm{S}} + \underline{\mu}_{st}^{\mathrm{S}} - \overline{\mu}_{st}^{\mathrm{S}} = 0, \quad \forall s \in \Omega^{\mathrm{S}}, t = 1, \ldots |\Omega^{\mathrm{T}}| - 1, \tag{4.16m}$$

$$\phi_{s|\Omega^{\mathrm{T}}|}^{\mathrm{S}} + \underline{\mu}_{s|\Omega^{\mathrm{T}}|}^{\mathrm{S}} - \overline{\mu}_{s|\Omega^{\mathrm{T}}|}^{\mathrm{S}} = 0, \quad \forall s \in \Omega^{\mathrm{S}}, \tag{4.16n}$$

$$\underline{\mu}_{ct}^{\mathrm{C}}, \overline{\mu}_{ct}^{\mathrm{C}} \geq 0, \quad \forall c \in \Omega^{\mathrm{C}}, \forall t \in \Omega^{\mathrm{T}}, \tag{4.16o}$$

$$\overline{\mu}_{rt}^{\mathrm{R}} \geq 0, \quad \forall r \in \Omega^{\mathrm{R}}, \forall t \in \Omega^{\mathrm{T}}, \tag{4.16p}$$

$$\overline{\mu}_{st}^{\mathrm{S,C}}, \overline{\mu}_{st}^{\mathrm{S,D}}, \underline{\mu}_{st}^{\mathrm{S}}, \overline{\mu}_{st}^{\mathrm{S}} \geq 0, \quad \forall s \in \Omega^{\mathrm{S}}, \forall t \in \Omega^{\mathrm{T}}, \tag{4.16q}$$

$$\underline{\mu}_{dt}^{\mathrm{D}}, \overline{\mu}_{dt}^{\mathrm{D}} \geq 0, \quad \forall d \in \Omega^{\mathrm{D}}, \forall t \in \Omega^{\mathrm{T}}, \tag{4.16r}$$

where variables in set $\Phi^{\mathrm{S}} = \Big\{ p_{rt}^{\mathrm{R,A}}, u_{rt}^{\mathrm{R},-}, u_{rt}^{\mathrm{R},+}, \forall r \in \Omega^{\mathrm{R}}, \forall t \in \Omega^{\mathrm{T}}; \lambda_t, \forall t \in \Omega^{\mathrm{T}}; \underline{\mu}_{ct}^{\mathrm{C}}, \overline{\mu}_{ct}^{\mathrm{C}}, \forall c \in \Omega^{\mathrm{C}}, \forall t \in \Omega^{\mathrm{T}}; \overline{\mu}_{rt}^{\mathrm{R}}, \forall r \in \Omega^{\mathrm{R}}, \forall t \in \Omega^{\mathrm{T}}; \overline{\mu}_{st}^{\mathrm{S,C}}, \overline{\mu}_{st}^{\mathrm{S,D}}, \phi_{st}^{\mathrm{S}}, \underline{\mu}_{st}^{\mathrm{S}}, \overline{\mu}_{st}^{\mathrm{S}}, \forall s \in \Omega^{\mathrm{S}}, \forall t \in \Omega^{\mathrm{T}}; \underline{\mu}_{dt}^{\mathrm{D}}, \overline{\mu}_{dt}^{\mathrm{D}}, \xi_d, \forall d \in \Omega^{\mathrm{D}}, \forall t \in \Omega^{\mathrm{T}} \Big\}$ are the optimization variables of problem (4.16). The dual lower-level objective function is minimized in (4.16a). Constraints (4.16b)–(4.16e) are associated with the uncertainty set while constraints (4.16f)–(4.16r) are the lower-level dual feasibility constraints.

Note that problem (4.16) includes some nonlinear terms in the objective function (4.16a), namely, terms $\overline{\mu}_{rt}^{\mathrm{R}} p_{rt}^{\mathrm{R,A}}$, $\forall r \in \Omega^{\mathrm{R}}, \forall t \in \Omega^{\mathrm{T}}$. These bilinear terms comprising the product of two continuous variables can be replaced by exact equivalent linear expressions as explained below. First, using constraint (4.16c):

$$\begin{aligned} \overline{\mu}_{rt}^{\mathrm{R}} p_{rt}^{\mathrm{R,A}} &= \overline{\mu}_{rt}^{\mathrm{R}} \left(\tilde{P}_{rt}^{\mathrm{R,A}} - u_{rt}^{\mathrm{R},-} \hat{P}_{rt}^{\mathrm{R,A}} + u_{rt}^{\mathrm{R},+} \hat{P}_{rt}^{\mathrm{R,A}} \right) \\ &= \overline{\mu}_{rt}^{\mathrm{R}} \tilde{P}_{rt}^{\mathrm{R,A}} - \overline{\mu}_{rt}^{\mathrm{R}} u_{rt}^{\mathrm{R},-} \hat{P}_{rt}^{\mathrm{R,A}} + \overline{\mu}_{rt}^{\mathrm{R}} u_{rt}^{\mathrm{R},+} \hat{P}_{rt}^{\mathrm{R,A}}, \quad \forall r \in \Omega^{\mathrm{R}}, \forall t \in \Omega^{\mathrm{T}}. \end{aligned} \tag{4.17}$$

Note that in Eq. (4.17) term $\overline{\mu}_{rt}^{\mathrm{R}} \tilde{P}_{rt}^{\mathrm{R,A}}$ is linear while terms $\overline{\mu}_{rt}^{\mathrm{R}} u_{rt}^{\mathrm{R},-} \hat{P}_{rt}^{\mathrm{R,A}}$ and $\overline{\mu}_{rt}^{\mathrm{R}} u_{rt}^{\mathrm{R},+} \hat{P}_{rt}^{\mathrm{R,A}}$ are still nonlinear terms comprising the product of a continuous variable, a binary variable, and a positive constant. As explained in [7], these nonlinear terms can be replaced by the following exact mixed-integer linear equations:

$$\overline{\mu}_{rt}^{\mathrm{R}} p_{rt}^{\mathrm{R,A}} = \overline{\mu}_{rt}^{\mathrm{R}} \tilde{P}_{rt}^{\mathrm{R,A}} - z_{rt}^{\mathrm{R},-} \hat{P}_{rt}^{\mathrm{R,A}} + z_{rt}^{\mathrm{R},+} \hat{P}_{rt}^{\mathrm{R,A}}, \quad \forall r \in \Omega^{\mathrm{R}}, \forall t \in \Omega^{\mathrm{T}}, \tag{4.18a}$$

$$0 \leq z_{rt}^{\mathrm{R},-} \leq M u_{rt}^{\mathrm{R},-}, \quad \forall r \in \Omega^{\mathrm{R}}, \forall t \in \Omega^{\mathrm{T}}, \tag{4.18b}$$

$$0 \leq \overline{\mu}_{rt}^{\mathrm{R}} - z_{rt}^{\mathrm{R},-} \leq M \left(1 - u_{rt}^{\mathrm{R},-} \right), \quad \forall r \in \Omega^{\mathrm{R}}, \forall t \in \Omega^{\mathrm{T}}, \tag{4.18c}$$

$$0 \leq z_{rt}^{\mathrm{R},+} \leq M u_{rt}^{\mathrm{R},+}, \quad \forall r \in \Omega^{\mathrm{R}}, \forall t \in \Omega^{\mathrm{T}}, \tag{4.18d}$$

$$0 \leq \overline{\mu}_{rt}^{\mathrm{R}} - z_{rt}^{\mathrm{R},+} \leq M \left(1 - u_{rt}^{\mathrm{R},+} \right), \quad \forall r \in \Omega^{\mathrm{R}}, \forall t \in \Omega^{\mathrm{T}}, \tag{4.18e}$$

where M is a large enough positive constant.

Therefore, using Eqs. (4.18), subproblem (4.16) is finally recast as an MILP problem [14].

Illustrative Example 4.7 Working of Eqs. (4.18)

Consider a problem with only one time period and one stochastic renewable generating unit. In such a case, nonlinear equation (4.17) is rewritten as follows:

$$\overline{\mu}^{\mathrm{R}} p^{\mathrm{R,A}} = \overline{\mu}^{\mathrm{R}} \tilde{P}^{\mathrm{R,A}} - \overline{\mu}^{\mathrm{R}} u^{\mathrm{R},-} \hat{P}^{\mathrm{R,A}} + \overline{\mu}^{\mathrm{R}} u^{\mathrm{R},+} \hat{P}^{\mathrm{R,A}},$$

where subscripts have been suppressed for the sake of clarity.

Recall that variables $u^{\mathrm{R},-}$ and $u^{\mathrm{R},+}$ are binary variables that indicate if the available generation level of the stochastic renewable unit is at its lower bound, its average level, or its upper bound. These three cases are analyzed below:

1. The available generation level of the stochastic renewable unit is at its lower bound ($\tilde{P}^{\mathrm{R,A}} - \hat{P}^{\mathrm{R,A}}$). In such a case, the values of binary variables are $u^{\mathrm{R},-} = 1$ and $u^{\mathrm{R},+} = 0$, and nonlinear equation (4.17) is recast as:

$$\overline{\mu}^{\mathrm{R}} p^{\mathrm{R,A}} = \overline{\mu}^{\mathrm{R}} \tilde{P}^{\mathrm{R,A}} - \overline{\mu}^{\mathrm{R}} \hat{P}^{\mathrm{R,A}}.$$

Instead, if the equivalent mixed-integer linear equations (4.18) are considered:

$$\begin{aligned}
&\overline{\mu}^{\mathrm{R}} p^{\mathrm{R,A}} = \overline{\mu}^{\mathrm{R}} \tilde{P}^{\mathrm{R,A}} - z^{\mathrm{R},-} \hat{P}^{\mathrm{R,A}} + z^{\mathrm{R},+} \hat{P}^{\mathrm{R,A}},\\
&0 \le z^{\mathrm{R},-} \le M,\\
&0 \le \overline{\mu}^{\mathrm{R}} - z^{\mathrm{R},-} \le 0,\\
&0 \le z^{\mathrm{R},+} \le 0,\\
&0 \le \overline{\mu}^{\mathrm{R}} - z^{\mathrm{R},+} \le M.
\end{aligned}$$

The third and fourth equations impose that $z^{\mathrm{R},-} = \overline{\mu}^{\mathrm{R}}$ and $z^{\mathrm{R},+} = 0$, respectively. Using this, both the second and fifth equations impose that $0 \le \overline{\mu}^{\mathrm{R}} \le M$. Note that M is a large enough positive constant and dual variable $\overline{\mu}^{\mathrm{R}}$ is nonnegative as per (4.16p). Thus, equation $0 \le \overline{\mu}^{\mathrm{R}} \le M$ is not bounding the solution. Finally, the first equation is recast as $\overline{\mu}^{\mathrm{R}} p^{\mathrm{R,A}} = \overline{\mu}^{\mathrm{R}} \tilde{P}^{\mathrm{R,A}} - \overline{\mu}^{\mathrm{R}} \hat{P}^{\mathrm{R,A}}$, which is identical to the one obtained using nonlinear equation (4.17).

2. The available generation level of the stochastic renewable unit is at its average level ($\tilde{P}^{\mathrm{R,A}}$). In such a case, the values of binary variables are $u^{\mathrm{R},-} = 0$ and $u^{\mathrm{R},+} = 0$, and nonlinear equation (4.17) is recast as:

$$\overline{\mu}^{\mathrm{R}} p^{\mathrm{R,A}} = \overline{\mu}^{\mathrm{R}} \tilde{P}^{\mathrm{R,A}}.$$

Instead, if the equivalent mixed-integer linear equations (4.18) are considered:

$$\begin{aligned}
&\overline{\mu}^{\mathrm{R}} p^{\mathrm{R,A}} = \overline{\mu}^{\mathrm{R}} \tilde{P}^{\mathrm{R,A}} - z^{\mathrm{R},-} \hat{P}^{\mathrm{R,A}} + z^{\mathrm{R},+} \hat{P}^{\mathrm{R,A}},\\
&0 \le z^{\mathrm{R},-} \le 0,\\
&0 \le \overline{\mu}^{\mathrm{R}} - z^{\mathrm{R},-} \le M,\\
&0 \le z^{\mathrm{R},+} \le 0,\\
&0 \le \overline{\mu}^{\mathrm{R}} - z^{\mathrm{R},+} \le M.
\end{aligned}$$

The second and fourth equations impose that $z^{\mathrm{R},-} = 0$ and $z^{\mathrm{R},+} = 0$, respectively. Using this, both the third and fifth equations impose that $0 \le \overline{\mu}^{\mathrm{R}} \le M$. Note that M is a large enough positive constant and dual variable $\overline{\mu}^{\mathrm{R}}$ is nonnegative as per (4.16p). Thus, equation $0 \le \overline{\mu}^{\mathrm{R}} \le M$ is not bounding the solution. Finally, the first equation is recast as $\overline{\mu}^{\mathrm{R}} p^{\mathrm{R,A}} = \overline{\mu}^{\mathrm{R}} \tilde{P}^{\mathrm{R,A}}$, which is identical to the one obtained using nonlinear equation (4.17).

3. The available generation level of the stochastic renewable unit is at its upper bound ($\tilde{P}^{\mathrm{R,A}} + \hat{P}^{\mathrm{R,A}}$). In such a case, the values of binary variables are $u^{\mathrm{R},-} = 0$ and $u^{\mathrm{R},+} = 1$, and nonlinear equation (4.17) is recast as:

$$\overline{\mu}^{\mathrm{R}} p^{\mathrm{R,A}} = \overline{\mu}^{\mathrm{R}} \tilde{P}^{\mathrm{R,A}} + \overline{\mu}^{\mathrm{R}} \hat{P}^{\mathrm{R,A}}.$$

Instead, if the equivalent mixed-integer linear equations (4.18) are considered:

$$\overline{\mu}^{\mathrm{R}} p^{\mathrm{R,A}} = \overline{\mu}^{\mathrm{R}} \tilde{p}^{\mathrm{R,A}} - z^{\mathrm{R,-}} \hat{p}^{\mathrm{R,A}} + z^{\mathrm{R,+}} \hat{p}^{\mathrm{R,A}},$$
$$0 \leq z^{\mathrm{R,-}} \leq 0,$$
$$0 \leq \overline{\mu}^{\mathrm{R}} - z^{\mathrm{R,-}} \leq M,$$
$$0 \leq z^{\mathrm{R,+}} \leq M,$$
$$0 \leq \overline{\mu}^{\mathrm{R}} - z^{\mathrm{R,+}} \leq 0.$$

The second and fifth equations impose that $z^{\mathrm{R,-}} = 0$ and $z^{\mathrm{R,+}} = \overline{\mu}^{\mathrm{R}}$, respectively. Using this, both the third and fourth equations impose that $0 \leq \overline{\mu}^{\mathrm{R}} \leq M$. Note that M is a large enough positive constant and dual variable $\overline{\mu}^{\mathrm{R}}$ is nonnegative as per (4.16p). Thus, equation $0 \leq \overline{\mu}^{\mathrm{R}} \leq M$ is not bounding the solution. Finally, the first equation is recast as $\overline{\mu}^{\mathrm{R}} p^{\mathrm{R,A}} = \overline{\mu}^{\mathrm{R}} \tilde{p}^{\mathrm{R,A}} + \overline{\mu}^{\mathrm{R}} \hat{p}^{\mathrm{R,A}}$, which is identical to the one obtained using nonlinear equation (4.17).

□

4.6.6.3 Solution Algorithm

The master problem (4.14) and the subproblem (4.16) are iteratively solved until convergence using the following algorithm:

1. Initialize the iteration counter ($\nu \leftarrow 0$), select the convergence tolerance (ϵ), and set the lower bound (LB) and upper bound (UB) to $-\infty$ and ∞, respectively.
2. Solve master problem (4.14) and obtain, among other variables, the optimal power traded by the VPP in the DA energy market $p_t^{\mathrm{DA}*}, \forall t \in \Omega^{\mathrm{T}}$, and the optimal status of the conventional power plants $u_{ct}^{\mathrm{C}*}, \forall c \in \Omega^{\mathrm{C}}, \forall t \in \Omega^{\mathrm{T}}$. Note that if $\nu = 0$, then constraints (4.14d)–(4.14m) are not included.
3. Update the upper bound using Eq. (4.19) below:

$$\mathrm{UB} = z^{\mathrm{M}*}, \tag{4.19}$$

   where $z^{\mathrm{M}*}$ is the optimal value of the objective function (4.14a) of the master problem.
4. Set $p_t^{\mathrm{DA}(\nu)} = p_t^{\mathrm{DA}*}, \forall t \in \Omega^{\mathrm{T}}$, and $u_{ct}^{\mathrm{C}(\nu)} = u_{ct}^{\mathrm{C}*}, \forall c \in \Omega^{\mathrm{C}}, \forall t \in \Omega^{\mathrm{T}}$, where $p_t^{\mathrm{DA}*}, \forall t \in \Omega^{\mathrm{T}}$, and $u_{ct}^{\mathrm{C}*}, \forall c \in \Omega^{\mathrm{C}}, \forall t \in \Omega^{\mathrm{T}}$, are obtained from the optimal solution of master problem (4.14) in step 2.
5. Solve subproblem (4.16) and obtain, among other variables, the worst realization of the available generation levels of the stochastic renewable generating units $p_{rt}^{\mathrm{R,A}*}, \forall r \in \Omega^{\mathrm{R}}, \forall t \in \Omega^{\mathrm{T}}$.
6. Update the lower bound using Eq. (4.20) below:

$$\mathrm{LB} = \max\left\{ \sum_{t \in \Omega^{\mathrm{T}}} \left(\lambda_t^{\mathrm{DA}} p_t^{\mathrm{DA}*} \Delta t - \sum_{c \in \Omega^{\mathrm{C}}} C_c^{\mathrm{C,F}} u_{ct}^{\mathrm{C}*} \right) + z^{\mathrm{S}*}, \mathrm{LB} \right\}, \tag{4.20}$$

   where $z^{\mathrm{S}*}$ is the optimal value of the objective function (4.16a) of the subproblem.
7. If $\mathrm{UB} - \mathrm{LB} \leq \epsilon$, the algorithm stops. The optimal power traded in the DA energy market and the optimal status of the conventional power plants are $p_t^{\mathrm{DA}*}, \forall t \in \Omega^{\mathrm{T}}$, and $u_{ct}^{\mathrm{C}*}, \forall c \in \Omega^{\mathrm{C}}, \forall t \in \Omega^{\mathrm{T}}$, respectively. Otherwise, go to step 8.
8. Update the iteration counter $\nu \leftarrow \nu + 1$.
9. Set $p_{rt}^{\mathrm{R,A}(\nu)} = p_{rt}^{\mathrm{R,A}*}, \forall r \in \Omega^{\mathrm{R}}, \forall t \in \Omega^{\mathrm{T}}$, where $p_{rt}^{\mathrm{R,A}*}, \forall r \in \Omega^{\mathrm{R}}, \forall t \in \Omega^{\mathrm{T}}$, is obtained from the optimal solution of master problem (4.16) in step 5.
10. Go to step 2.

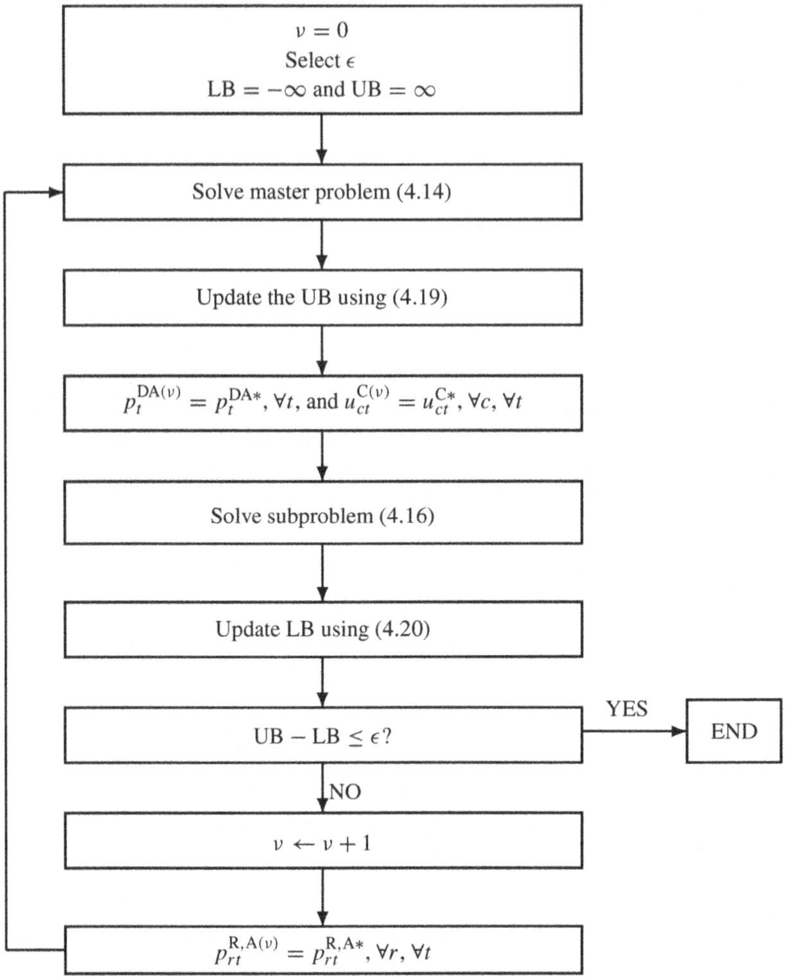

Fig. 4.12 Flowchart for the column-and-constraint generation algorithm

The algorithm flowchart is provided in Fig. 4.12 for the sake of clarity.

Illustrative Example 4.8 Adaptive Robust Self-Scheduling Problem

This examples considers the scheduling problem in the DA energy market of a VPP comprising a wind-power unit, an energy storage facility, and a flexible demand.

A 4-h planning horizon is considered. The DA energy market prices are provided in Fig. 4.13 and the maximum power traded (sold or bought) in the market is 40 MW. The average and fluctuation levels of the available wind-power generation are provided in Table 4.19. The technical data of the storage unit are provided in Table 4.20, being 20 MWh the minimum energy level in the storage unit in time period 4 to prevent this storage unit to be depleted at the end of the planning horizon. The lower and upper consumption bounds of the flexible demand are provided in Fig. 4.14, while the minimum energy consumption throughout the 4-h planning horizon is 60 MWh.

An uncertainty budget $\Gamma^{\text{R}} = 2$ is considered, which means that the available wind-power generation level is allowed to deviate from its average level in a maximum of two time periods.

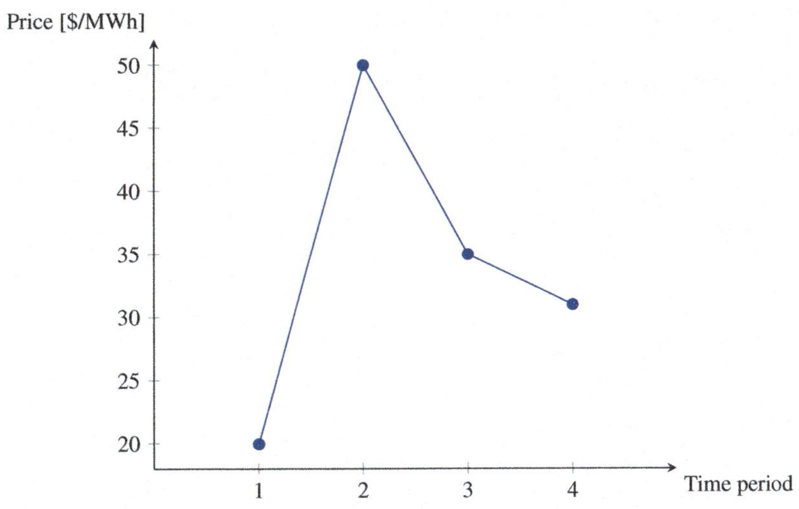

Fig. 4.13 Illustrative Example 4.8: DA energy market prices

Table 4.19 Illustrative Example 4.8: data of the wind-power unit

Time period	Average level [MW]	Fluctuation level [MW]
1	60	20
2	20	10
3	40	15
4	10	5

Table 4.20 Illustrative Example 4.8: data of the storage unit

Maximum charging power	20 MW
Maximum discharging power	20 MW
Maximum energy level	40 MWh
Minimum energy level	0
Initial energy level	20 MWh
Charging efficiency	0.9
Discharging efficiency	0.9

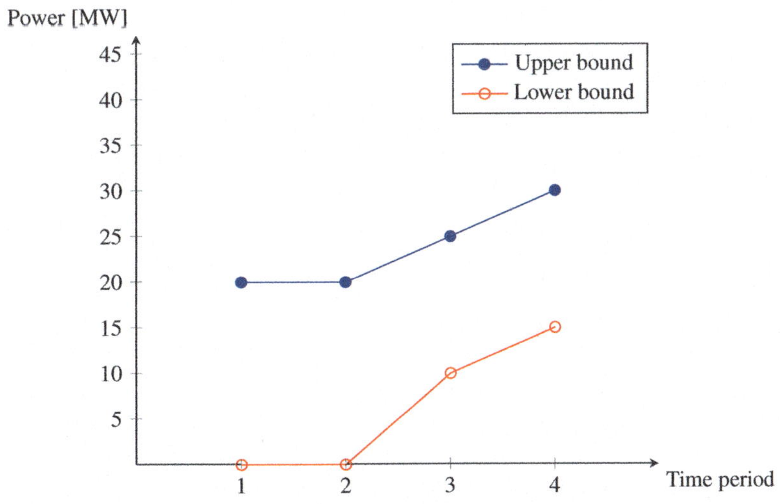

Fig. 4.14 Illustrative Example 4.8: lower and upper bounds of the power consumption

Considering these data, the problem is formulated as the following three-level optimization model:

$$\max_{\Phi^{\mathrm{UL}}} \min_{\Phi^{\mathrm{ML}}} \max_{\Phi^{\mathrm{LL}}} \; 20p_1^{\mathrm{DA}} + 50p_2^{\mathrm{DA}} + 35p_3^{\mathrm{DA}} + 31p_4^{\mathrm{DA}}$$

subject to

$$-40 \leq p_1^{\mathrm{DA}} \leq 40,$$

$$-40 \leq p_2^{\mathrm{DA}} \leq 40,$$

$$-40 \leq p_3^{\mathrm{DA}} \leq 40,$$

$$-40 \leq p_4^{\mathrm{DA}} \leq 40,$$

subject to

$$u_1^{\mathrm{R},-}, u_1^{\mathrm{R},+}, u_2^{\mathrm{R},-}, u_2^{\mathrm{R},+}, u_3^{\mathrm{R},-}, u_3^{\mathrm{R},+}, u_4^{\mathrm{R},-}, u_4^{\mathrm{R},+} \in \{0, 1\},$$

$$p_1^{\mathrm{R,A}} = 60 - 20u_1^{\mathrm{R},-} + 20u_1^{\mathrm{R},+},$$

$$p_2^{\mathrm{R,A}} = 20 - 10u_2^{\mathrm{R},-} + 10u_2^{\mathrm{R},+},$$

$$p_3^{\mathrm{R,A}} = 40 - 15u_3^{\mathrm{R},-} + 15u_3^{\mathrm{R},+},$$

$$p_4^{\mathrm{R,A}} = 10 - 5u_4^{\mathrm{R},-} + 5u_4^{\mathrm{R},+},$$

$$u_1^{\mathrm{R},-} + u_1^{\mathrm{R},+} \leq 1,$$

$$u_2^{\mathrm{R},-} + u_2^{\mathrm{R},+} \leq 1,$$

$$u_3^{\mathrm{R},-} + u_3^{\mathrm{R},+} \leq 1,$$

$$u_4^{\mathrm{R},-} + u_4^{\mathrm{R},+} \leq 1,$$

$$u_1^{\mathrm{R},-} + u_1^{\mathrm{R},+} + u_2^{\mathrm{R},-} + u_2^{\mathrm{R},+} + u_3^{\mathrm{R},-} + u_3^{\mathrm{R},+} + u_4^{\mathrm{R},-} + u_4^{\mathrm{R},+} \leq 2,$$

subject to

$$p_1^{\mathrm{DA}} = p_1^{\mathrm{R}} + p_1^{\mathrm{S,D}} - p_1^{\mathrm{S,C}} - p_1^{\mathrm{D}},$$

$$p_2^{\mathrm{DA}} = p_2^{\mathrm{R}} + p_2^{\mathrm{S,D}} - p_2^{\mathrm{S,C}} - p_2^{\mathrm{D}},$$

$$p_3^{\mathrm{DA}} = p_3^{\mathrm{R}} + p_3^{\mathrm{S,D}} - p_3^{\mathrm{S,C}} - p_3^{\mathrm{D}},$$

$$p_4^{\mathrm{DA}} = p_4^{\mathrm{R}} + p_4^{\mathrm{S,D}} - p_4^{\mathrm{S,C}} - p_4^{\mathrm{D}},$$

$$0 \leq p_1^{\mathrm{D}} \leq 20,$$

$$0 \leq p_2^{\mathrm{D}} \leq 20,$$

$$10 \leq p_3^{\mathrm{D}} \leq 25,$$

$$15 \leq p_4^{\mathrm{D}} \leq 30,$$

$$p_1^{\mathrm{D}} + p_2^{\mathrm{D}} + p_3^{\mathrm{D}} + p_4^{\mathrm{D}} \geq 60,$$

$$0 \leq p_1^{\mathrm{R}} \leq p_1^{\mathrm{R,A}},$$

$$0 \leq p_2^{\mathrm{R}} \leq p_2^{\mathrm{R,A}},$$

$$0 \leq p_3^{\mathrm{R}} \leq p_3^{\mathrm{R,A}},$$

$$0 \leq p_4^{\mathrm{R}} \leq p_4^{\mathrm{R,A}},$$

$$0 \leq p_1^{\mathrm{S,C}} \leq 20,$$

$$0 \leq p_2^{S,C} \leq 20,$$
$$0 \leq p_3^{S,C} \leq 20,$$
$$0 \leq p_4^{S,C} \leq 20,$$
$$0 \leq p_1^{S,D} \leq 20,$$
$$0 \leq p_2^{S,D} \leq 20,$$
$$0 \leq p_3^{S,D} \leq 20,$$
$$0 \leq p_4^{S,D} \leq 20,$$
$$e_1^S = 20 + 0.9p_1^{S,C} - \frac{p_1^{S,D}}{0.9},$$
$$e_2^S = e_1^S + 0.9p_2^{S,C} - \frac{p_2^{S,D}}{0.9},$$
$$e_3^S = e_2^S + 0.9p_3^{S,C} - \frac{p_3^{S,D}}{0.9},$$
$$e_4^S = e_3^S + 0.9p_4^{S,C} - \frac{p_4^{S,D}}{0.9},$$
$$0 \leq e_1^S \leq 40,$$
$$0 \leq e_2^S \leq 40,$$
$$0 \leq e_3^S \leq 40,$$
$$20 \leq e_4^S \leq 40,$$

where the optimization variables of the upper-, middle-, and lower-level problems are those included in sets $\Phi^{UL} = \left\{p_1^{DA}, p_2^{DA}, p_3^{DA}, p_4^{DA}\right\}$, $\Phi^{ML} = \left\{p_1^{R,A}, p_2^{R,A}, p_3^{R,A}, p_4^{R,A}\right\}$, and $\Phi^{LL} = \left\{p_1^{D}, p_2^{D}, p_3^{D}, p_4^{D}, p_1^{R}, p_2^{R}, p_3^{R}, p_4^{R}, e_1^{S}, e_2^{S}, e_3^{S}, e_4^{S}, p_1^{S,C}, p_2^{S,C}, p_3^{S,C}, p_4^{S,C}, p_1^{S,D}, p_2^{S,D}, p_3^{S,D}, p_4^{S,D}\right\}$, respectively. Note that there is only a single unit of each type in the VPP. Therefore, for the sake of clarity, optimization variables include just a single subscript denoting the time period.

The above three-level optimization problem is solved using the column-and-constraint generation algorithm described in Sect. 4.6.6.3 that comprises the following steps:

- Step 1. The iteration counter ($\nu \leftarrow 0$) is initialized, convergence tolerance is selected ($\epsilon = 0$), and the lower bound (LB) and upper bound (UB) are set to $-\infty$ and ∞, respectively.
- Step 2. The following master problem is solved:

$$\max_{\Phi^M} \ 20p_1^{DA} + 50p_2^{DA} + 35p_3^{DA} + 31p_4^{DA} + \theta$$

subject to

$$-40 \leq p_1^{DA} \leq 40,$$
$$-40 \leq p_2^{DA} \leq 40,$$
$$-40 \leq p_3^{DA} \leq 40,$$
$$-40 \leq p_4^{DA} \leq 40,$$

where $\Phi^M = \left\{p_1^{DA}, p_2^{DA}, p_3^{DA}, p_4^{DA}, \theta\right\}$. Note that as $\nu = 0$, constraints (4.14d)–(4.14m) are not included.

The optimal solution of this problem is:

$$p_1^{\mathrm{DA}*}, p_2^{\mathrm{DA}*}, p_3^{\mathrm{DA}*}, p_4^{\mathrm{DA}*} = 40,$$
$$\theta^* = \infty,$$

while the optimal value of the objective function is $z^{\mathrm{M}*} = \infty$.

- Step 3. The upper bound is updated using Eq. (4.19):

$$\mathrm{UB} = \infty.$$

- Step 4. Set $p_1^{\mathrm{DA}(1)} = 40$, $p_2^{\mathrm{DA}(1)} = 40$, $p_3^{\mathrm{DA}(1)} = 40$, and $p_4^{\mathrm{DA}(1)} = 40$.
- Step 5. The following subproblem is solved:

$$\begin{aligned}\min_{\Phi^{\mathrm{S}}} - \Big(& 40\lambda_1 + 40\lambda_2 + 40\lambda_3 + 40\lambda_4 + 0\underline{\mu}_1^{\mathrm{D}} + 0\underline{\mu}_2^{\mathrm{D}} + 10\underline{\mu}_3^{\mathrm{D}} + 15\underline{\mu}_4^{\mathrm{D}} \\ & - 20\overline{\mu}_1^{\mathrm{D}} - 20\overline{\mu}_2^{\mathrm{D}} - 25\overline{\mu}_3^{\mathrm{D}} - 30\overline{\mu}_4^{\mathrm{D}} - 60\overline{\mu}_1^{\mathrm{R}} + 20z_1^{\mathrm{R},-} - 20z_1^{\mathrm{R},+} \\ & - 20\overline{\mu}_2^{\mathrm{R}} + 10z_2^{\mathrm{R},-} - 10z_2^{\mathrm{R},+} - 40\overline{\mu}_3^{\mathrm{R}} + 15z_3^{\mathrm{R},-} - 15z_3^{\mathrm{R},+} - 10\overline{\mu}_4^{\mathrm{R}} \\ & + 5z_4^{\mathrm{R},-} - 5z_4^{\mathrm{R},+} - 20\overline{\mu}_1^{\mathrm{S,C}} - 20\overline{\mu}_2^{\mathrm{S,C}} - 20\overline{\mu}_3^{\mathrm{S,C}} - 20\overline{\mu}_4^{\mathrm{S,C}} - 20\overline{\mu}_1^{\mathrm{S,D}} \\ & - 20\overline{\mu}_2^{\mathrm{S,D}} - 20\overline{\mu}_3^{\mathrm{S,D}} - 20\overline{\mu}_4^{\mathrm{S,D}} + 0\underline{\mu}_1^{\mathrm{S}} + 0\underline{\mu}_2^{\mathrm{S}} + 0\underline{\mu}_3^{\mathrm{S}} + 20\underline{\mu}_4^{\mathrm{S}} \\ & - 40\overline{\mu}_1^{\mathrm{S}} - 40\overline{\mu}_2^{\mathrm{S}} - 40\overline{\mu}_3^{\mathrm{S}} - 40\overline{\mu}_4^{\mathrm{S}} + 60\xi + 20\phi_1^{\mathrm{S}} \Big)\end{aligned}$$

subject to

$$u_1^{\mathrm{R},-}, u_1^{\mathrm{R},+}, u_2^{\mathrm{R},-}, u_2^{\mathrm{R},+}, u_3^{\mathrm{R},-}, u_3^{\mathrm{R},+}, u_4^{\mathrm{R},-}, u_4^{\mathrm{R},+} \in \{0,1\},$$
$$p_1^{\mathrm{R,A}} = 60 - 20u_1^{\mathrm{R},-} + 20u_1^{\mathrm{R},+},$$
$$p_2^{\mathrm{R,A}} = 20 - 10u_2^{\mathrm{R},-} + 10u_2^{\mathrm{R},+},$$
$$p_3^{\mathrm{R,A}} = 40 - 15u_3^{\mathrm{R},-} + 15u_3^{\mathrm{R},+},$$
$$p_4^{\mathrm{R,A}} = 10 - 5u_4^{\mathrm{R},-} + 5u_4^{\mathrm{R},+},$$
$$u_1^{\mathrm{R},-} + u_1^{\mathrm{R},+} \le 1,$$
$$u_2^{\mathrm{R},-} + u_2^{\mathrm{R},+} \le 1,$$
$$u_3^{\mathrm{R},-} + u_3^{\mathrm{R},+} \le 1,$$
$$u_4^{\mathrm{R},-} + u_4^{\mathrm{R},+} \le 1,$$
$$u_1^{\mathrm{R},-} + u_1^{\mathrm{R},+} + u_2^{\mathrm{R},-} + u_2^{\mathrm{R},+} + u_3^{\mathrm{R},-} + u_3^{\mathrm{R},+} + u_4^{\mathrm{R},-} + u_4^{\mathrm{R},+} \le \Gamma^{\mathrm{R}},$$
$$0 \le z_1^{\mathrm{R},-} \le 1000u_1^{\mathrm{R},-},$$
$$0 \le z_2^{\mathrm{R},-} \le 1000u_2^{\mathrm{R},-},$$
$$0 \le z_3^{\mathrm{R},-} \le 1000u_3^{\mathrm{R},-},$$
$$0 \le z_4^{\mathrm{R},-} \le 1000u_4^{\mathrm{R},-},$$
$$0 \le \overline{\mu}_1^{\mathrm{R}} - z_1^{\mathrm{R},-} \le 1000\left(1 - u_1^{\mathrm{R},-}\right),$$
$$0 \le \overline{\mu}_2^{\mathrm{R}} - z_2^{\mathrm{R},-} \le 1000\left(1 - u_2^{\mathrm{R},-}\right),$$

$$0 \leq \overline{\mu}_3^{\mathrm{R}} - z_3^{\mathrm{R},-} \leq 1000\left(1 - u_3^{\mathrm{R},-}\right),$$

$$0 \leq \overline{\mu}_4^{\mathrm{R}} - z_4^{\mathrm{R},-} \leq 1000\left(1 - u_4^{\mathrm{R},-}\right),$$

$$0 \leq z_1^{\mathrm{R},+} \leq 1000u_1^{\mathrm{R},+},$$

$$0 \leq z_2^{\mathrm{R},+} \leq 1000u_2^{\mathrm{R},+},$$

$$0 \leq z_3^{\mathrm{R},+} \leq 1000u_3^{\mathrm{R},+},$$

$$0 \leq z_4^{\mathrm{R},+} \leq 1000u_4^{\mathrm{R},+},$$

$$0 \leq \overline{\mu}_1^{\mathrm{R}} - z_1^{\mathrm{R},+} \leq 1000\left(1 - u_1^{\mathrm{R},+}\right),$$

$$0 \leq \overline{\mu}_2^{\mathrm{R}} - z_2^{\mathrm{R},+} \leq 1000\left(1 - u_2^{\mathrm{R},+}\right),$$

$$0 \leq \overline{\mu}_3^{\mathrm{R}} - z_3^{\mathrm{R},+} \leq 1000\left(1 - u_3^{\mathrm{R},+}\right),$$

$$0 \leq \overline{\mu}_4^{\mathrm{R}} - z_4^{\mathrm{R},+} \leq 1000\left(1 - u_4^{\mathrm{R},+}\right),$$

$$\lambda_1 - \overline{\mu}_1^{\mathrm{R}} \leq 0,$$

$$\lambda_2 - \overline{\mu}_2^{\mathrm{R}} \leq 0,$$

$$\lambda_3 - \overline{\mu}_3^{\mathrm{R}} \leq 0,$$

$$\lambda_4 - \overline{\mu}_4^{\mathrm{R}} \leq 0,$$

$$\lambda_1 - \overline{\mu}_1^{\mathrm{S,D}} + \frac{1}{0.9}\phi_1^{\mathrm{S}} \leq 0,$$

$$\lambda_2 - \overline{\mu}_2^{\mathrm{S,D}} + \frac{1}{0.9}\phi_2^{\mathrm{S}} \leq 0,$$

$$\lambda_3 - \overline{\mu}_3^{\mathrm{S,D}} + \frac{1}{0.9}\phi_3^{\mathrm{S}} \leq 0,$$

$$\lambda_4 - \overline{\mu}_4^{\mathrm{S,D}} + \frac{1}{0.9}\phi_4^{\mathrm{S}} \leq 0,$$

$$-\lambda_1 - \overline{\mu}_1^{\mathrm{S,C}} + 0.9\phi_1^{\mathrm{S}} \leq 0,$$

$$-\lambda_2 - \overline{\mu}_2^{\mathrm{S,C}} + 0.9\phi_2^{\mathrm{S}} \leq 0,$$

$$-\lambda_3 - \overline{\mu}_3^{\mathrm{S,C}} + 0.9\phi_3^{\mathrm{S}} \leq 0,$$

$$-\lambda_4 - \overline{\mu}_4^{\mathrm{S,C}} + 0.9\phi_4^{\mathrm{S}} \leq 0,$$

$$-\lambda_1 + \underline{\mu}_1^{\mathrm{D}} - \overline{\mu}_1^{\mathrm{D}} + \xi = 0,$$

$$-\lambda_2 + \underline{\mu}_2^{\mathrm{D}} - \overline{\mu}_2^{\mathrm{D}} + \xi = 0,$$

$$-\lambda_3 + \underline{\mu}_3^{\mathrm{D}} - \overline{\mu}_3^{\mathrm{D}} + \xi = 0,$$

$$-\lambda_4 + \underline{\mu}_4^{\mathrm{D}} - \overline{\mu}_4^{\mathrm{D}} + \xi = 0,$$

$$\lambda_1, \lambda_2, \lambda_3, \lambda_4 \leq 1000,$$

$$-\lambda_1, -\lambda_2, -\lambda_3, -\lambda_4 \leq 1000,$$

$$\phi_1^{\mathrm{S}} - \phi_2^{\mathrm{S}} + \underline{\mu}_1^{\mathrm{S}} - \overline{\mu}_1^{\mathrm{S}} = 0,$$

$$\phi_2^{\mathrm{S}} - \phi_3^{\mathrm{S}} + \underline{\mu}_2^{\mathrm{S}} - \overline{\mu}_2^{\mathrm{S}} = 0,$$

$$\begin{aligned}
&\phi_3^{\mathrm{S}} - \phi_4^{\mathrm{S}} + \underline{\mu}_3^{\mathrm{S}} - \overline{\mu}_3^{\mathrm{S}} = 0,\\
&\phi_4^{\mathrm{S}} + \underline{\mu}_4^{\mathrm{S}} - \overline{\mu}_4^{\mathrm{S}} = 0,\\
&\overline{\mu}_1^{\mathrm{R}}, \overline{\mu}_2^{\mathrm{R}}, \overline{\mu}_3^{\mathrm{R}}, \overline{\mu}_4^{\mathrm{R}}, \overline{\mu}_1^{\mathrm{S,C}}, \overline{\mu}_2^{\mathrm{S,C}}, \overline{\mu}_3^{\mathrm{S,C}}, \overline{\mu}_4^{\mathrm{S,C}}, \overline{\mu}_1^{\mathrm{S,D}}, \overline{\mu}_2^{\mathrm{S,D}}, \overline{\mu}_3^{\mathrm{S,D}}, \overline{\mu}_4^{\mathrm{S,D}}, \underline{\mu}_1^{\mathrm{S}}, \underline{\mu}_2^{\mathrm{S}}, \underline{\mu}_3^{\mathrm{S}}, \underline{\mu}_4^{\mathrm{S}},\\
&\qquad \overline{\mu}_1^{\mathrm{S}}, \overline{\mu}_2^{\mathrm{S}}, \overline{\mu}_3^{\mathrm{S}}, \overline{\mu}_4^{\mathrm{S}}, \underline{\mu}_1^{\mathrm{D}}, \underline{\mu}_2^{\mathrm{D}}, \underline{\mu}_3^{\mathrm{D}}, \underline{\mu}_4^{\mathrm{D}}, \overline{\mu}_1^{\mathrm{D}}, \overline{\mu}_2^{\mathrm{D}}, \overline{\mu}_3^{\mathrm{D}}, \overline{\mu}_4^{\mathrm{D}} \geq 0,
\end{aligned}$$

where $\Phi^{\mathrm{S}} = \Big\{p_1^{\mathrm{R,A}}, p_2^{\mathrm{R,A}}, p_3^{\mathrm{R,A}}, p_4^{\mathrm{R,A}}, u_1^{\mathrm{R,-}}, u_2^{\mathrm{R,-}}, u_3^{\mathrm{R,-}}, u_4^{\mathrm{R,-}}, u_1^{\mathrm{R,+}}, u_2^{\mathrm{R,+}}, u_3^{\mathrm{R,+}}, u_4^{\mathrm{R,+}}, z_1^{\mathrm{R,-}}, z_2^{\mathrm{R,-}}, z_3^{\mathrm{R,-}}, z_4^{\mathrm{R,-}}, z_1^{\mathrm{R,+}}, z_2^{\mathrm{R,+}}, z_3^{\mathrm{R,+}}, z_4^{\mathrm{R,+}}, \lambda_1, \lambda_2, \lambda_3, \lambda_4, \overline{\mu}_1^{\mathrm{R}}, \overline{\mu}_2^{\mathrm{R}}, \overline{\mu}_3^{\mathrm{R}}, \overline{\mu}_4^{\mathrm{R}}, \overline{\mu}_1^{\mathrm{S,C}}, \overline{\mu}_2^{\mathrm{S,C}}, \overline{\mu}_3^{\mathrm{S,C}}, \overline{\mu}_4^{\mathrm{S,C}}, \overline{\mu}_1^{\mathrm{S,D}}, \overline{\mu}_2^{\mathrm{S,D}}, \overline{\mu}_3^{\mathrm{S,D}}, \overline{\mu}_4^{\mathrm{S,D}}, \phi_1^{\mathrm{S}}, \phi_2^{\mathrm{S}}, \phi_3^{\mathrm{S}}, \phi_4^{\mathrm{S}}, \underline{\mu}_1^{\mathrm{S}}, \underline{\mu}_2^{\mathrm{S}}, \underline{\mu}_3^{\mathrm{S}}, \underline{\mu}_4^{\mathrm{S}}, \overline{\mu}_1^{\mathrm{S}}, \overline{\mu}_2^{\mathrm{S}}, \overline{\mu}_3^{\mathrm{S}}, \overline{\mu}_4^{\mathrm{S}}, \underline{\mu}_1^{\mathrm{D}}, \underline{\mu}_2^{\mathrm{D}}, \underline{\mu}_3^{\mathrm{D}}, \underline{\mu}_4^{\mathrm{D}}, \overline{\mu}_1^{\mathrm{D}}, \overline{\mu}_2^{\mathrm{D}}, \overline{\mu}_3^{\mathrm{D}}, \overline{\mu}_4^{\mathrm{D}}, \xi\Big\}$.

The solution of this problem provides the worst uncertainty realization:

$$\begin{aligned}
p_1^{\mathrm{R,A}} &= 40,\\
p_2^{\mathrm{R,A}} &= 20,\\
p_3^{\mathrm{R,A}} &= 25,\\
p_4^{\mathrm{R,A}} &= 10.
\end{aligned}$$

Note that as $\Gamma^{\mathrm{R}} = 2$, the available wind-power generation level is allowed to deviate from its average value in a maximum of two time periods. In this case, given the power traded in the DA energy market obtained from the master problem in Step 2, the worst-case realization of the wind-power generation level is when it is at its lower bound in time periods 1 and 3, and at its average level in time periods 2 and 4. On the other hand, the optimal value of the objective function is $z^{\mathrm{S}*} = -1.250 \cdot 10^5$.

- Step 6. The lower bound is updated using Eq. (4.20):

$$\begin{aligned}
\mathrm{LB} &= \max\Big\{20 \cdot 40 + 50 \cdot 40 + 35 \cdot 40 + 31 \cdot 40 - 1.250 \cdot 10^5, -\infty\Big\}\\
&= -1.196 \cdot 10^5.
\end{aligned}$$

- Step 7. Compute $\mathrm{UB} - \mathrm{LB} = \infty - \left(-1.196 \cdot 10^5\right) = \infty$. As the difference between the upper and lower bounds is higher than the tolerance, continue with the following step.
- Step 8. The iteration counter is updated $\nu = 1$.
- Step 9. Set $p_1^{\mathrm{R,A}(1)} = 40$, $p_2^{\mathrm{R,A}(1)} = 20$, $p_3^{\mathrm{R,A}(1)} = 15$, and $p_4^{\mathrm{R,A}(1)} = 10$.
- Step 2. The following master problem is solved:

$$\max_{\Phi^{\mathrm{M}}} \; 20p_1^{\mathrm{DA}} + 50p_2^{\mathrm{DA}} + 35p_3^{\mathrm{DA}} + 31p_4^{\mathrm{DA}} + \theta$$

subject to

$$\begin{aligned}
&-40 \leq p_1^{\mathrm{DA}} \leq 40,\\
&-40 \leq p_2^{\mathrm{DA}} \leq 40,\\
&-40 \leq p_3^{\mathrm{DA}} \leq 40,\\
&-40 \leq p_4^{\mathrm{DA}} \leq 40,\\
&\theta \leq 0,\\
&p_1^{\mathrm{DA}} = p_{11}^{\mathrm{R}} + p_{11}^{\mathrm{S,D}} - p_{11}^{\mathrm{S,C}} - p_{11}^{\mathrm{D}},\\
&p_2^{\mathrm{DA}} = p_{21}^{\mathrm{R}} + p_{21}^{\mathrm{S,D}} - p_{21}^{\mathrm{S,C}} - p_{21}^{\mathrm{D}},\\
&p_3^{\mathrm{DA}} = p_{31}^{\mathrm{R}} + p_{31}^{\mathrm{S,D}} - p_{31}^{\mathrm{S,C}} - p_{31}^{\mathrm{D}},
\end{aligned}$$

$$p_4^{\mathrm{DA}} = p_{41}^{\mathrm{R}} + p_{41}^{\mathrm{S,D}} - p_{41}^{\mathrm{S,C}} - p_{41}^{\mathrm{D}},$$
$$0 \le p_{11}^{\mathrm{D}} \le 20,$$
$$0 \le p_{21}^{\mathrm{D}} \le 20,$$
$$10 \le p_{31}^{\mathrm{D}} \le 25,$$
$$15 \le p_{41}^{\mathrm{D}} \le 30,$$
$$p_{11}^{\mathrm{D}} + p_{21}^{\mathrm{D}} + p_{31}^{\mathrm{D}} + p_{41}^{\mathrm{D}} \ge 60,$$
$$0 \le p_{11}^{\mathrm{R}} \le 40,$$
$$0 \le p_{21}^{\mathrm{R}} \le 20,$$
$$0 \le p_{31}^{\mathrm{R}} \le 15,$$
$$0 \le p_{41}^{\mathrm{R}} \le 10,$$
$$0 \le p_{11}^{\mathrm{S,C}} \le 20,$$
$$0 \le p_{21}^{\mathrm{S,C}} \le 20,$$
$$0 \le p_{31}^{\mathrm{S,C}} \le 20,$$
$$0 \le p_{41}^{\mathrm{S,C}} \le 20,$$
$$0 \le p_{11}^{\mathrm{S,D}} \le 20,$$
$$0 \le p_{21}^{\mathrm{S,D}} \le 20,$$
$$0 \le p_{31}^{\mathrm{S,D}} \le 20,$$
$$0 \le p_{41}^{\mathrm{S,D}} \le 20,$$
$$e_{11}^{\mathrm{S}} = 20 + 0.9 p_{11}^{\mathrm{S,C}} - \frac{p_{11}^{\mathrm{S,D}}}{0.9},$$
$$e_{21}^{\mathrm{S}} = e_{11}^{\mathrm{S}} + 0.9 p_{21}^{\mathrm{S,C}} - \frac{p_{21}^{\mathrm{S,D}}}{0.9},$$
$$e_{31}^{\mathrm{S}} = e_{21}^{\mathrm{S}} + 0.9 p_{31}^{\mathrm{S,C}} - \frac{p_{31}^{\mathrm{S,D}}}{0.9},$$
$$e_{41}^{\mathrm{S}} = e_{31}^{\mathrm{S}} + 0.9 p_{41}^{\mathrm{S,C}} - \frac{p_{41}^{\mathrm{S,D}}}{0.9},$$
$$0 \le e_{11}^{\mathrm{S}} \le 40,$$
$$0 \le e_{21}^{\mathrm{S}} \le 40,$$
$$0 \le e_{31}^{\mathrm{S}} \le 40,$$
$$20 \le e_{41}^{\mathrm{S}} \le 40,$$

where $\Phi^{\mathrm{M}} = \left\{p_1^{\mathrm{DA}}, p_2^{\mathrm{DA}}, p_3^{\mathrm{DA}}, p_4^{\mathrm{DA}}, \theta, p_{11}^{\mathrm{D}}, p_{21}^{\mathrm{D}}, p_{31}^{\mathrm{D}}, p_{41}^{\mathrm{D}}, p_{11}^{\mathrm{R}}, p_{21}^{\mathrm{R}}, p_{31}^{\mathrm{R}}, p_{41}^{\mathrm{R}}, e_{11}^{\mathrm{S}}, e_{21}^{\mathrm{S}}, e_{31}^{\mathrm{S}}, e_{41}^{\mathrm{S}}, p_{11}^{\mathrm{S,C}}, p_{21}^{\mathrm{S,C}}, p_{31}^{\mathrm{S,C}}, p_{41}^{\mathrm{S,C}}, p_{11}^{\mathrm{S,D}}, p_{21}^{\mathrm{S,D}}, p_{31}^{\mathrm{S,D}}, p_{41}^{\mathrm{S,D}}\right\}$. Note that the operating variables include two subscripts. The first one indicates the time period while the second one denotes the iteration. As $\nu = 1$, in this case constraints (4.14d)–(4.14m) are included.

The optimal solution of this problem is:

$$p_1^{\mathrm{DA}*} = 20,$$
$$p_2^{\mathrm{DA}*} = 20,$$
$$p_3^{\mathrm{DA}*} = 15,$$
$$p_4^{\mathrm{DA}*} = -20,$$
$$\theta = 0.$$

- Step 3. The upper bound is updated using Eq. (4.19):

$$\mathrm{UB} = 1305.$$

- Step 4. Set $p_1^{\mathrm{DA}(1)} = 40$, $p_2^{\mathrm{DA}(1)} = 20$, $p_3^{\mathrm{DA}(1)} = 25$, and $p_4^{\mathrm{DA}(1)} = -20$.
- Step 5. The following subproblem is solved:

$$\begin{aligned}
\min_{\Phi^{\mathrm{S}}} -\Big(&40\lambda_1 + 20\lambda_2 + 15\lambda_3 - 20\lambda_4 + 0\underline{\mu}_1^{\mathrm{D}} + 0\underline{\mu}_2^{\mathrm{D}} + 10\underline{\mu}_3^{\mathrm{D}} + 15\underline{\mu}_4^{\mathrm{D}} \\
&- 20\overline{\mu}_1^{\mathrm{D}} - 20\overline{\mu}_2^{\mathrm{D}} - 25\overline{\mu}_3^{\mathrm{D}} - 30\overline{\mu}_4^{\mathrm{D}} - 60\overline{\mu}_1^{\mathrm{R}} + 20z_1^{\mathrm{R},-} - 20z_1^{\mathrm{R},+} \\
&- 20\overline{\mu}_2^{\mathrm{R}} + 10z_2^{\mathrm{R},-} - 10z_2^{\mathrm{R},+} - 40\overline{\mu}_3^{\mathrm{R}} + 15z_3^{\mathrm{R},-} - 15z_3^{\mathrm{R},+} - 10\overline{\mu}_4^{\mathrm{R}} \\
&+ 5z_4^{\mathrm{R},-} - 5z_4^{\mathrm{R},+} - 20\overline{\mu}_1^{\mathrm{S,C}} - 20\overline{\mu}_2^{\mathrm{S,C}} - 20\overline{\mu}_3^{\mathrm{S,C}} - 20\overline{\mu}_4^{\mathrm{S,C}} - 20\overline{\mu}_1^{\mathrm{S,D}} \\
&- 20\overline{\mu}_2^{\mathrm{S,D}} - 20\overline{\mu}_3^{\mathrm{S,D}} - 20\overline{\mu}_4^{\mathrm{S,D}} + 0\underline{\mu}_1^{\mathrm{S}} + 0\underline{\mu}_2^{\mathrm{S}} + 0\underline{\mu}_3^{\mathrm{S}} + 20\underline{\mu}_4^{\mathrm{S}} \\
&- 40\overline{\mu}_1^{\mathrm{S}} - 40\overline{\mu}_2^{\mathrm{S}} - 40\overline{\mu}_3^{\mathrm{S}} - 40\overline{\mu}_4^{\mathrm{S}} + 60\xi + 20\phi_1^{\mathrm{S}}\Big)
\end{aligned}$$

subject to

$$u_1^{\mathrm{R},-}, u_1^{\mathrm{R},+}, u_2^{\mathrm{R},-}, u_2^{\mathrm{R},+}, u_3^{\mathrm{R},-}, u_3^{\mathrm{R},+}, u_4^{\mathrm{R},-}, u_4^{\mathrm{R},+} \in \{0, 1\},$$
$$p_1^{\mathrm{R,A}} = 60 - 20u_1^{\mathrm{R},-} + 20u_1^{\mathrm{R},+},$$
$$p_2^{\mathrm{R,A}} = 20 - 10u_2^{\mathrm{R},-} + 10u_2^{\mathrm{R},+},$$
$$p_3^{\mathrm{R,A}} = 40 - 15u_3^{\mathrm{R},-} + 15u_3^{\mathrm{R},+},$$
$$p_4^{\mathrm{R,A}} = 10 - 5u_4^{\mathrm{R},-} + 5u_4^{\mathrm{R},+},$$
$$u_1^{\mathrm{R},-} + u_1^{\mathrm{R},+} \le 1,$$
$$u_2^{\mathrm{R},-} + u_2^{\mathrm{R},+} \le 1,$$
$$u_3^{\mathrm{R},-} + u_3^{\mathrm{R},+} \le 1,$$
$$u_4^{\mathrm{R},-} + u_4^{\mathrm{R},+} \le 1,$$
$$u_1^{\mathrm{R},-} + u_1^{\mathrm{R},+} + u_2^{\mathrm{R},-} + u_2^{\mathrm{R},+} + u_3^{\mathrm{R},-} + u_3^{\mathrm{R},+} + u_4^{\mathrm{R},-} + u_4^{\mathrm{R},+} \le \Gamma^{\mathrm{R}},$$
$$0 \le z_1^{\mathrm{R},-} \le 1000u_1^{\mathrm{R},-},$$
$$0 \le z_2^{\mathrm{R},-} \le 1000u_2^{\mathrm{R},-},$$
$$0 \le z_3^{\mathrm{R},-} \le 1000u_3^{\mathrm{R},-},$$
$$0 \le z_4^{\mathrm{R},-} \le 1000u_4^{\mathrm{R},-},$$

$$0 \leq \overline{\mu}_1^{\mathrm{R}} - z_1^{\mathrm{R},-} \leq 1000\left(1 - u_1^{\mathrm{R},-}\right),$$
$$0 \leq \overline{\mu}_2^{\mathrm{R}} - z_2^{\mathrm{R},-} \leq 1000\left(1 - u_2^{\mathrm{R},-}\right),$$
$$0 \leq \overline{\mu}_3^{\mathrm{R}} - z_3^{\mathrm{R},-} \leq 1000\left(1 - u_3^{\mathrm{R},-}\right),$$
$$0 \leq \overline{\mu}_4^{\mathrm{R}} - z_4^{\mathrm{R},-} \leq 1000\left(1 - u_4^{\mathrm{R},-}\right),$$
$$0 \leq z_1^{\mathrm{R},+} \leq 1000u_1^{\mathrm{R},+},$$
$$0 \leq z_2^{\mathrm{R},+} \leq 1000u_2^{\mathrm{R},+},$$
$$0 \leq z_3^{\mathrm{R},+} \leq 1000u_3^{\mathrm{R},+},$$
$$0 \leq z_4^{\mathrm{R},+} \leq 1000u_4^{\mathrm{R},+},$$
$$0 \leq \overline{\mu}_1^{\mathrm{R}} - z_1^{\mathrm{R},+} \leq 1000\left(1 - u_1^{\mathrm{R},+}\right),$$
$$0 \leq \overline{\mu}_2^{\mathrm{R}} - z_2^{\mathrm{R},+} \leq 1000\left(1 - u_2^{\mathrm{R},+}\right),$$
$$0 \leq \overline{\mu}_3^{\mathrm{R}} - z_3^{\mathrm{R},+} \leq 1000\left(1 - u_3^{\mathrm{R},+}\right),$$
$$0 \leq \overline{\mu}_4^{\mathrm{R}} - z_4^{\mathrm{R},+} \leq 1000\left(1 - u_4^{\mathrm{R},+}\right),$$
$$\lambda_1 - \overline{\mu}_1^{\mathrm{R}} \leq 0,$$
$$\lambda_2 - \overline{\mu}_2^{\mathrm{R}} \leq 0,$$
$$\lambda_3 - \overline{\mu}_3^{\mathrm{R}} \leq 0,$$
$$\lambda_4 - \overline{\mu}_4^{\mathrm{R}} \leq 0,$$
$$\lambda_1 - \overline{\mu}_1^{\mathrm{S,D}} + \frac{1}{0.9}\phi_1^{\mathrm{S}} \leq 0,$$
$$\lambda_2 - \overline{\mu}_2^{\mathrm{S,D}} + \frac{1}{0.9}\phi_2^{\mathrm{S}} \leq 0,$$
$$\lambda_3 - \overline{\mu}_3^{\mathrm{S,D}} + \frac{1}{0.9}\phi_3^{\mathrm{S}} \leq 0,$$
$$\lambda_4 - \overline{\mu}_4^{\mathrm{S,D}} + \frac{1}{0.9}\phi_4^{\mathrm{S}} \leq 0,$$
$$-\lambda_1 - \overline{\mu}_1^{\mathrm{S,C}} + 0.9\phi_1^{\mathrm{S}} \leq 0,$$
$$-\lambda_2 - \overline{\mu}_2^{\mathrm{S,C}} + 0.9\phi_2^{\mathrm{S}} \leq 0,$$
$$-\lambda_3 - \overline{\mu}_3^{\mathrm{S,C}} + 0.9\phi_3^{\mathrm{S}} \leq 0,$$
$$-\lambda_4 - \overline{\mu}_4^{\mathrm{S,C}} + 0.9\phi_4^{\mathrm{S}} \leq 0,$$
$$-\lambda_1 + \underline{\mu}_1^{\mathrm{D}} - \overline{\mu}_1^{\mathrm{D}} + \xi = 0,$$
$$-\lambda_2 + \underline{\mu}_2^{\mathrm{D}} - \overline{\mu}_2^{\mathrm{D}} + \xi = 0,$$
$$-\lambda_3 + \underline{\mu}_3^{\mathrm{D}} - \overline{\mu}_3^{\mathrm{D}} + \xi = 0,$$
$$-\lambda_4 + \underline{\mu}_4^{\mathrm{D}} - \overline{\mu}_4^{\mathrm{D}} + \xi = 0,$$
$$\lambda_1, \lambda_2, \lambda_3, \lambda_4 \leq 1000,$$

$$-\lambda_1, -\lambda_2, -\lambda_3, -\lambda_4 \leq 1000,$$
$$\phi_1^S - \phi_2^S + \underline{\mu}_1^S - \overline{\mu}_1^S = 0,$$
$$\phi_2^S - \phi_3^S + \underline{\mu}_2^S - \overline{\mu}_2^S = 0,$$
$$\phi_3^S - \phi_4^S + \underline{\mu}_3^S - \overline{\mu}_3^S = 0,$$
$$\phi_4^S + \underline{\mu}_4^S - \overline{\mu}_4^S = 0,$$
$$\overline{\mu}_1^R, \overline{\mu}_2^R, \overline{\mu}_3^R, \overline{\mu}_4^R, \overline{\mu}_1^{S,C}, \overline{\mu}_2^{S,C}, \overline{\mu}_3^{S,C}, \overline{\mu}_4^{S,C}, \overline{\mu}_1^{S,D}, \overline{\mu}_2^{S,D}, \overline{\mu}_3^{S,D}, \overline{\mu}_4^{S,D}, \underline{\mu}_1^S, \underline{\mu}_2^S, \underline{\mu}_3^S, \underline{\mu}_4^S,$$
$$\overline{\mu}_1^S, \overline{\mu}_2^S, \overline{\mu}_3^S, \overline{\mu}_4^S, \underline{\mu}_1^D, \underline{\mu}_2^D, \underline{\mu}_3^D, \underline{\mu}_4^D, \overline{\mu}_1^D, \overline{\mu}_2^D, \overline{\mu}_3^D, \overline{\mu}_4^D \geq 0,$$

where $\Phi^S = \Big\{p_1^{R,A}, p_2^{R,A}, p_3^{R,A}, p_4^{R,A}, u_1^{R,-}, u_2^{R,-}, u_3^{R,-}, u_4^{R,-}, u_1^{R,+}, u_2^{R,+}, u_3^{R,+}, u_4^{R,+}, z_1^{R,-}, z_2^{R,-}, z_3^{R,-}, z_4^{R,-}, z_1^{R,+}, z_2^{R,+}, z_3^{R,+}, z_4^{R,+}, \lambda_1, \lambda_2, \lambda_3, \lambda_4, \overline{\mu}_1^R, \overline{\mu}_2^R, \overline{\mu}_3^R, \overline{\mu}_4^R, \overline{\mu}_1^{S,C}, \overline{\mu}_2^{S,C}, \overline{\mu}_3^{S,C}, \overline{\mu}_4^{S,C}, \overline{\mu}_1^{S,D}, \overline{\mu}_2^{S,D}, \overline{\mu}_3^{S,D}, \overline{\mu}_4^{S,D}, \phi_1^S, \phi_2^S, \phi_3^S, \phi_4^S, \underline{\mu}_1^S, \underline{\mu}_2^S, \underline{\mu}_3^S, \underline{\mu}_4^S, \overline{\mu}_1^S, \overline{\mu}_2^S, \overline{\mu}_3^S, \overline{\mu}_4^S, \underline{\mu}_1^D, \underline{\mu}_2^D, \underline{\mu}_3^D, \underline{\mu}_4^D, \overline{\mu}_1^D, \overline{\mu}_2^D, \overline{\mu}_3^D, \overline{\mu}_4^D, \xi\Big\}$.

The worst uncertainty realization is:

$$p_1^{R,A} = 60,$$
$$p_2^{R,A} = 10,$$
$$p_3^{R,A} = 40,$$
$$p_4^{R,A} = 10.$$

Note that the optimal power traded in the DA energy market obtained from the solution of the master problem in iteration $\nu = 1$ is different to that obtained in iteration $\nu = 0$. Thus, the worst uncertainty realization is also different. In this case, the worst-case corresponds to the case in which the available wind-power generation level is at the lower bound in time period 2 and at its average level in time periods 1, 3, and 4. On the other hand, the optimal value of the objective function is $z^{S*} = -1.000 \cdot 10^4$.

- Step 6. The lower bound is updated using Eq. (4.20):

$$\text{LB} = \max\Big\{20 \cdot 20 + 50 \cdot 20 + 35 \cdot 15 + 31 \cdot (-20) - 1.00 \cdot 10^4, -1.1956 \cdot 10^5\Big\}$$
$$= -8695.$$

- Step 7. Compute UB − LB = 1305 − (−8695) = 10000. As the difference between the upper and lower bounds is higher than the tolerance, continue with the following step.
- Step 8. The iteration counter is updated $\nu = 2$.
- Step 9. Set $p_1^{R,A(2)} = 60$, $p_2^{R,A(2)} = 10$, $p_3^{R,A(2)} = 40$, and $p_4^{R,A(2)} = 10$.
- Step 2. The following master problem is solved:

$$\max_{\Phi^M} \; 20p_1^{DA} + 50p_2^{DA} + 35p_3^{DA} + 31p_4^{DA} + \theta$$

subject to

$$-40 \leq p_1^{DA} \leq 40,$$
$$-40 \leq p_2^{DA} \leq 40,$$
$$-40 \leq p_3^{DA} \leq 40,$$
$$-40 \leq p_4^{DA} \leq 40,$$
$$\theta \leq 0,$$

$$p_1^{\mathrm{DA}} = p_{11}^{\mathrm{R}} + p_{11}^{\mathrm{S,D}} - p_{11}^{\mathrm{S,C}} - p_{11}^{\mathrm{D}},$$
$$p_2^{\mathrm{DA}} = p_{21}^{\mathrm{R}} + p_{21}^{\mathrm{S,D}} - p_{21}^{\mathrm{S,C}} - p_{21}^{\mathrm{D}},$$
$$p_3^{\mathrm{DA}} = p_{31}^{\mathrm{R}} + p_{31}^{\mathrm{S,D}} - p_{31}^{\mathrm{S,C}} - p_{31}^{\mathrm{D}},$$
$$p_4^{\mathrm{DA}} = p_{41}^{\mathrm{R}} + p_{41}^{\mathrm{S,D}} - p_{41}^{\mathrm{S,C}} - p_{41}^{\mathrm{D}},$$
$$0 \le p_{11}^{\mathrm{D}} \le 20,$$
$$0 \le p_{21}^{\mathrm{D}} \le 20,$$
$$10 \le p_{31}^{\mathrm{D}} \le 25,$$
$$15 \le p_{41}^{\mathrm{D}} \le 30,$$
$$p_{11}^{\mathrm{D}} + p_{21}^{\mathrm{D}} + p_{31}^{\mathrm{D}} + p_{41}^{\mathrm{D}} \ge 60,$$
$$0 \le p_{11}^{\mathrm{R}} \le 40,$$
$$0 \le p_{21}^{\mathrm{R}} \le 20,$$
$$0 \le p_{31}^{\mathrm{R}} \le 25,$$
$$0 \le p_{41}^{\mathrm{R}} \le 10,$$
$$0 \le p_{11}^{\mathrm{S,C}} \le 20,$$
$$0 \le p_{21}^{\mathrm{S,C}} \le 20,$$
$$0 \le p_{31}^{\mathrm{S,C}} \le 20,$$
$$0 \le p_{41}^{\mathrm{S,C}} \le 20,$$
$$0 \le p_{11}^{\mathrm{S,D}} \le 20,$$
$$0 \le p_{21}^{\mathrm{S,D}} \le 20,$$
$$0 \le p_{31}^{\mathrm{S,D}} \le 20,$$
$$0 \le p_{41}^{\mathrm{S,D}} \le 20,$$
$$e_{11}^{\mathrm{S}} = 20 + 0.9 p_{11}^{\mathrm{S,C}} - \frac{p_{11}^{\mathrm{S,D}}}{0.9},$$
$$e_{21}^{\mathrm{S}} = e_{11}^{\mathrm{S}} + 0.9 p_{21}^{\mathrm{S,C}} - \frac{p_{21}^{\mathrm{S,D}}}{0.9},$$
$$e_{31}^{\mathrm{S}} = e_{21}^{\mathrm{S}} + 0.9 p_{31}^{\mathrm{S,C}} - \frac{p_{31}^{\mathrm{S,D}}}{0.9},$$
$$e_{41}^{\mathrm{S}} = e_{31}^{\mathrm{S}} + 0.9 p_{41}^{\mathrm{S,C}} - \frac{p_{41}^{\mathrm{S,D}}}{0.9},$$
$$0 \le e_{11}^{\mathrm{S}} \le 40,$$
$$0 \le e_{21}^{\mathrm{S}} \le 40,$$
$$0 \le e_{31}^{\mathrm{S}} \le 40,$$
$$20 \le e_{41}^{\mathrm{S}} \le 40,$$
$$\theta \le 0,$$
$$p_1^{\mathrm{DA}} = p_{12}^{\mathrm{R}} + p_{12}^{\mathrm{S,D}} - p_{12}^{\mathrm{S,C}} - p_{12}^{\mathrm{D}},$$

$$p_2^{\mathrm{DA}} = p_{22}^{\mathrm{R}} + p_{22}^{\mathrm{S,D}} - p_{22}^{\mathrm{S,C}} - p_{22}^{\mathrm{D}},$$
$$p_3^{\mathrm{DA}} = p_{32}^{\mathrm{R}} + p_{32}^{\mathrm{S,D}} - p_{32}^{\mathrm{S,C}} - p_{32}^{\mathrm{D}},$$
$$p_4^{\mathrm{DA}} = p_{42}^{\mathrm{R}} + p_{42}^{\mathrm{S,D}} - p_{42}^{\mathrm{S,C}} - p_{42}^{\mathrm{D}},$$
$$0 \le p_{12}^{\mathrm{D}} \le 20,$$
$$0 \le p_{22}^{\mathrm{D}} \le 20,$$
$$10 \le p_{32}^{\mathrm{D}} \le 25,$$
$$15 \le p_{42}^{\mathrm{D}} \le 30,$$
$$p_{12}^{\mathrm{D}} + p_{22}^{\mathrm{D}} + p_{32}^{\mathrm{D}} + p_{42}^{\mathrm{D}} \ge 60,$$
$$0 \le p_{12}^{\mathrm{R}} \le 60,$$
$$0 \le p_{22}^{\mathrm{R}} \le 10,$$
$$0 \le p_{32}^{\mathrm{R}} \le 40,$$
$$0 \le p_{42}^{\mathrm{R}} \le 10,$$
$$0 \le p_{12}^{\mathrm{S,C}} \le 20,$$
$$0 \le p_{22}^{\mathrm{S,C}} \le 20,$$
$$0 \le p_{32}^{\mathrm{S,C}} \le 20,$$
$$0 \le p_{42}^{\mathrm{S,C}} \le 20,$$
$$0 \le p_{12}^{\mathrm{S,D}} \le 20,$$
$$0 \le p_{22}^{\mathrm{S,D}} \le 20,$$
$$0 \le p_{32}^{\mathrm{S,D}} \le 20,$$
$$0 \le p_{42}^{\mathrm{S,D}} \le 20,$$
$$e_{12}^{\mathrm{S}} = 20 + 0.9 p_{12}^{\mathrm{S,C}} - \frac{p_{12}^{\mathrm{S,D}}}{0.9},$$
$$e_{22}^{\mathrm{S}} = e_{12}^{\mathrm{S}} + 0.9 p_{22}^{\mathrm{S,C}} - \frac{p_{22}^{\mathrm{S,D}}}{0.9},$$
$$e_{32}^{\mathrm{S}} = e_{22}^{\mathrm{S}} + 0.9 p_{32}^{\mathrm{S,C}} - \frac{p_{32}^{\mathrm{S,D}}}{0.9},$$
$$e_{42}^{\mathrm{S}} = e_{32}^{\mathrm{S}} + 0.9 p_{42}^{\mathrm{S,C}} - \frac{p_{42}^{\mathrm{S,D}}}{0.9},$$
$$0 \le e_{12}^{\mathrm{S}} \le 40,$$
$$0 \le e_{22}^{\mathrm{S}} \le 40,$$
$$0 \le e_{32}^{\mathrm{S}} \le 40,$$
$$20 \le e_{42}^{\mathrm{S}} \le 40,$$

where $\Phi^{\mathrm{M}} = \Big\{p_1^{\mathrm{DA}}, p_2^{\mathrm{DA}}, p_3^{\mathrm{DA}}, p_4^{\mathrm{DA}}, \theta, p_{11}^{\mathrm{D}}, p_{21}^{\mathrm{D}}, p_{31}^{\mathrm{D}}, p_{41}^{\mathrm{D}}, p_{11}^{\mathrm{R}}, p_{21}^{\mathrm{R}}, p_{31}^{\mathrm{R}}, p_{41}^{\mathrm{R}}, e_{11}^{\mathrm{S}}, e_{21}^{\mathrm{S}}, e_{31}^{\mathrm{S}}, e_{41}^{\mathrm{S}}, p_{11}^{\mathrm{S,C}}, p_{21}^{\mathrm{S,C}}, p_{31}^{\mathrm{S,C}}, p_{41}^{\mathrm{S,C}},$ $p_{11}^{\mathrm{S,D}}, p_{21}^{\mathrm{S,D}}, p_{31}^{\mathrm{S,D}}, p_{41}^{\mathrm{S,D}}, p_{12}^{\mathrm{D}}, p_{22}^{\mathrm{D}}, p_{32}^{\mathrm{D}}, p_{42}^{\mathrm{D}}, p_{12}^{\mathrm{R}}, p_{22}^{\mathrm{R}}, p_{32}^{\mathrm{R}}, p_{42}^{\mathrm{R}}, e_{12}^{\mathrm{S}}, e_{22}^{\mathrm{S}}, e_{32}^{\mathrm{S}}, e_{42}^{\mathrm{S}}, p_{12}^{\mathrm{S,C}}, p_{22}^{\mathrm{S,C}}, p_{32}^{\mathrm{S,C}}, p_{42}^{\mathrm{S,C}}, p_{12}^{\mathrm{S,D}}, p_{22}^{\mathrm{S,D}}, p_{32}^{\mathrm{S,D}},$ $p_{42}^{\mathrm{S,D}}\Big\}$.

The optimal solution of this problem is:

$$p_1^{\mathrm{DA}*} = 20,$$
$$p_2^{\mathrm{DA}*} = 10,$$
$$p_3^{\mathrm{DA}*} = 15,$$
$$p_4^{\mathrm{DA}*} = -10,$$
$$\theta = 0.$$

- Step 3. The upper bound is updated using Eq. (4.19):

$$\mathrm{UB} = 1115.$$

- Step 4. Set $p_1^{\mathrm{DA}(2)} = 20$, $p_2^{\mathrm{DA}(2)} = 10$, $p_3^{\mathrm{DA}(2)} = 15$, and $p_4^{\mathrm{DA}(2)} = -10$.
- Step 5. The following subproblem is solved:

$$\begin{aligned}\min_{\Phi^{\mathrm{S}}} - \Big(&20\lambda_1 + 10\lambda_2 + 15\lambda_3 - 10\lambda_4 + 0\underline{\mu}_1^{\mathrm{D}} + 0\underline{\mu}_2^{\mathrm{D}} + 10\underline{\mu}_3^{\mathrm{D}} + 15\underline{\mu}_4^{\mathrm{D}}\\
&- 20\overline{\mu}_1^{\mathrm{D}} - 20\overline{\mu}_2^{\mathrm{D}} - 25\overline{\mu}_3^{\mathrm{D}} - 30\overline{\mu}_4^{\mathrm{D}} - 60\overline{\mu}_1^{\mathrm{R}} + 20z_1^{\mathrm{R},-} - 20z_1^{\mathrm{R},+}\\
&- 20\overline{\mu}_2^{\mathrm{R}} + 10z_2^{\mathrm{R},-} - 10z_2^{\mathrm{R},+} - 40\overline{\mu}_3^{\mathrm{R}} + 15z_3^{\mathrm{R},-} - 15z_3^{\mathrm{R},+} - 10\overline{\mu}_4^{\mathrm{R}}\\
&+ 5z_4^{\mathrm{R},-} - 5z_4^{\mathrm{R},+} - 20\overline{\mu}_1^{\mathrm{S,C}} - 20\overline{\mu}_2^{\mathrm{S,C}} - 20\overline{\mu}_3^{\mathrm{S,C}} - 20\overline{\mu}_4^{\mathrm{S,C}} - 20\overline{\mu}_1^{\mathrm{S,D}}\\
&- 20\overline{\mu}_2^{\mathrm{S,D}} - 20\overline{\mu}_3^{\mathrm{S,D}} - 20\overline{\mu}_4^{\mathrm{S,D}} + 0\underline{\mu}_1^{\mathrm{S}} + 0\underline{\mu}_2^{\mathrm{S}} + 0\underline{\mu}_3^{\mathrm{S}} + 20\underline{\mu}_4^{\mathrm{S}}\\
&- 40\overline{\mu}_1^{\mathrm{S}} - 40\overline{\mu}_2^{\mathrm{S}} - 40\overline{\mu}_3^{\mathrm{S}} - 40\overline{\mu}_4^{\mathrm{S}} + 60\xi + 20\phi_1^{\mathrm{S}}\Big)\end{aligned}$$

subject to

$$u_1^{\mathrm{R},-}, u_1^{\mathrm{R},+}, u_2^{\mathrm{R},-}, u_2^{\mathrm{R},+}, u_3^{\mathrm{R},-}, u_3^{\mathrm{R},+}, u_4^{\mathrm{R},-}, u_4^{\mathrm{R},+} \in \{0, 1\},$$
$$p_1^{\mathrm{R,A}} = 60 - 20u_1^{\mathrm{R},-} + 20u_1^{\mathrm{R},+},$$
$$p_2^{\mathrm{R,A}} = 20 - 10u_2^{\mathrm{R},-} + 10u_2^{\mathrm{R},+},$$
$$p_3^{\mathrm{R,A}} = 40 - 15u_3^{\mathrm{R},-} + 15u_3^{\mathrm{R},+},$$
$$p_4^{\mathrm{R,A}} = 10 - 5u_4^{\mathrm{R},-} + 5u_4^{\mathrm{R},+},$$
$$u_1^{\mathrm{R},-} + u_1^{\mathrm{R},+} \le 1,$$
$$u_2^{\mathrm{R},-} + u_2^{\mathrm{R},+} \le 1,$$
$$u_3^{\mathrm{R},-} + u_3^{\mathrm{R},+} \le 1,$$
$$u_4^{\mathrm{R},-} + u_4^{\mathrm{R},+} \le 1,$$
$$u_1^{\mathrm{R},-} + u_1^{\mathrm{R},+} + u_2^{\mathrm{R},-} + u_2^{\mathrm{R},+} + u_3^{\mathrm{R},-} + u_3^{\mathrm{R},+} + u_4^{\mathrm{R},-} + u_4^{\mathrm{R},+} \le \Gamma^{\mathrm{R}},$$
$$0 \le z_1^{\mathrm{R},-} \le 1000u_1^{\mathrm{R},-},$$
$$0 \le z_2^{\mathrm{R},-} \le 1000u_2^{\mathrm{R},-},$$

$$0 \le z_3^{\mathrm{R},-} \le 1000 u_3^{\mathrm{R},-},$$
$$0 \le z_4^{\mathrm{R},-} \le 1000 u_4^{\mathrm{R},-},$$
$$0 \le \overline{\mu}_1^{\mathrm{R}} - z_1^{\mathrm{R},-} \le 1000\left(1 - u_1^{\mathrm{R},-}\right),$$
$$0 \le \overline{\mu}_2^{\mathrm{R}} - z_2^{\mathrm{R},-} \le 1000\left(1 - u_2^{\mathrm{R},-}\right),$$
$$0 \le \overline{\mu}_3^{\mathrm{R}} - z_3^{\mathrm{R},-} \le 1000\left(1 - u_3^{\mathrm{R},-}\right),$$
$$0 \le \overline{\mu}_4^{\mathrm{R}} - z_4^{\mathrm{R},-} \le 1000\left(1 - u_4^{\mathrm{R},-}\right),$$
$$0 \le z_1^{\mathrm{R},+} \le 1000 u_1^{\mathrm{R},+},$$
$$0 \le z_2^{\mathrm{R},+} \le 1000 u_2^{\mathrm{R},+},$$
$$0 \le z_3^{\mathrm{R},+} \le 1000 u_3^{\mathrm{R},+},$$
$$0 \le z_4^{\mathrm{R},+} \le 1000 u_4^{\mathrm{R},+},$$
$$0 \le \overline{\mu}_1^{\mathrm{R}} - z_1^{\mathrm{R},+} \le 1000\left(1 - u_1^{\mathrm{R},+}\right),$$
$$0 \le \overline{\mu}_2^{\mathrm{R}} - z_2^{\mathrm{R},+} \le 1000\left(1 - u_2^{\mathrm{R},+}\right),$$
$$0 \le \overline{\mu}_3^{\mathrm{R}} - z_3^{\mathrm{R},+} \le 1000\left(1 - u_3^{\mathrm{R},+}\right),$$
$$0 \le \overline{\mu}_4^{\mathrm{R}} - z_4^{\mathrm{R},+} \le 1000\left(1 - u_4^{\mathrm{R},+}\right),$$
$$\lambda_1 - \overline{\mu}_1^{\mathrm{R}} \le 0,$$
$$\lambda_2 - \overline{\mu}_2^{\mathrm{R}} \le 0,$$
$$\lambda_3 - \overline{\mu}_3^{\mathrm{R}} \le 0,$$
$$\lambda_4 - \overline{\mu}_4^{\mathrm{R}} \le 0,$$
$$\lambda_1 - \overline{\mu}_1^{\mathrm{S,D}} + \frac{1}{0.9}\phi_1^{\mathrm{S}} \le 0,$$
$$\lambda_2 - \overline{\mu}_2^{\mathrm{S,D}} + \frac{1}{0.9}\phi_2^{\mathrm{S}} \le 0,$$
$$\lambda_3 - \overline{\mu}_3^{\mathrm{S,D}} + \frac{1}{0.9}\phi_3^{\mathrm{S}} \le 0,$$
$$\lambda_4 - \overline{\mu}_4^{\mathrm{S,D}} + \frac{1}{0.9}\phi_4^{\mathrm{S}} \le 0,$$
$$-\lambda_1 - \overline{\mu}_1^{\mathrm{S,C}} + 0.9\phi_1^{\mathrm{S}} \le 0,$$
$$-\lambda_2 - \overline{\mu}_2^{\mathrm{S,C}} + 0.9\phi_2^{\mathrm{S}} \le 0,$$
$$-\lambda_3 - \overline{\mu}_3^{\mathrm{S,C}} + 0.9\phi_3^{\mathrm{S}} \le 0,$$
$$-\lambda_4 - \overline{\mu}_4^{\mathrm{S,C}} + 0.9\phi_4^{\mathrm{S}} \le 0,$$
$$-\lambda_1 + \underline{\mu}_1^{\mathrm{D}} - \overline{\mu}_1^{\mathrm{D}} + \xi = 0,$$
$$-\lambda_2 + \underline{\mu}_2^{\mathrm{D}} - \overline{\mu}_2^{\mathrm{D}} + \xi = 0,$$
$$-\lambda_3 + \underline{\mu}_3^{\mathrm{D}} - \overline{\mu}_3^{\mathrm{D}} + \xi = 0,$$

$$-\lambda_4+\underline{\mu}_4^{\mathrm{D}}-\overline{\mu}_4^{\mathrm{D}}+\xi=0,$$

$$\lambda_1,\lambda_2,\lambda_3,\lambda_4\leq 1000,$$

$$-\lambda_1,-\lambda_2,-\lambda_3,-\lambda_4\leq 1000,$$

$$\phi_1^{\mathrm{S}}-\phi_2^{\mathrm{S}}+\underline{\mu}_1^{\mathrm{S}}-\overline{\mu}_1^{\mathrm{S}}=0,$$

$$\phi_2^{\mathrm{S}}-\phi_3^{\mathrm{S}}+\underline{\mu}_2^{\mathrm{S}}-\overline{\mu}_2^{\mathrm{S}}=0,$$

$$\phi_3^{\mathrm{S}}-\phi_4^{\mathrm{S}}+\underline{\mu}_3^{\mathrm{S}}-\overline{\mu}_3^{\mathrm{S}}=0,$$

$$\phi_4^{\mathrm{S}}+\underline{\mu}_4^{\mathrm{S}}-\overline{\mu}_4^{\mathrm{S}}=0,$$

$$\overline{\mu}_1^{\mathrm{R}},\overline{\mu}_2^{\mathrm{R}},\overline{\mu}_3^{\mathrm{R}},\overline{\mu}_4^{\mathrm{R}},\overline{\mu}_1^{\mathrm{S,C}},\overline{\mu}_2^{\mathrm{S,C}},\overline{\mu}_3^{\mathrm{S,C}},\overline{\mu}_4^{\mathrm{S,C}},\overline{\mu}_1^{\mathrm{S,D}},\overline{\mu}_2^{\mathrm{S,D}},\overline{\mu}_3^{\mathrm{S,D}},\overline{\mu}_4^{\mathrm{S,D}},\underline{\mu}_1^{\mathrm{S}},\underline{\mu}_2^{\mathrm{S}},\underline{\mu}_3^{\mathrm{S}},\underline{\mu}_4^{\mathrm{S}},$$
$$\overline{\mu}_1^{\mathrm{S}},\overline{\mu}_2^{\mathrm{S}},\overline{\mu}_3^{\mathrm{S}},\overline{\mu}_4^{\mathrm{S}},\underline{\mu}_1^{\mathrm{D}},\underline{\mu}_2^{\mathrm{D}},\underline{\mu}_3^{\mathrm{D}},\underline{\mu}_4^{\mathrm{D}},\overline{\mu}_1^{\mathrm{D}},\overline{\mu}_2^{\mathrm{D}},\overline{\mu}_3^{\mathrm{D}},\overline{\mu}_4^{\mathrm{D}}\geq 0,$$

where $\Phi^{\mathrm{S}}=\Big\{p_1^{\mathrm{R,A}},p_2^{\mathrm{R,A}},p_3^{\mathrm{R,A}},p_4^{\mathrm{R,A}},u_1^{\mathrm{R,-}},u_2^{\mathrm{R,-}},u_3^{\mathrm{R,-}},u_4^{\mathrm{R,-}},u_1^{\mathrm{R,+}},u_2^{\mathrm{R,+}},u_3^{\mathrm{R,+}},u_4^{\mathrm{R,+}},z_1^{\mathrm{R,-}},z_2^{\mathrm{R,-}},z_3^{\mathrm{R,-}},z_4^{\mathrm{R,-}},z_1^{\mathrm{R,+}},$ $z_2^{\mathrm{R,+}},z_3^{\mathrm{R,+}},z_4^{\mathrm{R,+}},\lambda_1,\lambda_2,\lambda_3,\lambda_4,\overline{\mu}_1^{\mathrm{R}},\overline{\mu}_2^{\mathrm{R}},\overline{\mu}_3^{\mathrm{R}},\overline{\mu}_4^{\mathrm{R}},\overline{\mu}_1^{\mathrm{S,C}},\overline{\mu}_2^{\mathrm{S,C}},\overline{\mu}_3^{\mathrm{S,C}},\overline{\mu}_4^{\mathrm{S,C}},\overline{\mu}_1^{\mathrm{S,D}},\overline{\mu}_2^{\mathrm{S,D}},\overline{\mu}_3^{\mathrm{S,D}},\overline{\mu}_4^{\mathrm{S,D}},\phi_1^{\mathrm{S}},\phi_2^{\mathrm{S}},\phi_3^{\mathrm{S}},\phi_4^{\mathrm{S}},\underline{\mu}_1^{\mathrm{S}},$ $\underline{\mu}_2^{\mathrm{S}},\underline{\mu}_3^{\mathrm{S}},\underline{\mu}_4^{\mathrm{S}},\overline{\mu}_1^{\mathrm{S}},\overline{\mu}_2^{\mathrm{S}},\overline{\mu}_3^{\mathrm{S}},\overline{\mu}_4^{\mathrm{S}},\underline{\mu}_1^{\mathrm{D}},\underline{\mu}_2^{\mathrm{D}},\underline{\mu}_3^{\mathrm{D}},\underline{\mu}_4^{\mathrm{D}},\overline{\mu}_1^{\mathrm{D}},\overline{\mu}_2^{\mathrm{D}},\overline{\mu}_3^{\mathrm{D}},\overline{\mu}_4^{\mathrm{D}},\xi\Big\}$.

The worst uncertainty realization is:

$$p_1^{\mathrm{R,A}}=60,$$
$$p_2^{\mathrm{R,A}}=10,$$
$$p_3^{\mathrm{R,A}}=25,$$
$$p_4^{\mathrm{R,A}}=10.$$

On the other hand, the optimal value of the objective function is $z^{\mathrm{S}*}=-1.00\cdot 10^4$.

- Step 6. The lower bound is updated using Eq. (4.20):

$$\mathrm{LB}=\max\Big\{20\cdot 20+50\cdot 10+35\cdot 15+31\cdot(-10)-1.00\cdot 10^5,-8695\Big\}$$
$$=-8695.$$

- Step 7. Compute $\mathrm{UB}-\mathrm{LB}=1115-(-8695)=9810$. As the difference between the upper and lower bounds is higher than the tolerance, continue with the following step.
- Step 8. The iteration counter is updated $\nu=3$.
- Step 9. Set $p_1^{\mathrm{R,A}(3)}=60$, $p_2^{\mathrm{R,A}(3)}=10$, $p_3^{\mathrm{R,A}(3)}=25$, and $p_4^{\mathrm{R,A}(3)}=10$.
- Step 2. The following master problem is solved:

$$\max_{\Phi^{\mathrm{M}}}\ 20p_1^{\mathrm{DA}}+50p_2^{\mathrm{DA}}+35p_3^{\mathrm{DA}}+31p_4^{\mathrm{DA}}+\theta$$

subject to

$$-40\leq p_1^{\mathrm{DA}}\leq 40,$$
$$-40\leq p_2^{\mathrm{DA}}\leq 40,$$
$$-40\leq p_3^{\mathrm{DA}}\leq 40,$$
$$-40\leq p_4^{\mathrm{DA}}\leq 40,$$
$$\theta\leq 0,$$

$$p_1^{\text{DA}} = p_{11}^{\text{R}} + p_{11}^{\text{S,D}} - p_{11}^{\text{S,C}} - p_{11}^{\text{D}},$$
$$p_2^{\text{DA}} = p_{21}^{\text{R}} + p_{21}^{\text{S,D}} - p_{21}^{\text{S,C}} - p_{21}^{\text{D}},$$
$$p_3^{\text{DA}} = p_{31}^{\text{R}} + p_{31}^{\text{S,D}} - p_{31}^{\text{S,C}} - p_{31}^{\text{D}},$$
$$p_4^{\text{DA}} = p_{41}^{\text{R}} + p_{41}^{\text{S,D}} - p_{41}^{\text{S,C}} - p_{41}^{\text{D}},$$
$$0 \le p_{11}^{\text{D}} \le 20,$$
$$0 \le p_{21}^{\text{D}} \le 20,$$
$$10 \le p_{31}^{\text{D}} \le 25,$$
$$15 \le p_{41}^{\text{D}} \le 30,$$
$$p_{11}^{\text{D}} + p_{21}^{\text{D}} + p_{31}^{\text{D}} + p_{41}^{\text{D}} \ge 60,$$
$$0 \le p_{11}^{\text{R}} \le 40,$$
$$0 \le p_{21}^{\text{R}} \le 20,$$
$$0 \le p_{31}^{\text{R}} \le 25,$$
$$0 \le p_{41}^{\text{R}} \le 10,$$
$$0 \le p_{11}^{\text{S,C}} \le 20,$$
$$0 \le p_{21}^{\text{S,C}} \le 20,$$
$$0 \le p_{31}^{\text{S,C}} \le 20,$$
$$0 \le p_{41}^{\text{S,C}} \le 20,$$
$$0 \le p_{11}^{\text{S,D}} \le 20,$$
$$0 \le p_{21}^{\text{S,D}} \le 20,$$
$$0 \le p_{31}^{\text{S,D}} \le 20,$$
$$0 \le p_{41}^{\text{S,D}} \le 20,$$
$$e_{11}^{\text{S}} = 20 + 0.9 p_{11}^{\text{S,C}} - \frac{p_{11}^{\text{S,D}}}{0.9},$$
$$e_{21}^{\text{S}} = e_{11}^{\text{S}} + 0.9 p_{21}^{\text{S,C}} - \frac{p_{21}^{\text{S,D}}}{0.9},$$
$$e_{31}^{\text{S}} = e_{21}^{\text{S}} + 0.9 p_{31}^{\text{S,C}} - \frac{p_{31}^{\text{S,D}}}{0.9},$$
$$e_{41}^{\text{S}} = e_{3}^{\text{S}} + 0.9 p_{41}^{\text{S,C}} - \frac{p_{41}^{\text{S,D}}}{0.9},$$
$$0 \le e_{11}^{\text{S}} \le 40,$$
$$0 \le e_{21}^{\text{S}} \le 40,$$
$$0 \le e_{31}^{\text{S}} \le 40,$$
$$20 \le e_{41}^{\text{S}} \le 40,$$
$$\theta \le 0,$$
$$p_1^{\text{DA}} = p_{12}^{\text{R}} + p_{12}^{\text{S,D}} - p_{12}^{\text{S,C}} - p_{12}^{\text{D}},$$

$$p_2^{\mathrm{DA}} = p_{22}^{\mathrm{R}} + p_{22}^{\mathrm{S,D}} - p_{22}^{\mathrm{S,C}} - p_{22}^{\mathrm{D}},$$
$$p_3^{\mathrm{DA}} = p_{32}^{\mathrm{R}} + p_{32}^{\mathrm{S,D}} - p_{32}^{\mathrm{S,C}} - p_{32}^{\mathrm{D}},$$
$$p_4^{\mathrm{DA}} = p_{42}^{\mathrm{R}} + p_{42}^{\mathrm{S,D}} - p_{42}^{\mathrm{S,C}} - p_{42}^{\mathrm{D}},$$
$$0 \le p_{12}^{\mathrm{D}} \le 20,$$
$$0 \le p_{22}^{\mathrm{D}} \le 20,$$
$$10 \le p_{32}^{\mathrm{D}} \le 25,$$
$$15 \le p_{42}^{\mathrm{D}} \le 30,$$
$$p_{12}^{\mathrm{D}} + p_{22}^{\mathrm{D}} + p_{32}^{\mathrm{D}} + p_{42}^{\mathrm{D}} \ge 60,$$
$$0 \le p_{12}^{\mathrm{R}} \le 60,$$
$$0 \le p_{22}^{\mathrm{R}} \le 10,$$
$$0 \le p_{32}^{\mathrm{R}} \le 40,$$
$$0 \le p_{42}^{\mathrm{R}} \le 10,$$
$$0 \le p_{12}^{\mathrm{S,C}} \le 20,$$
$$0 \le p_{22}^{\mathrm{S,C}} \le 20,$$
$$0 \le p_{32}^{\mathrm{S,C}} \le 20,$$
$$0 \le p_{42}^{\mathrm{S,C}} \le 20,$$
$$0 \le p_{12}^{\mathrm{S,D}} \le 20,$$
$$0 \le p_{22}^{\mathrm{S,D}} \le 20,$$
$$0 \le p_{32}^{\mathrm{S,D}} \le 20,$$
$$0 \le p_{42}^{\mathrm{S,D}} \le 20,$$
$$e_{12}^{\mathrm{S}} = 20 + 0.9 p_{12}^{\mathrm{S,C}} - \frac{p_{12}^{\mathrm{S,D}}}{0.9},$$
$$e_{22}^{\mathrm{S}} = e_{12}^{\mathrm{S}} + 0.9 p_{22}^{\mathrm{S,C}} - \frac{p_{22}^{\mathrm{S,D}}}{0.9},$$
$$e_{32}^{\mathrm{S}} = e_{22}^{\mathrm{S}} + 0.9 p_{32}^{\mathrm{S,C}} - \frac{p_{32}^{\mathrm{S,D}}}{0.9},$$
$$e_{42}^{\mathrm{S}} = e_{32}^{\mathrm{S}} + 0.9 p_{42}^{\mathrm{S,C}} - \frac{p_{42}^{\mathrm{S,D}}}{0.9},$$
$$0 \le e_{12}^{\mathrm{S}} \le 40,$$
$$0 \le e_{22}^{\mathrm{S}} \le 40,$$
$$0 \le e_{32}^{\mathrm{S}} \le 40,$$
$$20 \le e_{42}^{\mathrm{S}} \le 40,$$
$$\theta \le 0,$$
$$p_1^{\mathrm{DA}} = p_{13}^{\mathrm{R}} + p_{13}^{\mathrm{S,D}} - p_{13}^{\mathrm{S,C}} - p_{13}^{\mathrm{D}},$$
$$p_2^{\mathrm{DA}} = p_{23}^{\mathrm{R}} + p_{23}^{\mathrm{S,D}} - p_{23}^{\mathrm{S,C}} - p_{23}^{\mathrm{D}},$$

$$
\begin{aligned}
& p_3^{\mathrm{DA}} = p_{33}^{\mathrm{R}} + p_{33}^{\mathrm{S,D}} - p_{33}^{\mathrm{S,C}} - p_{33}^{\mathrm{D}}, \\
& p_4^{\mathrm{DA}} = p_{43}^{\mathrm{R}} + p_{43}^{\mathrm{S,D}} - p_{43}^{\mathrm{S,C}} - p_{43}^{\mathrm{D}}, \\
& 0 \leq p_{13}^{\mathrm{D}} \leq 20, \\
& 0 \leq p_{23}^{\mathrm{D}} \leq 20, \\
& 10 \leq p_{33}^{\mathrm{D}} \leq 25, \\
& 15 \leq p_{43}^{\mathrm{D}} \leq 30, \\
& p_{13}^{\mathrm{D}} + p_{23}^{\mathrm{D}} + p_{33}^{\mathrm{D}} + p_{43}^{\mathrm{D}} \geq 60, \\
& 0 \leq p_{13}^{\mathrm{R}} \leq 60, \\
& 0 \leq p_{23}^{\mathrm{R}} \leq 10, \\
& 0 \leq p_{33}^{\mathrm{R}} \leq 25, \\
& 0 \leq p_{43}^{\mathrm{R}} \leq 10, \\
& 0 \leq p_{13}^{\mathrm{S,C}} \leq 20, \\
& 0 \leq p_{23}^{\mathrm{S,C}} \leq 20, \\
& 0 \leq p_{33}^{\mathrm{S,C}} \leq 20, \\
& 0 \leq p_{43}^{\mathrm{S,C}} \leq 20, \\
& 0 \leq p_{13}^{\mathrm{S,D}} \leq 20, \\
& 0 \leq p_{23}^{\mathrm{S,D}} \leq 20, \\
& 0 \leq p_{33}^{\mathrm{S,D}} \leq 20, \\
& 0 \leq p_{43}^{\mathrm{S,D}} \leq 20, \\
& e_{13}^{\mathrm{S}} = 20 + 0.9 p_{13}^{\mathrm{S,C}} - \frac{p_{13}^{\mathrm{S,D}}}{0.9}, \\
& e_{23}^{\mathrm{S}} = e_{13}^{\mathrm{S}} + 0.9 p_{23}^{\mathrm{S,C}} - \frac{p_{23}^{\mathrm{S,D}}}{0.9}, \\
& e_{33}^{\mathrm{S}} = e_{23}^{\mathrm{S}} + 0.9 p_{33}^{\mathrm{S,C}} - \frac{p_{33}^{\mathrm{S,D}}}{0.9}, \\
& e_{43}^{\mathrm{S}} = e_{33}^{\mathrm{S}} + 0.9 p_{43}^{\mathrm{S,C}} - \frac{p_{43}^{\mathrm{S,D}}}{0.9}, \\
& 0 \leq e_{13}^{\mathrm{S}} \leq 40, \\
& 0 \leq e_{23}^{\mathrm{S}} \leq 40, \\
& 0 \leq e_{33}^{\mathrm{S}} \leq 40, \\
& 20 \leq e_{43}^{\mathrm{S}} \leq 40,
\end{aligned}
$$

where $\Phi^{\mathrm{M}} = \Big\{ p_1^{\mathrm{DA}}, p_2^{\mathrm{DA}}, p_3^{\mathrm{DA}}, p_4^{\mathrm{DA}}, \theta, p_{11}^{\mathrm{D}}, p_{21}^{\mathrm{D}}, p_{31}^{\mathrm{D}}, p_{41}^{\mathrm{D}}, p_{11}^{\mathrm{R}}, p_{21}^{\mathrm{R}}, p_{31}^{\mathrm{R}}, p_{41}^{\mathrm{R}}, e_{11}^{\mathrm{S}}, e_{21}^{\mathrm{S}}, e_{31}^{\mathrm{S}}, e_{41}^{\mathrm{S}}, p_{11}^{\mathrm{S,C}}, p_{21}^{\mathrm{S,C}}, p_{31}^{\mathrm{S,C}}, p_{41}^{\mathrm{S,C}}, p_{11}^{\mathrm{S,D}}, p_{21}^{\mathrm{S,D}}, p_{31}^{\mathrm{S,D}}, p_{41}^{\mathrm{S,D}}, p_{12}^{\mathrm{D}}, p_{22}^{\mathrm{D}}, p_{32}^{\mathrm{D}}, p_{42}^{\mathrm{D}}, p_{12}^{\mathrm{R}}, p_{22}^{\mathrm{R}}, p_{32}^{\mathrm{R}}, p_{42}^{\mathrm{R}}, e_{12}^{\mathrm{S}}, e_{22}^{\mathrm{S}}, e_{32}^{\mathrm{S}}, e_{42}^{\mathrm{S}}, p_{12}^{\mathrm{S,C}}, p_{22}^{\mathrm{S,C}}, p_{32}^{\mathrm{S,C}}, p_{42}^{\mathrm{S,C}}, p_{12}^{\mathrm{S,D}}, p_{22}^{\mathrm{S,D}}, p_{32}^{\mathrm{S,D}}, p_{42}^{\mathrm{S,D}}, p_{13}^{\mathrm{D}}, p_{23}^{\mathrm{D}}, p_{33}^{\mathrm{D}}, p_{43}^{\mathrm{D}}, p_{13}^{\mathrm{R}}, p_{23}^{\mathrm{R}}, p_{33}^{\mathrm{R}}, p_{43}^{\mathrm{R}}, e_{13}^{\mathrm{S}}, e_{23}^{\mathrm{S}}, e_{33}^{\mathrm{S}}, e_{43}^{\mathrm{S}}, p_{13}^{\mathrm{S,C}}, p_{23}^{\mathrm{S,C}}, p_{33}^{\mathrm{S,C}}, p_{43}^{\mathrm{S,C}}, p_{13}^{\mathrm{S,D}}, p_{23}^{\mathrm{S,D}}, p_{33}^{\mathrm{S,D}}, p_{43}^{\mathrm{S,D}} \Big\}$.

The optimal solution of this problem is:

$$p_1^{\text{DA}*} = 30,$$
$$p_2^{\text{DA}*} = 10,$$
$$p_3^{\text{DA}*} = 15,$$
$$p_4^{\text{DA}*} = -20,$$
$$\theta = 0.$$

- Step 3. The upper bound is updated using Eq. (4.19):

$$\text{UB} = 1005.$$

- Step 4. Set $p_1^{\text{DA}(3)} = 30$, $p_3^{\text{DA}(3)} = 10$, $p_3^{\text{DA}(3)} = 15$, and $p_4^{\text{DA}(3)} = -20$.
- Step 5. The following subproblem is solved:

$$\begin{aligned}\min_{\Phi^{\text{S}}} - \Big(&30\lambda_1 + 10\lambda_2 + 15\lambda_3 - 20\lambda_4 + 0\underline{\mu}_1^{\text{D}} + 0\underline{\mu}_2^{\text{D}} + 10\underline{\mu}_3^{\text{D}} + 15\underline{\mu}_4^{\text{D}} \\ &- 20\overline{\mu}_1^{\text{D}} - 20\overline{\mu}_2^{\text{D}} - 25\overline{\mu}_3^{\text{D}} - 30\overline{\mu}_4^{\text{D}} - 60\overline{\mu}_1^{\text{R}} + 20z_1^{\text{R},-} - 20z_1^{\text{R},+} \\ &- 20\overline{\mu}_2^{\text{R}} + 10z_2^{\text{R},-} - 10z_2^{\text{R},+} - 40\overline{\mu}_3^{\text{R}} + 15z_3^{\text{R},-} - 15z_3^{\text{R},+} - 10\overline{\mu}_4^{\text{R}} \\ &+ 5z_4^{\text{R},-} - 5z_4^{\text{R},+} - 20\overline{\mu}_1^{\text{S,C}} - 20\overline{\mu}_2^{\text{S,C}} - 20\overline{\mu}_3^{\text{S,C}} - 20\overline{\mu}_4^{\text{S,C}} - 20\overline{\mu}_1^{\text{S,D}} \\ &- 20\overline{\mu}_2^{\text{S,D}} - 20\overline{\mu}_3^{\text{S,D}} - 20\overline{\mu}_4^{\text{S,D}} + 0\underline{\mu}_1^{\text{S}} + 0\underline{\mu}_2^{\text{S}} + 0\underline{\mu}_3^{\text{S}} + 20\underline{\mu}_4^{\text{S}} \\ &- 40\overline{\mu}_1^{\text{S}} - 40\overline{\mu}_2^{\text{S}} - 40\overline{\mu}_3^{\text{S}} - 40\overline{\mu}_4^{\text{S}} + 60\xi + 20\phi_1^{\text{S}}\Big)\end{aligned}$$

subject to

$$u_1^{\text{R},-}, u_1^{\text{R},+}, u_2^{\text{R},-}, u_2^{\text{R},+}, u_3^{\text{R},-}, u_3^{\text{R},+}, u_4^{\text{R},-}, u_4^{\text{R},+} \in \{0, 1\},$$
$$p_1^{\text{R,A}} = 60 - 20u_1^{\text{R},-} + 20u_1^{\text{R},+},$$
$$p_2^{\text{R,A}} = 20 - 10u_2^{\text{R},-} + 10u_2^{\text{R},+},$$
$$p_3^{\text{R,A}} = 40 - 15u_{3t}^{\text{R},-} + 15u_3^{\text{R},+},$$
$$p_4^{\text{R,A}} = 10 - 5u_4^{\text{R},-} + 5u_4^{\text{R},+},$$
$$u_1^{\text{R},-} + u_1^{\text{R},+} \leq 1,$$
$$u_2^{\text{R},-} + u_2^{\text{R},+} \leq 1,$$
$$u_3^{\text{R},-} + u_3^{\text{R},+} \leq 1,$$
$$u_4^{\text{R},-} + u_4^{\text{R},+} \leq 1,$$
$$u_1^{\text{R},-} + u_1^{\text{R},+} + u_2^{\text{R},-} + u_2^{\text{R},+} + u_3^{\text{R},-} + u_3^{\text{R},+} + u_4^{\text{R},-} + u_4^{\text{R},+} \leq \Gamma^{\text{R}},$$
$$0 \leq z_1^{\text{R},-} \leq 1000u_1^{\text{R},-},$$
$$0 \leq z_2^{\text{R},-} \leq 1000u_2^{\text{R},-},$$
$$0 \leq z_3^{\text{R},-} \leq 1000u_3^{\text{R},-},$$
$$0 \leq z_4^{\text{R},-} \leq 1000u_4^{\text{R},-},$$

$$0 \le \overline{\mu}_1^{\mathrm{R}} - z_1^{\mathrm{R},-} \le 1000\left(1 - u_1^{\mathrm{R},-}\right),$$
$$0 \le \overline{\mu}_2^{\mathrm{R}} - z_2^{\mathrm{R},-} \le 1000\left(1 - u_2^{\mathrm{R},-}\right),$$
$$0 \le \overline{\mu}_3^{\mathrm{R}} - z_3^{\mathrm{R},-} \le 1000\left(1 - u_3^{\mathrm{R},-}\right),$$
$$0 \le \overline{\mu}_4^{\mathrm{R}} - z_4^{\mathrm{R},-} \le 1000\left(1 - u_4^{\mathrm{R},-}\right),$$
$$0 \le z_1^{\mathrm{R},+} \le 1000 u_1^{\mathrm{R},+},$$
$$0 \le z_2^{\mathrm{R},+} \le 1000 u_2^{\mathrm{R},+},$$
$$0 \le z_3^{\mathrm{R},+} \le 1000 u_3^{\mathrm{R},+},$$
$$0 \le z_4^{\mathrm{R},+} \le 1000 u_4^{\mathrm{R},+},$$
$$0 \le \overline{\mu}_1^{\mathrm{R}} - z_1^{\mathrm{R},+} \le 1000\left(1 - u_1^{\mathrm{R},+}\right),$$
$$0 \le \overline{\mu}_2^{\mathrm{R}} - z_2^{\mathrm{R},+} \le 1000\left(1 - u_2^{\mathrm{R},+}\right),$$
$$0 \le \overline{\mu}_3^{\mathrm{R}} - z_3^{\mathrm{R},+} \le 1000\left(1 - u_3^{\mathrm{R},+}\right),$$
$$0 \le \overline{\mu}_4^{\mathrm{R}} - z_4^{\mathrm{R},+} \le 1000\left(1 - u_4^{\mathrm{R},+}\right),$$
$$\lambda_1 - \overline{\mu}_1^{\mathrm{R}} \le 0,$$
$$\lambda_2 - \overline{\mu}_2^{\mathrm{R}} \le 0,$$
$$\lambda_3 - \overline{\mu}_3^{\mathrm{R}} \le 0,$$
$$\lambda_4 - \overline{\mu}_4^{\mathrm{R}} \le 0,$$
$$\lambda_1 - \overline{\mu}_1^{\mathrm{S,D}} + \frac{1}{0.9}\phi_1^{\mathrm{S}} \le 0,$$
$$\lambda_2 - \overline{\mu}_2^{\mathrm{S,D}} + \frac{1}{0.9}\phi_2^{\mathrm{S}} \le 0,$$
$$\lambda_3 - \overline{\mu}_3^{\mathrm{S,D}} + \frac{1}{0.9}\phi_3^{\mathrm{S}} \le 0,$$
$$\lambda_4 - \overline{\mu}_4^{\mathrm{S,D}} + \frac{1}{0.9}\phi_4^{\mathrm{S}} \le 0,$$
$$-\lambda_1 - \overline{\mu}_1^{\mathrm{S,C}} + 0.9\phi_1^{\mathrm{S}} \le 0,$$
$$-\lambda_2 - \overline{\mu}_2^{\mathrm{S,C}} + 0.9\phi_2^{\mathrm{S}} \le 0,$$
$$-\lambda_3 - \overline{\mu}_3^{\mathrm{S,C}} + 0.9\phi_3^{\mathrm{S}} \le 0,$$
$$-\lambda_4 - \overline{\mu}_4^{\mathrm{S,C}} + 0.9\phi_4^{\mathrm{S}} \le 0,$$
$$-\lambda_1 + \underline{\mu}_1^{\mathrm{D}} - \overline{\mu}_1^{\mathrm{D}} + \xi = 0,$$
$$-\lambda_2 + \underline{\mu}_2^{\mathrm{D}} - \overline{\mu}_2^{\mathrm{D}} + \xi = 0,$$
$$-\lambda_3 + \underline{\mu}_3^{\mathrm{D}} - \overline{\mu}_3^{\mathrm{D}} + \xi = 0,$$
$$-\lambda_4 + \underline{\mu}_4^{\mathrm{D}} - \overline{\mu}_4^{\mathrm{D}} + \xi = 0,$$
$$\lambda_1, \lambda_2, \lambda_3, \lambda_4 \le 1000,$$

$$
\begin{aligned}
& -\lambda_1, -\lambda_2, -\lambda_3, -\lambda_4 \leq 1000, \\
& \phi_1^{\mathrm{S}} - \phi_2^{\mathrm{S}} + \underline{\mu}_1^{\mathrm{S}} - \overline{\mu}_1^{\mathrm{S}} = 0, \\
& \phi_2^{\mathrm{S}} - \phi_3^{\mathrm{S}} + \underline{\mu}_2^{\mathrm{S}} - \overline{\mu}_2^{\mathrm{S}} = 0, \\
& \phi_3^{\mathrm{S}} - \phi_4^{\mathrm{S}} + \underline{\mu}_3^{\mathrm{S}} - \overline{\mu}_3^{\mathrm{S}} = 0, \\
& \phi_4^{\mathrm{S}} + \underline{\mu}_4^{\mathrm{S}} - \overline{\mu}_4^{\mathrm{S}} = 0, \\
& \overline{\mu}_1^{\mathrm{R}}, \overline{\mu}_2^{\mathrm{R}}, \overline{\mu}_3^{\mathrm{R}}, \overline{\mu}_4^{\mathrm{R}}, \overline{\mu}_1^{\mathrm{S,C}}, \overline{\mu}_2^{\mathrm{S,C}}, \overline{\mu}_3^{\mathrm{S,C}}, \overline{\mu}_4^{\mathrm{S,C}}, \overline{\mu}_1^{\mathrm{S,D}}, \overline{\mu}_2^{\mathrm{S,D}}, \overline{\mu}_3^{\mathrm{S,D}}, \overline{\mu}_4^{\mathrm{S,D}}, \underline{\mu}_1^{\mathrm{S}}, \underline{\mu}_2^{\mathrm{S}}, \underline{\mu}_3^{\mathrm{S}}, \underline{\mu}_4^{\mathrm{S}}, \\
& \quad \overline{\mu}_1^{\mathrm{S}}, \overline{\mu}_2^{\mathrm{S}}, \overline{\mu}_3^{\mathrm{S}}, \overline{\mu}_4^{\mathrm{S}}, \underline{\mu}_1^{\mathrm{D}}, \underline{\mu}_2^{\mathrm{D}}, \underline{\mu}_3^{\mathrm{D}}, \underline{\mu}_4^{\mathrm{D}}, \overline{\mu}_1^{\mathrm{D}}, \overline{\mu}_2^{\mathrm{D}}, \overline{\mu}_3^{\mathrm{D}}, \overline{\mu}_4^{\mathrm{D}} \geq 0,
\end{aligned}
$$

where $\Phi^{\mathrm{S}} = \Big\{p_1^{\mathrm{R,A}}, p_2^{\mathrm{R,A}}, p_3^{\mathrm{R,A}}, p_4^{\mathrm{R,A}}, u_1^{\mathrm{R,-}}, u_2^{\mathrm{R,-}}, u_3^{\mathrm{R,-}}, u_4^{\mathrm{R,-}}, u_1^{\mathrm{R,+}}, u_2^{\mathrm{R,+}}, u_3^{\mathrm{R,+}}, u_4^{\mathrm{R,+}}, z_1^{\mathrm{R,-}}, z_2^{\mathrm{R,-}}, z_3^{\mathrm{R,-}}, z_4^{\mathrm{R,-}}, z_1^{\mathrm{R,+}}, z_2^{\mathrm{R,+}}, z_3^{\mathrm{R,+}}, z_4^{\mathrm{R,+}}, \lambda_1, \lambda_2, \lambda_3, \lambda_4, \overline{\mu}_1^{\mathrm{R}}, \overline{\mu}_2^{\mathrm{R}}, \overline{\mu}_3^{\mathrm{R}}, \overline{\mu}_4^{\mathrm{R}}, \overline{\mu}_1^{\mathrm{S,C}}, \overline{\mu}_2^{\mathrm{S,C}}, \overline{\mu}_3^{\mathrm{S,C}}, \overline{\mu}_4^{\mathrm{S,C}}, \overline{\mu}_1^{\mathrm{S,D}}, \overline{\mu}_2^{\mathrm{S,D}}, \overline{\mu}_3^{\mathrm{S,D}}, \overline{\mu}_4^{\mathrm{S,D}}, \phi_1^{\mathrm{S}}, \phi_2^{\mathrm{S}}, \phi_3^{\mathrm{S}}, \phi_4^{\mathrm{S}}, \underline{\mu}_1^{\mathrm{S}}, \underline{\mu}_2^{\mathrm{S}}, \underline{\mu}_3^{\mathrm{S}}, \underline{\mu}_4^{\mathrm{S}}, \overline{\mu}_1^{\mathrm{S}}, \overline{\mu}_2^{\mathrm{S}}, \overline{\mu}_3^{\mathrm{S}}, \overline{\mu}_4^{\mathrm{S}}, \underline{\mu}_1^{\mathrm{D}}, \underline{\mu}_2^{\mathrm{D}}, \underline{\mu}_3^{\mathrm{D}}, \underline{\mu}_4^{\mathrm{D}}, \overline{\mu}_1^{\mathrm{D}}, \overline{\mu}_2^{\mathrm{D}}, \overline{\mu}_3^{\mathrm{D}}, \overline{\mu}_4^{\mathrm{D}}, \xi\Big\}$.

The worst uncertainty realization is:

$$
\begin{aligned}
p_1^{\mathrm{R,A}} &= 60, \\
p_2^{\mathrm{R,A}} &= 20, \\
p_3^{\mathrm{R,A}} &= 40, \\
p_4^{\mathrm{R,A}} &= 10.
\end{aligned}
$$

On the other hand, the optimal value of the objective function is $z^{\mathrm{S}*} = 0$.

- Step 6. The lower bound is updated using Eq. (4.20):

$$
\mathrm{LB} = \max\{20 \cdot 30 + 50 \cdot 10 + 35 \cdot 15 + 31 \cdot (-20) + 0, -8695\} = 1005.
$$

- Step 7. Compute $\mathrm{UB} - \mathrm{LB} = 1005 - 1005 = 0$. As the difference between the upper and lower bounds is 0, the algorithm stops. The optimal power traded in the DA energy market is:

$$
\begin{aligned}
p_1^{\mathrm{DA}*} &= 30 \text{ MW}, \\
p_2^{\mathrm{DA}*} &= 10 \text{ MW}, \\
p_3^{\mathrm{DA}*} &= 15 \text{ MW}, \\
p_4^{\mathrm{DA}*} &= -20 \text{ MW}.
\end{aligned}
$$

This power traded in the DA energy market allows the VPP to maximize its worst-case expected profit provided that the available wind-power generation level can deviate from its average level in a maximum of two time periods. The maximum power is traded in the first time period, which despite having the lowest market price, it is the period with the highest available wind-power generation level. On the other hand, the VPP prefers buying from the DA energy market in the fourth time period in which the market price is lower than in periods 2 and 3, and the available wind-power level is low.

□

A GAMS [8] code for solving Illustrative Example 4.8 is provided in Sect. 4.8.

Illustrative Example 4.9 Impact of the Uncertainty Budget

The data of Illustrative Example 4.8 are considered again. Next, the impact of the uncertainty budget on the power traded by the VPP in the DA energy market is analyzed.

Results are provided in Fig. 4.15 that depict the optimal power traded in the market for different values of the uncertainty budget Γ^{R}. Note that the highest power levels correspond to the lowest values of this uncertainty budget. For example, if $\Gamma^{\mathrm{R}} = 0$, the VPP assumes that the available wind-power generation level is equal to the average level in all time periods,

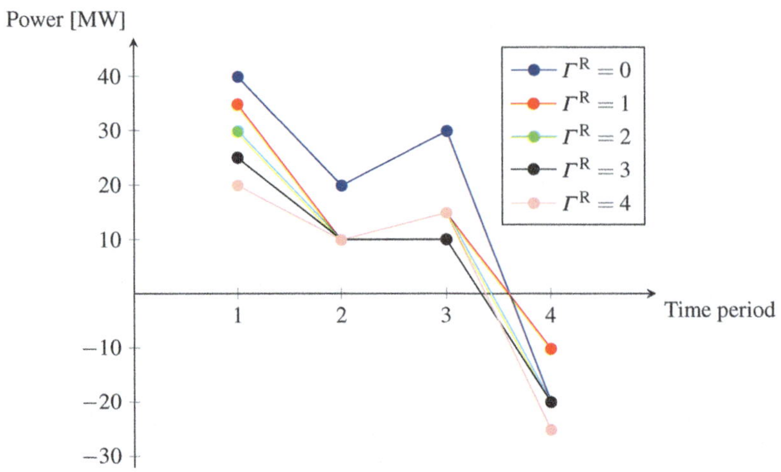

Fig. 4.15 Illustrative Example 4.9: power traded in the DA energy market

Table 4.21 Illustrative Example 4.9: worst-case profit for different values of the uncertainty budget

Γ^{R}	Objective function [$]
0	2230
1	1415
2	1005
3	730
4	650

and thus, it increases the power sold to the DA energy market. On the other hand, if $\Gamma^{R} = 4$, the VPP considers that the available wind-power generation level can deviate from the average level in all time periods and, then it reduces the power sold to the DA energy market.

Finally, the optimal value of the objective function for different uncertainty budgets is provided in Table 4.21. Note that the value of the objective function represents the worst-case profit achieved by the VPP. Thus, it decreases as the uncertainty budget increases since it is assumed that the available wind-power generation levels can be at their lower bounds in an increasing number of time periods.

□

Selecting the optimal value of the uncertainty budgets Γ_{r}^{R}, $\forall r \in \Omega^{R}$, is not trivial. If these parameters are fixed to their maximum values, i.e., to the number of time periods in the planning horizon, a robust solution is obtained since the VPP is protected against any deviation in the available wind-power generation level. However, the solution achieved in this case may be too conservative. On the other hand, if the uncertainty budgets are fixed to 0, the solution may be too risky since it assumes that the available wind-power generation level is at its average level. Any deviation from this average level may result in a power imbalance in the VPP for which an imbalance penalty may be charged to the VPP.

4.7 Summary and Further Reading

This chapter provides different methods for the scheduling problem of a risk-averse VPP participating in energy markets. These methods based on stochastic programming and robust optimization model the uncertainty in the decision-making problem and provide metrics to control the profit risk associated with the scheduling decisions.

Additional information about stochastic programming and robust optimization can be found in [6] and [3, 5], respectively. On the other hand, the interested readers are referred to the technical papers [1, 2, 11, 12] that provide further details about the problems described in this chapter, as well as more realistic case studies.

4.8 GAMS Codes

This section provides the GAMS files used to solve some of the illustrative examples of this chapter.

4.8.1 *Stochastic Programming Approach*

A GAMS code for solving Illustrative Example 4.1 is provided below:

```
* Stochastic scheduling of the risk-averse VPP in the day-ahead energy market
SET b 'Bus'                /b1*b3/;
SET l 'Line'               /l1*l3/;
SET t 'Time period'        /t1*t3/;
SET u 'Unit'               /u1*u5/;
SET o 'scenarios'          /o1*o2/;
SET g 'scenarios'          /g1*g2/;
SET i 'scenarios'          /i1/;
SET incDES(l,b)    'destination bus'        /l1.b2,l2.b3,l3.b3/;
SET incORI(l,b)    'origin bus'             /l1.b1,l2.b1,l3.b2/;
SET incD(u)        'demand'                 /u1*u2/;
SET incG(u)        'conventional power plant' /u3/;
SET incR(u)        'stochastic production facility' /u4/;
SET incS(u)        'storage unit'           /u5/;
SET incMB(b)       'buses connected to the main grid' /b1/;
SET incREF(b)      'reference bus'          /b1/;
SET incDB(u,b)     'demand at bus'          /u1.b2,u2.b3/;
SET incGB(u,b)     'conventional power plant at bus' /u3.b2/;
SET incRB(u,b)     'stochastic production facility at bus' /u4.b3/;
SET incSB(u,b)     'storage unit at bus' /u5.b3/;
SCALAR TT          'Number of trading hours in a day' /3/;
SCALAR delta       'Power-energy conversion factor' /1/;
SCALAR mup         'Output conversion efficiency of the storage unit' /0.9/;
SCALAR mun         'Input conversion efficiency of the storage unit' /0.9/;
SCALAR Sbase       'Base apparent power' /100/;
SCALAR Vbase       'Base voltage' /10/;
SCALAR PV_max      'Maximum power that can be produced by the VPP' /40/;
SCALAR PV_min      'Maximum power that can be consumed by the VPP' /-60/;
SCALAR beta        'Weight of CVaR' /1/;
SCALAR alpha       'Confidence level used for calculating CVaR' /0.8/;
***** Demands data
TABLE PD_max(u,t) 'Maximum load level for unit u in time period t'
         t1      t2      t3
u1       20      20      10
u2       30      30      15;
TABLE PD_min(u,t) 'Minimum load level for unit u in time period t'
         t1      t2      t3
u1       0       0       0
u2       5       5       5;
TABLE Utility(u,t) 'Utility of unit u for energy consumption in time period t'
         t1       t2       t3
u1       40       50       40
u2       50       60       50;
```

```
PARAMETER ED_min(u) 'Minimum daily energy consumption requested by unit i'
/u1        30
u2         50/;
PARAMETER RD_up(u) 'Load drop ramping limits for unit u'
/u1        15
u2         25/;
PARAMETER RD_do(u) 'Pick-up ramping limit for unit u'
/u1        15
u2         25/;
PARAMETER PD0(u) 'Initial load level for each unit'
/u1       10
u2        20/;
***** Conventional power plants data
PARAMETER PG_max(u) 'Maximum power production limit for unit u in time period t'
/u3        20/;
PARAMETER PG_min(u) 'Minimum power production limit for unit u in time period t'
/u3        0/;
PARAMETER RG_up(u) 'Ramping up limits for unit u'
/u3         20/;
PARAMETER RG_do(u) 'Ramping down limit for unit u'
/u3         20/;
PARAMETER CF(u) 'Fixed cost for unit u'
/u3       10/;
PARAMETER CG(u) 'Variable cost for unit u'
/u3       45/;
PARAMETER PG0(u) 'Initial power production of unit u'
/u3         0/;
PARAMETER V0(u) 'Initial status of unit u'
/u3         0/;
***** Storage units data
PARAMETER ES_max(u) 'Energy capacity of the storage unit u'
/u5       10/;
PARAMETER ES_min(u) 'Minimum energy capacity of the storage unit u'
/u5       5/;
PARAMETER PS_p(u) 'Maximum power that can be produced by unit u'
/u5       10/;
PARAMETER PS_n(u) 'Maximum power that can be transferred to unit u'
/u5       10/;
PARAMETER ES0(u) 'Initial energy stored in the storage unit'
/u5       5/;
***** Network data
PARAMETER PL_max(l) 'Capacity of line l'
/l1         50
l2          50
l3          50/;
PARAMETER BL(l) 'Susceptance of line l'
/l1         200
l2          200
l3          200/;
BL(l)=BL(l)/(Sbase/(Vbase**2));
```

```
PARAMETER PM_max(b) 'Maximum power that can be bought from the
    main grid'
/b1      100/;
***** Scenarios of uncertain parameters
TABLE PR_max 'Scenarios of available stochastic production in
    time period t'
                u4.t1    u4.t2    u4.t3
i1              9        7        1;
TABLE lambda_DA 'Scenarios of prices in the day-ahead energy
    market in time period t'
         t1        t2        t3
o1       20        40        30
o2       30        50        40;
TABLE lambda_RT 'Scenarios of prices in the real-time energy
    market in time period t'
        t1        t2        t3
o1.g1   25        50        25
o1.g2   10        45        28
o2.g1   35        60        35
o2.g2   20        55        38;
PARAMETER prob(o,g,i) 'Probability of scenarios'
/o1.g1.i1        0.16
o1.g2.i1        0.24
o2.g1.i1        0.24
o2.g2.i1        0.36/;
***** Actual data of the day-ahead market prices
PARAMETER lambda_DA_act(t) 'Actual day-ahead market price in time
     period t'
/t1       25
t2       50
t3       35/;
VARIABLES
pl(l,t,o,g,i)      'Power flow through line l in time period t'
pda(t)             'Power traded by the VPP in the day-ahead
    energy market in time period t'
prt(t,o,g,i)       'Power traded by the VPP in the real-time
    energy market in time period t'
pm(b,t,o,g,i)      'Power injection at bus b connected to the main
     grid in time period t'
teta(b,t,o,g,i)    'Voltage angle at bus b in time period t'
z                  'Objective'
zsc(o,g,i)         'Objective in each scenario'
p(u,t,o,g,i)       'Power dispatch of unit u in time period t'
rho                'Value at risk'
cvar               'Conditional value at risk metric'
ep                 'Expected profit';
POSITIVE VARIABLES
es(u,t,o,g,i)      'Energy stored at the end of time period t'
psp(u,t,o,g,i)     'Power produced by the storage unit in time
    period t'
psn(u,t,o,g,i)     'Power transferred to the storage unit in time
    period t'
mu(o,g,i)          'Auxiliary variable to calculate the
    conditional value at risk';
```

```
BINARY VARIABLES
v(u,t)              'Binary variable to state the commitment status
     of unit u in time period t';
*****************************************
** Equations definition *******
EQUATIONS
obj          'Objective function'
objscen      'Objective in each scenario'
expected_profit    'Expected profit'
cvar_calculation1 'Calculating the conditional value at risk'
cvar_calculation2 'Calculating the conditional value at risk'
nodal_balance_mg   'Power balance at the main grid buses'
nodal_balance      'Power balance at the remaining buses'
trade              'Power traded in the day-ahead and real-time
    markets'
line_power         'Power flow through lines'
line_max           'Upper bound of lines flows'
line_min           'Lower bound of lines flows'
voltage_angle_ref 'Voltage angle at reference bus'
voltage_angle_max 'Upper bound of voltage angles'
voltage_angle_min 'Lower bound of voltage angles'
tradeda_min        'Lower bound of power traded in the day-ahead
    market'
tradeda_max        'Upper bound of power traded in the day-ahead
    market'
tradedart_min      'Lower bound of power traded in the energy
    markets'
tradedart_max      'Upper bound of power traded in the energy
    markets'
trade_min          'Lower bound of power injection at each bus'
trade_max          'Upper bound of power injection at each bus'
dailyenergy_min    'Daily minimum energy consumption for each
    demand'
demand_max         'Upper bound of load level'
demand_min         'Lower bound of load level'
demand_ramp_down_initial      'Load ramp down limitation'
demand_ramp_pick_initial      'Load ramp up limitation'
demand_ramp_down   'Load ramp down limitation'
demand_ramp_pick   'Load ramp up limitation'
CPP_max            'Maximum power production limitation'
CPP_min            'Minimum power production limitation'
CPP_ramp_down_initial   'Ramp down production limitation'
CPP_ramp_up_initial     'Ramp up production limitation'
CCP_ramp_down      'Ramp down production limitation'
CCP_ramp_up        'Ramp up production limitation'
storage_balance_initial       'Power balance equation in the
    storage unit'
storage_balance    'Power balance equation in the storage unit'
storage_capacity   'Upper bound of energy stored in the storage
    unit'
storage_min        'Lower bound of energy stored in the storage
    unit'
storage_psp_max    'Maxmimum power that can be produced by storage
     unit'
```

```
storage_psn_max   'Maxmimum power that can be transferred to storage unit'
stochastic_max    'Maximum power production limitation';
*
obj..        z=E=ep+beta*CVAR;
objscen(o,g,i)..  zsc(o,g,i)=E=SUM(t,lambda_DA(o,t)*pda(t)*delta+lambda_RT(o,g,t)*prt(t,o,g,i)*delta+SUM(u$incD(u),Utility(u,t)*p(u,t,o,g,i)*delta)-SUM(u$incG(u),CF(u)*v(u,t)+CG(u)*p(u,t,o,g,i)*delta));
expected_profit.. ep=E=SUM((o,g,i),prob(o,g,i)*zsc(o,g,i));
cvar_calculation1..     CVAR=e=rho-(1/(1-alpha))*SUM((o,g,i),prob(o,g,i)*mu(o,g,i));
cvar_calculation2(o,g,i)..    -zsc(o,g,i)+rho-mu(o,g,i)=L=0;
nodal_balance_mg(b,t,o,g,i)$(incMB(b)).. SUM(u$incGB(u,b),p(u,t,o,g,i))+SUM(u$incRB(u,b),p(u,t,o,g,i))+SUM(u$incSB(u,b),psp(u,t,o,g,i)-psn(u,t,o,g,i))-SUM(l$incORI(l,b),pl(l,t,o,g,i))+SUM(l$incDES(l,b),pl(l,t,o,g,i))=E=pm(b,t,o,g,i)+SUM(u$incDB(u,b),p(u,t,o,g,i));
nodal_balance(b,t,o,g,i)$(NOT incMB(b)).. SUM(u$incGB(u,b),p(u,t,o,g,i))+SUM(u$incRB(u,b),p(u,t,o,g,i))+SUM(u$incSB(u,b),psp(u,t,o,g,i)-psn(u,t,o,g,i))-SUM(l$incORI(l,b),pl(l,t,o,g,i))+SUM(l$incDES(l,b),pl(l,t,o,g,i))=E=SUM(u$incDB(u,b),p(u,t,o,g,i));
trade(t,o,g,i)..  pda(t)+prt(t,o,g,i)=E=SUM(b$incMB(b),pm(b,t,o,g,i));
line_power(l,t,o,g,i).. pl(l,t,o,g,i)=E=Sbase*BL(l)*(SUM(b$incORI(l,b),teta(b,t,o,g,i))-SUM(b$incDES(l,b),teta(b,t,o,g,i)));
line_max(l,t,o,g,i)..   pl(l,t,o,g,i)=L=PL_max(l);
line_min(l,t,o,g,i)..   -PL_max(l)=L=pl(l,t,o,g,i);
voltage_angle_ref(b,t,o,g,i)$(incREF(b))..      teta(b,t,o,g,i)=E=0;
voltage_angle_max(b,t,o,g,i)..       teta(b,t,o,g,i)=L=Pi;
voltage_angle_min(b,t,o,g,i)..       -Pi=L=teta(b,t,o,g,i);
tradeda_min(t).. PV_min=L=pda(t);
tradeda_max(t).. pda(t)=L=PV_max;
tradedart_min(t,o,g,i)..       PV_min=L=pda(t)+prt(t,o,g,i);
tradedart_max(t,o,g,i)..       pda(t)+prt(t,o,g,i)=L=PV_max;
trade_min(b,t,o,g,i)$(incMB(b))..    -PM_max(b)=L=pm(b,t,o,g,i);
trade_max(b,t,o,g,i)$(incMB(b))..    pm(b,t,o,g,i)=L=PM_max(b);
dailyenergy_min(u,o,g,i)$(incD(u))..       SUM(t,p(u,t,o,g,i)*delta)=G=ED_min(u);
demand_max(u,t,o,g,i)$(incD(u))..    p(u,t,o,g,i)=L=PD_max(u,t);
demand_min(u,t,o,g,i)$(incD(u))..    PD_min(u,t)=L=p(u,t,o,g,i);
demand_ramp_down_initial(u,t,o,g,i)$((incD(u)) AND (ORD(t) EQ 1)) ..     -RD_do(u)=L=p(u,t,o,g,i)-PD0(u);
demand_ramp_pick_initial(u,t,o,g,i)$((incD(u)) AND (ORD(t) EQ 1)) ..      p(u,t,o,g,i)-PD0(u)=L=RD_up(u);
demand_ramp_down(u,t,o,g,i)$((incD(u)) AND (ORD(t) GE 2)).. -RD_do(u)=L=p(u,t,o,g,i)-p(u,t-1,o,g,i);
demand_ramp_pick(u,t,o,g,i)$((incD(u)) AND (ORD(t) GE 2)).. p(u,t,o,g,i)-p(u,t-1,o,g,i)=L=RD_up(u);
CPP_max(u,t,o,g,i)$(incG(u))..       p(u,t,o,g,i)=L=PG_max(u)*v(u,t);
CPP_min(u,t,o,g,i)$(incG(u))..       v(u,t)*PG_min(u)=L=p(u,t,o,g,i);
```

```
CPP_ramp_down_initial(u,t,o,g,i)$((incG(u)) AND (ORD(t) EQ 1))..
      -RG_do(u)=L=p(u,t,o,g,i)-PG0(u);
CPP_ramp_up_initial(u,t,o,g,i)$((incG(u)) AND (ORD(t) EQ 1))..
        p(u,t,o,g,i)-PG0(u)=L=RG_up(u);
CCP_ramp_down(u,t,o,g,i)$((incG(u)) AND (ORD(t) GE 2))..     -
    RG_do(u)=L=p(u,t,o,g,i)-p(u,t-1,o,g,i);
CCP_ramp_up(u,t,o,g,i)$((incG(u)) AND (ORD(t) GE 2))..      p(u,t
    ,o,g,i)-p(u,t-1,o,g,i)=L=RG_up(u);
storage_balance_initial(u,t,o,g,i)$((incS(u)) AND (ORD(t) EQ 1))
    ..      es(u,t,o,g,i)=E=ES0(u)+mun*psn(u,t,o,g,i)*delta-psp(u
    ,t,o,g,i)*delta/mup;
storage_balance(u,t,o,g,i)$((incS(u)) AND (ORD(t) GE 2)).. es(u,
    t,o,g,i)=E=es(u,t-1,o,g,i)+mun*psn(u,t,o,g,i)*delta-psp(u,t,o
    ,g,i)*delta/mup;
storage_capacity(u,t,o,g,i)$(incS(u))..    es(u,t,o,g,i)=L=ES_max(
    u);
storage_min(u,t,o,g,i)$(incS(u))..  es(u,t,o,g,i)=G=ES_min(u);
storage_psp_max(u,t,o,g,i)$(incS(u))..     psp(u,t,o,g,i)=L=PS_p(u
    );
storage_psn_max(u,t,o,g,i)$(incS(u))..     psn(u,t,o,g,i)=L=PS_n(u
    );
stochastic_max(u,t,o,g,i)$(incR(u))..      p(u,t,o,g,i)=L=PR_max(i
    ,u,t);
***********************************************
** Model definition **************
MODEL SSDA        /all/;
OPTION OPTCR=0;
OPTION OPTCA=0;
OPTION iterlim=1e8;
*************************************
*****Parameters used to save the results**********
PARAMETER state          'Convergence status';
PARAMETER pda_act        'Actual power traded in the day-ahead
    market';
PARAMETER revenueda_act 'Cost in the day-ahead market';
***********************************************
***** Solution ***********************
SOLVE SSDA USING MIP MAXIMIZING z;
state = SSDA.modelstat;
pda_act(t)=pda.l(t);
revenueda_act=SUM(t,pda_act(t)*lambda_DA_act(t));
*************************************
display state, pda_act, revenueda_act;
```

4.8.2 Robust Optimization Approach

A GAMS code for solving Illustrative Example 4.3 is provided below:

```
* Robust scheduling of the risk-averse VPP in the day-ahead
    energy market
SET b        'Bus'              /b1*b3/;
SET l        'Line'             /l1*l3/;
SET t        'Time period'      /t1*t3/;
SET u        'Unit'             /u1*u5/;
SET incDES(l,b)    'destination bus'        /l1.b2,l2.b3,l3.b3/;
SET incORI(l,b)    'origin bus'             /l1.b1,l2.b1,l3.b2/;
SET incD(u)        'demand'                              /u1*u2/;
SET incG(u)        'conventional power plant'            /u3/;
SET incR(u)        'stochastic production facility'      /u4/;
SET incS(u)        'storage unit'                        /u5/;
SET incMB(b)       'buses connected to the main grid'    /b1/;
SET incREF(b)      'reference bus'                       /b1/;
SET incDB(u,b)     'demand at bus'                /u1.b2,u2.b3/;
SET incGB(u,b)     'conventional power plant at bus'    /u3.b2/;
SET incRB(u,b) 'stochastic production facility at bus' /u4.b3/;
SET incSB(u,b)     'storage unit at bus'                 /u5.b3/;
SCALAR TT    'Number of trading hours in a day' /3/;
SCALAR delta       'Power-energy conversion factor' /1/;
SCALAR mup   'Output conversion efficiency of the storage unit'
    /0.9/;
SCALAR mun   'Input conversion efficiency of the storage unit'
    /0.9/;
SCALAR Sbase       'Base apparent power' /100/;
SCALAR Vbase       'Base voltage' /10/;
SCALAR PV_max      'Maximum power that can be produced by the VPP'
     /40/;
SCALAR PV_min      'Maximum power that can be consumed by the VPP'
     /-60/;
SCALAR GammaDA     'Parameter used to control the conservasim
    level of solution with respect to the uncertain prices
    constraints' /6/;
SCALAR GammaR      'Parameter used to control the protection level
     of constraints' /1/;
***** Demands data
TABLE PD_max(u,t) 'Maximum load level for unit u in time period t'

            t1       t2       t3
u1          20       20       10
u2          30       30       15;
TABLE PD_min(u,t) 'Minimum load level for unit u in time period t
    '
            t1       t2       t3
u1          0        0        0
u2          5        5        5;
TABLE Utility(u,t) 'Utility of unit u for energy consumption in
    time period t'
            t1        t2       t3
u1          40        50       40
u2          50        60       50;
PARAMETER ED_min(u) 'Minimum daily energy consumption requested
    by unit i'
/u1         30
u2          50/;
```

```
PARAMETER RD_up(u) 'Load drop ramping limits for unit u'
/u1       15
u2        25/;
PARAMETER RD_do(u) 'Pick-up ramping limit for unit u'
/u1       15
u2        25/;
PARAMETER PD0(u) 'Initial load level for each unit'
/u1      10
u2       20/;
***** Conventional power plants data
PARAMETER PG_max(u) 'Maximum power production limit for unit u in
     time period t'
/u3       20/;
PARAMETER PG_min(u) 'Minimum power production limit for unit u in
     time period t'
/u3       0/;
PARAMETER RG_up(u) 'Ramping up limits for unit u'
/u3        20/;
PARAMETER RG_do(u) 'Ramping down limit for unit u'
/u3        20/;
PARAMETER CF(u) 'Fixed cost for unit u'
/u3      10/;
PARAMETER CG(u) 'Variable cost for unit u'
/u3       45/;
PARAMETER PG0(u) 'Initial power production of unit u'
/u3        0/;
PARAMETER V0(u) 'Initial status of unit u'
/u3        0/;
***** Storage units data
PARAMETER ES_max(u) 'Energy capacity of the storage unit u'
/u5      10/;
PARAMETER ES_min(u) 'Minimum energy capacity of the storage unit u'
/u5      5/;
PARAMETER PS_p(u) 'Maximum power that can be produced by unit u'
/u5      10/;
PARAMETER PS_n(u) 'Maximum power that can be transferred to unit u'
/u5      10/;
PARAMETER ES0(u) 'Initial energy stored in the storage unit'
/u5      5/;
***** Network data
PARAMETER PL_max(l) 'Capacity of line l'
/l1        50
l2         50
l3         50/;
PARAMETER BL(l) 'Susceptance of line l'
/l1       200
l2        200
l3        200/;
BL(l)=BL(l)/(Sbase/(Vbase**2));
PARAMETER PM_max(b) 'Maximum power that can be bought from the
    main grid'
/b1      100/;
```

```
***** Scenarios of uncertain parameters
TABLE PR_max_n 'Nominal value of available stochastic production
    in time period t'
              t1       t2       t3
u4            6        7        5;
TABLE PR_max_d 'Deviation value of available stochastic
    production in time period t'
              t1       t2       t3
u4            2        3        4;
PARAMETER lambda_DA_n 'Normal value of prices in the day-ahead
    energy market in time period t'
/t1      26
t2       53
t3       31/;
PARAMETER lambda_DA_d 'Deviation value of prices in the day-ahead
     energy market in time period t'
/t1      3
t2       5
t3       6/;
PARAMETER r_RT_n 'Normal value of residual prices in the real-
    time energy market in time period t'
/t1      7
t2       5
t3       -4/;
PARAMETER r_RT_d 'Deviation value of residual prices in the real-
    time energy market in time period t'
/t1      4
t2       7
t3       8/;
***** Actual data of the day-ahead market prices
PARAMETER lambda_DA_act(t) 'Actual day-ahead market price in time
     period t'
/t1       25
t2        50
t3        35/;
VARIABLES
pl(l,t) 'Power flow through line l in time period t'
pda(t) 'Power traded by the VPP in the day-ahead energy market in
     time period t'
prt(t) 'Power traded by the VPP in the real-time energy market in
     time period t'
pm(b,t) 'Power injection at bus b connected to the main grid in
    time period t'
teta(b,t) 'Voltage angle at bus b in time period t'
z 'Objective'
zwr 'Objective without the robust part'
zr 'The robust part in the objective function'
p(u,t)  'Power dispatch of unit u in time period t';
POSITIVE VARIABLES
es(u,t)  'Energy stored at the end of time period t'
psp(u,t) 'Power produced by the storage unit in time period t'
psn(u,t) 'Power transferred to the storage unit in time period t'
kappaDA  'Dual variable used in the robust optimization model'
```

```
psiDA(t) 'Dual variable used in the robust optimization model'
yDA(t)   'Auxiliary variable used in the robust optimization model'
psiRT(t) 'Dual variable used in the robust optimization model'
yRT(t)   'Auxiliary variable used in the robust optimization model'
kappaR(u,t)'Dual variable used in the robust optimization model'
psiR(u,t) 'Dual variable used in the robust optimization model'
yR(u,t) 'Auxiliary variable used in the robust optimization model';
BINARY VARIABLES
v(u,t) 'Binary variable to state the commitment status of unit u in time period t';
***************************************
***** Equations definition ********************
EQUATIONS
obj 'Objective function'
objwr 'Objective without the robust part'
objr 'Robust part in the objective'
nodal_balance_mg 'Power balance at the main grid buses'
nodal_balance 'Power balance at the remaining buses'
trade 'Power traded in the day-ahead and real-time markets'
line_power 'Power flow through lines'
line_max 'Upper bound of lines flows'
line_min 'Lower bound of lines flows'
voltage_angle_ref 'Voltage angle at reference bus'
voltage_angle_max 'Upper bound of voltage angles'
voltage_angle_min 'Lower bound of voltage angles'
tradeda_min 'Lower bound of power traded in the day-ahead market'
tradeda_max 'Upper bound of power traded in the day-ahead market'
tradedart_min 'Lower bound of power traded in the energy markets'
tradedart_max 'Upper bound of power traded in the energy markets'
trade_min 'Lower bound of power injection at each bus'
trade_max 'Upper bound of power injection at each bus'
dailyenergy_min 'Daily minimum energy consumption for each demand'
demand_max 'Upper bound of load level'
demand_min 'Lower bound of load level'
demand_ramp_down_initial 'Load ramp down limitation'
demand_ramp_pick_initial 'Load ramp up limitation'
demand_ramp_down 'Load ramp down limitation'
demand_ramp_pick 'Load ramp up limitation'
CPP_max 'Maximum power production limitation'
CPP_min 'Minimum power production limitation'
CPP_ramp_down_initial 'Ramp down production limitation'
CPP_ramp_up_initial 'Ramp up production limitation'
CCP_ramp_down 'Ramp down production limitation'
CCP_ramp_up 'Ramp up production limitation'
storage_balance_initial 'Power balance equation in the storage unit'
storage_balance 'Power balance equation in the storage unit'
storage_capacity 'Upper bound of energy stored in the storage unit'
```

```
storage_min 'Lower bound of energy stored in the storage unit'
storage_psp_max 'Maximum power that can be produced by storage
    unit'
storage_psn_max 'Maximum power that can be transferred to storage
    unit'
robust_DA1 'Robust constraint to consider uncertain DA prices'
robust_DA2 'Robust constraint to consider uncertain DA prices'
robust_DA3 'Robust constraint to consider uncertain DA prices'
robust_RT1 'Robust constraint to consider uncertain residual
    prices'
robust_RT2 'Robust constraint to consider uncertain residual
    prices'
robust_RT3 'Robust constraint to consider uncertain residual
    prices'
stochastic_max1 'Maximum power production limitation of
    stochastic renewable units in robust form'
stochastic_max2 'Maximum power production limitation of
    stochastic renewable units in robust form'
stochastic_max3 'Maximum power production limitation of
    stochastic renewable units in robust form';
*
obj.. z=E=zwr-zr;
objwr..     zwr=E=SUM(t,lambda_DA_n(t)*(pda(t)*delta+prt(t)*delta
    )+r_RT_n(t)*prt(t)*delta+SUM(u$incD(u),Utility(u,t)*p(u,t)*
    delta)-SUM(u$incG(u),CF(u)*v(u,t)+CG(u)*p(u,t)*delta));
objr..      zr=E=GammaDA*kappaDA+SUM(t,psiDA(t)+psiRT(t));
nodal_balance_mg(b,t)$(incMB(b)).. SUM(u$incGB(u,b),p(u,t))+SUM(
    u$incRB(u,b),p(u,t))+SUM(u$incSB(u,b),psp(u,t)-psn(u,t))-SUM(
    l$incORI(l,b),pl(l,t))+SUM(l$incDES(l,b),pl(l,t))=E=pm(b,t)+
    SUM(u$incDB(u,b),p(u,t));
nodal_balance(b,t)$(NOT incMB(b))..      SUM(u$incGB(u,b),p(u,t))
    +SUM(u$incRB(u,b),p(u,t))+SUM(u$incSB(u,b),psp(u,t)-psn(u,t))
    -SUM(l$incORI(l,b),pl(l,t))+SUM(l$incDES(l,b),pl(l,t))=E=SUM(
    u$incDB(u,b),p(u,t));
trade(t).. pda(t)+prt(t)=E=SUM(b$incMB(b),pm(b,t));
line_power(l,t).. pl(l,t)=E=Sbase*BL(l)*(SUM(b$incORI(l,b),teta(b
    ,t))-SUM(b$incDES(l,b),teta(b,t)));
line_max(l,t)..   pl(l,t)=L=PL_max(l);
line_min(l,t)..   -PL_max(l)=L=pl(l,t);
voltage_angle_ref(b,t)$(incREF(b))..      teta(b,t)=E=0;
voltage_angle_max(b,t)..      teta(b,t)=L=Pi;
voltage_angle_min(b,t)..      -Pi=L=teta(b,t);
tradeda_min(t)..  PV_min=L=pda(t);
tradeda_max(t)..  pda(t)=L=PV_max;
tradedart_min(t)..      PV_min=L=pda(t)+prt(t);
tradedart_max(t)..      pda(t)+prt(t)=L=PV_max;
trade_min(b,t)$(incMB(b))..   -PM_max(b)=L=pm(b,t);
trade_max(b,t)$(incMB(b))..   pm(b,t)=L=PM_max(b);
dailyenergy_min(u)$(incD(u))..      SUM(t,p(u,t)*delta)=G=ED_min(
    u);
demand_max(u,t)$(incD(u))..   p(u,t)=L=PD_max(u,t);
demand_min(u,t)$(incD(u))..   PD_min(u,t)=L=p(u,t);
demand_ramp_down_initial(u,t)$((incD(u)) AND (ORD(t) EQ 1))..
         -RD_do(u)=L=p(u,t)-PD0(u);
```

```
demand_ramp_pick_initial(u,t)$((incD(u)) AND (ORD(t) EQ 1))..
         p(u,t)-PD0(u)=L=RD_up(u);
demand_ramp_down(u,t)$((incD(u)) AND (ORD(t) GE 2)).. -RD_do(u)=L
    =p(u,t)-p(u,t-1);
demand_ramp_pick(u,t)$((incD(u)) AND (ORD(t) GE 2)).. p(u,t)-p(u,
    t-1)=L=RD_up(u);
CPP_max(u,t)$(incG(u))..     p(u,t)=L=PG_max(u)*v(u,t);
CPP_min(u,t)$(incG(u))..     v(u,t)*PG_min(u)=L=p(u,t);
CPP_ramp_down_initial(u,t)$((incG(u)) AND (ORD(t) EQ 1)).. -
    RG_do(u)=L=p(u,t)-PG0(u);
CPP_ramp_up_initial(u,t)$((incG(u)) AND (ORD(t) EQ 1))..    p(u,t
    )-PG0(u)=L=RG_up(u);
CCP_ramp_down(u,t)$((incG(u)) AND (ORD(t) GE 2))..    -RG_do(u)=L
    =p(u,t)-p(u,t-1);
CCP_ramp_up(u,t)$((incG(u)) AND (ORD(t) GE 2))..      p(u,t)-p(u,
    t-1)=L=RG_up(u);
storage_balance_initial(u,t)$((incS(u)) AND (ORD(t) EQ 1))..
          es(u,t)=E=ES0(u)+mun*psn(u,t)*delta-psp(u,t)*delta/mup;
storage_balance(u,t)$((incS(u)) AND (ORD(t) GE 2))..          es(u,
    t)=E=es(u,t-1)+mun*psn(u,t)*delta-psp(u,t)*delta/mup;
storage_capacity(u,t)$(incS(u))..   es(u,t)=L=ES_max(u);
storage_min(u,t)$(incS(u)).. es(u,t)=G=ES_min(u);
storage_psp_max(u,t)$(incS(u))..    psp(u,t)=L=PS_p(u);
storage_psn_max(u,t)$(incS(u))..    psn(u,t)=L=PS_n(u);
robust_DA1(t)..   kappaDA+psiDA(t)=G=Lambda_DA_d(t)*yDA(t);
robust_DA2(t)..   pDA(t)*delta+pRT(t)*delta=L=yDA(t);
robust_DA3(t)..   pDA(t)*delta+pRT(t)*delta=G=-yDA(t);
robust_RT1(t)..   kappaDA+psiRT(t)=G=r_RT_d(t)*yRT(t);
robust_RT2(t)..   pRT(t)*delta=L=yRT(t);
robust_RT3(t)..   pRT(t)*delta=G=-yRT(t);
stochastic_max1(u,t)$(incR(u))..    p(u,t)+GammaR*KappaR(u,t)+
    PsiR(u,t)=L=PR_max_n(u,t);
stochastic_max2(u,t)$(incR(u))..    KappaR(u,t)+PsiR(u,t)=G=
    PR_max_d(u,t)*yR(u,t);
stochastic_max3(u,t)$(incR(u))..    yR(u,t)=G=1;
*******************************************
** Model definition ****************
MODEL RDA       /all/;
OPTION OPTCR=0;
OPTION OPTCA=0;
OPTION iterlim=1e8;
******************************
** Parameters used to save the results***************
PARAMETER state 'Convergence status';
PARAMETER pda_act 'Actual power traded in the day-ahead market';
PARAMETER revenueda_act 'Revenue in the day-ahead market';
********************************************
*** Solution ******************
SOLVE RDA USING MIP MAXIMIZING z;
state = RDA.modelstat;
pda_act(t)=pda.l(t);
revenueda_act=SUM(t,pda_act(t)*lambda_DA_act(t));
*************************************
display state, pda_act, revenueda_act;
```

4.8.3 Hybrid Stochastic-Robust Optimization Approach

A GAMS code for solving Illustrative Example 4.5 is provided below:

```
* Stochastic scheduling of the risk-neutral VPP in the day-ahead
    energy market
SET b          'Bus'                /b1*b3/;
SET l          'Line'               /l1*l3/;
SET t          'Time period'        /t1*t3/;
SET u          'Unit'               /u1*u5/;
SET o          'scenarios'          /o1*o2/;
SET g          'scenarios'          /g1*g2/;
SET incDES(l,b)     'destination bus' /l1.b2,l2.b3,l3.b3/;
SET incORI(l,b)     'origin bus' /l1.b1,l2.b1,l3.b2/;
SET incD(u)         'demand' /u1*u2/;
SET incG(u)         'conventional power plant' /u3/;
SET incR(u)         'stochastic production facility' /u4/;
SET incS(u)         'storage unit' /u5/;
SET incMB(b)        'buses connected to the main grid' /b1/;
SET incREF(b)       'reference bus' /b1/;
SET incDB(u,b)      'demand at bus' /u1.b2,u2.b3/;
SET incGB(u,b)      'conventional power plant at bus' /u3.b2/;
SET incRB(u,b)      'stochastic production facility at bus' /u4.b3/;
SET incSB(u,b)      'storage unit at bus' /u5.b3/;
SCALAR TT 'Number of trading hours in a day' /3/;
SCALAR delta 'Power-energy conversion factor' /1/;
SCALAR mup 'Output conversion efficiency of the storage unit'
    /0.9/;
SCALAR mun 'Input conversion efficiency of the storage unit'
    /0.9/;
SCALAR Sbase 'Base apparent power' /100/;
SCALAR Vbase 'Base voltage' /10/;
SCALAR PV_max 'Maximum power that can be produced by the VPP'/40/;
SCALAR PV_min 'Maximum power that can be consumed by the VPP'
    /-60/;
SCALAR beta 'Weight of CVaR' /1/;
SCALAR alpha 'Confidence level used for calculating CVaR' /0.8/;
SCALAR GammaR 'Parameter used to control the protection level of
    constraints' /1/;
***** Demands data
TABLE PD_max(u,t) 'Maximum load level for unit u in time period t'
          t1      t2      t3
u1        20      20      10
u2        30      30      15;
TABLE PD_min(u,t) 'Minimum load level for unit u in time period t'
          t1      t2      t3
u1        0       0       0
u2        5       5       5;
TABLE Utility(u,t) 'Utility of unit u for energy consumption in
    time period t'
          t1       t2       t3
u1        40       50       40
u2        50       60       50;
PARAMETER ED_min(u) 'Minimum daily energy consumption requested
    by unit i'
/u1        30
u2         50/;
```

```
PARAMETER RD_up(u) 'Load drop ramping limits for unit u'
/u1       15
u2        25/;
PARAMETER RD_do(u) 'Pick-up ramping limit for unit u'
/u1       15
u2        25/;
PARAMETER PD0(u) 'Initial load level for each unit'
/u1      10
u2       20/;
***** Conventional power plants data
PARAMETER PG_max(u) 'Maximum power production limit for unit u in
     time period t'
/u3       20/;
PARAMETER PG_min(u) 'Minimum power production limit for unit u in
     time period t'
/u3       0/;
PARAMETER RG_up(u) 'Ramping up limits for unit u'
/u3        20/;
PARAMETER RG_do(u) 'Ramping down limit for unit u'
/u3        20/;
PARAMETER CF(u) 'Fixed cost for unit u'
/u3      10/;
PARAMETER CG(u) 'Variable cost for unit u'
/u3      45/;
PARAMETER PG0(u) 'Initial power production of unit u'
/u3        0/;
PARAMETER V0(u) 'Initial status of unit u'
/u3        0/;
***** Storage units data
PARAMETER ES_max(u) 'Energy capacity of the storage unit u'
/u5      10/;
PARAMETER ES_min(u) 'Minimum energy capacity of the storage unit u'
/u5      5/;
PARAMETER PS_p(u) 'Maximum power that can be produced by unit u'
/u5      10/;
PARAMETER PS_n(u) 'Maximum power that can be transferred to unit u'
/u5      10/;
PARAMETER ES0(u) 'Initial energy stored in the storage unit'
/u5      5/;
***** Network data
PARAMETER PL_max(l) 'Capacity of line l'
/l1        50
l2         50
l3         50/;
PARAMETER BL(l) 'Susceptance of line l'
/l1       200
l2        200
l3        200/;
BL(l)=BL(l)/(Sbase/(Vbase**2));
PARAMETER PM_max(b) 'Maximum power that can be bought from the
     main grid'
/b1      100/;
```

```
***** Scenarios of uncertain parameters
TABLE PR_max_n 'Nominal value of available stochastic production in time period t'
              t1        t2        t3
u4            6         7         5;
TABLE PR_max_d 'Deviation value of available stochastic production in time period t'
              t1        t2        t3
u4            2         3         4;
TABLE lambda_DA 'Scenarios of prices in the day-ahead energy market in time period t'
         t1         t2         t3
o1       20         40         30
o2       30         50         40;
TABLE lambda_RT 'Scenarios of prices in the real-time energy market in time period t'
         t1         t2         t3
o1.g1    25         50         25
o1.g2    10         45         28
o2.g1    35         60         35
o2.g2    20         55         38;
PARAMETER prob(o,g) 'Probability of scenarios'
/o1.g1        0.16
o1.g2        0.24
o2.g1        0.24
o2.g2        0.36/;
***** Actual data of the day-ahead market prices
PARAMETER lambda_DA_act(t) 'Actual day-ahead market price in time period t'
/t1         25
t2         50
t3         35/;
VARIABLES
pl(l,t,o,g)         'Power flow through line l in time period t'
pda(t)              'Power traded by the VPP in the day-ahead energy market in time period t'
prt(t,o,g)          'Power traded by the VPP in the real-time energy market in time period t'
pm(b,t,o,g)         'Power injection at bus b connected to the main grid in time period t'
teta(b,t,o,g)       'Voltage angle at bus b in time period t'
z                   'Objective'
zsc(o,g)            'Objective in each scenario'
p(u,t,o,g)          'Power dispatch of unit u in time period t'
rho                 'Value at risk'
cvar                'Conditional value at risk metric'
ep                  'Expected profit';
POSITIVE VARIABLES
es(u,t,o,g)               'Energy stored at the end of time period t'
psp(u,t,o,g)              'Power produced by the storage unit in time period t'
psn(u,t,o,g)              'Power transferred to the storage unit in time period t'
```

```
mu(o,g)                  'Auxiliary variable to calculate the
    conditional value at risk'
KappaR(u,t,o,g)          'Dual variable used in the robust
    optimization model'
PsiR(u,t,o,g)            'Dual variable used in the robust
    optimization model'
yR(u,t,o,g)              'Auxiliary variable used in the robust
    optimization model';
BINARY VARIABLES
v(u,t)            'Binary variable to state the commitment status
     of unit u in time period t';
****************************************************
*** Equations definition *****
EQUATIONS
obj 'Objective function'
objscen 'Objective in each scenario'
expected_profit 'Expected profit'
cvar_calculation1 'Calculating the conditional value at risk'
cvar_calculation2 'Calculating the conditional value at risk'
nodal_balance_mg 'Power balance at the main grid buses'
nodal_balance 'Power balance at the remaining buses'
trade 'Power traded in the day-ahead and real-time markets'
line_power 'Power flow through lines'
line_max 'Upper bound of lines flows'
line_min 'Lower bound of lines flows'
voltage_angle_ref 'Voltage angle at reference bus'
voltage_angle_max 'Upper bound of voltage angles'
voltage_angle_min 'Lower bound of voltage angles'
tradeda_min 'Lower bound of power traded in the day-ahead market'
tradeda_max 'Upper bound of power traded in the day-ahead market'
tradedart_min 'Lower bound of power traded in the energy markets'
tradedart_max 'Upper bound of power traded in the energy markets'
trade_min 'Lower bound of power injection at each bus'
trade_max 'Upper bound of power injection at each bus'
dailyenergy_min 'Daily minimum energy consumption for each demand'
demand_max 'Upper bound of load level'
demand_min 'Lower bound of load level'
demand_ramp_down_initial 'Load ramp down limitation'
demand_ramp_pick_initial 'Load ramp up limitation'
demand_ramp_down 'Load ramp down limitation'
demand_ramp_pick 'Load ramp up limitation'
CPP_max 'Maximum power production limitation'
CPP_min 'Minimum power production limitation'
CPP_ramp_down_initial 'Ramp down production limitation'
CPP_ramp_up_initial 'Ramp up production limitation'
CCP_ramp_down 'Ramp down production limitation'
CCP_ramp_up 'Ramp up production limitation'
storage_balance_initial 'Power balance equation in the storage
    unit'
storage_balance 'Power balance equation in the storage unit'
storage_capacity 'Upper bound of energy stored in the storage
    unit'
storage_min 'Lower bound of energy stored in the storage unit'
```

```
storage_psp_max 'Maxmimum power that can be produced by storage
    unit'
storage_psn_max 'Maxmimum power that can be transferred to
    storage unit'
stochastic_max1 'Maximum power production limitation in robust
    form'
stochastic_max2 'Maximum power production limitation in robust
    form'
stochastic_max3 'Maximum power production limitation in robust
    form';
*
obj.. z=E=ep+beta*CVAR;
objscen(o,g)..    zsc(o,g)=E=SUM(t,lambda_DA(o,t)*pda(t)*delta+
    lambda_RT(o,g,t)*prt(t,o,g)*delta+SUM(u$incD(u),Utility(u,t)*
    p(u,t,o,g)*delta)-SUM(u$incG(u),CF(u)*v(u,t)+CG(u)*p(u,t,o,g)
    *delta));
expected_profit.. ep=E=SUM((o,g),prob(o,g)*zsc(o,g));
cvar_calculation1..     CVAR=e=rho-(1/(1-alpha))*SUM((o,g),prob(o
    ,g)*mu(o,g));
cvar_calculation2(o,g)..      -zsc(o,g)+rho-mu(o,g)=L=0;
nodal_balance_mg(b,t,o,g)$(incMB(b))..    SUM(u$incGB(u,b),p(u,t,
    o,g))+SUM(u$incRB(u,b),p(u,t,o,g))+SUM(u$incSB(u,b),psp(u,t,o
    ,g)-psn(u,t,o,g))-SUM(l$incORI(l,b),pl(l,t,o,g))+SUM(l$incDES
    (l,b),pl(l,t,o,g))=E=pm(b,t,o,g)+SUM(u$incDB(u,b),p(u,t,o,g));
nodal_balance(b,t,o,g)$(NOT incMB(b))..    SUM(u$incGB(u,b),p(u,t,
    o,g))+SUM(u$incRB(u,b),p(u,t,o,g))+SUM(u$incSB(u,b),psp(u,t,o
    ,g)-psn(u,t,o,g))-SUM(l$incORI(l,b),pl(l,t,o,g))+SUM(l$incDES
    (l,b),pl(l,t,o,g))=E=SUM(u$incDB(u,b),p(u,t,o,g));
trade(t,o,g)..    pda(t)+prt(t,o,g)=E=SUM(b$incMB(b),pm(b,t,o,g));
line_power(l,t,o,g)..   pl(l,t,o,g)=E=Sbase*BL(l)*(SUM(b$incORI(l
    ,b),teta(b,t,o,g))-SUM(b$incDES(l,b),teta(b,t,o,g)));
line_max(l,t,o,g)..     pl(l,t,o,g)=L=PL_max(l);
line_min(l,t,o,g)..     -PL_max(l)=L=pl(l,t,o,g);
voltage_angle_ref(b,t,o,g)$(incREF(b))..  teta(b,t,o,g)=E=0;
voltage_angle_max(b,t,o,g)..  teta(b,t,o,g)=L=Pi;
voltage_angle_min(b,t,o,g)..  -Pi=L=teta(b,t,o,g);
tradeda_min(t).. PV_min=L=pda(t);
tradeda_max(t).. pda(t)=L=PV_max;
tradedart_min(t,o,g)..  PV_min=L=pda(t)+prt(t,o,g);
tradedart_max(t,o,g)..  pda(t)+prt(t,o,g)=L=PV_max;
trade_min(b,t,o,g)$(incMB(b))..      -PM_max(b)=L=pm(b,t,o,g);
trade_max(b,t,o,g)$(incMB(b))..      pm(b,t,o,g)=L=PM_max(b);
dailyenergy_min(u,o,g)$(incD(u))..   SUM(t,p(u,t,o,g)*delta)=G=
    ED_min(u);
demand_max(u,t,o,g)$(incD(u))..      p(u,t,o,g)=L=PD_max(u,t);
demand_min(u,t,o,g)$(incD(u))..      PD_min(u,t)=L=p(u,t,o,g);
demand_ramp_down_initial(u,t,o,g)$((incD(u)) AND (ORD(t) EQ 1))..
     -RD_do(u)=L=p(u,t,o,g)-PD0(u);
demand_ramp_pick_initial(u,t,o,g)$((incD(u)) AND (ORD(t) EQ 1))..
     p(u,t,o,g)-PD0(u)=L=RD_up(u);
demand_ramp_down(u,t,o,g)$((incD(u)) AND (ORD(t) GE 2))..   -
    RD_do(u)=L=p(u,t,o,g)-p(u,t-1,o,g);
demand_ramp_pick(u,t,o,g)$((incD(u)) AND (ORD(t) GE 2))..   p(u,t
    ,o,g)-p(u,t-1,o,g)=L=RD_up(u);
```

```
CPP_max(u,t,o,g)$(incG(u)).. p(u,t,o,g)=L=PG_max(u)*v(u,t);
CPP_min(u,t,o,g)$(incG(u)).. v(u,t)*PG_min(u)=L=p(u,t,o,g);
CPP_ramp_down_initial(u,t,o,g)$((incG(u)) AND (ORD(t) EQ 1))..
        -RG_do(u)=L=p(u,t,o,g)-PG0(u);
CPP_ramp_up_initial(u,t,o,g)$((incG(u)) AND (ORD(t) EQ 1))..
          p(u,t,o,g)-PG0(u)=L=RG_up(u);
CCP_ramp_down(u,t,o,g)$((incG(u)) AND (ORD(t) GE 2))..        -
    RG_do(u)=L=p(u,t,o,g)-p(u,t-1,o,g);
CCP_ramp_up(u,t,o,g)$((incG(u)) AND (ORD(t) GE 2)).. p(u,t,o,g)-
    p(u,t-1,o,g)=L=RG_up(u);
storage_balance_initial(u,t,o,g)$((incS(u)) AND (ORD(t) EQ 1))..
      es(u,t,o,g)=E=ES0(u)+mun*psn(u,t,o,g)*delta-psp(u,t,o,g)*
    delta/mup;
storage_balance(u,t,o,g)$((incS(u)) AND (ORD(t) GE 2))..    es(u,
    t,o,g)=E=es(u,t-1,o,g)+mun*psn(u,t,o,g)*delta-psp(u,t,o,g)*
    delta/mup;
storage_capacity(u,t,o,g)$(incS(u))..     es(u,t,o,g)=L=ES_max(u)
    ;
storage_min(u,t,o,g)$(incS(u))..     es(u,t,o,g)=G=ES_min(u);
storage_psp_max(u,t,o,g)$(incS(u))..      psp(u,t,o,g)=L=PS_p(u);
storage_psn_max(u,t,o,g)$(incS(u))..      psn(u,t,o,g)=L=PS_n(u);
stochastic_max1(u,t,o,g)$(incR(u))..      p(u,t,o,g)+GammaR*
    KappaR(u,t,o,g)+PsiR(u,t,o,g)=L=PR_max_n(u,t);
stochastic_max2(u,t,o,g)$(incR(u))..      KappaR(u,t,o,g)+PsiR(u,
    t,o,g)=G=PR_max_d(u,t)*yR(u,t,o,g);
stochastic_max3(u,t,o,g)$(incR(u))..      yR(u,t,o,g)=G=1;
************************************************
***Model definition *****************
MODEL SSDA       /all/;
OPTION OPTCR=0;
OPTION OPTCA=0;
OPTION iterlim=1e8;
************************************************
*** Parameters used to save the results*******
PARAMETER state 'Convergence status';
PARAMETER pda_act 'Actual power traded in the day-ahead market';
PARAMETER revenueda_act 'Revenue in the day-ahead market';
******************************
*** Solution ********************
SOLVE SSDA USING MIP MAXIMIZING z;
state = SSDA.modelstat;
pda_act(t)=pda.l(t);
revenueda_act=SUM(t,pda_act(t)*lambda_DA_act(t));
*********************************************
display state, pda_act, revenueda_act;
```

4.8.4 *Adaptive Robust Optimization Approach*

A GAMS code for solving Illustrative Example 4.8 is provided below:

```
SETS
d         'demands'                          /d1*d1/
r         'stochastic renewable units'       /r1*r1/
s         'storage units'                    /s1*s1/
t         'time periods'                     /t1*t4/
it        'iteration counter'                /it1*it5/;
ALIAS(it,itt);
SCALAR PmarketMAX          'maximum power traded in the market'
    /40/;
PARAMETER price(t)         'energy market price'
/t1       20
t2        50
t3        35
t4        31/;
TABLE PRAv(r,t)            'average renewable generation level'
          t1       t2       t3       t4
r1        60       20       40       10;
TABLE PRFluc(r,t)          'renewable generation fluctuation level'
          t1       t2       t3       t4
r1        20       10       15       5;
TABLE STORAGEDATA(s,*)     'data of storage unit'
          PCmax    PDmax    EMAX     E0       EFFC     EFFD
s1        20       20       40       20       0.9      0.9;
TABLE EminS(s,t)           'minimum energy level of storage unit'
          t1       t2       t3       t4
s1        0        0        0        20;
TABLE PDmin(d,t)           'lower demand bound'
          t1       t2       t3       t4
d1        0        0        10       15;
TABLE PDmax(d,t)           'upper demand bound'
          t1       t2       t3       t4
d1        20       20       25       30;
PARAMETER EDmin(d)        'minimum energy consumption'
/d1       60/;
PARAMETER GammaR(r)        'uncertainty budget'
/r1       2/;
SCALAR    M        'large positive constant'          /1000/;
SCALAR ITAUX       'auxiliary parameter'              /0/;
PARAMETER
PE_AUX(t)          'power traded in the day-ahead energy market'
PRAAUX(r,t,it)     'available renewable generation';
PRAAUX(r,t,it)=0;
VARIABLES
Z_MP               'objective function of master problem'
Z_SP               'objective function of subproblem'
PE(t)              'power traded in the day-ahead energy market'
theta              'auxiliary variable'
lambda(t)          'dual variable'
phiS(s,t)          'dual variable'
xi(d)              'dual variable';
```

```
theta.up=1e6;
POSITIVE VARIABLES
ES(s,t,it)        'energy stored in storage c in time period t and
     iteration it'
PD(d,t,it)        'power consumption of demand d in time period t
     and iteration it'
PR(r,t,it)        'power production of renewable unit r in time
     period t and iteration it'
PRA(r,t)          'available power production of renewable unit r
     in time period t'
PSC(s,t,it)       'charging power of storage c in time period t
     and iteration it'
PSD(s,t,it)       'discharging power of storage c in time period t
     and iteration it'
muDL(d,t)         'dual variable'
muDU(d,t)         'dual variable'
muR(r,t)          'dual variable'
muSC(s,t)         'dual variable'
muSD(s,t)         'dual variable'
muSL(s,t)         'dual variable'
muSU(s,t)         'dual variable'
ZRP(r,t)          'auxiliary variable'
ZRN(r,t)          'auxiliary variable';
BINARY VARIABLES
URP(r,t)          'renewable generation at upper bound'
URN(r,t)          'renewable generation at lower bound';
EQUATIONS
MP_of     'Master problem: objective function'
MP_1      'Master problem: market limits I'
MP_2      'Master problem: market limits II'
MP_3      'Master problem: bounds for theta'
MP_4      'Master problem: balance in the VPP'
MP_5      'Master problem: demand limits I'
MP_6      'Master problem: demand limits II'
MP_7      'Master problem: minimum energy consumption'
MP_8      'Master problem: renewable generation limit'
MP_9      'Master problem: charging power limit'
MP_10     'Master problem: discharging power limit'
MP_11     'Master problem: energy evolution in the storage I'
MP_12     'Master problem: energy evolution in the storage II'
MP_13     'Master problem: energy limits in the storage I'
MP_14     'Master problem: energy limits in the storage II'
SP_of     'Subproblem: objective function'
SP_1      'Subproblem: available renewable generation'
SP_2      'Subproblem: uncertainty'
SP_3      'Subproblem: uncertainty budget'
SP_4      'Subproblem: auxiliary constraint'
SP_5      'Subproblem: auxiliary constraint'
SP_6      'Subproblem: auxiliary constraint'
SP_7      'Subproblem: auxiliary constraint'
SP_8      'Subproblem: auxiliary constraint'
SP_9      'Subproblem: auxiliary constraint'
SP_10     'Subproblem: dual constraint'
SP_11     'Subproblem: dual constraint'
SP_12     'Subproblem: dual constraint'
```

```
SP_13     'Subproblem: dual constraint'
SP_14     'Subproblem: dual constraint'
SP_15     'Subproblem: dual constraint'
SP_16     'Subproblem: dual constraint'
SP_17     'Subproblem: dual constraint'
balance   'power balance'
RPlim     'renewable production limits'
CPlim     'charging power limits'
DPlim     'discharging power limits'
ESbal1    'energy balance in the storage units I'
ESbal2    'energy balance in the storage units II'
ESlim1    'energy limits in the storage unit I'
ESlim2    'energy limits in the storage unit II';
MP_of..            Z_MP=e=sum(t,price(t)*PE(t))+theta;
MP_1(t)..    -PmarketMAX=l=PE(t);
MP_2(t)..    PE(t)=l=PmarketMAX;
MP_3(t,itt)$(ord(itt) LT (ITAUX+1))..        theta=l=0;
MP_4(t,itt)$(ord(itt) LT (ITAUX+1))..       PE(t)=E=sum(r,PR(r,t,
    itt))+sum(s,PSD(s,t,itt)-PSD(s,t,itt))-sum(d,PD(d,t,itt));
MP_5(d,t,itt)$(ord(itt) LT (ITAUX+1))..     PDmin(d,t)=l=PD(d,t,
    itt);
MP_6(d,t,itt)$(ord(itt) LT (ITAUX+1))..     PD(d,t,itt)=l=PDmax(d,
    t);
MP_7(d,itt)$(ord(itt) LT (ITAUX+1))..       sum(t,PD(d,t,itt))=g=
    EDmin(d);
MP_8(r,t,itt)$(ord(itt) LT (ITAUX+1))..     PR(r,t,itt)=l=PRAAUX(r
    ,t,itt);
MP_9(s,t,itt)$(ord(itt) LT (ITAUX+1))..     PSC(s,t,itt)=l=
    STORAGEDATA(s,'PCmax');
MP_10(s,t,itt)$(ord(itt) LT (ITAUX+1))..    PSD(s,t,itt)=l=
    STORAGEDATA(s,'PDmax');
MP_11(s,t,itt)$((ord(itt) LT (ITAUX+1)) AND (ord(t) GT 1))..
          ES(s,t,itt)=e=ES(s,t-1,itt)+PSC(s,t,itt)*STORAGEDATA(s,
    'EFFC')-PSD(s,t,itt)/STORAGEDATA(s,'EFFD');
MP_12(s,itt)$(ord(itt) LT (ITAUX+1))..      ES(s,'t1',itt)=e=
    STORAGEDATA(s,'E0')+PSC(s,'t1',itt)*STORAGEDATA(s,'EFFC')-PSD
    (s,'t1',itt)/STORAGEDATA(s,'EFFD');
MP_13(s,t,itt)$(ord(itt) LT (ITAUX+1))..    EminS(s,t)=l=ES(s,t,
    itt);
MP_14(s,t,itt)$(ord(itt) LT (ITAUX+1))..    ES(s,t,itt)=l=
    STORAGEDATA(s,'EMAX');
SP_of..                        Z_SP=e=-(sum(t,lambda(t)*PE_AUX(
    t)+sum(d,muDL(d,t)*PDmin(d,t)-muDU(d,t)*PDmax(d,t))-sum(r,muR
    (r,t)*PRAv(r,t)-ZRN(r,t)*PRFluc(r,t)+ZRP(r,t)*PRFluc(r,t))-
    sum(s,muSC(s,t)*STORAGEDATA(s,'PCmax')+muSD(s,t)*STORAGEDATA(
    s,'PDmax')-muSL(s,t)*EminS(s,t)+muSU(s,t)*STORAGEDATA(s,'Emax
    ')))+sum(d,xi(d)*EDmin(d))+sum(s,STORAGEDATA(s,'E0')*phiS(s,'
    t1')));
SP_1(r,t).. PRA(r,t)=e=PRAv(r,t)-URN(r,t)*PRfluc(r,t)+URP(r,t)*
    PRfluc(r,t);
SP_2(r,t).. URN(r,t)+URP(r,t)=l=1;
SP_3(r)..   sum(t,URN(r,t)+URP(r,t))=l=GammaR(r);
SP_4(r,t).. ZRN(r,t)=l=M*URN(r,t);
```

```
SP_5(r,t).. 0=l=MUR(r,t)-ZRN(r,t);
SP_6(r,t).. MUR(r,t)-ZRN(r,t)=l=M*(1-URN(r,t));
SP_7(r,t).. ZRP(r,t)=l=M*URP(r,t);
SP_8(r,t).. 0=l=MUR(r,t)-ZRP(r,t);
SP_9(r,t).. MUR(r,t)-ZRP(r,t)=l=M*(1-URP(r,t));
SP_10(r,t)..      lambda(t)-MUR(r,t)=l=0;
SP_11(s,t)..      lambda(t)-MUSD(s,t)+(1/STORAGEDATA(s,'EFFD'))*
    phiS(s,t)=l=0;
SP_12(s,t)..      -lambda(t)-MUSC(s,t)+STORAGEDATA(s,'EFFC')*phiS
    (s,t)=l=0;
SP_13(d,t)..      -lambda(t)+MUDL(d,t)-MUDU(d,t)+XI(d)=e=0;
SP_14(t).. lambda(t)=l=M;
SP_15(t).. -lambda(t)=l=M;
SP_16(s,t)$(ord(t) LT 4)..     phiS(s,t)-phiS(s,t+1)+MUSL(s,t)-
    MUSU(s,t)=e=0;
SP_17(s,t)$(ord(t) EQ 4)..     phiS(s,t)+MUSL(s,t)-MUSU(s,t)=e=0;
MODEL Master /MP_of,MP_1,MP_2,MP_3,MP_4,MP_5,MP_6,MP_7,MP_8,MP_9,
    MP_10,MP_11,MP_12,MP_13,MP_14/;
MODEL Subproblem /SP_of,SP_1,SP_2,SP_3,SP_4,SP_5,SP_6,SP_7,SP_8,
    SP_9,SP_10,SP_11,SP_12,SP_13,SP_14,SP_15,SP_16,SP_17/;
OPTION optcr=0;
************
***Step 1***
*Iteration counter it=0
SCALAR TOL       tolerance              /0/;
SCALAR LB        lower bound            /-1e10/;
SCALAR UB        upper bound            /1e10/;
************
***Step 2***
SOLVE Master using lp maximizing Z_MP;
DISPLAY PE.l, Z_MP.l, theta.l;
************
***Step 3***
UB=Z_MP.l;
DISPLAY UB;
************
***Step 4***
PE_AUX(t)=PE.l(t);
************
***Step 5***
SOLVE Subproblem using mip minimizing Z_SP;
DISPLAY PRA.l, Z_SP.l;
************
***Step 6***
LB=max(sum(t,price(t)*PE.l(t))+Z_SP.l,LB);
DISPLAY LB;
************
***Step 7***
***Step 8***
LOOP(it$((UB-LB) GT TOL),
************
```

```
***Step 9***
PRAAUX(r,t,it)=PRA.l(r,t);
ITAUX=ORD(IT);
DISPLAY ITAUX, PRAAUX;
************
***Step 2***
SOLVE Master using lp maximizing Z_MP;
DISPLAY PE.l, Z_MP.l, theta.l;
************
***Step 3***
UB=Z_MP.l;
DISPLAY UB;
************
***Step 4***
PE_AUX(t)=PE.l(t);
************
***Step 5***
SOLVE Subproblem using mip minimizing Z_SP;
DISPLAY PRA.l, Z_SP.l;
************
***Step 6***
LB=max(sum(t,price(t)*PE.l(t))+Z_SP.l,LB);
DISPLAY UB,LB;
);
DISPLAY PE.l;
```

References

1. Baringo, A., Baringo, L.: A stochastic adaptive robust optimization approach for the offering strategy of a virtual power plant. IEEE Trans. Power Syst. **32**(5), 3492–3504 (2017)
2. Baringo, A., Conejo, A.J.: Offering strategy via robust optimization. IEEE Trans. Power Syst. **26**(3), 1418–1425 (2011)
3. Ben-Tal, A., Goryashko, A., Guslitzer, E., Nemirovski, A.: Adjustable robust solutions of uncertain linear programs. Math. Program. **99**(2), 351–376 (2004)
4. Bertsimas, D., Sim, M.: Robust discrete optimization and network flows. Math. Program. **98**(1–3), 49–71 (2003)
5. Bertsimas, D., Sim, M.: The price of robustness. Oper. Res. **52**(1), 35–53 (2004)
6. Conejo, A.J., Carrión, M., Morales, J.M.: Decision Making Under Uncertainty in Electricity Markets. Springer, New York, US (2010)
7. Floudas, C.A.: Nonlinear and Mixed-Integer Optimization. Fundamentals and Applications. Oxford University Press, New York (1995)
8. GAMS.: Available at www.gams.com/ (2020)
9. ILOG CPLEX.: Available at www.ilog.com/products/cplex/ (2019)
10. Jiang, R., Zhang, M., Li, G., Guan, Y.: Two-stage network constrained robust unit commitment problem. Eur. J. Oper. Res. **234**(3), 751–762 (2014)
11. Rahimiyan, M., Baringo, L: Strategic bidding for a virtual power plant in the day-ahead and real-time markets: A price-taker robust optimization approach. IEEE Trans. Power Syst. **31**(4), 2676–2687 (2016)
12. Rahimiyan, M., Baringo, L: Real-time energy management of a smart virtual power plant. IET Gener. Transm. Distrib. **13**(11), 2015–2023 (2018)
13. Rockafellar, R.T., Uryasev, S.P.: Optimization of conditional value-at-risk. J. Risk **2**, 21–42 (2000)
14. Sioshansi, R., Conejo, A.J.: Optimization in Engineering. Models and Algorithms. Springer, New York, US (2017)
15. Zeng, B., Zhao, L.: Solving two-stage robust optimization problems using a column-and-constraint generation method. Oper. Res. Lett. **41**(5), 457–461 (2013)

Chapter 5
Optimal Scheduling of a Virtual Power Plant in Energy and Reserve Markets

This chapter analyzes the scheduling problem of a virtual power plant (VPP) participating in energy and reserve electricity markets. On one hand, the VPP sells or buys energy in the energy market with the aim of maximizing its profit while complying with the technical constraints of the different energy assets in the VPP. On the other hand, the VPP uses the flexibility of these energy resources to participate in the reserve electricity market, i.e., the VPP uses its flexibility to increase or decrease its total power production if requested by the system operator. This problem includes uncertainties in the market prices, the available production levels of stochastic renewable generating units, and the requests for reserve deployments. These uncertainties are modeled using both scenarios and confidence bounds, and the problem is formulated using different approaches based on stochastic programming and robust optimization.

5.1 Introduction

Virtual power plants (VPPs) aim at maximizing its expected profit by participating in different electricity markets. In this vein, Chaps. 3 and 4 provide different models to optimally decide the participation of VPPs in energy markets, including the day-ahead (DA), real-time (RT), and futures markets. In addition to energy markets, VPPs may use the flexibility of their energy resources to participate in other electricity markets such as reserve markets. The optimal participation of VPPs in both energy and reserve electricity markets is the problem analyzed in this chapter.

According to the offers submitted by participants in reserve markets, the system operator allocates some capacity to each of them. Then, in real time and depending on the system conditions, the system operator can request the participants in the reserve market to increase or decrease their production up to the scheduled capacity. The reserve market is divided into the up- and down-reserve markets depending on whether participants are committed to increase or decrease its production if requested, respectively.

For the sake of simplicity, this chapter analyzes only the DA energy and reserve electricity markets. This means that the VPP makes its scheduling decisions in these two markets one day in advance. Therefore, the VPP faces a number of uncertainties when it makes its decisions, including the uncertainties in the market prices, the available production levels of stochastic renewable generating units, as well as the requests of the system operator to deploy reserves. These uncertainties are modeled using both scenarios and confidence bounds so that the problem is formulated using stochastic programming and robust optimization models, respectively.

The remainder of this chapter is organized as follows. Section 5.2 defines the notation used in this chapter. Section 5.3 provides a deterministic formulation of the scheduling problem of a VPP in energy and reserve electricity markets, in which uncertain parameters are modeled using single forecasts. The deterministic model in Sect. 5.3 is extended in Sects. 5.4 and 5.5 using stochastic programming and adaptive robust optimization models, respectively. In these two models, uncertainties are characterized using scenarios and confidence bounds, respectively. Sections 5.3–5.5 include a number of clarifying examples. Section 5.6 summarizes the chapter and suggests some references for further reading. Finally, Sect. 5.7 provides the GAMS codes used to solve some of the examples described in Sects. 5.3–5.5.

L. Baringo, M. Rahimiyan, *Virtual Power Plants and Electricity Markets*, https://doi.org/10.1007/978-3-030-47602-1_5

5.2 Notation

The main notation used in this chapter is provided below for quick reference. Other symbols are defined as needed throughout the chapter. Note that if the symbols below include a subscript ω, it denotes their values in scenario ω.

5.2.1 Indexes

The indexes of the scheduling problem of a VPP in the energy and reserve markets are:

c Conventional power plants.
d Demands.
r Stochastic renewable generating units.
s Storage units.
t Time periods.
ω Scenarios.
ν Iterations of the column-and-constraint generation algorithm.

5.2.2 Sets

The sets of the scheduling problem of a VPP in the energy and reserve markets are:

Ω^{C} Conventional power plants.
Ω^{D} Demands.
Ω^{R} Stochastic renewable generating units.
Ω^{S} Storage units.
Ω^{T} Time periods.
Ω^{W} Scenarios.

5.2.3 Parameters

The parameters of the scheduling problem of a VPP in the energy and reserve markets are:

$C_c^{\mathrm{C,F}}$ Online cost of conventional power plant c [\$/h].
$C_c^{\mathrm{C,V}}$ Variable cost of conventional power plant c [\$/MWh].
$\underline{E}_d^{\mathrm{D}}$ Minimum energy consumption of demand d throughout the planning horizon [MWh].
$\underline{E}_{st}^{\mathrm{S}}$ Lower bound of the energy stored in storage unit s in time period t [MWh].
$\overline{E}_{st}^{\mathrm{S}}$ Upper bound of the energy stored in storage unit s in time period t [MWh].
$K_t^{\mathrm{R+}}$ Up-reserve request coefficient in time period t [pu].
$K_t^{\mathrm{R-}}$ Down-reserve request coefficient in time period t [pu].
$\underline{P}_{dt}^{\mathrm{D}}$ Lower bound of the power consumption of demand d in time period t [MW].
$\overline{P}_{dt}^{\mathrm{D}}$ Upper bound of the power consumption of demand d in time period t [MW].
$\underline{P}^{\mathrm{E}}$ Lower bound of the power traded in the energy market [MW].
$\overline{P}^{\mathrm{E}}$ Upper bound of the power traded in the energy market [MW].
$\overline{P}^{\mathrm{R+}}$ Upper bound of the power traded in the up-reserve market [MW].
$\overline{P}^{\mathrm{R-}}$ Upper bound of the power traded in the down-reserve market [MW].
$P_{rt}^{\mathrm{R,A}}$ Available generation level of stochastic renewable generating unit r in time period t [MW].
$\tilde{P}_{rt}^{\mathrm{R,A}}$ Average level of the available generation of stochastic renewable generating unit r in time period t [MW].

$\hat{P}^{\mathrm{R,A}}_{rt}$ Fluctuation level of the available generation of stochastic renewable generating unit r in time period t [MW].
$\overline{P}^{\mathrm{S,C}}_{s}$ Charging capacity of storage unit s [MW].
$\overline{P}^{\mathrm{S,D}}_{s}$ Discharging capacity of storage unit s [MW].
Δt Duration of time periods [h].
Γ^{K} Uncertainty budget of the reserve deployment request.
Γ^{R}_{r} Uncertainty budget of the stochastic renewable generating unit r.
λ^{E}_{t} Price in the energy market in time period t [\$/MWh].
$\lambda^{\mathrm{R+}}_{t}$ Up-reserve energy price in time period t [\$/MWh].
$\lambda^{\mathrm{R-}}_{t}$ Down-reserve energy price in time period t [\$/MWh].
$\tilde{\lambda}^{\mathrm{R+}}_{t}$ Capacity price in the up-reserve market in time period t [\$/MW].
$\tilde{\lambda}^{\mathrm{R-}}_{t}$ Capacity price in the down-reserve market in time period t [\$/MW].
π_{ω} Weight of scenario ω [pu].

5.2.4 Variables

The variables of the scheduling problem of a VPP in the energy and reserve markets are:

e^{S}_{st} Energy stored in storage unit s in time period t [MWh].
p^{C}_{ct} Power generation of conventional power plant c in time period t [MW].
p^{D}_{dt} Power consumption of demand d in time period t [MW].
p^{E}_{t} Power traded in the energy market in time period t [MW].
$p^{\mathrm{R+}}_{t}$ Power capacity traded in the up-reserve market in time period t [MW].
$p^{\mathrm{R-}}_{t}$ Power capacity traded in the down-reserve market in time period t [MW].
p^{R}_{rt} Production of stochastic renewable generating unit r in time period t [MW].
$p^{\mathrm{R,A}}_{rt}$ Available production level of stochastic renewable generating unit r in time period t [MW].
$p^{\mathrm{S,C}}_{st}$ Charging power level of storage unit s in time period t [MW].
$p^{\mathrm{S,D}}_{st}$ Discharging power level of storage unit s in time period t [MW].
u^{C}_{ct} Binary variable that is equal to 1 if conventional power plant c is generating electricity in time period t, being 0 otherwise.

5.3 Deterministic Approach

The optimal scheduling problem of a VPP participating in energy and reserve electricity markets can be formulated using the deterministic model below:

$$\max_{\Phi^{\mathrm{D}}} \sum_{t\in\Omega^{\mathrm{T}}} \Bigg[\lambda^{\mathrm{E}}_{t} p^{\mathrm{E}}_{t}\Delta t + \left(\tilde{\lambda}^{\mathrm{R+}}_{t} + K^{\mathrm{R+}}_{t}\lambda^{\mathrm{R+}}_{t}\Delta t\right) p^{\mathrm{R+}}_{t} + \left(\tilde{\lambda}^{\mathrm{R-}}_{t} - K^{\mathrm{R-}}_{t}\lambda^{\mathrm{R-}}_{t}\Delta t\right) p^{\mathrm{R-}}_{t}$$

$$- \sum_{c\in\Omega^{\mathrm{C}}} \left(C^{\mathrm{C,F}}_{c} u^{\mathrm{C}}_{ct} + C^{\mathrm{C,V}}_{c} p^{\mathrm{C}}_{ct}\Delta t\right)\Bigg] \quad (5.1a)$$

subject to

$$\underline{P}^{\mathrm{E}} \le p^{\mathrm{E}}_{t} \le \overline{P}^{\mathrm{E}}, \quad \forall t \in \Omega^{\mathrm{T}}, \quad (5.1b)$$

$$0 \le p^{\mathrm{R+}}_{t} \le \overline{P}^{\mathrm{R+}}, \quad \forall t \in \Omega^{\mathrm{T}}, \quad (5.1c)$$

$$0 \le p^{\mathrm{R-}}_{t} \le \overline{P}^{\mathrm{R-}}, \quad \forall t \in \Omega^{\mathrm{T}}, \quad (5.1d)$$

$$u^{\mathrm{C}}_{ct} \in \{0, 1\}, \quad \forall c \in \Omega^{\mathrm{C}}, \forall t \in \Omega^{\mathrm{T}}, \quad (5.1e)$$

$$p_t^{\mathrm{E}} + K_t^{\mathrm{R+}} p_t^{\mathrm{R+}} - K_t^{\mathrm{R-}} p_t^{\mathrm{R-}} = \sum_{c \in \Omega^{\mathrm{C}}} p_{ct}^{\mathrm{C}} + \sum_{r \in \Omega^{\mathrm{R}}} p_{rt}^{\mathrm{R}} + \sum_{s \in \Omega^{\mathrm{S}}} \left(p_{st}^{\mathrm{S,D}} - p_{st}^{\mathrm{S,C}} \right) - \sum_{d \in \Omega^{\mathrm{D}}} p_{dt}^{\mathrm{D}}, \quad \forall t \in \Omega^{\mathrm{T}}, \tag{5.1f}$$

$$\underline{P}_{dt}^{\mathrm{D}} \leq p_{dt}^{\mathrm{D}} \leq \overline{P}_{dt}^{\mathrm{D}}, \quad \forall d \in \Omega^{\mathrm{D}}, \forall t \in \Omega^{\mathrm{T}}, \tag{5.1g}$$

$$\sum_{t \in \Omega^{\mathrm{T}}} p_{dt}^{\mathrm{D}} \Delta t \geq \underline{E}_d^{\mathrm{D}}, \quad \forall d \in \Omega^{\mathrm{D}}, \tag{5.1h}$$

$$\underline{P}_{ct}^{\mathrm{C}} u_{ct}^{\mathrm{C}} \leq p_{ct}^{\mathrm{C}} \leq \overline{P}_{ct}^{\mathrm{C}} u_{ct}^{\mathrm{C}}, \quad \forall c \in \Omega^{\mathrm{C}}, \forall t \in \Omega^{\mathrm{T}}, \tag{5.1i}$$

$$0 \leq p_{rt}^{\mathrm{R}} \leq P_{rt}^{\mathrm{R,A}}, \quad \forall r \in \Omega^{\mathrm{R}}, \forall t \in \Omega^{\mathrm{T}}, \tag{5.1j}$$

$$0 \leq p_{st}^{\mathrm{S,C}} \leq \overline{P}_{st}^{\mathrm{S,C}}, \quad \forall s \in \Omega^{\mathrm{S}}, \forall t \in \Omega^{\mathrm{T}}, \tag{5.1k}$$

$$0 \leq p_{st}^{\mathrm{S,D}} \leq \overline{P}_{st}^{\mathrm{S,D}}, \quad \forall s \in \Omega^{\mathrm{S}}, \forall t \in \Omega^{\mathrm{T}}, \tag{5.1l}$$

$$e_{st}^{\mathrm{S}} = e_{s(t-1)}^{\mathrm{S}} + p_{st}^{\mathrm{S,C}} \Delta t \eta_s^{\mathrm{S,C}} - \frac{p_{st}^{\mathrm{S,D}} \Delta t}{\eta_s^{\mathrm{S,D}}}, \quad \forall s \in \Omega^{\mathrm{S}}, \forall t \in \Omega^{\mathrm{T}}, \tag{5.1m}$$

$$\underline{E}_{st}^{\mathrm{S}} \leq e_{st}^{\mathrm{S}} \leq \overline{E}_{st}^{\mathrm{S}}, \quad \forall s \in \Omega^{\mathrm{S}}, \forall t \in \Omega^{\mathrm{T}}, \tag{5.1n}$$

where set $\Phi^{\mathrm{D}} = \left\{ p_t^{\mathrm{E}}, p_t^{\mathrm{R+}}, p_t^{\mathrm{R-}}, \forall t \in \Omega^{\mathrm{T}}; u_{ct}^{\mathrm{C}}, p_{ct}^{\mathrm{C}}, \forall c \in \Omega^{\mathrm{C}}, \forall t \in \Omega^{\mathrm{T}}; p_{rt}^{\mathrm{R}}, \forall r \in \Omega^{\mathrm{R}}, \forall t \in \Omega^{\mathrm{T}}; p_{dt}^{\mathrm{D}}, \forall d \in \Omega^{\mathrm{D}}, \forall t \in \Omega^{\mathrm{T}}; e_{st}^{\mathrm{S}}, p_{st}^{\mathrm{S,C}}, p_{st}^{\mathrm{S,D}}, \forall s \in \Omega^{\mathrm{S}}, \forall t \in \Omega^{\mathrm{T}} \right\}$ includes the optimization variables of problem (5.1).

The objective function (5.1a) represents the profit of the VPP throughout the planning horizon and comprises the following terms:

1. Terms $\lambda_t^{\mathrm{E}} p_t^{\mathrm{E}} \Delta t, \forall t \in \Omega^{\mathrm{T}}$, are the revenues achieved by the VPP due to its participation in the energy market. Note that variables $p_t^{\mathrm{E}}, \forall t \in \Omega^{\mathrm{T}}$, can take both positive (in case of selling in the energy market) and negative (in case of buying) values. In the latter case, these terms represent the cost incurred by the VPP due to its participation in the energy market.
2. Terms $\left(\tilde{\lambda}_t^{\mathrm{R+}} + K_t^{\mathrm{R+}} \lambda_t^{\mathrm{R+}} \Delta t \right) p_t^{\mathrm{R+}}, \forall t \in \Omega^{\mathrm{T}}$, are the revenues achieved by the VPP due to its participation in the up-reserve market. These revenues are divided into capacity payments $\tilde{\lambda}_t^{\mathrm{R+}} p_t^{\mathrm{R+}}, \forall t \in \Omega^{\mathrm{T}}$, which do not depend on the requests of the system operator to deploy reserves, and energy payments $K_t^{\mathrm{R+}} \lambda_t^{\mathrm{R+}} p_t^{\mathrm{R+}} \Delta t, \forall t \in \Omega^{\mathrm{T}}$, which depend on the actual requirements of the system operator to deploy up reserves (represented by factors $K_t^{\mathrm{R+}}, \forall t \in \Omega^{\mathrm{T}}$).
3. Terms $\left(\tilde{\lambda}_t^{\mathrm{R-}} - K_t^{\mathrm{R-}} \lambda_t^{\mathrm{R-}} \Delta t \right) p_t^{\mathrm{R-}}, \forall t \in \Omega^{\mathrm{T}}$, are the revenues achieved by the VPP due to its participation in the down-reserve market. These revenues are divided into capacity payments $\tilde{\lambda}_t^{\mathrm{R-}} p_t^{\mathrm{R-}}, \forall t \in \Omega^{\mathrm{T}}$, which do not depend on the requests of the system operator to deploy reserves, and energy costs $K_t^{\mathrm{R-}} \lambda_t^{\mathrm{R-}} p_t^{\mathrm{R-}} \Delta t, \forall t \in \Omega^{\mathrm{T}}$, which depend on the actual requirements of the system operator to deploy down reserves (represented by factors $K_t^{\mathrm{R-}}, \forall t \in \Omega^{\mathrm{T}}$).
4. Terms $C_c^{\mathrm{C,F}} u_{ct}^{\mathrm{C}} + C_c^{\mathrm{C,V}} p_{ct}^{\mathrm{C}} \Delta t, \forall c \in \Omega^{\mathrm{C}}, \forall t \in \Omega^{\mathrm{T}}$, are the fixed and variable costs of the conventional power plants in the VPP.

Constraints (5.1b), (5.1c), and (5.1d) limit the power traded in the energy, up-reserve, and down-reserve markets, respectively. Constraints (5.1e) define the binary nature of variables $u_{ct}^{\mathrm{C}}, \forall c \in \Omega^{\mathrm{C}}, \forall t \in \Omega^{\mathrm{T}}$, representing the on/off status of conventional power plants. Equations (5.1f) impose the power balance in the VPP. The power traded in the energy market and the reserve deployment requests must be equal to the total net power (generation minus consumption) of the different assets in the VPP. Constraints (5.1g) and (5.1h) impose bounds on the power consumption of demands, while constraints (5.1i) and (5.1j) impose bounds on the power generated by conventional power plants and stochastic renewable generating units, respectively. Constraints (5.1k) and (5.1l) limit the charging and discharging power levels of storage units, respectively, while constraints (5.1m) define the energy level evolution in storage units. These energy levels are bounded using Eqs. (5.1n).

Problem (5.1) is a mixed-integer linear programming (MILP) problem [13].

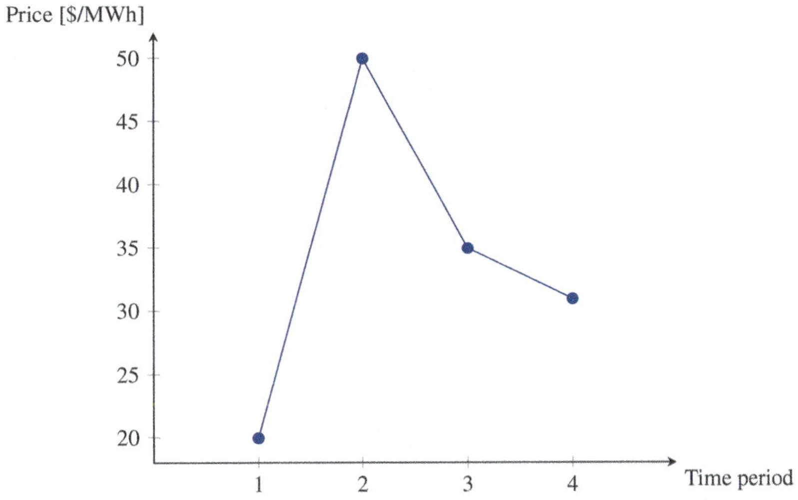

Fig. 5.1 Illustrative Example 5.1: energy market prices

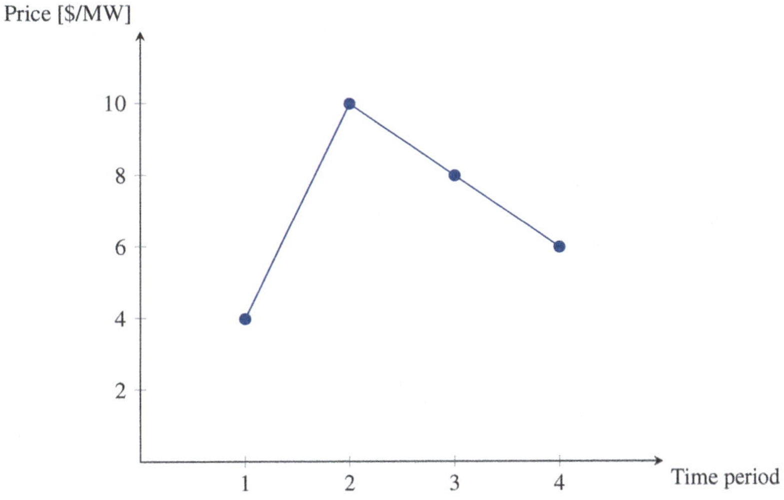

Fig. 5.2 Illustrative Example 5.1: up- and down-reserve market prices (capacity)

Illustrative Example 5.1 Deterministic Scheduling Problem in Energy and Reserve Markets

This example analyzes the scheduling problem in the DA energy and reserve electricity markets of a VPP comprising a conventional power plant, a wind-power unit, and a flexible demand.

A 4-h planning horizon is considered. The energy market prices are provided in Fig. 5.1 and the maximum power traded (sold or bought) in this market is 100 MW. The up- and down-reserve market prices for capacity and energy are respectively provided in Figs. 5.2 and 5.3. These prices are assumed to be the same for both the up- and down-reserve markets. The maximum power capacity that can be traded in these reserve markets is 50 MW.

The technical and economic data of the conventional power plant are provided in Table 5.1. The forecasts levels of the available wind-power generation are provided in Table 5.2. The lower and upper consumption bounds of the flexible demand are provided in Fig. 5.4, while the minimum energy consumption throughout the 4-h planning horizon is 60 MWh.

Finally, it is assumed that the VPP is requested to provide 50% of the scheduled capacity in the up-reserve market in time period 1, 80% of the scheduled capacity in the down-reserve market in time period 2, and 100% of the scheduled capacity in the up-reserve market in time period 3, while no reserve deployments are requested in time period 4.

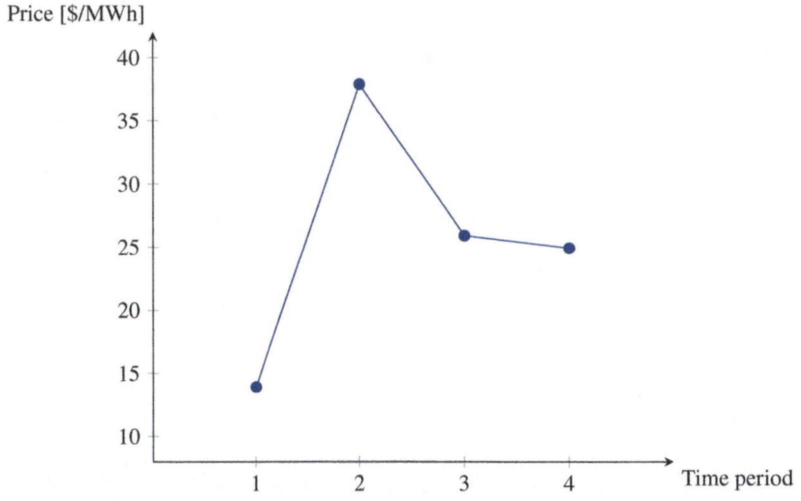

Fig. 5.3 Illustrative Example 5.1: up- and down-reserve market prices (energy)

Table 5.1 Illustrative Example 5.1: data of the conventional power plant

Capacity	60 MW
Minimum power output	10 MW
Online cost	$6/h
Variable cost	$33/MWh

Table 5.2 Illustrative Example 5.1: data of the wind-power unit

Time period	Forecast level [MW]
1	60
2	20
3	40
4	10

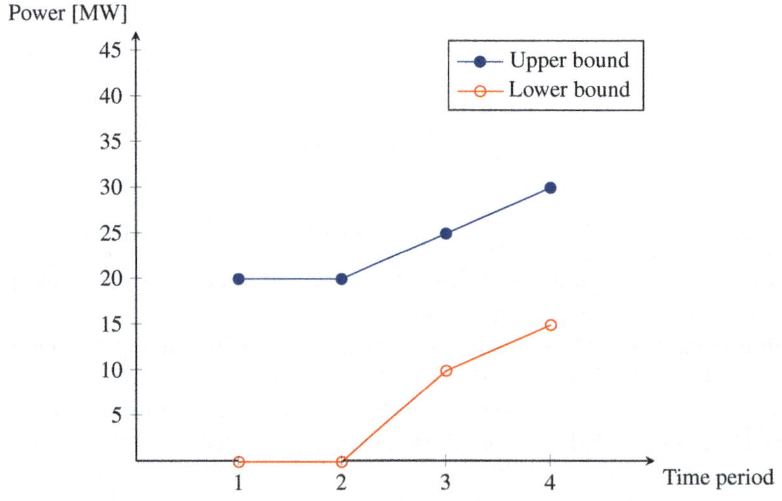

Fig. 5.4 Illustrative Example 5.1: lower and upper bounds of the power consumption

Considering these data, the problem is formulated below.

First, the objective function to be maximized is:

$$
\begin{aligned}
&20p_1^{\mathrm{E}} + 50p_2^{\mathrm{E}} + 35p_3^{\mathrm{E}} + 31p_4^{\mathrm{E}} \\
&\quad + (4 + 0.5 \cdot 14)\, p_1^{\mathrm{R+}} + (10 + 0 \cdot 38)\, p_2^{\mathrm{R+}} + (8 + 1 \cdot 26)\, p_3^{\mathrm{R+}} + (6 + 0 \cdot 25)\, p_4^{\mathrm{R+}} \\
&\quad + (4 - 0 \cdot 14)\, p_1^{\mathrm{R-}} + (10 - 0.8 \cdot 38)\, p_2^{\mathrm{R-}} + (8 - 0 \cdot 26)\, p_3^{\mathrm{R-}} + (6 - 0 \cdot 25)\, p_4^{\mathrm{R-}} \\
&\quad - \left(6u_1^{\mathrm{C}} + 6u_2^{\mathrm{C}} + 6u_3^{\mathrm{C}} + 6u_4^{\mathrm{C}} + 33p_1^{\mathrm{C}} + 33p_2^{\mathrm{C}} + 33p_3^{\mathrm{C}} + 33p_4^{\mathrm{C}}\right).
\end{aligned}
$$

The first line is the revenue achieved in the energy market. The second and third lines represent the revenues achieved in the up- and down-reserve markets, respectively, including both capacity and energy payments. Finally, the fourth line includes the fixed and variable costs of the conventional power plant.

Next the constraints of the problem are formulated:

- Limits on the power traded in the energy market:

$$
\begin{aligned}
-100 \le p_1^{\mathrm{E}} \le 100, \\
-100 \le p_2^{\mathrm{E}} \le 100, \\
-100 \le p_3^{\mathrm{E}} \le 100, \\
-100 \le p_4^{\mathrm{E}} \le 100.
\end{aligned}
$$

- Limits on the power traded in the up- and down-reserve markets:

$$
\begin{aligned}
0 \le p_1^{\mathrm{R+}} \le 50, \\
0 \le p_2^{\mathrm{R+}} \le 50, \\
0 \le p_3^{\mathrm{R+}} \le 50, \\
0 \le p_4^{\mathrm{R+}} \le 50, \\
0 \le p_1^{\mathrm{R-}} \le 50, \\
0 \le p_2^{\mathrm{R-}} \le 50, \\
0 \le p_3^{\mathrm{R-}} \le 50, \\
0 \le p_4^{\mathrm{R-}} \le 50.
\end{aligned}
$$

- Definition of binary variables:

$$
u_1^{\mathrm{C}}, u_2^{\mathrm{C}}, u_3^{\mathrm{C}}, u_4^{\mathrm{C}} \in \{0, 1\}.
$$

- Power balance in the VPP:

$$
\begin{aligned}
p_1^{\mathrm{E}} + 0.5 \cdot p_1^{\mathrm{R+}} - 0 \cdot p_1^{\mathrm{R-}} &= p_1^{\mathrm{C}} + p_1^{\mathrm{R}} - p_1^{\mathrm{D}}, \\
p_2^{\mathrm{E}} + 0 \cdot p_2^{\mathrm{R+}} - 0.8 \cdot p_2^{\mathrm{R-}} &= p_2^{\mathrm{C}} + p_2^{\mathrm{R}} - p_2^{\mathrm{D}}, \\
p_3^{\mathrm{E}} + 1 \cdot p_3^{\mathrm{R+}} - 0 \cdot p_3^{\mathrm{R-}} &= p_3^{\mathrm{C}} + p_3^{\mathrm{R}} - p_3^{\mathrm{D}}, \\
p_4^{\mathrm{E}} + 0 \cdot p_4^{\mathrm{R+}} - 0 \cdot p_4^{\mathrm{R-}} &= p_4^{\mathrm{C}} + p_4^{\mathrm{R}} - p_4^{\mathrm{D}}.
\end{aligned}
$$

- Limits on the power consumption of the demand:

$$0 \le p_1^{\rm D} \le 20,$$
$$0 \le p_2^{\rm D} \le 20,$$
$$10 \le p_3^{\rm D} \le 25,$$
$$15 \le p_4^{\rm D} \le 30.$$

- Minimum energy consumption of the demand throughout the planning horizon:

$$p_1^{\rm D} + p_2^{\rm D} + p_3^{\rm D} + p_4^{\rm D} \ge 60.$$

- Limits on the production of the conventional power plant:

$$10u_1^{\rm C} \le p_1^{\rm C} \le 60u_1^{\rm C},$$
$$10u_2^{\rm C} \le p_2^{\rm C} \le 60u_2^{\rm C},$$
$$10u_3^{\rm C} \le p_3^{\rm C} \le 60u_3^{\rm C},$$
$$10u_4^{\rm C} \le p_4^{\rm C} \le 60u_4^{\rm C}.$$

- Limits on the production of the wind-power unit:

$$0 \le p_1^{\rm R} \le 60,$$
$$0 \le p_2^{\rm R} \le 20,$$
$$0 \le p_3^{\rm R} \le 40,$$
$$0 \le p_4^{\rm R} \le 10.$$

Finally, the optimization variables of the problem are those included in set $\Phi^{\rm IE5.1} = \left\{p_1^{\rm E}, p_2^{\rm E}, p_3^{\rm E}, p_4^{\rm E}, p_1^{\rm R+}, p_2^{\rm R+}, p_3^{\rm R+}, p_4^{\rm R+}, p_1^{\rm R-}, p_2^{\rm R-}, p_3^{\rm R-}, p_4^{\rm R-}, u_1^{\rm C}, u_2^{\rm C}, u_3^{\rm C}, u_4^{\rm C}, p_1^{\rm C}, p_2^{\rm C}, p_3^{\rm C}, p_4^{\rm C}, p_1^{\rm R}, p_2^{\rm R}, p_3^{\rm R}, p_4^{\rm R}, p_1^{\rm D}, p_2^{\rm D}, p_3^{\rm D}, p_4^{\rm D}\right\}$. These variables include a single subscript that denotes the time period. Since there is just one unit of each type, the corresponding subscript has been suppressed for the sake of clarity.

The above problem is an MILP model that is solved using CPLEX [10] under GAMS [9]. The optimal power traded in the energy, up-reserve, and down-reserve markets is provided in Fig. 5.5, while the optimal power production and consumption levels of each element in the VPP are provided in Fig. 5.6.

On one hand, the VPP schedules 50 MW in both the up- and down-reserve markets in all time periods except in time period 3 for the up-reserve market. The reason is that in this deterministic problem it is assumed that the reserve deployment requests are known by the VPP in advance. Thus, the VPP knows that by providing the highest reserve levels it will be remunerated for the reserve capacity but it will have to provide reserves only in a limited number of time periods, e.g., in time period 3.

On the other hand, the scheduling of the VPP in the energy market depends on the prices in this market as well as on the technical and economic constraints of the energy assets. The VPP uses all the available wind-power production since it has no cost. Then, it turns on the conventional power plant in time periods 2 and 3 due to the comparatively high energy market prices. Similarly, it supplies the demand mainly in time periods 1 and 4, which have the lowest market prices. Note that the VPP sells energy in time periods 1, 2, and 3, while it prefers to buy from the energy market in time period 4.

Finally, the objective function indicates the total profit achieved by the VPP, which in this case is \$5748.

□

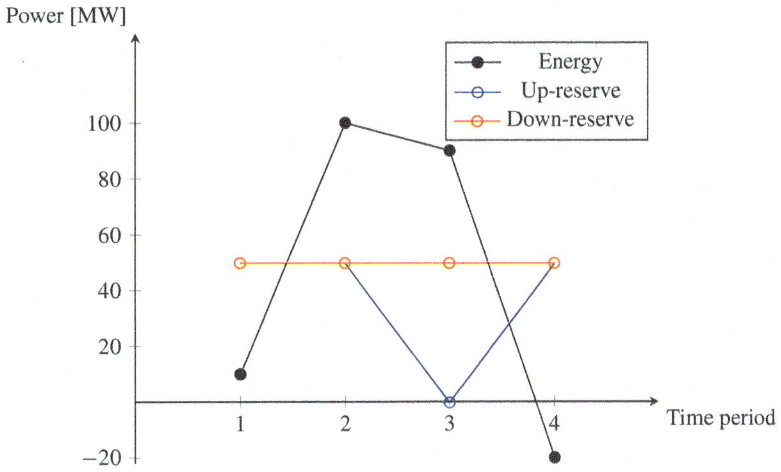

Fig. 5.5 Illustrative Example 5.1: power traded in the energy, up-reserve, and down-reserve markets

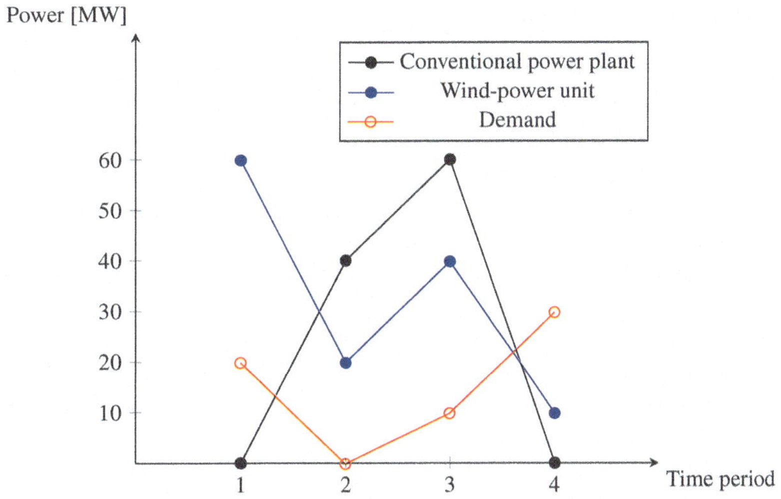

Fig. 5.6 Illustrative Example 5.1: power production and consumption of the energy assets in the VPP

A GAMS [9] code for solving Illustrative Example 5.1 is provided in Sect. 5.7.

Illustrative Example 5.2 Impact of Reserve Request Coefficients

The data and results of Illustrative Example 5.1 are considered again. The VPP uses these scheduling decisions in the energy and reserve markets. Then, in real-time, the conditions in the system change and thus also the reserve requests by the system operator. Instead of the reserve request considered in Illustrative Example 5.1, the VPP is requested to provide 100% of the up-reserve capacity in time periods 1 and 4, and 100% of the down-reserve capacity in time period 2, while no reserve deployment is requested in time period 3.

In such a case, note that the VPP must provide 60 MW in time period 1 (10 MW for the energy market and 50 MW due to the up-reserve request). In time period 1, the conventional power plant is not working. Since the on/off status of this unit cannot be changed in real time due to its dynamics, this means that the VPP would use the available 60 MW of wind-power production to comply the requirements in the electricity markets; however, the demand in the VPP would not be supplied in time period 1.

Then, in time period 2, the VPP must provide 150 MW (100 MW for the energy market and 50 MW due to the up-reserve request). There are 20 MW of available wind-power production; additionally, up to 60 MW can be used from the conventional power plant. However, note that this is not enough to provide the required 150 MW. This means that the VPP should participate in other markets such as balancing or RT markets to compensate these energy deviations, which may have a great impact on its profit.

□

Illustrative Example 5.2 shows that in order to obtain informed scheduling decisions in the energy and reserve electricity markets an accurate modeling of the uncertainties involved in the problem is necessary. This issue is addressed in the two following sections.

5.4 Stochastic Programming Approach

The deterministic model formulated in Sect. 5.3 is extended here to model the uncertainties in market prices, available production levels of stochastic renewable generating units, and requests of the system operator to deploy reserves. This is done using a stochastic programming approach. First, the uncertainty characterization is described and then the model formulation is provided.

5.4.1 Uncertainty Characterization and Decision Sequence

Scheduling decisions in the DA energy and reserve electricity markets are made for the following day one day in advance. At that time, the VPP faces a number of uncertainties, namely:

1. The market prices, including the energy market prices and the reserve market prices (for both capacity and energy).
2. The available production levels of stochastic renewable generating units.
3. The requests by the system operator to deploy reserves.

In order to obtain informed scheduling decisions it is necessary to model these uncertainties in the decision-making problem. An inaccurate modeling may result in loss of profit for the VPP and even in an infeasible operation of the generation and consumption assets.

Here, uncertainties are modeled using a predefined set of discrete scenario realization indexed by $\omega \in \Omega^{\mathrm{W}}$. Each scenario ω is characterized by parameters $\lambda^{\mathrm{E}}_{t\omega}$, $\tilde{\lambda}^{\mathrm{R+}}_{t\omega}$, $\lambda^{\mathrm{R+}}_{t\omega}$, $\tilde{\lambda}^{\mathrm{R-}}_{t\omega}$, $\lambda^{\mathrm{R-}}_{t\omega}$, $K^{\mathrm{R+}}_{t\omega}$, $K^{\mathrm{R-}}_{t\omega}$, and $P^{\mathrm{R,A}}_{rt\omega}$ that indicate the energy market price, the up-reserve market price (for capacity), the up-reserve market price (for energy), the down-reserve market price (for capacity), the down-reserve market price (for energy), the up-reserve request, the down-reserve request, and the available production levels of stochastic renewable generating units, respectively. Moreover, each scenario ω has a probability of occurrence π_{ω}, so that the sum of probabilities over all scenarios is equal to 1, i.e., $\sum_{\omega \in \Omega^{\mathrm{W}}} \pi_{\omega} = 1$.

Under such scenario characterization, the decision-making problem is modeled using two-stage stochastic programming [5]. In particular, the decision sequence is as follows:

1. The VPP makes its scheduling decisions in the DA energy and reserve electricity markets. It also decides the commitment status of the conventional power plants. These decisions are made one day in advance before knowing the actual values of the uncertain parameters. Therefore, these decisions do not depend on the scenario realization ω and are generally known as *here-and-now* decisions.
2. The DA energy and reserve markets are cleared and market prices are communicated to the VPP. Additionally, the VPP knows the actual available production level of the stochastic renewable generating units and is also informed about the actual reserve deployment requests.
3. Depending on the actual values of the uncertain parameters, the VPP decides the actual dispatch of the different generation and consumption assets. These decisions are made after knowing the actual scenario realization and thus they depend on the scenario realization ω. Therefore, these decisions are generally known as *wait-and-see* decisions.

The scenario tree associated with the above decision sequence is depicted in Fig. 5.7.

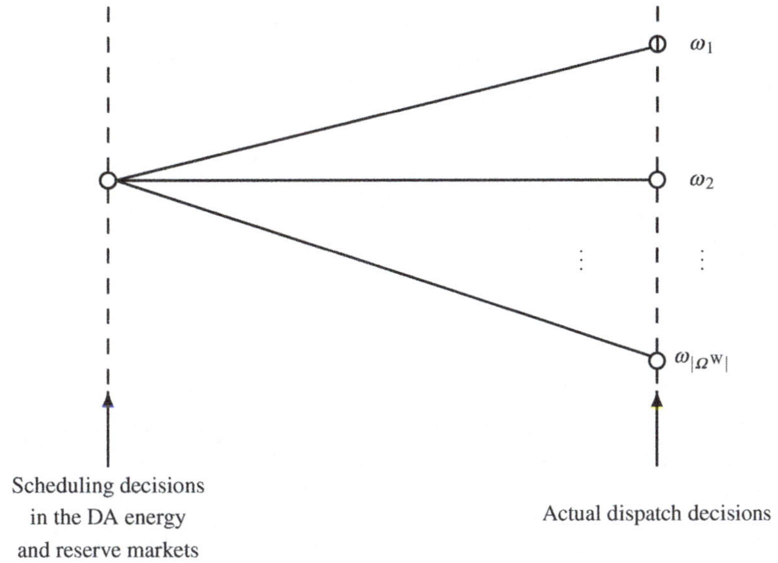

Fig. 5.7 Scheduling in the DA energy and reserve markets using a stochastic approach: scenario tree

5.4.2 Problem Formulation

Considering the uncertainty characterization and decision sequence described in Sect. 5.4.1, the scheduling problem of a VPP in the DA energy and reserve electricity markets is formulated as the following two-stage stochastic programming model:

$$\max_{\Phi^S} \sum_{\omega \in \Omega^W} \pi_\omega \Bigg\{ \sum_{t \in \Omega^T} \Bigg[\lambda^E_{t\omega} p^E_t \Delta t + \left(\tilde{\lambda}^{R+}_{t\omega} + K^{R+}_{t\omega} \lambda^{R+}_{t\omega} \Delta t \right) p^{R+}_t$$
$$+ \left(\tilde{\lambda}^{R-}_{t\omega} - K^{R-}_{t\omega} \lambda^{R-}_{t\omega} \Delta t \right) p^{R-}_t - \sum_{c \in \Omega^C} \left(C^{C,F}_c u^C_{ct} + C^{C,V}_c p^C_{ct\omega} \Delta t \right) \Bigg] \Bigg\} \quad (5.2a)$$

subject to

$$\underline{P}^E \le p^E_t \le \overline{P}^E, \quad \forall t \in \Omega^T, \quad (5.2b)$$

$$0 \le p^{R+}_t \le \overline{P}^{R+}, \quad \forall t \in \Omega^T, \quad (5.2c)$$

$$0 \le p^{R-}_t \le \overline{P}^{R-}, \quad \forall t \in \Omega^T, \quad (5.2d)$$

$$u^C_{ct} \in \{0, 1\}, \quad \forall c \in \Omega^C, \forall t \in \Omega^T, \quad (5.2e)$$

$$p^E_t + K^{R+}_{t\omega} p^{R+}_t - K^{R-}_{t\omega} p^{R-}_t = \sum_{c \in \Omega^C} p^C_{ct\omega} + \sum_{r \in \Omega^R} p^R_{rt\omega} + \sum_{s \in \Omega^S} \left(p^{S,D}_{st\omega} - p^{S,C}_{st\omega} \right)$$
$$- \sum_{d \in \Omega^D} p^D_{dt\omega}, \quad \forall t \in \Omega^T, \forall \omega \in \Omega^W, \quad (5.2f)$$

$$\underline{P}^D_{dt} \le p^D_{dt\omega} \le \overline{P}^D_{dt}, \quad \forall d \in \Omega^D, \forall t \in \Omega^T, \forall \omega \in \Omega^W, \quad (5.2g)$$

$$\sum_{t \in \Omega^T} p^D_{dt\omega} \Delta t \ge \underline{E}^D_d, \quad \forall d \in \Omega^D, \forall \omega \in \Omega^W, \quad (5.2h)$$

$$\underline{P}^C_{ct} u^C_{ct} \le p^C_{ct\omega} \le \overline{P}^C_{ct} u^C_{ct}, \quad \forall c \in \Omega^C, \forall t \in \Omega^T, \forall \omega \in \Omega^W, \quad (5.2i)$$

$$0 \le p^R_{rt\omega} \le P^{R,A}_{rt\omega}, \quad \forall r \in \Omega^R, \forall t \in \Omega^T, \forall \omega \in \Omega^W, \quad (5.2j)$$

$$0 \le p^{\mathrm{S,C}}_{st\omega} \le \overline{P}^{\mathrm{S,C}}_{st}, \quad \forall s \in \Omega^{\mathrm{S}}, \forall t \in \Omega^{\mathrm{T}}, \forall \omega \in \Omega^{\mathrm{W}}, \tag{5.2k}$$

$$0 \le p^{\mathrm{S,D}}_{st\omega} \le \overline{P}^{\mathrm{S,D}}_{st}, \quad \forall s \in \Omega^{\mathrm{S}}, \forall t \in \Omega^{\mathrm{T}}, \forall \omega \in \Omega^{\mathrm{W}}, \tag{5.2l}$$

$$e^{\mathrm{S}}_{st\omega} = e^{\mathrm{S}}_{s(t-1)\omega} + p^{\mathrm{S,C}}_{st\omega}\Delta t \eta^{\mathrm{S,C}}_{s} - \frac{p^{\mathrm{S,D}}_{st\omega}\Delta t}{\eta^{\mathrm{S,D}}_{s}}, \quad \forall s \in \Omega^{\mathrm{S}}, \forall t \in \Omega^{\mathrm{T}}, \forall \omega \in \Omega^{\mathrm{W}}, \tag{5.2m}$$

$$\underline{E}^{\mathrm{S}}_{st} \le e^{\mathrm{S}}_{st\omega} \le \overline{E}^{\mathrm{S}}_{st}, \quad \forall s \in \Omega^{\mathrm{S}}, \forall t \in \Omega^{\mathrm{T}}, \forall \omega \in \Omega^{\mathrm{W}}. \tag{5.2n}$$

where set $\Phi^{\mathrm{S}} = \Big\{p^{\mathrm{E}}_{t}, p^{\mathrm{R+}}_{t}, p^{\mathrm{R-}}_{t}, \forall t \in \Omega^{\mathrm{T}}; u^{\mathrm{C}}_{ct}, \forall c \in \Omega^{\mathrm{C}}, \forall t \in \Omega^{\mathrm{T}}; p^{\mathrm{C}}_{ct\omega}, \forall c \in \Omega^{\mathrm{C}}, \forall t \in \Omega^{\mathrm{T}}, \forall \omega \in \Omega^{\mathrm{W}}; p^{\mathrm{R}}_{rt\omega}, \forall r \in \Omega^{\mathrm{R}}, \forall t \in \Omega^{\mathrm{T}}, \forall \omega \in \Omega^{\mathrm{W}}; p^{\mathrm{D}}_{dt\omega}, \forall d \in \Omega^{\mathrm{D}}, \forall t \in \Omega^{\mathrm{T}}, \forall \omega \in \Omega^{\mathrm{W}}; e^{\mathrm{S}}_{st\omega}, p^{\mathrm{S,C}}_{st\omega}, p^{\mathrm{S,D}}_{st\omega}, \forall s \in \Omega^{\mathrm{S}}, \forall t \in \Omega^{\mathrm{T}}, \forall \omega \in \Omega^{\mathrm{W}}\Big\}$ includes the optimization variables of problem (5.2).

Constraints of problem (5.2) comprise constraints (5.2b)–(5.2e) that are identical to constraints (5.1b)–(5.1e) of the deterministic problem (5.1), as well as constraints (5.2f)–(5.2n), which are very similar to constraints (5.1f)–(5.1n) of the deterministic problem (5.1); however, in this case constraints (5.2f)–(5.2n) must be satisfied for all scenarios.

Illustrative Example 5.3 Stochastic Scheduling Problem in Energy and Reserve Markets

The VPP described in Illustrative Example 5.1 is considered again. The data of market prices, conventional power plant, and flexible demand are the same than those considered in Illustrative Example 5.1.

The uncertainty in the available wind-power production levels is modeled using two equiprobable scenarios, whose power profiles are depicted in Fig. 5.8. Additionally, the uncertainty in the reserve deployment requests is also modeled using also two equiprobable scenarios. The reserve request coefficients for these two scenarios are provided in Fig. 5.9, which depicts the difference between the up- and down-reserve request coefficients. Note that in each time period only one of these coefficients can be different from zero since the VPP is not required to provide up and down reserves simultaneously. This means that a positive value in Fig. 5.9 indicates that the down-reserve request coefficient is 0, while a negative value indicates that the up-reserve request coefficient is 0.

For the sake of simplicity, wind and reserve request scenarios are considered independent. Thus, all the possible combinations between scenarios are considered, which result in a total of 4 scenarios whose information is summarized in Table 5.3.

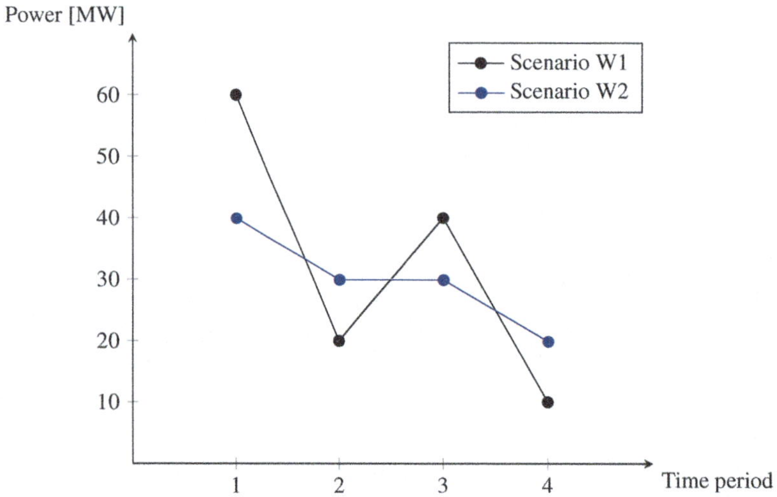

Fig. 5.8 Illustrative Example 5.3: scenarios for available wind generation levels

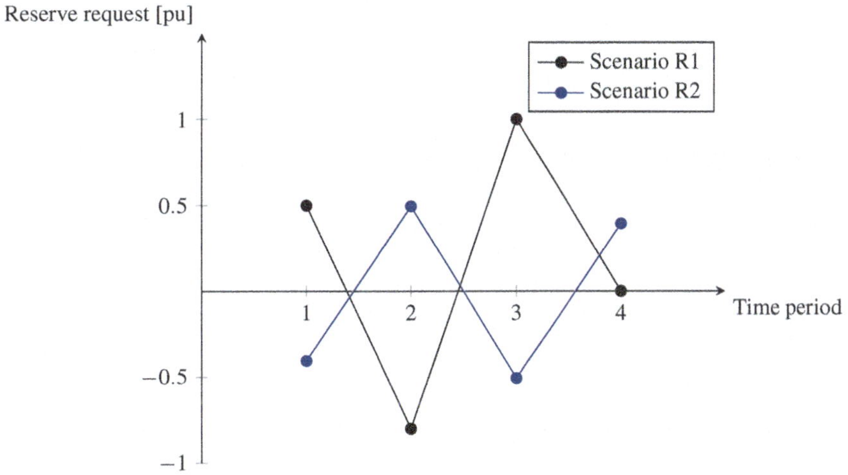

Fig. 5.9 Illustrative Example 5.3: scenarios for reserve request coefficients

Table 5.3 Illustrative Example 5.3: scenarios

Scenario	Wind scenario	Reserve scenario	Weight [pu]
1	W1	R1	$0.5 \times 0.5 = 0.25$
2	W1	R2	$0.5 \times 0.5 = 0.25$
3	W2	R1	$0.5 \times 0.5 = 0.25$
4	W2	R2	$0.5 \times 0.5 = 0.25$

Considering these data, the problem is formulated below.

First, the objective function to be maximized is:

$$
\begin{aligned}
&20p_1^{\text{E}} + 50p_2^{\text{E}} + 35p_3^{\text{E}} + 31p_4^{\text{E}} + 4p_1^{\text{R+}} + 10p_2^{\text{R+}} + 8p_3^{\text{R+}} + 6p_4^{\text{R+}} \\
&\quad + 4p_1^{\text{R-}} + 10p_2^{\text{R-}} + 8p_3^{\text{R-}} + 6p_4^{\text{R-}} - \left(6u_1^{\text{C}} + 6u_2^{\text{C}} + 6u_3^{\text{C}} + 6u_4^{\text{C}}\right) \\
&\quad + 0.25\Big[0.5 \cdot 14p_1^{\text{R+}} + 0 \cdot 38p_2^{\text{R+}} + 1 \cdot 26p_3^{\text{R+}} + 0 \cdot 25p_4^{\text{R+}} \\
&\qquad - \Big(0 \cdot 14p_1^{\text{R-}} + 0.8 \cdot 38p_2^{\text{R-}} + 0 \cdot 26p_3^{\text{R-}} + 0 \cdot 25p_4^{\text{R-}} \\
&\qquad\quad + 33p_{11}^{\text{C}} + 33p_{21}^{\text{C}} + 33p_{31}^{\text{C}} + 33p_{41}^{\text{C}}\Big)\Big] \\
&\quad + 0.25\Big[0 \cdot 14p_1^{\text{R+}} + 0.5 \cdot 38p_2^{\text{R+}} + 0 \cdot 26p_3^{\text{R+}} + 0.4 \cdot 25p_4^{\text{R+}} \\
&\qquad - \Big(0.4 \cdot 14p_1^{\text{R-}} + 0 \cdot 38p_2^{\text{R-}} + 0.5 \cdot 26p_3^{\text{R-}} + 0 \cdot 25p_4^{\text{R-}} \\
&\qquad\quad + 33p_{12}^{\text{C}} + 33p_{22}^{\text{C}} + 33p_{32}^{\text{C}} + 33p_{42}^{\text{C}}\Big)\Big] \\
&\quad + 0.25\Big[0.5 \cdot 14p_1^{\text{R+}} + 0 \cdot 38p_2^{\text{R+}} + 1 \cdot 26p_3^{\text{R+}} + 0 \cdot 25p_4^{\text{R+}} \\
&\qquad - \Big(0 \cdot 14p_1^{\text{R-}} + 0.8 \cdot 38p_2^{\text{R-}} + 0 \cdot 26p_3^{\text{R-}} + 0 \cdot 25p_4^{\text{R-}} \\
&\qquad\quad + 33p_{13}^{\text{C}} + 33p_{23}^{\text{C}} + 33p_{33}^{\text{C}} + 33p_{43}^{\text{C}}\Big)\Big] \\
&\quad + 0.25\Big[0 \cdot 14p_1^{\text{R+}} + 0.5 \cdot 38p_2^{\text{R+}} + 0 \cdot 26p_3^{\text{R+}} + 0.4 \cdot 25p_4^{\text{R+}}
\end{aligned}
$$

$$- \Big(0.4 \cdot 14 p_1^{\mathrm{R-}} + 0 \cdot 38 p_2^{\mathrm{R-}} + 0.5 \cdot 26 p_3^{\mathrm{R-}} + 0 \cdot 25 p_4^{\mathrm{R-}}$$

$$+33 p_{14}^{\mathrm{C}} + 33 p_{24}^{\mathrm{C}} + 33 p_{34}^{\mathrm{C}} + 33 p_{44}^{\mathrm{C}}\Big)\Big].$$

This objective function represents the expected profit achieved by the VPP. The first line is the revenue achieved in the energy market and the capacity revenue in the up-reserve market. The second line is the capacity revenue achieved in the down-reserve market minus the fixed cost of the conventional power plant. Note that the revenues and costs of these two lines do not depend on the scenario realization. The third, fourth, and fifth lines represent the energy revenues in the up-reserve market, the energy costs in the down-reserve market, and the variable cost of the conventional power plant in scenario 1. These terms are multiplied by the weight of the scenario (0.25). Similarly, lines 6–8, 9–11, and 12–14 include these same terms for scenarios 2, 3, and 4, respectively.

Next the constraints of the problem are formulated:

- Limits on the power traded in the DA energy market:

$$-100 \le p_1^{\mathrm{E}} \le 100,$$
$$-100 \le p_2^{\mathrm{E}} \le 100,$$
$$-100 \le p_3^{\mathrm{E}} \le 100,$$
$$-100 \le p_4^{\mathrm{E}} \le 100.$$

- Limits on the power traded in the up- and down-reserve markets:

$$0 \le p_1^{\mathrm{R+}} \le 50,$$
$$0 \le p_2^{\mathrm{R+}} \le 50,$$
$$0 \le p_3^{\mathrm{R+}} \le 50,$$
$$0 \le p_4^{\mathrm{R+}} \le 50,$$
$$0 \le p_1^{\mathrm{R-}} \le 50,$$
$$0 \le p_2^{\mathrm{R-}} \le 50,$$
$$0 \le p_3^{\mathrm{R-}} \le 50,$$
$$0 \le p_4^{\mathrm{R-}} \le 50.$$

- Definition of binary variables:

$$u_1^{\mathrm{C}}, u_2^{\mathrm{C}}, u_3^{\mathrm{C}}, u_4^{\mathrm{C}} \in \{0, 1\}.$$

- Power balance in the VPP:

$$p_1^{\mathrm{E}} + 0.5 \cdot p_1^{\mathrm{R+}} - 0 \cdot p_1^{\mathrm{R-}} = p_{11}^{\mathrm{C}} + p_{11}^{\mathrm{R}} - p_{11}^{\mathrm{D}},$$
$$p_2^{\mathrm{E}} + 0 \cdot p_2^{\mathrm{R+}} - 0.8 \cdot p_2^{\mathrm{R-}} = p_{21}^{\mathrm{C}} + p_{21}^{\mathrm{R}} - p_{21}^{\mathrm{D}},$$
$$p_3^{\mathrm{E}} + 1 \cdot p_3^{\mathrm{R+}} - 0 \cdot p_3^{\mathrm{R-}} = p_{31}^{\mathrm{C}} + p_{31}^{\mathrm{R}} - p_{31}^{\mathrm{D}},$$
$$p_4^{\mathrm{E}} + 0 \cdot p_4^{\mathrm{R+}} - 0 \cdot p_4^{\mathrm{R-}} = p_{41}^{\mathrm{C}} + p_{41}^{\mathrm{R}} - p_{41}^{\mathrm{D}},$$
$$p_1^{\mathrm{E}} + 0 \cdot p_1^{\mathrm{R+}} - 0.4 \cdot p_1^{\mathrm{R-}} = p_{12}^{\mathrm{C}} + p_{12}^{\mathrm{R}} - p_{12}^{\mathrm{D}},$$
$$p_2^{\mathrm{E}} + 0.5 \cdot p_2^{\mathrm{R+}} - 0 \cdot p_2^{\mathrm{R-}} = p_{22}^{\mathrm{C}} + p_{22}^{\mathrm{R}} - p_{22}^{\mathrm{D}},$$
$$p_3^{\mathrm{E}} + 0 \cdot p_3^{\mathrm{R+}} - 0.5 \cdot p_3^{\mathrm{R-}} = p_{32}^{\mathrm{C}} + p_{32}^{\mathrm{R}} - p_{32}^{\mathrm{D}},$$

$$p_4^{\mathrm{E}} + 0.4 \cdot p_4^{\mathrm{R}+} - 0 \cdot p_4^{\mathrm{R}-} = p_{42}^{\mathrm{C}} + p_{42}^{\mathrm{R}} - p_{42}^{\mathrm{D}},$$
$$p_1^{\mathrm{E}} + 0.5 \cdot p_1^{\mathrm{R}+} - 0 \cdot p_1^{\mathrm{R}-} = p_{13}^{\mathrm{C}} + p_{13}^{\mathrm{R}} - p_{13}^{\mathrm{D}},$$
$$p_2^{\mathrm{E}} + 0 \cdot p_2^{\mathrm{R}+} - 0.8 \cdot p_2^{\mathrm{R}-} = p_{23}^{\mathrm{C}} + p_{23}^{\mathrm{R}} - p_{23}^{\mathrm{D}},$$
$$p_3^{\mathrm{E}} + 1 \cdot p_3^{\mathrm{R}+} - 0 \cdot p_3^{\mathrm{R}-} = p_{33}^{\mathrm{C}} + p_{33}^{\mathrm{R}} - p_{33}^{\mathrm{D}},$$
$$p_4^{\mathrm{E}} + 0 \cdot p_4^{\mathrm{R}+} - 0 \cdot p_4^{\mathrm{R}-} = p_{43}^{\mathrm{C}} + p_{43}^{\mathrm{R}} - p_{43}^{\mathrm{D}},$$
$$p_1^{\mathrm{E}} + 0 \cdot p_1^{\mathrm{R}+} - 0.4 \cdot p_1^{\mathrm{R}-} = p_{14}^{\mathrm{C}} + p_{14}^{\mathrm{R}} - p_{14}^{\mathrm{D}},$$
$$p_2^{\mathrm{E}} + 0.5 \cdot p_2^{\mathrm{R}+} - 0 \cdot p_2^{\mathrm{R}-} = p_{24}^{\mathrm{C}} + p_{24}^{\mathrm{R}} - p_{24}^{\mathrm{D}},$$
$$p_3^{\mathrm{E}} + 0 \cdot p_3^{\mathrm{R}+} - 0.5 \cdot p_3^{\mathrm{R}-} = p_{34}^{\mathrm{C}} + p_{34}^{\mathrm{R}} - p_{34}^{\mathrm{D}},$$
$$p_4^{\mathrm{E}} + 0.4 \cdot p_4^{\mathrm{R}+} - 0 \cdot p_4^{\mathrm{R}-} = p_{44}^{\mathrm{C}} + p_{44}^{\mathrm{R}} - p_{44}^{\mathrm{D}}.$$

- Limits on the power consumption of the demand:

$$0 \leq p_{11}^{\mathrm{D}} \leq 20,$$
$$0 \leq p_{21}^{\mathrm{D}} \leq 20,$$
$$10 \leq p_{31}^{\mathrm{D}} \leq 25,$$
$$15 \leq p_{41}^{\mathrm{D}} \leq 30,$$
$$0 \leq p_{12}^{\mathrm{D}} \leq 20,$$
$$0 \leq p_{22}^{\mathrm{D}} \leq 20,$$
$$10 \leq p_{32}^{\mathrm{D}} \leq 25,$$
$$15 \leq p_{42}^{\mathrm{D}} \leq 30,$$
$$0 \leq p_{13}^{\mathrm{D}} \leq 20,$$
$$0 \leq p_{23}^{\mathrm{D}} \leq 20,$$
$$10 \leq p_{33}^{\mathrm{D}} \leq 25,$$
$$15 \leq p_{43}^{\mathrm{D}} \leq 30,$$
$$0 \leq p_{14}^{\mathrm{D}} \leq 20,$$
$$0 \leq p_{24}^{\mathrm{D}} \leq 20,$$
$$10 \leq p_{34}^{\mathrm{D}} \leq 25,$$
$$15 \leq p_{44}^{\mathrm{D}} \leq 30.$$

- Minimum energy consumption of the demand through the planning horizon:

$$p_{11}^{\mathrm{D}} + p_{21}^{\mathrm{D}} + p_{31}^{\mathrm{D}} + p_{41}^{\mathrm{D}} \geq 60,$$
$$p_{12}^{\mathrm{D}} + p_{22}^{\mathrm{D}} + p_{32}^{\mathrm{D}} + p_{42}^{\mathrm{D}} \geq 60,$$
$$p_{13}^{\mathrm{D}} + p_{23}^{\mathrm{D}} + p_{33}^{\mathrm{D}} + p_{43}^{\mathrm{D}} \geq 60,$$
$$p_{14}^{\mathrm{D}} + p_{24}^{\mathrm{D}} + p_{34}^{\mathrm{D}} + p_{44}^{\mathrm{D}} \geq 60.$$

- Limits on the production of the conventional power plant:

$$10u_1^{\mathrm{C}} \leq p_{11}^{\mathrm{C}} \leq 60u_1^{\mathrm{C}},$$
$$10u_2^{\mathrm{C}} \leq p_{21}^{\mathrm{C}} \leq 60u_2^{\mathrm{C}},$$
$$10u_3^{\mathrm{C}} \leq p_{31}^{\mathrm{C}} \leq 60u_3^{\mathrm{C}},$$
$$10u_4^{\mathrm{C}} \leq p_{41}^{\mathrm{C}} \leq 60u_4^{\mathrm{C}},$$
$$10u_1^{\mathrm{C}} \leq p_{12}^{\mathrm{C}} \leq 60u_1^{\mathrm{C}},$$
$$10u_2^{\mathrm{C}} \leq p_{22}^{\mathrm{C}} \leq 60u_2^{\mathrm{C}},$$
$$10u_3^{\mathrm{C}} \leq p_{32}^{\mathrm{C}} \leq 60u_3^{\mathrm{C}},$$
$$10u_4^{\mathrm{C}} \leq p_{42}^{\mathrm{C}} \leq 60u_4^{\mathrm{C}},$$
$$10u_1^{\mathrm{C}} \leq p_{13}^{\mathrm{C}} \leq 60u_1^{\mathrm{C}},$$
$$10u_2^{\mathrm{C}} \leq p_{23}^{\mathrm{C}} \leq 60u_2^{\mathrm{C}},$$
$$10u_3^{\mathrm{C}} \leq p_{33}^{\mathrm{C}} \leq 60u_3^{\mathrm{C}},$$
$$10u_4^{\mathrm{C}} \leq p_{43}^{\mathrm{C}} \leq 60u_4^{\mathrm{C}},$$
$$10u_1^{\mathrm{C}} \leq p_{14}^{\mathrm{C}} \leq 60u_1^{\mathrm{C}},$$
$$10u_2^{\mathrm{C}} \leq p_{24}^{\mathrm{C}} \leq 60u_2^{\mathrm{C}},$$
$$10u_3^{\mathrm{C}} \leq p_{34}^{\mathrm{C}} \leq 60u_3^{\mathrm{C}},$$
$$10u_4^{\mathrm{C}} \leq p_{44}^{\mathrm{C}} \leq 60u_4^{\mathrm{C}}.$$

- Limits on the production of the wind-power unit:

$$0 \leq p_{11}^{\mathrm{R}} \leq 60,$$
$$0 \leq p_{21}^{\mathrm{R}} \leq 20,$$
$$0 \leq p_{31}^{\mathrm{R}} \leq 40,$$
$$0 \leq p_{41}^{\mathrm{R}} \leq 10,$$
$$0 \leq p_{12}^{\mathrm{R}} \leq 60,$$
$$0 \leq p_{22}^{\mathrm{R}} \leq 20,$$
$$0 \leq p_{32}^{\mathrm{R}} \leq 40,$$
$$0 \leq p_{42}^{\mathrm{R}} \leq 10,$$
$$0 \leq p_{13}^{\mathrm{R}} \leq 40,$$
$$0 \leq p_{23}^{\mathrm{R}} \leq 30,$$
$$0 \leq p_{33}^{\mathrm{R}} \leq 30,$$
$$0 \leq p_{43}^{\mathrm{R}} \leq 20,$$
$$0 \leq p_{14}^{\mathrm{R}} \leq 40,$$
$$0 \leq p_{24}^{\mathrm{R}} \leq 30,$$

$$0 \le p_{34}^{\text{R}} \le 30,$$

$$0 \le p_{44}^{\text{R}} \le 20.$$

Finally, the optimization variables of the problem are those included in set $\Phi^{\text{IE5.3}} = \Big\{p_1^{\text{E}}, p_2^{\text{E}}, p_3^{\text{E}}, p_4^{\text{E}}, p_1^{\text{R+}}, p_2^{\text{R+}}, p_3^{\text{R+}}, p_4^{\text{R+}}, p_1^{\text{R-}}, p_2^{\text{R-}}, p_3^{\text{R-}}, p_4^{\text{R-}}, u_1^{\text{C}}, u_2^{\text{C}}, u_3^{\text{C}}, u_4^{\text{C}}, p_{11}^{\text{C}}, p_{21}^{\text{C}}, p_{31}^{\text{C}}, p_{41}^{\text{C}}, p_{12}^{\text{C}}, p_{22}^{\text{C}}, p_{32}^{\text{C}}, p_{42}^{\text{C}}, p_{13}^{\text{C}}, p_{23}^{\text{C}}, p_{33}^{\text{C}}, p_{43}^{\text{C}}, p_{14}^{\text{C}}, p_{24}^{\text{C}}, p_{34}^{\text{C}}, p_{44}^{\text{C}}, p_{11}^{\text{R}}, p_{21}^{\text{R}}, p_{31}^{\text{R}}, p_{41}^{\text{R}}, p_{12}^{\text{R}}, p_{22}^{\text{R}}, p_{32}^{\text{R}}, p_{42}^{\text{R}}, p_{13}^{\text{R}}, p_{23}^{\text{R}}, p_{33}^{\text{R}}, p_{43}^{\text{R}}, p_{14}^{\text{R}}, p_{24}^{\text{R}}, p_{34}^{\text{R}}, p_{44}^{\text{R}}, p_{11}^{\text{D}}, p_{21}^{\text{D}}, p_{31}^{\text{D}}, p_{41}^{\text{D}}, p_{12}^{\text{D}}, p_{22}^{\text{D}}, p_{32}^{\text{D}}, p_{42}^{\text{D}}, p_{13}^{\text{D}}, p_{23}^{\text{D}}, p_{33}^{\text{D}}, p_{43}^{\text{D}}, p_{14}^{\text{D}}, p_{24}^{\text{D}}, p_{34}^{\text{D}}, p_{44}^{\text{D}}\Big\}$. Some of these variables include a single subscript that denotes the time period. Other variables include two subscripts, the first one indicates the time period and the second one the scenario.

The above problem is a two-stage stochastic programming problem. In the first stage, the VPP decides its scheduling decisions in the energy and reserve markets, as well as the on/off status of the conventional power plant. These decisions are *here-and-now* decisions made before knowing the scenario realization, and thus they do not depend on the scenario realization. Then, in the second stage and after knowing the actual scenario realization, the VPP decides the optimal dispatch of the conventional power plant, the wind-power unit, and the flexible demand. The variables representing these *wait-and-see* dispatch decisions depend on scenario realization.

The resulting problem is an MILP model [13] that is solved using CPLEX [10] under GAMS [9].

The optimal power traded in the energy, up-reserve, and down-reserve markets is provided in Fig. 5.10.

The VPP sells in the energy market in the first three time periods due to the comparatively high market prices (in periods 2 and 3) and available wind-power generation levels (in period 1). Then, the VPP buys from the energy market in time period 4. This can be explained by the comparatively low market price and available wind-power generation level, as well as by the minimum consumption level in this time period. Regarding the participation in the up- and down-reserve markets, note that the VPP participates in the up-reserve market only in time period 4, while the participation in the down-reserve market is at the maximum levels most of the time.

After knowing the actual scenario realization, the VPP decides the actual dispatch of the different units. Figures 5.11 and 5.12 depict the demand consumption and wind-power production for different scenarios and time periods, respectively, while the conventional power plant is not turned on in this example. The profiles are different for different scenarios, depending on the reserve requests by the system operator and the available wind-power generation level. It is important to note that despite having a null cost, not all the available wind-power generation is used in all time periods and scenarios due to the uncertainties in the problem. To get a better use of the wind-power unit, the VPP may participate in other markets such as the balancing or RT markets. A storage unit may also allow a higher use of the wind-power unit.

Finally, the objective function indicates the expected profit achieved by the VPP, which in this case is $2697.5.

□

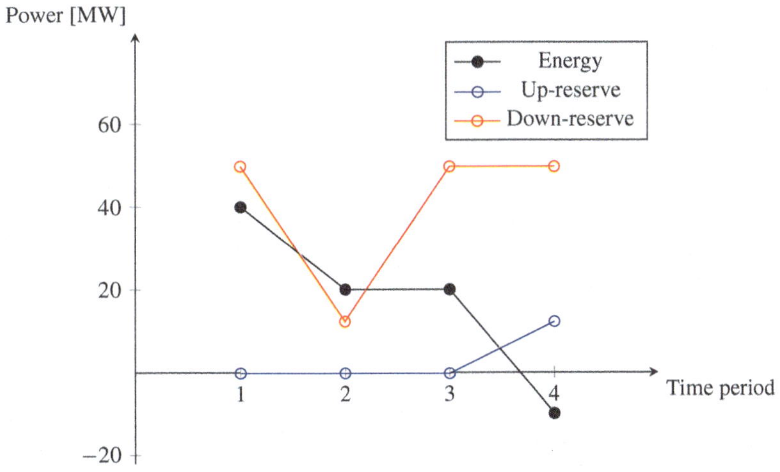

Fig. 5.10 Illustrative Example 5.3: power traded in the energy, up-reserve, and down-reserve markets

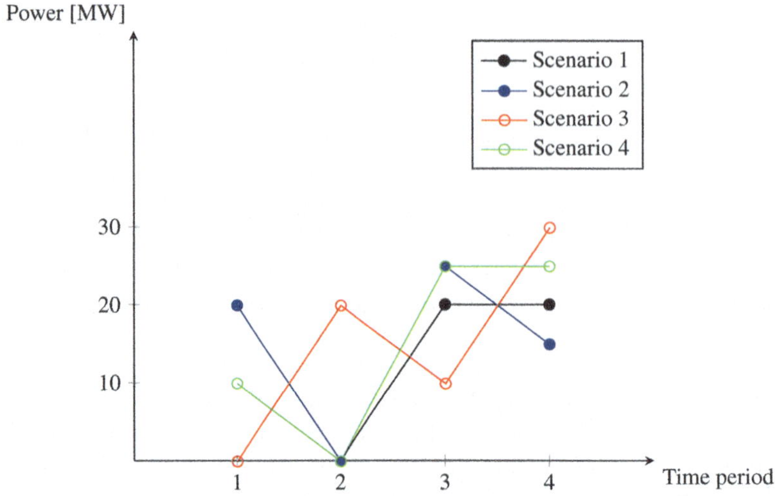

Fig. 5.11 Illustrative Example 5.3: power consumption in the VPP

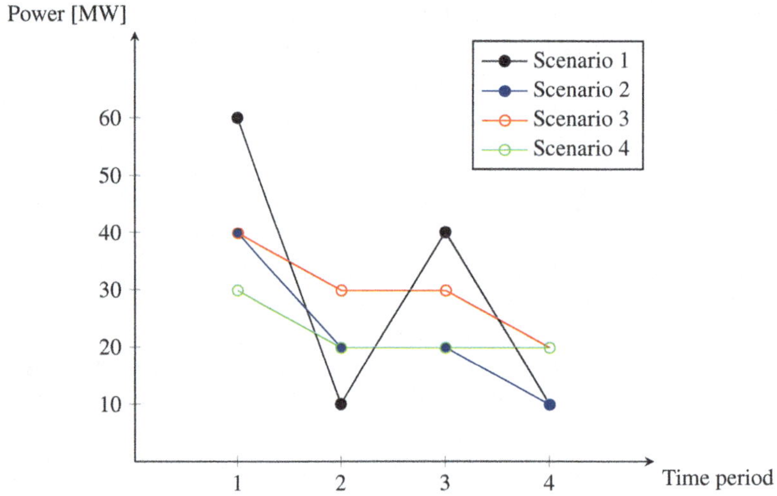

Fig. 5.12 Illustrative Example 5.3: wind-power production in the VPP

A GAMS [9] code for solving Illustrative Example 5.3 is provided in Sect. 5.7.

Illustrative Example 5.4 Value of the Stochastic Solution

The data of Illustrative Example 5.3 are considered again. However, in this case uncertain parameters are modeled using their forecast values, i.e., using a deterministic model such as the one described in Sect. 5.3. The forecast reserve request coefficients and available wind-power generation levels are depicted in Figs. 5.13 and 5.14, which are computed as the average values of the scenarios defined in Illustrative Example 5.3.

In this case, the optimal power traded in the energy, up-reserve, and down-reserve markets is provided in Fig. 5.15, 5.16, and 5.17, respectively. These figures compare the results with those obtained in Illustrative Example 5.3. Note that the solutions are completely different. Moreover, in this case the VPP decides to turn on the conventional power plant in time periods 2 and 3.

In order to further quantify the value of the stochastic solution, the problem formulated in Illustrative Example 5.3 is solved again with the values of first-stage decision variables fixed to those obtained using the deterministic model, i.e., the deterministic model is solved to determine the optimal power traded in the energy and reserve markets, as well as the on/off status of the conventional power plant, and then the operation of the VPP is determined for different scenarios. In such a case, the operation of the VPP is infeasible since it is not able to provide the reserve requests in some situations and at the same time to comply with the technical constraints of the different units in the VPP. This highlights the importance of an accurate modeling of the uncertainties in the problem.

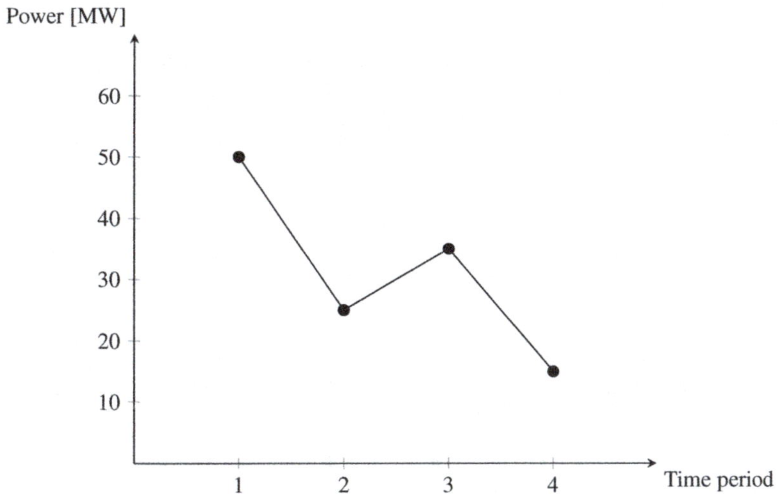

Fig. 5.13 Illustrative Example 5.6: forecast wind generation levels

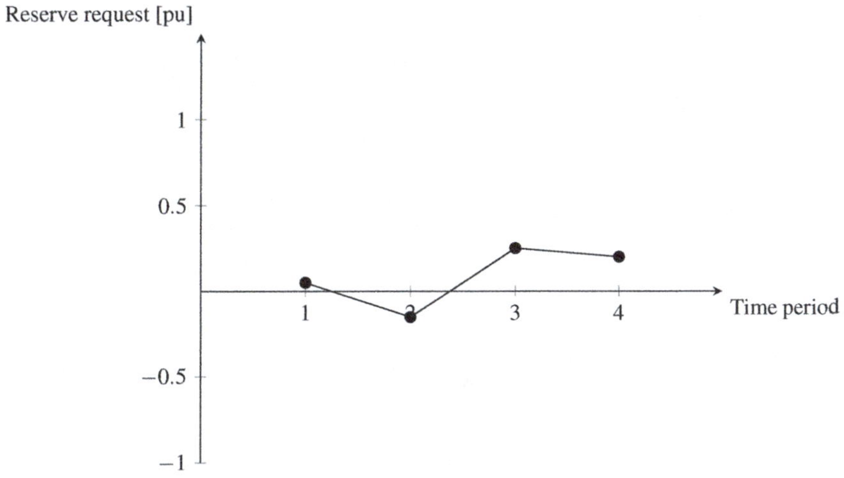

Fig. 5.14 Illustrative Example 5.3: forecast reserve request coefficients

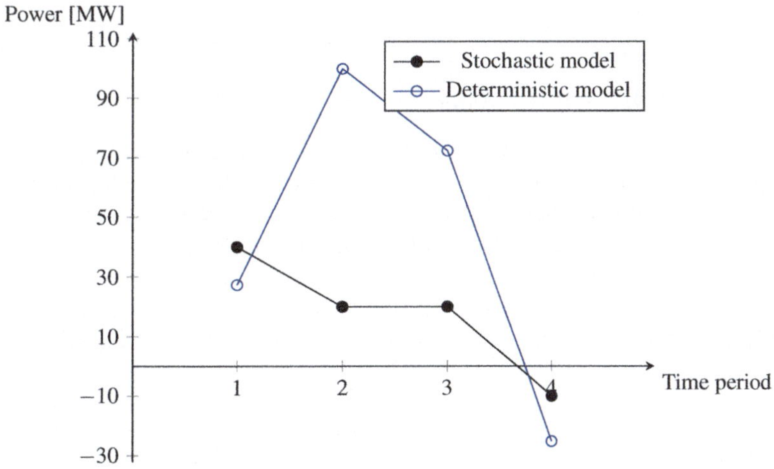

Fig. 5.15 Illustrative Example 5.6: power traded in the energy market

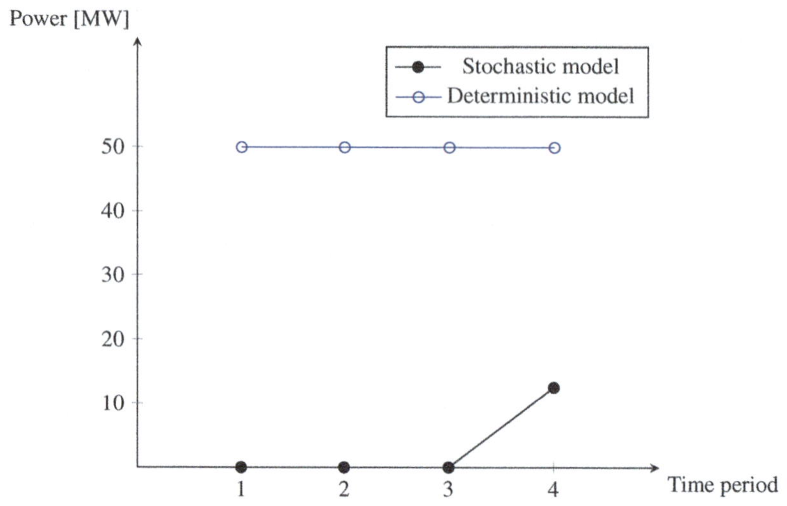

Fig. 5.16 Illustrative Example 5.6: power traded in the up-reserve market

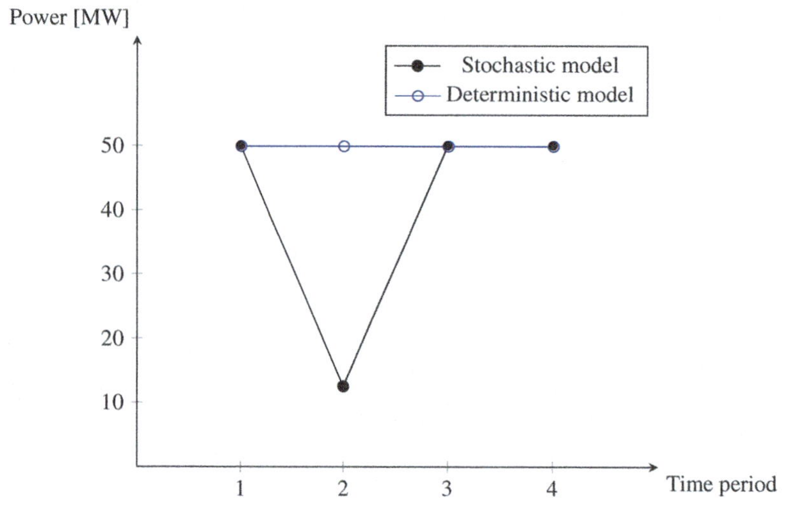

Fig. 5.17 Illustrative Example 5.6: power traded in the down-reserve market

□

Illustrative Example 5.5 Expected Value of Perfect Information

The data of Illustrative Example 5.3 are considered. In this case the value of having perfect information at the time of making the scheduling decisions in the energy and reserve markets is quantified. With this purpose, the problem formulated in Illustrative Example 5.3 is solved again; however, in this case the non-anticipativity constraints that impose that the scheduling decisions in the energy and reserve markets and the on/off status of the conventional power plant do not depend on the scenario realization are relaxed.

In such a case, the expected profit of the VPP would be \$4987, while the profit in Illustrative Example 5.3 was \$2697.5. This means that the VPP would be willing to pay up to \$2289.5 (\$4987−\$2697.5) for getting perfect information of the uncertain parameters in the problem.

□

5.5 Adaptive Robust Optimization Approach

The scheduling problem of a VPP in energy and reserve markets is analyzed in this section using an adaptive robust optimization approach.

5.5.1 Problem Description

As explained in Sect. 4.6.1 of Chap. 4, robust models allow making decisions by anticipating the worst-case uncertainty realizations. In the problem considered in this chapter, namely the scheduling of a VPP in energy and reserve markets, the decision sequence is as follows:

1. The VPP makes its scheduling decisions in the energy and reserve markets one day in advance. These decisions are made with the aim of maximizing the worst-case profit of the VPP.
2. These scheduling decisions are sought by anticipating that once they are made, the worst uncertainty will occur. In particular, the uncertainty in the reserve requests by the system operator and in the available production levels of stochastic renewable generating units is considered. For the sake of simplicity, market prices are considered known by the VPP in advance.
3. This worst-case uncertainty realization is determined by anticipating that after it occurs, the VPP can make corrective actions by optimally operating the different energy assets.

5.5.2 Formulation

The hierarchical structure described in the previous section can be represented using the following three-level optimization problem:

$$\max_{\Phi^{\text{UL}}} \sum_{t\in\Omega^{\text{T}}} \left(\lambda_t^{\text{E}} p_t^{\text{E}} \Delta t + \tilde{\lambda}_t^{\text{R+}} p_t^{\text{R+}} + \tilde{\lambda}_t^{\text{R-}} p_t^{\text{R-}} - \sum_{c\in\Omega^{\text{C}}} C_c^{\text{C,F}} u_{ct}^{\text{C}} \right)$$

$$+ \min_{\Phi^{\text{ML}}\in\Lambda} \max_{\Phi^{\text{LL}}\in\Theta} \sum_{t\in\Omega^{\text{T}}} \left(k_t^{\text{R+}} \lambda_t^{\text{R+}} p_t^{\text{R+}} - k_t^{\text{R-}} \lambda_t^{\text{R-}} p_t^{\text{R-}} - \sum_{c\in\Omega^{\text{C}}} C_c^{\text{C,V}} p_{ct}^{\text{C}} \right) \Delta t \tag{5.3a}$$

subject to

$$\underline{P}^{\text{E}} \le p_t^{\text{E}} \le \overline{P}^{\text{E}}, \quad \forall t \in \Omega^{\text{T}}, \tag{5.3b}$$

$$0 \le p_t^{\text{R+}} \le \overline{P}^{\text{R+}}, \quad \forall t \in \Omega^{\text{T}}, \tag{5.3c}$$

$$0 \le p_t^{\text{R-}} \le \overline{P}^{\text{R-}}, \quad \forall t \in \Omega^{\text{T}}, \tag{5.3d}$$

$$u_{ct}^{\text{C}} \in \{0, 1\}, \quad \forall c \in \Omega^{\text{C}}, \forall t \in \Omega^{\text{T}}, \tag{5.3e}$$

where the optimization variables of the upper-, middle-, and lower-level problems are those included in sets $\Phi^{\text{UL}} = \Big\{p_t^{\text{E}}, p_t^{\text{R+}}, p_t^{\text{R-}}, \forall t \in \Omega^{\text{T}}; u_{ct}^{\text{C}}, \forall c \in \Omega^{\text{C}}, \forall t \in \Omega^{\text{T}}\Big\}$, $\Phi^{\text{ML}} = \Big\{k_t^{\text{R+}}, k_t^{\text{R-}}, \forall t \in \Omega^{\text{T}}; p_{rt}^{\text{R,A}}, \forall r \in \Omega^{\text{R}}, \forall t \in \Omega^{\text{T}}\Big\}$, and $\Phi^{\text{LL}} = \Big\{p_{dt}^{\text{D}}, \forall d \in \Omega^{\text{D}}, \forall t; p_{ct}^{\text{C}}, \forall c \in \Omega^{\text{C}}, \forall t \in \Omega^{\text{T}}; p_{rt}^{\text{R}}, \forall r \in \Omega^{\text{R}}, \forall t \in \Omega^{\text{T}}; e_{st}^{\text{S,C}}, p_{st}^{\text{S,C}}, p_{st}^{\text{S,D}}, \forall s \in \Omega^{\text{S}}, \forall t \in \Omega^{\text{T}}\Big\}$, respectively.

Problem (5.3) is driven by the maximization of the worst-case profit of the VPP in the energy and reserve markets (5.3a). It comprises three optimization problems:

1. The upper-level problem determines the scheduling decisions of the VPP in the energy and reserve markets, as well as the scheduling of the conventional power plants. These decisions are made with the aim of maximizing the worst-case profit of the VPP and considering the limits on the power traded in the energy market (5.3b), the up-reserve market (5.3c), and the down-reserve market (5.3d), as well as the definition of binary variables to determine the on/off status of the conventional power plants (5.3e).
2. Given the upper-level decisions, the middle-level problem identifies the worst-case realization of reserve requests by the system operator and available stochastic renewable generation levels, i.e., the values of these variables that minimize the profit of the VPP. Note that the middle-level decision variables in set Φ^{ML} must be within the uncertainty set Λ, which is defined in Sect. 5.5.3.

3. Given upper- and middle-level decisions, the lower-level problem models the operation of the VPP maximizing its profit. Note that the lower-level decision variables in set Φ^{LL} must be within the feasibility set Θ, which is defined in Sect. 5.5.4.

5.5.3 Uncertainty Set

For the sake of simplicity, it is assumed that there is only uncertainty in the reserve requests by the system operator and in the available production levels of stochastic renewable generating units. These uncertainties are modeled using variables $k_t^{\mathrm{R+}}$ and $k_t^{\mathrm{R-}}$, $\forall t \in \Omega^{\mathrm{T}}$, which take values between 0 and 1 and model the fraction of the scheduled up- and down-reserve capacity requested by the system operator to be deployed, respectively, and variables $p_{rt}^{\mathrm{R,A}}$, $\forall r \in \Omega^{\mathrm{R}}, \forall t \in \Omega^{\mathrm{T}}$, which take values within predefined confidence bounds, i.e.:

$$k_t^{\mathrm{R+}} \in [0, 1], \quad \forall t \in \Omega^{\mathrm{T}}, \tag{5.4a}$$

$$k_t^{\mathrm{R-}} \in [0, 1], \quad \forall t \in \Omega^{\mathrm{T}}, \tag{5.4b}$$

$$p_{rt}^{\mathrm{R,A}} \in \left[\tilde{P}_{rt}^{\mathrm{R,A}} - \hat{P}_{rt}^{\mathrm{R,A}}, \tilde{P}_{rt}^{\mathrm{R,A}} + \hat{P}_{rt}^{\mathrm{R,A}}\right], \quad \forall r \in \Omega^{\mathrm{R}}, \forall t \in \Omega^{\mathrm{T}}. \tag{5.4c}$$

The so-called uncertainty budgets Γ^{K} and Γ_r^{R}, $\forall r \in \Omega^{\mathrm{R}}$ are also defined. Γ^{K} represents the maximum number of periods in which the VPP will be requested by the system operator to deploy reserves, while Γ_r^{R}, $\forall r \in \Omega^{\mathrm{R}}$, represent the maximum number of periods in which the available generation of stochastic renewable generating unit r experiences fluctuations with respect to its average value. These uncertainty budgets take values between zero and $|\Omega^{\mathrm{T}}|$ and allow controlling the conservativeness of the model:

1. If $\Gamma^{\mathrm{K}} = 0$, it means that the VPP is not requested to deploy request by the system operator at any time period, i.e., $k_t^{\mathrm{R+}} = k_t^{\mathrm{R-}} = 0$, $\forall t \in \Omega^{\mathrm{T}}$. This can be seen as a risky strategy.
2. If $\Gamma^{\mathrm{K}} = |\Omega^{\mathrm{T}}|$, it means that the system operator may request the VPP to deploy reserves at any time period. This can be seen as the most conservative strategy.
3. Intermediate values indicate that the VPP can be requested to deploy reserves in a maximum of Γ^{K} time periods.
4. In the case of Γ_r^{R}, the explanations are exactly the same as the ones provided in Sect. 4.6.3 of Chap. 4.

Note that a polyhedral uncertainty set is considered. As explained in Chap. 4 and shown in [11], in such a case the worst-case uncertainty realization corresponds to an extreme or vertex of the polyhedron representing the uncertainty sets in these cases. Therefore, it is possible to use the following binary-variable-based equivalent uncertainty set:

$$k_t^{\mathrm{R+}}, k_t^{\mathrm{R-}} \in \{0, 1\}, \quad \forall t \in \Omega^{\mathrm{T}}, \tag{5.5a}$$

$$k_t^{\mathrm{R+}} + k_t^{\mathrm{R-}} \leq 1, \quad \forall t \in \Omega^{\mathrm{T}}, \tag{5.5b}$$

$$\sum_{t \in \Omega^{\mathrm{T}}} \left(k_t^{\mathrm{R+}} + k_t^{\mathrm{R-}}\right) \leq \Gamma^{\mathrm{K}}, \tag{5.5c}$$

$$u_{rt}^{\mathrm{R,-}}, u_{rt}^{\mathrm{R,+}} \in \{0, 1\}, \quad \forall r \in \Omega^{\mathrm{R}}, \forall t \in \Omega^{\mathrm{T}}, \tag{5.5d}$$

$$p_{rt}^{\mathrm{R,A}} = \tilde{P}_{rt}^{\mathrm{R,A}} - u_{rt}^{\mathrm{R,-}} \hat{P}_{rt}^{\mathrm{R,A}} + u_{rt}^{\mathrm{R,+}} \hat{P}_{rt}^{\mathrm{R,A}}, \quad \forall r \in \Omega^{\mathrm{R}}, \forall t \in \Omega^{\mathrm{T}}, \tag{5.5e}$$

$$u_{rt}^{\mathrm{R,-}} + u_{rt}^{\mathrm{R,+}} \leq 1, \quad \forall r \in \Omega^{\mathrm{R}}, \forall t \in \Omega^{\mathrm{T}}, \tag{5.5f}$$

$$\sum_{t \in \Omega^{\mathrm{T}}} \left(u_{rt}^{\mathrm{R,-}} + u_{rt}^{\mathrm{R,+}}\right) \leq \Gamma_r^{\mathrm{R}}, \quad \forall r \in \Omega^{\mathrm{R}}. \tag{5.5g}$$

Binary variables $k_t^{\mathrm{R+}}$ and $k_t^{\mathrm{R+}}$ defined (5.5a) are associated with the worst-case reserve deployment requests. Equations (5.5b) guarantee that the VPP cannot be requested to provide both up- and down-reserves at the same time period. Constraints (5.5c) control the conservativeness of the model through the uncertainty budget Γ^{K}. Note that if the uncertainty budget Γ^{K} is 0, then $k_t^{\mathrm{R+}} = k_t^{\mathrm{R-}} = 0$, $\forall t \in \Omega^{\mathrm{T}}$. On the other hand, if $\Gamma^{\mathrm{K}} = |\Omega^{\mathrm{T}}|$, then $k_t^{\mathrm{R+}}$ and $k_t^{\mathrm{R-}}$ can be equal to 1 at any time period. Finally, the working of Eqs. (5.5d)–(5.5g) is exactly the same as Eqs. (4.11a)–(4.11d) of Chap. 4.

5.5.4 Feasibility of Operating Decision Variables

The feasibility of operating decision variables is modeled through set Θ as follows:

$$\Theta = \Big\{\Phi^{\mathrm{LL}} :$$

$$p_t^{\mathrm{E}} + k_t^{\mathrm{R+}} p_t^{\mathrm{R+}} - k_t^{\mathrm{R-}} p_t^{\mathrm{R-}} = \sum_{c\in\Omega^{\mathrm{C}}} p_{ct}^{\mathrm{C}} + \sum_{r\in\Omega^{\mathrm{R}}} p_{rt}^{\mathrm{R}} + \sum_{s\in\Omega^{\mathrm{S}}} \left(p_{st}^{\mathrm{S,D}} - p_{st}^{\mathrm{S,C}}\right) - \sum_{d\in\Omega^{\mathrm{D}}} p_{dt}^{\mathrm{D}}, \quad \forall t \in \Omega^{\mathrm{T}}, \tag{5.6a}$$

$$\underline{P}_{dt}^{\mathrm{D}} \le p_{dt}^{\mathrm{D}} \le \overline{P}_{dt}^{\mathrm{D}}, \quad \forall d \in \Omega^{\mathrm{D}}, \forall t \in \Omega^{\mathrm{T}}, \tag{5.6b}$$

$$\sum_{t\in\Omega^{\mathrm{T}}} p_{dt}^{\mathrm{D}} \Delta t \ge \underline{E}_d^{\mathrm{D}}, \quad \forall d \in \Omega^{\mathrm{D}}, \tag{5.6c}$$

$$\underline{P}_{ct}^{\mathrm{C}} u_{ct}^{\mathrm{C}} \le p_{ct}^{\mathrm{C}} \le \overline{P}_c^{\mathrm{C}} u_{ct}^{\mathrm{C}}, \quad \forall c \in \Omega^{\mathrm{C}}, \forall t \in \Omega^{\mathrm{T}}, \tag{5.6d}$$

$$0 \le p_{rt}^{\mathrm{R}} \le p_{rt}^{\mathrm{R,A}}, \quad \forall r \in \Omega^{\mathrm{R}}, \forall t \in \Omega^{\mathrm{T}}, \tag{5.6e}$$

$$0 \le p_{st}^{\mathrm{S,C}} \le \overline{P}_{st}^{\mathrm{S,C}}, \quad \forall s \in \Omega^{\mathrm{S}}, \forall t \in \Omega^{\mathrm{T}}, \tag{5.6f}$$

$$0 \le p_{st}^{\mathrm{S,D}} \le \overline{P}_{st}^{\mathrm{S,D}}, \quad \forall s \in \Omega^{\mathrm{S}}, \forall t \in \Omega^{\mathrm{T}}, \tag{5.6g}$$

$$e_{st}^{\mathrm{S}} = e_{s(t-1)}^{\mathrm{S}} + p_{st}^{\mathrm{S,C}} \Delta t \eta_s^{\mathrm{S,C}} - \frac{p_{st}^{\mathrm{S,D}} \Delta t}{\eta_s^{\mathrm{S,D}}}, \quad \forall s \in \Omega^{\mathrm{S}}, \forall t \in \Omega^{\mathrm{T}}, \tag{5.6h}$$

$$\underline{E}_{st}^{\mathrm{S}} \le e_{st}^{\mathrm{S}} \le \overline{E}_{st}^{\mathrm{S}}, \quad \forall s \in \Omega^{\mathrm{S}}, \forall t \in \Omega^{\mathrm{T}}\Big\}. \tag{5.6i}$$

Constraints (5.6a) define the power balance in the VPP. Constraints (5.6b) impose bounds on the power consumption of the demands, while constraints (5.6c) guarantee a minimum energy consumption throughout the planning horizon. Constraints (5.6d) and (5.6e) impose bounds on the production of the conventional power plant and the stochastic renewable generating units, respectively. Constraints (5.6f)–(5.6i) define the working of the storage facilities. Constraints (5.6f) and (5.6g) impose bounds on the charging and discharging power levels, respectively. Constraints (5.6h) define the energy level evolution of the storage units, while constraints (5.6i) impose bounds on the energy levels.

Note that the feasibility set Θ is parameterized in terms of upper-level decision variables in set Φ^{UL} and middle-level decision variables in set Φ^{ML}.

5.5.5 Detailed Formulation

For the sake of clarity, the detailed formulation of the adaptive robust optimization model for the scheduling problem of a VPP in the energy and reserve markets is provided below:

$$\max_{\Phi^{\mathrm{UL}}} \min_{\Phi^{\mathrm{ML}}} \max_{\Phi^{\mathrm{LL}}} \sum_{t\in\Omega^{\mathrm{T}}} \Bigg[\lambda_t^{\mathrm{E}} p_t^{\mathrm{E}} \Delta t + \left(\tilde{\lambda}_t^{\mathrm{R+}} + k_t^{\mathrm{R+}} \lambda_t^{\mathrm{R+}} \Delta t\right) p_t^{\mathrm{R+}} + \left(\tilde{\lambda}_t^{\mathrm{R-}} - k_t^{\mathrm{R-}} \lambda_t^{\mathrm{R-}} \Delta t\right) p_t^{\mathrm{R-}} - \sum_{c\in\Omega^{\mathrm{C}}} \left(C_c^{\mathrm{C,F}} u_{ct}^{\mathrm{C}} + C_c^{\mathrm{C,V}} p_{ct}^{\mathrm{C}} \Delta t\right)\Bigg] \tag{5.7a}$$

subject to

$$\underline{P}^{\mathrm{E}} \le p_t^{\mathrm{E}} \le \overline{P}^{\mathrm{E}}, \quad \forall t \in \Omega^{\mathrm{T}}, \tag{5.7b}$$

$$0 \le p_t^{\mathrm{R+}} \le \overline{P}^{\mathrm{R+}}, \quad \forall t \in \Omega^{\mathrm{T}}, \tag{5.7c}$$

$$0 \le p_t^{\mathrm{R}-} \le \overline{P}^{\mathrm{R}-}, \quad \forall t \in \Omega^{\mathrm{T}}, \tag{5.7d}$$

$$u_{ct}^{\mathrm{C}} \in \{0, 1\}, \quad \forall c \in \Omega^{\mathrm{C}}, \forall t \in \Omega^{\mathrm{T}}, \tag{5.7e}$$

subject to

$$k_t^{\mathrm{R}+}, k_t^{\mathrm{R}-} \in \{0, 1\}, \quad \forall t \in \Omega^{\mathrm{T}}, \tag{5.7f}$$

$$k_t^{\mathrm{R}+} + k_t^{\mathrm{R}-} \le 1, \quad \forall t \in \Omega^{\mathrm{T}}, \tag{5.7g}$$

$$\sum_{t \in \Omega^{\mathrm{T}}} \left(k_t^{\mathrm{R}+} + k_t^{\mathrm{R}-} \right) \le \Gamma^{\mathrm{K}}, \tag{5.7h}$$

$$u_{rt}^{\mathrm{R},-}, u_{rt}^{\mathrm{R},+} \in \{0, 1\}, \quad \forall r \in \Omega^{\mathrm{R}}, \forall t \in \Omega^{\mathrm{T}}, \tag{5.7i}$$

$$p_{rt}^{\mathrm{R,A}} = \tilde{P}_{rt}^{\mathrm{R,A}} - u_{rt}^{\mathrm{R},-} \hat{P}_{rt}^{\mathrm{R,A}} + u_{rt}^{\mathrm{R},+} \hat{P}_{rt}^{\mathrm{R,A}}, \quad \forall r \in \Omega^{\mathrm{R}}, \forall t \in \Omega^{\mathrm{T}}, \tag{5.7j}$$

$$u_{rt}^{\mathrm{R},-} + u_{rt}^{\mathrm{R},+} \le 1, \quad \forall r \in \Omega^{\mathrm{R}}, \forall t \in \Omega^{\mathrm{T}}, \tag{5.7k}$$

$$\sum_{t \in \Omega^{\mathrm{T}}} \left(u_{rt}^{\mathrm{R},-} + u_{rt}^{\mathrm{R},+} \right) \le \Gamma_r^{\mathrm{R}}, \quad \forall r \in \Omega^{\mathrm{R}} \tag{5.7l}$$

subject to

$$p_t^{\mathrm{E}} + k_t^{\mathrm{R}+} p_t^{\mathrm{R}+} - k_t^{\mathrm{R}-} p_t^{\mathrm{R}-} = \sum_{c \in \Omega^{\mathrm{C}}} p_{ct}^{\mathrm{C}} + \sum_{r \in \Omega^{\mathrm{R}}} p_{rt}^{\mathrm{R}} + \sum_{s \in \Omega^{\mathrm{S}}} \left(p_{st}^{\mathrm{S,D}} - p_{st}^{\mathrm{S,C}} \right) - \sum_{d \in \Omega^{\mathrm{D}}} p_{dt}^{\mathrm{D}}, \quad \forall t \in \Omega^{\mathrm{T}}, \tag{5.7m}$$

$$\underline{P}_{dt}^{\mathrm{D}} \le p_{dt}^{\mathrm{D}} \le \overline{P}_{dt}^{\mathrm{D}}, \quad \forall d \in \Omega^{\mathrm{D}}, \forall t \in \Gamma^{\mathrm{K}}, \tag{5.7n}$$

$$\sum_{t \in \Omega^{\mathrm{T}}} p_{dt}^{\mathrm{D}} \Delta t \ge \underline{E}_d^{\mathrm{D}}, \quad \forall d \in \Omega^{\mathrm{D}}, \tag{5.7o}$$

$$\underline{P}_{ct}^{\mathrm{C}} u_{ct}^{\mathrm{C}} \le p_{ct}^{\mathrm{C}} \le \overline{P}_{ct}^{\mathrm{C}} u_{ct}^{\mathrm{C}}, \quad \forall c \in \Omega^{\mathrm{C}}, \forall t \in \Omega^{\mathrm{T}}, \tag{5.7p}$$

$$0 \le p_{rt}^{\mathrm{R}} \le p_{rt}^{\mathrm{R,A}}, \quad \forall r \in \Omega^{\mathrm{R}}, \forall t \in \Omega^{\mathrm{T}}, \tag{5.7q}$$

$$0 \le p_{st}^{\mathrm{S,C}} \le \overline{P}_{st}^{\mathrm{S,C}}, \quad \forall s \in \Omega^{\mathrm{S}}, \forall t \in \Omega^{\mathrm{T}}, \tag{5.7r}$$

$$0 \le p_{st}^{\mathrm{S,D}} \le \overline{P}_{st}^{\mathrm{S,D}}, \quad \forall s \in \Omega^{\mathrm{S}}, \forall t \in \Omega^{\mathrm{T}}, \tag{5.7s}$$

$$e_{st}^{\mathrm{S}} = e_{s(t-1)}^{\mathrm{S}} + p_{st}^{\mathrm{S,C}} \Delta t \eta_s^{\mathrm{S,C}} - \frac{p_{st}^{\mathrm{S,D}} \Delta t}{\eta_s^{\mathrm{S,D}}}, \quad \forall s \in \Omega^{\mathrm{S}}, \forall t \in \Omega^{\mathrm{T}}, \tag{5.7t}$$

$$\underline{E}_{st}^{\mathrm{S}} \le e_{st}^{\mathrm{S}} \le \overline{E}_{st}^{\mathrm{S}}, \quad \forall s \in \Omega^{\mathrm{S}}, \forall t \in \Omega^{\mathrm{T}} \Big\}. \tag{5.7u}$$

Problem (5.7) aims at determining the scheduling decisions in the energy and reserve markets that maximize the worst-case profit (5.7a) of the VPP. The upper-level constraints (5.7b)–(5.7e) are described in Sect. 5.5.2. The middle-level constraints (5.7f)–(5.7l) define the uncertainty set. Finally, the lower-level constraints (5.7m)–(5.7u) describe the operation feasibility set.

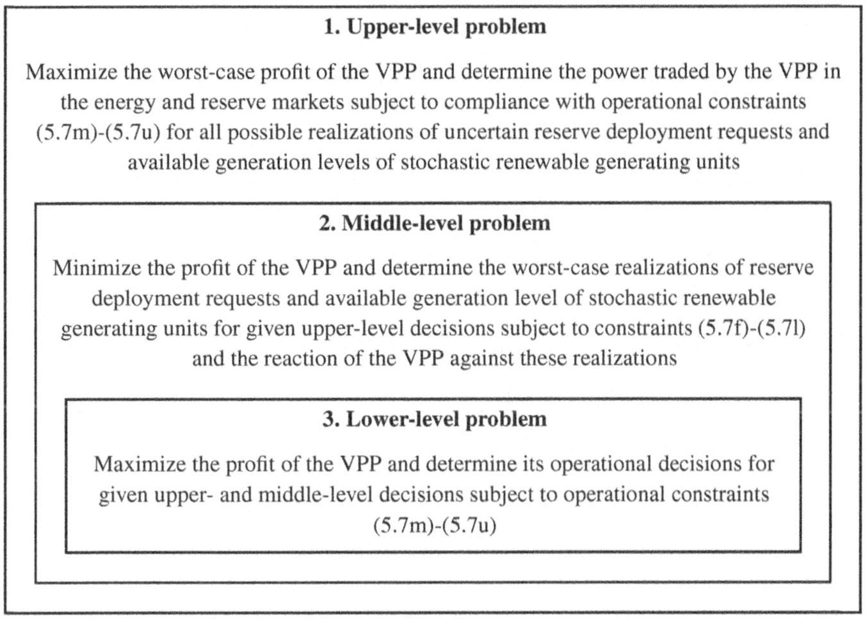

Fig. 5.18 Nested structure of the three-level optimization problem (5.7)

Figure 5.18 clarifies the nested structure of the three-level optimization problem (5.7).

5.5.6 *Solution Procedure*

The adaptive robust optimization problem (5.7) is a three-level optimization problem that can be solved using a column-and-constraint generation algorithm [14], similar to the one described in Sect. 4.6.6 of Chap. 4. The next sections provide the formulations of the master problem and the subproblem, as well as the iterative algorithm to achieve the optimal solution.

5.5.6.1 Master Problem

The master problem at iteration ν is formulated as follows:

$$\max_{\Phi^{\mathrm{M}}} \sum_{t\in\Omega^{\mathrm{T}}} \left(\lambda_t^{\mathrm{E}} p_t^{\mathrm{E}} \Delta t + \tilde{\lambda}_t^{\mathrm{R+}} p_t^{\mathrm{R+}} + \tilde{\lambda}_t^{\mathrm{R-}} p_t^{\mathrm{R-}} - \sum_{c\in\Omega^{\mathrm{C}}} C_c^{\mathrm{C,F}} u_{ct}^{\mathrm{C}} \right) + \theta \tag{5.8a}$$

subject to

$$\underline{P}^{\mathrm{E}} \le p_t^{\mathrm{E}} \le \overline{P}^{\mathrm{E}}, \quad \forall t \in \Omega^{\mathrm{T}}, \tag{5.8b}$$

$$0 \le p_t^{\mathrm{R+}} \le \overline{P}^{\mathrm{R+}}, \quad \forall t \in \Omega^{\mathrm{T}}, \tag{5.8c}$$

$$0 \le p_t^{\mathrm{R-}} \le \overline{P}^{\mathrm{R-}}, \quad \forall t \in \Omega^{\mathrm{T}}, \tag{5.8d}$$

$$u_{ct}^{\mathrm{C}} \in \{0, 1\}, \quad \forall c \in \Omega^{\mathrm{C}}, \forall t \in \Omega^{\mathrm{T}}, \tag{5.8e}$$

$$\theta \le \sum_{t\in\Omega^{\mathrm{T}}} \left(k_t^{\mathrm{R+}(\nu')} \lambda_t^{\mathrm{R+}} \Delta t p_t^{\mathrm{R+}} - k_t^{\mathrm{R-}(\nu')} \lambda_t^{\mathrm{R-}} \Delta t p_t^{\mathrm{R-}} - \sum_{c\in\Omega^{\mathrm{C}}} C_{ct}^{\mathrm{C,V}} p_{ct\nu'}^{\mathrm{C}} \Delta t \right),$$

$$\nu' = 1, \ldots, \nu, \tag{5.8f}$$

$$p_t^{\mathrm{E}} + k_t^{\mathrm{R+}(\nu')} p_t^{\mathrm{R+}} - k_t^{\mathrm{R-}(\nu')} p_t^{\mathrm{R-}} = \sum_{c \in \Omega^{\mathrm{C}}} p_{ct\nu'}^{\mathrm{C}} + \sum_{r \in \Omega^{\mathrm{R}}} p_{rt\nu'}^{\mathrm{R}} + \sum_{s \in \Omega^{\mathrm{S}}} \left(p_{st\nu'}^{\mathrm{S,D}} - p_{st\nu'}^{\mathrm{S,C}}\right)$$
$$- \sum_{d \in \Omega^{\mathrm{D}}} p_{dt\nu'}^{\mathrm{D}}, \quad \forall t \in \Omega^{\mathrm{T}}, \nu' = 1, \ldots, \nu, \tag{5.8g}$$

$$\underline{P}_{dt}^{\mathrm{D}} \leq p_{dt\nu'}^{\mathrm{D}} \leq \overline{P}_{dt}^{\mathrm{D}}, \quad \forall d \in \Omega^{\mathrm{D}}, \forall t \in \Omega^{\mathrm{T}}, \nu' = 1, \ldots, \nu, \tag{5.8h}$$

$$\sum_{t \in \Omega^{\mathrm{T}}} p_{dt\nu'}^{\mathrm{D}} \Delta t \geq \underline{E}_d^{\mathrm{D}}, \quad \forall d \in \Omega^{\mathrm{D}}, \nu' = 1, \ldots, \nu, \tag{5.8i}$$

$$\underline{P}_c^{\mathrm{C}} u_{ct}^{\mathrm{C}} \leq p_{ct\nu'}^{\mathrm{C}} \leq \overline{P}_c^{\mathrm{C}} u_{ct}^{\mathrm{C}}, \quad \forall c \in \Omega^{\mathrm{C}}, \forall t \in \Omega^{\mathrm{T}}, \nu' = 1, \ldots, \nu, \tag{5.8j}$$

$$0 \leq p_{rt\nu'}^{\mathrm{R}} \leq p_{rt}^{\mathrm{R,A}(\nu')}, \quad \forall r \in \Omega^{\mathrm{R}}, \forall t \in \Omega^{\mathrm{T}}, \nu' = 1, \ldots, \nu, \tag{5.8k}$$

$$0 \leq p_{st\nu'}^{\mathrm{S,C}} \leq \overline{P}_{st}^{\mathrm{S,C}}, \quad \forall s \in \Omega^{\mathrm{S}}, \forall t \in \Omega^{\mathrm{T}}, \nu' = 1, \ldots, \nu, \tag{5.8l}$$

$$0 \leq p_{st\nu'}^{\mathrm{S,D}} \leq \overline{P}_{st}^{\mathrm{S,D}}, \quad \forall s \in \Omega^{\mathrm{S}}, \forall t \in \Omega^{\mathrm{T}}, \nu' = 1, \ldots, \nu, \tag{5.8m}$$

$$e_{st\nu'}^{\mathrm{S}} = e_{s(t-1)\nu'}^{\mathrm{S}} + p_{st\nu'}^{\mathrm{S,C}} \Delta t \eta_s^{\mathrm{S,C}} - \frac{p_{st\nu'}^{\mathrm{S,D}} \Delta t}{\eta_s^{\mathrm{S,D}}}, \quad \forall s \in \Omega^{\mathrm{S}}, \forall t \in \Omega^{\mathrm{T}}, \nu' = 1, \ldots, \nu, \tag{5.8n}$$

$$\underline{E}_{st}^{\mathrm{S}} \leq e_{st\nu'}^{\mathrm{S}} \leq \overline{E}_{st}^{\mathrm{S}}, \quad \forall s \in \Omega^{\mathrm{S}}, \forall t \in \Omega^{\mathrm{T}}, \nu' = 1, \ldots, \nu, \tag{5.8o}$$

where variables in set $\Phi^{\mathrm{M}} = \Big\{\Phi^{\mathrm{UL}}; \theta; p_{dt\nu'}^{\mathrm{D}}, \forall d \in \Omega^{\mathrm{D}}, \forall t \in \Omega^{\mathrm{T}}, \nu' = 1, \ldots, \nu; p_{ct\nu'}^{\mathrm{C}}, \forall c \in \Omega^{\mathrm{C}}, \forall t \in \Omega^{\mathrm{T}}, \nu' = 1, \ldots, \nu;$ $p_{rt\nu'}^{\mathrm{R}}, \forall r \in \Omega^{\mathrm{R}}, \forall t \in \Omega^{\mathrm{T}}, \nu' = 1, \ldots, \nu; e_{st\nu'}^{\mathrm{S}}, p_{st\nu'}^{\mathrm{S,C}}, p_{st\nu'}^{\mathrm{S,D}}, \forall s \in \Omega^{\mathrm{S}}, \forall t \in \Omega^{\mathrm{T}}, \nu' = 1, \ldots, \nu\Big\}$ are the optimization variables of problem (5.8).

Master problem (5.8) is a relaxed version of the three-level optimization problem (5.7) wherein the auxiliary variable θ iteratively approximates the worst-case of the middle-level objective function. Thus, the size of master problem (5.8) increases with the iteration counter ν since a new set of constraints (5.8f)–(5.8o) is considered. Finally, note that parameters $k_t^{\mathrm{R+}(\nu')}, \forall t \in \Omega^{\mathrm{T}}$; $k_t^{\mathrm{R-}(\nu')}, \forall t \in \Omega^{\mathrm{T}}$; and $p_{rt}^{\mathrm{R,A}(\nu')}, \forall r \in \Omega^{\mathrm{R}}, \forall t \in \Omega^{\mathrm{T}}$, denote the optimal values for $k_t^{\mathrm{R+}}, \forall t \in \Omega^{\mathrm{T}}$; $k_t^{\mathrm{R-}}, \forall t \in \Omega^{\mathrm{T}}$; and $p_{rt}^{\mathrm{R,A}}, \forall r \in \Omega^{\mathrm{R}}, \forall t \in \Omega^{\mathrm{T}}$, respectively, obtained from the solution of the subproblem at iteration ν'.

5.5.6.2 Subproblem

The subproblem is a bi-level problem that corresponds to the two lowermost optimization levels of the original three-level problem (5.7) and is parameterized in terms of $p_t^{\mathrm{E}(\nu)}, p_t^{\mathrm{R+}(\nu)}, p_t^{\mathrm{R-}(\nu)}, \forall t \in \Omega^{\mathrm{T}}$; and $u_{ct}^{\mathrm{C}(\nu)}, \forall c \in \Omega^{\mathrm{C}}, \forall t \in \Omega^{\mathrm{T}}$, which are obtained from the solution of the master problem (5.8) at iteration ν. The formulation of the subproblem is provided below:

$$\min_{\Phi^{\mathrm{ML}}} \max_{\Phi^{\mathrm{LL}'}} \sum_{t \in \Omega^{\mathrm{T}}} \Bigg[k_t^{\mathrm{R+}} \lambda_t^{\mathrm{R+}} \Delta t p_t^{\mathrm{R+}(\nu)} - k_t^{\mathrm{R-}} \lambda_t^{\mathrm{R-}} \Delta t p_t^{\mathrm{R-}(\nu)}$$
$$- \sum_{c \in \Omega^{\mathrm{C}}} C_c^{\mathrm{C,V}} p_{ct}^{\mathrm{C}} \Delta t + M\left(h_t^{+} + h_t^{-}\right) \Bigg] \tag{5.9a}$$

subject to

$$k_t^{\mathrm{R+}}, k_t^{\mathrm{R-}} \in \{0,1\}, \quad \forall t \in \Omega^{\mathrm{T}}, \tag{5.9b}$$

$$k_t^{\mathrm{R+}} + k_t^{\mathrm{R-}} \leq 1, \quad \forall t \in \Omega^{\mathrm{T}}, \tag{5.9c}$$

$$\sum_{t \in \Omega^{\mathrm{T}}} \left(k_t^{\mathrm{R+}} + k_t^{\mathrm{R-}}\right) \leq \Gamma^{\mathrm{K}}, \tag{5.9d}$$

$$u_{rt}^{\mathrm{R,-}}, u_{rt}^{\mathrm{R,+}} \in \{0,1\}, \quad \forall r \in \Omega^{\mathrm{R}}, \forall t \in \Omega^{\mathrm{T}}, \tag{5.9e}$$

$$p_{rt}^{\text{R,A}} = \tilde{P}_{rt}^{\text{R,A}} - u_{rt}^{\text{R},-} \hat{P}_{rt}^{\text{R,A}} + u_{rt}^{\text{R},+} \hat{P}_{rt}^{\text{R,A}}, \quad \forall r \in \Omega^{\text{R}}, \forall t \in \Omega^{\text{T}}, \tag{5.9f}$$

$$u_{rt}^{\text{R},-} + u_{rt}^{\text{R},+} \leq 1, \quad \forall r \in \Omega^{\text{R}}, \forall t \in \Omega^{\text{T}}, \tag{5.9g}$$

$$\sum_{t \in \Omega^{\text{T}}} \left(u_{rt}^{\text{R},-} + u_{rt}^{\text{R},+} \right) \leq \Gamma_r^{\text{R}}, \quad \forall r \in \Omega^{\text{R}} \tag{5.9h}$$

subject to

$$p_t^{\text{E}(\nu)} + k_t^{\text{R}+} p_t^{\text{R}+(\nu)} - k_t^{\text{R}-} p_t^{\text{R}-(\nu)} = \sum_{c \in \Omega^{\text{C}}} p_{ct}^{\text{C}} + \sum_{r \in \Omega^{\text{R}}} p_{rt}^{\text{R}}$$

$$+ \sum_{s \in \Omega^{\text{S}}} \left(p_{st}^{\text{S,D}} - p_{st}^{\text{S,C}} \right) - \sum_{d \in \Omega^{\text{D}}} p_{dt}^{\text{D}} + h_t^+ - h_t^- \ : \lambda_t, \quad \forall t \in \Omega^{\text{T}}, \tag{5.9i}$$

$$\underline{P}_{dt}^{\text{D}} \leq p_{dt}^{\text{D}} \leq \overline{P}_{dt}^{\text{D}} \ : \underline{\mu}_{dt}^{\text{D}}, \overline{\mu}_{dt}^{\text{D}}, \quad \forall d \in \Omega^{\text{D}}, \forall t \in \Omega^{\text{T}}, \tag{5.9j}$$

$$\sum_{t \in \Omega^{\text{T}}} p_{dt}^{\text{D}} \Delta t \geq \underline{E}_d^{\text{D}} \ : \xi_d, \quad \forall d \in \Omega^{\text{D}}, \tag{5.9k}$$

$$\underline{P}_c^{\text{C}} u_{ct}^{\text{C}(\nu)} \leq p_{ct}^{\text{C}} \leq \overline{P}_c^{\text{C}} u_{ct}^{\text{C}(\nu)} \ : \underline{\mu}_{ct}^{\text{C}}, \overline{\mu}_{ct}^{\text{C}}, \quad \forall c \in \Omega^{\text{C}}, \forall t \in \Omega^{\text{T}}, \tag{5.9l}$$

$$0 \leq p_{rt}^{\text{R}} \leq p_{rt}^{\text{R,A}} \ : \overline{\mu}_{rt}^{\text{R}}, \quad \forall r \in \Omega^{\text{R}}, \forall t \in \Omega^{\text{T}}, \tag{5.9m}$$

$$0 \leq p_{st}^{\text{S,C}} \leq \overline{P}_{st}^{\text{S,C}} \ : \overline{\mu}_{st}^{\text{S,C}}, \quad \forall s \in \Omega^{\text{S}}, \forall t \in \Omega^{\text{T}}, \tag{5.9n}$$

$$0 \leq p_{st}^{\text{S,D}} \leq \overline{P}_{st}^{\text{S,D}} \ : \overline{\mu}_{st}^{\text{S,D}}, \quad \forall s \in \Omega^{\text{S}}, \forall t \in \Omega^{\text{T}}, \tag{5.9o}$$

$$e_{st}^{\text{S}} = e_{s(t-1)}^{\text{S}} + p_{st}^{\text{S,C}} \Delta t \eta_s^{\text{S,C}} - \frac{p_{st}^{\text{S,D}} \Delta t}{\eta_s^{\text{S,D}}} \ : \phi_{st}^{\text{S}}, \quad \forall s \in \Omega^{\text{S}}, \forall t \in \Omega^{\text{T}}, \tag{5.9p}$$

$$\underline{E}_{st}^{\text{S}} \leq e_{st}^{\text{S}} \leq \overline{E}_{st}^{\text{S}} \ : \underline{\mu}_{st}^{\text{S}}, \overline{\mu}_{st}^{\text{S}}, \quad \forall s \in \Omega^{\text{S}}, \forall t \in \Omega^{\text{T}}, \tag{5.9q}$$

$$h_t^+, h_t^- \geq 0, \quad \forall t \in \Omega^{\text{T}}, \tag{5.9r}$$

where $\Phi^{\text{LL}'} = \left\{ \Phi^{\text{LL}}; h_t^+, h_t^-, \forall t \in \Omega^{\text{T}} \right\}$, whereas the lower-level dual variables are provided following a colon.

At each iteration ν, subproblem (5.9) determines the worst-case reserve deployment requests and available generation levels of the stochastic renewable generating units, i.e., the values of variables in set Φ^{ML} that minimize the profit of the VPP for a given upper-level decision vector provided by the solution of the master problem (5.8). Constraints (5.9b)–(5.9h) represent the uncertainty set while constraints (5.9i)–(5.9r) represent the feasibility set. As the subproblem considers as input data the solution of the master problem, i.e., the power traded in the energy and reserve markets, as well as the status of the conventional power plants, the subproblem may be infeasible since the VPP may not be able to provide the required power in the markets and the requested reserves. In order to guarantee the feasibility of the subproblem along the iterative process, power balance constraints (5.9i) are relaxed with nonnegative slack variables h_t^+ and h_t^-, $\forall t$, that are penalized in the objective function (5.9a) using a sufficiently large positive constant M.

Subproblem (5.9) is a bi-level programming problem whose lower-level (5.9i)–(5.9r) problem is continuous and linear on its decision variables. Such a property allows using the duality theory of linear programming to convert bi-level problem (5.9) into an equivalent single-level problem [3]. To do so:

1. The lower-level problem in (5.9) is replaced by its dual constraints.
2. The objective function (5.9a) is replaced by the dual lower-level objective function.

Therefore, subproblem (5.9) is recast as the following single-level problem:

$$
\begin{aligned}
\min_{\Phi^{\mathrm{S}}} & \sum_{t \in \Omega^{\mathrm{T}}} \left[k_t^{\mathrm{R+}} \lambda_t^{\mathrm{R+}} \Delta t p_t^{\mathrm{R+}(\nu)} - k_t^{\mathrm{R-}} \lambda_t^{\mathrm{R-}} \Delta t p_t^{\mathrm{R-}(\nu)} \right] \\
& - \left[\sum_{t \in \Omega^{\mathrm{T}}} \left[\lambda_t \left(p_t^{\mathrm{E}(\nu)} + k_t^{\mathrm{R+}} p_t^{\mathrm{R+}(\nu)} - k_t^{\mathrm{R-}} p_t^{\mathrm{R-}(\nu)} \right) + \sum_{d \in \Omega^{\mathrm{D}}} \left(\underline{\mu}_{dt}^{\mathrm{D}} \underline{P}_{dt}^{\mathrm{D}} - \overline{\mu}_{dt}^{\mathrm{D}} \overline{P}_{dt}^{\mathrm{D}} \right) \right. \right. \\
& + \sum_{c \in \Omega^{\mathrm{C}}} \left(\underline{\mu}_{ct}^{\mathrm{C}} \underline{P}_c^{\mathrm{C}} u_{ct}^{\mathrm{C}(\nu)} - \overline{\mu}_{ct}^{\mathrm{C}} \overline{P}_c^{\mathrm{C}} u_{ct}^{\mathrm{C}(\nu)} \right) - \sum_{r \in \Omega^{\mathrm{R}}} \overline{\mu}_{rt}^{\mathrm{R}} p_{rt}^{\mathrm{R,A}} - \sum_{s \in \Omega^{\mathrm{S}}} \left(\overline{\mu}_{st}^{\mathrm{S,C}} \overline{P}_{st}^{\mathrm{S,C}} \right. \\
& \left. \left. + \overline{\mu}_{st}^{\mathrm{S,D}} \overline{P}_{st}^{\mathrm{S,D}} - \underline{\mu}_{st}^{\mathrm{S}} \underline{E}_{st}^{\mathrm{S}} + \overline{\mu}_{st}^{\mathrm{S}} \overline{E}_{st}^{\mathrm{S}} \right) \right] + \sum_{d \in \Omega^{\mathrm{D}}} \xi_d \underline{E}_d^{\mathrm{D}} + \left. \sum_{s \in \Omega^{\mathrm{S}}} \phi_{s1}^{\mathrm{S}} E_{s0}^{\mathrm{S}} \right]
\end{aligned}
\tag{5.10a}
$$

subject to

$$k_t^{\mathrm{R+}}, k_t^{\mathrm{R-}} \in \{0, 1\}, \quad \forall t \in \Omega^{\mathrm{T}}, \tag{5.10b}$$

$$k_t^{\mathrm{R+}} + k_t^{\mathrm{R-}} \leq 1, \quad \forall t \in \Omega^{\mathrm{T}}, \tag{5.10c}$$

$$\sum_{t \in \Omega^{\mathrm{T}}} \left(k_t^{\mathrm{R+}} + k_t^{\mathrm{R-}} \right) \leq \Gamma^{\mathrm{K}}, \tag{5.10d}$$

$$u_{rt}^{\mathrm{R,-}}, u_{rt}^{\mathrm{R,+}} \in \{0, 1\}, \quad \forall r \in \Omega^{\mathrm{R}}, \forall t \in \Omega^{\mathrm{T}}, \tag{5.10e}$$

$$p_{rt}^{\mathrm{R,A}} = \tilde{P}_{rt}^{\mathrm{R,A}} - u_{rt}^{\mathrm{R,-}} \hat{P}_{rt}^{\mathrm{R,A}} + u_{rt}^{\mathrm{R,+}} \hat{P}_{rt}^{\mathrm{R,A}}, \quad \forall r \in \Omega^{\mathrm{R}}, \forall t \in \Omega^{\mathrm{T}}, \tag{5.10f}$$

$$u_{rt}^{\mathrm{R,-}} + u_{rt}^{\mathrm{R,+}} \leq 1, \quad \forall r \in \Omega^{\mathrm{R}}, \forall t \in \Omega^{\mathrm{T}}, \tag{5.10g}$$

$$\sum_{t \in \Omega^{\mathrm{T}}} \left(u_{rt}^{\mathrm{R,-}} + u_{rt}^{\mathrm{R,+}} \right) \leq \Gamma_r^{\mathrm{R}}, \quad \forall r \in \Omega^{\mathrm{R}} \tag{5.10h}$$

$$\lambda_t + \underline{\mu}_{ct}^{\mathrm{C}} - \overline{\mu}_{ct}^{\mathrm{C}} = C_c^{\mathrm{C,V}}, \quad \forall c \in \Omega^{\mathrm{C}}, \forall t \in \Omega^{\mathrm{T}}, \tag{5.10i}$$

$$\lambda_t - \overline{\mu}_{rt}^{\mathrm{R}} \leq 0, \quad \forall r \in \Omega^{\mathrm{R}}, \forall t \in \Omega^{\mathrm{T}}, \tag{5.10j}$$

$$\lambda_t - \overline{\mu}_{st}^{\mathrm{S,D}} + \frac{\Delta t}{\eta_s^{\mathrm{S,D}}} \phi_{st}^{\mathrm{S}} \leq 0, \quad \forall s \in \Omega^{\mathrm{S}}, \forall t \in \Omega^{\mathrm{T}}, \tag{5.10k}$$

$$-\lambda_t - \overline{\mu}_{st}^{\mathrm{S,C}} + \eta_s^{\mathrm{S,C}} \Delta t \phi_{st}^{\mathrm{S}} \leq 0, \quad \forall s \in \Omega^{\mathrm{S}}, \forall t \in \Omega^{\mathrm{T}}, \tag{5.10l}$$

$$-\lambda_t + \underline{\mu}_{dt}^{\mathrm{D}} - \overline{\mu}_{dt}^{\mathrm{D}} + \Delta t \xi_d = 0, \quad \forall d \in \Omega^{\mathrm{D}}, \forall t \in \Omega^{\mathrm{T}}, \tag{5.10m}$$

$$\lambda_t \leq M, \quad \forall t \in \Omega^{\mathrm{T}}, \tag{5.10n}$$

$$-\lambda_t \leq M, \quad \forall t \in \Omega^{\mathrm{T}}, \tag{5.10o}$$

$$\phi_{st}^{\mathrm{S}} - \phi_{st+1}^{\mathrm{S}} + \underline{\mu}_{st}^{\mathrm{S}} - \overline{\mu}_{st}^{\mathrm{S}} = 0, \quad \forall s \in \Omega^{\mathrm{S}}, t = 1, \ldots |\Omega^{\mathrm{T}}| - 1, \tag{5.10p}$$

$$\phi_{s|\Omega^{\mathrm{T}}|}^{\mathrm{S}} + \underline{\mu}_{s|\Omega^{\mathrm{T}}|}^{\mathrm{S}} - \overline{\mu}_{s|\Omega^{\mathrm{T}}|}^{\mathrm{S}} = 0, \quad \forall s \in \Omega^{\mathrm{S}}, \tag{5.10q}$$

$$\underline{\mu}_{ct}^{\mathrm{C}}, \overline{\mu}_{ct}^{\mathrm{C}} \geq 0, \quad \forall c \in \Omega^{\mathrm{C}}, \forall t \in \Omega^{\mathrm{T}}, \tag{5.10r}$$

$$\overline{\mu}_{rt}^{\mathrm{R}} \geq 0, \quad \forall r \in \Omega^{\mathrm{R}}, \forall t \in \Omega^{\mathrm{T}}, \tag{5.10s}$$

$$\overline{\mu}_{st}^{\mathrm{S,C}}, \overline{\mu}_{st}^{\mathrm{S,D}}, \underline{\mu}_{st}^{\mathrm{S}}, \overline{\mu}_{st}^{\mathrm{S}} \geq 0, \quad \forall s \in \Omega^{\mathrm{S}}, \forall t \in \Omega^{\mathrm{T}}, \tag{5.10t}$$

$$\underline{\mu}_{dt}^{\mathrm{D}}, \overline{\mu}_{dt}^{\mathrm{D}} \geq 0, \quad \forall d \in \Omega^{\mathrm{D}}, \forall t \in \Omega^{\mathrm{T}}, \tag{5.10u}$$

where variables in set $\Phi^{\mathrm{S}} = \Big\{k_t^{\mathrm{R}+}, k_t^{\mathrm{R}-}, \forall t \in \Omega^{\mathrm{T}}; p_{rt}^{\mathrm{R,A}}, u_{rt}^{\mathrm{R},-}, u_{rt}^{\mathrm{R},+}, \forall r \in \Omega^{\mathrm{R}}, \forall t \in \Omega^{\mathrm{T}}; \lambda_t, \forall t \in \Omega^{\mathrm{T}}; \underline{\mu}_{ct}^{\mathrm{C}}, \overline{\mu}_{ct}^{\mathrm{C}}, \forall c \in \Omega^{\mathrm{C}}, \forall t \in \Omega^{\mathrm{T}}; \overline{\mu}_{rt}^{\mathrm{R}}, \forall r \in \Omega^{\mathrm{R}}, \forall t \in \Omega^{\mathrm{T}}; \overline{\mu}_{st}^{\mathrm{S,C}}, \overline{\mu}_{st}^{\mathrm{S,D}}, \phi_{st}^{\mathrm{S}}, \underline{\mu}_{st}^{\mathrm{S}}, \overline{\mu}_{st}^{\mathrm{S}}, \forall s \in \Omega^{\mathrm{S}}, \forall t \in \Omega^{\mathrm{T}}; \underline{\mu}_{dt}^{\mathrm{D}}, \overline{\mu}_{dt}^{\mathrm{D}}, \xi_d, \forall d \in \Omega^{\mathrm{D}}, \forall t \in \Omega^{\mathrm{T}}\Big\}$ are the optimization variables of problem (5.10). The dual lower-level objective function is minimized in (5.10a). Constraints (5.10b)–(5.10h) are associated with the uncertainty set while constraints (5.10i)–(5.10u) are the dual lower-level feasibility constraints.

Note that problem (5.10) includes some nonlinear terms in the objective function (5.10a), namely:

1. Terms $\lambda_t \left(k_t^{\mathrm{R}+} p_t^{\mathrm{R}+(\nu)} - k_t^{\mathrm{R}-} p_t^{\mathrm{R}-(\nu)}\right), \forall t \in \Omega^{\mathrm{T}}$. These bilinear terms comprise the product of a continuous variable, a binary variable, and a constant. As explained in [8], these nonlinear terms can be replaced by the following exact mixed-integer linear equations:

$$\lambda_t \left(k_t^{\mathrm{R}+} p_t^{\mathrm{R}+(\nu)} - k_t^{\mathrm{R}-} p_t^{\mathrm{R}-(\nu)}\right) = z_t^{\mathrm{K}+} p_t^{\mathrm{R}+(\nu)} - z_t^{\mathrm{K}-} p_t^{\mathrm{R}+(\nu)}, \quad \forall t \in \Omega^{\mathrm{T}}, \tag{5.11a}$$

$$- M k_t^{\mathrm{R}+} \le z_t^{\mathrm{K}+} \le M k_t^{\mathrm{R}+}, \quad \forall t \in \Omega^{\mathrm{T}}, \tag{5.11b}$$

$$- M \left(1 - k_t^{\mathrm{R}+}\right) \le \lambda_t - z_t^{\mathrm{K}+} \le M \left(1 - k_t^{\mathrm{R}+}\right), \quad \forall t \in \Omega^{\mathrm{T}}, \tag{5.11c}$$

$$- M k_t^{\mathrm{R}-} \le z_t^{\mathrm{K}-} \le M k_t^{\mathrm{R}-}, \quad \forall t \in \Omega^{\mathrm{T}}, \tag{5.11d}$$

$$- M \left(1 - k_t^{\mathrm{R}-}\right) \le \lambda_t - z_t^{\mathrm{K}-} \le M \left(1 - k_t^{\mathrm{R}-}\right), \quad \forall t \in \Omega^{\mathrm{T}}, \tag{5.11e}$$

where M is a large enough positive constant.
2. Terms $\overline{\mu}_{rt}^{\mathrm{R}} p_{rt}^{\mathrm{R,A}}, \forall r \in \Omega^{\mathrm{R}}, \forall t \in \Omega^{\mathrm{T}}$. These bilinear terms comprising the product of two continuous variables can be replaced by using Eqs. (4.18) as explained in Sect. 4.6.6.2 of Chap. 4.

Finally, subproblem (5.10) is recast as an MILP problem [13].

5.5.6.3 Solution Algorithm

The master problem (5.8) and the subproblem (5.10) are iteratively solved until convergence using the following algorithm:

1. Initialize the iteration counter ($\nu \leftarrow 0$), select the convergence tolerance (ϵ), and set the lower bound (LB) and upper bound (UB) to $-\infty$ and ∞, respectively.
2. Solve master problem (5.8) and obtain, among other variables, the optimal power traded by the VPP in the energy and reserve markets, i.e., variables $p_t^{\mathrm{E}*}, p_t^{\mathrm{R}+*}, p_t^{\mathrm{R}-*}, \forall t \in \Omega^{\mathrm{T}}$. Note that if $\nu = 0$, then constraints (5.8f)–(5.8o) are not included.
3. Update the upper bound using Eq. (5.12) below:

$$\mathrm{UB} = z^{\mathrm{M}*}, \tag{5.12}$$

where $z^{\mathrm{M}*}$ is the optimal value of the objective function (5.8a) of the master problem.
4. Set $p_t^{\mathrm{E}(\nu)} = p_t^{\mathrm{E}*}, \forall t \in \Omega^{\mathrm{T}}; p_t^{\mathrm{R}+(\nu)} = p_t^{\mathrm{R}+*}, \forall t \in \Omega^{\mathrm{T}}; p_t^{\mathrm{R}-(\nu)} = p_t^{\mathrm{R}-*}, \forall t \in \Omega^{\mathrm{T}}$; and $u_{ct}^{\mathrm{C}(\nu)} = u_{ct}^{\mathrm{C}*}, \forall c \in \Omega^{\mathrm{C}}, \forall t \in \Omega^{\mathrm{T}}$, where $p_t^{\mathrm{E}*}, \forall t \in \Omega^{\mathrm{T}}; p_t^{\mathrm{R}+*}, \forall t \in \Omega^{\mathrm{T}}; p_t^{\mathrm{R}-*}, \forall t \in \Omega^{\mathrm{T}}$; and $u_{ct}^{\mathrm{C}*}, \forall c \in \Omega^{\mathrm{C}}, \forall t \in \Omega^{\mathrm{T}}$, are obtained from the optimal solution of master problem (5.8) in step 2.
5. Solve subproblem (5.10) and obtain, among other variables, the worst realization of the reserve deployment requests and the available generation level of the stochastic renewable generating units, i.e., variables $k_t^{\mathrm{R}+*}, \forall t \in \Omega^{\mathrm{T}}; k_t^{\mathrm{R}-*}, \forall t \in \Omega^{\mathrm{T}}$; and $p_{rt}^{\mathrm{R,A}*}, \forall r \in \Omega^{\mathrm{R}}, \forall t \in \Omega^{\mathrm{T}}$.

6. Update the lower bound using Eq. (5.13) below:

$$\text{LB} = \max\Bigg\{ \sum_{t\in\Omega^{\text{T}}} \left(\lambda_t^{\text{E}} p_t^{\text{E}*} \Delta t + \tilde{\lambda}_t^{\text{R}+} p_t^{\text{R}+*} + \tilde{\lambda}_t^{\text{R}-} p_t^{\text{R}-*} - \sum_{c\in\Omega^{\text{C}}} C_c^{\text{C,F}} u_{ct}^{\text{C}*} \right) + z^{\text{S}*}, \text{LB} \Bigg\}, \tag{5.13}$$

where $z^{\text{S}*}$ is the optimal value of the objective function (5.10a) of the subproblem.
7. If UB − LB $\leq \epsilon$, the algorithm stops. The optimal power traded in the energy and reserve markets is $p_t^{\text{E}*}$, $p_t^{\text{R}+*}$, $p_t^{\text{R}-*}$, $\forall t \in \Omega^{\text{T}}$. Otherwise, go to step 8.
8. Update the iteration counter $\nu \leftarrow \nu + 1$.
9. Set $k_t^{\text{R}+(\nu)} = k_t^{\text{R}+*}, \forall t \in \Omega^{\text{T}}$; $k_t^{\text{R}-(\nu)} = k_t^{\text{R}-*}, \forall t \in \Omega^{\text{T}}$; and $p_{rt}^{\text{R,A}(\nu)} = p_{rt}^{\text{R,A}*}, \forall r \in \Omega^{\text{R}}, \forall t \in \Omega^{\text{T}}$, where $k_t^{\text{R}+*}, \forall t \in \Omega^{\text{T}}$; $k_t^{\text{R}-*}, \forall t \in \Omega^{\text{T}}$; and $p_{rt}^{\text{R,A}*}, \forall r \in \Omega^{\text{R}}, \forall t \in \Omega^{\text{T}}$, are obtained from the optimal solution of master problem (5.10) in step 5.
10. Go to step 2.

The algorithm flowchart is provided in Fig. 5.19 for the sake of clarity.

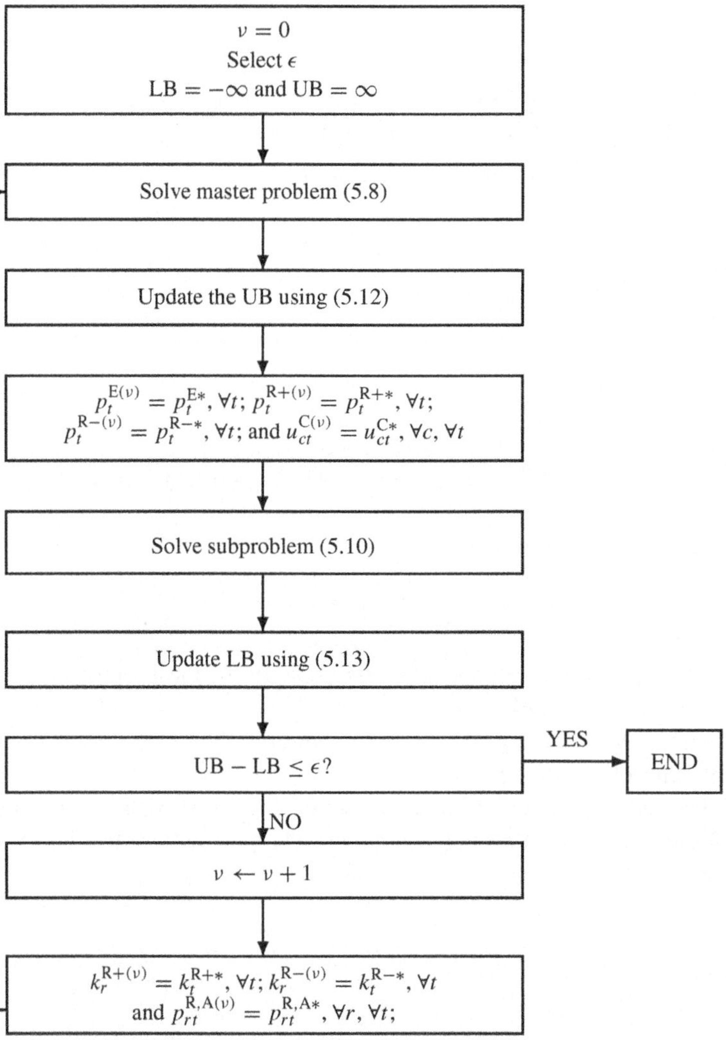

Fig. 5.19 Flowchart for the column-and-constraint generation algorithm

Illustrative Example 5.6 Robust Scheduling Problem in Energy and Reserve Markets

The VPP described in Illustrative Example 5.1 is considered again. The data of market prices, available wind-power generation levels, conventional power plant, and flexible demand are the same than those considered in Illustrative Example 5.1.

For the sake of simplicity, it is assumed that uncertainty affects only the reserve request coefficients, while the available wind-power generation levels are known.

An uncertainty budget $\Gamma^{\mathrm{K}} = 2$ is considered, which means that the VPP can be requested to deploy reserves in a maximum of 2 time periods.

Considering these data, the problem is formulated as the following three-level optimization model:

$$\begin{aligned}
&\max_{\Phi^{\mathrm{UL}}} \min_{\Phi^{\mathrm{ML}}} \max_{\Phi^{\mathrm{LL}}} \; 20p_1^{\mathrm{E}} + 50p_2^{\mathrm{E}} + 35p_3^{\mathrm{E}} + 31p_4^{\mathrm{E}} \\
&\quad + \left(4 + 14k_1^{\mathrm{R+}}\right) p_1^{\mathrm{R+}} + \left(10 + 38k_2^{\mathrm{R+}}\right) p_2^{\mathrm{R+}} + \left(8 + 26k_3^{\mathrm{R+}}\right) p_3^{\mathrm{R+}} + \left(6 + 25k_4^{\mathrm{R+}}\right) p_4^{\mathrm{R+}} \\
&\quad + \left(4 - 14k_1^{\mathrm{R-}}\right) p_1^{\mathrm{R-}} + \left(10 - 38k_2^{\mathrm{R-}}\right) p_2^{\mathrm{R-}} + \left(8 - 26k_3^{\mathrm{R-}}\right) p_3^{\mathrm{R-}} + \left(6 - 25k_4^{\mathrm{R-}}\right) p_4^{\mathrm{R-}} \\
&\quad - \left(6u_1^{\mathrm{C}} + 6u_2^{\mathrm{C}} + 6u_3^{\mathrm{C}} + 6u_4^{\mathrm{C}} + 33p_1^{\mathrm{C}} + 33p_2^{\mathrm{C}} + 33p_3^{\mathrm{C}} + 33p_4^{\mathrm{C}}\right)
\end{aligned}$$

subject to

$$\begin{aligned}
&-100 \le p_1^{\mathrm{E}} \le 100, \\
&-100 \le p_2^{\mathrm{E}} \le 100, \\
&-100 \le p_3^{\mathrm{E}} \le 100, \\
&-100 \le p_4^{\mathrm{E}} \le 100, \\
&0 \le p_1^{\mathrm{R+}} \le 50, \\
&0 \le p_2^{\mathrm{R+}} \le 50, \\
&0 \le p_3^{\mathrm{R+}} \le 50, \\
&0 \le p_4^{\mathrm{R+}} \le 50, \\
&0 \le p_1^{\mathrm{R-}} \le 50, \\
&0 \le p_2^{\mathrm{R-}} \le 50, \\
&0 \le p_3^{\mathrm{R-}} \le 50, \\
&0 \le p_4^{\mathrm{R-}} \le 50, \\
&u_1^{\mathrm{C}}, u_2^{\mathrm{C}}, u_3^{\mathrm{C}}, u_4^{\mathrm{C}} \in \{0, 1\},
\end{aligned}$$

subject to

$$\begin{aligned}
&k_1^{\mathrm{R+}}, k_2^{\mathrm{R+}}, k_3^{\mathrm{R+}}, k_4^{\mathrm{R+}}, k_1^{\mathrm{R-}}, k_2^{\mathrm{R-}}, k_3^{\mathrm{R-}}, k_4^{\mathrm{R-}} \in \{0, 1\}, \\
&k_1^{\mathrm{R+}} + k_1^{\mathrm{R-}} \le 1, \\
&k_2^{\mathrm{R+}} + k_2^{\mathrm{R-}} \le 1, \\
&k_3^{\mathrm{R+}} + k_3^{\mathrm{R-}} \le 1, \\
&k_4^{\mathrm{R+}} + k_4^{\mathrm{R-}} \le 1, \\
&k_1^{\mathrm{R+}} + k_1^{\mathrm{R-}} + k_2^{\mathrm{R+}} + k_2^{\mathrm{R-}} + k_3^{\mathrm{R+}} + k_3^{\mathrm{R-}} + k_4^{\mathrm{R+}} + k_4^{\mathrm{R-}} \le 2,
\end{aligned}$$

subject to

$$\begin{aligned}
&p_1^{\mathrm{E}} + k_1^{\mathrm{R+}} p_1^{\mathrm{R+}} - k_1^{\mathrm{R-}} p_1^{\mathrm{R-}} = p_1^{\mathrm{C}} + p_1^{\mathrm{R}} - p_1^{\mathrm{D}}, \\
&p_2^{\mathrm{E}} + k_2^{\mathrm{R+}} p_2^{\mathrm{R+}} - k_2^{\mathrm{R-}} p_2^{\mathrm{R-}} = p_2^{\mathrm{C}} + p_2^{\mathrm{R}} - p_2^{\mathrm{D}}, \\
&p_3^{\mathrm{E}} + k_3^{\mathrm{R+}} p_3^{\mathrm{R+}} - k_3^{\mathrm{R-}} p_3^{\mathrm{R-}} = p_3^{\mathrm{C}} + p_3^{\mathrm{R}} - p_3^{\mathrm{D}},
\end{aligned}$$

$$p_4^{\mathrm{E}} + k_4^{\mathrm{R}+} p_4^{\mathrm{R}+} - k_4^{\mathrm{R}-} p_4^{\mathrm{R}-} = p_4^{\mathrm{C}} + p_4^{\mathrm{R}} - p_4^{\mathrm{D}},$$
$$0 \le p_1^{\mathrm{D}} \le 20,$$
$$0 \le p_2^{\mathrm{D}} \le 20,$$
$$10 \le p_3^{\mathrm{D}} \le 25,$$
$$15 \le p_4^{\mathrm{D}} \le 30,$$
$$p_1^{\mathrm{D}} + p_2^{\mathrm{D}} + p_3^{\mathrm{D}} + p_4^{\mathrm{D}} \ge 60,$$
$$10u_1^{\mathrm{C}} \le p_1^{\mathrm{C}} \le 60u_1^{\mathrm{C}},$$
$$10u_2^{\mathrm{C}} \le p_2^{\mathrm{C}} \le 60u_2^{\mathrm{C}},$$
$$10u_3^{\mathrm{C}} \le p_3^{\mathrm{C}} \le 60u_3^{\mathrm{C}},$$
$$10u_4^{\mathrm{C}} \le p_4^{\mathrm{C}} \le 60u_4^{\mathrm{C}}$$
$$0 \le p_1^{\mathrm{R}} \le 60,$$
$$0 \le p_2^{\mathrm{R}} \le 20,$$
$$0 \le p_3^{\mathrm{R}} \le 40,$$
$$0 \le p_4^{\mathrm{R}} \le 10,$$

where the optimization variables of the upper-, middle-, and lower-level problems are those included in sets $\Phi^{\mathrm{UL}} = \Big\{p_1^{\mathrm{E}}, p_2^{\mathrm{E}}, p_3^{\mathrm{E}}, p_4^{\mathrm{E}}, p_1^{\mathrm{R}+}, p_2^{\mathrm{R}+}, p_3^{\mathrm{R}+}, p_4^{\mathrm{R}+}, p_1^{\mathrm{R}-}, p_2^{\mathrm{R}-}, p_3^{\mathrm{R}-}, p_4^{\mathrm{R}-}, u_1^{\mathrm{C}}, u_2^{\mathrm{C}}, u_3^{\mathrm{C}}, u_4^{\mathrm{C}}\Big\}$, $\Phi^{\mathrm{ML}} = \Big\{k_1^{\mathrm{R}+}, k_2^{\mathrm{R}+}, k_3^{\mathrm{R}+}, k_4^{\mathrm{R}+}, k_1^{\mathrm{R}-}, k_2^{\mathrm{R}-}, k_3^{\mathrm{R}-}, k_4^{\mathrm{R}-}\Big\}$, and $\Phi^{\mathrm{LL}} = \Big\{p_1^{\mathrm{C}}, p_2^{\mathrm{C}}, p_3^{\mathrm{C}}, p_4^{\mathrm{C}}, p_1^{\mathrm{R}}, p_2^{\mathrm{R}}, p_3^{\mathrm{R}}, p_4^{\mathrm{R}}, p_1^{\mathrm{D}}, p_2^{\mathrm{D}}, p_3^{\mathrm{D}}, p_4^{\mathrm{D}}\Big\}$, respectively.

The above three-level optimization problem is solved using the column-and-constraint generation algorithm described in Sect. 5.5.6.3 that comprises the following steps:

- Step 1. The iteration counter is initialized ($\nu \leftarrow 0$), the convergence tolerance is selected ($\epsilon = 0$), and the lower bound (LB) and upper bound (UB) are set to $-\infty$ and ∞, respectively.
- Step 2. The following master problem is solved:

$$\begin{aligned}\max_{\Phi^{\mathrm{M}}}\ & 20p_1^{\mathrm{E}} + 50p_2^{\mathrm{E}} + 35p_3^{\mathrm{E}} + 31p_4^{\mathrm{E}} + 4p_1^{\mathrm{R}+} + 10p_2^{\mathrm{R}+} + 8p_3^{\mathrm{R}+} + 6p_4^{\mathrm{R}+} \\ & + 4p_1^{\mathrm{R}-} + 10p_2^{\mathrm{R}-} + 8p_3^{\mathrm{R}-} + 6p_4^{\mathrm{R}-} - \left(6u_1^{\mathrm{C}} + 6u_2^{\mathrm{C}} + 6u_3^{\mathrm{C}} + 6u_4^{\mathrm{C}}\right) + \theta\end{aligned}$$

subject to

$$-100 \le p_1^{\mathrm{E}} \le 100,$$
$$-100 \le p_2^{\mathrm{E}} \le 100,$$
$$-100 \le p_3^{\mathrm{E}} \le 100,$$
$$-100 \le p_4^{\mathrm{E}} \le 100,$$
$$0 \le p_1^{\mathrm{R}+} \le 50,$$
$$0 \le p_2^{\mathrm{R}+} \le 50,$$
$$0 \le p_3^{\mathrm{R}+} \le 50,$$
$$0 \le p_4^{\mathrm{R}+} \le 50,$$
$$0 \le p_1^{\mathrm{R}-} \le 50,$$
$$0 \le p_2^{\mathrm{R}-} \le 50,$$

$$0 \le p_3^{\mathrm{R-}} \le 50,$$
$$0 \le p_4^{\mathrm{R-}} \le 50,$$
$$u_1^{\mathrm{C}}, u_2^{\mathrm{C}}, u_3^{\mathrm{C}}, u_4^{\mathrm{C}} \in \{0, 1\},$$

where $\Phi^{\mathrm{M}} = \left\{p_1^{\mathrm{E}}, p_2^{\mathrm{E}}, p_3^{\mathrm{E}}, p_4^{\mathrm{E}}, p_1^{\mathrm{R+}}, p_2^{\mathrm{R+}}, p_3^{\mathrm{R+}}, p_4^{\mathrm{R+}}, p_1^{\mathrm{R-}}, p_2^{\mathrm{R-}}, p_3^{\mathrm{R-}}, p_4^{\mathrm{R-}}, u_1^{\mathrm{C}}, u_2^{\mathrm{C}}, u_3^{\mathrm{C}}, u_4^{\mathrm{C}}, \theta\right\}$. Note that as $\nu = 0$, constraints (5.8f)–(5.8o) are not included.

The optimal solution of this problem is:

$$p_1^{\mathrm{E}*}, p_2^{\mathrm{E}*}, p_3^{\mathrm{E}*}, p_4^{\mathrm{E}*} = 100,$$
$$p_1^{\mathrm{R+}*}, p_2^{\mathrm{R+}*}, p_3^{\mathrm{R+}*}, p_4^{\mathrm{R+}*} = 50,$$
$$p_1^{\mathrm{R-}*}, p_2^{\mathrm{R-}*}, p_3^{\mathrm{R-}*}, p_4^{\mathrm{R-}*} = 50,$$
$$u_1^{\mathrm{C}*}, u_2^{\mathrm{C}*}, u_3^{\mathrm{C}*}, u_4^{\mathrm{C}*} = 0,$$
$$\theta^* = \infty,$$

while the optimal value of the objective function is $z^{\mathrm{M}*} = \infty$.

- Step 3. The upper bound is updated using Eq. (5.12):

$$\mathrm{UB} = \infty.$$

- Step 4. Set $p_1^{\mathrm{E}(0)} = 100$, $p_2^{\mathrm{E}(0)} = 100$, $p_3^{\mathrm{E}(0)} = 100$, $p_4^{\mathrm{E}(0)} = 100$, $p_1^{\mathrm{R+}(0)} = 50$, $p_2^{\mathrm{R+}(0)} = 50$, $p_3^{\mathrm{R+}(0)} = 50$, $p_4^{\mathrm{R+}(0)} = 50$, $p_1^{\mathrm{R-}(0)} = 50$, $p_2^{\mathrm{R-}(0)} = 50$, $p_3^{\mathrm{R-}(0)} = 50$, $p_4^{\mathrm{R-}(0)} = 50$, $u_1^{\mathrm{C}(0)} = 0$, $u_2^{\mathrm{C}(0)} = 0$, $u_3^{\mathrm{C}(0)} = 0$, and $u_4^{\mathrm{C}(0)} = 0$.
- Step 5. The following subproblem is solved:

$$\begin{aligned}\min_{\Phi^{\mathrm{S}}} & 14k_1^{\mathrm{R+}} \cdot 50 + 38k_2^{\mathrm{R+}} \cdot 50 + 26k_3^{\mathrm{R+}} \cdot 50 + 25k_4^{\mathrm{R+}} \cdot 50 - \Big(14k_1^{\mathrm{R-}} \cdot 50 \\ & +38k_2^{\mathrm{R-}} \cdot 50 + 26k_3^{\mathrm{R-}} \cdot 50 + 25k_4^{\mathrm{R-}} \cdot 50\Big) - \Big[100\lambda_1 + 100\lambda_2 + 100\lambda_3 \\ & + 100\lambda_4 + 50z_1^{\mathrm{K+}} + 50z_2^{\mathrm{K+}} + 50z_3^{\mathrm{K+}} + 50z_4^{\mathrm{K+}} - 50z_1^{\mathrm{K-}} - 50z_2^{\mathrm{K-}} \\ & - 50z_3^{\mathrm{K-}} - 50z_4^{\mathrm{K-}} + 0\underline{\mu}_1^{\mathrm{D}} + 0\underline{\mu}_2^{\mathrm{D}} + 10\underline{\mu}_3^{\mathrm{D}} + 15\underline{\mu}_4^{\mathrm{D}} - 20\overline{\mu}_1^{\mathrm{D}} - 20\overline{\mu}_2^{\mathrm{D}} \\ & - 25\overline{\mu}_3^{\mathrm{D}} - 30\overline{\mu}_4^{\mathrm{D}} + 10 \cdot 0\underline{\mu}_1^{\mathrm{C}} + 10 \cdot 0\underline{\mu}_2^{\mathrm{C}} + 10 \cdot 0\underline{\mu}_3^{\mathrm{C}} + 10 \cdot 0\underline{\mu}_4^{\mathrm{C}} - 60 \cdot 0\overline{\mu}_1^{\mathrm{C}} \\ & - 60 \cdot 0\overline{\mu}_2^{\mathrm{C}} - 60 \cdot 0\overline{\mu}_3^{\mathrm{C}} - 60 \cdot 0\overline{\mu}_4^{\mathrm{C}} - 60\overline{\mu}_1^{\mathrm{R}} - 20\overline{\mu}_2^{\mathrm{R}} - 40\overline{\mu}_3^{\mathrm{R}} - 10\overline{\mu}_4^{\mathrm{R}}\Big]\end{aligned}$$

subject to

$$k_1^{\mathrm{R+}}, k_2^{\mathrm{R+}}, k_3^{\mathrm{R+}}, k_4^{\mathrm{R+}}, k_1^{\mathrm{R-}}, k_2^{\mathrm{R-}}, k_3^{\mathrm{R-}}, k_4^{\mathrm{R-}} \in \{0, 1\},$$
$$k_1^{\mathrm{R+}} + k_1^{\mathrm{R-}} \le 1,$$
$$k_2^{\mathrm{R+}} + k_2^{\mathrm{R-}} \le 1,$$
$$k_3^{\mathrm{R+}} + k_3^{\mathrm{R-}} \le 1,$$
$$k_4^{\mathrm{R+}} + k_4^{\mathrm{R-}} \le 1,$$
$$k_1^{\mathrm{R+}} + k_1^{\mathrm{R-}} + k_2^{\mathrm{R+}} + k_2^{\mathrm{R-}} + k_3^{\mathrm{R+}} + k_3^{\mathrm{R-}} + k_4^{\mathrm{R+}} + k_4^{\mathrm{R-}} \le 2,$$
$$\lambda_1 + \underline{\mu}_1^{\mathrm{C}} - \overline{\mu}_1^{\mathrm{C}} = 33,$$
$$\lambda_2 + \underline{\mu}_2^{\mathrm{C}} - \overline{\mu}_2^{\mathrm{C}} = 33,$$

$$
\begin{aligned}
&\lambda_3 + \underline{\mu}_3^{\mathrm{C}} - \overline{\mu}_3^{\mathrm{C}} = 33,\\
&\lambda_4 + \underline{\mu}_4^{\mathrm{C}} - \overline{\mu}_4^{\mathrm{C}} = 33,\\
&-1000k_1^{\mathrm{R+}} \le z_1^{\mathrm{K+}} \le 1000k_1^{\mathrm{R+}},\\
&-1000\left(1 - k_1^{\mathrm{R+}}\right) \le \lambda_1 - z_1^{\mathrm{K+}} \le 1000\left(1 - k_1^{\mathrm{R+}}\right),\\
&-1000k_1^{\mathrm{R-}} \le z_1^{\mathrm{K-}} \le 1000k_1^{\mathrm{R-}},\\
&-1000\left(1 - k_1^{\mathrm{R-}}\right) \le \lambda_1 - z_1^{\mathrm{K-}} \le 1000\left(1 - k_1^{\mathrm{R-}}\right),\\
&-1000k_2^{\mathrm{R+}} \le z_2^{\mathrm{K+}} \le 1000k_2^{\mathrm{R+}},\\
&-1000\left(1 - k_2^{\mathrm{R+}}\right) \le \lambda_2 - z_2^{\mathrm{K+}} \le 1000\left(1 - k_2^{\mathrm{R+}}\right),\\
&-1000k_2^{\mathrm{R-}} \le z_2^{\mathrm{K-}} \le 1000k_2^{\mathrm{R-}},\\
&-1000\left(1 - k_2^{\mathrm{R-}}\right) \le \lambda_1 - z_2^{\mathrm{K-}} \le 1000\left(1 - k_2^{\mathrm{R-}}\right),\\
&-1000k_3^{\mathrm{R+}} \le z_3^{\mathrm{K+}} \le 1000k_3^{\mathrm{R+}},\\
&-1000\left(1 - k_3^{\mathrm{R+}}\right) \le \lambda_3 - z_3^{\mathrm{K+}} \le 1000\left(1 - k_3^{\mathrm{R+}}\right),\\
&-1000k_3^{\mathrm{R-}} \le z_3^{\mathrm{K-}} \le 1000k_3^{\mathrm{R-}},\\
&-1000\left(1 - k_3^{\mathrm{R-}}\right) \le \lambda_3 - z_3^{\mathrm{K-}} \le 1000\left(1 - k_3^{\mathrm{R-}}\right),\\
&-1000k_4^{\mathrm{R+}} \le z_4^{\mathrm{K+}} \le 1000k_4^{\mathrm{R+}},\\
&-1000\left(1 - k_4^{\mathrm{R+}}\right) \le \lambda_4 - z_4^{\mathrm{K+}} \le 1000\left(1 - k_4^{\mathrm{R+}}\right),\\
&-1000k_4^{\mathrm{R-}} \le z_4^{\mathrm{K-}} \le 1000k_4^{\mathrm{R-}},\\
&-1000\left(1 - k_4^{\mathrm{R-}}\right) \le \lambda_4 - z_4^{\mathrm{K-}} \le 1000\left(1 - k_4^{\mathrm{R-}}\right),\\
&\lambda_1 - \overline{\mu}_1^{\mathrm{R}} \le 0,\\
&\lambda_2 - \overline{\mu}_2^{\mathrm{R}} \le 0,\\
&\lambda_3 - \overline{\mu}_3^{\mathrm{R}} \le 0,\\
&\lambda_4 - \overline{\mu}_4^{\mathrm{R}} \le 0,\\
&-\lambda_1 + \underline{\mu}_1^{\mathrm{D}} - \overline{\mu}_1^{\mathrm{D}} + \xi = 0,\\
&-\lambda_2 + \underline{\mu}_2^{\mathrm{D}} - \overline{\mu}_2^{\mathrm{D}} + \xi = 0,\\
&-\lambda_3 + \underline{\mu}_3^{\mathrm{D}} - \overline{\mu}_3^{\mathrm{D}} + \xi = 0,\\
&-\lambda_4 + \underline{\mu}_4^{\mathrm{D}} - \overline{\mu}_4^{\mathrm{D}} + \xi = 0,\\
&\lambda_1, \lambda_2, \lambda_3, \lambda_4 \le 1000,\\
&-\lambda_1, -\lambda_2, -\lambda_3, -\lambda_4 \le 1000,\\
&\underline{\mu}_1^{\mathrm{C}}, \overline{\mu}_1^{\mathrm{C}}, \underline{\mu}_2^{\mathrm{C}}, \overline{\mu}_2^{\mathrm{C}}, \underline{\mu}_3^{\mathrm{C}}, \overline{\mu}_3^{\mathrm{C}}, \underline{\mu}_4^{\mathrm{C}}, \overline{\mu}_4^{\mathrm{C}}, \overline{\mu}_1^{\mathrm{R}}, \overline{\mu}_2^{\mathrm{R}}, \overline{\mu}_3^{\mathrm{R}}, \overline{\mu}_4^{\mathrm{R}}, \underline{\mu}_1^{\mathrm{D}}, \overline{\mu}_1^{\mathrm{D}}, \underline{\mu}_2^{\mathrm{D}}, \overline{\mu}_2^{\mathrm{D}}, \underline{\mu}_3^{\mathrm{D}}, \overline{\mu}_3^{\mathrm{D}},\\
&\qquad \underline{\mu}_4^{\mathrm{D}}, \overline{\mu}_4^{\mathrm{D}} \ge 0,
\end{aligned}
$$

where $\Phi^{\mathrm{S}} = \left\{k_1^{\mathrm{R}+}, k_2^{\mathrm{R}+}, k_3^{\mathrm{R}+}, k_4^{\mathrm{R}+}, k_1^{\mathrm{R}-}, k_2^{\mathrm{R}-}, k_3^{\mathrm{R}-}, k_4^{\mathrm{R}-}, z_1^{\mathrm{K}+}, z_2^{\mathrm{K}+}, z_3^{\mathrm{K}+}, z_4^{\mathrm{K}+}, z_1^{\mathrm{K}-}, z_2^{\mathrm{K}-}, z_3^{\mathrm{K}-}, z_4^{\mathrm{K}-}, \lambda_1, \lambda_2, \lambda_3, \lambda_4, \underline{\mu}_1^{\mathrm{C}}, \underline{\mu}_2^{\mathrm{C}}, \underline{\mu}_3^{\mathrm{C}}, \underline{\mu}_4^{\mathrm{C}}, \overline{\mu}_1^{\mathrm{C}}, \overline{\mu}_2^{\mathrm{C}}, \overline{\mu}_3^{\mathrm{C}}, \overline{\mu}_4^{\mathrm{C}}, \overline{\mu}_1^{\mathrm{R}}, \overline{\mu}_2^{\mathrm{R}}, \overline{\mu}_3^{\mathrm{R}}, \overline{\mu}_4^{\mathrm{R}}, \underline{\mu}_1^{\mathrm{D}}, \underline{\mu}_2^{\mathrm{D}}, \underline{\mu}_3^{\mathrm{D}}, \underline{\mu}_4^{\mathrm{D}}, \overline{\mu}_1^{\mathrm{D}}, \overline{\mu}_2^{\mathrm{D}}, \overline{\mu}_3^{\mathrm{D}}, \overline{\mu}_4^{\mathrm{D}}, \xi\right\}$.

The worst uncertainty realization is:

$$k_2^{\mathrm{R}+}, k_3^{\mathrm{R}+}, k_1^{\mathrm{R}-}, k_2^{\mathrm{R}-}, k_3^{\mathrm{R}-}, k_4^{\mathrm{R}-} = 0,$$
$$k_1^{\mathrm{R}+}, k_4^{\mathrm{R}+} = 1.$$

Note that $\Gamma^{\mathrm{K}} = 2$ and thus the VPP can be requested to provide reserves in a maximum of two time periods. In this case, it is found that given the power traded in the energy and reserve markets obtained from the master problem in step 2, the worst case realization is that the VPP is requested to provide up reserves in time periods 1 and 4. On the other hand, the optimal value of the objective function is $z^{\mathrm{S}*} = -4.33 \cdot 10^5$.

- Step 6. The lower bound is updated using Eq. (5.13):

$$\mathrm{LB} = -4.166 \cdot 10^5.$$

- Step 7. Compute $\mathrm{UB} - \mathrm{LB} = \infty - \left(-4.166 \cdot 10^5\right) = \infty$. As the difference between the upper and lower bounds is higher than the tolerance, continue with the following step.
- Step 8. The iteration counter is updated $\nu = 1$.
- Step 9. Set $k_1^{\mathrm{R}+(1)} = k_4^{\mathrm{R}+(1)} = 1$ and $k_2^{\mathrm{R}+(1)} = k_3^{\mathrm{R}+(1)} = k_1^{\mathrm{R}-(1)} = k_2^{\mathrm{R}-(1)} = k_3^{\mathrm{R}-(1)} = k_4^{\mathrm{R}-(1)} = 0$.
- Step 2. The following master problem is solved:

$$\max_{\Phi^{\mathrm{M}}} \; 20p_1^{\mathrm{E}} + 50p_2^{\mathrm{E}} + 35p_3^{\mathrm{E}} + 31p_4^{\mathrm{E}} + 4p_1^{\mathrm{R}+} + 10p_2^{\mathrm{R}+} + 8p_3^{\mathrm{R}+} + 6p_4^{\mathrm{R}+}$$
$$+ 4p_1^{\mathrm{R}-} + 10p_2^{\mathrm{R}-} + 8p_3^{\mathrm{R}-} + 6p_4^{\mathrm{R}-} - \left(6u_1^{\mathrm{C}} + 6u_2^{\mathrm{C}} + 6u_3^{\mathrm{C}} + 6u_4^{\mathrm{C}}\right) + \theta$$

subject to

$$-100 \le p_1^{\mathrm{E}} \le 100,$$
$$-100 \le p_2^{\mathrm{E}} \le 100,$$
$$-100 \le p_3^{\mathrm{E}} \le 100,$$
$$-100 \le p_4^{\mathrm{E}} \le 100,$$
$$0 \le p_1^{\mathrm{R}+} \le 50,$$
$$0 \le p_2^{\mathrm{R}+} \le 50,$$
$$0 \le p_3^{\mathrm{R}+} \le 50,$$
$$0 \le p_4^{\mathrm{R}+} \le 50,$$
$$0 \le p_1^{\mathrm{R}-} \le 50,$$
$$0 \le p_2^{\mathrm{R}-} \le 50,$$
$$0 \le p_3^{\mathrm{R}-} \le 50,$$
$$0 \le p_4^{\mathrm{R}-} \le 50,$$
$$u_1^{\mathrm{C}}, u_2^{\mathrm{C}}, u_3^{\mathrm{C}}, u_4^{\mathrm{C}} \in \{0, 1\},$$
$$\theta \le 1 \cdot 14p_1^{\mathrm{R}+} + 0 \cdot 38p_2^{\mathrm{R}+} + 0 \cdot 26p_3^{\mathrm{R}+} + 1 \cdot 25p_4^{\mathrm{R}+} - \left(0 \cdot 14p_1^{\mathrm{R}-} + 0 \cdot 38p_2^{\mathrm{R}-}\right.$$
$$\left. +0 \cdot 26p_3^{\mathrm{R}-} + 0 \cdot 25p_4^{\mathrm{R}-}\right) - \left(33p_{11}^{\mathrm{C}} + 33p_{21}^{\mathrm{C}} + 33p_{31}^{\mathrm{C}} + 33p_{41}^{\mathrm{C}}\right),$$

$$p_1^{\mathrm{E}} + 1 \cdot p_1^{\mathrm{R+}} - 0 \cdot p_1^{\mathrm{R-}} = p_{11}^{\mathrm{C}} + p_{11}^{\mathrm{R}} - p_{11}^{\mathrm{D}},$$
$$p_2^{\mathrm{E}} + 0 \cdot p_2^{\mathrm{R+}} - 0 \cdot p_2^{\mathrm{R-}} = p_{21}^{\mathrm{C}} + p_{21}^{\mathrm{R}} - p_{21}^{\mathrm{D}},$$
$$p_3^{\mathrm{E}} + 0 \cdot p_3^{\mathrm{R+}} - 0 \cdot p_3^{\mathrm{R-}} = p_{31}^{\mathrm{C}} + p_{31}^{\mathrm{R}} - p_{31}^{\mathrm{D}},$$
$$p_4^{\mathrm{E}} + 1 \cdot p_4^{\mathrm{R+}} - 0 \cdot p_4^{\mathrm{R-}} = p_{41}^{\mathrm{C}} + p_{41}^{\mathrm{R}} - p_{41}^{\mathrm{D}},$$
$$10u_1^{\mathrm{C}} \le p_{11}^{\mathrm{C}} \le 60u_1^{\mathrm{C}},$$
$$10u_2^{\mathrm{C}} \le p_{21}^{\mathrm{C}} \le 60u_2^{\mathrm{C}},$$
$$10u_3^{\mathrm{C}} \le p_{31}^{\mathrm{C}} \le 60u_3^{\mathrm{C}},$$
$$10u_4^{\mathrm{C}} \le p_{41}^{\mathrm{C}} \le 60u_4^{\mathrm{C}},$$
$$0 \le p_{11}^{\mathrm{D}} \le 20,$$
$$0 \le p_{21}^{\mathrm{D}} \le 20,$$
$$10 \le p_{31}^{\mathrm{D}} \le 25,$$
$$15 \le p_{41}^{\mathrm{D}} \le 30,$$
$$p_{11}^{\mathrm{D}} + p_{21}^{\mathrm{D}} + p_{31}^{\mathrm{D}} + p_{41}^{\mathrm{D}} \ge 60,$$
$$0 \le p_{11}^{\mathrm{R}} \le 60,$$
$$0 \le p_{21}^{\mathrm{R}} \le 40,$$
$$0 \le p_{31}^{\mathrm{R}} \le 20,$$
$$0 \le p_{41}^{\mathrm{R}} \le 10,$$

where $\Phi^{\mathrm{M}} = \Big\{p_1^{\mathrm{E}}, p_2^{\mathrm{E}}, p_3^{\mathrm{E}}, p_4^{\mathrm{E}}, p_1^{\mathrm{R+}}, p_2^{\mathrm{R+}}, p_3^{\mathrm{R+}}, p_4^{\mathrm{R+}}, p_1^{\mathrm{R-}}, p_2^{\mathrm{R-}}, p_3^{\mathrm{R-}}, p_4^{\mathrm{R-}}, u_1^{\mathrm{C}}, u_2^{\mathrm{C}}, u_3^{\mathrm{C}}, u_4^{\mathrm{C}}, \theta, p_{11}^{\mathrm{D}}, p_{21}^{\mathrm{D}}, p_{31}^{\mathrm{D}}, p_{41}^{\mathrm{D}}, p_{11}^{\mathrm{R}}, p_{21}^{\mathrm{R}}, p_{31}^{\mathrm{R}}, p_{41}^{\mathrm{R}}, p_{11}^{\mathrm{C}}, p_{21}^{\mathrm{C}}, p_{31}^{\mathrm{C}}, p_{41}^{\mathrm{C}}\Big\}$. Note that the operating variables include two subscripts. The first one indicates the time period while the second one denotes the iteration. As $\nu = 1$, in this case constraints (5.8f)–(5.8o) are included.

The optimal solution of this problem is:

$$p_1^{\mathrm{E}*} = 30,$$
$$p_2^{\mathrm{E}*} = p_3^{\mathrm{E}*} = 85,$$
$$p_4^{\mathrm{E}*} = -15,$$
$$p_1^{\mathrm{R+}*}, p_4^{\mathrm{R+}*} = 0,$$
$$p_3^{\mathrm{R+}*}, p_4^{\mathrm{R+}*} = 50,$$
$$p_1^{\mathrm{R-}*}, p_2^{\mathrm{R-}*}, p_3^{\mathrm{R-}*}, p_4^{\mathrm{R-}*} = 50,$$
$$u_1^{\mathrm{C}*}, u_4^{\mathrm{C}*} = 0$$
$$u_2^{\mathrm{C}*}, u_3^{\mathrm{C}*} = 1,$$
$$\theta^* = -3960,$$

while the optimal value of the objective function is $z^{\mathrm{M}*} = 5688$.

- Step 3. The upper bound is updated using Eq. (5.12):

$$\mathrm{UB} = 5688.$$

- Step 4. Set $p_1^{\mathrm{E}(1)} = 30$, $p_2^{\mathrm{E}(1)} = 85$, $p_3^{\mathrm{E}(1)} = 85$, $p_4^{\mathrm{E}(1)} = -15$, $p_1^{\mathrm{R}+(1)} = 0$, $p_2^{\mathrm{R}+(1)} = 50$, $p_3^{\mathrm{R}+(1)} = 50$, $p_4^{\mathrm{R}+(1)} = 0$, $p_1^{\mathrm{R}-(1)} = 50$, $p_2^{\mathrm{R}-(1)} = 50$, $p_3^{\mathrm{R}-(1)} = 50$, $p_4^{\mathrm{R}-(1)} = 50$, $u_1^{\mathrm{C}(1)} = 0$, $u_2^{\mathrm{C}(1)} = 1$, $u_3^{\mathrm{C}(1)} = 1$, and $u_4^{\mathrm{C}(1)} = 0$.
- Step 5. The following subproblem is solved:

$$
\begin{aligned}
\min_{\Phi^{\mathrm{S}}} \, & 14k_1^{\mathrm{R}+} \cdot 0 + 38k_2^{\mathrm{R}+} \cdot 50 + 26k_3^{\mathrm{R}+} \cdot 50 + 25k_4^{\mathrm{R}+} \cdot 0 - \Big(14k_1^{\mathrm{R}-} \cdot 50 \\
& +38k_2^{\mathrm{R}-} \cdot 50 + 26k_3^{\mathrm{R}-} \cdot 50 + 25k_4^{\mathrm{R}-} \cdot 50\Big) - \Big[30\lambda_1 + 85\lambda_2 + 85\lambda_3 \\
& - 15\lambda_4 + 0z_1^{\mathrm{K}+} + 50z_2^{\mathrm{K}+} + 50z_3^{\mathrm{K}+} + 0z_4^{\mathrm{K}+} - 50z_1^{\mathrm{K}-} - 50z_2^{\mathrm{K}-} - 50z_3^{\mathrm{K}-} \\
& - 50z_4^{\mathrm{K}-} + 0\underline{\mu}_1^{\mathrm{D}} + 0\underline{\mu}_2^{\mathrm{D}} + 10\underline{\mu}_3^{\mathrm{D}} + 15\underline{\mu}_4^{\mathrm{D}} - 20\overline{\mu}_1^{\mathrm{D}} - 20\overline{\mu}_2^{\mathrm{D}} - 25\overline{\mu}_3^{\mathrm{D}} \\
& - 30\overline{\mu}_4^{\mathrm{D}} + 10 \cdot 0\underline{\mu}_1^{\mathrm{C}} + 10 \cdot 1\underline{\mu}_2^{\mathrm{C}} + 10 \cdot 1\underline{\mu}_3^{\mathrm{C}} + 10 \cdot 0\underline{\mu}_4^{\mathrm{C}} - 60 \cdot 0\overline{\mu}_1^{\mathrm{C}} - 60 \cdot 1\overline{\mu}_2^{\mathrm{C}} \\
& - 60 \cdot 1\overline{\mu}_3^{\mathrm{C}} - 60 \cdot 0\overline{\mu}_4^{\mathrm{C}} - 60\overline{\mu}_1^{\mathrm{R}} - 20\overline{\mu}_2^{\mathrm{R}} - 40\overline{\mu}_3^{\mathrm{R}} - 10\overline{\mu}_4^{\mathrm{R}}\Big]
\end{aligned}
$$

subject to

$$
\begin{aligned}
& k_1^{\mathrm{R}+}, k_2^{\mathrm{R}+}, k_3^{\mathrm{R}+}, k_4^{\mathrm{R}+}, k_1^{\mathrm{R}-}, k_2^{\mathrm{R}-}, k_3^{\mathrm{R}-}, k_4^{\mathrm{R}-} \in \{0, 1\}, \\
& k_1^{\mathrm{R}+} + k_1^{\mathrm{R}-} \le 1, \\
& k_2^{\mathrm{R}+} + k_2^{\mathrm{R}-} \le 1, \\
& k_3^{\mathrm{R}+} + k_3^{\mathrm{R}-} \le 1, \\
& k_4^{\mathrm{R}+} + k_4^{\mathrm{R}-} \le 1, \\
& k_1^{\mathrm{R}+} + k_1^{\mathrm{R}-} + k_2^{\mathrm{R}+} + k_2^{\mathrm{R}-} + k_3^{\mathrm{R}+} + k_3^{\mathrm{R}-} + k_4^{\mathrm{R}+} + k_4^{\mathrm{R}-} \le 2, \\
& \lambda_1 + \underline{\mu}_1^{\mathrm{C}} - \overline{\mu}_1^{\mathrm{C}} = 33, \\
& \lambda_2 + \underline{\mu}_2^{\mathrm{C}} - \overline{\mu}_2^{\mathrm{C}} = 33, \\
& \lambda_3 + \underline{\mu}_3^{\mathrm{C}} - \overline{\mu}_3^{\mathrm{C}} = 33, \\
& \lambda_4 + \underline{\mu}_4^{\mathrm{C}} - \overline{\mu}_4^{\mathrm{C}} = 33, \\
& \quad 1000k_1^{\mathrm{R}+} \le z_1^{\mathrm{K}+} \le 1000k_1^{\mathrm{R}+}, \\
& -1000\left(1 - k_1^{\mathrm{R}+}\right) \le \lambda_1 - z_1^{\mathrm{K}+} \le 1000\left(1 - k_1^{\mathrm{R}+}\right), \\
& -1000k_1^{\mathrm{R}-} \le z_1^{\mathrm{K}-} \le 1000k_1^{\mathrm{R}-}, \\
& -1000\left(1 - k_1^{\mathrm{R}-}\right) \le \lambda_1 - z_1^{\mathrm{K}-} \le 1000\left(1 - k_1^{\mathrm{R}-}\right), \\
& -1000k_2^{\mathrm{R}+} \le z_2^{\mathrm{K}+} \le 1000k_2^{\mathrm{R}+}, \\
& -1000\left(1 - k_2^{\mathrm{R}+}\right) \le \lambda_2 - z_2^{\mathrm{K}+} \le 1000\left(1 - k_2^{\mathrm{R}+}\right), \\
& -1000k_2^{\mathrm{R}-} \le z_2^{\mathrm{K}-} \le 1000k_2^{\mathrm{R}-}, \\
& -1000\left(1 - k_2^{\mathrm{R}-}\right) \le \lambda_1 - z_2^{\mathrm{K}-} \le 1000\left(1 - k_2^{\mathrm{R}-}\right), \\
& -1000k_3^{\mathrm{R}+} \le z_3^{\mathrm{K}+} \le 1000k_3^{\mathrm{R}+}, \\
& -1000\left(1 - k_3^{\mathrm{R}+}\right) \le \lambda_3 - z_3^{\mathrm{K}+} \le 1000\left(1 - k_3^{\mathrm{R}+}\right), \\
& -1000k_3^{\mathrm{R}-} \le z_3^{\mathrm{K}-} \le 1000k_3^{\mathrm{R}-},
\end{aligned}
$$

$$
\begin{aligned}
& -1000\left(1-k_3^{\mathrm{R}-}\right) \leq \lambda_3 - z_3^{\mathrm{K}-} \leq 1000\left(1-k_3^{\mathrm{R}-}\right),\\
& -1000k_4^{\mathrm{R}+} \leq z_4^{\mathrm{K}+} \leq 1000k_4^{\mathrm{R}+},\\
& -1000\left(1-k_4^{\mathrm{R}+}\right) \leq \lambda_4 - z_4^{\mathrm{K}+} \leq 1000\left(1-k_4^{\mathrm{R}+}\right),\\
& -1000k_4^{\mathrm{R}-} \leq z_4^{\mathrm{K}-} \leq 1000k_4^{\mathrm{R}-},\\
& -1000\left(1-k_4^{\mathrm{R}-}\right) \leq \lambda_4 - z_4^{\mathrm{K}-} \leq 1000\left(1-k_4^{\mathrm{R}-}\right),\\
& \lambda_1 - \overline{\mu}_1^{\mathrm{R}} \leq 0,\\
& \lambda_2 - \overline{\mu}_2^{\mathrm{R}} \leq 0,\\
& \lambda_3 - \overline{\mu}_3^{\mathrm{R}} \leq 0,\\
& \lambda_4 - \overline{\mu}_4^{\mathrm{R}} \leq 0,\\
& -\lambda_1 + \underline{\mu}_1^{\mathrm{D}} - \overline{\mu}_1^{\mathrm{D}} + \xi = 0,\\
& -\lambda_2 + \underline{\mu}_2^{\mathrm{D}} - \overline{\mu}_2^{\mathrm{D}} + \xi = 0,\\
& -\lambda_3 + \underline{\mu}_3^{\mathrm{D}} - \overline{\mu}_3^{\mathrm{D}} + \xi = 0,\\
& -\lambda_4 + \underline{\mu}_4^{\mathrm{D}} - \overline{\mu}_4^{\mathrm{D}} + \xi = 0,\\
& \lambda_1, \lambda_2, \lambda_3, \lambda_4 \leq 1000,\\
& -\lambda_1, -\lambda_2, -\lambda_3, -\lambda_4 \leq 1000,\\
& \underline{\mu}_1^{\mathrm{C}}, \overline{\mu}_1^{\mathrm{C}}, \underline{\mu}_2^{\mathrm{C}}, \overline{\mu}_2^{\mathrm{C}}, \underline{\mu}_3^{\mathrm{C}}, \overline{\mu}_3^{\mathrm{C}}, \underline{\mu}_4^{\mathrm{C}}, \overline{\mu}_4^{\mathrm{C}}, \overline{\mu}_1^{\mathrm{R}}, \overline{\mu}_2^{\mathrm{R}}, \overline{\mu}_3^{\mathrm{R}}, \overline{\mu}_4^{\mathrm{R}}, \underline{\mu}_1^{\mathrm{D}}, \overline{\mu}_1^{\mathrm{D}}, \underline{\mu}_2^{\mathrm{D}}, \overline{\mu}_2^{\mathrm{D}}, \underline{\mu}_3^{\mathrm{D}}, \overline{\mu}_3^{\mathrm{D}},\\
& \quad \underline{\mu}_4^{\mathrm{D}}, \overline{\mu}_4^{\mathrm{D}} \geq 0,
\end{aligned}
$$

where $\Phi^{\mathrm{S}} = \Big\{k_1^{\mathrm{R}+}, k_2^{\mathrm{R}+}, k_3^{\mathrm{R}+}, k_4^{\mathrm{R}+}, k_1^{\mathrm{R}-}, k_2^{\mathrm{R}-}, k_3^{\mathrm{R}-}, k_4^{\mathrm{R}-}, z_1^{\mathrm{K}+}, z_2^{\mathrm{K}+}, z_3^{\mathrm{K}+}, z_4^{\mathrm{K}+}, z_1^{\mathrm{K}-}, z_2^{\mathrm{K}-}, z_3^{\mathrm{K}-}, z_4^{\mathrm{K}-}, \lambda_1, \lambda_2, \lambda_3, \lambda_4, \underline{\mu}_1^{\mathrm{C}}, \underline{\mu}_2^{\mathrm{C}}, \underline{\mu}_3^{\mathrm{C}}, \underline{\mu}_4^{\mathrm{C}}, \overline{\mu}_1^{\mathrm{C}}, \overline{\mu}_2^{\mathrm{C}}, \overline{\mu}_3^{\mathrm{C}}, \overline{\mu}_4^{\mathrm{C}}, \overline{\mu}_1^{\mathrm{R}}, \overline{\mu}_2^{\mathrm{R}}, \overline{\mu}_3^{\mathrm{R}}, \overline{\mu}_4^{\mathrm{R}}, \underline{\mu}_1^{\mathrm{D}}, \underline{\mu}_2^{\mathrm{D}}, \underline{\mu}_3^{\mathrm{D}}, \underline{\mu}_4^{\mathrm{D}}, \overline{\mu}_1^{\mathrm{D}}, \overline{\mu}_2^{\mathrm{D}}, \overline{\mu}_3^{\mathrm{D}}, \overline{\mu}_4^{\mathrm{D}}, \xi\Big\}$.

In this case, the worst uncertainty realization is:

$$
\begin{aligned}
& k_1^{\mathrm{R}+}, k_4^{\mathrm{R}+}, k_1^{\mathrm{R}-}, k_2^{\mathrm{R}-}, k_3^{\mathrm{R}-}, k_4^{\mathrm{R}-} = 0,\\
& k_2^{\mathrm{R}+}, k_3^{\mathrm{R}+} = 1.
\end{aligned}
$$

In this second iteration, it is found that given the power traded in the energy and reserve markets obtained from the master problem in step 2, the worst case realization is that the VPP is requested to provide up reserves in time periods 2 and 3. On the other hand, the optimal value of the objective function is $z^{\mathrm{S}*} = -1.008 \cdot 10^5$.

- Step 6. The lower bound is updated using Eq. (5.13):

$$
\mathrm{LB} = -91112.
$$

- Step 7. Compute $\mathrm{UB} - \mathrm{LB} = 5688 - (-91112) = 96800$. As the difference between the upper and lower bounds is higher than the tolerance, continue with the following step.
- Step 8. The iteration counter $\nu = 2$ is updated.

- Step 9. Set $k_2^{\mathrm{R+(2)}} = k_3^{\mathrm{R+(2)}} = 1$ and $k_1^{\mathrm{R+(2)}} = k_4^{\mathrm{R+(2)}} = k_1^{\mathrm{R-(2)}} = k_2^{\mathrm{R-(2)}} = k_3^{\mathrm{R-(2)}} = k_4^{\mathrm{R-(2)}} = 0$.
- Step 2. The following master problem is solved:

$$\begin{aligned}\max_{\Phi^{\mathrm{M}}}\ & 20p_1^{\mathrm{E}} + 50p_2^{\mathrm{E}} + 35p_3^{\mathrm{E}} + 31p_4^{\mathrm{E}} + 4p_1^{\mathrm{R+}} + 10p_2^{\mathrm{R+}} + 8p_3^{\mathrm{R+}} + 6p_4^{\mathrm{R+}} \\ & + 4p_1^{\mathrm{R-}} + 10p_2^{\mathrm{R-}} + 8p_3^{\mathrm{R-}} + 6p_4^{\mathrm{R-}} - \left(6u_1^{\mathrm{C}} + 6u_2^{\mathrm{C}} + 6u_3^{\mathrm{C}} + 6u_4^{\mathrm{C}}\right) + \theta\end{aligned}$$

subject to

$$\begin{aligned}
& -100 \le p_1^{\mathrm{E}} \le 100, \\
& -100 \le p_2^{\mathrm{E}} \le 100, \\
& -100 \le p_3^{\mathrm{E}} \le 100, \\
& -100 \le p_4^{\mathrm{E}} \le 100, \\
& 0 \le p_1^{\mathrm{R+}} \le 50, \\
& 0 \le p_2^{\mathrm{R+}} \le 50, \\
& 0 \le p_3^{\mathrm{R+}} \le 50, \\
& 0 \le p_4^{\mathrm{R+}} \le 50, \\
& 0 \le p_1^{\mathrm{R-}} \le 50, \\
& 0 \le p_2^{\mathrm{R-}} \le 50, \\
& 0 \le p_3^{\mathrm{R-}} \le 50, \\
& 0 \le p_4^{\mathrm{R-}} \le 50, \\
& u_1^{\mathrm{C}}, u_2^{\mathrm{C}}, u_3^{\mathrm{C}}, u_4^{\mathrm{C}} \in \{0, 1\}, \\
& \theta \le 1 \cdot 14p_1^{\mathrm{R+}} + 0 \cdot 38p_2^{\mathrm{R+}} + 0 \cdot 26p_3^{\mathrm{R+}} + 1 \cdot 25p_4^{\mathrm{R+}} - \Big(0 \cdot 14p_1^{\mathrm{R-}} + 0 \cdot 38p_2^{\mathrm{R-}} \\
& \qquad + 0 \cdot 26p_3^{\mathrm{R-}} + 0 \cdot 25p_4^{\mathrm{R-}}\Big) - \left(33p_{11}^{\mathrm{C}} + 33p_{21}^{\mathrm{C}} + 33p_{31}^{\mathrm{C}} + 33p_{41}^{\mathrm{C}}\right), \\
& p_1^{\mathrm{E}} + 1 \cdot p_1^{\mathrm{R+}} - 0 \cdot p_1^{\mathrm{R-}} = p_{11}^{\mathrm{C}} + p_{11}^{\mathrm{R}} - p_{11}^{\mathrm{D}}, \\
& p_2^{\mathrm{E}} + 0 \cdot p_2^{\mathrm{R+}} - 0 \cdot p_2^{\mathrm{R-}} = p_{21}^{\mathrm{C}} + p_{21}^{\mathrm{R}} - p_{21}^{\mathrm{D}}, \\
& p_3^{\mathrm{E}} + 0 \cdot p_3^{\mathrm{R+}} - 0 \cdot p_3^{\mathrm{R-}} = p_{31}^{\mathrm{C}} + p_{31}^{\mathrm{R}} - p_{31}^{\mathrm{D}}, \\
& p_4^{\mathrm{E}} + 1 \cdot p_4^{\mathrm{R+}} - 0 \cdot p_4^{\mathrm{R-}} = p_{41}^{\mathrm{C}} + p_{41}^{\mathrm{R}} - p_{41}^{\mathrm{D}}, \\
& 10u_1^{\mathrm{C}} \le p_{11}^{\mathrm{C}} \le 60u_1^{\mathrm{C}}, \\
& 10u_2^{\mathrm{C}} \le p_{21}^{\mathrm{C}} \le 60u_2^{\mathrm{C}}, \\
& 10u_3^{\mathrm{C}} \le p_{31}^{\mathrm{C}} \le 60u_3^{\mathrm{C}}, \\
& 10u_4^{\mathrm{C}} \le p_{41}^{\mathrm{C}} \le 60u_4^{\mathrm{C}},
\end{aligned}$$

$$0 \leq p_{11}^{\mathrm{D}} \leq 20,$$
$$0 \leq p_{21}^{\mathrm{D}} \leq 20,$$
$$10 \leq p_{31}^{\mathrm{D}} \leq 25,$$
$$15 \leq p_{41}^{\mathrm{D}} \leq 30,$$
$$p_{11}^{\mathrm{D}} + p_{21}^{\mathrm{D}} + p_{31}^{\mathrm{D}} + p_{41}^{\mathrm{D}} \geq 60,$$
$$0 \leq p_{11}^{\mathrm{R}} \leq 60,$$
$$0 \leq p_{21}^{\mathrm{R}} \leq 40,$$
$$0 \leq p_{31}^{\mathrm{R}} \leq 20,$$
$$0 \leq p_{41}^{\mathrm{R}} \leq 10,$$
$$\theta \leq 0 \cdot 14 p_1^{\mathrm{R+}} + 1 \cdot 38 p_2^{\mathrm{R+}} + 1 \cdot 26 p_3^{\mathrm{R+}} + 0 \cdot 25 p_4^{\mathrm{R+}} - \Big(0 \cdot 14 p_1^{\mathrm{R-}} + 0 \cdot 38 p_2^{\mathrm{R-}}$$
$$+ 0 \cdot 26 p_3^{\mathrm{R-}} + 0 \cdot 25 p_4^{\mathrm{R-}}\Big) - \Big(33 p_{12}^{\mathrm{C}} + 33 p_{22}^{\mathrm{C}} + 33 p_{32}^{\mathrm{C}} + 33 p_{42}^{\mathrm{C}}\Big),$$
$$p_1^{\mathrm{E}} + 0 \cdot p_1^{\mathrm{R+}} - 0 \cdot p_1^{\mathrm{R-}} = p_{12}^{\mathrm{C}} + p_{12}^{\mathrm{R}} - p_{12}^{\mathrm{D}},$$
$$p_2^{\mathrm{E}} + 1 \cdot p_2^{\mathrm{R+}} - 0 \cdot p_2^{\mathrm{R-}} = p_{22}^{\mathrm{C}} + p_{22}^{\mathrm{R}} - p_{22}^{\mathrm{D}},$$
$$p_3^{\mathrm{E}} + 1 \cdot p_3^{\mathrm{R+}} - 0 \cdot p_3^{\mathrm{R-}} = p_{32}^{\mathrm{C}} + p_{32}^{\mathrm{R}} - p_{32}^{\mathrm{D}},$$
$$p_4^{\mathrm{E}} + 0 \cdot p_4^{\mathrm{R+}} - 0 \cdot p_4^{\mathrm{R-}} = p_{42}^{\mathrm{C}} + p_{42}^{\mathrm{R}} - p_{42}^{\mathrm{D}},$$
$$10 u_1^{\mathrm{C}} \leq p_{12}^{\mathrm{C}} \leq 60 u_1^{\mathrm{C}},$$
$$10 u_2^{\mathrm{C}} \leq p_{22}^{\mathrm{C}} \leq 60 u_2^{\mathrm{C}},$$
$$10 u_3^{\mathrm{C}} \leq p_{32}^{\mathrm{C}} \leq 60 u_3^{\mathrm{C}},$$
$$10 u_4^{\mathrm{C}} \leq p_{42}^{\mathrm{C}} \leq 60 u_4^{\mathrm{C}},$$
$$0 \leq p_{12}^{\mathrm{D}} \leq 20,$$
$$0 \leq p_{22}^{\mathrm{D}} \leq 20,$$
$$10 \leq p_{32}^{\mathrm{D}} \leq 25,$$
$$15 \leq p_{42}^{\mathrm{D}} \leq 30,$$
$$p_{12}^{\mathrm{D}} + p_{22}^{\mathrm{D}} + p_{32}^{\mathrm{D}} + p_{42}^{\mathrm{D}} \geq 60,$$
$$0 \leq p_{12}^{\mathrm{R}} \leq 60,$$
$$0 \leq p_{22}^{\mathrm{R}} \leq 40,$$
$$0 \leq p_{32}^{\mathrm{R}} \leq 20,$$
$$0 \leq p_{42}^{\mathrm{R}} \leq 10,$$

where $\Phi^{\mathrm{M}} = \Big\{p_1^{\mathrm{E}}, p_2^{\mathrm{E}}, p_3^{\mathrm{E}}, p_4^{\mathrm{E}}, p_1^{\mathrm{R}+}, p_2^{\mathrm{R}+}, p_3^{\mathrm{R}+}, p_4^{\mathrm{R}+}, p_1^{\mathrm{R}-}, p_2^{\mathrm{R}-}, p_3^{\mathrm{R}-}, p_4^{\mathrm{R}-}, u_1^{\mathrm{C}}, u_2^{\mathrm{C}}, u_3^{\mathrm{C}}, u_4^{\mathrm{C}}, \theta, p_{11}^{\mathrm{D}}, p_{21}^{\mathrm{D}}, p_{31}^{\mathrm{D}}, p_{41}^{\mathrm{D}}, p_{11}^{\mathrm{R}},$ $p_{21}^{\mathrm{R}}, p_{31}^{\mathrm{R}}, p_{41}^{\mathrm{R}}, p_{11}^{\mathrm{C}}, p_{21}^{\mathrm{C}}, p_{31}^{\mathrm{C}}, p_{41}^{\mathrm{C}}, p_{12}^{\mathrm{D}}, p_{22}^{\mathrm{D}}, p_{32}^{\mathrm{D}}, p_{42}^{\mathrm{D}}, p_{12}^{\mathrm{R}}, p_{22}^{\mathrm{R}}, p_{32}^{\mathrm{R}}, p_{42}^{\mathrm{R}}, p_{12}^{\mathrm{C}}, p_{22}^{\mathrm{C}}, p_{32}^{\mathrm{C}}, p_{42}^{\mathrm{C}}\Big\}$.

The optimal solution of this problem is:

$$
\begin{aligned}
&p_1^{\mathrm{E}*} = 30,\\
&p_2^{\mathrm{E}*} = 85,\\
&p_3^{\mathrm{E}*} = 35,\\
&p_4^{\mathrm{E}*} = -5,\\
&p_1^{\mathrm{R}+*}, p_2^{\mathrm{R}+*} = 0\\
&p_3^{\mathrm{R}+*} = 50,\\
&p_3^{\mathrm{R}+*} = 43.75,\\
&p_1^{\mathrm{R}-*}, p_2^{\mathrm{R}-*}, p_3^{\mathrm{R}-*}, p_4^{\mathrm{R}-*} = 50,\\
&u_1^{\mathrm{C}*} = 0,\\
&u_2^{\mathrm{C}*}, u_3^{\mathrm{C}*}, u_4^{\mathrm{C}*} = 1,\\
&\theta^* = -2999,
\end{aligned}
$$

while the optimal value of the objective function is $z^{\mathrm{M}*} = 4974.5$.

- Step 3. The upper bound using Eq. (5.12):

$$\mathrm{UB} = 4974.5.$$

- Step 4. Set $p_1^{\mathrm{E}(2)} = 30$, $p_2^{\mathrm{E}(2)} = 85$, $p_3^{\mathrm{E}(2)} = 35$, $p_4^{\mathrm{E}(2)} = -5$, $p_1^{\mathrm{R}+(2)} = 0$, $p_2^{\mathrm{R}+(2)} = 0$, $p_3^{\mathrm{R}+(2)} = 50$, $p_4^{\mathrm{R}+(2)} = 43.75$, $p_1^{\mathrm{R}-(2)} = 50$, $p_2^{\mathrm{R}-(2)} = 50$, $p_3^{\mathrm{R}-(2)} = 50$, $p_4^{\mathrm{R}-(2)} = 50$, $u_1^{\mathrm{C}(2)} = 0$, $u_2^{\mathrm{C}(2)} = 1$, $u_3^{\mathrm{C}(2)} = 1$, and $u_4^{\mathrm{C}(2)} = 1$.
- Step 5. The following subproblem is solved:

$$
\begin{aligned}
\min_{\Phi^{\mathrm{S}}} & 14k_1^{\mathrm{R}+} \cdot 0 + 38k_2^{\mathrm{R}+} \cdot 0 + 26k_3^{\mathrm{R}+} \cdot 50 + 25k_4^{\mathrm{R}+} \cdot 43.75 - \Big(14k_1^{\mathrm{R}-} \cdot 50\\
&+38k_2^{\mathrm{R}-} \cdot 50 + 26k_3^{\mathrm{R}-} \cdot 50 + 25k_4^{\mathrm{R}-} \cdot 50\Big) - \Big[30\lambda_1 + 85\lambda_2 + 35\lambda_3 - 5\lambda_4\\
&+ 0z_1^{\mathrm{K}+} + 0z_2^{\mathrm{K}+} + 50z_3^{\mathrm{K}+} + 43.75z_4^{\mathrm{K}+} - 50z_1^{\mathrm{K}-} - 50z_2^{\mathrm{K}-} - 50z_3^{\mathrm{K}-} - 50z_4^{\mathrm{K}-}\\
&+ 0\underline{\mu}_1^{\mathrm{D}} + 0\underline{\mu}_2^{\mathrm{D}} + 10\underline{\mu}_3^{\mathrm{D}} + 15\underline{\mu}_4^{\mathrm{D}} - 20\overline{\mu}_1^{\mathrm{D}} - 20\overline{\mu}_2^{\mathrm{D}} - 25\overline{\mu}_3^{\mathrm{D}} - 30\overline{\mu}_4^{\mathrm{D}}\\
&+ 10 \cdot 0\underline{\mu}_1^{\mathrm{C}} + 10 \cdot 1\underline{\mu}_2^{\mathrm{C}} + 10 \cdot 1\underline{\mu}_3^{\mathrm{C}} + 10 \cdot 1\underline{\mu}_4^{\mathrm{C}} - 60 \cdot 0\overline{\mu}_1^{\mathrm{C}} - 60 \cdot 1\overline{\mu}_2^{\mathrm{C}}\\
&- 60 \cdot 1\overline{\mu}_3^{\mathrm{C}} - 60 \cdot 1\overline{\mu}_4^{\mathrm{C}} - 60\overline{\mu}_1^{\mathrm{R}} - 20\overline{\mu}_2^{\mathrm{R}} - 40\overline{\mu}_3^{\mathrm{R}} - 10\overline{\mu}_4^{\mathrm{R}}\Big]
\end{aligned}
$$

subject to

$$
\begin{aligned}
&k_1^{\mathrm{R}+}, k_2^{\mathrm{R}+}, k_3^{\mathrm{R}+}, k_4^{\mathrm{R}+}, k_1^{\mathrm{R}-}, k_2^{\mathrm{R}-}, k_3^{\mathrm{R}-}, k_4^{\mathrm{R}-} \in \{0, 1\},\\
&k_1^{\mathrm{R}+} + k_1^{\mathrm{R}-} \le 1,\\
&k_2^{\mathrm{R}+} + k_2^{\mathrm{R}-} \le 1,\\
&k_3^{\mathrm{R}+} + k_3^{\mathrm{R}-} \le 1,\\
&k_4^{\mathrm{R}+} + k_4^{\mathrm{R}-} \le 1,
\end{aligned}
$$

$$k_1^{R+} + k_1^{R-} + k_2^{R+} + k_2^{R-} + k_3^{R+} + k_3^{R-} + k_4^{R+} + k_4^{R-} \leq 2,$$
$$\lambda_1 + \underline{\mu}_1^C - \overline{\mu}_1^C = 33,$$
$$\lambda_2 + \underline{\mu}_2^C - \overline{\mu}_2^C = 33,$$
$$\lambda_3 + \underline{\mu}_3^C - \overline{\mu}_3^C = 33,$$
$$\lambda_4 + \underline{\mu}_4^C - \overline{\mu}_4^C = 33,$$
$$-1000k_1^{R+} \leq z_1^{K+} \leq 1000k_1^{R+},$$
$$-1000\left(1 - k_1^{R+}\right) \leq \lambda_1 - z_1^{K+} \leq 1000\left(1 - k_1^{R+}\right),$$
$$-1000k_1^{R-} \leq z_1^{K-} \leq 1000k_1^{R-},$$
$$-1000\left(1 - k_1^{R-}\right) \leq \lambda_1 - z_1^{K-} \leq 1000\left(1 - k_1^{R-}\right),$$
$$-1000k_2^{R+} \leq z_2^{K+} \leq 1000k_2^{R+},$$
$$-1000\left(1 - k_2^{R+}\right) \leq \lambda_2 - z_2^{K+} \leq 1000\left(1 - k_2^{R+}\right),$$
$$-1000k_2^{R-} \leq z_2^{K-} \leq 1000k_2^{R-},$$
$$-1000\left(1 - k_2^{R-}\right) \leq \lambda_1 - z_2^{K-} \leq 1000\left(1 - k_2^{R-}\right),$$
$$-1000k_3^{R+} \leq z_3^{K+} \leq 1000k_3^{R+},$$
$$-1000\left(1 - k_3^{R+}\right) \leq \lambda_3 - z_3^{K+} \leq 1000\left(1 - k_3^{R+}\right),$$
$$-1000k_3^{R-} \leq z_3^{K-} \leq 1000k_3^{R-},$$
$$-1000\left(1 - k_3^{R-}\right) \leq \lambda_3 - z_3^{K-} \leq 1000\left(1 - k_3^{R-}\right),$$
$$-1000k_4^{R+} \leq z_4^{K+} \leq 1000k_4^{R+},$$
$$-1000\left(1 - k_4^{R+}\right) \leq \lambda_4 - z_4^{K+} \leq 1000\left(1 - k_4^{R+}\right),$$
$$-1000k_4^{R-} \leq z_4^{K-} \leq 1000k_4^{R-},$$
$$-1000\left(1 - k_4^{R-}\right) \leq \lambda_4 - z_4^{K-} \leq 1000\left(1 - k_4^{R-}\right),$$
$$\lambda_1 - \overline{\mu}_1^R \leq 0,$$
$$\lambda_2 - \overline{\mu}_2^R \leq 0,$$
$$\lambda_3 - \overline{\mu}_3^R \leq 0,$$
$$\lambda_4 - \overline{\mu}_4^R \leq 0,$$
$$-\lambda_1 + \underline{\mu}_1^D - \overline{\mu}_1^D + \xi = 0,$$
$$-\lambda_2 + \underline{\mu}_2^D - \overline{\mu}_2^D + \xi = 0,$$
$$-\lambda_3 + \underline{\mu}_3^D - \overline{\mu}_3^D + \xi = 0,$$
$$-\lambda_4 + \underline{\mu}_4^D - \overline{\mu}_4^D + \xi = 0,$$
$$\lambda_1, \lambda_2, \lambda_3, \lambda_4 \leq 1000,$$
$$-\lambda_1, -\lambda_2, -\lambda_3, -\lambda_4 \leq 1000,$$

$$\underline{\mu}_1^{\rm C}, \overline{\mu}_1^{\rm C}, \underline{\mu}_2^{\rm C}, \overline{\mu}_2^{\rm C}, \underline{\mu}_3^{\rm C}, \overline{\mu}_3^{\rm C}, \underline{\mu}_4^{\rm C}, \overline{\mu}_4^{\rm C}, \overline{\mu}_1^{\rm R}, \overline{\mu}_2^{\rm R}, \overline{\mu}_3^{\rm R}, \overline{\mu}_4^{\rm R}, \underline{\mu}_1^{\rm D}, \overline{\mu}_1^{\rm D}, \underline{\mu}_2^{\rm D}, \overline{\mu}_2^{\rm D}, \underline{\mu}_3^{\rm D}, \overline{\mu}_3^{\rm D},$$
$$\underline{\mu}_4^{\rm D}, \overline{\mu}_4^{\rm D} \geq 0,$$

where $\Phi^{\rm S} = \Big\{k_1^{\rm R+}, k_2^{\rm R+}, k_3^{\rm R+}, k_4^{\rm R+}, k_1^{\rm R-}, k_2^{\rm R-}, k_3^{\rm R-}, k_4^{\rm R-}, z_1^{\rm K+}, z_2^{\rm K+}, z_3^{\rm K+}, z_4^{\rm K+}, z_1^{\rm K-}, z_2^{\rm K-}, z_3^{\rm K-}, z_4^{\rm K-}, \lambda_1, \lambda_2, \lambda_3, \lambda_4, \underline{\mu}_1^{\rm C},$ $\underline{\mu}_2^{\rm C}, \underline{\mu}_3^{\rm C}, \underline{\mu}_4^{\rm C}, \overline{\mu}_1^{\rm C}, \overline{\mu}_2^{\rm C}, \overline{\mu}_3^{\rm C}, \overline{\mu}_4^{\rm C}, \overline{\mu}_1^{\rm R}, \overline{\mu}_2^{\rm R}, \overline{\mu}_3^{\rm R}, \overline{\mu}_4^{\rm R}, \underline{\mu}_1^{\rm D}, \underline{\mu}_2^{\rm D}, \underline{\mu}_3^{\rm D}, \underline{\mu}_4^{\rm D}, \overline{\mu}_1^{\rm D}, \overline{\mu}_2^{\rm D}, \overline{\mu}_3^{\rm D}, \overline{\mu}_4^{\rm D}, \xi\Big\}$.

In this case, the worst uncertainty realization is:

$$k_1^{\rm R+}, k_2^{\rm R+}, k_3^{\rm R+}, k_4^{\rm R+}, k_1^{\rm R-}, k_2^{\rm R-} = 0,$$
$$k_3^{\rm R-}, k_4^{\rm R-} = 1.$$

On the other hand, the optimal value of the objective function is $z^{\rm S*} = -5190$.

- Step 6. The lower bound is updated using Eq. (5.13):

$$\text{LB} = 2274.5.$$

- Step 7. Compute UB − LB = 4974.5 − 2274.5 = 2200. As the difference between the upper and lower bounds is higher than the tolerance, continue with the following step.

The above steps are repeated until the LB and UB are equal and the algorithm stops. After convergence, it is obtained that the optimal power traded in the energy and reserve markets is:

$$p_1^{\rm E*} = 30 \text{ MW},$$
$$p_2^{\rm E*} = 85 \text{ MW},$$
$$p_3^{\rm E*} = 61.56 \text{ MW},$$
$$p_4^{\rm E*} = 34.27 \text{ MW},$$
$$p_1^{\rm R+*} = 5.21 \text{ MW},$$
$$p_2^{\rm R+*} = 0,$$
$$p_3^{\rm R+*} = 18.22 \text{ MW},$$
$$p_4^{\rm R+*} = 15.95 \text{ MW},$$
$$p_1^{\rm R-*} = 9.11 \text{ MW},$$
$$p_2^{\rm R-*} = p_3^{\rm R-*} = p_4^{\rm R-*} = 50 \text{ MW}.$$

This solution allows the VPP to maximize its worst-case expected profit provided that it can be requested to provide reserves in a maximum of two time periods.

□

A GAMS [9] code for solving Illustrative Example 5.6 is provided in Sect. 5.7.

Illustrative Example 5.7 Impact of the Uncertainty Budget

The data of Illustrative Example 5.6 are considered again. Next, the impact of the uncertainty budget on the power traded by the VPP in the energy and reserve markets is analyzed.

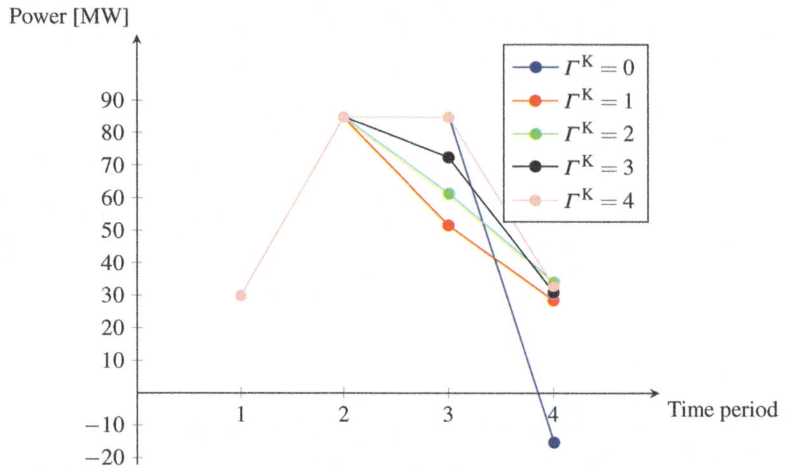

Fig. 5.20 Illustrative Example 5.7: power traded in the energy market

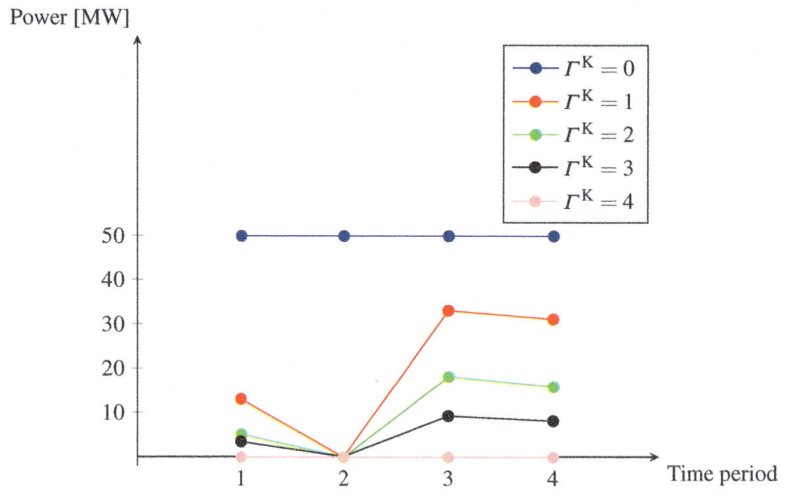

Fig. 5.21 Illustrative Example 5.7: power traded in the up-reserve market

Results are provided in Figs. 5.20, 5.21, and 5.22 that respectively depict the optimal power traded in the energy, up-reserve, and down-reserve markets for different values of the uncertainty budget Γ^{K}. The power traded in the energy market generally increases as the uncertainty budget is increased. Moreover, the VPP only buys from the energy market in time period 4 and in the case of an uncertainty budget equal to 0. Regarding the power traded in the up- and down-reserve markets, note that if the uncertainty budget is equal to 0, the VPP trades the maximum 50 MW in both the up- and down-reserve markets since this case considers that the VPP is not requested to provide reserves in any time period. However, as the level of uncertainty increases the power traded in the reserve markets decreases. For example, in the more conservative case ($\Gamma^{K} = 4$) the VPP does not participate in the up-reserve market. Note also that the participation in the down-reserve market is higher than in the up-reserve market. This can be explained by the fact that it is generally easier to decrease the production (or to increase the consumption of the demands) than increasing it (or reducing the demand consumption).

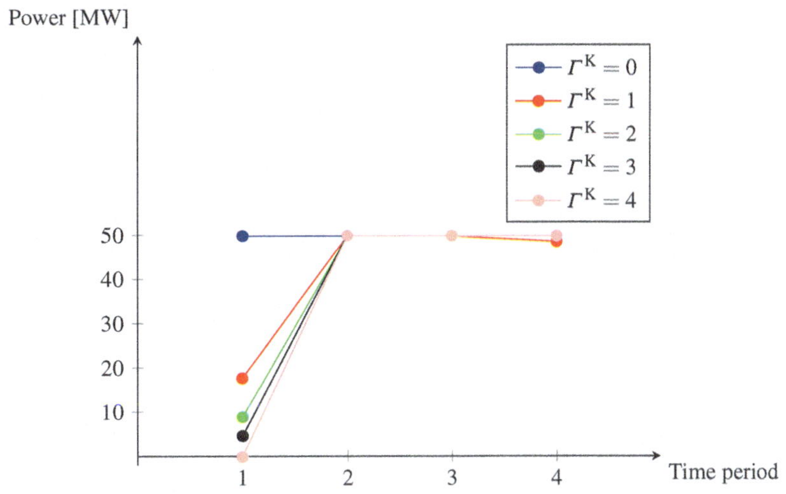

Fig. 5.22 Illustrative Example 5.7: power traded in the down-reserve market

Table 5.4 Illustrative Example 5.7: results

Γ^{R}	Objective function [$]
0	6188.0
1	4747.7
2	4357.8
3	4240.0
4	4236.2

Finally, the optimal value of the objective function for different uncertainty budgets is provided in Table 5.4. Note that the value of the objective function represents the worst-case profit achieved by the VPP. Thus, it decreases as the uncertainty budget is increased since the VPP can be requested to provide reserves in an increasing number of time periods.

□

5.6 Summary and Further Reading

This chapter provides different methods for the scheduling problem of a VPP participating in energy and reserve markets. First, a deterministic model that assumes that the VPP has all the information available at the time of making its decisions. Then, this model is extended to account for the uncertainty in the decision-making problem. With this purpose, two models based on stochastic programming and robust optimization are described.

Additional information about stochastic programming and robust optimization can be found in [5, 6] and [2, 4], respectively. On the other hand, the interested readers are referred to the technical paper by Mashhour and Moghaddas-Tafreshi [12] that proposes a nonequilibrium model based on the deterministic price-based unit commitment for the bidding strategy of a VPP participating in energy and spinning reserve markets, the paper by Dabbagh and Sheikh-El-Eslami [7] that describes a stochastic model for the offering strategy of a VPP in energy and reserve markets considering the risk associated with the offering decisions, and the paper by Baringo et al. [1] that provides further details about the adaptive robust optimization problem described in Sect. 5.5 of this chapter, as well as more realistic case studies.

5.7 GAMS Codes

This section provides the GAMS codes used for solving some of the illustrative examples of this chapter.

5.7.1 *Deterministic Approach*

An input GAMS [9] code for solving Illustrative Example 5.1 is provided below:

```
SETS
c         'conventional power plants'       /c1*c1/
d         'demands'                         /d1*d1/
r         'stochastic renewable units'      /r1*r1/
t         'time periods'                    /t1*t4/;
SCALAR PEMAX          'maximum power traded in the energy market'
    /100/;
SCALAR PURMAX         'maximum power traded in the up-reserve
    market' /50/;
SCALAR PDRMAX         'maximum power traded in the down-reserve
    market' /50/;
PARAMETER priceE(t)        'energy market price'
/t1       20
t2        50
t3        35
t4        31/;
PARAMETER priceUR(t)       'up-reserve market price (capacity)'
/t1       4
t2        10
t3        8
t4        6/;
PARAMETER priceDR(t)       'down-reserve market price (capacity)'
/t1       4
t2        10
t3        8
t4        6/;
PARAMETER priceURE(t)      'up-reserve market price (energy)'
/t1       14
t2        38
t3        26
t4        25/;
PARAMETER priceDRE(t)      'down-reserve market price (energy)'
/t1       14
t2        38
t3        26
t4        25/;
PARAMETER KUR(t)           'up-reserve requirements'
/t1       0.5
t2        0
t3        1
t4        0/;
PARAMETER KDR(t)           'down-reserve requirements'
/t1       0
t2        0.8
t3        0
t4        0/;
TABLE GENDATA(c,*)         'conventional generating unit data'
          Pmin     Pmax     VC       FC
c1        10       60       33       6;
TABLE PRA(r,t)             'forecast renewable generation level'
          t1       t2       t3       t4
r1        60       20       40       10;
```

```
TABLE PDmin(d,t)          'lower demand bound'
          t1       t2       t3       t4
d1        0        0        10       15;
TABLE PDmax(d,t)          'upper demand bound'
          t1       t2       t3       t4
d1        20       20       25       30;
PARAMETER EDmin(d)      'minimum energy consumption'
/d1       60/;
VARIABLES
Z          'objective function'
PE(t)      'power traded in the energy market in time period t';
POSITIVE VARIABLES
PCPP(c,t) 'power production of conventional power plant c in time
     period t'
PD(d,t)    'power consumption of demand d in time period t'
PR(r,t)    'power production of renewable unit r in time period t'
PRU(t)     'power traded in the up-reserve market in time period t'
PRD(t)     'power traded in the up-reserve market in time period t';
BINARY VARIABLES
UC(c,t)    'on-off status of conventional power plant c in time
    period t';
EQUATIONS
OF        'objective function'
EMLim1    'energy market limits I'
EMLim2    'energy market limits II'
URLim     'up-reserve market limits'
DRLim     'down-reserve market limits'
Bal       'power balance in the VPP'
DLim1     'power demand limits I'
DLim2     'power demand limits II'
EDlim     'minimum energy consumption in the day'
Clim1     'conventional power plant production limits I'
Clim2     'conventional power plant production limits II'
Rlim      'stochastic renewable generating unit limits';
*
OF..                z=e=sum(t,priceE(t)*PE(t)+(priceUR(t)+KUR(t)*
    priceURE(t))*PRU(t)+(priceDR(t)-KDR(t)*priceDRE(t))*PRD(t))-
    sum((c,t),GENDATA(c,'FC')*UC(c,t)+GENDATA(c,'VC')*PCPP(c,t));
EMLim1(t)..         -PEMAX=l=PE(t);
EMLim2(t)..         PE(t)=l=PEMAX;
URLim(t)..          PRU(t)=l=PURMAX;
DRLim(t)..          PRD(t)=l=PDRMAX;
Bal(t)..            PE(t)+KUR(t)*PRU(t)-KDR(t)*PRD(t)=E=sum(c,PCPP(c
    ,t))+sum(r,PR(r,t))-sum(d,PD(d,t));
DLim1(d,t)..        PDmin(d,t)=l=PD(d,t);
DLim2(d,t)..        PD(d,t)=l=PDmax(d,t);
EDlim(d)..          sum(t,PD(d,t))=g=EDmin(d);
Clim1(c,t)..        GENDATA(c,'Pmin')*UC(c,t)=l=PCPP(c,t);
Clim2(c,t)..        PCPP(c,t)=l=GENDATA(c,'Pmax')*UC(c,t);
Rlim(r,t)..         PR(r,t)=l=PRA(r,t);
MODEL Deterministic /all/;
OPTION optcr=0;
SOLVE Deterministic using mip maximizing z;
DISPLAY z.l,PE.l,PRU.l,PRD.l,PCPP.l,PR.l,PD.l;
```

5.7.2 Stochastic Programming Approach

An input GAMS [9] code for solving Illustrative Example 5.3 is provided below:

```
SETS
c         'conventional power plants'        /c1*c1/
d         'demands'                          /d1*d1/
r         'stochastic renewable units'       /r1*r1/
t         'time periods'                     /t1*t4/
w         'scenarios'                        /w1*w4/;
SCALAR PEMAX        'maximum power traded in the energy market'
    /100/;
SCALAR PURMAX       'maximum power traded in the up-reserve
    market' /50/;
SCALAR PDRMAX       'maximum power traded in the down-reserve
    market' /50/;
PARAMETER priceE(t)      'energy market price'
/t1       20
t2        50
t3        35
t4        31/;
PARAMETER priceUR(t)      'up-reserve market price (capacity)'
/t1       4
t2        10
t3        8
t4        6/;
PARAMETER priceDR(t)      'down-reserve market price (capacity)'
/t1       4
t2        10
t3        8
t4        6/;
PARAMETER priceURE(t)     'up-reserve market price (energy)'
/t1       14
t2        38
t3        26
t4        25/;
PARAMETER priceDRE(t)     'down-reserve market price (energy)'
/t1       14
t2        38
t3        26
t4        25/;
PARAMETER weigth(w)       'weight of scenarios'
/w1       0.25
w2        0.25
w3        0.25
w4        0.25/;
TABLE KUR(t,w)          'up-reserve requirements'
          w1       w2       w3       w4
t1        0.5      0        0.5      0
t2        0        0.5      0        0.5
t3        1        0        1        0
t4        0        0.4      0        0.4;
```

```
TABLE KDR(t,w)          'down-reserve requirements'
          w1       w2       w3       w4
t1        0        0.4      0        0.4
t2        0.8      0        0.8      0
t3        0        0.5      0        0.5
t4        0        0        0        0;
TABLE GENDATA(c,*)        'conventional generating unit data'
          Pmin     Pmax     VC       FC
c1        10       60       33       6;
TABLE PRA(r,t,w)          'forecast renewable generation level'
          w1       w2       w3       w4
r1.t1     60       60       40       40
r1.t2     20       20       30       30
r1.t3     40       40       30       30
r1.t4     10       10       20       20;
TABLE PDmin(d,t)          'lower demand bound'
          t1       t2       t3       t4
d1        0        0        10       15;
TABLE PDmax(d,t)          'upper demand bound'
          t1       t2       t3       t4
d1        20       20       25       30;
PARAMETER EDmin(d)        'minimum energy consumption'
/d1       60/;
VARIABLES
Z            'objective function'
PE(t)        'power traded in the energy market in time period t';
POSITIVE VARIABLES
PCPP(c,t,w) 'power production of conventional power plant c in
    time period t'
PD(d,t,w)    'power consumption of demand d in time period t'
PR(r,t,w)    'power production of renewable unit r in time period t'
PRU(t)       'power traded in the up-reserve market in time period t'
PRD(t)       'power traded in the up-reserve market in time period t';
BINARY VARIABLES
UC(c,t)      'on-off status of conventional power plant c in time
    period t';
EQUATIONS
OF         'objective function'
EMLim1     'energy market limits I'
EMLim2     'energy market limits II'
URLim      'up-reserve market limits'
DRLim      'down-reserve market limits'
Bal        'power balance in the VPP'
DLim1      'power demand limits I'
DLim2      'power demand limits II'
EDlim      'minimum energy consumption in the day'
Clim1      'conventional power plant production limits I'
Clim2      'conventional power plant production limits II'
Rlim       'stochastic renewable generating unit limits';
*
OF..              z=e=sum(w,weigth(w)*sum(t,priceE(t)*PE(t)+(
    priceUR(t)+KUR(t,w)*priceURE(t))*PRU(t)+(priceDR(t)-KDR(t,w)*
    priceDRE(t))*PRD(t))-sum((c,t),GENDATA(c,'FC')*UC(c,t)+
    GENDATA(c,'VC')*PCPP(c,t,w)));
```

```
EMLim1(t)..        -PEMAX=l=PE(t);
EMLim2(t)..        PE(t)=l=PEMAX;
URLim(t)..         PRU(t)=l=PURMAX;
DRLim(t)..         PRD(t)=l=PDRMAX;
Bal(t,w)..         PE(t)+KUR(t,w)*PRU(t)-KDR(t,w)*PRD(t)=E=sum(c,
    PCPP(c,t,w))+sum(r,PR(r,t,w))-sum(d,PD(d,t,w));
DLim1(d,t,w)..     PDmin(d,t)=l=PD(d,t,w);
DLim2(d,t,w)..     PD(d,t,w)=l=PDmax(d,t);
EDlim(d,w)..       sum(t,PD(d,t,w))=g=EDmin(d);
Clim1(c,t,w)..     GENDATA(c,'Pmin')*UC(c,t)=l=PCPP(c,t,w);
Clim2(c,t,w)..     PCPP(c,t,w)=l=GENDATA(c,'Pmax')*UC(c,t);
Rlim(r,t,w)..      PR(r,t,w)=l=PRA(r,t,w);
MODEL Stochastic /all/;
OPTION optcr=0;
SOLVE Stochastic using mip maximizing z;
DISPLAY z.l,PE.l,PRU.l,PRD.l,PCPP.l,PR.l,PD.l;
```

5.7.3 *Adaptive Robust Optimization Approach*

An input GAMS [9] code for solving Illustrative Example 5.6 is provided below:

```
SETS
c         'conventional power plants'        /c1*c1/
d         'demands'                          /d1*d1/
r         'stochastic renewable units'       /r1*r1/
t         'time periods'                     /t1*t4/
it        'iteration counter'                /it1*it15/;
ALIAS(it,itt);
SCALAR PEMAX          'maximum power traded in the energy market'
  /100/;
SCALAR PURMAX         'maximum power traded in the up-reserve
    market' /50/;
SCALAR PDRMAX         'maximum power traded in the down-reserve
    market' /50/;
PARAMETER priceE(t)        'energy market price'
/t1       20
t2        50
t3        35
t4        31/;
PARAMETER priceUR(t)       'up-reserve market price (capacity)'
/t1       4
t2        10
t3        8
t4        6/;
PARAMETER priceDR(t)       'down-reserve market price (capacity)'
/t1       4
t2        10
t3        8
t4        6/;
PARAMETER priceURE(t)      'up-reserve market price (energy)'
/t1       14
t2        38
t3        26
t4        25/;
PARAMETER priceDRE(t)      'down-reserve market price (energy)'
/t1       14
t2        38
t3        26
t4        25/;
TABLE GENDATA(c,*)         'conventional generating unit data'
          Pmin     Pmax     VC       FC
c1        10       60       33       6;
TABLE PRA(r,t)             'forecast renewable generation level'
          t1       t2       t3       t4
r1        50       25       35       15;
TABLE PDmin(d,t)           'lower demand bound'
          t1       t2       t3       t4
d1        0        0        10       15;
TABLE PDmax(d,t)           'upper demand bound'
          t1       t2       t3       t4
d1        20       20       25       30;
```

```
PARAMETER EDmin(d)
/d1      60/;
SCALAR GammaK     'uncertainty budget'              /2/;
SCALAR   M        'large positive constant'         /1000/;
SCALAR ITAUX      'auxiliary parameter'             /0/;
PARAMETERS
PE_AUX(t)         'power traded in the energy market'
PRU_AUX(t)        'power traded in the up-reserve market'
PRD_AUX(t)        'power traded in the down-reserve market'
UC_AUX(c,t)       'on-off status of conventional power plant c in
    time period t'
KUAUX(t,it)       'up-reserve request'
KDAUX(t,it)       'down-reserve request';
KUAUX(t,it)=0;
KDAUX(t,it)=0;
VARIABLES
Z_MP              'objective function of master problem'
Z_SP              'objective function of subproblem'
PE(t)             'power traded in the energy market'
theta             'auxiliary variable'
lambda(t)         'dual variable'
xi(d)             'dual variable';
theta.up=1e6;
POSITIVE VARIABLES
PRU(t)            'power traded in the up-reserve market'
PRD(t)            'power traded in the down-reserve market'
PCPP(c,t,it)      'power production of conventional power plant c
    in time period t and iteration it'
PD(d,t,it)        'power consumption of demand d in time period t
    and iteration it'
PR(r,t,it)        'power production of renewable unit r in time
    period t and iteration it'
muCL(c,t)         'dual variable'
muCU(c,t)         'dual variable'
muDL(d,t)         'dual variable'
muDU(d,t)         'dual variable'
muR(r,t)          'dual variable'
ZKP(t)            'auxiliary variable'
ZKN(t)            'auxiliary variable';
BINARY VARIABLES
KU(t)             'up-reserve requests'
KD(t)             'down-reserve requests'
UC(c,t)           'on-off status of conventional power plant c in
    time period t';
EQUATIONS
MP_of      'Master problem: objective function'
MP_1       'Master problem: energy market limits I'
MP_2       'Master problem: energy market limits II'
MP_3       'Master problem: up-reserve market limits'
MP_4       'Master problem: up-reserve market limits'
MP_5       'Master problem: bounds for theta'
MP_6       'Master problem: balance in the VPP'
MP_7       'Master problem: demand limits I'
MP_8       'Master problem: demand limits II'
```

```
MP_9      'Master problem: minimum energy consumption'
MP_10     'Master problem: renewable generation limit'
MP_11     'Master problem: conventional generation limit I'
MP_12     'Master problem: conventional generation limit II'
SP_of     'Subproblem: objective function'
SP_1      'Subproblem: uncertainty'
SP_2      'Subproblem: uncertainty budget'
SP_3      'Subproblem: auxiliary constraint'
SP_4      'Subproblem: auxiliary constraint'
SP_5      'Subproblem: auxiliary constraint'
SP_6      'Subproblem: auxiliary constraint'
SP_7      'Subproblem: auxiliary constraint'
SP_8      'Subproblem: auxiliary constraint'
SP_9      'Subproblem: auxiliary constraint'
SP_10     'Subproblem: auxiliary constraint'
SP_11     'Subproblem: dual constraint'
SP_12     'Subproblem: dual constraint'
SP_13     'Subproblem: dual constraint'
SP_14     'Subproblem: dual constraint'
SP_15     'Subproblem: dual constraint';
*
MP_of..              Z_MP=e=sum(t,priceE(t)*PE(t)+priceUR(t)*PRU(
    t)+priceDR(t)*PRD(t)-sum(c,GENDATA(c,'FC')*UC(c,t)))+theta;
MP_1(t)..            -PEMAX=l=PE(t);
MP_2(t)..            PE(t)=l=PEMAX;
MP_3(t)..            PRU(t)=l=PURMAX;
MP_4(t)..            PRD(t)=l=PDRMAX;
MP_5(itt)$(ord(itt) LT (ITAUX+1))..      theta=l=sum(t,KUAUX(t,
    itt)*priceURE(t)*PRU(t)-KDAUX(t,itt)*priceDRE(t)*PRD(t)-sum(c
    ,GENDATA(c,'VC')*PC(c,t,itt)));
MP_6(t,itt)$(ord(itt) LT (ITAUX+1))..    PE(t)+KUAUX(t,itt)*PRU(t
    )-KDAUX(t,itt)*PRD(t)=E=sum(c,PCPP(c,t,itt))+sum(r,PR(r,t,itt
    ))-sum(d,PD(d,t,itt));
MP_7(d,t,itt)$(ord(itt) LT (ITAUX+1))..  PDmin(d,t)=l=PD(d,t,itt)
    ;
MP_8(d,t,itt)$(ord(itt) LT (ITAUX+1))..  PD(d,t,itt)=l=PDmax(d,t)
    ;
MP_9(d,itt)$(ord(itt) LT (ITAUX+1))..    sum(t,PD(d,t,itt))=g=
    EDmin(d);
MP_10(r,t,itt)$(ord(itt) LT (ITAUX+1)).. PR(r,t,itt)=l=PRA(r,t);
MP_11(c,t,itt)$(ord(itt) LT (ITAUX+1)).. GENDATA(c,'Pmin')*UC(c,t
    )=l=PCPP(c,t,itt);
MP_12(c,t,itt)$(ord(itt) LT (ITAUX+1)).. PCPP(c,t,itt)=l=GENDATA(
    c,'Pmax')*UC(c,t);
SP_of..            Z_SP=e=sum(t,KU(t)*priceURE(t)*PRU_AUX(t)-KD(t)
    *priceDRE(t)*PRD_AUX(t))-(sum(t,lambda(t)*PE_AUX(t)+ZKP(t)*
    PRU_AUX(t)-ZKN(t)*PRD_AUX(t)+sum(d,muDL(d,t)*PDmin(d,t)-muDU(
    d,t)*PDmax(d,t))+sum(c,muCL(c,t)*GENDATA(c,'Pmin')*UC_AUX(c,t
    )-muCU(c,t)*GENDATA(c,'Pmax')*UC_AUX(c,t))-sum(r,muR(r,t)*PRA
    (r,t)))+sum(d,xi(d)*EDmin(d)));
SP_1(t)..          KU(t)+KD(t)=l=1;
SP_2..              sum(t,KU(t)+KD(t))=l=GammaK;
SP_3(t)..          -M*KU(t)=l=ZKP(t);
SP_4(t)..          ZKP(t)=l=M*KU(t);
```

```
SP_5(t)..          -M*(1-KU(t))=l=lambda(t)-ZKP(t);
SP_6(t)..          lambda(t)-ZKP(t)=l=M*(1-KU(t));
SP_7(t)..          -M*KD(t)=l=ZKN(t);
SP_8(t)..          ZKN(t)=l=M*KD(t);
SP_9(t)..          -M*(1-KD(t))=l=lambda(t)-ZKN(t);
SP_10(t)..         lambda(t)-ZKN(t)=l=M*(1-KD(t));
SP_11(c,t)..             lambda(t)+MUCL(c,t)-MUCU(c,t)=e=GENDATA(c
    ,'VC');
SP_12(r,t)..             lambda(t)-MUR(r,t)=l=0;
SP_13(d,t)..             -lambda(t)+MUDL(d,t)-MUDU(d,t)+XI(d)=e=0;
SP_14(t)..         lambda(t)=l=M;
SP_15(t)..         -lambda(t)=l=M;
MODEL Master /MP_of,MP_1,MP_2,MP_3,MP_4,MP_5,MP_6,MP_7,MP_8,MP_9,
    MP_10,MP_11,MP_12/;
MODEL Subproblem /SP_of,SP_1,SP_2,SP_3,SP_4,SP_5,SP_6,SP_7,SP_8,
    SP_9,SP_10,SP_11,SP_12,SP_13,SP_14,SP_15/;
OPTION optcr=0;
************
***Step 1***
*Iteration counter it=0
SCALAR TOL        'tolerance'             /0/;
SCALAR LB         'lower bound'           /-1e10/;
SCALAR UB         'upper bound'           /1e10/;
************
***Step 2***
SOLVE Master using mip maximizing Z_MP;
DISPLAY PE.l, PRU.l, PRD.l, Z_MP.l, theta.l;
************
***Step 3***
UB=Z_MP.l;
DISPLAY UB;
************
***Step 4***
PE_AUX(t)=PE.l(t);
PRU_AUX(t)=PRU.l(t);
PRD_AUX(t)=PRD.l(t);
UC_AUX(c,t)=UC.l(c,t);
************
***Step 5***
SOLVE Subproblem using mip minimizing Z_SP;
DISPLAY KU.l, KD.l, Z_SP.l;
************
***Step 6***
LB=max(sum(t,priceE(t)*PE.l(t)+priceUR(t)*PRU.l(t)+priceDR(t)*PRD
    .l(t)-sum(c,GENDATA(c,'FC')*UC.l(c,t)))+Z_SP.l,LB);
DISPLAY LB;
************
***Step 7***
***Step 8***
LOOP(it$((UB-LB) GT TOL),
************
***Step 9***
KUAUX(t,it)=KU.l(t);
KDAUX(t,it)=KD.l(t);
```

```
ITAUX=ORD(IT);
DISPLAY ITAUX, KUAUX, KDAUX;
************
***Step 2***
SOLVE Master using mip maximizing Z_MP;
DISPLAY PE.l, PRU.l, PRD.l, Z_MP.l, theta.l;
************
***Step 3***
UB=Z_MP.l;
DISPLAY UB;
************
***Step 4***
PE_AUX(t)=PE.l(t);
PRU_AUX(t)=PRU.l(t);
PRD_AUX(t)=PRD.l(t);
UC_AUX(c,t)=UC.l(c,t);
************
***Step 5***
SOLVE Subproblem using mip minimizing Z_SP;
DISPLAY KU.l, KD.l, Z_SP.l;
************
***Step 6***
LB=max(sum(t,priceE(t)*PE.l(t)+priceUR(t)*PRU.l(t)+priceDR(t)*PRD
    .l(t)-sum(c,GENDATA(c,'FC')*UC.l(c,t)))+Z_SP.l,LB);
DISPLAY UB,LB;
);
DISPLAY PE.l, PRU.l, PRD.l;
```

References

1. Baringo, A., Baringo, L, Arroyo, J.M.: Day-ahead self-scheduling of a virtual power plant in energy and reserve electricity markets under uncertainty. IEEE Trans. Power Syst. **34**(3), 1881–1894 (2019)
2. Ben-Tal, A., Goryashko, A., Guslitzer, E., Nemirovski, A.: Adjustable robust solutions of uncertain linear programs. Math. Program. **99**(2), 351–376 (2004)
3. Bertsimas, D., Sim, M.: Robust discrete optimization and network flows. Math. Program. **98**(1–3), 49–71 (2003)
4. Bertsimas, D., Sim, M.: The price of robustness. Oper. Res. **52**(1), 35–53 (2004)
5. Birge, J.R., Louveaux, F: Introduction to Stochastic Programming, 2nd edn. Springer, New York, US (2011)
6. Conejo, A.J., Carrión, M., Morales, J.M.: Decision Making Under Uncertainty in Electricity Markets. Springer, New York, US (2010)
7. Dabbagh, S.R., Sheikh-El-Eslami, M.K.: Risk assessment of virtual power plants offering in energy and reserve markets. IEEE Trans. Power Syst. **31**(5), 3572–3582 (2016)
8. Floudas, C.A.: Nonlinear and Mixed-Integer Optimization. Fundamentals and Applications. Oxford University Press, New York (1995)
9. GAMS.: Available at www.gams.com/ (2020)
10. ILOG CPLEX.: Available at www.ilog.com/products/cplex/ (2020)
11. Jiang, R., Zhang, M., Li, G., Guan, Y.: Two-stage network constrained robust unit commitment problem. Eur. J. Oper. Res. **234**(3), 751–762 (2014)
12. Mashhour, E., Moghaddas-Tafreshi, S.M.: Bidding strategy of virtual power plant for participating in energy and spinning reserve markets–Part I: Problem formulation. IEEE Trans. Power Syst. **26**(2), 949–956 (2011)
13. Sioshansi, R., Conejo, A.J.: Optimization in Engineering. Models and Algorithms. Springer, New York, US (2017)
14. Zeng, B., Zhao, L.: Solving two-stage robust optimization problems using a column-and-constraint generation method. Oper. Res. Lett. **41**(5), 457–461 (2013)

Chapter 6
Price-Maker Virtual Power Plants

This chapter provides equilibrium-based mathematical models that can be used for scheduling price-maker virtual power plants (VPPs) participating in energy and reserve electricity markets. These mathematical models allow price-maker VPPs to influence the prices in these markets to their own benefit, i.e., the proposed models allow VPPs to exercise market power. Uncertainties in market prices, production levels of stochastic renewable generating units, and reserve deployment requests are addressed in the problem using two different methods based on deterministic and stochastic programming approaches.

6.1 Introduction

Previous chapters focus on the scheduling problem of a price-taker virtual power plant (VPP), i.e., a VPP that is not able to alter market prices. However, VPPs may comprise large enough electricity assets that make them susceptible to exercise market power. Exercising market power means that the VPP is able to alter market prices to its own benefit, e.g., to achieve higher profits, by strategically selecting its offering decisions in the electricity markets.

Price-maker VPPs may exercise market power through both financial and physical capacity withholding strategies. Through the former strategy, VPPs consider an offer/bid price above/below the marginal production cost/consumption utility of their generation/consumption assets. Alternatively, considering the latter, the VPP may use a power offer/bid quantity lower than the maximum possible value.

Bi-level models are generally considered in the technical literature to represent the market power of market agents. These models include two problems, namely an upper-level problem that represents the profit-maximization objective of the market agent and a lower-level problem representing the market clearing. This way it is possible to endogenously represent in the decision-making problem the formation of market prices.

These bi-level models are generally reformulated as mathematical programs with equilibrium constraints (MPECs) in which the profit-maximization problem of a market agent is formulated subject to market equilibrium conditions. These MPECs have been widely used to model the price-maker behavior of different market agents in energy markets [3].

The VPP under consideration comprises a combination of electricity assets that include demands, conventional power plants, stochastic renewable generating units, and storage units. The VPP participates in both the day-ahead (DA) energy and reserve electricity markets. Thus, the VPP decides the offer/bid quantity-price in the energy market, as well as the offer quantity-price in the reserve market for up and down reserves.

The scheduling decisions are made by the price-maker VPP one day in advance. This means that decisions are made under uncertainty, which affects market prices and available production levels of stochastic renewable generating units, as well as the up/down reserve requirements that will be called on by the system operator to provide. To handle these uncertainties, two different approaches are described based on deterministic and stochastic programming models. In the deterministic approach, the VPP seeks to maximize its profit under a single scenario realization of uncertainties for the scheduling horizon. However, in the stochastic programming approach the VPP maximizes its expected profit over a predefined scenario set.

The remainder of this chapter is organized as follows. Section 6.2 describes the scheduling problem of a price-maker VPP. Section 6.3 formulates the scheduling problem in the DA energy market of a price-maker VPP using a deterministic bi-level model. Section 6.4 extends the formulation described in Sect. 6.3 considering a stochastic programming approach. Section 6.5 extends the stochastic approach in Sect. 6.4 for a price-maker VPP participating in the DA energy and reserve markets. Section 6.6 provides a brief summary of the chapter, as well as the main conclusions and some suggestions for further reading. Finally, Sect. 6.7 provides the GAMS codes used to solve some of the examples of the chapter.

L. Baringo, M. Rahimiyan, *Virtual Power Plants and Electricity Markets*, https://doi.org/10.1007/978-3-030-47602-1_6

6.2 Problem Description

This section provides a brief description of the main characteristics of the price-maker VPP models described in this chapter.

6.2.1 Notation

The main notation used in this chapter is provided below for quick reference. Other symbols are defined as needed throughout the chapter. Note that if the symbols below include a subscript γ, it denotes their values in scenario γ.

6.2.1.1 Indexes

The indexes of the price-maker VPP model are:

c Conventional power plants in the VPP.
d Demands in the VPP.
g Producers other than the VPP.
o Market agents other than the VPP in the reserve market.
q Demands other than the VPP.
r Stochastic renewable generating units in the VPP.
s Storage units in the VPP.
t Time periods.
γ Scenarios.

6.2.1.2 Sets

The sets of the price-maker VPP model are:

Ω^{C} Conventional power plants in the VPP.
Ω^{D} Demands in the VPP.
Ω^{G} Producers other than the VPP.
Ω^{O} Market agents in the reserve market other than the VPP.
Ω^{Q} Demands other than the VPP.
Ω^{R} Stochastic renewable generating units in the VPP.
Ω^{S} Storage units in the VPP.
Ω^{T} Time periods.

6.2.1.3 Parameters

The parameters of the price-maker VPP model are:

$C_c^{\mathrm{C,F}}$ Online cost of conventional power plant c [\$/h].
$C_c^{\mathrm{C,V}}$ Variable cost of conventional power plant c [\$/MWh].
$\underline{E}_d^{\mathrm{D}}$ Minimum energy consumption of demand d throughout the planning horizon [MWh].
$\underline{E}_{st}^{\mathrm{S}}$ Lower bound of the energy stored in storage unit s in time period t [MWh].
$\overline{E}_{st}^{\mathrm{S}}$ Upper bound of the energy stored in storage unit s in time period t [MWh].
$K_t^{\mathrm{R+}}$ Up-reserve deployment request by the system operator in time period t [pu].
$K_t^{\mathrm{R-}}$ Down-reserve deployment request by the system operator in time period t [pu].
$\underline{P}_c^{\mathrm{C}}$ Minimum power production of conventional generating unit c [MW].
$\overline{P}_c^{\mathrm{C}}$ Maximum power production of conventional generating unit c [MW].
$\underline{P}_{dt}^{\mathrm{D}}$ Lower bound of the power consumption of demand d in time period t [MW].

$\overline{P}_{dt}^{\mathrm{D}}$ Upper bound of the power consumption of demand d in time period t [MW].
$\overline{P}_{gt}^{\mathrm{G,E}}$ Power offer quantity of producer g in the energy market in time period t [MW].
$\overline{P}_{ot}^{\mathrm{O+}}$ Up-reserve offer capacity of market agent o in the reserve market in time period t [MW].
$\overline{P}_{ot}^{\mathrm{O-}}$ Down-reserve offer capacity of market agent o in the reserve market in time period t [MW].
$\overline{P}_{qt}^{\mathrm{Q,E}}$ Power bid quantity of demand q in the energy market in time period t [MW].
$P_{rt}^{\mathrm{R,A}}$ Available generation level of the stochastic renewable generating unit r in time period t [MW].
$P_{t}^{\mathrm{R+}}$ Up-reserve capacity required in the reserve market in time period t [MW].
$P_{t}^{\mathrm{R-}}$ Down-reserve capacity required in the reserve market in time period t [MW].
$\overline{P}_{s}^{\mathrm{S,C}}$ Charging capacity of storage unit s [MW].
$\overline{P}_{s}^{\mathrm{S,D}}$ Discharging capacity of storage unit s [MW].
$\overline{P}^{\mathrm{V+}}$ Power capacity of generation assets in the VPP [MW].
$\overline{P}^{\mathrm{V-}}$ Power capacity of consumption assets in the VPP [MW].
U_{dt}^{D} Utility of demand d in time period t [\$/MWh].
$\chi_{gt}^{\mathrm{G,E}}$ Offer price of producer g in the energy market in time period t [\$/MWh].
$\chi_{ot}^{\mathrm{O+}}$ Up-reserve offer price of market agent o in the reserve market in time period t [\$/MW].
$\chi_{ot}^{\mathrm{O-}}$ Down-reserve bid price of market agent o in the reserve market in time period t [\$/MW].
$\chi_{qt}^{\mathrm{Q,E}}$ Bid price of demand q in the energy market in time period t [\$/MWh].
Δt Duration of time periods [h].
$\eta_{s}^{\mathrm{S,C}}$ Charging efficiency of storage unit s [%].
$\eta_{s}^{\mathrm{S,D}}$ Discharging efficiency of storage unit s [%].

6.2.1.4 Variables

The variables of the price-maker VPP model are:

e_{st}^{S} Energy stored in storage unit s in time period t [MWh].
p_{ct}^{C} Power generation of conventional power plant c in time period t [MW].
p_{dt}^{D} Power consumption of demand d in time period t [MW].
$p_{t}^{\mathrm{E+}}$ Power sold by the VPP to the energy market in time period t [MW].
$p_{t}^{\mathrm{E-}}$ Power bought by the VPP from the energy market in time period t [MW].
$\overline{p}_{t}^{\mathrm{E+}}$ Power offer quantity submitted by the VPP to the energy market in time period t [MW].
$\overline{p}_{t}^{\mathrm{E-}}$ Power bid quantity submitted by the VPP to the energy market in time period t [MW].
$p_{gt}^{\mathrm{G,E}}$ Power produced by producer g in the energy market in time period t [MW].
$p_{ot}^{\mathrm{O+}}$ Up-reserve capacity sold by market agent o in the reserve market in time period t [MW].
$p_{ot}^{\mathrm{O-}}$ Down-reserve capacity sold by market agent o in the reserve market in time period t [MW].
$p_{qt}^{\mathrm{Q,E}}$ Power consumed by demand q in the energy market in time period t [MW].
p_{rt}^{R} Production of stochastic renewable generating unit r in time period t [MW].
$p_{t}^{\mathrm{R+}}$ Up-reserve capacity sold by the VPP in the reserve market in time period t [MW].
$p_{t}^{\mathrm{R-}}$ Down-reserve capacity sold by the VPP in the reserve market in time period t [MW].
$\overline{p}_{t}^{\mathrm{R+}}$ Up-reserve offer capacity submitted by the VPP to the reserve market in time period t [MW].
$\overline{p}_{t}^{\mathrm{R-}}$ Down-reserve offer capacity submitted by the VPP to the reserve market in time period t [MW].
$p_{st}^{\mathrm{S,C}}$ Charging power level of storage unit s in time period t [MW].
$p_{st}^{\mathrm{S,D}}$ Discharging power level of storage unit s in time period t [MW].
u_{ct}^{C} Binary variable that is equal to 1 if conventional power plant c is generating electricity in time period t, being 0 otherwise.
u_{st}^{S} Binary variable that is used to prevent the simultaneous charging and discharging of storage unit s in time period t.
λ_{t}^{E} Energy market price in time period t [\$/MWh].
$\lambda_{t}^{\mathrm{R+}}$ Up-reserve market price in time period t [\$/MW].
$\lambda_{t}^{\mathrm{R-}}$ Down-reserve market price in time period t [\$/MW].

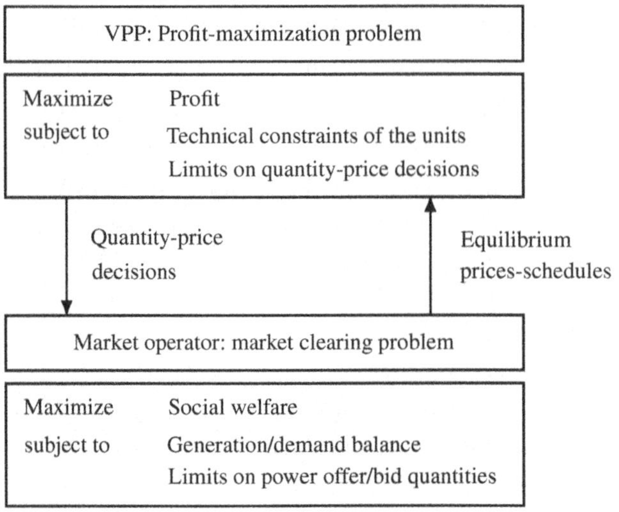

Fig. 6.1 Structure of the bi-level model for the scheduling of a price-maker VPP in the DA energy market

ν_t^{E+} Offer price submitted by the VPP to the energy market in time period t [$/MWh].
ν_t^{E-} Bid price submitted by the VPP to the energy market in time period t [$/MWh].
ν_t^{R+} Up-reserve offer price submitted by the VPP to the reserve market in time period t [$/MW].
ν_t^{R-} Down-reserve offer price submitted by the VPP to the reserve market in time period t [$/MW].

6.2.2 Scheduling Problem

A price-maker VPP that participates in the DA energy and reserve electricity markets is considered. For the sake of clarity, the description below focuses on the participation of the VPP in the energy market; however, it can be easily extended to the reserve market for which additional details are provided in Sect. 6.5.

The main characteristic of a price-maker VPP is its ability to alter market prices to its own benefit by strategically selecting the offer quantity and/or price. In order to take this into account, it is necessary to include in the decision-making problem the clearing of the market. This can be done using bi-level models in which the price-maker VPP and the market operator participate in a Stackelberg game as explained below:

1. In the upper-level problem, the price-maker VPP seeks to maximize its profit achieved in the energy market considering the technical and economic constraints of the units. Thus, the price-maker VPP behaves as the *leader* of the game and decides the offer/bid quantity-price in the energy market.
2. In the lower-level problem, the market operator maximizes the social welfare in the energy market considering the offers and bids submitted by the price-maker VPP and other market participants. Thus, the market operator acts as the *follower* of the game and reacts to the decisions of the price-maker VPP, providing the market prices and the schedules of market participants.

Figure 6.1 clarifies the structure of this bi-level model, usually also referred to as complementarity model, used to derive the optimal scheduling of a price-maker VPP in the DA energy market.

6.3 Deterministic Approach

This section provides the detailed formulation of a deterministic bi-level model for the optimal scheduling of a price-maker VPP in the DA energy market. In this deterministic model, all uncertain parameters are represented using single-point forecasts.

6.3.1 Bi-Level Model

The scheduling problem of a price-maker VPP in the DA energy market is formulated through the following bi-level model:

$$\max_{\Phi,\Phi_t^{\mathrm{E}}} \sum_{t\in\Omega^{\mathrm{T}}} \Big[\lambda_t^{\mathrm{E}}(p_t^{\mathrm{E+}}\Delta t - p_t^{\mathrm{E-}}\Delta t) + \sum_{d\in\Omega^{\mathrm{D}}} U_{dt}^{\mathrm{D}} p_{dt}^{\mathrm{D}}\Delta t - \sum_{c\in\Omega^{\mathrm{C}}} (C_c^{\mathrm{C,F}} u_{ct}^{\mathrm{C}} + C_c^{\mathrm{C,V}} p_{ct}^{\mathrm{C}}\Delta t)\Big] \quad (6.1a)$$

subject to

$$\sum_{c\in\Omega^{\mathrm{C}}} p_{ct}^{\mathrm{C}} + \sum_{r\in\Omega^{\mathrm{R}}} p_{rt}^{\mathrm{R}} + \sum_{s\in\Omega^{\mathrm{S}}} (p_{st}^{\mathrm{S,D}} - p_{st}^{\mathrm{S,C}}) - \sum_{d\in\Omega^{\mathrm{D}}} p_{dt}^{\mathrm{D}} = p_t^{\mathrm{E+}} - p_t^{\mathrm{E-}}, \quad \forall t\in\Omega^{\mathrm{T}}, \quad (6.1b)$$

$$\sum_{t\in\Omega^{\mathrm{T}}} p_{dt}^{\mathrm{D}}\Delta t \geq \underline{E}_d^{\mathrm{D}}, \quad \forall d\in\Omega^{\mathrm{D}}, \quad (6.1c)$$

$$\underline{P}_{dt}^{\mathrm{D}} \leq p_{dt}^{\mathrm{D}} \leq \overline{P}_{dt}^{\mathrm{D}}, \quad \forall d\in\Omega^{\mathrm{D}}, \forall t\in\Omega^{\mathrm{T}}, \quad (6.1d)$$

$$\underline{P}_{ct}^{\mathrm{C}} u_{ct}^{\mathrm{C}} \leq p_{ct}^{\mathrm{C}} \leq \overline{P}_{ct}^{\mathrm{C}} u_{ct}^{\mathrm{C}}, \quad \forall c\in\Omega^{\mathrm{C}}, \forall t\in\Omega^{\mathrm{T}}, \quad (6.1e)$$

$$u_{ct}^{\mathrm{C}} \in \{0,1\}, \quad \forall c\in\Omega^{\mathrm{C}}, \forall t\in\Omega^{\mathrm{T}}, \quad (6.1f)$$

$$e_{st}^{\mathrm{S}} = e_{s(t-1)}^{\mathrm{S}} + \eta_s^{\mathrm{S,C}} p_{st}^{\mathrm{S,C}}\Delta t - \frac{p_{st}^{\mathrm{S,D}}}{\eta_s^{\mathrm{S,D}}}\Delta t, \quad \forall s\in\Omega^{\mathrm{S}}, \forall t\in\Omega^{\mathrm{T}}, \quad (6.1g)$$

$$\underline{E}_{st}^{\mathrm{S}} \leq e_{st}^{\mathrm{S}} \leq \overline{E}_{st}^{\mathrm{S}}, \quad \forall s\in\Omega^{\mathrm{S}}, \forall t\in\Omega^{\mathrm{T}}, \quad (6.1h)$$

$$0 \leq p_{st}^{\mathrm{S,C}} \leq \overline{P}_s^{\mathrm{S,C}} u_{st}^{\mathrm{S}}, \quad \forall s\in\Omega^{\mathrm{S}}, \forall t\in\Omega^{\mathrm{T}}, \quad (6.1i)$$

$$0 \leq p_{st}^{\mathrm{S,D}} \leq \overline{P}_s^{\mathrm{S,D}} (1 - u_{st}^{\mathrm{S}}), \quad \forall s\in\Omega^{\mathrm{S}}, \forall t\in\Omega^{\mathrm{T}}, \quad (6.1j)$$

$$u_{st}^{\mathrm{S}} \in \{0,1\}, \quad \forall s\in\Omega^{\mathrm{S}}, \forall t\in\Omega^{\mathrm{T}}, \quad (6.1k)$$

$$0 \leq p_{rt}^{\mathrm{R}} \leq P_{rt}^{\mathrm{R,A}}, \quad \forall r\in\Omega^{\mathrm{R}}, \forall t\in\Omega^{\mathrm{T}}, \quad (6.1l)$$

$$\overline{p}_t^{\mathrm{E+}} \leq \overline{P}_t^{\mathrm{V+}} \quad \forall t\in\Omega^{\mathrm{T}} \quad (6.1m)$$

$$\overline{p}_t^{\mathrm{E-}} \leq \overline{P}_t^{\mathrm{V-}}, \quad \forall t\in\Omega^{\mathrm{T}}, \quad (6.1n)$$

$$\overline{p}_t^{\mathrm{E+}}, \overline{p}_t^{\mathrm{E-}} \geq 0, \quad \forall t\in\Omega^{\mathrm{T}}, \quad (6.1o)$$

$$\nu_t^{\mathrm{E+}}, \nu_t^{\mathrm{E-}} \geq 0, \quad \forall t\in\Omega^{\mathrm{T}}, \quad (6.1p)$$

$$\lambda_t^{\mathrm{E}}, p_t^{\mathrm{E+}}, p_t^{\mathrm{E-}} \in \Psi_t^{\mathrm{E}}(\cdot), \quad \forall t\in\Omega^{\mathrm{T}}, \quad (6.1q)$$

where variables in sets $\Phi = \Big\{p_{dt}^{\mathrm{D}}, \forall d\in\Omega^{\mathrm{D}}, \forall t\in\Omega^{\mathrm{T}}; e_{st}^{\mathrm{S}}, p_{st}^{\mathrm{S,C}}, p_{st}^{\mathrm{S,D}}, u_{st}^{\mathrm{S}}, \forall s\in\Omega^{\mathrm{S}}, \forall t\in\Omega^{\mathrm{T}}; p_{ct}^{\mathrm{C}}, u_{ct}^{\mathrm{C}}, \forall c\in\Omega^{\mathrm{C}}, \forall t\in\Omega^{\mathrm{T}}; p_{rt}^{\mathrm{R}}, \forall r\in\Omega^{\mathrm{R}}, \forall t\in\Omega^{\mathrm{T}}; \overline{p}_t^{\mathrm{E+}}, \overline{p}_t^{\mathrm{E-}}, \nu_t^{\mathrm{E+}}, \nu_t^{\mathrm{E-}}, \forall t\in\Omega^{\mathrm{T}}\Big\}$ and $\Phi_t^{\mathrm{E}} = \Big\{p_{gt}^{\mathrm{G,E}}, \forall g\in\Omega^{\mathrm{G}}; p_{qt}^{\mathrm{Q,E}}, \forall q\in\Omega^{\mathrm{Q}}; p_t^{\mathrm{E+}}, p_t^{\mathrm{E-}}\Big\}$, $\forall t\in\Omega^{\mathrm{T}}$, are the optimization variables of problem (6.1).

The aim of bi-level model (6.1) is to determine the offer/bid quantities ($\overline{p}_t^{\mathrm{E+}}, \overline{p}_t^{\mathrm{E-}}$) and prices ($\nu_t^{\mathrm{E+}}, \nu_t^{\mathrm{E-}}$) that maximize the profit achieved by the VPP. The objective function (6.1a) to be maximized is the profit of the VPP and includes the revenues achieved in the energy market, the utility of the flexible demands, and the operating costs of the conventional power plants in the VPP. Constraints (6.1b)–(6.1l) define the feasible operating region of the units in the VPP. Constraints (6.1b) impose the power balance in the VPP, considering the power traded in the energy market throughout the scheduling

Table 6.1 Illustrative Example 6.1: data of the conventional power plant

Fixed production cost	\$0/h
Variable production cost	\$45/MWh
Maximum power production limit	40 MW
Minimum power production limit	0 MWh

Table 6.2 Illustrative Example 6.1: data of the demand

Minimum energy consumption	20 MWh
Maximum power load	30 MW
Minimum power load	0 MW
Utility	\$60/MWh

horizon. Constraints (6.1c) impose a lower bound on the daily energy consumption of the demands, while constraints (6.1d) limit the power consumption of demands in each time period. Constraints (6.1e) impose limits on the power production of conventional power plants, while constraints (6.1f) define binary variables u^{C}_{ct}. Constraints (6.1g)–(6.1k) model the working of storage units, including the energy balance (6.1g), energy limits (6.1h), charging power limits (6.1i), discharging power limits (6.1j), and definition of binary variables u^{S}_{st} in (6.1k) to avoid the simultaneous charging and discharging. Constraints (6.1l) impose that the power production of stochastic renewable generating units must be lower than or equal to the available production level. Constraints (6.1m) and (6.1n) impose upper bounds on the power quantity that can be offered and bid by the VPP, respectively. These bounds can be fixed equal to the available generation and consumption capacities of the energy assets in the VPP. Constraints (6.1o) and (6.1p) define the quantity and price decisions as nonnegative variables.

Finally, constraints (6.1q) state that both the market prices and the power scheduled of the VPP are the result of the market-clearing problem represented by set Ψ^{E}_t, which will be formulated in Sect. 6.3.2.

Illustrative Example 6.1 Deterministic Bi-level Model for the Scheduling of a Price-Maker VPP in the DA Energy Market

A price-maker VPP that includes one conventional power plant, one demand, and one photovoltaic (PV) generating unit is considered. The VPP participates in the DA energy market. For the sake of simplicity, a 1-h scheduling horizon is considered. The data of the conventional power plant and the demand are given in Tables 6.1 and 6.2, respectively. The installed capacity of the PV generating unit is 10 MW, being 8 MW the available production level. Thus, the generation capacity in the VPP is 50 MW, while the maximum demand consumption is 30 MW. These two quantities will be used to bound the power quantities offered and bid by the VPP.

Based on the above data, the scheduling problem of the price-maker VPP in the DA energy market is formulated as the following bi-level model.

First, the objective function of the problem to be maximized is:

$$\lambda^{\mathrm{E}}(p^{\mathrm{E+}} - p^{\mathrm{E-}}) + (60p^{\mathrm{D}} - 45p^{\mathrm{C}}).$$

This objective function represents the profit achieved by the VPP. The first term is the revenue minus the energy procurement cost in the market. The second term is the utility of the demand minus the production cost of the conventional power plant in the VPP.

Next, the constraints of the problem are formulated.

- Power balance equality:

$$p^{\mathrm{C}} + p^{\mathrm{R}} - p^{\mathrm{D}} = p^{\mathrm{E+}} - p^{\mathrm{E-}}.$$

- Technical limits of the demand in the VPP:

$$p^{\mathrm{D}} \geq 20,$$
$$0 \leq p^{\mathrm{D}} \leq 30.$$

- Technical limits of the conventional power plant in the VPP:

$$0 \leq p^{\mathrm{C}} \leq 40u^{\mathrm{C}}.$$

- Technical limits of the PV generating unit in the VPP:

$$0 \leq p^{\mathrm{R}} \leq 8.$$

- Limits on the offering and bidding decisions:

$$\overline{p}^{\mathrm{E+}} \leq 50,$$
$$\overline{p}^{\mathrm{E-}} \leq 30,$$
$$\overline{p}^{\mathrm{E+}}, \overline{p}^{\mathrm{E-}} \geq 0,$$
$$\nu^{\mathrm{E+}}, \nu^{\mathrm{E-}} \geq 0.$$

- Market-clearing conditions:

$$\lambda^{\mathrm{E}}, p^{\mathrm{E+}}, p^{\mathrm{E-}} \in \Psi^{\mathrm{E}}.$$

Finally, note that the optimization variables of the problem are those included in set $\Phi^{\mathrm{IE6.1}} = \left\{p^{\mathrm{D}}, p^{\mathrm{C}}, u^{\mathrm{C}}, p^{\mathrm{R}}, \overline{p}^{\mathrm{E+}}, \overline{p}^{\mathrm{E-}}, \nu^{\mathrm{E+}}, \nu^{\mathrm{E-}}, \lambda^{\mathrm{E}}, p^{\mathrm{E+}}, p^{\mathrm{E-}}\right\}$. For the sake of clarity, subscripts in these variables are suppressed since there is a single time period and just one unit of each type.

□

6.3.2 Market-Clearing Problem

As explained in Sect. 6.3.1, both the market prices and power quantities scheduled in the energy market and used in problem (6.1) are obtained from the market-clearing problem represented by sets Ψ_t^{E}, $\forall t \in \Omega^{\mathrm{T}}$. In each time period t, the market-clearing problem can be formulated as the following linear programming (LP) problem:

$$\Psi_t^{\mathrm{E}}(\cdot) = \Bigg\{$$

$$\min_{\Phi_t^{\mathrm{E}}} \quad \nu_t^{\mathrm{E+}} p_t^{\mathrm{E+}} + \sum_{g \in \Omega^{\mathrm{G}}} \chi_{gt}^{\mathrm{G,E}} p_{gt}^{\mathrm{G,E}} - \nu_t^{\mathrm{E-}} p_t^{\mathrm{E-}} - \sum_{q \in \Omega^{\mathrm{Q}}} \chi_{qt}^{\mathrm{Q,E}} p_{qt}^{\mathrm{Q,E}} \tag{6.2a}$$

subject to

$$p_t^{\mathrm{E+}} + \sum_{g \in \Omega^{\mathrm{G}}} p_{gt}^{\mathrm{G,E}} = p_t^{\mathrm{E-}} + \sum_{q \in \Omega^{\mathrm{Q}}} p_{qt}^{\mathrm{Q,E}} \ : \ \lambda_t^{\mathrm{E}}, \tag{6.2b}$$

$$0 \leq p_{gt}^{\mathrm{G,E}} \leq \overline{P}_{gt}^{\mathrm{G,E}} \ : \ \underline{\mu}_{gt}^{\mathrm{G,E}}, \overline{\mu}_{gt}^{\mathrm{G,E}}, \quad \forall g \in \Omega^{\mathrm{G}}, \tag{6.2c}$$

$$0 \leq p_{qt}^{\mathrm{Q,E}} \leq \overline{P}_{qt}^{\mathrm{Q,E}} \ : \ \underline{\mu}_{qt}^{\mathrm{Q,E}}, \overline{\mu}_{qt}^{\mathrm{Q,E}}, \quad \forall q \in \Omega^{\mathrm{Q}}, \tag{6.2d}$$

$$0 \leq p_t^{\mathrm{E+}} \leq \overline{p}_t^{\mathrm{E+}} \ : \ \underline{\mu}_t^{\mathrm{E+}}, \overline{\mu}_t^{\mathrm{E+}}, \tag{6.2e}$$

$$0 \leq p_t^{\mathrm{E-}} \leq \overline{p}_t^{\mathrm{E-}} \ : \ \underline{\mu}_t^{\mathrm{E-}}, \overline{\mu}_t^{\mathrm{E-}}, \tag{6.2f}$$

$\Big\}$, $\forall t \in \Omega^{\mathrm{T}}$, where variables in set Φ_t^{E} are the optimization variables of problem (6.2) and dual variables associated with the constraints are provided following a colon.

The objective function (6.2a) to be minimized is the minus social welfare, which is equivalent to maximize the social welfare. It includes the four terms below:

1. Term $\nu_t^{\mathrm{E+}} p_t^{\mathrm{E+}}$ is the cost of the power production of the VPP.
2. Term $\sum_{g \in \Omega^{\mathrm{G}}} \chi_{gt}^{\mathrm{G,E}} p_{gt}^{\mathrm{G,E}}$ is the cost of the power production of producers other than the VPP.
3. Term $\nu_t^{\mathrm{E-}} p_t^{\mathrm{E-}}$ is the utility of the power consumption of the VPP.
4. Term $\sum_{q \in \Omega^{\mathrm{Q}}} \chi_{qt}^{\mathrm{Q,E}} p_{qt}^{\mathrm{Q,E}}$ is the utility of the power consumption of demands other than the VPP.

Constraints of the market-clearing problem include equality (6.2b) that imposes the power balance of generation and consumption; constraints (6.2c)–(6.2f) that determine the feasible operating region of producers and demands considering their offer and bid quantities; constraints (6.2c) that limit the power production of producers other than the VPP; constraints (6.2d) that impose upper bounds on the power consumption of demands other than the VPP; as well as constraints (6.2e) and (6.2f) that limit the power production and consumption of the VPP, respectively.

The market operator determines the power dispatch of producers and consumers by solving the above market-clearing problem (6.2) for all time periods in the scheduling horizon.

Illustrative Example 6.2 Energy Market-Clearing Problem

The VPP described in Illustrative Example 6.1 is considered again. This VPP decides a pair of quantity-price ($\nu^{\mathrm{E+}}$, $\overline{p}^{\mathrm{E+}}$)=(30 MW,\$40/MWh) to sell energy in the DA energy market. Additionally, there are two producers and two demands participating in this market, whose offering and bidding data are presented in Tables 6.3 and 6.4, respectively.

The market operator solves the market-clearing problem of the DA energy market to determine the market price as well as the power schedules of the VPP, producers, and demands.

First, the objective function to be minimized is:

$$40p^{\mathrm{E+}} + 30p_1^{\mathrm{G,E}} + 50p_2^{\mathrm{G,E}} - 50p_1^{\mathrm{Q,E}} - 70p_1^{\mathrm{Q,E}}.$$

This objective function is the minus social welfare. The first term is the production cost of the VPP, the next two terms are the production costs of the other two producers, while the remaining terms are the minus utilities of the two demands.

The market-clearing problem is subject to the following constraints:

- Power balance:

$$p^{\mathrm{E+}} + p_1^{\mathrm{G,E}} + p_2^{\mathrm{G,E}} = p_1^{\mathrm{Q,E}} + p_2^{\mathrm{Q,E}} \ : \ \lambda^{\mathrm{E}}.$$

- Limits on the power production of the two producers:

$$0 \le p_1^{\mathrm{G,E}} \le 50 \ : \ \underline{\mu}_1^{\mathrm{G,E}}, \overline{\mu}_1^{\mathrm{G,E}},$$
$$0 \le p_2^{\mathrm{G,E}} \le 50 \ : \ \underline{\mu}_2^{\mathrm{G,E}}, \overline{\mu}_2^{\mathrm{G,E}}.$$

- Limits on the power consumption of the two demands:

$$0 \le p_1^{\mathrm{Q,E}} \le 50 \ : \ \underline{\mu}_1^{\mathrm{Q,E}}, \overline{\mu}_1^{\mathrm{Q,E}},$$
$$0 \le p_2^{\mathrm{Q,E}} \le 50 \ : \ \underline{\mu}_2^{\mathrm{Q,E}}, \overline{\mu}_2^{\mathrm{Q,E}}.$$

Table 6.3 Illustrative Example 6.2: data of two producers other than the VPP

Parameter	Producer 1	Producer 2
Offer price	\$30/MWh	\$50/MWh
Power offer quantity	50 MW	50 MW

Table 6.4 Illustrative Example 6.2: data of two demands other than the VPP

Parameter	Demand 1	Demand 2
Bid price	\$50/MWh	\$70/MWh
Power bid quantity	50 MW	50 MW

Table 6.5 Illustrative Example 6.2: schedules of two producers, two demands, and the VPP in the DA energy market

	Producer 1	Producer 2	Demand 1	Demand 2	VPP
Power traded [MW]	50	0	30	50	30

- Limits on the power production of the VPP:

$$0 \leq p^{\mathrm{E+}} \leq 30 \; : \; \underline{\mu}^{\mathrm{E+}}, \overline{\mu}^{\mathrm{E+}}.$$

Finally, note that the optimization variables of the problem are those included in set $\Phi^{\mathrm{IE6.2}} = \left\{p_1^{\mathrm{G,E}}, p_2^{\mathrm{G,E}}, p_1^{\mathrm{Q,E}}, p_2^{\mathrm{Q,E}}, p^{\mathrm{E+}}\right\}$.

The above problem is an LP model that is solved using CPLEX [5] under GAMS [4]. The optimal schedules of the market participants are provided in Table 6.5.

The total power production is 80 MW, which is allocated to producer 1 (50 MW) and the VPP (30 MW) since they have the lowest offer prices ($30/MWh and $40/MWh, respectively). On the other hand, the bid price of demand 2 is higher than that of demand 1 ($70/MWh versus $50/MWh), and thus the power scheduled of demand 2 is 50 MW, its power bid quantity, while only 30 MW is scheduled for demand 1.

The resulting market-clearing price is $50/MWh, being the optimal value of the social welfare equal to $2300.

□

An input GAMS code for solving Illustrative Example 6.2 is provided in Sect. 6.7.

Illustrative Example 6.3 Impact of Offering and Bidding Decisions on the Market-Clearing Problem

Illustrative Example 6.2 is considered again. Next, the impact of the offering and bidding decisions submitted by the market participants on the market-clearing price is analyzed. To do this, the bid price of demand 1 is changed to $45/MWh and the market-clearing problem is solved again. The optimal power schedules of market participants are the same than those obtained in Illustrative Example 6.2; however, the optimal value of the social welfare decreases to $2150 and the market-clearing price changes to $45/MWh.

□

6.3.2.1 KKT Conditions of the Market-Clearing Problem

The market-clearing problem (6.2) is continuous and linear on its decision variables, i.e., it is a convex optimization problem. Thus, the Karush-Kuhn-Tucker (KKT) optimality conditions provide both necessary and sufficient conditions for global optimal solution [1]. This means that the clearing problem of the DA energy market can be replaced by the following KKT optimality conditions:

$$\Psi_t^{\mathrm{E}}(\cdot) = \Bigg\{$$

$$p_t^{\mathrm{E+}} + \sum_{g \in \Omega^{\mathrm{G}}} p_{gt}^{\mathrm{G,E}} = p_t^{\mathrm{E-}} + \sum_{q \in \Omega^{\mathrm{Q}}} p_{qt}^{\mathrm{Q,E}}, \tag{6.3a}$$

$$\chi_{gt}^{\mathrm{G,E}} - \lambda_t^{\mathrm{E}} + \overline{\mu}_{gt}^{\mathrm{G,E}} - \underline{\mu}_{gt}^{\mathrm{G,E}} = 0, \quad \forall g \in \Omega^{\mathrm{G}}, \tag{6.3b}$$

$$-\chi_{qt}^{\mathrm{Q,E}} + \lambda_t^{\mathrm{E}} + \overline{\mu}_{qt}^{\mathrm{Q,E}} - \underline{\mu}_{qt}^{\mathrm{Q,E}} = 0, \quad \forall q \in \Omega^{\mathrm{Q}}, \tag{6.3c}$$

$$\nu_t^{\mathrm{E+}} - \lambda_t^{\mathrm{E}} + \overline{\mu}_t^{\mathrm{E+}} - \underline{\mu}_t^{\mathrm{E+}} = 0, \tag{6.3d}$$

$$-\nu_t^{\mathrm{E-}} + \lambda_t^{\mathrm{E}} + \overline{\mu}_t^{\mathrm{E-}} - \underline{\mu}_t^{\mathrm{E-}} = 0, \tag{6.3e}$$

$$0 \leq \underline{\mu}_{gt}^{\mathrm{G,E}} \perp p_{gt}^{\mathrm{G,E}} \geq 0, \quad \forall g \in \Omega^{\mathrm{G}}, \tag{6.3f}$$

$$0 \leq \overline{\mu}_{gt}^{\mathrm{G,E}} \perp \overline{P}_{gt}^{\mathrm{G,E}} - p_{gt}^{\mathrm{G,E}} \geq 0, \quad \forall g \in \Omega^{\mathrm{G}}, \tag{6.3g}$$

$$0 \leq \underline{\mu}_{qt}^{\mathrm{Q,E}} \perp p_{qt}^{\mathrm{Q,E}} \geq 0, \quad \forall q \in \Omega^{\mathrm{Q}}, \tag{6.3h}$$

$$0 \le \overline{\mu}_{qt}^{\mathrm{Q,E}} \perp \overline{P}_{qt}^{\mathrm{Q,E}} - p_{qt}^{\mathrm{Q,E}} \ge 0, \quad \forall q \in \Omega^{\mathrm{Q}}, \tag{6.3i}$$

$$0 \le \underline{\mu}_{t}^{\mathrm{E+}} \perp p_{t}^{\mathrm{E+}} \ge 0, \tag{6.3j}$$

$$0 \le \overline{\mu}_{t}^{\mathrm{E+}} \perp \overline{p}_{t}^{\mathrm{E+}} - p_{t}^{\mathrm{E+}} \ge 0, \tag{6.3k}$$

$$0 \le \underline{\mu}_{t}^{\mathrm{E-}} \perp p_{t}^{\mathrm{E-}} \ge 0, \tag{6.3l}$$

$$0 \le \overline{\mu}_{t}^{\mathrm{E-}} \perp \overline{p}_{t}^{\mathrm{E-}} - p_{t}^{\mathrm{E-}} \ge 0 \tag{6.3m}$$

$\Big\}, \forall t \in \Omega^{\mathrm{T}}$.

Equalities (6.3a)–(6.3e) are obtained by taking the derivatives of the Lagrange function of the market-clearing problem (6.2) with respect to the dual variable λ_t^{E} and the primal variables $p_{gt}^{\mathrm{G,E}}, \forall g \in \Omega^{\mathrm{G}}$; $p_{qt}^{\mathrm{Q,E}}, \forall q \in \Omega^{\mathrm{Q}}$; $p_t^{\mathrm{E+}}$, $p_t^{\mathrm{E-}}$. Constraints (6.3f)–(6.3m) are the complementarity constraints regarding inequalities (6.2c)–(6.2f) in the market-clearing problem (6.2). Note that complementarity constraints have the form $0 \le x \perp y \ge 0$, which is equivalent to $x, y \ge 0$ and the slackness condition $x \cdot y = 0$. Appendix A shows how to formulate the Lagrange function of an LP problem, as well as how to derive the KKT optimality conditions.

Illustrative Example 6.4 Energy Market-Clearing Problem: KKT Conditions

The market-clearing problem described in Illustrative Example 6.2 is considered again. Here, the KKT conditions of this problem are provided.

First, the Lagrange function is formulated as:

$$\begin{aligned}
&L(p^{\mathrm{E+}}, p_1^{\mathrm{G,E}}, p_2^{\mathrm{G,E}}, p_1^{\mathrm{Q,E}}, p_2^{\mathrm{Q,E}}, \lambda^{\mathrm{E}}, \overline{\mu}_1^{\mathrm{G,E}}, \underline{\mu}_1^{\mathrm{G,E}}, \overline{\mu}_1^{\mathrm{Q,E}}, \underline{\mu}_2^{\mathrm{Q,E}}, \overline{\mu}^{\mathrm{E+}}, \underline{\mu}^{\mathrm{E+}}) \\
&= (40p^{\mathrm{E+}} + 30p_1^{\mathrm{G,E}} + 50p_2^{\mathrm{G,E}} - 50p_1^{\mathrm{Q,E}} - 70p_2^{\mathrm{Q,E}}) \\
&+ \lambda^{\mathrm{E}}(p_1^{\mathrm{Q,E}} + p_2^{\mathrm{Q,E}} - p^{\mathrm{E+}} - p_1^{\mathrm{G,E}} - p_2^{\mathrm{G,E}}) \\
&+ \underline{\mu}_1^{\mathrm{G,E}}(-p_1^{\mathrm{G,E}}) + \underline{\mu}_2^{\mathrm{G,E}}(-p_2^{\mathrm{G,E}}) + \overline{\mu}_1^{\mathrm{G,E}}(-50 + p_1^{\mathrm{G,E}}) + \overline{\mu}_2^{\mathrm{G,E}}(-50 + p_2^{\mathrm{G,E}}) \\
&+ \underline{\mu}_1^{\mathrm{Q,E}}(-p_1^{\mathrm{Q,E}}) + \underline{\mu}_2^{\mathrm{Q,E}}(-p_2^{\mathrm{Q,E}}) + \overline{\mu}_1^{\mathrm{Q,E}}(-50 + p_1^{\mathrm{Q,E}}) + \overline{\mu}_2^{\mathrm{Q,E}}(-50 + p_2^{\mathrm{Q,E}}) \\
&+ \underline{\mu}^{\mathrm{E+}}(-p^{\mathrm{E+}}) + \overline{\mu}^{\mathrm{E+}}(-30 + p^{\mathrm{E+}}).
\end{aligned}$$

Next, the KKT conditions are provided:

- Derivative of the Lagrange function with respect to variable λ^{E}:

$$p_1^{\mathrm{Q,E}} + p_2^{\mathrm{Q,E}} - p^{\mathrm{E+}} - p_1^{\mathrm{G,E}} - p_2^{\mathrm{G,E}} = 0.$$

- Derivative of the Lagrange function with respect to variables $p_1^{\mathrm{G,E}}$, $p_2^{\mathrm{G,E}}$, $p_1^{\mathrm{Q,E}}$, $p_2^{\mathrm{Q,E}}$, and $p^{\mathrm{E+}}$:

$$\begin{aligned}
&30 - \lambda^{\mathrm{E}} + \overline{\mu}_1^{\mathrm{G,E}} - \underline{\mu}_1^{\mathrm{G,E}} = 0, \\
&50 - \lambda^{\mathrm{E}} + \overline{\mu}_2^{\mathrm{G,E}} - \underline{\mu}_2^{\mathrm{G,E}} = 0, \\
&-50 + \lambda^{\mathrm{E}} + \overline{\mu}_1^{\mathrm{Q,E}} - \underline{\mu}_1^{\mathrm{Q,E}} = 0, \\
&-70 + \lambda^{\mathrm{E}} + \overline{\mu}_2^{\mathrm{Q,E}} - \underline{\mu}_2^{\mathrm{Q,E}} = 0, \\
&40 - \lambda^{\mathrm{E}} + \overline{\mu}^{\mathrm{E+}} - \underline{\mu}^{\mathrm{E+}} = 0.
\end{aligned}$$

- Complementarity constraints:

$$0 \leq \underline{\mu}_1^{\mathrm{G,E}} \perp p_1^{\mathrm{G,E}} \geq 0,$$
$$0 \leq \underline{\mu}_2^{\mathrm{G,E}} \perp p_2^{\mathrm{G,E}} \geq 0,$$
$$0 \leq \overline{\mu}_1^{\mathrm{G,E}} \perp 50 - p_1^{\mathrm{G,E}} \geq 0,$$
$$0 \leq \overline{\mu}_2^{\mathrm{G,E}} \perp 50 - p_2^{\mathrm{G,E}} \geq 0,$$
$$0 \leq \underline{\mu}_1^{\mathrm{Q,E}} \perp p_1^{\mathrm{Q,E}} \geq 0,$$
$$0 \leq \underline{\mu}_2^{\mathrm{Q,E}} \perp p_2^{\mathrm{Q,E}} \geq 0,$$
$$0 \leq \overline{\mu}_1^{\mathrm{Q,E}} \perp 50 - p_1^{\mathrm{Q,E}} \geq 0,$$
$$0 \leq \overline{\mu}_2^{\mathrm{Q,E}} \perp 50 - p_2^{\mathrm{Q,E}} \geq 0,$$
$$0 \leq \underline{\mu}^{\mathrm{E+}} \perp p^{\mathrm{E+}} \geq 0,$$
$$0 \leq \overline{\mu}^{\mathrm{E+}} \perp 30 - p^{\mathrm{E+}} \geq 0.$$

The above equations provide a set of nonlinear equations that are jointly solved. Solving these equations provide the solution of the market-clearing problem, including both the scheduled power of the different market participants and the market-clearing price.

□

6.3.2.2 Linear Complementarity Constraints

Nonlinear complementarity constraints have the form $x \cdot y = 0$ with $x, y \geq 0$. These nonlinear terms can be linearized through the big-M method using the following expressions:

$$0 \leq x \leq M^{\mathrm{X}} u, \tag{6.4a}$$
$$0 \leq y \leq M^{\mathrm{Y}}(1 - u), \tag{6.4b}$$
$$u \in \{0, 1\}, \tag{6.4c}$$

where M^{X} and M^{Y} are large enough positive constants.

Therefore, complementarity constraints (6.3f)–(6.3m) can be replaced by the following mixed-integer linear constraints:

$$\Big\{ 0 \leq \underline{\mu}_{gt}^{\mathrm{G,E}} \leq \underline{N}_{gt}^{\mathrm{G,E}} \underline{u}_{gt}^{\mathrm{G,E}}, \quad \forall g \in \Omega^{\mathrm{G}}, \tag{6.5a}$$
$$0 \leq p_{gt}^{\mathrm{G,E}} \leq \underline{M}_{gt}^{\mathrm{G,E}}(1 - \underline{u}_{gt}^{\mathrm{G,E}}), \quad \forall g \in \Omega^{\mathrm{G}}, \tag{6.5b}$$
$$0 \leq \overline{\mu}_{gt}^{\mathrm{G,E}} \leq \overline{N}_{gt}^{\mathrm{G,E}} \overline{u}_{gt}^{\mathrm{G,E}}, \quad \forall g \in \Omega^{\mathrm{G}}, \tag{6.5c}$$
$$0 \leq \overline{P}_{gt}^{\mathrm{G,E}} - p_{gt}^{\mathrm{G,E}} \leq \overline{M}_{gt}^{\mathrm{G,E}}(1 - \overline{u}_{gt}^{\mathrm{G,E}}), \quad \forall g \in \Omega^{\mathrm{G}}, \tag{6.5d}$$
$$\underline{u}_{gt}^{\mathrm{G,E}}, \overline{u}_{gt}^{\mathrm{G,E}} \in \{0, 1\}, \quad \forall g \in \Omega^{\mathrm{G}}, \tag{6.5e}$$
$$0 \leq \underline{\mu}_{qt}^{\mathrm{Q,E}} \leq \underline{N}_{qt}^{\mathrm{Q,E}} \underline{u}_{qt}^{\mathrm{Q,E}}, \quad \forall q \in \Omega^{\mathrm{Q}}, \tag{6.5f}$$
$$0 \leq p_{qt}^{\mathrm{Q,E}} \leq \underline{M}_{qt}^{\mathrm{Q,E}}(1 - \underline{u}_{qt}^{\mathrm{Q,E}}), \quad \forall q \in \Omega^{\mathrm{Q}}, \tag{6.5g}$$
$$0 \leq \overline{\mu}_{qt}^{\mathrm{Q,E}} \leq \overline{N}_{qt}^{\mathrm{Q,E}} \overline{u}_{qt}^{\mathrm{Q,E}}, \quad \forall q \in \Omega^{\mathrm{Q}}, \tag{6.5h}$$
$$0 \leq \overline{P}_{qt}^{\mathrm{Q,E}} - p_{qt}^{\mathrm{Q,E}} \leq \overline{M}_{qt}^{\mathrm{Q,E}}(1 - \overline{u}_{qt}^{\mathrm{Q,E}}), \quad \forall q \in \Omega^{\mathrm{Q}}, \tag{6.5i}$$

$$\underline{u}_{qt}^{\mathrm{Q,E}}, \overline{u}_{qt}^{\mathrm{Q,E}} \in \{0, 1\}, \quad \forall q \in \Omega^{\mathrm{Q}}, \tag{6.5j}$$

$$0 \le \underline{\mu}_t^{\mathrm{E+}} \le \underline{N}_t^{\mathrm{E+}} \underline{u}_t^{\mathrm{E+}}, \tag{6.5k}$$

$$0 \le p_t^{\mathrm{E+}} \le \underline{M}_t^{\mathrm{E+}}(1 - \underline{u}_t^{\mathrm{E+}}), \tag{6.5l}$$

$$0 \le \overline{\mu}_t^{\mathrm{E+}} \le \overline{N}_t^{\mathrm{E+}} \overline{u}_t^{\mathrm{E+}}, \tag{6.5m}$$

$$0 \le \overline{p}_t^{\mathrm{E+}} - p_t^{\mathrm{E+}} \le \overline{M}_t^{\mathrm{E+}}(1 - \overline{u}_t^{\mathrm{E+}}) \tag{6.5n}$$

$$0 \le \underline{\mu}_t^{\mathrm{E-}} \le \underline{N}_t^{\mathrm{E-}} \underline{u}_t^{\mathrm{E-}}, \tag{6.5o}$$

$$0 \le p_t^{\mathrm{E-}} \le \underline{M}_t^{\mathrm{E-}}(1 - \underline{u}_t^{\mathrm{E-}}), \tag{6.5p}$$

$$0 \le \overline{\mu}_t^{\mathrm{E-}} \le \overline{N}_t^{\mathrm{E-}} \overline{u}_t^{\mathrm{E-}}, \tag{6.5q}$$

$$0 \le \overline{p}_t^{\mathrm{E-}} - p_t^{\mathrm{E-}} \le \overline{M}_t^{\mathrm{E-}}(1 - \overline{u}_t^{\mathrm{E-}}), \tag{6.5r}$$

$$\underline{u}_t^{\mathrm{E+}}, \overline{u}_t^{\mathrm{E+}}, \underline{u}_t^{\mathrm{E-}}, \overline{u}_t^{\mathrm{E-}} \in \{0, 1\} \tag{6.5s}$$

$\Big\}$, $\forall t \in \Omega^{\mathrm{T}}$, where $\underline{N}_{gt}^{\mathrm{G,E}}$, $\underline{M}_{gt}^{\mathrm{G,E}}$, $\overline{N}_{gt}^{\mathrm{G,E}}$, $\overline{M}_{gt}^{\mathrm{G,E}}$, $\forall g \in \Omega^{\mathrm{G}}$, $\forall t \in \Omega^{\mathrm{T}}$; $\underline{N}_{qt}^{\mathrm{Q,E}}$, $\underline{M}_{qt}^{\mathrm{Q,E}}$, $\overline{N}_{qt}^{\mathrm{Q,E}}$, $\overline{M}_{qt}^{\mathrm{Q,E}}$, $\forall q \in \Omega^{\mathrm{Q}}$, $\forall t \in \Omega^{\mathrm{T}}$; $\underline{N}_t^{\mathrm{E+}}$, $\underline{M}_t^{\mathrm{E+}}$, $\overline{N}_t^{\mathrm{E+}}$, $\overline{M}_t^{\mathrm{E+}}$, $\underline{N}_t^{\mathrm{E-}}$, $\underline{M}_t^{\mathrm{E-}}$, $\overline{N}_t^{\mathrm{E-}}$, $\overline{M}_t^{\mathrm{E-}}$, $\forall t \in \Omega^{\mathrm{T}}$, are large enough positive constants.

Illustrative Example 6.5 Energy Market-Clearing Problem: Linearized KKT Conditions

Illustrative Example 6.4 is considered again. Next, complementarity constraints are replaced by the following exact mixed-integer linear constraints:

$$0 \le \underline{\mu}_1^{\mathrm{G,E}} \le \underline{N}_1^{\mathrm{G,E}} \underline{u}_1^{\mathrm{G,E}},$$

$$0 \le p_1^{\mathrm{G,E}} \le \underline{M}_1^{\mathrm{G,E}}(1 - \underline{u}_1^{\mathrm{G,E}}),$$

$$0 \le \underline{\mu}_2^{\mathrm{G,E}} \le \underline{N}_2^{\mathrm{G,E}} \underline{u}_2^{\mathrm{G,E}},$$

$$0 \le p_2^{\mathrm{G,E}} \le \underline{M}_2^{\mathrm{G,E}}(1 - \underline{u}_2^{\mathrm{G,E}}),$$

$$0 \le \overline{\mu}_1^{\mathrm{G,E}} \le \overline{N}_1^{\mathrm{G,E}} \overline{u}_1^{\mathrm{G,E}},$$

$$0 \le 50 - p_1^{\mathrm{G,E}} \le \overline{M}_1^{\mathrm{G,E}}(1 - \overline{u}_1^{\mathrm{G,E}}),$$

$$0 \le \overline{\mu}_2^{\mathrm{G,E}} \le \overline{N}_2^{\mathrm{G,E}} \overline{u}_2^{\mathrm{G,E}},$$

$$0 \le 50 - p_2^{\mathrm{G,E}} \le \overline{M}_2^{\mathrm{G,E}}(1 - \overline{u}_2^{\mathrm{G,E}}),$$

$$0 \le \underline{\mu}_1^{\mathrm{Q,E}} \le \underline{N}_1^{\mathrm{Q,E}} \underline{u}_1^{\mathrm{Q,E}},$$

$$0 \le p_1^{\mathrm{Q,E}} \le \underline{M}_1^{\mathrm{Q,E}}(1 - \underline{u}_1^{\mathrm{Q,E}}),$$

$$0 \le \underline{\mu}_2^{\mathrm{Q,E}} \le \underline{N}_2^{\mathrm{Q,E}} \underline{u}_2^{\mathrm{Q,E}},$$

$$0 \le p_2^{\mathrm{Q,E}} \le \underline{M}_2^{\mathrm{Q,E}}(1 - \underline{u}_2^{\mathrm{Q,E}}),$$

$$0 \le \overline{\mu}_1^{\mathrm{Q,E}} \le \overline{N}_1^{\mathrm{Q,E}} \overline{u}_1^{\mathrm{Q,E}},$$

$$0 \le 50 - p_1^{\mathrm{Q,E}} \le \overline{M}_1^{\mathrm{Q,E}}(1 - \overline{u}_1^{\mathrm{Q,E}}),$$

$$0 \le \overline{\mu}_2^{\mathrm{Q,E}} \le \overline{N}_2^{\mathrm{Q,E}} \overline{u}_2^{\mathrm{Q,E}},$$

$$0 \le 50 - p_2^{\mathrm{Q,E}} \le \overline{M}_2^{\mathrm{Q,E}}(1 - \overline{u}_2^{\mathrm{Q,E}}),$$

Table 6.6 Illustrative Example 6.5: optimal values of the primal variables

Primal variable	$p_1^{\mathrm{G,E}}$	$p_2^{\mathrm{G,E}}$	$p_1^{\mathrm{Q,E}}$	$p_2^{\mathrm{Q,E}}$	$p^{\mathrm{E+}}$
Value [MW]	50	0	30	50	30

Table 6.7 Illustrative Example 6.5: optimal values of the dual variables

Dual variable	λ^{E}	$\underline{\mu}_1^{\mathrm{G,E}}$	$\underline{\mu}_2^{\mathrm{G,E}}$	$\overline{\mu}_1^{\mathrm{G,E}}$	$\overline{\mu}_2^{\mathrm{G,E}}$	$\underline{\mu}_1^{\mathrm{Q,E}}$	$\underline{\mu}_2^{\mathrm{Q,E}}$	$\overline{\mu}_1^{\mathrm{Q,E}}$	$\overline{\mu}_2^{\mathrm{Q,E}}$	$\underline{\mu}^{\mathrm{E+}}$	$\overline{\mu}^{\mathrm{E+}}$
Value [\$/MWh]	50	0	0	20	0	0	0	0	20	0	10

$$0 \leq \underline{\mu}^{\mathrm{E+}} \leq \underline{N}^{\mathrm{E+}} \underline{u}^{\mathrm{E+}},$$

$$0 \leq p^{\mathrm{E+}} \leq \underline{M}^{\mathrm{E+}} (1 - \underline{u}^{\mathrm{E+}}),$$

$$0 \leq \overline{\mu}^{\mathrm{E+}} \leq \overline{N}^{\mathrm{E+}} \overline{u}^{\mathrm{E+}},$$

$$0 \leq 30 - p^{\mathrm{E+}} \leq \overline{M}^{\mathrm{E+}} (1 - \overline{u}^{\mathrm{E+}}),$$

$$\underline{u}_1^{\mathrm{G,E}}, \underline{u}_2^{\mathrm{G,E}}, \overline{u}_1^{\mathrm{G,E}}, \overline{u}_2^{\mathrm{G,E}}, \underline{u}_1^{\mathrm{Q,E}}, \underline{u}_2^{\mathrm{Q,E}}, \overline{u}_1^{\mathrm{Q,E}}, \overline{u}_2^{\mathrm{Q,E}}, \underline{u}^{\mathrm{E+}}, \overline{u}^{\mathrm{E+}} \in \{0, 1\} .$$

Large enough values of parameters N and M are considered, e.g., 10000 and 100, respectively. Then, the system of equations is solved to obtain the primal and dual variables provided in Tables 6.6 and 6.7, respectively. Observe that the results are the same than those obtained in Illustrative Example 6.2.

□

An input GAMS code for solving Illustrative Example 6.5 is provided in Sect. 6.7.

6.3.2.3 Dual Formulation of the Market-Clearing Problem

The market-clearing problem (6.2) is an LP whose dual optimization problem can be formulated as well as an LP problem, whose formulation is provided below:

$$\max_{\Upsilon_t^{\mathrm{E}}} \quad - \overline{P}_t^{\mathrm{E+}} \overline{\mu}_t^{\mathrm{E+}} - \sum_{g \in \Omega^{\mathrm{G}}} \overline{P}_{gt}^{\mathrm{G,E}} \overline{\mu}_{gt}^{\mathrm{G,E}} - \overline{P}_t^{\mathrm{E-}} \overline{\mu}_t^{\mathrm{E-}} - \sum_{q \in \Omega^{\mathrm{Q}}} \overline{P}_{qt}^{\mathrm{Q,E}} \overline{\mu}_{qt}^{\mathrm{Q,E}} \tag{6.6a}$$

subject to

$$\chi_{gt}^{\mathrm{G,E}} - \lambda_t^{\mathrm{E}} + \overline{\mu}_{gt}^{\mathrm{G,E}} - \underline{\mu}_{gt}^{\mathrm{G,E}} = 0, \quad \forall g \in \Omega^{\mathrm{G,}} \tag{6.6b}$$

$$- \chi_{qt}^{\mathrm{Q,E}} + \lambda_t^{\mathrm{E}} + \overline{\mu}_{qt}^{\mathrm{Q,E}} - \underline{\mu}_{qt}^{\mathrm{Q,E}} = 0, \quad \forall q \in \Omega^{\mathrm{Q}}, \tag{6.6c}$$

$$\nu_t^{\mathrm{E+}} - \lambda_t^{\mathrm{E}} + \overline{\mu}_t^{\mathrm{E+}} - \underline{\mu}_t^{\mathrm{E+}} = 0, \tag{6.6d}$$

$$- \nu_t^{\mathrm{E-}} + \lambda_t^{\mathrm{E}} + \overline{\mu}_t^{\mathrm{E-}} - \underline{\mu}_t^{\mathrm{E-}} = 0, \tag{6.6e}$$

$$\overline{\mu}_{gt}^{\mathrm{G,E}}, \underline{\mu}_{gt}^{\mathrm{G,E}} \geq 0, \quad \forall g \in \Omega^{\mathrm{G}}, \tag{6.6f}$$

$$\overline{\mu}_{qt}^{\mathrm{Q,E}}, \underline{\mu}_{qt}^{\mathrm{Q,E}} \geq 0, \quad \forall q \in \Omega^{\mathrm{Q}}, \tag{6.6g}$$

$$\overline{\mu}_t^{\mathrm{E+}}, \underline{\mu}_t^{\mathrm{E+}}, \overline{\mu}_t^{\mathrm{E-}}, \underline{\mu}_t^{\mathrm{E-}} \geq 0, \tag{6.6h}$$

where dual variables in set $\Upsilon_t^{\mathrm{E}} = \left\{\lambda_t^{\mathrm{E}}; \underline{\mu}_{gt}^{\mathrm{G,E}}, \overline{\mu}_{gt}^{\mathrm{G,E}}, \forall g \in \Omega^{\mathrm{G}}; \underline{\mu}_{qt}^{\mathrm{Q,E}}, \overline{\mu}_{qt}^{\mathrm{Q,E}}, \forall q \in \Omega^{\mathrm{Q}}; \underline{\mu}_t^{\mathrm{E+}}, \overline{\mu}_t^{\mathrm{E+}}, \underline{\mu}_t^{\mathrm{E-}}, \overline{\mu}_t^{\mathrm{E-}}\right\}$, $\forall t \in \Omega^{\mathrm{T}}$, are the optimization variables of dual problem (6.6). Detailed information about how to obtain the dual problem from a standard LP problem is provided in Appendix A.

The market-clearing problem (6.2) is an LP problem. Therefore, the strong duality holds [1]. This means that the duality gap is zero at the optimal solution. In other words, the optimal values of the objective functions of the primal optimization problem (6.2) and its equivalent dual optimization problem (6.6) are equal.

Illustrative Example 6.6 Energy Market-Clearing Problem: Dual Optimization Problem

The market-clearing problem analyzed in Illustrative Example 6.2 is considered again. Here, the dual optimization problem is formulated:

$$\max_{\Upsilon^{\mathrm{IE6.6}}} \quad -30\overline{\mu}^{\mathrm{E+}} - 50\overline{\mu}_1^{\mathrm{G,E}} - 50\overline{\mu}_2^{\mathrm{G,E}} - 50\overline{\mu}_1^{\mathrm{Q,E}} - 50\overline{\mu}_2^{\mathrm{Q,E}}$$

subject to

$$30 - \lambda^{\mathrm{E}} + \overline{\mu}_1^{\mathrm{G,E}} - \underline{\mu}_1^{\mathrm{G,E}} = 0,$$

$$50 - \lambda^{\mathrm{E}} + \overline{\mu}_2^{\mathrm{G,E}} - \underline{\mu}_2^{\mathrm{G,E}} = 0,$$

$$-50 + \lambda^{\mathrm{E}} + \overline{\mu}_1^{\mathrm{Q,E}} - \underline{\mu}_1^{\mathrm{Q,E}} = 0,$$

$$-70 + \lambda^{\mathrm{E}} + \overline{\mu}_2^{\mathrm{Q,E}} - \underline{\mu}_2^{\mathrm{Q,E}} = 0,$$

$$40 - \lambda^{\mathrm{E}} + \overline{\mu}^{\mathrm{E+}} - \underline{\mu}^{\mathrm{E+}} = 0,$$

$$\overline{\mu}_1^{\mathrm{G,E}}, \overline{\mu}_2^{\mathrm{G,E}}, \underline{\mu}_1^{\mathrm{G,E}}, \underline{\mu}_2^{\mathrm{G,E}} \geq 0,$$

$$\overline{\mu}_1^{\mathrm{Q,E}}, \overline{\mu}_2^{\mathrm{Q,E}}, \underline{\mu}_1^{\mathrm{Q,E}}, \underline{\mu}_2^{\mathrm{Q,E}} \geq 0,$$

$$\overline{\mu}^{\mathrm{E+}}, \underline{\mu}^{\mathrm{E+}} \geq 0,$$

where $\Upsilon^{\mathrm{E}} = \left\{\lambda^{\mathrm{E6.6}}, \overline{\mu}_1^{\mathrm{G,E}}, \overline{\mu}_2^{\mathrm{G,E}}, \underline{\mu}_1^{\mathrm{G,E}}, \underline{\mu}_2^{\mathrm{G,E}}, \overline{\mu}_1^{\mathrm{Q,E}}, \overline{\mu}_2^{\mathrm{Q,E}}, \underline{\mu}_1^{\mathrm{Q,E}}, \underline{\mu}_2^{\mathrm{Q,E}}, \overline{\mu}^{\mathrm{E+}}, \underline{\mu}^{\mathrm{E+}}\right\}$.

The above dual problem is an LP problem that can be solved using CPLEX [5] under GAMS [4]. The optimal values of the dual variables are provided in Table 6.8, which are equal to the values obtained through the KKT conditions in Illustrative Example 6.5. Additionally, the optimal value of the objective function is −\$2300, which is equal to the objective function of the energy market-clearing problem solved in Illustrative Example 6.2. As previously explained, the strong duality holds for the market-clearing problem.

□

An input GAMS code for solving Illustrative Example 6.6 is provided in Sect. 6.7.

6.3.3 MPEC Model

The scheduling problem of a price-maker VPP is formulated using bi-level problem (6.1). To solve this problem, it is necessary to reformulate it as a single-level problem. To do so, the lower-level problem, i.e., the market-clearing problem (6.2), is replaced by its KKT conditions rendering the following single-level problem:

$$\max_{\Phi, \Phi^{\mathrm{E}}, \Upsilon^{\mathrm{E}}} \quad \sum_{t \in \Omega^{\mathrm{T}}} \Big[\lambda_t^{\mathrm{E}}(p_t^{\mathrm{E+}}\Delta t - p_t^{\mathrm{E-}}\Delta t)$$
$$+ \sum_{d \in \Omega^{\mathrm{D}}} U_{dt}^{\mathrm{D}} p_{dt}^{\mathrm{D}} \Delta t - \sum_{c \in \Omega^{\mathrm{C}}} (C_c^{\mathrm{C,F}} u_{ct}^{\mathrm{C}} + C_c^{\mathrm{C,V}} p_{ct}^{\mathrm{C}} \Delta t)\Big] \qquad (6.7a)$$

Table 6.8 Illustrative Example 6.6: optimal values of the dual variables

Dual variable	λ^{E}	$\underline{\mu}_1^{\mathrm{G,E}}$	$\underline{\mu}_2^{\mathrm{G,E}}$	$\overline{\mu}_1^{\mathrm{G,E}}$	$\overline{\mu}_2^{\mathrm{G,E}}$	$\underline{\mu}_1^{\mathrm{Q,E}}$	$\underline{\mu}_2^{\mathrm{Q,E}}$	$\overline{\mu}_1^{\mathrm{Q,E}}$	$\overline{\mu}_2^{\mathrm{Q,E}}$	$\underline{\mu}^{\mathrm{E+}}$	$\overline{\mu}^{\mathrm{E+}}$
Value [\$/MWh]	50	0	0	20	0	0	0	0	20	0	10

subject to

$$\text{Constraints (6.1b)–(6.1p)}, \tag{6.7b}$$

$$\text{Constraints (6.3)}, \quad \forall t \in \Omega^{\mathrm{T}}. \tag{6.7c}$$

This problem is generally known as an MPEC.

Illustrative Example 6.7 Deterministic MPEC for the Scheduling in the DA Energy Market

The data of Illustrative Examples 6.1 and 6.2 are considered again, including the data of the VPP, and the offering and bidding decisions of market participants. Next, the MPEC for the scheduling problem of this price-maker VPP in the DA energy market is formulated:

$$\max_{\Phi,\Phi^{\mathrm{E}},\Upsilon^{\mathrm{E}}} \quad \left[\lambda^{\mathrm{E}}(p^{\mathrm{E+}} - p^{\mathrm{E-}}) + (60p^{\mathrm{D}} - 45p^{\mathrm{C}})\right]$$

subject to

$$p^{\mathrm{C}} + p^{\mathrm{R}} - p^{\mathrm{D}} = p^{\mathrm{E+}} - p^{\mathrm{E-}},$$

$$20 \le p^{\mathrm{D}},$$

$$0 \le p^{\mathrm{D}} \le 30,$$

$$0 \le p^{\mathrm{C}} \le 40u^{\mathrm{C}},$$

$$0 \le p^{\mathrm{R}} \le 8,$$

$$\overline{p}^{\mathrm{E+}} \le 50,$$

$$\overline{p}^{\mathrm{E-}} \le 30,$$

$$\overline{p}^{\mathrm{E+}}, \overline{p}^{\mathrm{E-}} \ge 0,$$

$$\nu^{\mathrm{E+}}, \nu^{\mathrm{E-}} \ge 0.$$

$$p_1^{\mathrm{Q,E}} + p_2^{\mathrm{Q,E}} + p^{\mathrm{E-}} - p^{\mathrm{E+}} - p_1^{\mathrm{G,E}} - p_2^{\mathrm{G,E}} = 0,$$

$$30 - \lambda^{\mathrm{E}} + \overline{\mu}_1^{\mathrm{G,E}} - \underline{\mu}_1^{\mathrm{G,E}} = 0,$$

$$50 - \lambda^{\mathrm{E}} + \overline{\mu}_2^{\mathrm{G,E}} - \underline{\mu}_2^{\mathrm{G,E}} = 0,$$

$$-50 + \lambda^{\mathrm{E}} + \overline{\mu}_1^{\mathrm{Q,E}} - \underline{\mu}_1^{\mathrm{Q,E}} = 0,$$

$$-70 + \lambda^{\mathrm{E}} + \overline{\mu}_2^{\mathrm{Q,E}} - \underline{\mu}_2^{\mathrm{Q,E}} = 0,$$

$$\nu^{\mathrm{E+}} - \lambda^{\mathrm{E}} + \overline{\mu}^{\mathrm{E+}} - \underline{\mu}^{\mathrm{E+}} = 0,$$

$$-\nu^{\mathrm{E-}} + \lambda^{\mathrm{E}} + \overline{\mu}^{\mathrm{E-}} - \underline{\mu}^{\mathrm{E-}} = 0,$$

$$0 \le \underline{\mu}_1^{\mathrm{G,E}} \perp p_1^{\mathrm{G,E}} \ge 0,$$

$$0 \le \underline{\mu}_2^{\mathrm{G,E}} \perp p_2^{\mathrm{G,E}} \ge 0,$$

$$0 \le \overline{\mu}_1^{\mathrm{G,E}} \perp 50 - p_1^{\mathrm{G,E}} \ge 0,$$

$$0 \le \overline{\mu}_2^{\mathrm{G,E}} \perp 50 - p_2^{\mathrm{G,E}} \ge 0,$$

$$0 \le \underline{\mu}_1^{\mathrm{Q,E}} \perp p_1^{\mathrm{Q,E}} \ge 0,$$

$$0 \le \underline{\mu}_2^{\mathrm{Q,E}} \perp p_2^{\mathrm{Q,E}} \ge 0,$$

$$0 \le \overline{\mu}_1^{\rm Q,E} \perp 50 - p_1^{\rm Q,E} \ge 0,$$
$$0 \le \overline{\mu}_2^{\rm Q,E} \perp 50 - p_2^{\rm Q,E} \ge 0,$$
$$0 \le \underline{\mu}^{\rm E+} \perp p^{\rm E+} \ge 0,$$
$$0 \le \overline{\mu}^{\rm E+} \perp \overline{p}^{\rm E+} - p^{\rm E+} \ge 0,$$
$$0 \le \underline{\mu}^{\rm E-} \perp p^{\rm E-} \ge 0,$$
$$0 \le \overline{\mu}^{\rm E-} \perp \overline{p}^{\rm E-} - p^{\rm E-} \ge 0.$$

where $\Phi = \left\{p^{\rm D}, p^{\rm C}, u^{\rm C}, p^{\rm R}, \overline{p}^{\rm E+}, \overline{p}^{\rm E-}, \nu^{\rm E+}, \nu^{\rm E-}\right\}$, $\Phi^{\rm E} = \left\{p_1^{\rm G,E}, p_2^{\rm G,E}, p_1^{\rm Q,E}, p_2^{\rm Q,E}, p^{\rm E+}, p^{\rm E-}\right\}$, and $\Upsilon^{\rm E} = \left\{\lambda^{\rm E}, \underline{\mu}_1^{\rm G,E}, \underline{\mu}_2^{\rm G,E}, \overline{\mu}_1^{\rm G,E}, \overline{\mu}_2^{\rm G,E}, \underline{\mu}_1^{\rm Q,E}, \underline{\mu}_2^{\rm Q,E}, \overline{\mu}_1^{\rm Q,E}, \overline{\mu}_2^{\rm Q,E}, \underline{\mu}^{\rm E+}, \overline{\mu}^{\rm E+}, \underline{\mu}^{\rm E-}, \overline{\mu}^{\rm E-}\right\}$.

Note that the objective function of the above MPEC is the same than that considered in Illustrative Example 6.1, while constraints include the constraints modeling the feasible operating region of the VPP (as in Illustrative Example 6.1) and the KKT conditions of the market-clearing problem (as in Illustrative Example 6.4).

□

6.3.4 MILP Model

MPEC (6.7) is a single-level nonlinear problem since it includes the following nonlinear terms:

1. Terms $\lambda_t^{\rm E}(p_t^{\rm E+}\Delta t - p_t^{\rm E-}\Delta t)$, $\forall t \in \Omega^{\rm T}$, in the objective function (6.7a).
2. Complementarity constraints (6.3f)–(6.3m).

However, it is possible to replace these nonlinear terms by exact equivalent mixed-integer linear expressions as explained below.

On one hand, in order to achieve an equivalent linear formulation for the nonlinear terms $\lambda_t^{\rm E}(p_t^{\rm E+} - p_t^{\rm E-})$, $\forall t \in \Omega^{\rm T}$, first, the KKT conditions (6.3d) and (6.3e) are multiplied by $p_t^{\rm E+}$ and $p_t^{\rm E-}$, respectively, and then summed, obtaining:

$$\begin{aligned}&\nu_t^{\rm E+}p_t^{\rm E+} - \lambda_t^{\rm E}p_t^{\rm E+} + \overline{\mu}_t^{\rm E+}p_t^{\rm E+} - \underline{\mu}_t^{\rm E+}p_t^{\rm E+}\\&\quad - \nu_t^{\rm E-}p_t^{\rm E-} + \lambda_t^{\rm E}p_t^{\rm E-} + \overline{\mu}_t^{\rm E-}p_t^{\rm E-} - \underline{\mu}_t^{\rm E-}p_t^{\rm E-} = 0, \quad \forall t \in \Omega^{\rm T}.\end{aligned} \tag{6.8}$$

Note that the two terms $\underline{\mu}_t^{\rm E+}p_t^{\rm E+}$ and $\underline{\mu}_t^{\rm E-}p_t^{\rm E-}$ are equal to zero based on the complementarity constraints (6.3j) and (6.3l), respectively. Additionally, the two terms $\overline{\mu}_t^{\rm E+}p_t^{\rm E+}$ and $\overline{\mu}_t^{\rm E-}p_t^{\rm E-}$ can be replaced by $\overline{\mu}_t^{\rm E+}\overline{p}_t^{\rm E+}$ and $\overline{\mu}_t^{\rm E-}\overline{p}_t^{\rm E-}$, respectively, based on the complementarity constraints (6.3k) and (6.3m). Thus, Eqs. (6.8) can be rewritten as:

$$\begin{aligned}&\lambda_t^{\rm E}p_t^{\rm E+} - \lambda_t^{\rm E}p_t^{\rm E-}\\&\quad = \nu_t^{\rm E+}p_t^{\rm E+} + \overline{\mu}_t^{\rm E+}\overline{p}_t^{\rm E+} - \nu_t^{\rm E-}p_t^{\rm E-} + \overline{\mu}_t^{\rm E-}\overline{p}_t^{\rm E-}, \quad \forall t \in \Omega^{\rm T}.\end{aligned} \tag{6.9}$$

Note that the left-hand side of Eqs. (6.9) is the nonlinear term $\lambda_t^{\rm E}(p_t^{\rm E+} - p_t^{\rm E-})$ in the objective function (6.7a).

Next, the strong duality equality is imposed, which states that, at the optimum, the values of the objective function of the primal (6.2) and dual problems (6.6) related to the energy market-clearing problem are the same, i.e.:

$$\begin{aligned}&\nu_t^{\rm E+}p_t^{\rm E+} + \sum_{g\in\Omega^{\rm G}}\chi_{gt}^{\rm G,E}p_{gt}^{\rm G,E} - \nu_t^{\rm E-}p_t^{\rm E-} - \sum_{q\in\Omega^{\rm Q}}\chi_{qt}^{\rm Q,E}p_{qt}^{\rm Q,E}\\&\quad = -\overline{p}_t^{\rm E+}\overline{\mu}_t^{\rm E+} - \sum_{g\in\Omega^{\rm G}}\overline{P}_{gt}^{\rm G,E}\overline{\mu}_{gt}^{\rm G,E} - \overline{p}_t^{\rm E-}\overline{\mu}_t^{\rm E-} - \sum_{q\in\Omega^{\rm Q}}\overline{P}_{qt}^{\rm Q,E}\overline{\mu}_{qt}^{\rm Q,E}, \quad \forall t \in \Omega^{\rm T}.\end{aligned} \tag{6.10}$$

Using the strong duality (6.10), the right-hand side of Eqs. (6.9) can be rewritten as follows:

$$\lambda_t^{\mathrm{E}}(p_t^{\mathrm{E}+} - p_t^{\mathrm{E}-}) = \sum_{q\in\Omega^{\mathrm{Q}}} \chi_{qt}^{\mathrm{Q,E}} p_{qt}^{\mathrm{Q,E}} - \sum_{g\in\Omega^{\mathrm{G}}} \chi_{gt}^{\mathrm{G,E}} p_{gt}^{\mathrm{G,E}}$$

$$- \sum_{g\in\Omega^{\mathrm{G}}} \overline{P}_{gt}^{\mathrm{G,E}} \overline{\mu}_{gt}^{\mathrm{G,E}} - \sum_{q\in\Omega^{\mathrm{Q}}} \overline{P}_{qt}^{\mathrm{Q,E}} \overline{\mu}_{qt}^{\mathrm{Q,E}}, \quad \forall t \in \Omega^{\mathrm{T}}. \tag{6.11}$$

Note that the right-hand side of Eqs. (6.11) is linear. Therefore, nonlinear terms $\lambda_t^{\mathrm{E}}(p_t^{\mathrm{E}+} - p_t^{\mathrm{E}-})$, $\forall t \in \Omega^{\mathrm{T}}$, in the objective function (6.7a) can be replaced by the right-hand side of Eqs. (6.11).

On the other hand, nonlinear complementarity constraints can be replaced by equivalent mixed-integer linear expressions as explained in Sect. 6.3.2.2.

As a result, nonlinear MPEC (6.7) can be finally recast as the following mixed-integer linear programming (MILP) model:

$$\max_{\Phi,\Phi^{\mathrm{E}},\Upsilon^{\mathrm{E}},\Phi^{\mathrm{M,E}}} \sum_{t\in\Omega^{\mathrm{T}}} \Big[\sum_{q\in\Omega^{\mathrm{Q}}} \chi_{qt}^{\mathrm{Q,E}} p_{qt}^{\mathrm{Q,E}} \Delta t - \sum_{g\in\Omega^{\mathrm{G}}} \chi_{gt}^{\mathrm{G,E}} p_{gt}^{\mathrm{G,E}} \Delta t$$

$$- \sum_{g\in\Omega^{\mathrm{G}}} \overline{P}_{gt}^{\mathrm{G,E}} \overline{\mu}_{gt}^{\mathrm{G,E}} \Delta t - \sum_{q\in\Omega^{\mathrm{Q}}} \overline{P}_{qt}^{\mathrm{Q,E}} \overline{\mu}_{qt}^{\mathrm{Q,E}} \Delta t$$

$$+ \sum_{d\in\Omega^{\mathrm{D}}} U_{dt}^{\mathrm{D}} p_{dt}^{\mathrm{D}} \Delta t - \sum_{c\in\Omega^{\mathrm{C}}} (C_c^{\mathrm{C,F}} u_{ct}^{\mathrm{C}} + C_c^{\mathrm{C,V}} p_{ct}^{\mathrm{C}} \Delta t) \Big] \tag{6.12a}$$

subject to

$$\text{Constraints (6.1b)–(6.1p)}, \tag{6.12b}$$

$$\text{Constraints (6.3a)–(6.3e)}, \quad \forall t \in \Omega^{\mathrm{T}}, \tag{6.12c}$$

$$\text{Constraints (6.5)}, \quad \forall t \in \Omega^{\mathrm{T}}, \tag{6.12d}$$

where $\Phi^{\mathrm{M,E}} = \left\{ \underline{u}_{gt}^{\mathrm{G,E}}, \overline{u}_{gt}^{\mathrm{G,E}}, \forall g \in \Omega^{\mathrm{G}}, \forall t \in \Omega^{\mathrm{T}}; \underline{u}_{qt}^{\mathrm{Q,E}}, \overline{u}_{qt}^{\mathrm{Q,E}}, \forall q \in \Omega^{\mathrm{Q}}, \forall t \in \Omega^{\mathrm{T}}; \underline{u}_t^{\mathrm{E}+}, \overline{u}_t^{\mathrm{E}+}, \underline{u}_t^{\mathrm{E}-}, \overline{u}_t^{\mathrm{E}-}, \forall t \in \Omega^{\mathrm{T}} \right\}$.

Illustrative Example 6.8 Deterministic MILP Model for the Scheduling in the DA Energy Market

The nonlinear MPEC formulated in Illustrative Example 6.7 is recast as the following MILP model:

$$\max_{\Phi,\Phi^{\mathrm{E}},\Upsilon^{\mathrm{E}},\Phi^{\mathrm{M,E}}} \Big[50 p_1^{\mathrm{Q,E}} \Delta t + 70 p_2^{\mathrm{Q,E}} \Delta t - 30 p_1^{\mathrm{G,E}} \Delta t - 50 p_2^{\mathrm{G,E}} \Delta t$$

$$- 50 \overline{\mu}_1^{\mathrm{G,E}} \Delta t - 50 \overline{\mu}_2^{\mathrm{G,E}} \Delta t - 50 \overline{\mu}_1^{\mathrm{Q,E}} \Delta t - 50 \overline{\mu}_2^{\mathrm{Q,E}} \Delta t \Big]$$

$$+ \Big[(60 p^{\mathrm{D}} - 45 p^{\mathrm{C}}) \Big]$$

subject to

$$p^{\mathrm{C}} + p^{\mathrm{R}} - p^{\mathrm{D}} = p^{\mathrm{E}+} - p^{\mathrm{E}-},$$

$$20 \le p^{\mathrm{D}},$$

$$0 \le p^{\mathrm{D}} \le 30,$$

$$0 \le p^{\mathrm{C}} \le 40 u^{\mathrm{C}},$$

$$0 \le p^{\mathrm{R}} \le 8,$$

$$\overline{p}^{\mathrm{E}+} \le 50,$$

$$\overline{p}^{\mathrm{E}-} \leq 30,$$

$$\overline{p}^{\mathrm{E}+}, \overline{p}^{\mathrm{E}-} \geq 0,$$

$$\nu^{\mathrm{E}+}, \nu^{\mathrm{E}-} \geq 0,$$

$$p_1^{\mathrm{Q,E}} + p_2^{\mathrm{Q,E}} + p^{\mathrm{E}-} - p^{\mathrm{E}+} - p_1^{\mathrm{G,E}} - p_2^{\mathrm{G,E}} = 0,$$

$$30 - \lambda^{\mathrm{E}} + \overline{\mu}_1^{\mathrm{G,E}} - \underline{\mu}_1^{\mathrm{G,E}} = 0,$$

$$50 - \lambda^{\mathrm{E}} + \overline{\mu}_2^{\mathrm{G,E}} - \underline{\mu}_2^{\mathrm{G,E}} = 0,$$

$$-50 + \lambda^{\mathrm{E}} + \overline{\mu}_1^{\mathrm{Q,E}} - \underline{\mu}_1^{\mathrm{Q,E}} = 0,$$

$$-70 + \lambda^{\mathrm{E}} + \overline{\mu}_2^{\mathrm{Q,E}} - \underline{\mu}_2^{\mathrm{Q,E}} = 0,$$

$$\nu^{\mathrm{E}+} - \lambda^{\mathrm{E}} + \overline{\mu}^{\mathrm{E}+} - \underline{\mu}^{\mathrm{E}+} = 0,$$

$$-\nu^{\mathrm{E}-} + \lambda^{\mathrm{E}} + \overline{\mu}^{\mathrm{E}-} - \underline{\mu}^{\mathrm{E}-} = 0,$$

$$0 \leq \underline{\mu}_1^{\mathrm{G,E}} \leq \underline{N}_1^{\mathrm{G,E}} \underline{u}_1^{\mathrm{G,E}},$$

$$0 \leq p_1^{\mathrm{G,E}} \leq \underline{M}_1^{\mathrm{G,E}} (1 - \underline{u}_1^{\mathrm{G,E}}),$$

$$0 \leq \underline{\mu}_2^{\mathrm{G,E}} \leq \underline{N}_2^{\mathrm{G,E}} \underline{u}_2^{\mathrm{G,E}},$$

$$0 \leq p_2^{\mathrm{G,E}} \leq \underline{M}_2^{\mathrm{G,E}} (1 - \underline{u}_2^{\mathrm{G,E}}),$$

$$0 \leq \overline{\mu}_1^{\mathrm{G,E}} \leq \overline{N}_1^{\mathrm{G,E}} \overline{u}_1^{\mathrm{G,E}},$$

$$0 \leq 50 - p_1^{\mathrm{G,E}} \leq \overline{M}_1^{\mathrm{G,E}} (1 - \overline{u}_1^{\mathrm{G,E}}),$$

$$0 \leq \overline{\mu}_2^{\mathrm{G,E}} \leq \overline{N}_2^{\mathrm{G,E}} \overline{u}_2^{\mathrm{G,E}},$$

$$0 \leq 50 - p_2^{\mathrm{G,E}} \leq \overline{M}_2^{\mathrm{G,E}} (1 - \overline{u}_2^{\mathrm{G,E}}),$$

$$0 \leq \underline{\mu}_1^{\mathrm{Q,E}} \leq \underline{N}_1^{\mathrm{Q,E}} \underline{u}_1^{\mathrm{Q,E}},$$

$$0 \leq p_1^{\mathrm{Q,E}} \leq \underline{M}_1^{\mathrm{Q,E}} (1 - \underline{u}_1^{\mathrm{Q,E}}),$$

$$0 \leq \underline{\mu}_2^{\mathrm{Q,E}} \leq \underline{N}_2^{\mathrm{Q,E}} \underline{u}_2^{\mathrm{Q,E}},$$

$$0 \leq p_2^{\mathrm{Q,E}} \leq \underline{M}_2^{\mathrm{Q,E}} (1 - \underline{u}_2^{\mathrm{Q,E}}),$$

$$0 \leq \overline{\mu}_1^{\mathrm{Q,E}} \leq \overline{N}_1^{\mathrm{Q,E}} \overline{u}_1^{\mathrm{Q,E}},$$

$$0 \leq 50 - p_1^{\mathrm{Q,E}} \leq \overline{M}_1^{\mathrm{Q,E}} (1 - \overline{u}_1^{\mathrm{Q,E}}),$$

$$0 \leq \overline{\mu}_2^{\mathrm{Q,E}} \leq \overline{N}_2^{\mathrm{Q,E}} \overline{u}_2^{\mathrm{Q,E}},$$

$$0 \leq 50 - p_2^{\mathrm{Q,E}} \leq \overline{M}_2^{\mathrm{Q,E}} (1 - \overline{u}_2^{\mathrm{Q,E}}),$$

$$0 \leq \underline{\mu}^{\mathrm{E}+} \leq \underline{N}^{\mathrm{E}+} \underline{u}^{\mathrm{E}+},$$

$$0 \leq p^{\mathrm{E}+} \leq \underline{M}^{\mathrm{E}+} (1 - \underline{u}^{\mathrm{E}+}),$$

$$0 \leq \overline{\mu}^{\mathrm{E}+} \leq \overline{N}^{\mathrm{E}+} \overline{u}^{\mathrm{E}+},$$

$$0 \leq \overline{p}^{\mathrm{E}+} - p^{\mathrm{E}+} \leq \overline{M}^{\mathrm{E}+} (1 - \overline{u}^{\mathrm{E}+}),$$

$$0 \leq \underline{\mu}^{\mathrm{E}-} \leq \underline{N}^{\mathrm{E}-} \underline{u}^{\mathrm{E}-},$$

Table 6.9 Illustrative Example 6.8: scheduling decisions by the VPP in the DA energy market

Variable	$\overline{p}^{\mathrm{E+}}$ [MW]	$\nu^{\mathrm{E+}}$ [\$/MWh]	$p^{\mathrm{E+}}$ [MW]	u^{C}
Value	18	50	18	1

Table 6.10 Illustrative Example 6.8: power dispatch of the units in the VPP

Variable	p^{C}	p^{D}	p^{R}
Value [MW]	40	30	8

Table 6.11 Illustrative Example 6.8: optimal schedules of market participants

Variable	$p_1^{\mathrm{G,E}}$	$p_2^{\mathrm{G,E}}$	$p_1^{\mathrm{Q,E}}$	$p_2^{\mathrm{Q,E}}$
Value [MW]	50	32	50	50

$$0 \le p^{\mathrm{E-}} \le \underline{M}^{\mathrm{E-}}(1 - \underline{u}^{\mathrm{E-}}),$$
$$0 \le \overline{\mu}^{\mathrm{E-}} \le \overline{N}^{\mathrm{E-}}\overline{u}^{\mathrm{E-}},$$
$$0 \le \overline{p}^{\mathrm{E-}} - p^{\mathrm{E-}} \le \overline{M}^{\mathrm{E-}}(1 - \overline{u}^{\mathrm{E-}}),$$
$$\underline{u}_1^{\mathrm{G,E}}, \underline{u}_2^{\mathrm{G,E}}, \overline{u}_1^{\mathrm{G,E}}, \overline{u}_2^{\mathrm{G,E}}, \underline{u}_1^{\mathrm{Q,E}}, \underline{u}_2^{\mathrm{Q,E}}, \overline{u}_1^{\mathrm{Q,E}}, \overline{u}_2^{\mathrm{Q,E}}, \underline{u}^{\mathrm{E+}}, \overline{u}^{\mathrm{E+}}, \underline{u}^{\mathrm{E-}}, \overline{u}^{\mathrm{E-}} \in \{0, 1\},$$

where $\Phi = \left\{p^{\mathrm{D}}, p^{\mathrm{C}}, u^{\mathrm{C}}, p^{\mathrm{R}}, \overline{p}^{\mathrm{E+}}, \overline{p}^{\mathrm{E-}}, \nu^{\mathrm{E+}}, \nu^{\mathrm{E-}}\right\}$, $\Phi^{\mathrm{E}} = \left\{p_1^{\mathrm{G,E}}, p_2^{\mathrm{G,E}}, p_1^{\mathrm{Q,E}}, p_2^{\mathrm{Q,E}}, p^{\mathrm{E+}}, p^{\mathrm{E-}}\right\}$, $\Upsilon^{\mathrm{E}} = \left\{\lambda^{\mathrm{E}}, \underline{\mu}_1^{\mathrm{G,E}}, \underline{\mu}_2^{\mathrm{G,E}}, \overline{\mu}_1^{\mathrm{G,E}}, \overline{\mu}_2^{\mathrm{G,E}}, \underline{\mu}_1^{\mathrm{Q,E}}, \underline{\mu}_2^{\mathrm{Q,E}}, \overline{\mu}_1^{\mathrm{Q,E}}, \overline{\mu}_2^{\mathrm{Q,E}}, \underline{\mu}^{\mathrm{E+}}, \overline{\mu}^{\mathrm{E+}}, \underline{\mu}^{\mathrm{E-}}, \overline{\mu}^{\mathrm{E-}}\right\}$, and $\Phi^{\mathrm{M,E}} = \left\{\underline{u}_1^{\mathrm{G,E}}, \underline{u}_2^{\mathrm{G,E}}, \overline{u}_1^{\mathrm{G,E}}, \overline{u}_2^{\mathrm{G,E}}, \underline{u}_1^{\mathrm{Q,E}}, \underline{u}_2^{\mathrm{Q,E}}, \overline{u}_1^{\mathrm{Q,E}}, \overline{u}_2^{\mathrm{Q,E}}, \underline{u}^{\mathrm{E+}}, \overline{u}^{\mathrm{E+}}, \underline{u}^{\mathrm{E-}}, \overline{u}^{\mathrm{E-}}\right\}$.

Large enough values of parameters N and M are considered, e.g., 10000 and 100, respectively. Then, the above MILP model is solved using CPLEX [5] under GAMS [4].

Table 6.9 provides the results. The second and third columns present the offer quantity and price, respectively. The fourth column identifies the power scheduled to be sold to the DA energy market. The last column shows the status of conventional power plant in the VPP. Accordingly, the VPP is willing to sell energy in the DA energy market.

The market-clearing price is equal to \$50/MWh. This means that the VPP receives \$900 (50×18) by selling energy in the DA energy market.

The dispatch decisions of the different components of the VPP are provided in Table 6.10. The second, third, and fourth columns provide the power production of conventional power plant, the power consumption of demand, and the power production of PV generating unit, respectively. Accordingly, the power consumption utility and production cost are equal to \$1800 and \$1800, respectively.

Finally, the optimal schedules of the remaining market participants are provided in Table 6.11. The second and third columns provide the power sold by the two producers, while the fourth and fifth columns present the power bought by the two demands.

□

An input GAMS code for solving Illustrative Example 6.8 is provided in Sect. 6.7.

6.4 Stochastic Programming Approach

This section describes the scheduling problem of a price-maker VPP in the DA energy market considering a stochastic programming approach.

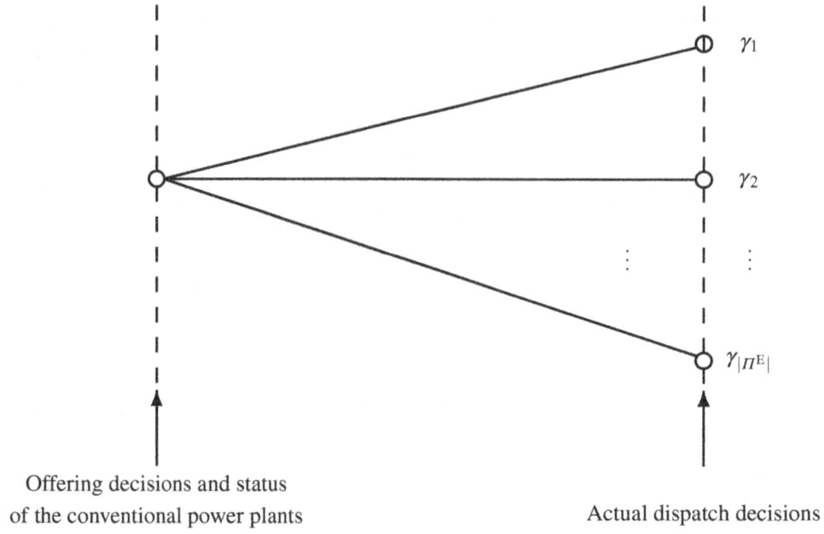

Fig. 6.2 Stochastic complementarity model for scheduling a price-maker VPP in the DA energy market: scenario tree

6.4.1 Uncertainty Model

Section 6.3 presents a detailed formulation of a deterministic complementarity model for scheduling a price-maker VPP in the DA energy market. This model is used to decide a pair of quantity-price, while modeling uncertain parameters using single-point forecasts during the scheduling horizon. This deterministic model is easy to implement; however, the uncertainty model is not accurate enough. In practice, the VPP faces the actual realization of uncertainties that is highly probable to be different from the forecast one. Under such a condition, the decision may be neither optimal nor feasible. To avoid these issues, the stochastic programming approach described below models uncertainties considering a predefined set of scenarios denoted by Π^{E} and indexed by γ.

The decision sequence of the problem can be summarized in the following steps:

1. The VPP decides its offering decisions in the DA energy market, as well as the status of its own conventional power plants to achieve maximum possible expected profit over all scenario realizations. These decisions are made one day in advance, i.e., before knowing the actual scenario realization. Therefore, these decisions are *here-and-now* since they do not depend on the scenario realization.
2. Once a scenario γ is realized, the VPP is informed about the actual realization of uncertain parameters, e.g., the actual available production levels of the stochastic renewable generating units in the VPP, the actual offering and bidding decisions of other market participants, and the actual market-clearing outputs.
3. The VPP dispatches its units. These decisions are *wait-and-see* decisions since they depend on the actual scenario realization.

Figure 6.2 provides the scenario tree associated with the above decision sequence problem.

For the sake of clarity, in the following it is assumed that uncertainty only affects the available production levels of stochastic renewable generating units. However, note that considering additional uncertain parameters such as the offering and bidding decisions of other market participants is straightforward.

6.4.2 Formulation

This section provides the formulation of a stochastic programming model for the scheduling of a price-maker VPP in the DA energy market.

6.4.2.1 Bi-Level Model

Considering the uncertainty characterization described in Sect. 6.4.1, the scheduling problem of a price-maker VPP in the DA energy market is formulated as the following bi-level model:

$$\max_{\Phi,\Phi^{\mathrm{E}},\Upsilon^{\mathrm{E}}} \sum_{t\in\Omega^{\mathrm{T}}} \lambda_t^{\mathrm{E}}(p_t^{\mathrm{E}+}\Delta t - p_t^{\mathrm{E}-}\Delta t) + \sum_{\gamma\in\Pi^{\mathrm{E}}} \pi_\gamma \sum_{t\in\Omega^{\mathrm{T}}} \Big[\sum_{d\in\Omega^{\mathrm{D}}} U_{dt}^{\mathrm{D}} p_{dt\gamma}^{\mathrm{D}}\Delta t - \sum_{c\in\Omega^{\mathrm{C}}} (C_c^{\mathrm{C,F}} u_{ct}^{\mathrm{C}} + C_c^{\mathrm{C,V}} p_{ct\gamma}^{\mathrm{C}}\Delta t)\Big] \quad (6.13a)$$

subject to

$$\sum_{c\in\Omega^{\mathrm{C}}} p_{ct\gamma}^{\mathrm{C}} + \sum_{r\in\Omega^{\mathrm{R}}} p_{rt\gamma}^{\mathrm{R}} + \sum_{s\in\Omega^{\mathrm{S}}} (p_{st\gamma}^{\mathrm{S,D}} - p_{st\gamma}^{\mathrm{S,C}}) - \sum_{d\in\Omega^{\mathrm{D}}} p_{dt\gamma}^{\mathrm{D}} = p_t^{\mathrm{E}+} - p_t^{\mathrm{E}-}, \quad \forall t\in\Omega^{\mathrm{T}}, \forall\gamma\in\Pi^{\mathrm{E}}, \quad (6.13b)$$

$$\sum_{t\in\Omega^{\mathrm{T}}} p_{dt\gamma}^{\mathrm{D}}\Delta t \ge \underline{E}_d^{\mathrm{D}}, \quad \forall d\in\Omega^{\mathrm{D}}, \forall\gamma\in\Pi^{\mathrm{E}}, \quad (6.13c)$$

$$\underline{P}_{dt}^{\mathrm{D}} \le p_{dt\gamma}^{\mathrm{D}} \le \overline{P}_{dt}^{\mathrm{D}}, \quad \forall d\in\Omega^{\mathrm{D}}, \forall t\in\Omega^{\mathrm{T}}, \forall\gamma\in\Pi^{\mathrm{E}}, \quad (6.13d)$$

$$\underline{P}_{ct}^{\mathrm{C}} u_{ct}^{\mathrm{C}} \le p_{ct\gamma}^{\mathrm{C}} \le \overline{P}_{ct}^{\mathrm{C}} u_{ct}^{\mathrm{C}}, \quad \forall c\in\Omega^{\mathrm{C}}, \forall t\in\Omega^{\mathrm{T}}, \forall\gamma\in\Pi^{\mathrm{E}}, \quad (6.13e)$$

$$e_{st\gamma}^{\mathrm{S}} = e_{s(t-1)\gamma}^{\mathrm{S}} + \eta_s^{\mathrm{S,C}} p_{st\gamma}^{\mathrm{S,C}}\Delta t - \frac{p_{st\gamma}^{\mathrm{S,D}}}{\eta_s^{\mathrm{S,D}}}\Delta t, \quad \forall s\in\Omega^{\mathrm{S}}, \forall t\in\Omega^{\mathrm{T}}, \forall\gamma\in\Pi^{\mathrm{E}}, \quad (6.13f)$$

$$\underline{E}_s^{\mathrm{S}} \le e_{st\gamma}^{\mathrm{S}} \le \overline{E}_s^{\mathrm{S}}, \quad \forall s\in\Omega^{\mathrm{S}}, \forall t\in\Omega^{\mathrm{T}}, \forall\gamma\in\Pi^{\mathrm{E}}, \quad (6.13g)$$

$$0 \le p_{st\gamma}^{\mathrm{S,C}} \le \overline{P}_s^{\mathrm{S,C}} u_{st\gamma}^{\mathrm{S}}, \quad \forall s\in\Omega^{\mathrm{S}}, \forall t\in\Omega^{\mathrm{T}}, \forall\gamma\in\Pi^{\mathrm{E}}, \quad (6.13h)$$

$$0 \le p_{st\gamma}^{\mathrm{S,D}} \le \overline{P}_s^{\mathrm{S,D}}(1 - u_{st\gamma}^{\mathrm{S}}), \quad \forall s\in\Omega^{\mathrm{S}}, \forall t\in\Omega^{\mathrm{T}}, \forall\gamma\in\Pi^{\mathrm{E}}, \quad (6.13i)$$

$$0 \le p_{rt\gamma}^{\mathrm{R}} \le P_{rt\gamma}^{\mathrm{R,A}}, \quad \forall r\in\Omega^{\mathrm{R}}, \forall t\in\Omega^{\mathrm{T}}, \forall\gamma\in\Pi^{\mathrm{E}}, \quad (6.13j)$$

$$\overline{p}_t^{\mathrm{E}+} \le \overline{P}_t^{\mathrm{V}+}, \quad \forall t\in\Omega^{\mathrm{T}}, \quad (6.13k)$$

$$\overline{p}_t^{\mathrm{E}-} \le \overline{P}_t^{\mathrm{V}-}, \quad \forall t\in\Omega^{\mathrm{T}}, \quad (6.13l)$$

$$\overline{p}_t^{\mathrm{E}+}, \overline{p}_t^{\mathrm{E}-} \ge 0, \quad \forall t\in\Omega^{\mathrm{T}}, \quad (6.13m)$$

$$\nu_t^{\mathrm{E}+}, \nu_t^{\mathrm{E}-} \ge 0, \quad \forall t\in\Omega^{\mathrm{T}}, \quad (6.13n)$$

$$\lambda_t^{\mathrm{E}}, p_t^{\mathrm{E}+}, p_t^{\mathrm{E}-} \in \Psi_t^{\mathrm{E}}(\cdot) = \Big\{$$

$$\underset{\Phi^{\mathrm{E}}}{\text{Minimize}} \quad \nu_t^{\mathrm{E}+} p_t^{\mathrm{E}+} + \sum_{g\in\Omega^{\mathrm{G}}} \chi_{gt}^{\mathrm{G,E}} p_{gt}^{\mathrm{G,E}} - \nu_t^{\mathrm{E}-} p_t^{\mathrm{E}-} - \sum_{q\in\Omega^{\mathrm{Q}}} \chi_{qt}^{\mathrm{Q,E}} p_{qt}^{\mathrm{Q,E}} \quad (6.13o)$$

subject to

$$p_t^{\mathrm{E}+} + \sum_{g\in\Omega^{\mathrm{G}}} p_{gt}^{\mathrm{G,E}} = p_t^{\mathrm{E}-} + \sum_{q\in\Omega^{\mathrm{Q}}} p_{qt}^{\mathrm{Q,E}} \;:\; \lambda_t^{\mathrm{E}}, \quad (6.13p)$$

$$0 \le p_{gt}^{\mathrm{G,E}} \le \overline{P}_{gt}^{\mathrm{G,E}} \;:\; \underline{\mu}_{gt}^{\mathrm{G,E}}, \overline{\mu}_{gt}^{\mathrm{G,E}}, \quad \forall g\in\Omega^{\mathrm{G}}, \quad (6.13q)$$

$$0 \le p_{qt}^{\mathrm{Q,E}} \le \overline{P}_{qt}^{\mathrm{Q,E}} \;:\; \underline{\mu}_{qt}^{\mathrm{Q,E}}, \overline{\mu}_{qt}^{\mathrm{Q,E}}, \quad \forall q\in\Omega^{\mathrm{Q}}, \quad (6.13r)$$

$$0 \le p_t^{\mathrm{E+}} \le \overline{p}_t^{\mathrm{E+}} \ : \ \underline{\mu}_t^{\mathrm{E+}}, \overline{\mu}_t^{\mathrm{E+}}, \tag{6.13s}$$

$$0 \le p_t^{\mathrm{E-}} \le \overline{p}_t^{\mathrm{E-}} \ : \ \underline{\mu}_t^{\mathrm{E-}}, \overline{\mu}_t^{\mathrm{E-}}, \tag{6.13t}$$

$\Big\}$, $\forall t \in \Omega^{\mathrm{T}}$, where variables in sets $\Phi = \Big\{p_{dt\gamma}^{\mathrm{D}}, \forall d \in \Omega^{\mathrm{D}}, \forall t \in \Omega^{\mathrm{T}}, \forall \gamma \in \Pi^{\mathrm{E}}; e_{st\gamma}^{\mathrm{S}}, p_{st\gamma}^{\mathrm{S,C}}, p_{st\gamma}^{\mathrm{S,D}}, u_{st\gamma}^{\mathrm{S}}, \forall s \in \Omega^{\mathrm{S}}, \forall t \in \Omega^{\mathrm{T}}, \forall \gamma \in \Pi^{\mathrm{E}}; p_{ct\gamma}^{\mathrm{C}}, \forall c \in \Omega^{\mathrm{C}}, \forall t \in \Omega^{\mathrm{T}}, \forall \gamma \in \Pi^{\mathrm{E}}; u_{ct}^{\mathrm{C}}, \forall c \in \Omega^{\mathrm{C}}, \forall t \in \Omega^{\mathrm{T}}; p_{rt\gamma}^{\mathrm{R}}, \forall r \in \Omega^{\mathrm{R}}, \forall t \in \Omega^{\mathrm{T}}, \forall \gamma \in \Pi^{\mathrm{E}}; \overline{p}_t^{\mathrm{E+}}, \overline{p}_t^{\mathrm{E-}}, v_t^{\mathrm{E+}}, v_t^{\mathrm{E-}}, \forall t \in \Omega^{\mathrm{T}}\Big\}$, $\Phi^{\mathrm{E}} = \Big\{p_{gt}^{\mathrm{G,E}}, \forall g \in \Omega^{\mathrm{G}}, \forall t \in \Omega^{\mathrm{T}}; p_{qt}^{\mathrm{Q,E}}, \forall q \in \Omega^{\mathrm{Q}}, \forall t \in \Omega^{\mathrm{T}}; p_t^{\mathrm{E+}}, p_t^{\mathrm{E-}}, \forall t \in \Omega^{\mathrm{T}}\Big\}$, and $\Upsilon^{\mathrm{E}} = \Big\{\lambda_t^{\mathrm{E}}, \forall t \in \Omega^{\mathrm{T}}; \underline{\mu}_{gt}^{\mathrm{G,E}}, \overline{\mu}_{gt}^{\mathrm{G,E}}, g \in \Omega^{\mathrm{G}}, \forall t \in \Omega^{\mathrm{T}}; \underline{\mu}_{qt}^{\mathrm{Q,E}}, \overline{\mu}_{qt}^{\mathrm{Q,E}}, q \in \Omega^{\mathrm{Q}}, \forall t \in \Omega^{\mathrm{T}}; \underline{\mu}_t^{\mathrm{E+}}, \overline{\mu}_t^{\mathrm{E+}}, \underline{\mu}_t^{\mathrm{E-}}, \overline{\mu}_t^{\mathrm{E-}}, \forall t \in \Omega^{\mathrm{T}}\Big\}$ are the optimization variables of problem (6.13).

Bi-level model (6.13) comprises the upper-level problem (6.13a)–(6.13n), representing the VPP operating and offering decisions, and the lower-level problem (6.13o)–(6.13t), representing the clearing of the DA energy market.

The VPP decides its offering decisions in the DA energy market that maximize its expected profit represented by objective function (6.13a), which includes the following two terms:

1. The revenue/cost related to the participation (selling/purchasing) in the DA energy market, i.e., terms $\lambda_t^{\mathrm{E}}(p_t^{\mathrm{E+}}\Delta t - p_t^{\mathrm{E-}}\Delta t), \forall t \in \Omega^{\mathrm{T}}$.
2. The expected profit related to the utility of demands in the VPP minus the production cost of conventional power plants in the VPP, i.e., terms $\sum_{d \in \Omega^{\mathrm{D}}} U_{dt}^{\mathrm{D}} p_{dt\gamma}^{\mathrm{D}} \Delta t - \sum_{c \in \Omega^{\mathrm{C}}} (C_c^{\mathrm{C,F}} u_{ct}^{\mathrm{C}} + C_c^{\mathrm{C,V}} p_{ct\gamma}^{\mathrm{C}} \Delta t), \forall t \in \Omega^{\mathrm{T}}, \forall \gamma \in \Pi^{\mathrm{E}}$. Note that each of these terms is multiplied by the weight of the corresponding scenario.

Constraints (6.13b)–(6.13j) define the feasible operating region of the units in the VPP for all scenarios, while constraints (6.13k)–(6.13n) impose limits on the power quantities and prices submitted to the DA energy market. Regarding the market-clearing problem (6.13o)–(6.13t), it is equivalent to the problem formulated in Sect. 6.3.2.

6.4.2.2 MPEC

Lower-level problem (6.13o)–(6.13t) is continuous and linear on its decision variables, i.e., it is a convex optimization problem. Thus, the lower-level problem (6.13o)–(6.13t) can be replaced by its KKT conditions as equilibrium constraints, and the stochastic bi-level optimization problem (6.13) can be recast as the following stochastic MPEC:

$$\max_{\Phi, \Phi^{\mathrm{E}}, \Upsilon^{\mathrm{E}}} \quad \sum_{t \in \Omega^{\mathrm{T}}} \lambda_t^{\mathrm{E}}(p_t^{\mathrm{E+}}\Delta t - p_t^{\mathrm{E-}}\Delta t)$$
$$+ \sum_{\gamma \in \Pi^{\mathrm{E}}} \pi_\gamma \sum_{t \in \Omega^{\mathrm{T}}} \Big[\sum_{d \in \Omega^{\mathrm{D}}} U_{dt}^{\mathrm{D}} p_{dt\gamma}^{\mathrm{D}} \Delta t - \sum_{c \in \Omega^{\mathrm{C}}} (C_c^{\mathrm{C,F}} u_{ct}^{\mathrm{C}} + C_c^{\mathrm{C,V}} p_{ct\gamma}^{\mathrm{C}} \Delta t)\Big] \tag{6.14a}$$

subject to

$$\text{Constraints (6.13b)–(6.13n)}, \tag{6.14b}$$

$$\text{Constraints (6.3)}, \quad \forall t \in \Omega^{\mathrm{T}}. \tag{6.14c}$$

6.4.2.3 MILP Model

Note that the stochastic MPEC model (6.14) is nonlinear due to nonlinearities in the objective function (6.14a) and the complementarity constraints in (6.14c). The nonlinear MPEC model (6.14) can be transformed into the equivalent MILP model below:

$$\max_{\Phi, \Phi^{\mathrm{E}}, \Upsilon^{\mathrm{E}}, \Phi^{\mathrm{M,E}}} \quad \sum_{t \in \Omega^{\mathrm{T}}} \Big[\sum_{q \in \Omega^{\mathrm{Q}}} \chi_{qt}^{\mathrm{Q,E}} p_{qt}^{\mathrm{Q,E}} \Delta t - \sum_{g \in \Omega^{\mathrm{G}}} \chi_{gt}^{\mathrm{G,E}} p_{gt}^{\mathrm{G,E}} \Delta t$$

Table 6.12 Illustrative Example 6.9: data of available production by the PV generating unit

Scenario	1	2
Production [MW]	10	5
Probability	0.6	0.4

$$- \sum_{g \in \Omega^{\mathrm{G}}} \overline{P}_{gt}^{\mathrm{G,E}} \overline{\mu}_{gt}^{\mathrm{G,E}} \Delta t - \sum_{q \in \Omega^{\mathrm{Q}}} \overline{P}_{qt}^{\mathrm{Q,E}} \overline{\mu}_{qt}^{\mathrm{Q,E}} \Delta t \Big]$$

$$+ \sum_{\gamma \in \Pi^{\mathrm{E}}} \pi_\gamma \sum_{t \in \Omega^{\mathrm{T}}} \Big[\sum_{d \in \Omega^{\mathrm{D}}} U_{dt}^{\mathrm{D}} p_{dt\gamma}^{\mathrm{D}} \Delta t - \sum_{c \in \Omega^{\mathrm{C}}} (C_c^{\mathrm{C,F}} u_{ct}^{\mathrm{C}} + C_c^{\mathrm{C,V}} p_{ct\gamma}^{\mathrm{C}} \Delta t) \Big] \tag{6.15a}$$

subject to

$$\text{Constraints (6.13b)–(6.13n)}, \tag{6.15b}$$

$$\text{Constraints (6.3a)–(6.3e)}, \quad \forall t \in \Omega^{\mathrm{T}}, \tag{6.15c}$$

$$\text{Constraints (6.5)}, \quad \forall t \in \Omega^{\mathrm{T}}, \tag{6.15d}$$

where Φ, Φ^{E}, Υ^{E}, and $\Phi^{\mathrm{M,E}} = \Big\{ \underline{u}_{gt}^{\mathrm{G,E}}, \overline{u}_{gt}^{\mathrm{G,E}}, \forall g \in \Omega^{\mathrm{G}}, \forall t \in \Omega^{\mathrm{T}}; \underline{u}_{qt}^{\mathrm{Q,E}}, \overline{u}_{qt}^{\mathrm{Q,E}}, \forall q \in \Omega^{\mathrm{Q}}, \forall t \in \Omega^{\mathrm{T}}; \underline{u}_t^{\mathrm{E+}}, \overline{u}_t^{\mathrm{E+}}, \underline{u}_t^{\mathrm{E-}}, \overline{u}_t^{\mathrm{E-}}, \forall t \in \Omega^{\mathrm{T}} \Big\}$, are the optimization variables of the stochastic MILP problem (6.15).

Illustrative Example 6.9 Scheduling the Price-Maker VPP in the DA Energy Market: Stochastic MILP Model

The data of the VPP in Illustrative Examples 6.1 and the data of the DA market in Illustrative Example 6.2 are considered again. Next, this illustrative example takes into account the uncertainty in the available production level of the PV generating unit at the time of making the offering and bidding decisions by the price-maker VPP in the DA energy market. The available production level of the PV generating unit is represented using 2 scenarios as presented in Table 6.12.

Using the above data, the scheduling problem of the price-maker VPP in the DA energy market is formulated as the following stochastic MILP model:

$$\max_{\Phi, \Phi^{\mathrm{E}}, \Upsilon^{\mathrm{E}}} \Big[50 p_1^{\mathrm{Q,E}} \Delta t + 70 p_2^{\mathrm{Q,E}} \Delta t - 30 p_1^{\mathrm{G,E}} \Delta t - 50 p_2^{\mathrm{G,E}} \Delta t$$

$$- 50 \overline{\mu}_1^{\mathrm{G,E}} \Delta t - 50 \overline{\mu}_2^{\mathrm{G,E}} \Delta t - 50 \overline{\mu}_1^{\mathrm{Q,E}} \Delta t - 50 \overline{\mu}_2^{\mathrm{Q,E}} \Delta t \Big]$$

$$+ \Big[0.6 (60 p_1^{\mathrm{D}} - 45 p_1^{\mathrm{C}}) + 0.4 (60 p_2^{\mathrm{D}} - 45 p_2^{\mathrm{C}}) \Big]$$

subject to

$$p_1^{\mathrm{C}} + p_1^{\mathrm{R}} - p_1^{\mathrm{D}} = p^{\mathrm{E+}} - p^{\mathrm{E-}},$$

$$p_2^{\mathrm{C}} + p_2^{\mathrm{R}} - p_2^{\mathrm{D}} = p^{\mathrm{E+}} - p^{\mathrm{E-}},$$

$$20 \le p_1^{\mathrm{D}},$$

$$20 \le p_2^{\mathrm{D}},$$

$$0 \le p_1^{\mathrm{D}} \le 30,$$

$$0 \le p_2^{\mathrm{D}} \le 30,$$

$$0 \le p_1^{\mathrm{C}} \le 40 u^{\mathrm{C}},$$

$$0 \le p_2^{\mathrm{C}} \le 40 u^{\mathrm{C}},$$

$$0 \le p_1^{\mathrm{R}} \le 10,$$

$$0 \le p_2^{\mathrm{R}} \le 5,$$

$$
\begin{aligned}
&\overline{p}^{\mathrm{E}+} \leq 50,\\
&\overline{p}^{\mathrm{E}-} \leq 30,\\
&\overline{p}^{\mathrm{E}+}, \overline{p}^{\mathrm{E}-} \geq 0,\\
&\nu^{\mathrm{E}+}, \nu^{\mathrm{E}-} \geq 0,\\
&p_1^{\mathrm{Q,E}} + p_2^{\mathrm{Q,E}} + p^{\mathrm{E}-} - p^{\mathrm{E}+} - p_1^{\mathrm{G,E}} - p_2^{\mathrm{G,E}} = 0,\\
&30 - \lambda^{\mathrm{E}} + \overline{\mu}_1^{\mathrm{G,E}} - \underline{\mu}_1^{\mathrm{G,E}} = 0,\\
&50 - \lambda^{\mathrm{E}} + \overline{\mu}_2^{\mathrm{G,E}} - \underline{\mu}_2^{\mathrm{G,E}} = 0,\\
&-50 + \lambda^{\mathrm{E}} + \overline{\mu}_1^{\mathrm{Q,E}} - \underline{\mu}_1^{\mathrm{Q,E}} = 0,\\
&-70 + \lambda^{\mathrm{E}} + \overline{\mu}_2^{\mathrm{Q,E}} - \underline{\mu}_2^{\mathrm{Q,E}} = 0,\\
&\nu^{\mathrm{E}+} - \lambda^{\mathrm{E}} + \overline{\mu}^{\mathrm{E}+} - \underline{\mu}^{\mathrm{E}+} = 0,\\
&-\nu^{\mathrm{E}-} + \lambda^{\mathrm{E}} + \overline{\mu}^{\mathrm{E}-} - \underline{\mu}^{\mathrm{E}-} = 0,\\
&0 \leq \underline{\mu}_1^{\mathrm{G,E}} \leq \underline{N}_1^{\mathrm{G,E}} \underline{u}_1^{\mathrm{G,E}},\\
&0 \leq p_1^{\mathrm{G,E}} \leq \underline{M}_1^{\mathrm{G,E}} (1 - \underline{u}_1^{\mathrm{G,E}}),\\
&0 \leq \underline{\mu}_2^{\mathrm{G,E}} \leq \underline{N}_2^{\mathrm{G,E}} \underline{u}_2^{\mathrm{G,E}},\\
&0 \leq p_2^{\mathrm{G,E}} \leq \underline{M}_2^{\mathrm{G,E}} (1 - \underline{u}_2^{\mathrm{G,E}}),\\
&0 \leq \overline{\mu}_1^{\mathrm{G,E}} \leq \overline{N}_1^{\mathrm{G,E}} \overline{u}_1^{\mathrm{G,E}},\\
&0 \leq 50 - p_1^{\mathrm{G,E}} \leq \overline{M}_1^{\mathrm{G,E}} (1 - \overline{u}_1^{\mathrm{G,E}}),\\
&0 \leq \overline{\mu}_2^{\mathrm{G,E}} \leq \overline{N}_2^{\mathrm{G,E}} \overline{u}_2^{\mathrm{G,E}},\\
&0 \leq 50 - p_2^{\mathrm{G,E}} \leq \overline{M}_2^{\mathrm{G,E}} (1 - \overline{u}_2^{\mathrm{G,E}}),\\
&0 \leq \underline{\mu}_1^{\mathrm{Q,E}} \leq \underline{N}_1^{\mathrm{Q,E}} \underline{u}_1^{\mathrm{Q,E}},\\
&0 \leq p_1^{\mathrm{Q,E}} \leq \underline{M}_1^{\mathrm{Q,E}} (1 - \underline{u}_1^{\mathrm{Q,E}}),\\
&0 \leq \underline{\mu}_2^{\mathrm{Q,E}} \leq \underline{N}_2^{\mathrm{Q,E}} \underline{u}_2^{\mathrm{Q,E}},\\
&0 \leq p_2^{\mathrm{Q,E}} \leq \underline{M}_2^{\mathrm{Q,E}} (1 - \underline{u}_2^{\mathrm{Q,E}}),\\
&0 \leq \overline{\mu}_1^{\mathrm{Q,E}} \leq \overline{N}_1^{\mathrm{Q,E}} \overline{u}_1^{\mathrm{Q,E}},\\
&0 \leq 50 - p_1^{\mathrm{Q,E}} \leq \overline{M}_1^{\mathrm{Q,E}} (1 - \overline{u}_1^{\mathrm{Q,E}}),\\
&0 \leq \overline{\mu}_2^{\mathrm{Q,E}} \leq \overline{N}_2^{\mathrm{Q,E}} \overline{u}_2^{\mathrm{Q,E}},\\
&0 \leq 50 - p_2^{\mathrm{Q,E}} \leq \overline{M}_2^{\mathrm{Q,E}} (1 - \overline{u}_2^{\mathrm{Q,E}}),\\
&0 \leq \underline{\mu}^{\mathrm{E}+} \leq \underline{N}^{\mathrm{E}+} \underline{u}^{\mathrm{E}+},\\
&0 \leq p^{\mathrm{E}+} \leq \underline{M}^{\mathrm{E}+} (1 - \underline{u}^{\mathrm{E}+}),\\
&0 \leq \overline{\mu}^{\mathrm{E}+} \leq \overline{N}^{\mathrm{E}+} \overline{u}^{\mathrm{E}+},\\
&0 \leq \overline{p}^{\mathrm{E}+} - p^{\mathrm{E}+} \leq \overline{M}^{\mathrm{E}+} (1 - \overline{u}^{\mathrm{E}+}),
\end{aligned}
$$

Table 6.13 Illustrative Example 6.9: scheduling decisions of the VPP in the DA energy market

Variable	$\overline{p}^{\mathrm{E+}}$ [MW]	$\nu^{\mathrm{E+}}$ [\$/MWh]	$p^{\mathrm{E+}}$ [MW]	u^{C}
Value	15	50	15	1

Table 6.14 Illustrative Example 6.9: power dispatch of the units in the VPP

Variable	p^{C}	p^{D}	p^{R}
Value at scenario 1 [MW]	35	30	10
Value at scenario 2 [MW]	40	30	5

Table 6.15 Illustrative Example 6.9: optimal schedules of two producers and two demands in the DA energy market

Variable	$p_1^{\mathrm{G,E}}$	$p_2^{\mathrm{G,E}}$	$p_1^{\mathrm{Q,E}}$	$p_2^{\mathrm{Q,E}}$
Value [MW]	50	35	50	50

$$0 \leq \underline{\mu}^{\mathrm{E-}} \leq \underline{N}^{\mathrm{E-}} \underline{u}^{\mathrm{E-}},$$

$$0 \leq p^{\mathrm{E-}} \leq \underline{M}^{\mathrm{E-}}(1 - \underline{u}^{\mathrm{E-}}),$$

$$0 \leq \overline{\mu}^{\mathrm{E-}} \leq \overline{N}^{\mathrm{E-}} \overline{u}^{\mathrm{E-}},$$

$$0 \leq \overline{p}^{\mathrm{E-}} - p^{\mathrm{E-}} \leq \overline{M}^{\mathrm{E-}}(1 - \overline{u}^{\mathrm{E-}}),$$

where $\Phi = \left\{p_1^{\mathrm{D}}, p_2^{\mathrm{D}}, p_1^{\mathrm{C}}, p_2^{\mathrm{C}}, u^{\mathrm{C}}, p_1^{\mathrm{R}}, p_2^{\mathrm{R}}, \overline{p}^{\mathrm{E+}}, \overline{p}^{\mathrm{E-}}, \nu^{\mathrm{E+}}, \nu^{\mathrm{E-}}\right\}$, $\Phi^{\mathrm{E}} = \left\{p_1^{\mathrm{G,E}}, p_2^{\mathrm{G,E}}, p_1^{\mathrm{Q,E}}, p_2^{\mathrm{Q,E}}, p^{\mathrm{E+}}, p^{\mathrm{E-}}\right\}$, $\Upsilon^{\mathrm{E}} = \left\{\lambda^{\mathrm{E}}, \underline{\mu}_1^{\mathrm{G,E}}, \underline{\mu}_2^{\mathrm{G,E}}, \overline{\mu}_1^{\mathrm{G,E}}, \overline{\mu}_2^{\mathrm{G,E}}, \underline{\mu}_1^{\mathrm{Q,E}}, \underline{\mu}_2^{\mathrm{Q,E}}, \overline{\mu}_1^{\mathrm{Q,E}}, \overline{\mu}_2^{\mathrm{Q,E}}, \underline{\mu}^{\mathrm{E+}}, \overline{\mu}^{\mathrm{E+}}, \underline{\mu}^{\mathrm{E-}}, \overline{\mu}^{\mathrm{E-}}\right\}$, and $\Phi^{\mathrm{M,E}} = \left\{\underline{u}_1^{\mathrm{G,E}}, \underline{u}_2^{\mathrm{G,E}}, \overline{u}_1^{\mathrm{G,E}}, \overline{u}_2^{\mathrm{G,E}}, \underline{u}_1^{\mathrm{Q,E}}, \underline{u}_2^{\mathrm{Q,E}}, \overline{u}_1^{\mathrm{Q,E}}, \overline{u}_2^{\mathrm{Q,E}}, \underline{u}^{\mathrm{E+}}, \overline{u}^{\mathrm{E+}}, \underline{u}^{\mathrm{E-}}, \overline{u}^{\mathrm{E-}}\right\}$.

Parameters N and M are considered equal to 10000 and 100, respectively. Then, the above MILP model is solved using CPLEX [5] under GAMS [4]. Table 6.13 provides the offering and bidding decisions of the VPP, as well as the power traded in the DA energy market. The second and third columns present the offer quantity and price of the VPP, respectively. The fourth column determines the amount of power that is sold to the DA energy market by the VPP. The last column provides the status of the conventional power plant in the VPP. The results show that the VPP decides to sell energy in the DA energy market.

Note that the VPP is willing to sell 15 MW at \$50/MWh in the DA energy market. Considering the offering and bidding decisions of other market participants, the VPP can sell 15 MW in the DA energy market, being the resulting market-clearing price equal to \$50/MWh. Therefore, the VPP achieves \$750 by selling energy in the DA energy market.

The dispatch decisions by the VPP for each scenario realization are also presented in Table 6.14. The second, third, and fourth columns provide the power production of the conventional power plant, the power consumption of the demand, and the power production of the PV generating unit for each scenario realization, respectively. In scenario 1, the utility of power consumption and production cost are equal to \$1800 and \$1575, respectively. Thus, the profit obtained by dispatch decisions in scenario 1 is equal to \$225. However, the utility minus production cost is zero in scenario 2. This yields an expected profit of \$135 to be obtained from the dispatch decisions.

The objective function of the stochastic MPEC model is equal to \$885 as the summation of the profit in the DA energy market and the expected profit related to dispatch decisions. Note that only the offering and bidding decisions by the VPP and the status of conventional power plant are implemented when participating in the DA energy market, while the dispatch decisions depend on the actual uncertainty realizations, which may be different to the scenarios considered.

The optimal schedules of the two producers and the two demands in the DA energy market are provided in Table 6.15. The second and third columns provide the amount of power to be sold by the two producers in the DA energy market. The fourth and fifth columns present the amount of power to be bought by the two demands in the DA energy market.

□

An input GAMS code for solving Illustrative Example 6.9 is provided in Sect. 6.7.

Illustrative Example 6.10 Value of the Stochastic Solution

The data of Illustrative Example 6.9 are considered again. Next, the performance of the stochastic scheduling compared to the deterministic one is analyzed through the concept of the value of the stochastic solution (VSS). To calculate the VSS, a deterministic form of the problem in Illustrative Example 6.9 in which the uncertain parameter of the available stochastic

PV production level is replaced by its average value is considered. In other words, deterministic problem introduced in Illustrative Example 6.8 is solved. Then, the stochastic problem in Illustrative Example 6.9 is solved again but with the first-stage scheduling decisions fixed to their values obtained from solving the deterministic problem. Accordingly, the VSS can be calculated as follows:

$$\mathrm{VSS} = z^{\mathrm{S}*} - z^{\mathrm{D}*},$$

where $z^{\mathrm{S}*}$ is the optimal value of the objective function of the stochastic problem in Illustrative Example 6.9 and $z^{\mathrm{D}*}$ is the optimal value of the objective function of the modified stochastic problem, i.e., that fixing the first-stage scheduling decisions to their values obtained from solving the deterministic problem.

Solving the modified stochastic problem, the optimal value of the objective function is \$882. Considering the optimal value of the objective function in Illustrative Example 6.9, the VSS is calculated as:

$$\mathrm{VSS} = \$885 - \$882 = \$3.$$

□

Illustrative Example 6.11 Expected Value of Perfect Information

The data of Illustrative Example 6.9 are considered again. This example is intended for measuring the value of perfect information using the so-called expected value of perfect information (EVPI) metric. In this case, a modified stochastic problem in which all scheduling and dispatch decisions are allowed to be made after knowing the actual realization of scenarios is solved. Then, the EVPI can be computed as:

$$\mathrm{EVPI} = z^{\mathrm{P}*} - z^{\mathrm{S}*},$$

where $z^{\mathrm{P}*}$ is the optimal value of the objective function of the modified stochastic problem in which all decisions are scenario dependent and $z^{\mathrm{S}*}$ is the optimal value of the objective function of the stochastic problem in Illustrative Example 6.9.

The modified stochastic problem is solved and the optimal value of the objective function is \$900. Then, considering the optimal value of the objective function in Illustrative Example 6.9, the EVPI is calculated as:

$$\mathrm{EVPI} = \$900 - \$885 = \$15.$$

□

6.5 Participation of Virtual Power Plants in Energy and Reserve Markets

This section describes the scheduling problem of a price-maker VPP in the DA energy and reserve electricity markets using a stochastic programming approach.

6.5.1 Problem Description

Sections 6.3 and 6.4 formulate the scheduling problem of the price-maker VPP in the DA energy market using deterministic and stochastic programming approaches, respectively. The VPP submits the quantity-price decisions in the DA energy market for selling and buying energy. In addition to the energy market, the VPP can also take the advantage of using the flexibility of its energy assets to participate in the reserve market to increase its profit. Therefore, the VPP submits the quantity-price decisions in the reserve market for up and down reserves required for maintaining the system security.

The price-maker VPP can influence the prices in the DA energy and reserve markets through its offering and bidding decisions. To consider the potential impact of the price-maker VPP on the prices in the DA energy and reserve markets, a complementarity model similar to the ones described in Sects. 6.3 and 6.4 is presented. This way, the scheduling problem of the price-maker VPP in the DA energy and reserve markets is modeled as a bi-level optimization problem. The price-maker

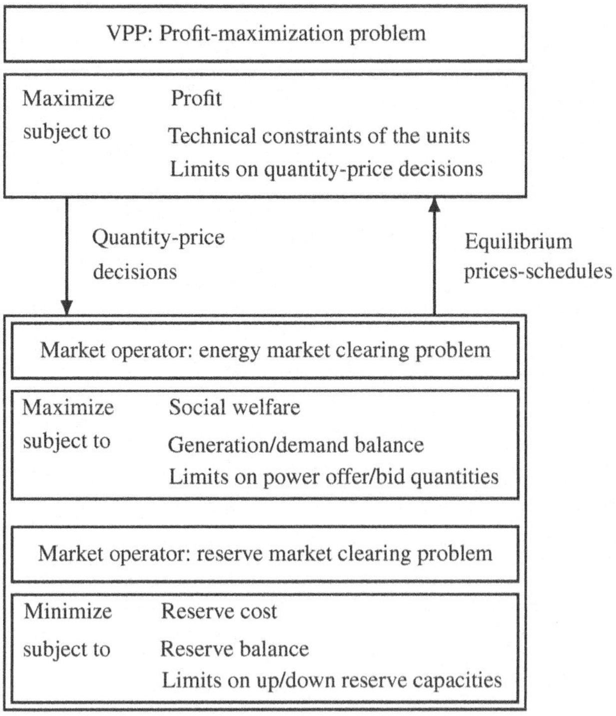

Fig. 6.3 Structure of the bi-level model for the scheduling of a price-maker VPP in the DA energy and reserve markets

VPP acts as the *leader*, while the market operator in charge of the energy and reserve markets acts as the *follower* of the Stackelberg game as explained below:

1. In the upper-level problem, the price-maker VPP submits the quantity-price decisions in the DA energy market, as well as the quantity-price decisions in the reserve market for up and down reserves. Through its offering and bidding decisions, the price-maker VPP tries to influence the prices in the DA energy and reserve markets to achieve the maximum profit.
2. In the lower-level problem, the market operator clears the DA energy market in response to the submitted decisions by the price-maker VPP, as well as by the remaining market participants. The market operator determines the schedules of the price-maker VPP and other market participants that maximize the social welfare in the DA energy market.
3. As an additional lower level problem, the market operator clears the reserve market in response to the offering decisions for up and down reserves submitted by the price-maker VPP and the rest market agents. The market operator allocates the up and down reserves to the market participants, while minimizing the total cost of reserve capacity required.

Figure 6.3 clarifies the interaction between the price-maker VPP and the market operator of the DA energy and reserve markets in the complementarity model.

6.5.2 Uncertainty Model

The price-maker VPP faces a number of uncertainties at the time of deciding the offering and bidding decisions in the DA energy and reserve markets, namely:

1. The available production levels of stochastic renewable generating units in the VPP.
2. The up- and down-reserve deployments requested by the system operator.

Uncertainties are modeling using a number of scenarios for each uncertain parameter $P_{rt}^{\mathrm{R,A}}$, $K_t^{\mathrm{R+}}$, $K_t^{\mathrm{R-}}$. These scenarios are used to build the scenario set $\Pi^{\mathrm{E,R}}$.

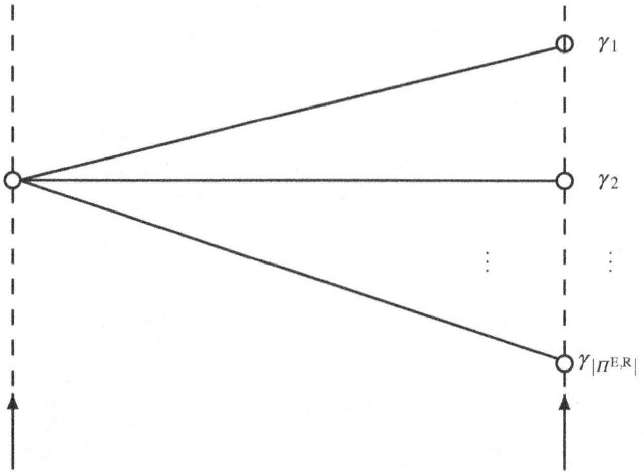

Fig. 6.4 Stochastic complementarity model for scheduling a price-maker VPP in the DA energy and reserve markets: scenario tree

Considering the above framework, the decision sequence for scheduling the price-maker VPP in the DA energy and reserve markets is as follows:

1. The price-maker VPP decides the quantity-price decisions in the DA energy and reserve markets. Additionally, the VPP determines the commitment status of its conventional power plants. The VPP seeks to maximize the expected profit over the scenario set.
2. Once a scenario γ occurs, the VPP knows the available production levels of its stochastic renewable generating units, as well as the up- and down-reserve deployment requests by the system operator. Since the offering and bidding decisions, as well as the status of conventional power plants in the VPP, are determined before the actual realization of scenario γ, they are *here-and-now* decisions with respect to the scenarios.
3. The price-maker VPP dispatches its own conventional power plants, the demands, the stochastic renewable generating units, and the storage units. Thus, these dispatch decisions are *wait-and-see* decisions with respect to the scenarios.

Figure 6.4 presents the scenario tree used in the stochastic complementarity model for scheduling a price-maker VPP in the DA energy and reserve markets. Note also that without loss of generality, additional uncertainties such as the offering and bidding decisions of other market participants can be considered.

6.5.3 Formulation

This section provides the formulation of the stochastic model for the optimal participation of a price-maker VPP in the energy and reserve electricity markets.

6.5.3.1 Bi-Level Model

The scheduling problem of a price-maker VPP in the DA energy and reserve markets is formulated as the following stochastic bi-level model:

$$\max_{\Phi,\Phi^{E},\Phi^{R}} \quad \sum_{t\in\Omega^{T}} \left[\lambda_t^{E}(p_t^{E+}\Delta t - p_t^{E-}\Delta t) + (\lambda_t^{R+}p_t^{R+} + \lambda_t^{R-}p_t^{R-})\right]$$

$$+ \sum_{\gamma\in\Pi^{E,R}} \pi_\gamma \sum_{t\in\Omega^{T}} \Big[\sum_{d\in\Omega^{D}} U_{dt}^{D} p_{dt\gamma}^{D}\Delta t - \sum_{c\in\Omega^{C}} (C_c^{C,F}u_{ct}^{C} + C_c^{C,V}p_{ct\gamma}^{C}\Delta t)\Big] \tag{6.16a}$$

subject to

$$\sum_{c \in \Omega^{\mathrm{C}}} p^{\mathrm{C}}_{ct\gamma} + \sum_{r \in \Omega^{\mathrm{R}}} p^{\mathrm{R}}_{rt\gamma} + \sum_{s \in \Omega^{\mathrm{S}}} (p^{\mathrm{S,D}}_{st\gamma} - p^{\mathrm{S,C}}_{st\gamma}) - \sum_{d \in \Omega^{\mathrm{D}}} p^{\mathrm{D}}_{dt\gamma}$$
$$= p^{\mathrm{E+}}_t - p^{\mathrm{E-}}_t + K^{\mathrm{R+}}_{t\gamma} p^{\mathrm{R+}}_t - K^{\mathrm{R-}}_{t\gamma} p^{\mathrm{R-}}_t, \quad \forall t \in \Omega^{\mathrm{T}}, \forall \gamma \in \Pi^{\mathrm{E,R}}, \tag{6.16b}$$

$$\sum_{t \in \Omega^{\mathrm{T}}} p^{\mathrm{D}}_{dt\gamma} \Delta t \geq \underline{E}^{\mathrm{D}}_d, \quad \forall d \in \Omega^{\mathrm{D}}, \forall \gamma \in \Pi^{\mathrm{E,R}}, \tag{6.16c}$$

$$\underline{P}^{\mathrm{D}}_{dt} \leq p^{\mathrm{D}}_{dt\gamma} \leq \overline{P}^{\mathrm{D}}_{dt}, \quad \forall d \in \Omega^{\mathrm{D}}, \forall t \in \Omega^{\mathrm{T}}, \forall \gamma \in \Pi^{\mathrm{E,R}}, \tag{6.16d}$$

$$\underline{P}^{\mathrm{C}}_c u^{\mathrm{C}}_{ct} \leq p^{\mathrm{C}}_{ct\gamma} \leq \overline{P}^{\mathrm{C}}_c u^{\mathrm{C}}_{ct}, \quad \forall c \in \Omega^{\mathrm{C}}, \forall t \in \Omega^{\mathrm{T}}, \forall \gamma \in \Pi^{\mathrm{E,R}}, \tag{6.16e}$$

$$e^{\mathrm{S}}_{st\gamma} = e^{\mathrm{S}}_{s(t-1)\gamma} + \eta^{\mathrm{S,C}}_s p^{\mathrm{S,C}}_{st\gamma} \Delta t - \frac{p^{\mathrm{S,D}}_{st\gamma}}{\eta^{\mathrm{S,D}}_s} \Delta t, \quad \forall s \in \Omega^{\mathrm{S}}, \forall t \in \Omega^{\mathrm{T}}, \forall \gamma \in \Pi^{\mathrm{E,R}}, \tag{6.16f}$$

$$\underline{E}^{\mathrm{S}}_s \leq e^{\mathrm{S}}_{st\gamma} \leq \overline{E}^{\mathrm{S}}_s, \quad \forall s \in \Omega^{\mathrm{S}}, \forall t \in \Omega^{\mathrm{T}}, \forall \gamma \in \Pi^{\mathrm{E,R}}, \tag{6.16g}$$

$$0 \leq p^{\mathrm{S,C}}_{st\gamma} \leq \overline{P}^{\mathrm{S,C}}_s u^{\mathrm{S}}_{st\gamma}, \quad \forall s \in \Omega^{\mathrm{S}}, \forall t \in \Omega^{\mathrm{T}}, \forall \gamma \in \Pi^{\mathrm{E,R}}, \tag{6.16h}$$

$$0 \leq p^{\mathrm{S,D}}_{st\gamma} \leq \overline{P}^{\mathrm{S,D}}_s (1 - u^{\mathrm{S}}_{st\gamma}), \quad \forall s \in \Omega^{\mathrm{S}}, \forall t \in \Omega^{\mathrm{T}}, \forall \gamma \in \Pi^{\mathrm{E,R}}, \tag{6.16i}$$

$$0 \leq p^{\mathrm{R}}_{rt\gamma} \leq P^{\mathrm{R,A}}_{rt\gamma}, \quad \forall r \in \Omega^{\mathrm{R}}, \forall t \in \Omega^{\mathrm{T}}, \forall \gamma \in \Pi^{\mathrm{E,R}}, \tag{6.16j}$$

$$\overline{p}^{\mathrm{E+}}_t, \overline{p}^{\mathrm{R+}}_t \leq \overline{P}^{\mathrm{V+}}_t, \quad \forall t \in \Omega^{\mathrm{T}}, \tag{6.16k}$$

$$\overline{p}^{\mathrm{E-}}_t, \overline{p}^{\mathrm{R-}}_t \leq \overline{P}^{\mathrm{V-}}_t, \quad \forall t \in \Omega^{\mathrm{T}}, \tag{6.16l}$$

$$\overline{p}^{\mathrm{E+}}_t + \overline{p}^{\mathrm{R+}}_t \leq \overline{P}^{\mathrm{V+}}_t, \quad \forall t \in \Omega^{\mathrm{T}}, \tag{6.16m}$$

$$\overline{p}^{\mathrm{E-}}_t + \overline{p}^{\mathrm{R-}}_t \leq \overline{P}^{\mathrm{V-}}_t, \quad \forall t \in \Omega^{\mathrm{T}}, \tag{6.16n}$$

$$\overline{p}^{\mathrm{E+}}_t, \overline{p}^{\mathrm{E-}}_t, \overline{p}^{\mathrm{R+}}_t, \overline{p}^{\mathrm{R-}}_t \geq 0, \quad \forall t \in \Omega^{\mathrm{T}}, \tag{6.16o}$$

$$\nu^{\mathrm{E+}}_t, \nu^{\mathrm{E-}}_t, \nu^{\mathrm{R+}}_t, \nu^{\mathrm{R-}}_t \geq 0, \quad \forall t \in \Omega^{\mathrm{T}}, \tag{6.16p}$$

$$\lambda^{\mathrm{E}}_t, p^{\mathrm{E+}}_t, p^{\mathrm{E-}}_t \in \Psi^{\mathrm{E}}_t, \quad \forall t \in \Omega^{\mathrm{T}}, \tag{6.16q}$$

$$\lambda^{\mathrm{R+}}_t, \lambda^{\mathrm{R-}}_t, p^{\mathrm{R+}}_t, p^{\mathrm{R-}}_t \in \Psi^{\mathrm{R}}_t, \quad \forall t \in \Omega^{\mathrm{T}}, \tag{6.16r}$$

where variables in sets $\Phi = \Big\{ p^{\mathrm{D}}_{dt\gamma}, \forall d \in \Omega^{\mathrm{D}}, \forall t \in \Omega^{\mathrm{T}}, \forall \gamma \in \Pi^{\mathrm{E,R}}; e^{\mathrm{S}}_{st\gamma}, p^{\mathrm{S,C}}_{st\gamma}, p^{\mathrm{S,D}}_{st\gamma}, u^{\mathrm{S}}_{st\gamma}, \forall s \in \Omega^{\mathrm{S}}, \forall t \in \Omega^{\mathrm{T}}, \forall \gamma \in \Pi^{\mathrm{E,R}};$ $p^{\mathrm{C}}_{ct\gamma}, u^{\mathrm{C}}_{ct}, \forall c \in \Omega^{\mathrm{C}}, \forall t \in \Omega^{\mathrm{T}}, \forall \gamma \in \Pi^{\mathrm{E,R}}; p^{\mathrm{R}}_{rt\gamma}, \forall r \in \Omega^{\mathrm{R}}, \forall t \in \Omega^{\mathrm{T}}, \forall \gamma \in \Pi^{\mathrm{E,R}}; \overline{p}^{\mathrm{E+}}_t, \overline{p}^{\mathrm{E-}}_t, \nu^{\mathrm{E+}}_t, \nu^{\mathrm{E-}}_t, \overline{p}^{\mathrm{R+}}_t, \overline{p}^{\mathrm{R-}}_t, \nu^{\mathrm{R+}}_t,$ $\nu^{\mathrm{R-}}_t, \forall t \in \Omega^{\mathrm{T}} \Big\}$, $\Phi^{\mathrm{E}}_t, \forall t \in \Omega^{\mathrm{T}}$, and $\Phi^{\mathrm{R}}_t = \Big\{ p^{\mathrm{O+}}_{ot}, p^{\mathrm{O-}}_{ot}, \forall o \in \Omega^{\mathrm{O}}; p^{\mathrm{R+}}_t, p^{\mathrm{R-}}_t \Big\}$, $\forall t \in \Omega^{\mathrm{T}}$, are the optimization variables of problem (6.16).

Bi-level model (6.16) comprises the upper-level problem (6.16a)–(6.16p) that determines the offering and bidding decisions of the VPP in the DA energy and reserve markets, as well as the commitment status of conventional power plants in the VPP, while maximizing the expected profit over the scenario set, and the lower-level problems represented by sets Ψ^{E}_t and Ψ^{R}_t in constraints (6.16q) and (6.16r), which are related to the DA energy market-clearing problem and the reserve market-clearing problem, respectively.

6.5.3.2 Market-Clearing Problems

The formulation of the DA energy market-clearing problem is identical to problem (6.2) provided in Sect. 6.3.2, while the reserve market-clearing problem can be also formulated as the following LP problem:

$$\min_{\Phi_t^{\mathrm{R}}} \quad \nu_t^{\mathrm{R+}} p_t^{\mathrm{R+}} + \sum_{o \in \Omega^{\mathrm{O}}} \chi_{ot}^{\mathrm{O+}} p_{ot}^{\mathrm{O+}} + \nu_t^{\mathrm{R-}} p_t^{\mathrm{R-}} + \sum_{o \in \Omega^{\mathrm{O}}} \chi_{ot}^{\mathrm{O-}} p_{ot}^{\mathrm{O-}} \tag{6.17a}$$

subject to

$$p_t^{\mathrm{R+}} + \sum_{o \in \Omega^{\mathrm{O}}} p_{ot}^{\mathrm{O+}} = P_t^{\mathrm{R+}} \ : \lambda_t^{\mathrm{R+}}, \tag{6.17b}$$

$$p_t^{\mathrm{R-}} + \sum_{o \in \Omega^{\mathrm{O}}} p_{ot}^{\mathrm{O-}} = P_t^{\mathrm{R-}} \ : \lambda_t^{\mathrm{R-}}, \tag{6.17c}$$

$$0 \le p_{ot}^{\mathrm{O+}} \le \overline{P}_{ot}^{\mathrm{O+}} \ : \underline{\mu}_{ot}^{\mathrm{O+}}, \overline{\mu}_{ot}^{\mathrm{O+}} \quad \forall o \in \Omega^{\mathrm{O}}, \tag{6.17d}$$

$$0 \le p_{ot}^{\mathrm{O-}} \le \overline{P}_{ot}^{\mathrm{O-}} \ : \underline{\mu}_{ot}^{\mathrm{O-}}, \overline{\mu}_{ot}^{\mathrm{O-}}, \quad \forall o \in \Omega^{\mathrm{O}}, \tag{6.17e}$$

$$0 \le p_t^{\mathrm{R+}} \le \overline{p}_t^{\mathrm{R+}} \ : \underline{\mu}_t^{\mathrm{R+}}, \overline{\mu}_t^{\mathrm{R+}}, \tag{6.17f}$$

$$0 \le p_t^{\mathrm{R-}} \le \overline{p}_t^{\mathrm{R-}} \ : \underline{\mu}_t^{\mathrm{R-}}, \overline{\mu}_t^{\mathrm{R-}}, \tag{6.17g}$$

where set Φ_t^{R} includes the optimization variables of the reserve market-clearing problem (6.17), while variables in set $\Upsilon_t^{\mathrm{R}} = \left\{\lambda_t^{\mathrm{R+}}, \lambda_t^{\mathrm{R-}}; \underline{\mu}_{ot}^{\mathrm{O+}}, \overline{\mu}_{ot}^{\mathrm{O+}}, \underline{\mu}_{ot}^{\mathrm{O-}}, \overline{\mu}_{ot}^{\mathrm{O-}}, o \in \Omega^{\mathrm{O}}; \underline{\mu}_t^{\mathrm{R+}}, \overline{\mu}_t^{\mathrm{R+}}, \underline{\mu}_t^{\mathrm{R-}}, \overline{\mu}_t^{\mathrm{R-}}\right\}$, $\forall t \in \Omega^{\mathrm{T}}$, are the optimization variables of its equivalent dual problem.

6.5.3.3 MPEC

Note that the market-clearing problems (6.2) and (6.17) are continuous and linear. Thus, each lower-level problem can be equivalently replaced with its corresponding KKT conditions. This way, the stochastic bi-level model (6.16) can be converted into the stochastic MPEC model below:

$$\begin{aligned} \max_{\Phi, \Phi^{\mathrm{E}}, \Upsilon^{\mathrm{E}}, \Phi^{\mathrm{R}}, \Upsilon^{\mathrm{R}}} \quad & \sum_{t \in \Omega^{\mathrm{T}}} \left[\lambda_t^{\mathrm{E}}(p_t^{\mathrm{E+}} \Delta t - p_t^{\mathrm{E-}} \Delta t) + (\lambda_t^{\mathrm{R+}} p_t^{\mathrm{R+}} + \lambda_t^{\mathrm{R-}} p_t^{\mathrm{R-}})\right] \\ + \sum_{\gamma \in \Pi^{\mathrm{E,R}}} \pi_\gamma \sum_{t \in \Omega^{\mathrm{T}}} & \Big[\sum_{d \in \Omega^{\mathrm{D}}} U_{dt}^{\mathrm{D}} p_{dt\gamma}^{\mathrm{D}} \Delta t - \sum_{c \in \Omega^{\mathrm{C}}} (C_c^{\mathrm{C,F}} u_{ct}^{\mathrm{C}} + C_c^{\mathrm{C,V}} p_{ct\gamma}^{\mathrm{C}} \Delta t)\Big] \end{aligned} \tag{6.18a}$$

subject to

$$\text{Constraints (6.16b)–(6.16p)}, \tag{6.18b}$$

$$\Big\{\text{Constraints (6.3)}, \tag{6.18c}$$

$$p_t^{\mathrm{R+}} + \sum_{o \in \Omega^{\mathrm{O}}} p_{ot}^{\mathrm{O+}} = P_t^{\mathrm{R+}}, \tag{6.18d}$$

$$p_t^{\mathrm{R-}} + \sum_{o \in \Omega^{\mathrm{O}}} p_{ot}^{\mathrm{O-}} = P_t^{\mathrm{R-}}, \tag{6.18e}$$

$$\chi_{ot}^{\mathrm{O+}} - \lambda_t^{\mathrm{R+}} + \overline{\mu}_{ot}^{\mathrm{O+}} - \underline{\mu}_{ot}^{\mathrm{O+}} = 0, \quad \forall o \in \Omega^{\mathrm{O}}, \tag{6.18f}$$

$$\chi_{ot}^{\mathrm{O-}} - \lambda_t^{\mathrm{R-}} + \overline{\mu}_{ot}^{\mathrm{O-}} - \underline{\mu}_{ot}^{\mathrm{O-}} = 0, \quad \forall o \in \Omega^{\mathrm{O}}, \tag{6.18g}$$

$$\nu_t^{\mathrm{R+}} - \lambda_t^{\mathrm{R+}} + \overline{\mu}_t^{\mathrm{R+}} - \underline{\mu}_t^{\mathrm{R+}} = 0, \tag{6.18h}$$

$$\nu_t^{\mathrm{R-}} - \lambda_t^{\mathrm{R-}} + \overline{\mu}_t^{\mathrm{R-}} - \underline{\mu}_t^{\mathrm{R-}} = 0, \tag{6.18i}$$

$$0 \leq \underline{\mu}_{ot}^{\mathrm{O+}} \perp p_{ot}^{\mathrm{O+}} \geq 0, \quad \forall o \in \Omega^{\mathrm{O}}, \tag{6.18j}$$

$$0 \leq \overline{\mu}_{ot}^{\mathrm{O+}} \perp \overline{P}_{ot}^{\mathrm{O+}} - p_{ot}^{\mathrm{O+}} \geq 0, \quad \forall o \in \Omega^{\mathrm{O}}, \tag{6.18k}$$

$$0 \leq \underline{\mu}_{ot}^{\mathrm{O-}} \perp p_{ot}^{\mathrm{O-}} \geq 0, \quad \forall o \in \Omega^{\mathrm{O}}, \tag{6.18l}$$

$$0 \leq \overline{\mu}_{ot}^{\mathrm{O-}} \perp \overline{P}_{ot}^{\mathrm{O-}} - p_{ot}^{\mathrm{O-}} \geq 0, \quad \forall o \in \Omega^{\mathrm{O}}, \tag{6.18m}$$

$$0 \leq \underline{\mu}_t^{\mathrm{E+}} \perp p_t^{\mathrm{E+}} \geq 0, \tag{6.18n}$$

$$0 \leq \overline{\mu}_t^{\mathrm{E+}} \perp \overline{p}_t^{\mathrm{E+}} - p_t^{\mathrm{E+}} \geq 0, \tag{6.18o}$$

$$0 \leq \underline{\mu}_t^{\mathrm{E-}} \perp p_t^{\mathrm{E-}} \geq 0, \tag{6.18p}$$

$$0 \leq \overline{\mu}_t^{\mathrm{E-}} \perp \overline{p}_t^{\mathrm{E-}} - p_t^{\mathrm{E-}} \geq 0 \tag{6.18q}$$

$\}$, $\forall t \in \Omega^{\mathrm{T}}$, where sets Φ, Φ^{E}, Υ^{E}, Φ^{R}, and Υ^{R} include the optimization variables of the stochastic MPEC problem (6.18).

The stochastic MPEC model (6.18) includes the upper-level problem (6.18a)–(6.18b), the KKT conditions (6.18c) corresponding to the DA energy market-clearing problem, and the KKT conditions (6.18d)–(6.18q) corresponding to the reserve market-clearing problem.

The stochastic MPEC model (6.18) is nonlinear due to two terms of $\lambda_t^{\mathrm{E}}(p_t^{\mathrm{E+}}\Delta t - p_t^{\mathrm{E-}}\Delta t)$ and $(\lambda_t^{\mathrm{R+}} p_t^{\mathrm{R+}} + \lambda_t^{\mathrm{R-}} p_t^{\mathrm{R-}})$ in the objective function (6.18a), and the slackness conditions in the complementarity constraints (6.18c) and (6.18j)–(6.18q).

6.5.3.4 MILP Model

As explained in Sect. 6.3.4, nonlinear terms in the stochastic MPEC model (6.18) can be replaced by equivalent mixed-integer linear expressions, rendering the following MILP problem:

$$\begin{aligned}
\max_{\Phi,\Phi^{\mathrm{E}},\Upsilon^{\mathrm{E}},\Phi^{\mathrm{M,E}},\Phi^{\mathrm{R}},\Upsilon^{\mathrm{R}},\Phi^{\mathrm{M,R}}} \quad & \sum_{t\in\Omega^{\mathrm{T}}} \Big[\sum_{q\in\Omega^{\mathrm{Q}}} \chi_{qt}^{\mathrm{Q,E}} p_{qt}^{\mathrm{Q,E}} \Delta t - \sum_{g\in\Omega^{\mathrm{G}}} \chi_{gt}^{\mathrm{G,E}} p_{gt}^{\mathrm{G,E}} \Delta t \\
& - \sum_{g\in\Omega^{\mathrm{G}}} \overline{P}_{gt}^{\mathrm{G,E}} \overline{\mu}_{gt}^{\mathrm{G,E}} \Delta t - \sum_{q\in\Omega^{\mathrm{Q}}} \overline{P}_{qt}^{\mathrm{Q,E}} \overline{\mu}_{qt}^{\mathrm{Q,E}} \Delta t \Big] + \sum_{t\in\Omega^{\mathrm{T}}} \Big[p_t^{\mathrm{R+}} \lambda_t^{\mathrm{R+}} + p_t^{\mathrm{R-}} \lambda_t^{\mathrm{R-}} \\
& - \sum_{o\in\Omega^{\mathrm{O}}} (\overline{P}_{ot}^{\mathrm{O+}} \overline{\mu}_{ot}^{\mathrm{O+}} + \overline{P}_{ot}^{\mathrm{O-}} \overline{\mu}_{ot}^{\mathrm{O-}} + \chi_{ot}^{\mathrm{O+}} p_{ot}^{\mathrm{O+}} + \chi_{ot}^{\mathrm{O-}} p_{ot}^{\mathrm{O-}}) \Big] \\
& + \sum_{\gamma\in\Pi^{\mathrm{E,R}}} \pi_\gamma \sum_{t\in\Omega^{\mathrm{T}}} \Big[\sum_{d\in\Omega^{\mathrm{D}}} U_{dt}^{\mathrm{D}} p_{dt\gamma}^{\mathrm{D}} \Delta t - \sum_{c\in\Omega^{\mathrm{C}}} (C_c^{\mathrm{C,F}} u_{ct}^{\mathrm{C}} + C_c^{\mathrm{C,V}} p_{ct\gamma}^{\mathrm{C}} \Delta t) \Big]
\end{aligned} \tag{6.19a}$$

subject to

$$\text{Constraints (6.16b)–(6.16p)}, \tag{6.19b}$$

$$\Big\{ \text{Constraints (6.3a)–(6.3e) and (6.5)}, \tag{6.19c}$$

$$\text{Constraints (6.18d)–(6.18i)} \tag{6.19d}$$

$$0 \leq \underline{\mu}_{ot}^{\mathrm{O+}} \leq \underline{N}_{ot}^{\mathrm{O+}} \underline{u}_{ot}^{\mathrm{O+}}, \quad \forall o \in \Omega^{\mathrm{O}}, \tag{6.19e}$$

$$0 \leq p_{ot}^{\mathrm{O+}} \leq \underline{M}_{ot}^{\mathrm{O+}} (1 - \underline{u}_{ot}^{\mathrm{O+}}), \quad \forall o \in \Omega^{\mathrm{O}}, \tag{6.19f}$$

$$0 \leq \overline{\mu}_{ot}^{\mathrm{O+}} \leq \overline{N}_{ot}^{\mathrm{O+}} \overline{u}_{ot}^{\mathrm{O+}}, \quad \forall o \in \Omega^{\mathrm{O}}, \tag{6.19g}$$

$$0 \le \overline{P}_{ot}^{\mathrm{O}+} - p_{ot}^{\mathrm{O}+} \le \overline{M}_{ot}^{\mathrm{O}+}(1 - \overline{u}_{ot}^{\mathrm{O}+}), \quad \forall o \in \Omega^{\mathrm{O}}, \tag{6.19h}$$

$$0 \le \underline{\mu}_{ot}^{\mathrm{O}-} \le \underline{N}_{ot}^{\mathrm{O}-}\underline{u}_{ot}^{\mathrm{O}-}, \quad \forall o \in \Omega^{\mathrm{O}}, \tag{6.19i}$$

$$0 \le p_{ot}^{\mathrm{O}-} \le \underline{M}_{ot}^{\mathrm{O}-}(1 - \underline{u}_{ot}^{\mathrm{O}-}), \quad \forall o \in \Omega^{\mathrm{O}}, \tag{6.19j}$$

$$0 \le \overline{\mu}_{ot}^{\mathrm{O}-} \le \overline{N}_{ot}^{\mathrm{O}-}\overline{u}_{ot}^{\mathrm{O}-}, \quad \forall o \in \Omega^{\mathrm{O}}, \tag{6.19k}$$

$$0 \le \overline{P}_{ot}^{\mathrm{O}-} - p_{ot}^{\mathrm{O}-} \le \overline{M}_{ot}^{\mathrm{O}-}(1 - \overline{u}_{ot}^{\mathrm{O}-}), \quad \forall o \in \Omega^{\mathrm{O}}, \tag{6.19l}$$

$$0 \le \underline{\mu}_{t}^{\mathrm{R}+} \le \underline{N}_{t}^{\mathrm{R}+}\underline{u}_{t}^{\mathrm{R}+}, \tag{6.19m}$$

$$0 \le p_{t}^{\mathrm{R}+} \le \underline{M}_{t}^{\mathrm{R}+}(1 - \underline{u}_{t}^{\mathrm{R}+}), \tag{6.19n}$$

$$0 \le \overline{\mu}_{t}^{\mathrm{R}+} \le \overline{N}_{t}^{\mathrm{R}+}\overline{u}_{t}^{\mathrm{R}+}, \tag{6.19o}$$

$$0 \le \overline{p}_{t}^{\mathrm{R}+} - p_{t}^{\mathrm{R}+} \le \overline{M}_{t}^{\mathrm{R}+}(1 - \overline{u}_{t}^{\mathrm{R}+}), \tag{6.19p}$$

$$0 \le \underline{\mu}_{t}^{\mathrm{R}-} \le \underline{N}_{t}^{\mathrm{R}-}\underline{u}_{t}^{\mathrm{R}-}, \tag{6.19q}$$

$$0 \le p_{t}^{\mathrm{R}-} \le \underline{M}_{t}^{\mathrm{R}-}(1 - \underline{u}_{t}^{\mathrm{R}-}), \tag{6.19r}$$

$$0 \le \overline{\mu}_{t}^{\mathrm{R}-} \le \overline{N}_{t}^{\mathrm{R}-}\overline{u}_{t}^{\mathrm{R}-}, \tag{6.19s}$$

$$0 \le \overline{p}_{t}^{\mathrm{R}-} - p_{t}^{\mathrm{R}-} \le \overline{M}_{t}^{\mathrm{R}-}(1 - \overline{u}_{t}^{\mathrm{R}-}) \tag{6.19t}$$

$\Big\}$, $\forall t \in \Omega^{\mathrm{T}}$, where sets Φ, Φ^{E}, Υ^{E}, $\Phi^{\mathrm{M,E}} = \Big\{\underline{u}_{gt}^{\mathrm{G,E}}, \overline{u}_{gt}^{\mathrm{G,E}}, \forall g \in \Omega^{\mathrm{G}}, \forall t \in \Omega^{\mathrm{T}}; \underline{u}_{qt}^{\mathrm{Q,E}}, \overline{u}_{qt}^{\mathrm{Q,E}}, \forall q \in \Omega^{\mathrm{Q}}, \forall t \in \Omega^{\mathrm{T}}; \underline{u}_{t}^{\mathrm{E}+}, \overline{u}_{t}^{\mathrm{E}+}, \underline{u}_{t}^{\mathrm{E}-}, \overline{u}_{t}^{\mathrm{E}-}, \forall t \in \Omega^{\mathrm{T}}\Big\}$, Φ^{R}, Υ^{R}, and $\Phi^{\mathrm{M,R}} = \Big\{\underline{u}_{ot}^{\mathrm{O}+}, \overline{u}_{ot}^{\mathrm{O}+}, \forall o \in \Omega^{\mathrm{O}}, \forall t \in \Omega^{\mathrm{T}}; \underline{u}_{ot}^{\mathrm{O}-}, \overline{u}_{ot}^{\mathrm{O}-}, \forall o \in \Omega^{\mathrm{O}}, \forall t \in \Omega^{\mathrm{T}}; \underline{u}_{t}^{\mathrm{R}+}, \overline{u}_{t}^{\mathrm{R}+}, \underline{u}_{t}^{\mathrm{R}-}, \overline{u}_{t}^{\mathrm{R}-}, \forall t \in \Omega^{\mathrm{T}}\Big\}$, include the optimization variables of the stochastic MILP problem (6.19).

Illustrative Example 6.12 Scheduling the Price-Maker VPP in the DA Energy and Reserve Markets: Stochastic MPEC Model

This illustrative example considers the VPP described in Illustrative Example 6.1, considering that it participates in the DA energy market (offering and bidding data of other participants provided in Illustrative Example 6.2) and also in the reserve market. Offering data of participants in the up- and down-reserve market are provided in Table 6.16. The reserve capacity required in the reserve market is 30 MW for both up and down reserves.

Uncertainty in the reserve deployment requests by the system operator is modeled considering two scenarios whose data are presented in Table 6.17.

Considering the above data, the bi-level model used to schedule the considered price-maker VPP in both the DA energy and reserve markets is formulated below.

Table 6.16 Illustrative Example 6.12: data of the reserve market

	Up-reserve		Down-reserve	
Parameter	Unit 1	Unit 2	Unit 1	Unit 2
Price [$/MW]	10	20	10	20
Power quantity [MW]	20	20	20	20

Table 6.17 Illustrative Example 6.12: data of reserve deployment by the system operator

Scenario	1	2
Up-reserve [pu]	0.6	0
Down-reserve [pu]	0	0.8
Probability	0.6	0.4

First, the upper-level problem is formulated as follows:

$$\max_{\Phi,\Phi^{\mathrm{E}},\Phi^{\mathrm{R}}} \Big[50p_1^{\mathrm{Q,E}}\Delta t + 70p_2^{\mathrm{Q,E}}\Delta t - 30p_1^{\mathrm{G,E}}\Delta t - 50p_2^{\mathrm{G,E}}\Delta t$$
$$- 50\overline{\mu}_1^{\mathrm{G,E}}\Delta t - 50\overline{\mu}_2^{\mathrm{G,E}}\Delta t - 50\overline{\mu}_1^{\mathrm{Q,E}}\Delta t - 50\overline{\mu}_2^{\mathrm{Q,E}}\Delta t\Big]$$
$$+ \Big[30\lambda^{\mathrm{R+}} + 30\lambda^{\mathrm{R-}} - (20\overline{\mu}_1^{\mathrm{O+}} + 20\overline{\mu}_2^{\mathrm{O+}}$$
$$+ 20\overline{\mu}_1^{\mathrm{O-}} + 20\overline{\mu}_2^{\mathrm{O-}} + 10p_1^{\mathrm{O+}} + 20p_2^{\mathrm{O+}} + 10p_1^{\mathrm{O-}} + 20p_2^{\mathrm{O-}})\Big]$$
$$+ \Big[0.6(60p_1^{\mathrm{D}} - 45p_1^{\mathrm{C}}) + 0.4(60p_2^{\mathrm{D}} - 45p_2^{\mathrm{C}})\Big]$$

subject to

$$p_1^{\mathrm{C}} + p_1^{\mathrm{R}} - p_1^{\mathrm{D}} = p^{\mathrm{E+}} - p^{\mathrm{E-}} + 0.6p^{\mathrm{R+}},$$
$$p_2^{\mathrm{C}} + p_2^{\mathrm{R}} - p_2^{\mathrm{D}} = p^{\mathrm{E+}} - p^{\mathrm{E-}} - 0.8p^{\mathrm{R-}},$$
$$20 \le p_1^{\mathrm{D}},$$
$$20 \le p_2^{\mathrm{D}},$$
$$0 \le p_1^{\mathrm{D}} \le 30,$$
$$0 \le p_2^{\mathrm{D}} \le 30,$$
$$0 \le p_1^{\mathrm{C}} \le 40u^{\mathrm{C}},$$
$$0 \le p_2^{\mathrm{C}} \le 40u^{\mathrm{C}},$$
$$0 \le p_1^{\mathrm{R}} \le 8,$$
$$0 \le p_2^{\mathrm{R}} \le 8,$$
$$\overline{p}^{\mathrm{E+}} \le 50,$$
$$\overline{p}^{\mathrm{E-}} \le 30,$$
$$\overline{p}^{\mathrm{R+}} \le 50,$$
$$\overline{p}^{\mathrm{R-}} \le 30,$$
$$\overline{p}^{\mathrm{E+}} + \overline{p}^{\mathrm{R+}} \le 50,$$
$$\overline{p}^{\mathrm{E-}} + \overline{p}^{\mathrm{R-}} \le 30,$$
$$\overline{p}^{\mathrm{E+}}, \overline{p}^{\mathrm{E-}}, \overline{p}^{\mathrm{R+}}, \overline{p}^{\mathrm{R-}} \ge 0,$$
$$\nu^{\mathrm{E+}}, \nu^{\mathrm{E-}}, \nu^{\mathrm{R+}}, \nu^{\mathrm{R-}} \ge 0.$$

Next, the lower-level problems comprise the clearing of the DA energy and reserve markets. On one hand, the equilibrium constraints corresponding to the DA energy market are the same than those considered in Illustrative Examples 6.4 and 6.5. On the other hand, the equilibrium constraints corresponding to the DA reserve markets are formulated as follows:

$$p^{\mathrm{R+}} + p_1^{\mathrm{O+}} + p_2^{\mathrm{O+}} = 30,$$
$$p^{\mathrm{R-}} + p_1^{\mathrm{O-}} + p_2^{\mathrm{O-}} = 30,$$
$$10 - \lambda^{\mathrm{R+}} + \overline{\mu}_1^{\mathrm{O+}} - \underline{\mu}_1^{\mathrm{O+}} = 0,$$
$$20 - \lambda^{\mathrm{R+}} + \overline{\mu}_2^{\mathrm{O+}} - \underline{\mu}_2^{\mathrm{O+}} = 0,$$

$$
\begin{aligned}
& 10 - \lambda^{\text{R}-} + \overline{\mu}_1^{\text{O}-} - \underline{\mu}_1^{\text{O}-} = 0, \\
& 20 - \lambda^{\text{R}-} + \overline{\mu}_2^{\text{O}-} - \underline{\mu}_2^{\text{O}-} = 0, \\
& \nu^{\text{R}+} - \lambda^{\text{R}+} + \overline{\mu}^{\text{R}+} - \underline{\mu}^{\text{R}+} = 0, \\
& \nu^{\text{R}-} - \lambda^{\text{R}-} + \overline{\mu}^{\text{R}-} - \underline{\mu}^{\text{R}-} = 0, \\
& 0 \leq \underline{\mu}_1^{\text{O}+} \leq \underline{N}_1^{\text{O}+} \underline{u}_1^{\text{O}+}, \\
& 0 \leq p_1^{\text{O}+} \leq \underline{M}_1^{\text{O}+} (1 - \underline{u}_1^{\text{O}+}), \\
& 0 \leq \underline{\mu}_2^{\text{O}+} \leq \underline{N}_2^{\text{O}+} \underline{u}_2^{\text{O}+}, \\
& 0 \leq p_2^{\text{O}+} \leq \underline{M}_2^{\text{O}+} (1 - \underline{u}_2^{\text{O}+}), \\
& 0 \leq \overline{\mu}_1^{\text{O}+} \leq \overline{N}_1^{\text{O}+} \overline{u}_1^{\text{O}+}, \\
& 0 \leq 20 - p_1^{\text{O}+} \leq \overline{M}_1^{\text{O}+} (1 - \overline{u}_1^{\text{O}+}), \\
& 0 \leq \overline{\mu}_2^{\text{O}+} \leq \overline{N}_2^{\text{O}+} \overline{u}_2^{\text{O}+}, \\
& 0 \leq 20 - p_2^{\text{O}+} \leq \overline{M}_2^{\text{O}+} (1 - \overline{u}_2^{\text{O}+}), \\
& 0 \leq \underline{\mu}_1^{\text{O}-} \leq \underline{N}_1^{\text{O}-} \underline{u}_1^{\text{O}-}, \\
& 0 \leq p_1^{\text{O}} \leq \underline{M}_1^{\text{O}-} (1 - \underline{u}_1^{\text{O}-}), \\
& 0 \leq \underline{\mu}_2^{\text{O}-} \leq \underline{N}_2^{\text{O}-} \underline{u}_2^{\text{O}-}, \\
& 0 \leq p_2^{\text{O}-} \leq \underline{M}_2^{\text{O}-} (1 - \underline{u}_2^{\text{O}-}), \\
& 0 \leq \overline{\mu}_1^{\text{O}-} \leq \overline{N}_1^{\text{O}-} \overline{u}_1^{\text{O}-}, \\
& 0 \leq 20 - p_1^{\text{O}-} \leq \overline{M}_1^{\text{O}-} (1 - \overline{u}_1^{\text{O}-}), \\
& 0 \leq \overline{\mu}_2^{\text{O}-} \leq \overline{N}_2^{\text{O}-} \overline{u}_2^{\text{O}-}, \\
& 0 \leq 20 - p_2^{\text{O}-} \leq \overline{M}_2^{\text{O}-} (1 - \overline{u}_2^{\text{O}-}), \\
& 0 \leq \underline{\mu}^{\text{R}+} \leq \underline{N}^{\text{R}+} \underline{u}^{\text{R}+}, \\
& 0 \leq p^{\text{R}+} \leq \underline{M}^{\text{R}+} (1 - \underline{u}^{\text{R}+}) \\
& 0 \leq \overline{\mu}^{\text{R}+} \leq \overline{N}^{\text{R}+} \overline{u}^{\text{R}+}, \\
& 0 \leq \overline{p}^{\text{R}+} - p^{\text{R}+} \leq \overline{M}^{\text{R}+} (1 - \overline{u}^{\text{R}+}) \\
& 0 \leq \underline{\mu}^{\text{R}-} \leq \underline{N}^{\text{R}-} \underline{u}^{\text{R}-}, \\
& 0 \leq p^{\text{R}-} \leq \underline{M}^{\text{R}-} (1 - \underline{u}^{\text{R}-}) \\
& 0 \leq \overline{\mu}^{\text{R}-} \leq \overline{N}^{\text{R}-} \overline{u}^{\text{R}-}, \\
& 0 \leq \overline{p}^{\text{R}-} - p^{\text{R}-} \leq \overline{M}^{\text{R}-} (1 - \overline{u}^{\text{R}-}),
\end{aligned}
$$

Table 6.18 Illustrative Example 6.12: scheduling decisions of the VPP in the DA energy market

Variable	$\overline{p}^{\text{E+}}$ [MW]	$\nu^{\text{E+}}$ [$/MWh]	$p^{\text{E+}}$ [MW]	u^{C}
Value	12	50	12	1

Table 6.19 Illustrative Example 6.12: scheduling decisions by the VPP in the reserve market

Variable	$\overline{p}^{\text{R+}}$ [MW]	$\nu^{\text{R+}}$ [$/MW]	$\overline{p}^{\text{R}-}$ [MW]	$\nu^{\text{R}-}$ [$/MW]	$p^{\text{R+}}$ [MW]	$p^{\text{R}-}$ [MW]
Value	10	20	30	10	10	30

Table 6.20 Illustrative Example 6.12: power dispatch of the units in the VPP

Variable	p^{C}	p^{D}	p^{R}
Value at scenario 1 [MW]	40	30	8
Value at scenario 2 [MW]	10	30	8

where sets $\Phi = \left\{p_1^{\text{D}}, p_2^{\text{D}}, p_1^{\text{C}}, p_2^{\text{C}}, u^{\text{C}}, p_1^{\text{R}}, p_2^{\text{R}}, \overline{p}^{\text{E+}}, \overline{p}^{\text{E}-}, \overline{p}^{\text{R+}}, \overline{p}^{\text{R}-}, \nu^{\text{E+}}, \nu^{\text{E}-}, \nu^{\text{R+}}, \nu^{\text{R}-}\right\}$, $\Phi^{\text{E}} = \left\{p_1^{\text{G,E}}, p_2^{\text{G,E}}, p_1^{\text{Q,E}}, p_2^{\text{Q,E}}, p^{\text{E+}}, p^{\text{E}-}\right\}$, $\Upsilon^{\text{E}} = \left\{\lambda^{\text{E}}, \underline{\mu}_1^{\text{G,E}}, \underline{\mu}_2^{\text{G,E}}, \overline{\mu}_1^{\text{G,E}}, \overline{\mu}_2^{\text{G,E}}, \underline{\mu}_1^{\text{Q,E}}, \underline{\mu}_2^{\text{Q,E}}, \overline{\mu}_1^{\text{Q,E}}, \overline{\mu}_2^{\text{Q,E}}, \underline{\mu}^{\text{E+}}, \overline{\mu}^{\text{E+}}, \underline{\mu}^{\text{E}-}, \overline{\mu}^{\text{E}-}\right\}$, $\Phi^{\text{M,E}} = \left\{\underline{u}_1^{\text{G,E}}, \underline{u}_2^{\text{G,E}}, \overline{u}_1^{\text{G,E}}, \overline{u}_2^{\text{G,E}}, \underline{u}_1^{\text{Q,E}}, \underline{u}_2^{\text{Q,E}}, \overline{u}_1^{\text{Q,E}}, \overline{u}_2^{\text{Q,E}}, \underline{u}^{\text{E+}}, \overline{u}^{\text{E+}}, \underline{u}^{\text{E}-}, \overline{u}^{\text{E}-}\right\}$, $\Phi^{\text{R}} = \left\{p_1^{\text{O+}}, p_2^{\text{O+}}, p_1^{\text{O}-}, p_2^{\text{O}-}, p^{\text{R+}}, p^{\text{R}-}\right\}$, $\Upsilon^{\text{R}} = \left\{\lambda^{\text{R+}}, \lambda^{\text{R}-}, \underline{\mu}_1^{\text{O+}}, \underline{\mu}_2^{\text{O+}}, \overline{\mu}_1^{\text{O+}}, \overline{\mu}_2^{\text{O+}}, \underline{\mu}_1^{\text{O}-}, \underline{\mu}_2^{\text{O}-}, \overline{\mu}_1^{\text{O}-}, \overline{\mu}_2^{\text{O}-}, \underline{\mu}^{\text{R+}}, \overline{\mu}^{\text{R+}}, \underline{\mu}^{\text{R}-}, \overline{\mu}^{\text{R}-}\right\}$, and $\Phi^{\text{M,R}} = \left\{\underline{u}_1^{\text{O+}}, \underline{u}_2^{\text{O+}}, \overline{u}_1^{\text{O+}}, \overline{u}_2^{\text{O+}}, \underline{u}_1^{\text{O}-}, \underline{u}_2^{\text{O}-}, \overline{u}_1^{\text{O}-}, \overline{u}_2^{\text{O}-}, \underline{u}^{\text{R+}}, \overline{u}^{\text{R+}}, \underline{u}^{\text{R}-}, \overline{u}^{\text{R}-}\right\}$ include the optimization variables of the MILP model.

Parameters N and M are considered to be equal to 10000 and 100, respectively. Then, the above MILP model is solved using CPLEX [5] under GAMS [4].

Table 6.18 provides the offering and bidding decisions of the VPP in the DA energy market, as well as the power traded in this market. The second and third columns present the offer quantity and price, respectively. The fourth column determines the amount of power that is sold to the DA energy market. The last column provides the status of the conventional power plant in the VPP. The VPP is willing to offer 12 MW at $50/MWh. Considering the offering and bidding decisions of the remaining market participants, the VPP sells 12 MW in the DA energy market, being the price equal to $50/MWh. Therefore, the VPP achieves $600 by selling energy in the DA energy market.

Table 6.19 presents the offering and bidding decisions for up and down reserves in the reserve market. The second and third columns provide the capacity and the price for up reserve, respectively. The fourth and fifth columns present the capacity and the price for down reserve, respectively. The sixth and seventh columns determine the up- and down-reserve capacities scheduled to the VPP, respectively. Observe that the VPP decides to sell both up- and down-reserve capacities. The prices for up- and down-reserves are $20/MW and $10/MW, respectively. Therefore, the VPP achieves $500 due to its participation in the reserve market.

The dispatch decisions by the VPP for each scenario realization are presented in Table 6.20. The second, third, and fourth columns provide the power production of the conventional power plant, the power consumption of the demand, and the power production of the PV generating unit for each scenario realization, respectively. In scenario 1, the power consumption utility and production cost are equal to $1800 and $1800, respectively. Thus, the profit obtained due to the dispatch decisions in scenario 1 is zero. However, the power consumption utility and the production cost are $1800 and $450 in scenario 2, and thus the profit is $1350. This yields an expected profit of $540 due to the dispatch decisions.

The objective function of the stochastic MPEC model is equal to $1640 as the summation of three terms, namely the profit in the DA energy market, the profit in the reserve market, and the expected profit due to the dispatch decisions.

Table 6.21 Illustrative Example 6.12: optimal schedules of two producers and two demands in the DA energy market

Variable	$p_1^{\mathrm{G,E}}$	$p_2^{\mathrm{G,E}}$	$p_1^{\mathrm{Q,E}}$	$p_2^{\mathrm{Q,E}}$
Value [MW]	50	0	12	50

Table 6.22 Illustrative Example 6.12: optimal schedules of units in the reserve market

Variable	$p_1^{\mathrm{O+}}$	$p_2^{\mathrm{O+}}$	$p_1^{\mathrm{O-}}$	$p_2^{\mathrm{O-}}$
Value [MW]	20	0	0	0

The optimal schedules of the other two producers and two demands in the energy market are provided in Table 6.21. The second and third columns provide the amount of power to be sold by the two producers. The fourth and fifth columns present the amount of power to be bought by the two demands. Additionally, the schedules of the units other than the VPP in the reserve market are provided in Table 6.22. The second and third columns present the up-reserve capacity allocated to units 1 and 2, respectively, while the fourth and fifth columns present the down-reserve capacity allocated to units 1 and 2, respectively.

□

An input GAMS code for solving Illustrative Example 6.12 is provided in Sect. 6.7.

6.6 Summary and Further Reading

This chapter describes the scheduling problem of a price-maker VPP participating in the DA energy and reserve markets. The problem is formulated as a bi-level model that considers the technical and economical constraints of the units in the VPP in the upper level, as well as the market-clearing problems of the energy and reserve markets in the lower level. The market-clearing problems are formulated using the corresponding equilibrium constraints, rendering an MPEC. This equilibrium-based model allows the VPP to alter the prices in the DA energy and reserve markets to its own benefit. This is done by strategic offering and bidding decisions, comprising both the price and power quantities offered in the markets. To handle uncertainties involved, the scheduling problem is formulated using both a deterministic and a stochastic programming approach.

Additional information on the fundamentals of the complementarity models presented in this chapter can be found in [3]. Moreover, the basis of the theory and formulation of stochastic programming models can be found in [2]. Finally, the interested readers are referred to the technical papers [6] and [7] to find further details on the mathematical formulation of stochastic complementarity models for scheduling price-maker VPPs in energy markets, as well as to its application in more realistic case studies.

6.7 GAMS Codes

This section provides the GAMS codes used to solve some of the illustrative examples analyzed through the chapter.

6.7.1 Energy Market-Clearing Problem: Primal Problem

An input GAMS code [4] used to solve the energy market-clearing problem in Illustrative Example 6.2 is presented below:

```
SETS
q         'demands other than the VPP'                  /q1*q2/
g         'generating units other than the VPP'         /g1*g2/
t         'time periods'                                /t1/;
********************** VPP Data*********************************
PARAMETER NU_E_POS(t)        'Offer price of the VPP  [$/MWh]'
/t1       40/;
PARAMETER NU_E_NEG(t)        'Bid price of the VPP  [$/MWh]'
/t1       0/;
PARAMETER P_E_POS_MAX(t)      'Power offer quantity of the VPP [MW]'
/t1       30/;
PARAMETER P_E_NEG_MAX(t)      'Power bid quantity of the VPP [MW]'
/t1       0/;
********************** Generating Unit Data**********************
TABLE CHI_G_E(g,t)      'Offer price of producers [$/MWh]'
          t1
g1        30
g2        50
TABLE P_G_E_MAX(g,t)    'Power offer quantity of generating units [MW]'
          t1
g1        50
g2        50
********************** Demand Data*******************************
TABLE CHI_Q_E(q,t)       'Bid price of demands   [$/MWh]'
          t1
q1        50
q2        70
TABLE P_Q_E_MAX(q,t)      'Power bid quantity of demands [MW]'
          t1
q1        50
q2        50;
VARIABLES
Z              'Value of the objective function';
POSITIVE VARIABLES
P_E_POS(t)  'Power scheduled to be produced by the VPP [MW]'
P_E_NEG(t)  'Power scheduled to be consumed by the VPP [MW]'
P_G_E(g,t)  'Power scheduled to be produced by generating units [MW]'

P_Q_E(q,t)  'Power scheduled to be consumed by demands [MW]';
EQUATIONS
DAMCPobj, DAMCP1, DAMCP2, DAMCP3, DAMCP4, DAMCP5;
DAMCPobj..                   Z=E=SUM(t,NU_E_POS(t)*P_E_POS(t)+SUM
    (g,CHI_G_E(g,t)*P_G_E(g,t))-NU_E_NEG(t)*P_E_NEG(t)-SUM(q,
    CHI_Q_E(q,t)*P_Q_E(q,t)));
DAMCP1(t)..                  P_E_NEG(t)+SUM(q,P_Q_E(q,t))-P_E_POS
    (t)-SUM(g,P_G_E(g,t))=E=0;
DAMCP2(g,t)..                P_G_E(g,t)=L=P_G_E_MAX(g,t);
DAMCP3(q,t)..                P_Q_E(q,t)=L=P_Q_E_MAX(q,t);
DAMCP4(t)..                  P_E_POS(t)=L=P_E_POS_MAX(t);
DAMCP5(t)..                  P_E_NEG(t)=L=P_E_NEG_MAX(t);
MODEL DAMCP /ALL/;
PARAMETERS state;
OPTION OPTCA=0;
OPTION OPTCR=0;
SOLVE DAMCP USING LP MINIMIZING Z;
state=DAMCP.modelstat;
DISPLAY state, DAMCP1.m;
```

6.7.2 Energy Market-Clearing Problem: KKT Conditions

An input GAMS code [5] used to solve the KKT conditions corresponding to the energy market-clearing problem described in Illustrative Example 6.5 is provided below:

```
SETS
q        'demands other than the VPP'                  /q1*q2/
g        'generating units other than the VPP'         /g1*g2/
t        'time periods'                                /t1/;
********************* VPP Data**********************************
PARAMETER NU_E_POS(t)        'Offer price of the VPP  [$/MWh]'
/t1      40/;
PARAMETER NU_E_NEG(t)        'Bid price of the VPP  [$/MWh]'
/t1      0/;
PARAMETER P_E_POS_MAX(t)     'Power offer quantity of the VPP [MW]'
/t1      30/;
PARAMETER P_E_NEG_MAX(t)      'Power bid quantity of the VPP [MW]'
/t1      0/;
********************* Generating Unit Data**********************
TABLE CHI_G_E(g,t)       'Offer price of producers [$/MWh]'
         t1
g1       30
g2       50
TABLE P_G_E_MAX(g,t)      'Power offer quantity of generating units [MW]'
         t1
g1       50
g2       50
********************* Demand Data*******************************
TABLE CHI_Q_E(q,t)       'Bid price of demands   [$/MWh]'
         t1
q1       50
q2       70
TABLE P_Q_E_MAX(q,t)      'Power bid quantity of demands [MW]'
         t1
q1       50
q2       50;
SCALARS
N_G_E_MIN         /10000/
N_G_E_MAX         /10000/
M_G_E_MIN         /100/
M_G_E_MAX         /100/
N_Q_E_MIN         /10000/
N_Q_E_MAX         /10000/
M_Q_E_MIN         /100/
M_Q_E_MAX         /100/
N_E_POS_MIN       /10000/
N_E_POS_MAX       /10000/
M_E_POS_MIN       /100/
M_E_POS_MAX       /100/
N_E_NEG_MIN       /10000/
N_E_NEG_MAX       /10000/
M_E_NEG_MIN       /100/
M_E_NEG_MAX       /100/;
```

```
VARIABLES
Z                          'Value of the objective function'
LAMBDA_E(t)                'Energy market price  [$/MWh]';
POSITIVE VARIABLES
P_E_POS(t)  'Power scheduled to be produced by the VPP [MW]'
P_E_NEG(t)  'Power scheduled to be consumed by the VPP [MW]'
P_G_E(g,t)  'Power scheduled to be produced by generating
            units [MW]'
P_Q_E(q,t)  'Power scheduled to be consumed by demands [MW]'
MU_E_POS_MAX(t)             'Dual variable'
MU_E_POS_MIN(t)             'Dual variable'
MU_E_NEG_MAX(t)             'Dual variable'
MU_E_NEG_MIN(t)             'Dual variable'
MU_G_E_MAX(g,t)             'Dual variable'
MU_G_E_MIN(g,t)             'Dual variable'
MU_Q_E_MAX(q,t)             'Dual variable'
MU_Q_E_MIN(q,t)             'Dual variable';
BINARY VARIABLES
U_G_E_MIN(g,t)
U_G_E_MAX(g,t)
U_Q_E_MIN(q,t)
U_Q_E_MAX(q,t)
U_E_POS_MIN(t)
U_E_POS_MAX(t)
U_E_NEG_MIN(t)
U_E_NEG_MAX(t);
EQUATIONS
DAMCPobj,DAMCP2, DAMCP3, DAMCP4, DAMCP5,
DAMCP_KKT1, DAMCP_KKT2, DAMCP_KKT3, DAMCP_KKT4, DAMCP_KKT5,
DAMCP_KKT6a, DAMCP_KKT6b, DAMCP_KKT7a, DAMCP_KKT7b, DAMCP_KKT8a,
    DAMCP_KKT8b, DAMCP_KKT9a, DAMCP_KKT9b,
DAMCP_KKT10a, DAMCP_KKT10b, DAMCP_KKT11a, DAMCP_KKT11b,
    DAMCP_KKT12a, DAMCP_KKT12b, DAMCP_KKT13a, DAMCP_KKT13b;
DAMCPobj..                 Z=E=1;
DAMCP2(g,t)..      P_G_E(g,t)=L=P_G_E_MAX(g,t);
DAMCP3(q,t)..      P_Q_E(q,t)=L=P_Q_E_MAX(q,t);
DAMCP4(t)..        P_E_POS(t)=L=P_E_POS_MAX(t);
DAMCP5(t)..        P_E_NEG(t)=L=P_E_NEG_MAX(t);
DAMCP_KKT1(t)..    P_E_NEG(t)+SUM(q,P_Q_E(q,t))-P_E_POS(t)-SUM(g,
    P_G_E(g,t))=E=0;
DAMCP_KKT2(g,t).. CHI_G_E(g,t)-LAMBDA_E(t)+MU_G_E_MAX(g,t)-
    MU_G_E_MIN(g,t)=E=0;
DAMCP_KKT3(q,t).. -CHI_Q_E(q,t)+LAMBDA_E(t)+MU_Q_E_MAX(q,t)-
    MU_Q_E_MIN(q,t)=E=0;
DAMCP_KKT4(t)..    NU_E_POS(t)-LAMBDA_E(t)+MU_E_POS_MAX(t)-
    MU_E_POS_MIN(t)=E=0;
DAMCP_KKT5(t)..    -NU_E_NEG(t)+LAMBDA_E(t)+MU_E_NEG_MAX(t)-
    MU_E_NEG_MIN(t)=E=0;
DAMCP_KKT6a(g,t).. MU_G_E_MIN(g,t)=L=N_G_E_MIN*U_G_E_MIN(g,t);
DAMCP_KKT6b(g,t).. P_G_E(g,t)=L=M_G_E_MIN*(1-U_G_E_MIN(g,t));
```

```
DAMCP_KKT7a(g,t).. MU_G_E_MAX(g,t)=L=N_G_E_MAX*U_G_E_MAX(g,t);
DAMCP_KKT7b(g,t).. P_G_E_MAX(g,t)-P_G_E(g,t)=L=M_G_E_MAX*(1-
    U_G_E_MAX(g,t));
DAMCP_KKT8a(q,t).. MU_Q_E_MIN(q,t)=L=N_Q_E_MIN*U_Q_E_MIN(q,t);
DAMCP_KKT8b(q,t).. P_Q_E(q,t)=L=M_Q_E_MIN*(1-U_Q_E_MIN(q,t));
DAMCP_KKT9a(q,t).. MU_Q_E_MAX(q,t)=L=N_Q_E_MAX*U_Q_E_MAX(q,t);
DAMCP_KKT9b(q,t).. P_Q_E_MAX(q,t)-P_Q_E(q,t)=L=M_Q_E_MAX*(1-
    U_Q_E_MAX(q,t));
DAMCP_KKT10a(t)..  MU_E_POS_MIN(t)=L=N_E_POS_MIN*U_E_POS_MIN(t);
DAMCP_KKT10b(t)..  P_E_POS(t)=L=M_E_POS_MIN*(1-U_E_POS_MIN(t));
DAMCP_KKT11a(t)..  MU_E_POS_MAX(t)=L=N_E_POS_MAX*U_E_POS_MAX(t);
DAMCP_KKT11b(t)..  P_E_POS_MAX(t)-P_E_POS(t)=L=M_E_POS_MAX*(1-
    U_E_POS_MAX(t));
DAMCP_KKT12a(t)..  MU_E_NEG_MIN(t)=L=N_E_NEG_MIN*U_E_NEG_MIN(t);
DAMCP_KKT12b(t)..  P_E_NEG(t)=L=M_E_NEG_MIN*(1-U_E_NEG_MIN(t));
DAMCP_KKT13a(t)..   MU_E_NEG_MAX(t)=L=N_E_NEG_MAX*U_E_NEG_MAX(t);
DAMCP_KKT13b(t)..  P_E_NEG_MAX(t)-P_E_NEG(t)=L=M_E_NEG_MAX*(1-
    U_E_NEG_MAX(t));
MODEL DAMCPKKT /ALL/;
PARAMETERS state, obj;
OPTION OPTCA=0;
OPTION OPTCR=0;
SOLVE DAMCPKKT USING MIP MAXIMIZING Z;
state=DAMCPKKT.modelstat;
obj=SUM(t,NU_E_POS(t)*P_E_POS.l(t)+SUM(g,CHI_G_E(g,t)*P_G_E.l(g,t
    ))-NU_E_NEG(t)*P_E_NEG.l(t)-SUM(q,CHI_Q_E(q,t)*P_Q_E.l(q,t)))
    ;
DISPLAY state, obj;
```

6.7.3 Energy Market-Clearing Problem: Dual Problem

An input GAMS code [5] that is used to solve the dual problem related to the energy market-clearing problem analyzed in Illustrative Example 6.6 is presented below:

```
SETS
q          'demands other than the VPP'                    /q1*q2/
g          'generating units other than the VPP'           /g1*g2/
t          'time periods'                                  /t1/;
********************* VPP Data*********************************
PARAMETER NU_E_POS(t)         'Offer price of the VPP   [$/MWh]'
/t1        40/;
PARAMETER NU_E_NEG(t)         'Bid price of the VPP   [$/MWh]'
/t1        0/;
PARAMETER P_E_POS_MAX(t)       'Power offer quantity of the VPP [MW
    ]'
/t1        30/;
PARAMETER P_E_NEG_MAX(t)       'Power bid quantity of the VPP [MW]'
/t1        0/;
********************* Generating Unit Data**********************
```

```
TABLE CHI_G_E(g,t)       'Offer price of producers [$/MWh]'
          t1
g1        30
g2        50
TABLE P_G_E_MAX(g,t)      'Power offer quantity of generating units [MW]'
          t1
g1        50
g2        50
********************* Demand Data********************************
TABLE CHI_Q_E(q,t)       'Bid price of demands  [$/MWh]'
          t1
q1        50
q2        70
TABLE P_Q_E_MAX(q,t)      'Power bid quantity of demands [MW]'
          t1
q1        50
q2        50;
VARIABLES
Z                         'Value of the objective function'
LAMBDA_E(t)               'Dual variable: energy market price [$/MWh]';
POSITIVE VARIABLES
MU_E_POS_MAX(t)           'Dual variable'
MU_E_POS_MIN(t)           'Dual variable'
MU_E_NEG_MAX(t)           'Dual variable'
MU_E_NEG_MIN(t)           'Dual variable'
MU_G_E_MAX(g,t)           'Dual variable'
MU_G_E_MIN(g,t)           'Dual variable'
MU_Q_E_MAX(q,t)           'Dual variable'
MU_Q_E_MIN(q,t)           'Dual variable';
EQUATIONS
DAMCPobj, DAMCP_DUAL1, DAMCP_DUAL2, DAMCP_DUAL3, DAMCP_DUAL4;
DAMCPobj..          Z=E=SUM(t,-P_E_POS_MAX(t)*MU_E_POS_MAX(t)-SUM(g,P_G_E_MAX(g,t)*MU_G_E_MAX(g,t))-P_E_NEG_MAX(t)*MU_E_NEG_MAX(t)-SUM(q,P_Q_E_MAX(q,t)*MU_Q_E_MAX(q,t)));
DAMCP_DUAL1(g,t).. CHI_G_E(g,t)-LAMBDA_E(t)+MU_G_E_MAX(g,t)-MU_G_E_MIN(g,t)=E=0;
DAMCP_DUAL2(q,t).. -CHI_Q_E(q,t)+LAMBDA_E(t)+MU_Q_E_MAX(q,t)-MU_Q_E_MIN(q,t)=E=0;
DAMCP_DUAL3(t)..   NU_E_POS(t)-LAMBDA_E(t)+MU_E_POS_MAX(t)-MU_E_POS_MIN(t)=E=0;
DAMCP_DUAL4(t)..   -NU_E_NEG(t)+LAMBDA_E(t)+MU_E_NEG_MAX(t)-MU_E_NEG_MIN(t)=E=0;
MODEL DAMCP_DUAL /ALL/;
PARAMETERS state;
OPTION OPTCA=0;
OPTION OPTCR=0;
SOLVE DAMCP_DUAL USING LP MAXIMIZING Z;
state=DAMCP_DUAL.modelstat;
DISPLAY state;
```

6.7.4 Deterministic Scheduling in the Energy Market

An input GAMS code [4] used to solve the deterministic MILP model for the scheduling of a price-maker VPP in the DA energy market described in Illustrative Example 6.8 is provided below:

```
SETS
c         'conventional power plants in the VPP'              /c1/
d         'demands in the VPP'                                /d1/
r         'stochastic renewable generating unit in the VPP' /r1/
q         'demands other than the VPP'                        /q1*q2 /
g         'generating units other than the VPP'               /g1*g2 /
t         'time periods'                                      /t1/;
********************* VPP Data**********************************
PARAMETER C_C_F(c)    'Fixed cost of conventional generating units [$]'
/c1       0/;
PARAMETER C_C_V(c)    'Variable cost of conventional generating units [$/MWh]'
/c1       45/;
PARAMETER P_C_MAX(c) 'Maximum power production of conventional generating units [MW]'
/c1        40/;
PARAMETER P_C_MIN(c) 'Minimum power production of conventional generating units [MW]'
/c1        0/;
TABLE U_D(d,t)        'Utility of demands [$/MWh]'
          t1
d1        60;
PARAMETER E_D_MIN(d) 'Minimum daily energy consumption of demands [MWh]'
/d1       20/;
TABLE P_D_MAX(d,t)    'Upper bound of the power consumption of demands [MW]'
          t1
d1        30;
TABLE P_D_MIN(d,t)    'Lower bound of the power consumption of demands [MW]'
          t1
d1        0;
TABLE P_R_A(r,t)      'Available generation of  stochastic renewable generating units [MW]'
          t1
r1        8;
PARAMETER P_V_POS_MAX 'Power capacity of generation assets available by the VPP [MW]' /50/;
PARAMETER P_V_NEG_MAX 'Power capacity of consumption assets available by the VPP [MW]' /30/;
SCALAR delta           'Duration of time periods [h]' /1/;
********************* Generating Unit Data**********************
TABLE CHI_G_E(g,t)       'Offer price of producers [$/MWh]'
          t1
g1        30
g2        50
```

```
TABLE P_G_E_MAX(g,t)      'Power offer quantity of generating units [MW]'
          t1
g1        50
g2        50
********************* Demand Data*********************************
TABLE CHI_Q_E(q,t)       'Bid price of demands  [$/MWh]'
          t1
q1        50
q2        70
TABLE P_Q_E_MAX(q,t)      'Power bid quantity of demands [MW]'
          t1
q1        50
q2        50;
SCALARS
N_G_E_MIN         /10000/
N_G_E_MAX        /10000/
M_G_E_MIN        /100/
M_G_E_MAX        /100/
N_Q_E_MIN        /10000/
N_Q_E_MAX        /10000/
M_Q_E_MIN        /100/
M_Q_E_MAX        /100/
N_E_POS_MIN      /10000/
N_E_POS_MAX      /10000/
M_E_POS_MIN      /100/
M_E_POS_MAX      /100/
N_E_NEG_MIN      /10000/
N_E_NEG_MAX      /10000/
M_E_NEG_MIN      /100/
M_E_NEG_MAX      /100/;
VARIABLES
Z               'Value of the objective function [$]'
LAMBDA_E(t)     'Energy market price  [$/MWh]'
OBJ1            'Utility minus production cost [$]'
OBJ2            'Revenue minus energy procurement cost in the DA energy market [$]';
POSITIVE VARIABLES
P_C(c,t)        'Power generation of conventional power plants [MW]'
P_D(d,t)        'Power consumption of demands [MW]'
P_R(r,t)        'Production of renewable generating unitst [MW]'
NU_E_POS(t)     'Offer price decided by the VPP [$/MWh]'
NU_E_NEG(t)     'Bid price decided by the VPP [$/MWh]'
P_E_POS_MAX(t) 'Power offer quantity by the VPP [MW]'
P_E_NEG_MAX(t) 'Power bid quantity by the VPP [MW]'
P_E_POS(t)   'Total power scheduled to be produced by the VPP [MW]'
P_E_NEG(t)   'Total power scheduled to be consumed by the VPP [MW]'
P_G_E(g,t)   'Power scheduled to be produced by generating units [MW]'
P_Q_E(q,t)   'Power scheduled to be consumed by demands [MW]'
```

```
MU_E_POS_MAX(t) 'Dual variable'
MU_E_POS_MIN(t) 'Dual variable'
MU_E_NEG_MAX(t) 'Dual variable'
MU_E_NEG_MIN(t) 'Dual variable'
MU_G_E_MAX(g,t) 'Dual variable'
MU_G_E_MIN(g,t) 'Dual variable'
MU_Q_E_MAX(q,t) 'Dual variable'
MU_Q_E_MIN(q,t) 'Dual variable';
BINARY VARIABLES
U_C(c,t)          'Status of conventional power plants in VPP'
U_G_E_MIN(g,t)    'Auxiliary variable'
U_G_E_MAX(g,t)    'Auxiliary variable'
U_Q_E_MIN(q,t)    'Auxiliary variable'
U_Q_E_MAX(q,t)    'Auxiliary variable'
U_E_POS_MIN(t)    'Auxiliary variable'
U_E_POS_MAX(t)    'Auxiliary variable'
U_E_NEG_MIN(t)    'Auxiliary variable'
U_E_NEG_MAX(t)    'Auxiliary variable';
EQUATIONS
VPPDAobj     'Total profit',
VPPDAOBJ1    'Utility minus production cost',
VPPDAOBJ2    'Revenue minus energy procurement cost in the DA
    energy market',
VPPDA1       'Power balance constraint',
VPPDA2       'Minimum daily energy consumption of demands in the
    VPP',
VPPDA_MIN3   'Lower bound of power consumption of demands in the
    VPP',
VPPDA_MAX3   'Upper bound of power consumption of demands in the
    VPP',
VPPDA_MIN4   'Lower bound of power production of conventional
    power plants in the VPP',
VPPDA_MAX4   'Upper bound of power production of conventional
    power plants in the VPP',
VPPDA5       'Upper bound of power production of stochastic
    renewable generating unis in the VPP',
VPPDA6       'Maximum power that can be sold by the VPP',
VPPDA7       'Maximum power that can be bought by the VPP',
DAMCP2, DAMCP3, DAMCP4, DAMCP5 Constraints of the energy market
    clearing problem,
DAMCP_KKT1, DAMCP_KKT2, DAMCP_KKT3, DAMCP_KKT4, DAMCP_KKT5 KKT,
    DAMCP_KKT6a, DAMCP_KKT6b, DAMCP_KKT7a, DAMCP_KKT7b,
    DAMCP_KKT8a, DAMCP_KKT8b, DAMCP_KKT9a, DAMCP_KKT9b,
    DAMCP_KKT10a, DAMCP_KKT10b, DAMCP_KKT11a, DAMCP_KKT11b,
    DAMCP_KKT12a, DAMCP_KKT12b, DAMCP_KKT13a, DAMCP_KKT13b KKT
    conditions;
VPPDAobj..        Z=E=OBJ1+OBJ2;
VPPDAOBJ1..       OBJ1=E=SUM(t,SUM(d,U_D(d,t)*P_D(d,t)*delta) -
    SUM(c,C_C_F(c)*U_C(c,t)+C_C_V(c)*P_C(c,t)*delta));
VPPDAOBJ2..       OBJ2=E=SUM(t,SUM(q,CHI_Q_E(q,t)*P_Q_E(q,t)) -
    SUM(g,CHI_G_E(g,t)*P_G_E(g,t)) - SUM(g,P_G_E_MAX(g,t)*
    MU_G_E_MAX(g,t)) - SUM(q,P_Q_E_MAX(q,t)*MU_Q_E_MAX(q,t)));
VPPDA1(t)..       SUM(c,P_C(c,t))+SUM(r,P_R(r,t))-SUM(d,P_D(d,t))
    =E=P_E_POS(t)-P_E_NEG(t);
```

```
VPPDA2(d)..         SUM(t,P_D(d,t)*delta)=G=E_D_MIN(d);
VPPDA_MIN3(d,t).. P_D_MIN(d,t)=L=P_D(d,t);
VPPDA_MAX3(d,t).. P_D(d,t)=L=P_D_MAX(d,t);
VPPDA_MIN4(c,t).. P_C_MIN(c)*U_C(c,t)=L=P_C(c,t);
VPPDA_MAX4(c,t).. P_C(c,t)=l=P_C_MAX(c)*U_C(c,t);
VPPDA5(r,t)..       P_R(r,t)=L=P_R_A(r,t);
VPPDA6(t)..         P_E_POS_MAX(t)=L=P_V_POS_MAX;
VPPDA7(t)..         P_E_NEG_MAX(t)=L=P_V_NEG_MAX;
DAMCP2(g,t)..       P_G_E(g,t)=L=P_G_E_MAX(g,t);
DAMCP3(q,t)..       P_Q_E(q,t)=L=P_Q_E_MAX(q,t);
DAMCP4(t)..         P_E_POS(t)=L=P_E_POS_MAX(t);
DAMCP5(t)..         P_E_NEG(t)=L=P_E_NEG_MAX(t);
DAMCP_KKT1(t)..     P_E_NEG(t)+SUM(q,P_Q_E(q,t))-P_E_POS(t)-SUM(g,
    P_G_E(g,t))=E=0;
DAMCP_KKT2(g,t).. CHI_G_E(g,t)-LAMBDA_E(t)+MU_G_E_MAX(g,t)-
    MU_G_E_MIN(g,t)=E=0;
DAMCP_KKT3(q,t).. -CHI_Q_E(q,t)+LAMBDA_E(t)+MU_Q_E_MAX(q,t)-
    MU_Q_E_MIN(q,t)=E=0;
DAMCP_KKT4(t)..     NU_E_POS(t)-LAMBDA_E(t)+MU_E_POS_MAX(t)-
    MU_E_POS_MIN(t)=E=0;
DAMCP_KKT5(t)..     -NU_E_NEG(t)+LAMBDA_E(t)+MU_E_NEG_MAX(t)-
    MU_E_NEG_MIN(t)=E=0;
DAMCP_KKT6a(g,t).. MU_G_E_MIN(g,t)=L=N_G_E_MIN*U_G_E_MIN(g,t);
DAMCP_KKT6b(g,t).. P_G_E(g,t)=L=M_G_E_MIN*(1-U_G_E_MIN(g,t));
DAMCP_KKT7a(g,t).. MU_G_E_MAX(g,t)=L=N_G_E_MAX*U_G_E_MAX(g,t);
DAMCP_KKT7b(g,t).. P_G_E_MAX(g,t)-P_G_E(g,t)=L=M_G_E_MAX*(1-
    U_G_E_MAX(g,t));
DAMCP_KKT8a(q,t).. MU_Q_E_MIN(q,t)=L=N_Q_E_MIN*U_Q_E_MIN(q,t);
DAMCP_KKT8b(q,t).. P_Q_E(q,t)=L=M_Q_E_MIN*(1-U_Q_E_MIN(q,t));
DAMCP_KKT9a(q,t).. MU_Q_E_MAX(q,t)=L=N_Q_E_MAX*U_Q_E_MAX(q,t);
DAMCP_KKT9b(q,t).. P_Q_E_MAX(q,t)-P_Q_E(q,t)=L=M_Q_E_MAX*(1-
    U_Q_E_MAX(q,t));
DAMCP_KKT10a(t).. MU_E_POS_MIN(t)=L=N_E_POS_MIN*U_E_POS_MIN(t);
DAMCP_KKT10b(t).. P_E_POS(t)=L=M_E_POS_MIN*(1-U_E_POS_MIN(t));
DAMCP_KKT11a(t).. MU_E_POS_MAX(t)=L=N_E_POS_MAX*U_E_POS_MAX(t);
DAMCP_KKT11b(t).. P_E_POS_MAX(t)-P_E_POS(t)=L=M_E_POS_MAX*(1-
    U_E_POS_MAX(t));
DAMCP_KKT12a(t).. MU_E_NEG_MIN(t)=L=N_E_NEG_MIN*U_E_NEG_MIN(t);
DAMCP_KKT12b(t).. P_E_NEG(t)=L=M_E_NEG_MIN*(1-U_E_NEG_MIN(t));
DAMCP_KKT13a(t).. MU_E_NEG_MAX(t)=L=N_E_NEG_MAX*U_E_NEG_MAX(t);
DAMCP_KKT13b(t).. P_E_NEG_MAX(t)-P_E_NEG(t)=L=M_E_NEG_MAX*(1-
    U_E_NEG_MAX(t));
MODEL DA_MPEC /ALL/;
PARAMETERS state, obj;
OPTION OPTCA=0;
OPTION OPTCR=0;
SOLVE DA_MPEC USING MIP MAXIMIZING Z;
state=DA_MPEC.modelstat;
obj=OBJ1.l+SUM(t,LAMBDA_E.l(t)*(P_E_POS.l(t)-P_E_NEG.l(t)));
DISPLAY state, obj;
```

6.7.5 Stochastic Scheduling in the Energy Market

An input GAMS code [4] that is used to solve the stochastic MPEC model for scheduling a price-maker VPP in the DA energy market in the Illustrative Example 6.9 is provided below:

```
SETS
c     'conventional power plants in the VPP'                 /c1/
d     'demands in the VPP'                                   /d1/
r     'stochastic renewable generating unit in the VPP'      /r1/
q     'demands other than the VPP'                           /q1*q2/
g     'generating units other than the VPP'                  /g1*g2/
t     'time periods'                                         /t1/
ga    'scenarios'/;                                          /ga1*ga2
***************** VPP Data***************************
PARAMETER C_C_F(c)   'Fixed cost of conventional generating unit c [$]'
/c1      0/;
PARAMETER C_C_V(c)   'Variable cost of conventional generating unit c [$/MWh]'
/c1      45/;
PARAMETER P_C_MAX(c) 'Maximum power production of conventional generating unit c [MW]'
/c1       40/;
PARAMETER P_C_MIN(c) 'Minimum power production of conventional generating unit c [MW]'
/c1       0/;
TABLE U_D(d,t)       'Utility of demand d in time period [$/MWh]'
         t1
d1       60;
PARAMETER E_D_MIN(d) 'Minimum daily energy consumption of demand d [MWh]'
/d1      20/;
TABLE P_D_MAX(d,t)   'Upper bound of the power consumption of demand d in time period t [MW]'
         t1
d1       30;
TABLE P_D_MIN(d,t)   'Lower bound of the power consumption of demand d in time period t [MW]'
         t1
d1       0;
TABLE P_R_A(r,t,ga)  'Available generation of the stochastic renewable generating unit r in time period t in scenario ga [MW]'
         ga1     ga2
r1.t1    10      5;
PARAMETER P_V_POS_MAX       'Power capacity of generation assets available by the VPP [MW] /50/';
PARAMETER P_V_NEG_MAX       'Power capacity of consumption assets available by the VPP [MW] /30/';
SCALAR delta                'Duration of time periods [h] /1/';
********************** Generating Units Data ***********************
TABLE CHI_G_E(g,t)      'Offer price by producer g in the DA energy market in time period t [$/MWh]'
         t1
g1       30
g2       50
```

```
TABLE P_G_E_MAX(g,t)      'Power offer quantity by producer g in
    the DA energy market in time' period t[MW]
          t1
g1        50
g2        50;
********************* Demands Data*******************************
TABLE CHI_Q_E(q,t)        'Bid price by demand q in the DA energy
    market in time period t [$/MWh]'
          t1
q1        50
q2        70
TABLE P_Q_E_MAX(q,t)      'Power bid quantity by demand q in the
    DA energy market in time period t[MW]'
          t1
q1        50
q2        50;
******************** probability of scenarios********************
PARAMETER PI(ga)          ' Probability of scenario ga'
/ga1      0.6
ga2       0.4/;
SCALARS
N_G_E_MIN         /10000/
N_G_E_MAX         /10000/
M_G_E_MIN         /100/
M_G_E_MAX         /100/
N_Q_E_MIN         /10000/
N_Q_E_MAX         /10000/
M_Q_E_MIN         /100/
M_Q_E_MAX         /100/
N_E_POS_MIN       /10000/
N_E_POS_MAX       /10000/
M_E_POS_MIN       /100/
M_E_POS_MAX       /100/
N_E_NEG_MIN       /10000/
N_E_NEG_MAX       /10000/
M_E_NEG_MIN       /100/
M_E_NEG_MAX       /100/;
VARIABLES
Z                    'Value of the objective function [$]'
LAMBDA_E(t)         'Energy market price in time period t [$/MWh]'
OBJ1                 'Utility minus production cost [$]'
OBJ2                 'Revenue minus energy procurement cost in the
     DA energy market[$]';
POSITIVE VARIABLES
P_C(c,t,ga)                'Power generation of conventional power
    plant c in time period t in scenario ga [MW]'
P_D(d,t,ga)                'Power consumption of demand d in time
    period t in scenario ga [MW]'
P_R(r,t,ga)                'Production of renewable generating unit
     r in time period t in scenario ga [MW]'
NU_E_POS(t)                'Offer price decided by the VPP in the
    DA energy market in time period t[$/MWh]'
NU_E_NEG(t)                'Bid price decided by the VPP in the DA
    energy market in time period t[$/MWh]'
```

```
P_E_POS_MAX(t)            'Power offer quantity by the VPP in the
    DA energy market in time period t[MW]'
P_E_NEG_MAX(t)            'Power bid quantity by the VPP in the DA
     energy market in time period t[MW]'
P_E_POS(t)               'Total power scheduled to be produced by
    the VPP in the DA energy market [MW]'
P_E_NEG(t)               'Total power scheduled to be consumed by
    the VPP in the DA energy market [MW]'
P_G_E(g,t)               'Power scheduled to be produced by the
    gth generating unit in the DA energy market [MW]'
P_Q_E(q,t)               'Power scheduled to be consumed by the
    dth demand in the DA energy market [MW]'
MU_E_POS_MAX(t)           'Dual variable'
MU_E_POS_MIN(t)           'Dual variable'
MU_E_NEG_MAX(t)           'Dual variable'
MU_E_NEG_MIN(t)           'Dual variable'
MU_G_E_MAX(g,t)           'Dual variable'
MU_G_E_MIN(g,t)           'Dual variable'
MU_Q_E_MAX(q,t)           'Dual variable'
MU_Q_E_MIN(q,t)           'Dual variable';
BINARY VARIABLES
U_C(c,t)         'Binary variable equal to 1 if conventional
    power plant c is generating electricity in time period t,
    being 0 otherwise'
U_G_E_MIN(g,t)
U_G_E_MAX(g,t)
U_Q_E_MIN(q,t)
U_Q_E_MAX(q,t)
U_E_POS_MIN(t)
U_E_POS_MAX(t)
U_E_NEG_MIN(t)
U_E_NEG_MAX(t);
EQUATIONS
VPPDAobj         'Total profit',
VPPDAOBJ1        'Utility minus production cost',
VPPDAOBJ2        'Revenue minus energy procurement cost in the DA
     energy market',
VPPDA1           'Power balance constraint',
VPPDA2           'Minimum dailiy energy consumption of demands in
     the VPP',
VPPDA_MIN3       'Lower bound of power consumption by each demand
     in the VPP',
VPPDA_MAX3       'Upper bound of power consumption by each demand
     in the VPP',
VPPDA_MIN4       'Lower bound of power production by each
    conventional power plant in the VPP',
VPPDA_MAX4       'Upper bound of power production by each
    conventional power plant in the VPP',
VPPDA5           'Upper bound of power production by each
    stochastic renewable generating unit in the VPP',
VPPDA6           'Offer quantity bounded by maximum power that
    can be sold by the VPP in the DA energy market',
VPPDA7           'Bid quantity bounded by maximum power that can
    be bought from the VPP in the DA energy market',
```

```
DAMCP2, DAMCP3, DAMCP4, DAMCP5 'Constraints of the DA energy market clearing problem',
DAMCP_KKT1, DAMCP_KKT2, DAMCP_KKT3, DAMCP_KKT4, DAMCP_KKT5 'KKT conditions',
DAMCP_KKT6a, DAMCP_KKT6b, DAMCP_KKT7a, DAMCP_KKT7b, DAMCP_KKT8a, DAMCP_KKT8b, DAMCP_KKT9a, DAMCP_KKT9b 'KKT conditions',
DAMCP_KKT10a, DAMCP_KKT10b, DAMCP_KKT11a, DAMCP_KKT11b, DAMCP_KKT12a, DAMCP_KKT12b, DAMCP_KKT13a, DAMCP_KKT13b 'KKT conditions';
VPPDAobj..                Z=E=OBJ1+OBJ2;
VPPDAOBJ1..               OBJ1=E=SUM(ga,PI(ga)*SUM(t,SUM(d,U_D(d,t)*P_D(d,t,ga)*delta) - SUM(c,C_C_F(c)*U_C(c,t)+C_C_V(c)*P_C(c,t,ga)*delta)));
VPPDAOBJ2..               OBJ2=E=SUM(t,SUM(q,CHI_Q_E(q,t)*P_Q_E(q,t)) - SUM(g,CHI_G_E(g,t)*P_G_E(g,t)) - SUM(g,P_G_E_MAX(g,t)*MU_G_E_MAX(g,t)) - SUM(q,P_Q_E_MAX(q,t)*MU_Q_E_MAX(q,t)));
VPPDA1(t,ga)..            SUM(c,P_C(c,t,ga))+SUM(r,P_R(r,t,ga))-SUM(d,P_D(d,t,ga))=E=P_E_POS(t)-P_E_NEG(t);
VPPDA2(d,ga)..            SUM(t,P_D(d,t,ga)*delta)=G=E_D_MIN(d);
VPPDA_MIN3(d,t,ga)..      P_D_MIN(d,t)=L=P_D(d,t,ga);
VPPDA_MAX3(d,t,ga)..      P_D(d,t,ga)=L=P_D_MAX(d,t);
VPPDA_MIN4(c,t,ga)..      P_C_MIN(c)*U_C(c,t)=L=P_C(c,t,ga);
VPPDA_MAX4(c,t,ga)..      P_C(c,t,ga)=l=P_C_MAX(c)*U_C(c,t);
VPPDA5(r,t,ga)..          P_R(r,t,ga)=L=P_R_A(r,t,ga);
VPPDA6(t)..               P_E_POS_MAX(t)=L=P_V_POS_MAX;
VPPDA7(t)..               P_E_NEG_MAX(t)=L=P_V_NEG_MAX;
DAMCP2(g,t)..             P_G_E(g,t)=L=P_G_E_MAX(g,t);
DAMCP3(q,t)..             P_Q_E(q,t)=L=P_Q_E_MAX(q,t);
DAMCP4(t)..               P_E_POS(t)=L=P_E_POS_MAX(t);
DAMCP5(t)..               P_E_NEG(t)=L=P_E_NEG_MAX(t);
DAMCP_KKT1(t)..           P_E_NEG(t)+SUM(q,P_Q_E(q,t))-P_E_POS(t)-SUM(g,P_G_E(g,t))=E=0;
DAMCP_KKT2(g,t)..         CHI_G_E(g,t)-LAMBDA_E(t)+MU_G_E_MAX(g,t)-MU_G_E_MIN(g,t)=E=0;
DAMCP_KKT3(q,t)..         -CHI_Q_E(q,t)+LAMBDA_E(t)+MU_Q_E_MAX(q,t)-MU_Q_E_MIN(q,t)=E=0;
DAMCP_KKT4(t)..           NU_E_POS(t)-LAMBDA_E(t)+MU_E_POS_MAX(t)-MU_E_POS_MIN(t)=E=0;
DAMCP_KKT5(t)..           -NU_E_NEG(t)+LAMBDA_E(t)+MU_E_NEG_MAX(t)-MU_E_NEG_MIN(t)=E=0;
DAMCP_KKT6a(g,t)..         MU_G_E_MIN(g,t)=L=N_G_E_MIN*U_G_E_MIN(g,t);
DAMCP_KKT6b(g,t)..         P_G_E(g,t)=L=M_G_E_MIN*(1-U_G_E_MIN(g,t));
DAMCP_KKT7a(g,t)..         MU_G_E_MAX(g,t)=L=N_G_E_MAX*U_G_E_MAX(g,t);
DAMCP_KKT7b(g,t)..         P_G_E_MAX(g,t)-P_G_E(g,t)=L=M_G_E_MAX*(1-U_G_E_MAX(g,t));
DAMCP_KKT8a(q,t)..         MU_Q_E_MIN(q,t)=L=N_Q_E_MIN*U_Q_E_MIN(q,t);
DAMCP_KKT8b(q,t)..         P_Q_E(q,t)=L=M_Q_E_MIN*(1-U_Q_E_MIN(q,t));
DAMCP_KKT9a(q,t)..         MU_Q_E_MAX(q,t)=L=N_Q_E_MAX*U_Q_E_MAX(q,t);
```

```
DAMCP_KKT9b(q,t)..          P_Q_E_MAX(q,t)-P_Q_E(q,t)=L=M_Q_E_MAX
    *(1-U_Q_E_MAX(q,t));
DAMCP_KKT10a(t)..           MU_E_POS_MIN(t)=L=N_E_POS_MIN*
    U_E_POS_MIN(t);
DAMCP_KKT10b(t)..           P_E_POS(t)=L=M_E_POS_MIN*(1-
    U_E_POS_MIN(t));
DAMCP_KKT11a(t)..           MU_E_POS_MAX(t)=L=N_E_POS_MAX*
    U_E_POS_MAX(t);
DAMCP_KKT11b(t)..           P_E_POS_MAX(t)-P_E_POS(t)=L=
    M_E_POS_MAX*(1-U_E_POS_MAX(t));
DAMCP_KKT12a(t)..           MU_E_NEG_MIN(t)=L=N_E_NEG_MIN*
    U_E_NEG_MIN(t);
DAMCP_KKT12b(t)..           P_E_NEG(t)=L=M_E_NEG_MIN*(1-
    U_E_NEG_MIN(t));
DAMCP_KKT13a(t)..           MU_E_NEG_MAX(t)=L=N_E_NEG_MAX*
    U_E_NEG_MAX(t);
DAMCP_KKT13b(t)..           P_E_NEG_MAX(t)-P_E_NEG(t)=L=
    M_E_NEG_MAX*(1-U_E_NEG_MAX(t));
MODEL DA_MPEC /ALL/;
PARAMETERS state, obj;
OPTION OPTCA=0;
OPTION OPTCR=0;
SOLVE DA_MPEC USING MIP MAXIMIZING Z;
state=DA_MPEC.modelstat;
obj=OBJ1.l+SUM(t,LAMBDA_E.l(t)*(P_E_POS.l(t)-P_E_NEG.l(t)));
DISPLAY state, obj;
```

6.7.6 *Stochastic Scheduling in the Energy and Reserve Markets*

An input GAMS code [4] that is used to solve the stochastic MPEC model for scheduling a price-maker VPP in the DA energy and reserve markets in the Illustrative Example 6.12 is provided below:

```
SETS
c        'conventional power plants in the VPP'              /c1/
d        'demands in the VPP'                                /d1/
r        'stochastic renewable generating unit in the VPP'/r1/
q        'demands other than the VPP'                        /q1*q2/
g        'generating units other than the VPP'               /g1*g2/
o        'units other than the VPP in the reserve market'    /o1*o2/
t        'time periods'                                      /t1/
ga        'scenarios'                                         /ga1*
    ga2/;
********************* VPP Data
    *************************************************
PARAMETER C_C_F(c)       'Fixed cost of conventional generating
    unit c [$]'
/c1        0/;
PARAMETER C_C_V(c)       'Variable cost of conventional generating
     unit c [$/MWh]'
/c1        45/;
PARAMETER P_C_MAX(c)     'Maximum power production of conventional
     generating unit c [MW]'
/c1         40/;
PARAMETER P_C_MIN(c)     'Minimum power production of conventional
     generating unit c [MW]'
```

```
/c1        0/;
TABLE U_D(d,t)            'Utility of demand d in time period [$/MWh]'
           t1
d1         60;
PARAMETER E_D_MIN(d)       'Minimum daily energy consumption of demand d [MWh]'
/d1        20/;
TABLE P_D_MAX(d,t)         'Upper bound of the power consumption of demand d in time period t [MW]'
           t1
d1         30;
TABLE P_D_MIN(d,t)         'Lower bound of the power consumption of demand d in time period t [MW]'
           t1
d1         0;
TABLE P_R_A(r,t)         'Available generation of the stochastic renewable generating unit r in time period t [MW]'
           t1
r1         8;
PARAMETER P_V_POS_MAX       'Power capacity of generation assets available by the VPP [MW]' /50/;
PARAMETER P_V_NEG_MAX       'Power capacity of consumption assets available by the VPP [MW]' /30/;
SCALAR delta                'Duration of time periods [h]' /1/;
********************* Generating Units Data *************************************
TABLE CHI_G_E(g,t)         'Offer price by producer g in the DA energy market in time period t [$/MWh]'
           t1
g1         30
g2         50
TABLE P_G_E_MAX(g,t)       'Power offer quantity by producer g in the DA energy market in time period t[MW]'
           t1
g1         50
g2         50;
********************* Demands Data ************************************************
TABLE CHI_Q_E(q,t)         'Bid price by demand q in the DA energy market in time period t [$/MWh]'
           t1
q1         50
q2         70
TABLE P_Q_E_MAX(q,t)       'Power bid quantity by demand q in the DA energy market in time period t[MW]'
           t1
q1         50
q2         50;
********************* Up-Reserve by other units *********************************
```

```
TABLE CHI_O_POS(o,t)       'Offer price by unit o in the reserve
    market for up-reserve in time period t [$/MW]'
          t1
o1        10
o2        20
TABLE P_O_POS_MAX(o,t)     'Power offer quantity by unit o in the
    reserve market for up-reserve in time period t[MW]'
          t1
o1        20
o2        20;
********************* Up-Reserve by other units
    *********************************
TABLE CHI_O_NEG(o,t)       'Bid price by unit o in the reserve
    market for down-reserve in time period t [$/MW]'
          t1
o1        10
o2        20
TABLE P_O_NEG_MAX(o,t)     'Power bid quantity by unit o in the
    reserve market for down-reserve in time period t[MW]'
          t1
o1        20
o2        20;
********************* Up- and down-reserve required in the
    reserve market*******
PARAMETER P_R_POS_REQ(t)     'Up-reserve capacity required in the
    reserve market in time period t [MW]'
/t1        30/;
PARAMETER P_R_NEG_REQ(t)     'Down-reserve capacity required in
    the reserve market in time period t [MW]'
/t1        30/;
********************* scenarios
    *************************************************
TABLE K_R_POS(t,ga)          'Up-reserve deployment requested by
    the system operator in time period t [pu]'
           ga1     ga2
t1        0.6      0;
TABLE K_R_NEG(t,ga)          'Down-reserve deployment requested by
     the system operator in time period t [pu]'
           ga1     ga2
t1        0        0.8;
PARAMETER PI(ga)           'Probability of scenario ga'
/ga1      0.6
ga2       0.4/;
SCALARS
N_G_E_MIN        /10000/
N_G_E_MAX        /10000/
M_G_E_MIN        /100/
M_G_E_MAX        /100/
N_Q_E_MIN        /10000/
N_Q_E_MAX        /10000/
M_Q_E_MIN        /100/
```

```
M_Q_E_MAX          /100/
N_E_POS_MIN        /10000/
N_E_POS_MAX        /10000/
M_E_POS_MIN        /100/
M_E_POS_MAX        /100/
N_E_NEG_MIN        /10000/
N_E_NEG_MAX        /10000/
M_E_NEG_MIN        /100/
M_E_NEG_MAX        /100/
N_O_POS_MIN        /10000/
N_O_POS_MAX        /10000/
M_O_POS_MIN        /100/
M_O_POS_MAX        /100/
N_O_NEG_MIN        /10000/
N_O_NEG_MAX        /10000/
M_O_NEG_MIN        /100/
M_O_NEG_MAX        /100/
N_R_POS_MIN        /10000/
N_R_POS_MAX        /10000/
M_R_POS_MIN        /100/
M_R_POS_MAX        /100/
N_R_NEG_MIN        /10000/
N_R_NEG_MAX        /10000/
M_R_NEG_MIN        /100/
M_R_NEG_MAX        /100/;
VARIABLES
Z                          'Value of the objective function [$]'
LAMBDA_E(t)              'Energy market price in time period t [$/
    MWh]'
LAMBDA_R_POS(t)            'Up-reserve capacity price in the
    reserve market [$/MW]'
LAMBDA_R_NEG(t)            'Down-reserve capacity price in the
    reserve market [$/MW]'
OBJ1                       'Utility minus production cost [$]'
OBJ2                       'Revenue minus energy procurement cost
    in the DA energy market[$]'
OBJ3                       'Revenue by selling up- and down-reserve
     capacities in the reserve market [$]'
ZZ;
POSITIVE VARIABLES
P_C(c,t,ga)                'Power generation of conventional power
    plant c in time period t in scenario ga [MW]'
P_D(d,t,ga)                'Power consumption of demand d in time
    period t in scenario ga [MW]'
P_R(r,t,ga)                'Production of renewable generating unit
     r in time period t in scenario ga [MW]'
NU_E_POS(t)                'Offer price decided by the VPP in the
    DA energy market in time period t[$/MWh]'
NU_E_NEG(t)                'Bid price decided by the VPP in the DA
    energy market in time period t[$/MWh]'
NU_R_POS(t)                'Offer price decided by the VPP in the
    reserve market for up-reserve capacity in time period t[$/MW]
    '
```

```
NU_R_NEG(t)              'Bid price decided by the VPP in the reserve market for down-reserve capacity in time period t[$/MW]'
P_E_POS_MAX(t)           'Power offer quantity by the VPP in the DA energy market in time period t[MW]'
P_E_NEG_MAX(t)           'Power bid quantity by the VPP in the DA energy market in time period t[MW]'
P_R_POS_MAX(t)           'Capacity offer quantity by the VPP in the DA energy market in time period t[MW]'
P_R_NEG_MAX(t)           'Capacity bid quantity by the VPP in the DA energy market in time period t[MW]'
P_E_POS(t)             'Total power scheduled to be produced by the VPP in the DA energy market [MW]'
P_E_NEG(t)             'Total power scheduled to be consumed by the VPP in the DA energy market [MW]'
P_R_POS(t)             'Up-reserve capacity scheduled to be sold by the VPP in the reserve market [MW]'
P_R_NEG(t)             'Down-reserve capacity scheduled to be sold by the VPP in the reserve market [MW]'
P_G_E(g,t)             'Power scheduled to be produced by the gth generating unit in the DA energy market [MW]'
P_Q_E(q,t)             'Power scheduled to be consumed by the dth demand in the DA energy market [MW]'
P_O_POS(o,t)           'Up-reserve capacity scheduled to be sold by the gth generating unit in the reserve market [MW]'
P_O_NEG(o,t)           'Down-reserve capacity scheduled to be sold by the dth demand in the reserve market [MW]'
MU_E_POS_MAX(t)          'Dual variable'
MU_E_POS_MIN(t)          'Dual variable'
MU_E_NEG_MAX(t)          'Dual variable'
MU_E_NEG_MIN(t)          'Dual variable'
MU_G_E_MAX(g,t)          'Dual variable'
MU_G_E_MIN(g,t)          'Dual variable'
MU_Q_E_MAX(q,t)          'Dual variable'
MU_Q_E_MIN(q,t)          'Dual variable'
MU_R_POS_MAX(t)          'Dual variable'
MU_R_POS_MIN(t)          'Dual variable'
MU_R_NEG_MAX(t)          'Dual variable'
MU_R_NEG_MIN(t)          'Dual variable'
MU_O_POS_MAX(o,t)        'Dual variable'
MU_O_POS_MIN(o,t)        'Dual variable'
MU_O_NEG_MAX(o,t)        'Dual variable'
MU_O_NEG_MIN(o,t)        'Dual variable';
BINARY VARIABLES
U_C(c,t)         'Binary variable equal to 1 if conventional power plant c is generating electricity in time period t, being 0 otherwise'
U_G_E_MIN(g,t)
U_G_E_MAX(g,t)
U_Q_E_MIN(q,t)
U_Q_E_MAX(q,t)
U_E_POS_MIN(t)
U_E_POS_MAX(t)
U_E_NEG_MIN(t)
U_E_NEG_MAX(t)
```

```
U_R_POS_MAX(t)
U_R_POS_MIN(t)
U_R_NEG_MAX(t)
U_R_NEG_MIN(t)
U_O_POS_MAX(o,t)
U_O_POS_MIN(o,t)
U_O_NEG_MAX(o,t)
U_O_NEG_MIN(o,t);
EQUATIONS
VPPDAobj          'Total profit',
VPPDAOBJ1         'Utility minus production cost',
VPPDAOBJ2         'Revenue minus energy procurement cost in the DA
     energy market',
VPPROBJ           'Revenue by selling up- and down-reserve
    capacities in the reserve market',
VPPDA1            'Power balance constraint',
VPPDA2            'Minimum dailiy energy consumption of demands in
     the VPP',
VPPDA_MIN3        'Lower bound of power consumption by each demand
     in the VPP',
VPPDA_MAX3        'Upper bound of power consumption by each demand
     in the VPP',
VPPDA_MIN4        'Lower bound of power production by each
    conventional power plant in the VPP',
VPPDA_MAX4        'Upper bound of power production by each
    conventional power plant in the VPP',
VPPDA5            'Upper bound of power production by each
    stochastic renewable generating unit in the VPP',
VPPDA6            'Offer quantity bounded by maximum power that
    can be sold by the VPP in the DA energy market',
VPPDA7            'Bid quantity bounded by maximum power that can
    be bought from the VPP in the DA energy market',
VPPR6             'Up-reserve capacity bounded by maximum power
    that can be sold by the VPP in the DA energy market',
VPPR7             'Down-reserve capacity bounded by maximum power
    that can be bought from the VPP in the DA energy market',
VPPDAR6           'Suummation of offer quantity and up-reserve
    capacity bounded by maximum power that can be sold by the VPP
     in the DA energy market',
VPPDAR7           'Summation of bid quantity and down-reserve
    capacity bounded by maximum power that can be bought from the
     VPP in the DA energy market',
DAMCP2, DAMCP3, DAMCP4, DAMCP5 'Inequality constraints of the DA
    energy market clearing problem',
DAMCP_KKT1, DAMCP_KKT2, DAMCP_KKT3, DAMCP_KKT4, DAMCP_KKT5 'KKT
    conditions for the DA energy market clearing problem',
DAMCP_KKT6a, DAMCP_KKT6b, DAMCP_KKT7a, DAMCP_KKT7b, DAMCP_KKT8a,
    DAMCP_KKT8b, DAMCP_KKT9a, DAMCP_KKT9b 'KKT conditions for the
     DA energy market clearing problem',
DAMCP_KKT10a, DAMCP_KKT10b, DAMCP_KKT11a, DAMCP_KKT11b,
    DAMCP_KKT12a, DAMCP_KKT12b, DAMCP_KKT13a, DAMCP_KKT13b 'KKT
    conditions for the DA energy market clearing problem',
RMCP3, RMCP4, RMCP5, RMCP6 'Inequality onstraints of the reserve
    market clearing problem',
```

```
RMCP_KKT1, RMCP_KKT2, RMCP_KKT3, RMCP_KKT4, RMCP_KKT5, RMCP_KKT6
    'kkt conditions for the reserve market clearing problem',
RMCP_KKT7a, RMCP_KKT7b, RMCP_KKT8a, RMCP_KKT8b, RMCP_KKT9a,
    RMCP_KKT9b, RMCP_KKT10a, RMCP_KKT10b 'kkt conditions for the
    reserve market clearing problem',
RMCP_KKT11a, RMCP_KKT11b, RMCP_KKT12a, RMCP_KKT12b, RMCP_KKT13a,
    RMCP_KKT13b, RMCP_KKT14a, RMCP_KKT14b 'kkt conditions for the
     reserve market clearing problem'
;
VPPDAobj..                   Z=E=OBJ1+OBJ2+OBJ3;
VPPDAOBJ1..                  OBJ1=E=SUM(ga,PI(ga)*SUM(t,SUM(d,U_D(d
    ,t)*P_D(d,t,ga)*delta) - SUM(c,C_C_F(c)*U_C(c,t)+C_C_V(c)*P_C
    (c,t,ga)*delta)));
VPPDAOBJ2..                  OBJ2=E=SUM(t,SUM(q,CHI_Q_E(q,t)*P_Q_E(
    q,t)) - SUM(g,CHI_G_E(g,t)*P_G_E(g,t)) - SUM(g,P_G_E_MAX(g,t)
    *MU_G_E_MAX(g,t)) - SUM(q,P_Q_E_MAX(q,t)*MU_Q_E_MAX(q,t)));
VPPROBJ..                    OBJ3=E=SUM(t,P_R_POS_REQ(t)*
    LAMBDA_R_POS(t)+P_R_NEG_REQ(t)*LAMBDA_R_NEG(t)-SUM(o,
    P_O_POS_MAX(o,t)*MU_O_POS_MAX(o,t)+P_O_NEG_MAX(o,t)*
    MU_O_NEG_MAX(o,t)+CHI_O_POS(o,t)*P_O_POS(o,t)+CHI_O_NEG(o,t)*
    P_O_NEG(o,t)));
VPPDA1(t,ga)..               SUM(c,P_C(c,t,ga))+SUM(r,P_R(r,t,ga))-
    SUM(d,P_D(d,t,ga))=E=P_E_POS(t)-P_E_NEG(t)+K_R_POS(t,ga)*
    P_R_POS(t)-K_R_NEG(t,ga)*P_R_NEG(t);
VPPDA2(d,ga)..               SUM(t,P_D(d,t,ga)*delta)=G=E_D_MIN(d);
VPPDA_MIN3(d,t,ga)..         P_D_MIN(d,t)=L=P_D(d,t,ga);
VPPDA_MAX3(d,t,ga)..         P_D(d,t,ga)=L=P_D_MAX(d,t);
VPPDA_MIN4(c,t,ga)..         P_C_MIN(c)*U_C(c,t)=L=P_C(c,t,ga);
VPPDA_MAX4(c,t,ga)..         P_C(c,t,ga)=l=P_C_MAX(c)*U_C(c,t);
VPPDA5(r,t,ga)..             P_R(r,t,ga)=L=P_R_A(r,t);
VPPDA6(t)..                  P_E_POS_MAX(t)=L=P_V_POS_MAX;
VPPDA7(t)..                  P_E_NEG_MAX(t)=L=P_V_NEG_MAX;
VPPR6(t)..                   P_R_POS_MAX(t)=L=P_V_POS_MAX;
VPPR7(t)..                   P_R_NEG_MAX(t)=L=P_V_NEG_MAX;
VPPDAR6(t)..                 P_E_POS_MAX(t)+P_R_POS_MAX(t)=L=
    P_V_POS_MAX;
VPPDAR7(t)..                 P_E_NEG_MAX(t)+P_R_NEG_MAX(t)=L=
    P_V_NEG_MAX;
DAMCP2(g,t)..                P_G_E(g,t)=L=P_G_E_MAX(g,t);
DAMCP3(q,t)..                P_Q_E(q,t)=L=P_Q_E_MAX(q,t);
DAMCP4(t)..                  P_E_POS(t)=L=P_E_POS_MAX(t);
DAMCP5(t)..                  P_E_NEG(t)=L=P_E_NEG_MAX(t);
DAMCP_KKT1(t)..              P_E_NEG(t)+SUM(q,P_Q_E(q,t))-P_E_POS(t
    )-SUM(g,P_G_E(g,t))=E=0;
DAMCP_KKT2(g,t)..            CHI_G_E(g,t)-LAMBDA_E(t)+MU_G_E_MAX(g,
    t)-MU_G_E_MIN(g,t)=E=0;
DAMCP_KKT3(q,t)..            -CHI_Q_E(q,t)+LAMBDA_E(t)+MU_Q_E_MAX(q
    ,t)-MU_Q_E_MIN(q,t)=E=0;
DAMCP_KKT4(t)..              NU_E_POS(t)-LAMBDA_E(t)+MU_E_POS_MAX(t
    )-MU_E_POS_MIN(t)=E=0;
DAMCP_KKT5(t)..              -NU_E_NEG(t)+LAMBDA_E(t)+MU_E_NEG_MAX(
    t)-MU_E_NEG_MIN(t)=E=0;
DAMCP_KKT6a(g,t)..            MU_G_E_MIN(g,t)=L=N_G_E_MIN*U_G_E_MIN
    (g,t);
DAMCP_KKT6b(g,t)..            P_G_E(g,t)=L=M_G_E_MIN*(1-U_G_E_MIN(g
    ,t));
```

```
DAMCP_KKT7a(g,t)..            MU_G_E_MAX(g,t)=L=N_G_E_MAX*U_G_E_MAX
     (g,t);
DAMCP_KKT7b(g,t)..            P_G_E_MAX(g,t)-P_G_E(g,t)=L=M_G_E_MAX
     *(1-U_G_E_MAX(g,t));
DAMCP_KKT8a(q,t)..            MU_Q_E_MIN(q,t)=L=N_Q_E_MIN*U_Q_E_MIN
     (q,t);
DAMCP_KKT8b(q,t)..            P_Q_E(q,t)=L=M_Q_E_MIN*(1-U_Q_E_MIN(q
     ,t));
DAMCP_KKT9a(q,t)..            MU_Q_E_MAX(q,t)=L=N_Q_E_MAX*U_Q_E_MAX
     (q,t);
DAMCP_KKT9b(q,t)..            P_Q_E_MAX(q,t)-P_Q_E(q,t)=L=M_Q_E_MAX
     *(1-U_Q_E_MAX(q,t));
DAMCP_KKT10a(t)..             MU_E_POS_MIN(t)=L=N_E_POS_MIN*
     U_E_POS_MIN(t);
DAMCP_KKT10b(t)..             P_E_POS(t)=L=M_E_POS_MIN*(1-
     U_E_POS_MIN(t));
DAMCP_KKT11a(t)..             MU_E_POS_MAX(t)=L=N_E_POS_MAX*
     U_E_POS_MAX(t);
DAMCP_KKT11b(t)..             P_E_POS_MAX(t)-P_E_POS(t)=L=
     M_E_POS_MAX*(1-U_E_POS_MAX(t));
DAMCP_KKT12a(t)..             MU_E_NEG_MIN(t)=L=N_E_NEG_MIN*
     U_E_NEG_MIN(t);
DAMCP_KKT12b(t)..             P_E_NEG(t)=L=M_E_NEG_MIN*(1-
     U_E_NEG_MIN(t));
DAMCP_KKT13a(t)..             MU_E_NEG_MAX(t)=L=N_E_NEG_MAX*
     U_E_NEG_MAX(t);
DAMCP_KKT13b(t)..             P_E_NEG_MAX(t)-P_E_NEG(t)=L=
     M_E_NEG_MAX*(1-U_E_NEG_MAX(t));
RMCP3(o,t)..                P_O_POS(o,t)=L=P_O_POS_MAX(o,t);
RMCP4(o,t)..                P_O_NEG(o,t)=L=P_O_NEG_MAX(o,t);
RMCP5(t)..                  P_R_POS(t)=L=P_R_POS_MAX(t);
RMCP6(t)..                  P_R_NEG(t)=L=P_R_NEG_MAX(t);
RMCP_KKT1(t)..              P_R_POS(t)+SUM(o,P_O_POS(o,t))-
    P_R_POS_REQ(t)=E=0;
RMCP_KKT2(t)..              P_R_NEG(t)+SUM(o,P_O_NEG(o,t))-
    P_R_NEG_REQ(t)=E=0;
RMCP_KKT3(o,t)..            CHI_O_POS(o,t)-LAMBDA_R_POS(t)+
    MU_O_POS_MAX(o,t)-MU_O_POS_MIN(o,t)=E=0;
RMCP_KKT4(o,t)..            CHI_O_NEG(o,t)-LAMBDA_R_NEG(t)+
    MU_O_NEG_MAX(o,t)-MU_O_NEG_MIN(o,t)=E=0;
RMCP_KKT5(t)..              NU_R_POS(t)-LAMBDA_R_POS(t)+
    MU_R_POS_MAX(t)-MU_R_POS_MIN(t)=E=0;
RMCP_KKT6(t)..              NU_R_NEG(t)-LAMBDA_R_NEG(t)+
    MU_R_NEG_MAX(t)-MU_R_NEG_MIN(t)=E=0;
RMCP_KKT7a(o,t)..           MU_O_POS_MIN(o,t)=L=N_O_POS_MIN*
    U_O_POS_MIN(o,t);
RMCP_KKT7b(o,t)..           P_O_POS(o,t)=L=M_O_POS_MIN*(1-
    U_O_POS_MIN(o,t));
RMCP_KKT8a(o,t)..           MU_O_POS_MAX(o,t)=L=N_O_POS_MAX*
    U_O_POS_MAX(o,t);
RMCP_KKT8b(o,t)..           P_O_POS_MAX(o,t)-P_O_POS(o,t)=L=
    M_O_POS_MAX*(1-U_O_POS_MAX(o,t));
RMCP_KKT9a(o,t)..           MU_O_NEG_MIN(o,t)=L=N_O_NEG_MIN*
    U_O_NEG_MIN(o,t);
RMCP_KKT9b(o,t)..           P_O_NEG(o,t)=L=M_O_NEG_MIN*(1-
    U_O_NEG_MIN(o,t));
```

```
RMCP_KKT10a(o,t)..          MU_O_NEG_MAX(o,t)=L=N_O_NEG_MAX*
    U_O_NEG_MAX(o,t);
RMCP_KKT10b(o,t)..          P_O_NEG_MAX(o,t)-P_O_NEG(o,t)=L=
    M_O_NEG_MAX*(1-U_O_NEG_MAX(o,t));
RMCP_KKT11a(t)..            MU_R_POS_MIN(t)=L=N_R_POS_MIN*
    U_R_POS_MIN(t);
RMCP_KKT11b(t)..            P_R_POS(t)=L=M_R_POS_MIN*(1-
    U_R_POS_MIN(t));
RMCP_KKT12a(t)..            MU_R_POS_MAX(t)=L=N_R_POS_MAX*
    U_R_POS_MAX(t);
RMCP_KKT12b(t)..            P_R_POS_MAX(t)-P_R_POS(t)=L=
    M_R_POS_MAX*(1-U_R_POS_MAX(t));
RMCP_KKT13a(t)..            MU_R_NEG_MIN(t)=L=N_R_NEG_MIN*
    U_R_NEG_MIN(t);
RMCP_KKT13b(t)..            P_R_NEG(t)=L=M_R_NEG_MIN*(1-
    U_R_NEG_MIN(t));
RMCP_KKT14a(t)..            MU_R_NEG_MAX(t)=L=N_R_NEG_MAX*
    U_R_NEG_MAX(t);
RMCP_KKT14b(t)..            P_R_NEG_MAX(t)-P_R_NEG(t)=L=
    M_R_NEG_MAX*(1-U_R_NEG_MAX(t));
MODEL DAR_MPEC /ALL/;
PARAMETERS state, obj;
OPTION OPTCA=0;
OPTION OPTCR=0;
SOLVE DAR_MPEC USING MIP MAXIMIZING Z;
state=DAR_MPEC.modelstat;
obj=OBJ1.l+SUM(t,LAMBDA_E.l(t)*(P_E_POS.l(t)-P_E_NEG.l(t))) + SUM
    (t,LAMBDA_R_POS.l(t)*P_R_POS.l(t)+LAMBDA_R_NEG.l(t)*P_R_NEG.l
    (t));
DISPLAY state, obj;
```

References

1. Boyd, S., Vandenberghe, L.: Convex Optimization. Cambridge, University Press (2009)
2. Conejo, A.J., Carrión, M., Morales, J.M.: Decision Making Under Uncertainty in Electricity Markets. Springer, New York, US (2010)
3. Gabriel, S.A., Conejo, A.J., Fuller, J.D., Hobbs, B.F., Ruiz, C.: Complementarity Modeling in Energy Markets. Springer, New York, US (2013)
4. GAMS.: Available at www.gams.com (2020)
5. ILOG CPLEX.: Available at www.ilog.com/products/cplex/ (2020)
6. Kardakos, E.G., Simoglou, C.K., Bakirtzis, A.G.: Optimal offering strategy of a virtual power plant: a stochastic bi-level approach. IEEE Trans. Power Syst. **7**(2), 794–806 (2016)
7. Pourghaderi, N., Fotuhi-Firuzabad, M., Moeini-Aghtaie, M., Kabirifar, M.: Commercial demand response programs in bidding of a technical virtual power plant. IEEE Trans. Ind. Inf. **14**(11), 5100–5111 (2018)

Chapter 7
Expansion Planning of Virtual Power Plants

This chapter describes the expansion planning problem of a virtual power plant (VPP) trading energy in electricity markets. The VPP comprises generating units, including both conventional and stochastic renewable units, storage facilities, and flexible demands. With the aim of maximizing its profit, it has the possibility of building new conventional, stochastic renewable, and storage units. The resulting model is first formulated using a deterministic model that disregards uncertainties. This deterministic model is extended to account for uncertainties using a scenario-based two-stage stochastic programming model. Finally, the profit risk associated with the investment decisions is modeled using the conditional-value-at-risk.

7.1 Introduction

The aim of a virtual power plant (VPP) is the maximization of its profit by participating in different electricity markets. This problem is analyzed in Chaps. 3–6 of this book. However, the models described in these chapters consider that the fleet of assets managed by the VPP is fixed. Nevertheless, the VPP can also increase its profit by building new generating units. This chapter analyzes this problem, i.e., the optimal expansion planning of a VPP.

Expansion planning problems in power systems are tackled using two different approaches [2]. The first one is using a centralized framework in which the planner decides the expansion plans that minimize the total cost based on a social welfare criterion. This approach is useful, for example, for the transmission system operator making decisions on the reinforcements of the transmission network [5], or the transmission and storage facilities [8] to be constructed in a power system. However, this framework is not suitable for a profit-oriented VPP. The second option is using a decentralized or market framework that considers an investor that makes the expansion decisions with the aim of maximizing its profit, which is the approach used in this chapter.

Expansion decisions are generally made for a medium- or long-term planning horizon. Thus, these decisions are made within an uncertain framework and, therefore, an accurate modeling of the uncertainty sources is important in order to obtain informed expansion decisions. In particular, this chapter describes a model based on two-stage stochastic programming [1].

In addition, expansion decisions involve an important profit risk that should be considered in the decision-making problem. To do this, the conditional-value-at-risk (CVaR) is used in this chapter.

The remaining parts of this chapter are organized as follows. Section 7.2 describes the main characteristics of the expansion planning problem of a VPP. This problem is analyzed using a deterministic approach that disregards uncertainty in Sect. 7.3. Then, Sect. 7.4 models the impact of uncertainties on expansion planning decisions using an approach based on stochastic programming, while Sect. 7.5 incorporates in this problem the CVaR metric, which allows to account for the profit risk associated with the investment decisions. Section 7.6 summarizes the chapter and suggests some references for further reading. Finally, Sect. 7.7 provides the GAMS codes used to solve some of the illustrative examples provided in the chapter.

7.2 Problem Description

This chapter describes and formulates the expansion planning problem of a VPP considering different approaches. All these methods have some common characteristics that are described in this section for the sake of clarity.

L. Baringo, M. Rahimiyan, *Virtual Power Plants and Electricity Markets*, https://doi.org/10.1007/978-3-030-47602-1_7

7.2.1 Notation

The main notation used in this chapter is provided below for quick reference. Other symbols are defined as needed throughout the chapter. Note that if the symbols below include a subscript ω it denotes their values in scenario ω.

7.2.1.1 Indexes

The indexes of the expansion planning problem of a VPP are:

c Conventional power plants.
d Demands.
r Stochastic renewable generating units.
s Storage units.
t Time periods.
ω Scenarios.

7.2.1.2 Sets

The sets of the expansion planning problem of a VPP are:

Ω^{C} Conventional power plants.
$\tilde{\Omega}^{\text{C}}$ Candidate conventional power plants.
Ω^{D} Demands.
Ω^{R} Stochastic renewable generating units.
$\tilde{\Omega}^{\text{R}}$ Candidate stochastic renewable generating units.
Ω^{S} Storage units.
$\tilde{\Omega}^{\text{S}}$ Candidate storage units.
Ω^{T} Time periods.
Ω^{W} Scenarios.

7.2.1.3 Parameters

The parameters of the expansion planning problem of a VPP are:

$C_c^{\text{C,I}}$ Investment cost of conventional power plant c [\$].
$C_c^{\text{C,V}}$ Variable cost of conventional generating unit c [\$/MWh].
$\overline{C}^{\text{I}}$ Investment budget [\$].
$C_r^{\text{R,I}}$ Investment cost of stochastic renewable generating unit r [\$].
$C_s^{\text{S,I}}$ Investment cost of storage unit s [\$].
$\underline{E}_d^{\text{D}}$ Minimum energy consumption of demand d [MWh].
$\underline{E}_{st}^{\text{S}}$ Lower bound of the energy stored in storage unit s in time period t [MWh].
$\overline{E}_{st}^{\text{S}}$ Upper bound of the energy stored in storage unit s in time period t [MWh].
$\overline{P}_c^{\text{C}}$ Upper bound of the power production of conventional power plant c [MW].
$\underline{P}_{dt}^{\text{D}}$ Lower bound of the power consumption of demand d in time period t [MW].
$\overline{P}_{dt}^{\text{D}}$ Upper bound of the power consumption of demand d in time period t [MW].
$\underline{P}^{\text{E}}$ Lower bound of the power traded in the market [MW].
$\overline{P}^{\text{E}}$ Upper bound of the power traded in the market [MW].
$P_{rt}^{\text{R,A}}$ Available generation of the stochastic renewable generating unit r in time period t [MW].
$\overline{P}_{st}^{\text{S,C}}$ Charging capacity of storage unit s [MW].
$\overline{P}_{st}^{\text{S,D}}$ Discharging capacity of storage unit s [MW].
β Weighting parameter to model the trade-off between expected profit and profit risk [pu].

π_ω Weight of scenario ω [pu].
σ Factor to make investment and operation cost comparable [pu].

7.2.1.4 Variables

The variables of the expansion planning problem of a VPP are:

e^{S}_{st} Energy stored in storage unit s in time period t [MWh].
p^{C}_{ct} Power production of conventional power plant c in time period t [MW].
p^{D}_{dt} Power consumption of demand d in time period t [MW].
p^{E}_{t} Power traded in the energy market in time period t [MW].
p^{R}_{rt} Production of stochastic renewable generating unit r in time period t [MW].
$p^{\mathrm{S,C}}_{st}$ Charging power level of storage unit s in time period t [MW].
$p^{\mathrm{S,D}}_{st}$ Discharging power level of storage unit s in time period t [MW].
$u^{\mathrm{C,I}}_{c}$ Binary variable that is equal to 1 if conventional power plant c is built, being 0 otherwise.
$u^{\mathrm{R,I}}_{r}$ Binary variable that is equal to 1 if stochastic renewable generating unit r is built, being 0 otherwise.
$u^{\mathrm{S,I}}_{s}$ Binary variable that is equal to 1 if storage unit s is built, being 0 otherwise.
τ Auxiliary CVaR-related variable [\$].
ϑ Value-at-risk [\$].

7.2.2 *Time Framework*

Expansion planning problems are usually considered for a long-term planning horizon, e.g., 30 years. In this kind of problems, there are two expansion planning strategies. The first is to make the expansion planning decisions at a single point in time, generally at the beginning of the planning horizon. The resulting problem is generally known as a single-stage or static model. The second option is to consider that expansion planning decisions are made at different points in time. These problems are generally known as multi-stage or dynamic models. Figure 7.1 illustrates the differences between these two models.

The main advantage of static or single-stage models is their simplicity. However, they require a large investment budget since all expansion planning decisions are made at a single point in time. Moreover, note that expansion planning decisions are made for a long-term planning horizon. In this sense, it is difficult to forecast the needs of the VPP for the following 20 or 30 years. Therefore, the expansion planning decisions made at the beginning of the planning horizon may be sub-optimal if the needs of the VPP change in the future. These two disadvantages of static models are solved by considering a dynamic

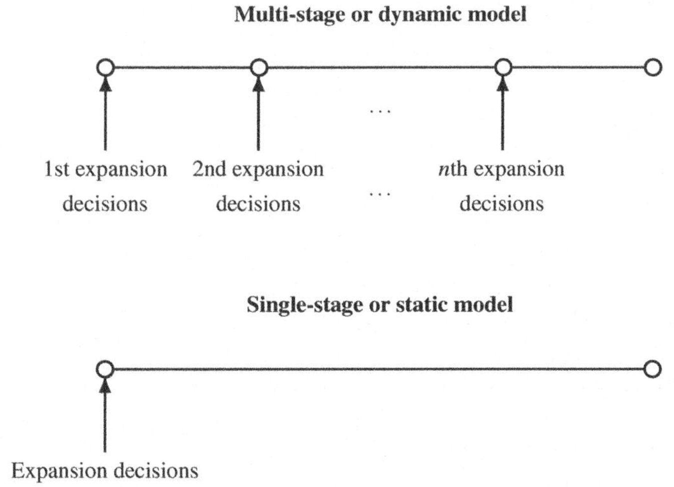

Fig. 7.1 Static and dynamic expansion models

or multi-stage model at the cost of solving a comparatively more complex problem, at least from a computational point of view.

For the sake of simplicity, the remainder of this chapter considers a static approach for the expansion planning problem of a VPP. In this model, all decisions are made at the beginning of the planning horizon, which is represented using different time periods indexed by t that models different system conditions, e.g., different demand or stochastic renewable production conditions.

7.3 Deterministic Approach

Considering the problem description in the previous section, the expansion planning problem of a VPP is first formulated using a simple deterministic model. In this model, all parameters involved in the decision-making problem are modeled using forecasts, i.e., uncertainty is not considered. The resulting model is formulated below:

$$\max_{\Phi^{\mathrm{D}}} \quad \sigma \sum_{t \in \Omega^{\mathrm{T}}} \left(\lambda_t^{\mathrm{E}} p_t^{\mathrm{E}} \Delta t - \sum_{c \in \Omega^{\mathrm{C}}} C_c^{\mathrm{C,V}} p_{ct}^{\mathrm{C}} \Delta t \right) - \left(\sum_{c \in \tilde{\Omega}^{\mathrm{C}}} C_c^{\mathrm{C,I}} u_c^{\mathrm{C,I}} + \sum_{r \in \tilde{\Omega}^{\mathrm{R}}} C_r^{\mathrm{R,I}} u_r^{\mathrm{R,I}} + \sum_{s \in \tilde{\Omega}^{\mathrm{S}}} C_s^{\mathrm{S,I}} u_s^{\mathrm{S,I}} \right) \tag{7.1a}$$

subject to

$$\sum_{c \in \tilde{\Omega}^{\mathrm{C}}} C_c^{\mathrm{C,I}} u_c^{\mathrm{C,I}} + \sum_{r \in \tilde{\Omega}^{\mathrm{R}}} C_r^{\mathrm{R,I}} u_r^{\mathrm{R,I}} + \sum_{s \in \tilde{\Omega}^{\mathrm{S}}} C_s^{\mathrm{S,I}} u_s^{\mathrm{S,I}} \le \overline{C}^{\mathrm{I}}, \tag{7.1b}$$

$$u_c^{\mathrm{C,I}} \in \{0, 1\}, \quad \forall c \in \tilde{\Omega}^{\mathrm{C}}, \tag{7.1c}$$

$$u_r^{\mathrm{R,I}} \in \{0, 1\}, \quad \forall r \in \tilde{\Omega}^{\mathrm{R}}, \tag{7.1d}$$

$$u_s^{\mathrm{S,I}} \in \{0, 1\}, \quad \forall s \in \tilde{\Omega}^{\mathrm{S}}, \tag{7.1e}$$

$$\underline{P}^{\mathrm{E}} \le p_t^{\mathrm{E}} \le \overline{P}^{\mathrm{E}}, \quad \forall t \in \Omega^{\mathrm{T}}, \tag{7.1f}$$

$$p_t^{\mathrm{E}} = \sum_{c \in \Omega^{\mathrm{C}}} p_{ct}^{\mathrm{C}} + \sum_{r \in \Omega^{\mathrm{R}}} p_{rt}^{\mathrm{R}} + \sum_{s \in \Omega^{\mathrm{S}}} \left(p_{st}^{\mathrm{S,D}} - p_{st}^{\mathrm{S,C}} \right) - \sum_{d \in \Omega^{\mathrm{D}}} p_{dt}^{\mathrm{D}}, \quad \forall t \in \Omega^{\mathrm{T}}, \tag{7.1g}$$

$$\underline{P}_{dt}^{\mathrm{D}} \le p_{dt}^{\mathrm{D}} \le \overline{P}_{dt}^{\mathrm{D}}, \quad \forall d \in \Omega^{\mathrm{D}}, \forall t \in \Omega^{\mathrm{T}}, \tag{7.1h}$$

$$\sum_{t \in \Omega^{\mathrm{T}}} p_{dt}^{\mathrm{D}} \Delta t \ge \underline{E}_d^{\mathrm{D}}, \quad \forall d \in \Omega^{\mathrm{D}}, \tag{7.1i}$$

$$0 \le p_{ct}^{\mathrm{C}} \le \overline{P}_t^{\mathrm{C}}, \quad \forall c \in \Omega^{\mathrm{C}} \setminus c \in \tilde{\Omega}^{\mathrm{C}}, \forall t \in \Omega^{\mathrm{T}}, \tag{7.1j}$$

$$0 \le p_{ct}^{\mathrm{C}} \le \overline{P}_t^{\mathrm{C}} u_c^{\mathrm{C,I}}, \quad \forall c \in \tilde{\Omega}^{\mathrm{C}}, \forall t \in \Omega^{\mathrm{T}}, \tag{7.1k}$$

$$0 \le p_{rt}^{\mathrm{R}} \le P_{rt}^{\mathrm{R,A}}, \quad \forall r \in \Omega^{\mathrm{R}} \setminus r \in \tilde{\Omega}^{\mathrm{R}}, \forall t \in \Omega^{\mathrm{T}}, \tag{7.1l}$$

$$0 \le p_{rt}^{\mathrm{R}} \le P_{rt}^{\mathrm{R,A}} u_r^{\mathrm{R,I}}, \quad \forall r \in \tilde{\Omega}^{\mathrm{R}}, \forall t \in \Omega^{\mathrm{T}}, \tag{7.1m}$$

$$0 \le p_{st}^{\mathrm{S,C}} \le \overline{P}_s^{\mathrm{S,C}}, \quad \forall s \in \Omega^{\mathrm{S}} \setminus s \in \tilde{\Omega}^{\mathrm{S}}, \forall t \in \Omega^{\mathrm{T}}, \tag{7.1n}$$

$$0 \le p_{st}^{\mathrm{S,D}} \le \overline{P}_s^{\mathrm{S,D}}, \quad \forall s \in \Omega^{\mathrm{S}} \setminus s \in \tilde{\Omega}^{\mathrm{S}}, \forall t \in \Omega^{\mathrm{T}}, \tag{7.1o}$$

$$e_{st}^{\mathrm{S}} = e_{s(t-1)}^{\mathrm{S}} + p_{st}^{\mathrm{S,C}} \Delta t \eta_s^{\mathrm{S,C}} - \frac{p_{st}^{\mathrm{S,D}} \Delta t}{\eta_s^{\mathrm{S,D}}}, \quad \forall s \in \Omega^{\mathrm{S}}, \forall t \in \Omega^{\mathrm{T}}, \tag{7.1p}$$

$$\underline{E}_{st}^{\mathrm{S}} \le e_{st}^{\mathrm{S}} \le \overline{E}_{st}^{\mathrm{S}}, \quad \forall s \in \Omega^{\mathrm{S}}, \forall t \in \Omega^{\mathrm{T}}, \tag{7.1q}$$

$$0 \le p_{st}^{\mathrm{S,C}} \le \overline{P}_{s}^{\mathrm{S,C}} u_{s}^{\mathrm{S,I}}, \quad \forall s \in \tilde{\Omega}^{\mathrm{S}}, \forall t \in \Omega^{\mathrm{T}}, \tag{7.1r}$$

$$0 \le p_{st}^{\mathrm{S,D}} \le \overline{P}_{s}^{\mathrm{S,D}} u_{s}^{\mathrm{S,I}}, \quad \forall s \in \tilde{\Omega}^{\mathrm{S}}, \forall t \in \Omega^{\mathrm{T}}, \tag{7.1s}$$

where set $\Phi^{\mathrm{D}} = \Big\{u_c^{\mathrm{C,I}}, \forall c \in \tilde{\Omega}^{\mathrm{C}}; u_r^{\mathrm{R,I}}, \forall r \in \tilde{\Omega}^{\mathrm{R}}; u_s^{\mathrm{S,I}}, \forall s \in \tilde{\Omega}^{\mathrm{S}}; p_t^{\mathrm{E}}, \forall t \in \Omega^{\mathrm{T}}; p_{ct}^{\mathrm{C}}, \forall c \in \Omega^{\mathrm{C}}, \forall t \in \Omega^{\mathrm{T}}; p_{rt}^{\mathrm{R}}, \forall r \in \Omega^{\mathrm{R}}, \forall t \in \Omega^{\mathrm{T}}; p_{dt}^{\mathrm{D}}, \forall d \in \Omega^{\mathrm{D}}, \forall t \in \Omega^{\mathrm{T}}; e_{st}^{\mathrm{S}}, p_{st}^{\mathrm{S,C}}, p_{st}^{\mathrm{S,D}}, \forall s \in \Omega^{\mathrm{S}}, \forall t \in \Omega^{\mathrm{T}}\Big\}$ includes the optimization variables of problem (7.1).

The objective function (7.1a) represents the profit achieved by the VPP and comprises the following terms:

1. Terms $\lambda_t^{\mathrm{E}} p_t^{\mathrm{E}} \Delta t, \forall t \in \Omega^{\mathrm{T}}$, are the operating revenues achieved by the VPP due to its participation in the market. Note that variables $p_t^{\mathrm{E}}, \forall t \in \Omega^{\mathrm{T}}$, can take both positive (in case of selling) and negative (in case of buying) values. In the later case, these terms represent the minus cost incurred by the VPP.
2. Terms $C_c^{\mathrm{C,V}} p_{ct}^{\mathrm{C}} \Delta t, \forall c \in \Omega^{\mathrm{C}}, \forall t \in \Omega^{\mathrm{T}}$, are the operating costs of the conventional power plants in the VPP.
3. Term $\sum_{c \in \tilde{\Omega}^{\mathrm{C}}} C_c^{\mathrm{C,I}} u_c^{\mathrm{C,I}} + \sum_{r \in \tilde{\Omega}^{\mathrm{R}}} C_r^{\mathrm{R,I}} u_c^{\mathrm{R,I}} + \sum_{s \in \tilde{\Omega}^{\mathrm{S}}} C_s^{\mathrm{S,I}} u_s^{\mathrm{S,I}}$ is the investment cost in new conventional, stochastic renewable, and storage units.

Note that the operating revenues and costs in items 1 and 2 are multiplied by factor σ in order to make operating revenues and costs comparable with investment costs.

Constraint (7.1b) imposes bounds on the investment budget. Constraints (7.1c), (7.1d), and (7.1e) define binary variables that indicate if candidate conventional, stochastic renewable, and storage units are built, respectively. Constraints (7.1f) limit the power traded in the market. Equations (7.1g) impose the power balance in the VPP. Constraints (7.1h) and (7.1i) impose bounds on the power consumption of demands. Constraints (7.1j) and (7.1k) impose bounds on the power generated by existing and candidate conventional power plants, respectively. Note that in the latter the capacity of the conventional power plant is multiplied by binary variable $u_c^{\mathrm{C,I}}$ whose value is 1 only if the corresponding unit is built. Similarly, constraints (7.1l) and (7.1m) impose bounds on the power generated by existing and candidate stochastic renewable generating units, respectively. Constraints (7.1n) and (7.1o) limit the charging and discharging power levels of existing and candidate storage units, respectively, while constraints (7.1p) define the energy level evolution in these units. These energy levels are bounded using Eqs. (7.1q). Finally, constraints (7.1r) and (7.1s) limit the charging and discharging power levels of candidate storage units, respectively.

Problem (7.1) is a mixed-integer linear programming (MILP) problem [7].

Illustrative Example 7.1 Deterministic Expansion Planning Problem

This example illustrates the expansion planning problem of a VPP in which expansion decisions are made for a planning horizon that is represented using just 4 time periods for the sake of simplicity.

The VPP comprises a conventional power plant ($c = 1$), a wind-power unit ($r = 1$), and a flexible demand (without subscript for the sake of simplicity). The technical and economic data of the conventional power plant are provided in Table 7.1. The forecasts levels of the available wind-power generation levels are provided in Table 7.2. The lower and upper consumption bounds of the flexible demand are provided in Fig. 7.2, while the minimum energy consumption throughout the 4-h planning horizon is 60 MWh.

The VPP can build additional generating units to maximize its profit. Prospective generating assets include two conventional power plants ($c = 2, 3$), two stochastic solar-power units ($r = 2, 3$), and a storage facility ($s = 1$). Data of prospective conventional power plants are provided in Table 7.3. The forecast levels of the available generation of candidate

Table 7.1 Illustrative Example 7.1: data of the existing conventional power plant

Capacity	60 MW
Variable cost	$33/MWh

Table 7.2 Illustrative Example 7.1: data of the existing wind-power unit

Time period	Forecast level [MW]
1	60
2	20
3	40
4	10

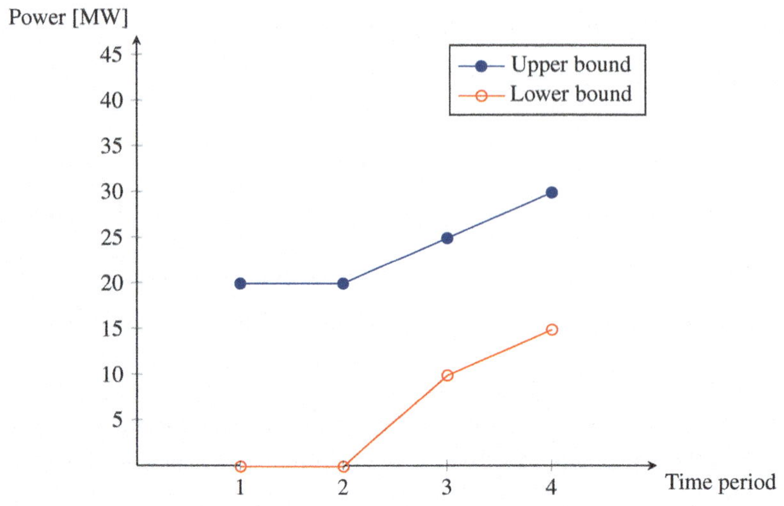

Fig. 7.2 Illustrative Example 7.1: lower and upper bounds of the power consumption

Table 7.3 Illustrative Example 7.1: data of the prospective conventional power plants

Unit	Capacity	Variable cost	Investment cost
2	40	$22/MWh	$2,000,000
3	80	$12/MWh	$7,000,000

Table 7.4 Illustrative Example 7.1: forecast levels of the available generation of the candidate solar power units

Unit	Time period 1	Time period 2	Time period 3	Time period 4
2	0	80 MW	60 MW	0
3	90 MW	70 MW	80 MW	50 MW

Table 7.5 Illustrative Example 7.1: investment costs of the candidate solar-power units

Unit	Investment cost [$]
2	7,000,000
3	10,000,000

Table 7.6 Illustrative Example 7.1: data of the candidate storage facility

Charging capacity	20 MW
Discharging capacity	20 MW
Minimum energy level 1	0
Maximum energy level 1	40
Charging efficiency	0.9
Discharging efficiency	0.9
Initial energy level	0
Investment cost	$500,000

solar-power units are provided in Table 7.4, while the corresponding investment costs are given in Table 7.5. Data of the candidate storage unit are provided in Table 7.6.

The VPP trades energy in a market whose prices are provided in Fig. 7.3, being the maximum power traded (sold or bought) in the market equal to 200 MW.

Finally, an investment budget equal to $100,000,000 is considered, while the value of parameter σ to make operating costs/revenues and investment costs comparable is 2190. This factor can be explained as follows. The planning horizon is represented by a year divided in four time periods. All investment cost are annualized values; therefore factor σ is obtained as 8760 (the number of hours of a year) divided by four (the considered time periods).

Considering these data, the problem is formulated below.

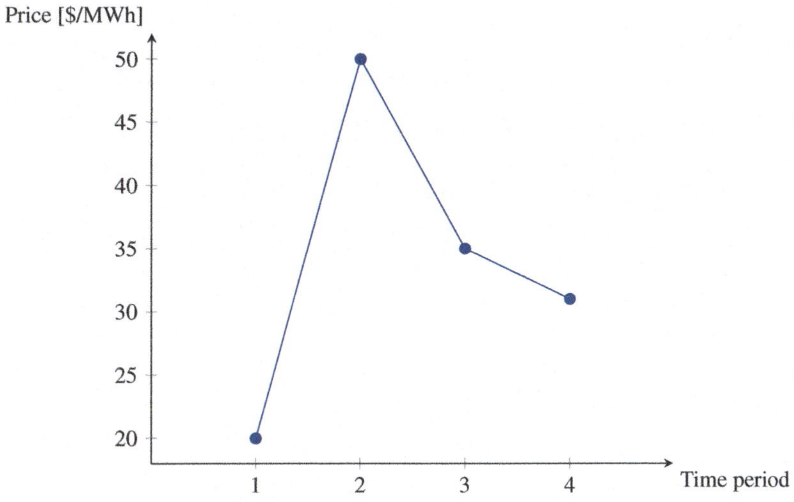

Fig. 7.3 Illustrative Example 7.1: energy market prices

First, the objective function to be maximized is:

$$
\begin{aligned}
&2190\Big[20p_1^{\mathrm{E}} + 50p_2^{\mathrm{E}} + 35p_3^{\mathrm{E}} + 31p_4^{\mathrm{E}} - \left(33p_{11}^{\mathrm{C}} + 33p_{12}^{\mathrm{C}} + 33p_{13}^{\mathrm{C}} + 33p_{14}^{\mathrm{C}}\right) \\
&\quad - \left(22p_{21}^{\mathrm{C}} + 22p_{22}^{\mathrm{C}} + 22p_{23}^{\mathrm{C}} + 22p_{24}^{\mathrm{C}}\right) - \left(12p_{31}^{\mathrm{C}} + 12p_{32}^{\mathrm{C}} + 12p_{33}^{\mathrm{C}} + 12p_{34}^{\mathrm{C}}\right)\Big] \\
&\quad - \left(2000000u_2^{\mathrm{C,I}} + 7000000u_3^{\mathrm{C,I}} + 7000000u_2^{\mathrm{R,I}} + 10000000u_3^{\mathrm{R,I}} + 500000u_1^{\mathrm{S,I}}\right).
\end{aligned}
$$

The first line is the revenue achieved by the VPP due to its participation in the electricity market minus the variable costs of the existing power plant. The second line represents the variable costs of the candidate conventional power plants. The third line includes the investment costs in candidate conventional power plants, stochastic renewable units, and storage units. Finally, note that the terms in the first and second lines are multiplied by factor 2190 in order to make operating revenues and costs comparable with investment costs.

Next, the constraints of the problem are formulated:

- Investment budget:

$$
\begin{aligned}
&2000000u_2^{\mathrm{C,I}} + 7000000u_3^{\mathrm{C,I}} + 7000000u_2^{\mathrm{R,I}} + 10000000u_3^{\mathrm{R,I}} \\
&\quad + 500000u_1^{\mathrm{S,I}} \le 1000000000.
\end{aligned}
$$

- Definition of binary variables:

$$
u_2^{\mathrm{C,I}}, u_3^{\mathrm{C,I}}, u_2^{\mathrm{R,I}}, u_3^{\mathrm{R,I}}, u_1^{\mathrm{S,I}} \in \{0, 1\}.
$$

- Limits on the power traded in the market:

$$
\begin{aligned}
&-200 \le p_1^{\mathrm{E}} \le 200, \\
&-200 \le p_2^{\mathrm{E}} \le 200, \\
&-200 \le p_3^{\mathrm{E}} \le 200, \\
&-200 \le p_4^{\mathrm{E}} \le 200.
\end{aligned}
$$

- Power balance in the VPP:

$$p_1^{\mathrm{E}} = p_{11}^{\mathrm{C}} + p_{21}^{\mathrm{C}} + p_{31}^{\mathrm{C}} + p_{11}^{\mathrm{R}} + p_{21}^{\mathrm{R}} + p_{31}^{\mathrm{R}} + p_{11}^{\mathrm{S,D}} - p_{11}^{\mathrm{S,C}} - p_1^{\mathrm{D}},$$
$$p_2^{\mathrm{E}} = p_{12}^{\mathrm{C}} + p_{22}^{\mathrm{C}} + p_{32}^{\mathrm{C}} + p_{12}^{\mathrm{R}} + p_{22}^{\mathrm{R}} + p_{32}^{\mathrm{R}} + p_{12}^{\mathrm{S,D}} - p_{12}^{\mathrm{S,C}} - p_2^{\mathrm{D}},$$
$$p_3^{\mathrm{E}} = p_{13}^{\mathrm{C}} + p_{23}^{\mathrm{C}} + p_{33}^{\mathrm{C}} + p_{13}^{\mathrm{R}} + p_{23}^{\mathrm{R}} + p_{33}^{\mathrm{R}} + p_{13}^{\mathrm{S,D}} - p_{13}^{\mathrm{S,C}} - p_3^{\mathrm{D}},$$
$$p_4^{\mathrm{E}} = p_{14}^{\mathrm{C}} + p_{24}^{\mathrm{C}} + p_{34}^{\mathrm{C}} + p_{14}^{\mathrm{R}} + p_{24}^{\mathrm{R}} + p_{34}^{\mathrm{R}} + p_{14}^{\mathrm{S,D}} - p_{14}^{\mathrm{S,C}} - p_4^{\mathrm{D}}.$$

- Limits on the power consumption of the demand:

$$0 \le p_1^{\mathrm{D}} \le 20,$$
$$0 \le p_2^{\mathrm{D}} \le 20,$$
$$10 \le p_3^{\mathrm{D}} \le 25,$$
$$15 \le p_4^{\mathrm{D}} \le 30.$$

- Minimum energy consumption of the demand through the planning horizon:

$$p_1^{\mathrm{D}} + p_2^{\mathrm{D}} + p_3^{\mathrm{D}} + p_4^{\mathrm{D}} \ge 60.$$

- Limits on the production of the existing conventional power plant:

$$0 \le p_{11}^{\mathrm{C}} \le 60,$$
$$0 \le p_{12}^{\mathrm{C}} \le 60,$$
$$0 \le p_{13}^{\mathrm{C}} \le 60,$$
$$0 \le p_{14}^{\mathrm{C}} \le 60.$$

- Limits on the production of the candidate conventional power plants:

$$0 \le p_{21}^{\mathrm{C}} \le 40u_2^{\mathrm{C,I}},$$
$$0 \le p_{22}^{\mathrm{C}} \le 40u_2^{\mathrm{C,I}},$$
$$0 \le p_{23}^{\mathrm{C}} \le 40u_2^{\mathrm{C,I}},$$
$$0 \le p_{24}^{\mathrm{C}} \le 40u_2^{\mathrm{C,I}},$$
$$0 \le p_{31}^{\mathrm{C}} \le 80u_3^{\mathrm{C,I}},$$
$$0 \le p_{32}^{\mathrm{C}} \le 80u_3^{\mathrm{C,I}},$$
$$0 \le p_{33}^{\mathrm{C}} \le 80u_3^{\mathrm{C,I}},$$
$$0 \le p_{34}^{\mathrm{C}} \le 80u_3^{\mathrm{C,I}}.$$

- Limits on the production of the existing stochastic renewable generating unit:

$$0 \le p_{11}^{\mathrm{R}} \le 60,$$
$$0 \le p_{12}^{\mathrm{R}} \le 20,$$
$$0 \le p_{13}^{\mathrm{R}} \le 40,$$
$$0 \le p_{14}^{\mathrm{R}} \le 10.$$

- Limits on the production of the candidate stochastic renewable generating units:

$$0 \le p_{21}^{\mathrm{R}} \le 0u_2^{\mathrm{R,I}},$$
$$0 \le p_{22}^{\mathrm{R}} \le 80u_2^{\mathrm{R,I}},$$
$$0 \le p_{23}^{\mathrm{R}} \le 60u_2^{\mathrm{R,I}},$$
$$0 \le p_{24}^{\mathrm{R}} \le 0u_2^{\mathrm{R,I}},$$
$$0 \le p_{31}^{\mathrm{R}} \le 90u_3^{\mathrm{R,I}},$$
$$0 \le p_{32}^{\mathrm{R}} \le 70u_3^{\mathrm{R,I}},$$
$$0 \le p_{33}^{\mathrm{R}} \le 80u_3^{\mathrm{R,I}},$$
$$0 \le p_{34}^{\mathrm{R}} \le 50u_3^{\mathrm{R,I}}.$$

- Charging power limits of the candidate storage unit:

$$0 \le p_{11}^{\mathrm{S,C}} \le 20u_1^{\mathrm{S,I}},$$
$$0 \le p_{12}^{\mathrm{S,C}} \le 20u_1^{\mathrm{S,I}},$$
$$0 \le p_{13}^{\mathrm{S,C}} \le 20u_1^{\mathrm{S,I}},$$
$$0 \le p_{14}^{\mathrm{S,C}} \le 20u_1^{\mathrm{S,I}}.$$

- Discharging power limits of the candidate storage unit:

$$0 \le p_{11}^{\mathrm{S,D}} \le 20u_1^{\mathrm{S,I}},$$
$$0 \le p_{12}^{\mathrm{S,D}} \le 20u_1^{\mathrm{S,I}},$$
$$0 \le p_{13}^{\mathrm{S,D}} \le 20u_1^{\mathrm{S,I}},$$
$$0 \le p_{14}^{\mathrm{S,D}} \le 20u_1^{\mathrm{S,I}}.$$

- Energy balance in the candidate storage unit:

$$e_{11}^{\mathrm{S}} = 0 + 0.9p_{11}^{\mathrm{S,C}} - \frac{p_{11}^{\mathrm{S,D}}}{0.9},$$
$$e_{12}^{\mathrm{S}} = e_{11}^{\mathrm{S}} + 0.9p_{12}^{\mathrm{S,C}} - \frac{p_{12}^{\mathrm{S,D}}}{0.9},$$
$$e_{13}^{\mathrm{S}} = e_{12}^{\mathrm{S}} + 0.9p_{13}^{\mathrm{S,C}} - \frac{p_{13}^{\mathrm{S,D}}}{0.9},$$
$$e_{14}^{\mathrm{S}} = e_{13}^{\mathrm{S}} + 0.9p_{14}^{\mathrm{S,C}} - \frac{p_{14}^{\mathrm{S,D}}}{0.9}.$$

- Energy stored limits:

$$0 \le e_{11}^{\mathrm{S}} \le 40,$$
$$0 \le e_{12}^{\mathrm{S}} \le 40,$$
$$0 \le e_{13}^{\mathrm{S}} \le 40,$$
$$0 \le e_{14}^{\mathrm{S}} \le 40.$$

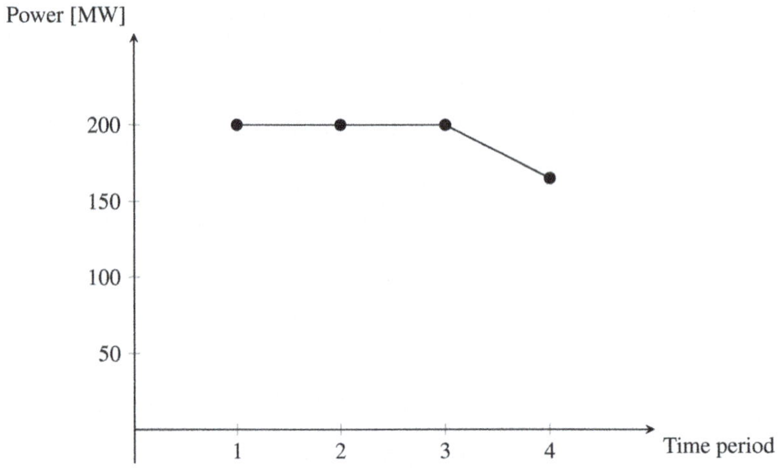

Fig. 7.4 Illustrative Example 7.1: power traded in the market

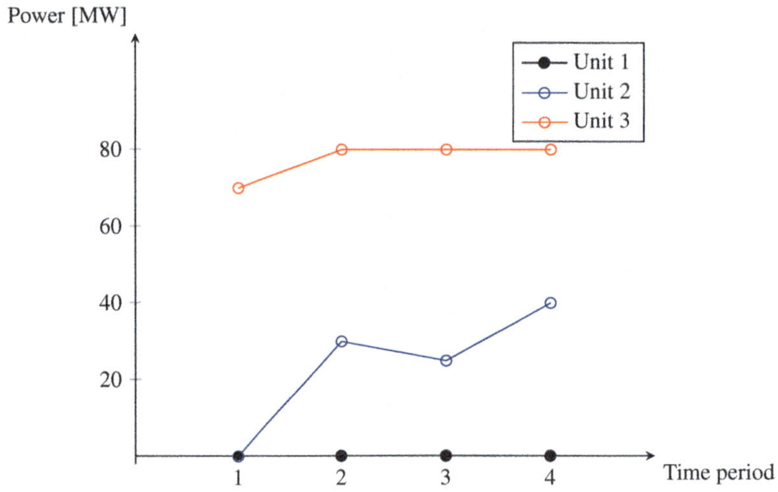

Fig. 7.5 Illustrative Example 7.1: production of the conventional power plants

Finally, the optimization variables of the problem are those included in set $\Phi^{\mathrm{IE7.1}} = \Big\{u_2^{\mathrm{C,I}}, u_3^{\mathrm{C,I}}, u_2^{\mathrm{R,I}}, u_3^{\mathrm{R,I}}, u_1^{\mathrm{S,I}}, p_1^{\mathrm{E}}, p_2^{\mathrm{E}}, p_3^{\mathrm{E}}, p_4^{\mathrm{E}}, p_{11}^{\mathrm{C}}, p_{12}^{\mathrm{C}}, p_{13}^{\mathrm{C}}, p_{14}^{\mathrm{C}}, p_{21}^{\mathrm{C}}, p_{22}^{\mathrm{C}}, p_{23}^{\mathrm{C}}, p_{24}^{\mathrm{C}}, p_{31}^{\mathrm{C}}, p_{32}^{\mathrm{C}}, p_{33}^{\mathrm{C}}, p_{34}^{\mathrm{C}}, p_{11}^{\mathrm{R}}, p_{12}^{\mathrm{R}}, p_{13}^{\mathrm{R}}, p_{14}^{\mathrm{R}}, p_{21}^{\mathrm{R}}, p_{22}^{\mathrm{R}}, p_{23}^{\mathrm{R}}, p_{24}^{\mathrm{R}}, p_{31}^{\mathrm{R}}, p_{32}^{\mathrm{R}}, p_{33}^{\mathrm{R}}, p_{34}^{\mathrm{R}}, p_{11}^{\mathrm{S,C}}, p_{12}^{\mathrm{S,C}}, p_{13}^{\mathrm{S,C}}, p_{14}^{\mathrm{S,C}}, p_{11}^{\mathrm{S,D}}, p_{12}^{\mathrm{S,D}}, p_{13}^{\mathrm{S,D}}, p_{14}^{\mathrm{S,D}}, e_{11}^{\mathrm{S}}, e_{12}^{\mathrm{S}}, e_{13}^{\mathrm{S}}, e_{14}^{\mathrm{S}}, p_1^{\mathrm{D}}, p_2^{\mathrm{D}}, p_3^{\mathrm{D}}, p_4^{\mathrm{D}}\Big\}$.

The optimal expansion plan consists in building the two conventional power plants ($c = 2, 3$) and the candidate solar power unit ($r = 3$).

The power traded by the VPP in the market using both the existing and candidate units is provided in Fig. 7.4.

The production of the conventional power plants is provided in Fig. 7.5. Note that the VPP does not use the existing conventional power plant (unit 1) due to its comparable high variable costs.

The production of the stochastic renewable units is provided in Fig. 7.6, which shows that the VPP uses all the available renewable production in all time periods.

The demand consumption is provided in Fig. 7.7. Note that the demand is supplied in the first and fourth time periods, which have the comparatively lowest market prices, as well as in the third time period due to the high renewable production level.

Finally, note that the optimal value of the objective function, i.e., the profit achieved by the VPP is \$25.468 million. □

A GAMS [4] code for solving Illustrative Example 7.1 is provided in Sect. 7.7.

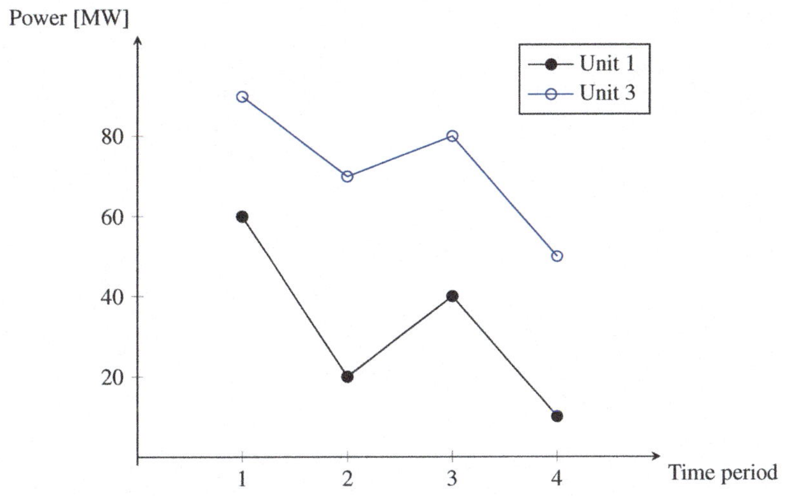

Fig. 7.6 Illustrative Example 7.1: production of the stochastic renewable generating units

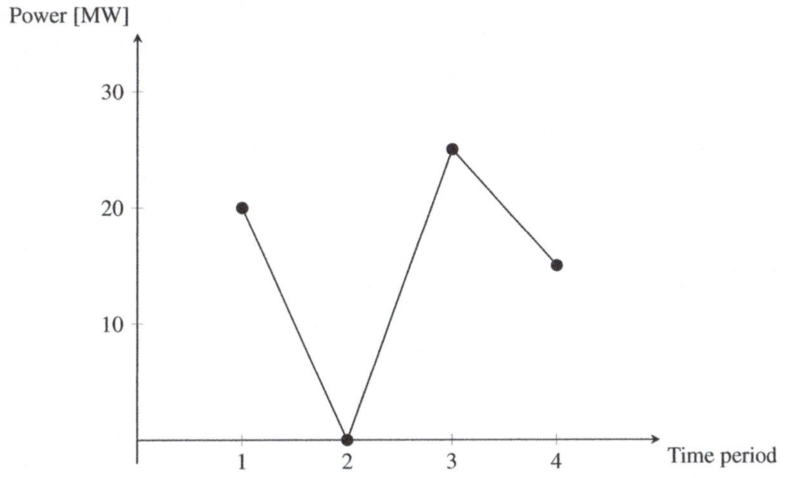

Fig. 7.7 Illustrative Example 7.1: demand consumption

Illustrative Example 7.2 Impact of the Investment Budget

The expansion planning problem of the VPP described in Illustrative Example 7.1 is considered again. Next the impact of the investment budget on the expansion decisions and the profit of the VPP is analyzed. With this purpose, the problem described in Illustrative Example 7.1 is solved for different values of the investment budget.

Figure 7.8 shows that the profit achieved by the VPP increases as the investment budget increases. When the investment budget is null, i.e., when the VPP cannot expand, the profit is \$7.380 million. By building new facilities the VPP increases its profit up to \$30.794 million, which means that the VPP can increase its profit by 317% by expanding its generating assets. Note also that the profit does not change for investment budgets larger than \$20 million. Note that Fig. 7.8 shows also how the profit splits between operating revenues and investment costs.

Table 7.7 provides the candidate units built for different values of the investment budget. Observe that the expansion planning decisions change as the investment budget is increased. For low investment budgets, the VPP builds the storage facility, which is the cheapest unit. However, when the investment budget increases, the VPP prefers building the conventional and stochastic renewable generating units. Note also that for investment budgets larger than \$20 million the expansion decisions do not change despite the fact that the VPP is able to build additional units. This indicates that the expansion planning problem faced by a VPP is not a trivial problem and appropriate expansion tools are needed.

□

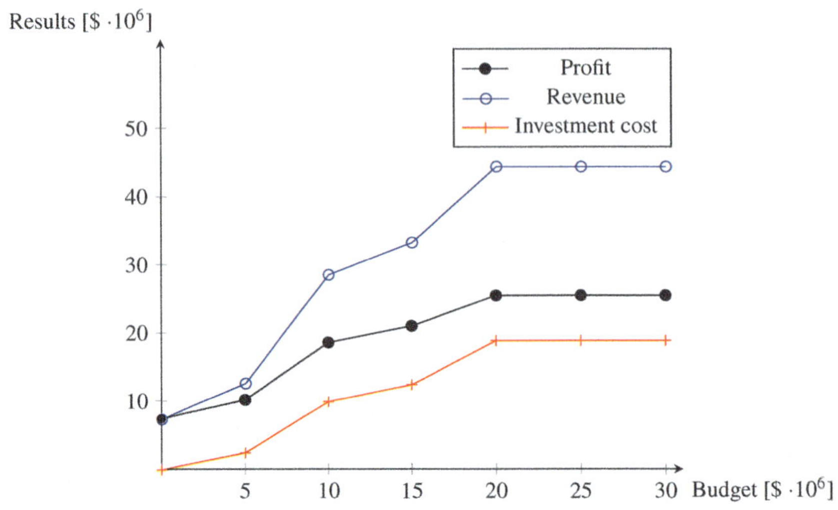

Fig. 7.8 Illustrative Example 7.2: profit achieved by the VPP for different investment budgets

Table 7.7 Illustrative Example 7.2: candidate units built for different investment budgets

Investment budget [$ $\cdot 10^6$]	Candidate units built
0	–
5	Conventional power plant 1
	Storage unit 1
10	Stochastic renewable generating unit 3
15	Conventional power plant 2
	Stochastic renewable generating unit 3
	Storage unit 1
20	Conventional power plants 2 and 3
	Stochastic renewable generating unit 3
25	Conventional power plants 2 and 3
	Stochastic renewable generating unit 3
30	Conventional power plants 2 and 3
	Stochastic renewable generating unit 3

Illustrative Example 7.3 Impact of the Limit of the Power Traded in the Market

The expansion planning problem of the VPP described in Illustrative Example 7.1 is considered again. Here the impact of the limit of the power traded in the market on the expansion decisions and the profit of the VPP is analyzed.

Figure 7.9 shows that the profit achieved by the VPP increases as the limit of the power traded with the market increases up to a value of 400 MW. Note that if this limit is equal to 0, it means that the VPP does not participate in the market and it uses its generation assets only to supply the demand in the VPP. In this case, the profit achieved by the VPP is equal to −$0.361 million.

Table 7.8 provides the candidate units built for different limits of the power traded in the market. When the VPP does not participate in the market, no candidate units are built. However, as the power limit is increased, additional units are built. Note that if this limit is equal or higher than 400 MW, then the VPP builds all the candidate generating assets.

□

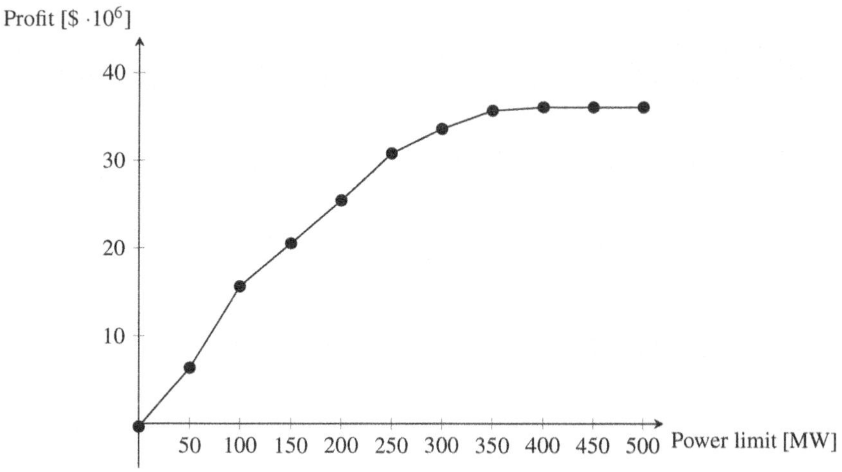

Fig. 7.9 Illustrative Example 7.2: profit achieved by the VPP for different limits of the power traded in the market

Table 7.8 Illustrative Example 7.2: candidate units built for different limits of the power traded in the market

Power limit [MW]	Candidate units built
0	–
50	Conventional power plant 2
100	Stochastic renewable generating unit 3
	Storage unit 1
150	Stochastic renewable generating units 2 and 3
200	Conventional power plants 2 and 3
	Stochastic renewable generating unit 3
250	Conventional power plant 3
	Stochastic renewable generating units 2 and 3
300	Conventional power plants 2 and 3
	Stochastic renewable generating units 2 and 3
350	Conventional power plants 2 and 3
	Stochastic renewable generating units 2 and 3
400	Conventional power plants 2 and 3
	Stochastic renewable generating units 2 and 3
	Storage unit 1

7.4 Stochastic Programming Approach

Expansion planning decisions are made for a medium- or long-term planning horizon. Therefore, at the time of making these decisions, there is a number of uncertainties that must be considered. For example, the VPP does not know its demand needs in the future nor the operating cost of the units, which may condition the expansion decisions made.

In order to account for these uncertainties, a set of scenarios that model different realizations of the uncertain parameters is considered. Moreover, a probability of occurrence to each of these scenarios is assigned. This uncertainty representation allows formulating the optimal expansion planning problem of a VPP as the following stochastic programming model:

$$\max_{\Phi^{\mathrm{S}}} \sum_{\omega \in \Omega^{\mathrm{W}}} \pi_{\omega} \sigma \sum_{t \in \Omega^{\mathrm{T}}} \left(\lambda^{\mathrm{E}}_{t\omega} p^{\mathrm{E}}_{t\omega} \Delta t - \sum_{c \in \Omega^{\mathrm{C}}} C^{\mathrm{C,V}}_{c\omega} p^{\mathrm{C}}_{ct\omega} \Delta t \right)$$
$$- \left(\sum_{c \in \tilde{\Omega}^{\mathrm{C}}} C^{\mathrm{C,I}}_{c} u^{\mathrm{C,I}}_{c} + \sum_{r \in \tilde{\Omega}^{\mathrm{R}}} C^{\mathrm{R,I}}_{r} u^{\mathrm{R,I}}_{r} + \sum_{s \in \tilde{\Omega}^{\mathrm{S}}} C^{\mathrm{S,I}}_{s} u^{\mathrm{S,I}}_{s} \right) \quad (7.2\mathrm{a})$$

subject to

$$\sum_{c \in \tilde{\Omega}^{\mathrm{C}}} C^{\mathrm{C,I}}_{c} u^{\mathrm{C,I}}_{c} + \sum_{r \in \tilde{\Omega}^{\mathrm{R}}} C^{\mathrm{R,I}}_{r} u^{\mathrm{R,I}}_{r} + \sum_{s \in \tilde{\Omega}^{\mathrm{S}}} C^{\mathrm{S,I}}_{s} u^{\mathrm{S,I}}_{s} \leq \overline{C}^{\mathrm{I}}, \quad (7.2\mathrm{b})$$

$$u^{\mathrm{C,I}}_{c} \in \{0, 1\}, \quad \forall c \in \tilde{\Omega}^{\mathrm{C}}, \quad (7.2\mathrm{c})$$

$$u^{\mathrm{R,I}}_{r} \in \{0, 1\}, \quad \forall r \in \tilde{\Omega}^{\mathrm{R}}, \quad (7.2\mathrm{d})$$

$$u^{\mathrm{S,I}}_{s} \in \{0, 1\}, \quad \forall s \in \tilde{\Omega}^{\mathrm{S}}, \quad (7.2\mathrm{e})$$

$$\underline{P}^{\mathrm{E}} \leq p^{\mathrm{E}}_{t\omega} \leq \overline{P}^{\mathrm{E}}, \quad \forall t \in \Omega^{\mathrm{T}}, \forall \omega \in \Omega^{\mathrm{W}}, \quad (7.2\mathrm{f})$$

$$p^{\mathrm{E}}_{t\omega} = \sum_{c \in \Omega^{\mathrm{C}}} p^{\mathrm{C}}_{ct\omega} + \sum_{r \in \Omega^{\mathrm{R}}} p^{\mathrm{R}}_{rt\omega} + \sum_{s \in \Omega^{\mathrm{S}}} \left(p^{\mathrm{S,D}}_{st\omega} - p^{\mathrm{S,C}}_{st\omega} \right)$$
$$- \sum_{d \in \Omega^{\mathrm{D}}} p^{\mathrm{D}}_{dt\omega}, \quad \forall t \in \Omega^{\mathrm{T}}, \forall \omega \in \Omega^{\mathrm{W}}, \quad (7.2\mathrm{g})$$

$$\underline{P}^{\mathrm{D}}_{dt\omega} \leq p^{\mathrm{D}}_{dt\omega} \leq \overline{P}^{\mathrm{D}}_{dt\omega}, \quad \forall d \in \Omega^{\mathrm{D}}, \forall t \in \Omega^{\mathrm{T}}, \forall \omega \in \Omega^{\mathrm{W}}, \quad (7.2\mathrm{h})$$

$$\sum_{t \in \Omega^{\mathrm{T}}} p^{\mathrm{D}}_{dt\omega} \Delta t \geq \underline{E}^{\mathrm{D}}_{d\omega}, \quad \forall d \in \Omega^{\mathrm{D}}, \forall \omega \in \Omega^{\mathrm{W}}, \quad (7.2\mathrm{i})$$

$$0 \leq p^{\mathrm{C}}_{ct\omega} \leq \overline{P}^{\mathrm{C}}_{c}, \quad \forall c \in \Omega^{\mathrm{C}} \setminus c \in \tilde{\Omega}^{\mathrm{C}}, \forall t \in \Omega^{\mathrm{T}}, \forall \omega \in \Omega^{\mathrm{W}}, \quad (7.2\mathrm{j})$$

$$0 \leq p^{\mathrm{C}}_{ct\omega} \leq \overline{P}^{\mathrm{C}}_{c} u^{\mathrm{C,I}}_{c}, \quad \forall c \in \tilde{\Omega}^{\mathrm{C}}, \forall t \in \Omega^{\mathrm{T}}, \forall \omega \in \Omega^{\mathrm{W}}, \quad (7.2\mathrm{k})$$

$$0 \leq p^{\mathrm{R}}_{rt\omega} \leq P^{\mathrm{R,A}}_{rt}, \quad \forall r \in \Omega^{\mathrm{R}} \setminus r \in \tilde{\Omega}^{\mathrm{R}}, \forall t \in \Omega^{\mathrm{T}}, \forall \omega \in \Omega^{\mathrm{W}}, \quad (7.2\mathrm{l})$$

$$0 \leq p^{\mathrm{R}}_{rt\omega} \leq P^{\mathrm{R,A}}_{rt} u^{\mathrm{R,I}}_{r}, \quad \forall r \in \tilde{\Omega}^{\mathrm{R}}, \forall t \in \Omega^{\mathrm{T}}, \forall \omega \in \Omega^{\mathrm{W}}, \quad (7.2\mathrm{m})$$

$$0 \leq p^{\mathrm{S,C}}_{st\omega} \leq \overline{P}^{\mathrm{S,C}}_{s}, \quad \forall s \in \Omega^{\mathrm{S}} \setminus s \in \tilde{\Omega}^{\mathrm{S}}, \forall t \in \Omega^{\mathrm{T}}, \forall \omega \in \Omega^{\mathrm{W}}, \quad (7.2\mathrm{n})$$

$$0 \leq p^{\mathrm{S,D}}_{st\omega} \leq \overline{P}^{\mathrm{S,D}}_{s}, \quad \forall s \in \Omega^{\mathrm{S}} \setminus s \in \tilde{\Omega}^{\mathrm{S}}, \forall t \in \Omega^{\mathrm{T}}, \forall \omega \in \Omega^{\mathrm{W}}, \quad (7.2\mathrm{o})$$

$$e^{\mathrm{S}}_{st\omega} = e^{\mathrm{S}}_{s(t-1)\omega} + p^{\mathrm{S,C}}_{st\omega} \Delta t \eta^{\mathrm{S,C}}_{s} - \frac{p^{\mathrm{S,D}}_{st\omega} \Delta t}{\eta^{\mathrm{S,D}}_{s}}, \quad \forall s \in \Omega^{\mathrm{S}}, \forall t \in \Omega^{\mathrm{T}}, \forall \omega \in \Omega^{\mathrm{W}}, \quad (7.2\mathrm{p})$$

$$\underline{E}^{\mathrm{S}}_{st} \leq e^{\mathrm{S}}_{st\omega} \leq \overline{E}^{\mathrm{S}}_{st}, \quad \forall s \in \Omega^{\mathrm{S}}, \forall t \in \Omega^{\mathrm{T}}, \forall \omega \in \Omega^{\mathrm{W}}, \quad (7.2\mathrm{q})$$

$$0 \leq p^{\mathrm{S,C}}_{st\omega} \leq \overline{P}^{\mathrm{S,C}}_{s} u^{\mathrm{S,I}}_{s}, \quad \forall s \in \tilde{\Omega}^{\mathrm{S}}, \forall t \in \Omega^{\mathrm{T}}, \forall \omega \in \Omega^{\mathrm{W}}, \quad (7.2\mathrm{r})$$

$$0 \leq p^{\mathrm{S,D}}_{st\omega} \leq \overline{P}^{\mathrm{S,D}}_{s} u^{\mathrm{S,I}}_{s}, \quad \forall s \in \tilde{\Omega}^{\mathrm{S}}, \forall t \in \Omega^{\mathrm{T}}, \forall \omega \in \Omega^{\mathrm{W}}, \quad (7.2\mathrm{s})$$

where set $\Phi^{\mathrm{S}} = \left\{u_c^{\mathrm{C,I}}, \forall c \in \tilde{\Omega}^{\mathrm{C}}; u_r^{\mathrm{R,I}}, \forall r \in \tilde{\Omega}^{\mathrm{R}}; u_s^{\mathrm{S,I}}, \forall s \in \tilde{\Omega}^{\mathrm{S}}; p_{t\omega}^{\mathrm{E}}, \forall t \in \Omega^{\mathrm{T}}, \forall \omega \in \Omega^{\mathrm{W}}; p_{ct\omega}^{\mathrm{C}}, \forall c \in \Omega^{\mathrm{C}}, \forall t \in \Omega^{\mathrm{T}}, \forall \omega \in \Omega^{\mathrm{W}}; p_{rt\omega}^{\mathrm{R}}, \forall r \in \Omega^{\mathrm{R}}, \forall t \in \Omega^{\mathrm{T}}, \forall \omega \in \Omega^{\mathrm{W}}; p_{dt\omega}^{\mathrm{D}}, \forall d \in \Omega^{\mathrm{D}}, \forall t \in \Omega^{\mathrm{T}}, \forall \omega \in \Omega^{\mathrm{W}}; e_{st\omega}^{\mathrm{S}}, p_{st\omega}^{\mathrm{S,C}}, p_{st\omega}^{\mathrm{S,D}}, \forall s \in \Omega^{\mathrm{S}}, \forall t \in \Omega^{\mathrm{T}}, \forall \omega \in \Omega^{\mathrm{W}}\right\}$ includes the optimization variables of problem (7.2).

The formulation of the stochastic problem (7.2) is similar to the deterministic one (7.1) formulated in the previous section. The main differences between these two problems are summarized below:

1. The objective function (7.2a) represents the expected profit achieved by the VPP. Therefore, the terms in the first line of (7.2a) representing the operating profits are multiplied by the weight of the corresponding scenario π_ω.
2. Constraints (7.2f)–(7.2s) must be satisfied for all scenarios ω.

Two types of decisions can be distinguished in the problem. On one hand, expansion planning decisions, i.e., variables $u_c^{\mathrm{C,I}}, \forall c \in \tilde{\Omega}^{\mathrm{C}}; u_r^{\mathrm{R,I}}, \forall r \in \tilde{\Omega}^{\mathrm{R}}; u_s^{\mathrm{S,I}}, \forall s \in \tilde{\Omega}^{\mathrm{S}}$, which are made before knowing the actual scenario realization. Therefore, these variables do not depend on scenario ω and are usually known as *here-and-now* or first-stage decisions. On the other hand, operating decisions, i.e., variables $p_{t\omega}^{\mathrm{E}}, \forall t \in \Omega^{\mathrm{T}}, \forall \omega \in \Omega^{\mathrm{W}}; p_{ct\omega}^{\mathrm{C}}, \forall c \in \Omega^{\mathrm{C}}, \forall t \in \Omega^{\mathrm{T}}, \forall \omega \in \Omega^{\mathrm{W}}; p_{rt\omega}^{\mathrm{R}}, \forall r \in \Omega^{\mathrm{R}}, \forall t \in \Omega^{\mathrm{T}}, \forall \omega \in \Omega^{\mathrm{W}}; p_{dt\omega}^{\mathrm{D}}, \forall d \in \Omega^{\mathrm{D}}, \forall t \in \Omega^{\mathrm{T}}, \forall \omega \in \Omega^{\mathrm{W}}; e_{st\omega}^{\mathrm{S}}, p_{st\omega}^{\mathrm{S,C}}, p_{st\omega}^{\mathrm{S,D}}, \forall s \in \Omega^{\mathrm{S}}, \forall t \in \Omega^{\mathrm{T}}, \forall \omega \in \Omega^{\mathrm{W}}$, which are made after knowing the actual scenario realization, and thus they depend on scenario ω. These decisions are generally known as *wait-and-see* or second-stage decisions. Figure 7.10 depicts the scenario tree summarizing the decision-making process.

Model (7.2) is an MILP problem [7].

Illustrative Example 7.4 Stochastic Expansion Planning Problem

The VPP and expansion planning alternatives described in Illustrative Example 7.1 are considered again. However, in this case it is assumed that there are some uncertain parameters at the time of making the expansion decisions. In particular, the demand requirements in the VPP (both demand limits and minimum energy consumption) can be 50% lower than, equal to, and 50% higher than those considered in Illustrative Example 7.1, with probabilities of 0.3, 0.4, and 0.3, respectively. Moreover, the operating cost of conventional power plants can be 30% lower than, equal to, and 30% higher than those considered in Illustrative Example 7.1, with probabilities of 0.2, 0.6, and 0.2, respectively. Both uncertain parameters are considered to be independent, and thus a total of 9 scenarios are considered in the problem (3 × 3) whose data are summarized in Table 7.9.

Using these data, next the stochastic expansion planning problem is formulated.

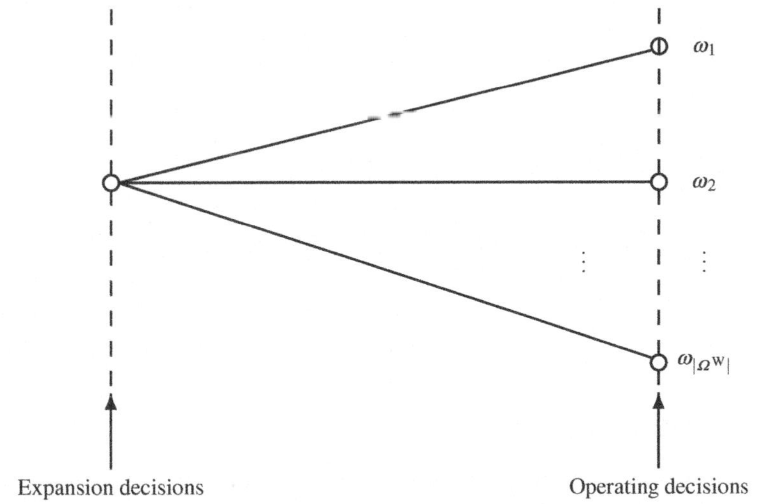

Fig. 7.10 Stochastic expansion planning of a VPP: scenario tree

Table 7.9 Illustrative Example 7.4: scenario data

Scenario	Demand	Costs	Probability
1	Low (−50%)	Low (−30%)	$0.3 \times 0.2 = 0.06$
2	Low (−50%)	Medium (=)	$0.3 \times 0.6 = 0.18$
3	Low (−50%)	High (+30%)	$0.3 \times 0.2 = 0.06$
4	Medium (=)	Low (−30%)	$0.4 \times 0.2 = 0.08$
5	Medium (=)	Medium (=)	$0.4 \times 0.6 = 0.24$
6	Medium (=)	High (+30%)	$0.4 \times 0.2 = 0.08$
7	High (+50%)	Low (−30%)	$0.3 \times 0.2 = 0.06$
8	High (+50%)	Medium (=)	$0.3 \times 0.6 = 0.18$
9	High (+50%)	High (+30%)	$0.3 \times 0.2 = 0.06$

First, the objective function to be maximized:

$$
\begin{aligned}
2190\Big[\, & 0.06\Big(20p^{\mathrm{E}}_{11} + 50p^{\mathrm{E}}_{21} + 35p^{\mathrm{E}}_{31} + 31p^{\mathrm{E}}_{41} - \Big(23.1p^{\mathrm{C}}_{111} + 23.1p^{\mathrm{C}}_{121} + 23.1p^{\mathrm{C}}_{131} \\
& + 23.1p^{\mathrm{C}}_{141} + 15.4p^{\mathrm{C}}_{211} + 15.4p^{\mathrm{C}}_{221} + 15.4p^{\mathrm{C}}_{231} + 15.4p^{\mathrm{C}}_{241} + 8.4p^{\mathrm{C}}_{311} \\
& + 8.4p^{\mathrm{C}}_{321} + 8.4p^{\mathrm{C}}_{331} + 8.4p^{\mathrm{C}}_{341}\Big)\Big) \\
& + 0.18\Big(20p^{\mathrm{E}}_{12} + 50p^{\mathrm{E}}_{22} + 35p^{\mathrm{E}}_{32} + 31p^{\mathrm{E}}_{42} - \Big(33p^{\mathrm{C}}_{112} + 33p^{\mathrm{C}}_{122} + 33p^{\mathrm{C}}_{132} \\
& + 33p^{\mathrm{C}}_{142} + 22p^{\mathrm{C}}_{212} + 22p^{\mathrm{C}}_{222} + 22p^{\mathrm{C}}_{232} + 22p^{\mathrm{C}}_{242} + 12p^{\mathrm{C}}_{312} \\
& + 12p^{\mathrm{C}}_{322} + 12p^{\mathrm{C}}_{332} + 12p^{\mathrm{C}}_{342}\Big)\Big) \\
& + 0.06\Big(20p^{\mathrm{E}}_{13} + 50p^{\mathrm{E}}_{23} + 35p^{\mathrm{E}}_{33} + 31p^{\mathrm{E}}_{43} - \Big(42.9p^{\mathrm{C}}_{113} + 42.9p^{\mathrm{C}}_{123} + 42.9p^{\mathrm{C}}_{133} \\
& + 42.9p^{\mathrm{C}}_{143} + 28.6p^{\mathrm{C}}_{213} + 28.6p^{\mathrm{C}}_{223} + 28.6p^{\mathrm{C}}_{233} + 28.6p^{\mathrm{C}}_{243} + 15.6p^{\mathrm{C}}_{313} \\
& + 15.6p^{\mathrm{C}}_{323} + 15.6p^{\mathrm{C}}_{333} + 15.6p^{\mathrm{C}}_{343}\Big)\Big) \\
& + 0.08\Big(20p^{\mathrm{E}}_{14} + 50p^{\mathrm{E}}_{24} + 35p^{\mathrm{E}}_{34} + 31p^{\mathrm{E}}_{44} - \Big(23.1p^{\mathrm{C}}_{114} + 23.1p^{\mathrm{C}}_{124} + 23.1p^{\mathrm{C}}_{134} \\
& + 23.1p^{\mathrm{C}}_{144} + 15.4p^{\mathrm{C}}_{214} + 15.4p^{\mathrm{C}}_{224} + 15.4p^{\mathrm{C}}_{234} + 15.4p^{\mathrm{C}}_{244} + 8.4p^{\mathrm{C}}_{314} \\
& + 8.4p^{\mathrm{C}}_{324} + 8.4p^{\mathrm{C}}_{334} + 8.4p^{\mathrm{C}}_{344}\Big)\Big) \\
& + 0.24\Big(20p^{\mathrm{E}}_{15} + 50p^{\mathrm{E}}_{25} + 35p^{\mathrm{E}}_{35} + 31p^{\mathrm{E}}_{45} - \Big(33p^{\mathrm{C}}_{115} + 33p^{\mathrm{C}}_{125} + 33p^{\mathrm{C}}_{135} \\
& + 33p^{\mathrm{C}}_{145} + 22p^{\mathrm{C}}_{215} + 22p^{\mathrm{C}}_{225} + 22p^{\mathrm{C}}_{235} + 22p^{\mathrm{C}}_{245} + 12p^{\mathrm{C}}_{315} \\
& + 12p^{\mathrm{C}}_{325} + 12p^{\mathrm{C}}_{335} + 12p^{\mathrm{C}}_{345}\Big)\Big) \\
& + 0.08\Big(20p^{\mathrm{E}}_{16} + 50p^{\mathrm{E}}_{26} + 35p^{\mathrm{E}}_{36} + 31p^{\mathrm{E}}_{46} - \Big(42.9p^{\mathrm{C}}_{116} + 42.9p^{\mathrm{C}}_{126} + 42.9p^{\mathrm{C}}_{136} \\
& + 42.9p^{\mathrm{C}}_{146} + 28.6p^{\mathrm{C}}_{216} + 28.6p^{\mathrm{C}}_{226} + 28.6p^{\mathrm{C}}_{236} + 28.6p^{\mathrm{C}}_{246} + 15.6p^{\mathrm{C}}_{316} \\
& + 15.6p^{\mathrm{C}}_{326} + 15.6p^{\mathrm{C}}_{336} + 15.6p^{\mathrm{C}}_{346}\Big)\Big) \\
& + 0.06\Big(20p^{\mathrm{E}}_{17} + 50p^{\mathrm{E}}_{27} + 35p^{\mathrm{E}}_{37} + 31p^{\mathrm{E}}_{47} - \Big(23.1p^{\mathrm{C}}_{117} + 23.1p^{\mathrm{C}}_{127} + 23.1p^{\mathrm{C}}_{137} \\
& + 23.1p^{\mathrm{C}}_{147} + 15.4p^{\mathrm{C}}_{217} + 15.4p^{\mathrm{C}}_{227} + 15.4p^{\mathrm{C}}_{237} + 15.4p^{\mathrm{C}}_{247} + 8.4p^{\mathrm{C}}_{317} \\
& + 8.4p^{\mathrm{C}}_{327} + 8.4p^{\mathrm{C}}_{337} + 8.4p^{\mathrm{C}}_{347}\Big)\Big)
\end{aligned}
$$

$$
\begin{aligned}
&+0.18\Big(20p^{\mathrm{E}}_{18}+50p^{\mathrm{E}}_{28}+35p^{\mathrm{E}}_{38}+31p^{\mathrm{E}}_{48}-\Big(33p^{\mathrm{C}}_{118}+33p^{\mathrm{C}}_{128}+33p^{\mathrm{C}}_{138}\\
&\quad+33p^{\mathrm{C}}_{148}+22p^{\mathrm{C}}_{218}+22p^{\mathrm{C}}_{228}+22p^{\mathrm{C}}_{238}+22p^{\mathrm{C}}_{248}+12p^{\mathrm{C}}_{318}\\
&\quad+12p^{\mathrm{C}}_{328}+12p^{\mathrm{C}}_{338}+12p^{\mathrm{C}}_{348}\Big)\Big)\\
&+0.06\Big(20p^{\mathrm{E}}_{19}+50p^{\mathrm{E}}_{29}+35p^{\mathrm{E}}_{39}+31p^{\mathrm{E}}_{49}-\Big(42.9p^{\mathrm{C}}_{119}+42.9p^{\mathrm{C}}_{129}+42.9p^{\mathrm{C}}_{139}\\
&\quad+42.9p^{\mathrm{C}}_{149}+28.6p^{\mathrm{C}}_{219}+28.6p^{\mathrm{C}}_{229}+28.6p^{\mathrm{C}}_{239}+28.6p^{\mathrm{C}}_{249}+15.6p^{\mathrm{C}}_{319}\\
&\quad+15.6p^{\mathrm{C}}_{329}+15.6p^{\mathrm{C}}_{339}+15.6p^{\mathrm{C}}_{349}\Big)\Big)\Big]\\
&-\Big(2000000u^{\mathrm{C,I}}_{2}+7000000u^{\mathrm{C,I}}_{3}+7000000u^{\mathrm{R,I}}_{2}+10000000u^{\mathrm{R,I}}_{3}+500000u^{\mathrm{S,I}}_{1}\Big)\,.
\end{aligned}
$$

Next the constraints of the problem are formulated.

On one hand, the investment budget constraint and the definition of binary variables are exactly the same than those formulated in Illustrative Example 7.1. On the other hand, constraints related to the power balance in the VPP, as well as those related to the limits on the production of existing and candidate conventional power plants, stochastic renewable generating units, and storage units are also the same; however, these constraints must be satisfied for all scenarios $\omega = 1, \ldots, 9$.

Finally, it is necessary to modify the constraints related to the demand consumption in the VPP as follows:

- Limits on the power consumption of the demand:

$$
\begin{aligned}
0 &\le p^{\mathrm{D}}_{11} \le 10,\\
0 &\le p^{\mathrm{D}}_{21} \le 10,\\
5 &\le p^{\mathrm{D}}_{31} \le 12.5,\\
7.5 &\le p^{\mathrm{D}}_{41} \le 15,\\
0 &\le p^{\mathrm{D}}_{12} \le 10,\\
0 &\le p^{\mathrm{D}}_{22} \le 10,\\
5 &\le p^{\mathrm{D}}_{32} \le 12.5,\\
7.5 &\le p^{\mathrm{D}}_{42} \le 15,\\
0 &\le p^{\mathrm{D}}_{13} \le 10,\\
0 &\le p^{\mathrm{D}}_{23} \le 10,\\
5 &\le p^{\mathrm{D}}_{33} \le 12.5,\\
7.5 &\le p^{\mathrm{D}}_{43} \le 15,\\
0 &\le p^{\mathrm{D}}_{14} \le 20,\\
0 &\le p^{\mathrm{D}}_{24} \le 20,\\
10 &\le p^{\mathrm{D}}_{34} \le 25,\\
15 &\le p^{\mathrm{D}}_{44} \le 30,\\
0 &\le p^{\mathrm{D}}_{15} \le 20,\\
0 &\le p^{\mathrm{D}}_{25} \le 20,\\
10 &\le p^{\mathrm{D}}_{35} \le 25,
\end{aligned}
$$

$$15 \le p^{\rm D}_{45} \le 30,$$
$$0 \le p^{\rm D}_{16} \le 20,$$
$$0 \le p^{\rm D}_{26} \le 20,$$
$$10 \le p^{\rm D}_{36} \le 25,$$
$$15 \le p^{\rm D}_{46} \le 30,$$
$$0 \le p^{\rm D}_{17} \le 30,$$
$$0 \le p^{\rm D}_{27} \le 30,$$
$$15 \le p^{\rm D}_{37} \le 37.5,$$
$$22.5 \le p^{\rm D}_{47} \le 45,$$
$$0 \le p^{\rm D}_{18} \le 30,$$
$$0 \le p^{\rm D}_{28} \le 30,$$
$$15 \le p^{\rm D}_{38} \le 37.5,$$
$$22.5 \le p^{\rm D}_{48} \le 45,$$
$$0 \le p^{\rm D}_{19} \le 30,$$
$$0 \le p^{\rm D}_{29} \le 30,$$
$$15 \le p^{\rm D}_{39} \le 37.5,$$
$$22.5 \le p^{\rm D}_{49} \le 45.$$

- Minimum energy consumption of the demand through the planning horizon:

$$p^{\rm D}_{11} + p^{\rm D}_{21} + p^{\rm D}_{31} + p^{\rm D}_{41} \ge 30,$$
$$p^{\rm D}_{12} + p^{\rm D}_{22} + p^{\rm D}_{32} + p^{\rm D}_{42} \ge 30,$$
$$p^{\rm D}_{13} + p^{\rm D}_{23} + p^{\rm D}_{33} + p^{\rm D}_{43} \ge 30,$$
$$p^{\rm D}_{14} + p^{\rm D}_{24} + p^{\rm D}_{34} + p^{\rm D}_{44} \ge 60,$$
$$p^{\rm D}_{15} + p^{\rm D}_{25} + p^{\rm D}_{35} + p^{\rm D}_{45} \ge 60,$$
$$p^{\rm D}_{16} + p^{\rm D}_{26} + p^{\rm D}_{36} + p^{\rm D}_{46} \ge 60,$$
$$p^{\rm D}_{17} + p^{\rm D}_{27} + p^{\rm D}_{37} + p^{\rm D}_{47} \ge 90,$$
$$p^{\rm D}_{18} + p^{\rm D}_{28} + p^{\rm D}_{38} + p^{\rm D}_{48} \ge 90,$$
$$p^{\rm D}_{19} + p^{\rm D}_{29} + p^{\rm D}_{39} + p^{\rm D}_{49} \ge 90.$$

The problem is solved, obtaining that it is optimal to build candidate conventional power plant 3, stochastic renewable unit 3, and storage unit 1. Note that although the expected values of the uncertain parameters are the same than those considered in Illustrative Example 7.1, the expansion decisions are different, which highlights the importance of an accurate modeling of the uncertainty in the problem.

The optimal value of the objective function, i.e., the expected profit of the VPP including both operating profits and investment costs, is $25.731 million.

□

A GAMS [4] code for solving Illustrative Example 7.4 is provided in Sect. 7.7.

Illustrative Example 7.5 Value of the Stochastic Solution

The data of Illustrative Example 7.4 are considered again. In order to justify the use of stochastic programming over the deterministic problem described in Sect. 7.3, the value of the stochastic solution (VSS) metric is computed below. The VSS quantifies the advantage of using stochastic programming over a deterministic approach. To compute the VSS, a deterministic instance of the problem formulated in Illustrative Example 7.4 is solved, for which uncertain parameters are set equal to their average values, i.e., the problem formulated in Illustrative Example 7.1. Then, the stochastic programming problem formulated in Illustrative Example 7.4 is solved again but in this case fixing the expansion decisions to those provided by the deterministic problem. Finally, the VSS is computed as:

$$\mathrm{VSS} = z^{\mathrm{S}*} - z^{\mathrm{D}*},$$

where $z^{\mathrm{S}*}$ is the optimal value of the objective function of the stochastic programming problem, while $z^{\mathrm{D}*}$ is the optimal value of the objective function of the modified stochastic programming problem, i.e., that with investment decisions fixed to those obtained in the deterministic problem.

Considering the data of Illustrative Example 7.4, the VSS is:

$$\mathrm{VSS} = \$25.731 \cdot 10^6 - \$25.589 \cdot 10^6 = \$0.142 \cdot 10^6.$$

□

Illustrative Example 7.6 Expected Value of Perfect Information

The data of Illustrative Example 7.4 are considered again. In this case, another relevant metric is computed, namely the so-called expected value of perfect information (EVPI). The EVPI represents how much the VPP is willing to pay for obtaining perfect information about the future. To compute this metric, a modified version of the stochastic problem formulated in Illustrative Example 7.4 is solved, in which the non-anticipativity constraints are relaxed, i.e., all decision variables are scenario dependent. In this case, expansion decisions are made with perfect information. Then, the EVPI is computed as:

$$\mathrm{EVPI} = z^{\mathrm{P}*} - z^{\mathrm{S}*},$$

where $z^{\mathrm{P}*}$ is the optimal value of the objective function of such a modified stochastic programming problem with perfect information and $z^{\mathrm{S}*}$ is the optimal value of the objective function of the original stochastic programming problem.

Considering the data of Illustrative Example 7.4, the EVPI is:

$$\mathrm{EVPI} = \$25.990 \cdot 10^6 - \$25.731 \cdot 10^6 = \$0.259 \cdot 10^6.$$

□

7.5 Stochastic Risk-Constrained Approach

Making expansion decisions generally requires investing a large amount of money. Moreover, these decisions are made for a medium- or long-term planning horizon, and thus these decisions are usually made under uncertainty. This means that making investment decisions for a VPP may be risky if uncertainty is not adequately addressed in the decision-making problem. However, even if uncertainty is well represented, the investment plan may be risky if certain scenarios are realized. This is shown in the following example, which illustrates the concept of risk in the VPP expansion planning problem.

Illustrative Example 7.7 Risk Modeling in the VPP Expansion Planning Problem

A VPP is considering two possible expansion plans to be carried out, namely *Plan I* and *Plan II*. Uncertain parameters in the problem are modeled using four scenarios with the same probability (0.25). The VPP computes the profit (revenues minus costs) for each expansion plan and scenario. These profits are provided in Table 7.10.

Table 7.10 Illustrative Example 7.7: profits of expansion plans I and II for different scenarios [$\$\cdot 10^6$]

Scenario	1	2	3	4
Expansion plan I	−20	40	48	20
Expansion plan II	10	24	30	16

With the data of Table 7.10, the VPP computes the expected profit associated with each expansion plan:

$$E(\pi_I) = (-20\cdot 0.25 + 40\cdot 0.25 + 48\cdot 0.25 + 20\cdot 0.25)\cdot 10^6 = 22\cdot 10^6.$$

$$E(\pi_{II}) = (10\cdot 0.25 + 24\cdot 0.25 + 30\cdot 0.25 + 16\cdot 0.25)\cdot 10^6 = 20\cdot 10^6.$$

Following the criterion considered in Sect. 7.4, i.e., selecting the expansion plan that maximizes the expected profit, it is clear that the VPP would select expansion plan I since it has the highest expected profit. However, observe the profits per scenario provided in Table 7.10 and note that if expansion plan I is selected and scenario 1 is finally realized, the VPP would have to face a negative profit, i.e., the VPP would have losses, which it may be not willing to assume.

On the other hand, if expansion plan II is selected, the profit achieved by the VPP will be likely lower (the profits are lower in 3 of the 4 scenarios in comparison with expansion plan I). However, the VPP knows that if expansion plan II is carried out, the profit in all scenarios will be positive. In other words, making expansion plan I is a risky strategy since the profit of the VPP will be negative with a probability of 25%, while making expansion plan II is a more conservative strategy since the expected profit is lower but the VPP knows that its profit will be always positive.

□

The above example highlights the importance of modeling the profit risk associated with the expansion decisions. One option is using the CVaR metric, which is defined in a profit maximization problem as the expected profit of the $(1-\alpha)$ 100% scenarios yielding the lowest profits for a given α confidence level. This risk metric is coherent, linear, and easy to incorporate in optimization problems [6].

Taking into account the previous considerations, the expansion planning problem of a VPP considering a stochastic risk-constrained approach is formulated as the following MILP problem:

$$\begin{aligned} \max_{\Phi^{\text{SR}}} \quad & \sum_{\omega\in\Omega^{\text{W}}} \pi_\omega \sigma \sum_{t\in\Omega^{\text{T}}} \left(\lambda^{\text{E}}_{t\omega} p^{\text{E}}_{t\omega}\Delta t - \sum_{c\in\Omega^{\text{C}}} C^{\text{C,V}}_{c\omega} p^{\text{C}}_{ct\omega}\Delta t\right) \\ & - \left(\sum_{c\in\tilde{\Omega}^{\text{C}}} C^{\text{C,I}}_c u^{\text{C,I}}_c + \sum_{r\in\tilde{\Omega}^{\text{R}}} C^{\text{R,I}}_r u^{\text{R,I}}_r + \sum_{s\in\tilde{\Omega}^{\text{S}}} C^{\text{S,I}}_s u^{\text{S,I}}_s\right) \\ & + \beta\left(\vartheta - \frac{1}{1-\alpha}\sum_{\omega\in\Omega^{\text{W}}} \pi_\omega \tau_\omega\right) \end{aligned} \tag{7.3a}$$

subject to

$$\sum_{c\in\tilde{\Omega}^{\text{C}}} C^{\text{C,I}}_c u^{\text{C,I}}_c + \sum_{r\in\tilde{\Omega}^{\text{R}}} C^{\text{R,I}}_r u^{\text{R,I}}_r + \sum_{s\in\tilde{\Omega}^{\text{S}}} C^{\text{S,I}}_s u^{\text{S,I}}_s \le \overline{C}^{\text{I}}, \tag{7.3b}$$

$$u^{\text{C,I}}_c \in \{0,1\}, \quad \forall c \in \tilde{\Omega}^{\text{C}}, \tag{7.3c}$$

$$u^{\text{R,I}}_r \in \{0,1\}, \quad \forall r \in \tilde{\Omega}^{\text{R}}, \tag{7.3d}$$

$$u^{\text{S,I}}_s \in \{0,1\}, \quad \forall s \in \tilde{\Omega}^{\text{S}}, \tag{7.3e}$$

$$\begin{aligned} \vartheta - \Bigg[\sigma & \sum_{t\in\Omega^{\text{T}}} \left(\lambda^{\text{E}}_{t\omega} p^{\text{E}}_{t\omega}\Delta t - \sum_{c\in\Omega^{\text{C}}} C^{\text{C,V}}_{c\omega} p^{\text{C}}_{ct\omega}\Delta t\right) \\ & - \left(\sum_{c\in\tilde{\Omega}^{\text{C}}} C^{\text{C,I}}_c u^{\text{C,I}}_c + \sum_{r\in\tilde{\Omega}^{\text{R}}} C^{\text{R,I}}_r u^{\text{R,I}}_r + \sum_{s\in\tilde{\Omega}^{\text{S}}} C^{\text{S,I}}_s u^{\text{S,I}}_s\right)\Bigg] \le \tau_\omega, \quad \forall \omega \in \Omega^{\text{W}}, \end{aligned} \tag{7.3f}$$

$$\tau_\omega \ge 0, \quad \forall \omega \in \Omega^{\text{W}}, \tag{7.3g}$$

$$\underline{P}^{\mathrm{E}} \leq p^{\mathrm{E}}_{t\omega} \leq \overline{P}^{\mathrm{E}}, \quad \forall t \in \Omega^{\mathrm{T}}, \forall \omega \in \Omega^{\mathrm{W}}, \tag{7.3h}$$

$$p^{\mathrm{E}}_{t\omega} = \sum_{c \in \Omega^{\mathrm{C}}} p^{\mathrm{C}}_{ct\omega} + \sum_{r \in \Omega^{\mathrm{R}}} p^{\mathrm{R}}_{rt\omega} + \sum_{s \in \Omega^{\mathrm{S}}} \left(p^{\mathrm{S,D}}_{st\omega} - p^{\mathrm{S,C}}_{st\omega} \right) - \sum_{d \in \Omega^{\mathrm{D}}} p^{\mathrm{D}}_{dt\omega}, \quad \forall t \in \Omega^{\mathrm{T}}, \forall \omega \in \Omega^{\mathrm{W}}, \tag{7.3i}$$

$$\underline{P}^{\mathrm{D}}_{dt\omega} \leq p^{\mathrm{D}}_{dt\omega} \leq \overline{P}^{\mathrm{D}}_{dt\omega}, \quad \forall d \in \Omega^{\mathrm{D}}, \forall t \in \Omega^{\mathrm{T}}, \forall \omega \in \Omega^{\mathrm{W}}, \tag{7.3j}$$

$$\sum_{t \in \Omega^{\mathrm{T}}} p^{\mathrm{D}}_{dt\omega} \Delta t \geq \underline{E}^{\mathrm{D}}_{d\omega}, \quad \forall d \in \Omega^{\mathrm{D}}, \forall \omega \in \Omega^{\mathrm{W}}, \tag{7.3k}$$

$$0 \leq p^{\mathrm{C}}_{ct\omega} \leq \overline{P}^{\mathrm{C}}_{c}, \quad \forall c \in \Omega^{\mathrm{C}} \setminus c \in \tilde{\Omega}^{\mathrm{C}}, \forall t \in \Omega^{\mathrm{T}}, \forall \omega \in \Omega^{\mathrm{W}}, \tag{7.3l}$$

$$0 \leq p^{\mathrm{C}}_{ct\omega} \leq \overline{P}^{\mathrm{C}}_{c} u^{\mathrm{C,I}}_{c}, \quad \forall c \in \tilde{\Omega}^{\mathrm{C}}, \forall t \in \Omega^{\mathrm{T}}, \forall \omega \in \Omega^{\mathrm{W}}, \tag{7.3m}$$

$$0 \leq p^{\mathrm{R}}_{rt\omega} \leq P^{\mathrm{R,A}}_{rt}, \quad \forall r \in \Omega^{\mathrm{R}} \setminus r \in \tilde{\Omega}^{\mathrm{R}}, \forall t \in \Omega^{\mathrm{T}}, \forall \omega \in \Omega^{\mathrm{W}}, \tag{7.3n}$$

$$0 \leq p^{\mathrm{R}}_{rt\omega} \leq P^{\mathrm{R,A}}_{rt} u^{\mathrm{R,I}}_{r}, \quad \forall r \in \tilde{\Omega}^{\mathrm{R}}, \forall t \in \Omega^{\mathrm{T}}, \forall \omega \in \Omega^{\mathrm{W}}, \tag{7.3o}$$

$$0 \leq p^{\mathrm{S,C}}_{st\omega} \leq \overline{P}^{\mathrm{S,C}}_{s}, \quad \forall s \in \Omega^{\mathrm{S}} \setminus s \in \tilde{\Omega}^{\mathrm{S}}, \forall t \in \Omega^{\mathrm{T}}, \forall \omega \in \Omega^{\mathrm{W}}, \tag{7.3p}$$

$$0 \leq p^{\mathrm{S,D}}_{st\omega} \leq \overline{P}^{\mathrm{S,D}}_{s}, \quad \forall s \in \Omega^{\mathrm{S}} \setminus s \in \tilde{\Omega}^{\mathrm{S}}, \forall t \in \Omega^{\mathrm{T}}, \forall \omega \in \Omega^{\mathrm{W}}, \tag{7.3q}$$

$$e^{\mathrm{S}}_{st\omega} = e^{\mathrm{S}}_{s(t-1)\omega} + p^{\mathrm{S,C}}_{st\omega} \Delta t \eta^{\mathrm{S,C}}_{s} - \frac{p^{\mathrm{S,D}}_{st\omega} \Delta t}{\eta^{\mathrm{S,D}}_{s}}, \quad \forall s \in \Omega^{\mathrm{S}}, \forall t \in \Omega^{\mathrm{T}}, \forall \omega \in \Omega^{\mathrm{W}}, \tag{7.3r}$$

$$\underline{E}^{\mathrm{S}}_{st} \leq e^{\mathrm{S}}_{st\omega} \leq \overline{E}^{\mathrm{S}}_{st}, \quad \forall s \in \Omega^{\mathrm{S}}, \forall t \in \Omega^{\mathrm{T}}, \forall \omega \in \Omega^{\mathrm{W}}, \tag{7.3s}$$

$$0 \leq p^{\mathrm{S,C}}_{st\omega} \leq \overline{P}^{\mathrm{S,C}}_{s} u^{\mathrm{S,I}}_{s}, \quad \forall s \in \tilde{\Omega}^{\mathrm{S}}, \forall t \in \Omega^{\mathrm{T}}, \forall \omega \in \Omega^{\mathrm{W}}, \tag{7.3t}$$

$$0 \leq p^{\mathrm{S,D}}_{st\omega} \leq \overline{P}^{\mathrm{S,D}}_{s} u^{\mathrm{S,I}}_{s}, \quad \forall s \in \tilde{\Omega}^{\mathrm{S}}, \forall t \in \Omega^{\mathrm{T}}, \forall \omega \in \Omega^{\mathrm{W}}, \tag{7.3u}$$

where set $\Phi^{\mathrm{SR}} = \Big\{ u^{\mathrm{C,I}}_{c}, \forall c \in \tilde{\Omega}^{\mathrm{C}};\ u^{\mathrm{R,I}}_{r}, \forall r \in \tilde{\Omega}^{\mathrm{R}};\ u^{\mathrm{S,I}}_{s}, \forall s \in \tilde{\Omega}^{\mathrm{S}};\ \vartheta;\ \tau_{\omega}, \forall \omega \in \Omega^{\mathrm{W}};\ p^{\mathrm{E}}_{t\omega}, \forall t \in \Omega^{\mathrm{T}}, \forall \omega \in \Omega^{\mathrm{W}};\ p^{\mathrm{C}}_{ct\omega}, \forall c \in \Omega^{\mathrm{C}}, \forall t \in \Omega^{\mathrm{T}}, \forall \omega;\ p^{\mathrm{R}}_{rt\omega}, \forall r \in \Omega^{\mathrm{R}}, \forall t \in \Omega^{\mathrm{T}}, \forall \omega \in \Omega^{\mathrm{W}};\ p^{\mathrm{D}}_{dt\omega}, \forall d \in \Omega^{\mathrm{D}}, \forall t \Omega^{\mathrm{T}}, \forall \omega \in \Omega^{\mathrm{W}};\ e^{\mathrm{S}}_{st\omega}, p^{\mathrm{S,C}}_{st\omega}, p^{\mathrm{S,D}}_{st\omega}, \forall s \in \Omega^{\mathrm{S}}, \forall t \in \Omega^{\mathrm{T}}, \forall \omega \in \Omega^{\mathrm{W}} \Big\}$ includes the optimization variables of problem (7.3).

The formulation of the stochastic risk-constrained problem (7.3) is very similar to the stochastic one (7.2) formulated in the previous section. The main differences between these two problems are summarized below:

1. The objective function (7.3a) includes an additional term $\vartheta - \frac{1}{1-\alpha} \sum_{\omega \in \Omega^{\mathrm{W}}} \pi_{\omega} \tau_{\omega}$ representing the CVaR.
2. Constraints (7.3f)–(7.3g) are included.

Note that the CVaR in the objective function is multiplied by a weighting parameter β that models the trade-off between the expected profit achieved by the VPP and the CVaR. If $\beta = 0$, then the VPP behaves as a risk-neutral agent, i.e., the VPP maximizes its expected profit and neglects the profit risk associated with its expansion decisions. In such a case, problems (7.3) and (7.2) are equivalent. On the other hand, large enough values of β allow representing a risk-averse VPP that maximizes not only its expected profit but also the CVaR. Finally, note that constraints (7.3f) and (7.3g) allow incorporating the CVaR risk metric as a set of linear constraints as explained in [6].

Illustrative Example 7.8 Stochastic Risk-Constrained Expansion Planning Problem

The data of Illustrative Example 7.4 are considered again. In this case, the problem is solved using the stochastic risk-constrained model (7.3) described above. To do so, a confidence level $\alpha = 0.95$ is considered, as well as two values of the weighting parameter β, namely $\beta = 0$ and $\beta = 1$, which respectively represent a risk-neutral and a risk-averse VPP.

Regarding the formulation of the problem, note that the following modifications are needed in the formulation of the problem provided in Illustrative Example 7.4:

- The objective function is replaced by the following one that includes also the CVaR:

$$
\begin{aligned}
2190\Big[&0.06\big(20p^{\mathrm{E}}_{11} + 50p^{\mathrm{E}}_{21} + 35p^{\mathrm{E}}_{31} + 31p^{\mathrm{E}}_{41} - \big(23.1p^{\mathrm{C}}_{111} + 23.1p^{\mathrm{C}}_{121} + 23.1p^{\mathrm{C}}_{131}\\
&+23.1p^{\mathrm{C}}_{141} + 15.4p^{\mathrm{C}}_{211} + 15.4p^{\mathrm{C}}_{221} + 15.4p^{\mathrm{C}}_{231} + 15.4p^{\mathrm{C}}_{241} + 8.4p^{\mathrm{C}}_{311}\\
&+8.4p^{\mathrm{C}}_{321} + 8.4p^{\mathrm{C}}_{331} + 8.4p^{\mathrm{C}}_{341}\big)\big)\\
&+0.18\big(20p^{\mathrm{E}}_{12} + 50p^{\mathrm{E}}_{22} + 35p^{\mathrm{E}}_{32} + 31p^{\mathrm{E}}_{42} - \big(33p^{\mathrm{C}}_{112} + 33p^{\mathrm{C}}_{122} + 33p^{\mathrm{C}}_{132}\\
&+33p^{\mathrm{C}}_{142} + 22p^{\mathrm{C}}_{212} + 22p^{\mathrm{C}}_{222} + 22p^{\mathrm{C}}_{232} + 22p^{\mathrm{C}}_{242} + 12p^{\mathrm{C}}_{312}\\
&+12p^{\mathrm{C}}_{322} + 12p^{\mathrm{C}}_{332} + 12p^{\mathrm{C}}_{342}\big)\big)\\
&+0.06\big(20p^{\mathrm{E}}_{13} + 50p^{\mathrm{E}}_{23} + 35p^{\mathrm{E}}_{33} + 31p^{\mathrm{E}}_{43} - \big(42.9p^{\mathrm{C}}_{113} + 42.9p^{\mathrm{C}}_{123} + 42.9p^{\mathrm{C}}_{133}\\
&+42.9p^{\mathrm{C}}_{143} + 28.6p^{\mathrm{C}}_{213} + 28.6p^{\mathrm{C}}_{223} + 28.6p^{\mathrm{C}}_{233} + 28.6p^{\mathrm{C}}_{243} + 15.6p^{\mathrm{C}}_{313}\\
&+15.6p^{\mathrm{C}}_{323} + 15.6p^{\mathrm{C}}_{333} + 15.6p^{\mathrm{C}}_{343}\big)\big)\\
&+0.08\big(20p^{\mathrm{E}}_{14} + 50p^{\mathrm{E}}_{24} + 35p^{\mathrm{E}}_{34} + 31p^{\mathrm{E}}_{44} - \big(23.1p^{\mathrm{C}}_{114} + 23.1p^{\mathrm{C}}_{124} + 23.1p^{\mathrm{C}}_{134}\\
&+23.1p^{\mathrm{C}}_{144} + 15.4p^{\mathrm{C}}_{214} + 15.4p^{\mathrm{C}}_{224} + 15.4p^{\mathrm{C}}_{234} + 15.4p^{\mathrm{C}}_{244} + 8.4p^{\mathrm{C}}_{314}\\
&+8.4p^{\mathrm{C}}_{324} + 8.4p^{\mathrm{C}}_{334} + 8.4p^{\mathrm{C}}_{344}\big)\big)\\
&+0.24\big(20p^{\mathrm{E}}_{15} + 50p^{\mathrm{E}}_{25} + 35p^{\mathrm{E}}_{35} + 31p^{\mathrm{E}}_{45} - \big(33p^{\mathrm{C}}_{115} + 33p^{\mathrm{C}}_{125} + 33p^{\mathrm{C}}_{135}\\
&+33p^{\mathrm{C}}_{145} + 22p^{\mathrm{C}}_{215} + 22p^{\mathrm{C}}_{225} + 22p^{\mathrm{C}}_{235} + 22p^{\mathrm{C}}_{245} + 12p^{\mathrm{C}}_{315}\\
&+12p^{\mathrm{C}}_{325} + 12p^{\mathrm{C}}_{335} + 12p^{\mathrm{C}}_{345}\big)\big)\\
&+0.08\big(20p^{\mathrm{E}}_{16} + 50p^{\mathrm{E}}_{26} + 35p^{\mathrm{E}}_{36} + 31p^{\mathrm{E}}_{46} - \big(42.9p^{\mathrm{C}}_{116} + 42.9p^{\mathrm{C}}_{126} + 42.9p^{\mathrm{C}}_{136}\\
&+42.9p^{\mathrm{C}}_{146} + 28.6p^{\mathrm{C}}_{216} + 28.6p^{\mathrm{C}}_{226} + 28.6p^{\mathrm{C}}_{236} + 28.6p^{\mathrm{C}}_{246} + 15.6p^{\mathrm{C}}_{316}\\
&+15.6p^{\mathrm{C}}_{326} + 15.6p^{\mathrm{C}}_{336} + 15.6p^{\mathrm{C}}_{346}\big)\big)\\
&+0.06\big(20p^{\mathrm{E}}_{17} + 50p^{\mathrm{E}}_{27} + 35p^{\mathrm{E}}_{37} + 31p^{\mathrm{E}}_{47} - \big(23.1p^{\mathrm{C}}_{117} + 23.1p^{\mathrm{C}}_{127} + 23.1p^{\mathrm{C}}_{137}\\
&+23.1p^{\mathrm{C}}_{147} + 15.4p^{\mathrm{C}}_{217} + 15.4p^{\mathrm{C}}_{227} + 15.4p^{\mathrm{C}}_{237} + 15.4p^{\mathrm{C}}_{247} + 8.4p^{\mathrm{C}}_{317}\\
&+8.4p^{\mathrm{C}}_{327} + 8.4p^{\mathrm{C}}_{337} + 8.4p^{\mathrm{C}}_{347}\big)\big)\\
&+0.18\big(20p^{\mathrm{E}}_{18} + 50p^{\mathrm{E}}_{28} + 35p^{\mathrm{E}}_{38} + 31p^{\mathrm{E}}_{48} - \big(33p^{\mathrm{C}}_{118} + 33p^{\mathrm{C}}_{128} + 33p^{\mathrm{C}}_{138}\\
&+33p^{\mathrm{C}}_{148} + 22p^{\mathrm{C}}_{218} + 22p^{\mathrm{C}}_{228} + 22p^{\mathrm{C}}_{238} + 22p^{\mathrm{C}}_{248} + 12p^{\mathrm{C}}_{318}\\
&+12p^{\mathrm{C}}_{328} + 12p^{\mathrm{C}}_{338} + 12p^{\mathrm{C}}_{348}\big)\big)
\end{aligned}
$$

$$+\,0.06\Big(20p^{E}_{19}+50p^{E}_{29}+35p^{E}_{39}+31p^{E}_{49}-\Big(42.9p^{C}_{119}+42.9p^{C}_{129}+42.9p^{C}_{139}$$
$$+42.9p^{C}_{149}+28.6p^{C}_{219}+28.6p^{C}_{229}+28.6p^{C}_{239}+28.6p^{C}_{249}+15.6p^{C}_{319}$$
$$+15.6p^{C}_{329}+15.6p^{C}_{339}+15.6p^{C}_{349}\Big)\Big)\Big]$$
$$-\Big(2000000u^{C,I}_{2}+7000000u^{C,I}_{3}+7000000u^{R,I}_{2}+10000000u^{R,I}_{3}+500000u^{S,I}_{1}\Big)$$
$$+\,\beta\Big(\vartheta-\frac{1}{1-0.95}\,(0.06\tau_1+0.18\tau_2+0.06\tau_3+0.08\tau_4+0.24\tau_5$$
$$+0.08\tau_6+0.06\tau_7+0.18\tau_8+0.06\tau_9)\Big).$$

- The following additional constraints are included:

$$\vartheta-2190\Big[20p^{E}_{11}+50p^{E}_{21}+35p^{E}_{31}+31p^{E}_{41}-\Big(23.1p^{C}_{111}+23.1p^{C}_{121}+23.1p^{C}_{131}$$
$$+23.1p^{C}_{141}+15.4p^{C}_{211}+15.4p^{C}_{221}+15.4p^{C}_{231}+15.4p^{C}_{241}+8.4p^{C}_{311}$$
$$+8.4p^{C}_{321}+8.4p^{C}_{331}+8.4p^{C}_{341}\Big)\Big]-\Big(2000000u^{C,I}_{2}+7000000u^{C,I}_{3}$$
$$+7000000u^{R,I}_{2}+10000000u^{R,I}_{3}+500000u^{S,I}_{1}\Big)\le\tau_1,$$

$$\vartheta-2190\Big[20p^{E}_{12}+50p^{E}_{22}+35p^{E}_{32}+31p^{E}_{42}-\Big(33p^{C}_{112}+33p^{C}_{122}+33p^{C}_{132}$$
$$+33p^{C}_{142}+22p^{C}_{212}+22p^{C}_{222}+22p^{C}_{232}+22p^{C}_{242}+12p^{C}_{312}$$
$$+12p^{C}_{322}+12p^{C}_{332}+12p^{C}_{342}\Big)\Big]-\Big(2000000u^{C,I}_{2}+7000000u^{C,I}_{3}$$
$$+7000000u^{R,I}_{2}+10000000u^{R,I}_{3}+500000u^{S,I}_{1}\Big)\le\tau_2,$$

$$\vartheta-2190\Big[20p^{E}_{13}+50p^{E}_{23}+35p^{E}_{33}+31p^{E}_{43}-\Big(42.9p^{C}_{113}+42.9p^{C}_{123}+42.9p^{C}_{133}$$
$$+42.9p^{C}_{143}+28.6p^{C}_{213}+28.6p^{C}_{223}+28.6p^{C}_{233}+28.6p^{C}_{243}+15.6p^{C}_{313}$$
$$+15.6p^{C}_{323}+15.6p^{C}_{333}+15.6p^{C}_{343}\Big)\Big]-\Big(2000000u^{C,I}_{2}+7000000u^{C,I}_{3}$$
$$+7000000u^{R,I}_{2}+10000000u^{R,I}_{3}+500000u^{S,I}_{1}\Big)\le\tau_3,$$

$$\vartheta-2190\Big[20p^{E}_{14}+50p^{E}_{24}+35p^{E}_{34}+31p^{E}_{44}-\Big(23.1p^{C}_{114}+23.1p^{C}_{124}+23.1p^{C}_{134}$$
$$+23.1p^{C}_{144}+15.4p^{C}_{214}+15.4p^{C}_{224}+15.4p^{C}_{234}+15.4p^{C}_{244}+8.4p^{C}_{314}$$

$$+8.4p^{\mathrm{C}}_{324}+8.4p^{\mathrm{C}}_{334}+8.4p^{\mathrm{C}}_{344}\Big)\Big]-\Big(2000000u^{\mathrm{C,I}}_{2}+7000000u^{\mathrm{C,I}}_{3}$$

$$+7000000u^{\mathrm{R,I}}_{2}+10000000u^{\mathrm{R,I}}_{3}+500000u^{\mathrm{S,I}}_{1}\Big)\leq\tau_4,$$

$$\vartheta-2190\Big[20p^{\mathrm{E}}_{15}+50p^{\mathrm{E}}_{25}+35p^{\mathrm{E}}_{35}+31p^{\mathrm{E}}_{45}-\Big(33p^{\mathrm{C}}_{115}+33p^{\mathrm{C}}_{125}+33p^{\mathrm{C}}_{135}$$

$$+33p^{\mathrm{C}}_{145}+22p^{\mathrm{C}}_{215}+22p^{\mathrm{C}}_{225}+22p^{\mathrm{C}}_{235}+22p^{\mathrm{C}}_{245}+12p^{\mathrm{C}}_{315}$$

$$+12p^{\mathrm{C}}_{325}+12p^{\mathrm{C}}_{335}+12p^{\mathrm{C}}_{345}\Big)\Big]-\Big(2000000u^{\mathrm{C,I}}_{2}+7000000u^{\mathrm{C,I}}_{3}$$

$$+7000000u^{\mathrm{R,I}}_{2}+10000000u^{\mathrm{R,I}}_{3}+500000u^{\mathrm{S,I}}_{1}\Big)\leq\tau_5,$$

$$\vartheta-2190\Big[20p^{\mathrm{E}}_{16}+50p^{\mathrm{E}}_{26}+35p^{\mathrm{E}}_{36}+31p^{\mathrm{E}}_{46}-\Big(42.9p^{\mathrm{C}}_{116}+42.9p^{\mathrm{C}}_{126}+42.9p^{\mathrm{C}}_{136}$$

$$+42.9p^{\mathrm{C}}_{146}+28.6p^{\mathrm{C}}_{216}+28.6p^{\mathrm{C}}_{226}+28.6p^{\mathrm{C}}_{236}+28.6p^{\mathrm{C}}_{246}+15.6p^{\mathrm{C}}_{316}$$

$$+15.6p^{\mathrm{C}}_{326}+15.6p^{\mathrm{C}}_{336}+15.6p^{\mathrm{C}}_{346}\Big)\Big]-\Big(2000000u^{\mathrm{C,I}}_{2}+7000000u^{\mathrm{C,I}}_{3}$$

$$+7000000u^{\mathrm{R,I}}_{2}+10000000u^{\mathrm{R,I}}_{3}+500000u^{\mathrm{S,I}}_{1}\Big)\leq\tau_6,$$

$$\vartheta-2190\Big[20p^{\mathrm{E}}_{17}+50p^{\mathrm{E}}_{27}+35p^{\mathrm{E}}_{37}+31p^{\mathrm{E}}_{47}-\Big(23.1p^{\mathrm{C}}_{117}+23.1p^{\mathrm{C}}_{127}+23.1p^{\mathrm{C}}_{137}$$

$$+23.1p^{\mathrm{C}}_{147}+15.4p^{\mathrm{C}}_{217}+15.4p^{\mathrm{C}}_{227}+15.4p^{\mathrm{C}}_{237}+15.4p^{\mathrm{C}}_{247}+8.4p^{\mathrm{C}}_{317}$$

$$+8.4p^{\mathrm{C}}_{327}+8.4p^{\mathrm{C}}_{337}+8.4p^{\mathrm{C}}_{347}\Big)\Big]-\Big(2000000u^{\mathrm{C,I}}_{2}+7000000u^{\mathrm{C,I}}_{3}$$

$$+7000000u^{\mathrm{R,I}}_{2}+10000000u^{\mathrm{R,I}}_{3}+500000u^{\mathrm{S,I}}_{1}\Big)\leq\tau_7,$$

$$\vartheta-2190\Big[20p^{\mathrm{E}}_{18}+50p^{\mathrm{E}}_{28}+35p^{\mathrm{E}}_{38}+31p^{\mathrm{E}}_{48}-\Big(33p^{\mathrm{C}}_{118}+33p^{\mathrm{C}}_{128}+33p^{\mathrm{C}}_{138}$$

$$+33p^{\mathrm{C}}_{148}+22p^{\mathrm{C}}_{218}+22p^{\mathrm{C}}_{228}+22p^{\mathrm{C}}_{238}+22p^{\mathrm{C}}_{248}+12p^{\mathrm{C}}_{318}$$

$$+12p^{\mathrm{C}}_{328}+12p^{\mathrm{C}}_{338}+12p^{\mathrm{C}}_{348}\Big)\Big]-\Big(2000000u^{\mathrm{C,I}}_{2}+7000000u^{\mathrm{C,I}}_{3}$$

$$+7000000u^{\mathrm{R,I}}_{2}+10000000u^{\mathrm{R,I}}_{3}+500000u^{\mathrm{S,I}}_{1}\Big)\leq\tau_8,$$

$$\vartheta-2190\Big[20p^{\mathrm{E}}_{19}+50p^{\mathrm{E}}_{29}+35p^{\mathrm{E}}_{39}+31p^{\mathrm{E}}_{49}-\Big(42.9p^{\mathrm{C}}_{119}+42.9p^{\mathrm{C}}_{129}+42.9p^{\mathrm{C}}_{139}$$

$$+42.9p^{\mathrm{C}}_{149}+28.6p^{\mathrm{C}}_{219}+28.6p^{\mathrm{C}}_{229}+28.6p^{\mathrm{C}}_{239}+28.6p^{\mathrm{C}}_{249}+15.6p^{\mathrm{C}}_{319}$$

$$+15.6p^{\mathrm{C}}_{329}+15.6p^{\mathrm{C}}_{339}+15.6p^{\mathrm{C}}_{349}\Big)\Big]-\Big(2000000u^{\mathrm{C,I}}_{2}+7000000u^{\mathrm{C,I}}_{3}$$

$$+7000000u^{\mathrm{R,I}}_{2}+10000000u^{\mathrm{R,I}}_{3}+500000u^{\mathrm{S,I}}_{1}\Big)\leq\tau_9,$$

$$\tau_1,\tau_2,\tau_3,\tau_4,\tau_5,\tau_6,\tau_7,\tau_8,\tau_9\geq 0.$$

Table 7.11 Illustrative Example 7.8: results

β	Candidate units built	Expected profit	CVaR
0	Conventional power plant 3	$25.731·10^6$	$20.685·10^6$
	Stochastic renewable generating unit 3		
	Storage unit 1		
1	Conventional power plant 3	$24.960·10^6$	$21.526·10^6$
	Stochastic renewable generating units 2 and 3		
	Storage unit 1		

The optimal solution achieved for the two considered values of the weighting parameter β is provided in Table 7.11. Note that the difference between these two solutions is on the building of stochastic renewable generating unit 2 in the case of the risk-averse VPP ($\beta = 1$). This can be explained by the fact that uncertainty affects the variable costs of conventional power plants and demand in the VPP. A risk-averse VPP prefers building an additional stochastic renewable generating unit. This results in a lower expected profit. However, if scenarios with high variable costs and/or high demand are realized, the VPP can use this additional renewable generating unit whose variable cost is 0, while the risk-neutral VPP ($\beta = 0$) should use the comparatively expensive conventional power plants.

□

A GAMS [4] code for solving Illustrative Example 7.8 is provided in Sect. 7.7.

7.6 Summary and Further Reading

This chapter describes the expansion planning problem of a VPP that invests in new generation assets with the aim of maximizing its profit. The problem is first formulated using a deterministic model, which is then extended to account for uncertainties using a stochastic programming approach. Finally, the profit risk associated with the expansion decisions is modeled using the CVaR metric.

Further information about investment models can be found in the book by Conejo et al. [2], while alternative ways of incorporating risk measures in stochastic programming problems can be found in [3].

7.7 GAMS Codes

This section provides the GAMS codes used to solve some of the illustrative examples of this chapter.

7.7.1 Deterministic Expansion Planning

An input GAMS [4] code for solving Illustrative Example 7.1 is provided below:

```
SETS
c          'conventional power plants'                  /c1*c3/
candC(c)   'candidate conventional power plants'        /c2*c3/
d          'demands'                                    /d1*d1/
r          'stochastic renewable units'                 /r1*r3/
candR(r)   'candidate stochastic renewable units'       /r2*r3/
s          'storage units'                              /s1*s1/
candS(s)   'candidate storage units'                    /s1/
t          'time periods'                               /t1*t4/;
SCALAR PEMAX      'maximum power traded in the market'       /250/;
SCALAR sigma      'factor to make costs comparable' /2190/;
SCALAR IB         'investment budget'                        /1e8/;
PARAMETER priceE(t)      'market price'
/t1       20
t2        50
t3        35
t4        31/;
TABLE GENDATA(c,*)       'conventional generating unit data'
          Pmax    VC     IC
c1        60      33     0
c2        40      22     2e6
c3        80      12     7e6;
TABLE PRA(r,t)           'forecast renewable generation level'
          t1      t2     t3      t4
r1        60      20     40      10
r2        0       80     60      0
r3        90      70     80      50;
PARAMETER ICR(r)         'investment cost of stochastic renewable units'
/r2       7e6
r3        1e7/;
TABLE PDmin(d,t)         'lower demand bound'
          t1      t2     t3      t4
d1        0       0      10      15;
TABLE PDmax(d,t)         'upper demand bound'
          t1      t2     t3      t4
d1        20      20     25      30;
PARAMETER EDmin(d)       'minimum energy consumption'
/d1       60/;
TABLE STORAGEDATA(s,*)         'storage data'
      PCmax    PDmax    Emin    Emax    EFFC    EFFD    E0     IC
s1    20       20       0       40      0.9     0.9     0      5e5;
VARIABLES
Z           'objective function'
PE(t)       'power traded in the market in time period t';
POSITIVE VARIABLES
ES(s,t)     'energy stored in storage c in time period t'
PCPP(c,t)   'power production of conventional power plant c in time period t'
PD(d,t)     'power consumption of demand d in time period t'
PR(r,t)     'power production of renewable unit r in time period t'
```

```
PSC(s,t)  'charging power of storage s in time period t'
PSD(s,t)  'discharging power of storage s in time period t';
BINARY VARIABLES
UCI(c)     'investment in conventional power plant c'
URI(r)     'investment in stochastic renewable unit r'
USI(s)     'investment in storage unit unit s';
EQUATIONS
OF        'objective function'
Budget    'investment budget'
EMLim1    'market limits I'
EMLim2    'market limits II'
Bal       'power balance in the VPP'
DLim1     'power demand limits I'
DLim2     'power demand limits II'
EDlim     'minimum energy consumption in the day'
Clim1     'existing conventional power plant production limits'
Clim2     'candidate conventional power plant production limits'
Rlim1     'existing stochastic renewable generating unit limits'
Rlim2     'candidate stochastic renewable generating unit limits'
ESbal1    'energy balance in the storage units I'
ESbal2    'energy balance in the storage units II'
ESlim1    'energy limits in the storage unit I'
ESlim2    'energy limits in the storage unit II'
CPlim1    'charging power limits of existing storage units'
CPlim2    'charging power limits of candidate storage units'
DPlim1    'discharging power limits of existing storage units'
DPlim2    'discharging power limits of candidate storage units';
OF..              z=e=sigma*(sum(t,priceE(t)*PE(t))-sum((c,t),
    GENDATA(c,'VC')*PCPP(c,t)))-sum(c$candC(c),GENDATA(c,'IC')*
    UCI(c))-sum(r$candR(r),ICR(r)*URI(r))-sum(s$candS(s),
    STORAGEDATA(s,'IC')*USI(s));
Budget..          sum(c$candC(c),GENDATA(c,'IC')*UCI(c))+sum(
    r$candR(r),ICR(r)*URI(r))+sum(s$candS(s),STORAGEDATA(s,'IC')*
    USI(s))=l=IB;
EMLim1(t)..       -PEMAX=l=PE(t);
EMLim2(t)..       PE(t)=l=PEMAX;
Bal(t)..          PE(t)=E=sum(c,PC(c,t))+sum(r,PR(r,t))+sum(s,PSD(
    s,t)-PSC(s,t))-sum(d,PD(d,t));
DLim1(d,t)..      PDmin(d,t)=l=PD(d,t);
DLim2(d,t)..      PD(d,t)=l=PDmax(d,t);
EDlim(d)..        sum(t,PD(d,t))=g=EDmin(d);
Clim1(c,t)$(not candC(c))..      PCPP(c,t)=l=GENDATA(c,'Pmax');
Clim2(c,t)$(candC(c))..          PCPP(c,t)=l=GENDATA(c,'Pmax')*
    UCI(c);
Rlim1(r,t)$(not candR(r))..      PR(r,t)=l=PRA(r,t);
Rlim2(r,t)$(candR(r))..          PR(r,t)=l=PRA(r,t)*URI(r);
ESbal1(s)..                      ES(s,'t1')=e=STORAGEDATA(s,'E0')
    +STORAGEDATA(s,'EFFC')*PSC(s,'t1')-PSD(s,'t1')/STORAGEDATA(s,
    'EFFD');

ESbal2(s,t)$(ord(t) GT 1)..      ES(s,t)=e=ES(s,t-1)+STORAGEDATA(
    s,'EFFC')*PSC(s,t)-PSD(s,t)/STORAGEDATA(s,'EFFD');
ESlim1(s,t)..                    ES(s,t)=l=STORAGEDATA(s,'Emax');
ESlim2(s,t)..                    ES(s,t)=g=STORAGEDATA(s,'Emin');
CPlim1(s,t)$(not candS(s))..     PSC(s,t)=l=STORAGEDATA(s,'PCmax'
    );
CPlim2(s,t)$(candS(s))..         PSC(s,t)=l=STORAGEDATA(s,'PCmax'
    )*USI(s);
DPlim1(s,t)$(not candS(s))..     PSD(s,t)=l=STORAGEDATA(s,'PDmax'
    );
DPlim2(s,t)$(candS(s))..         PSD(s,t)=l=STORAGEDATA(s,'PDmax'
    )*USI(s);
MODEL Deterministic /all/;
OPTION optcr=0;
SOLVE Deterministic using mip maximizing z;
DISPLAY z.l,UCI.l,URI.l,USI.l,PE.l,PCPP.l,PR.l,PD.l;
```

7.7.2 *Stochastic Expansion Planning*

An input GAMS [4] code for solving Illustrative Example 7.4 is provided below:

```
SETS
c        'conventional power plants'                /c1*c3/
candC(c) 'candidate conventional power plants'      /c2*c3/
d        'demands'                                  /d1*d1/
r        'stochastic renewable units'               /r1*r3/
candR(r) 'candidate stochastic renewable units'     /r2*r3/
s        'storage units'                            /s1*s1/
candS(s) 'candidate storage units'                  /s1/
t        'time periods'                             /t1*t4/
w        'scenarios'                                /w1*w9/;
SCALAR PEMAX     'maximum power traded in the market'      /200/;
SCALAR sigma     'factor to make costs comparable'
    /2190/;
SCALAR IB        'investment budget'                       /100e6
    /;
PARAMETER priceE(t)     'market price'
/t1      20
t2       50
t3       35
t4       31/;
TABLE GENDATA(c,*)              'conventional generating unit data'
         Pmax     VC       IC
c1       60       33       0
c2       40       22       2e6
c3       80       12       7e6;
PARAMETER GenVC(c,w)     'variable costs conventional power
    plants';
GenVC(c,'w1')=GENDATA(c,'VC')*0.7;
GenVC(c,'w2')=GENDATA(c,'VC')*1;
GenVC(c,'w3')=GENDATA(c,'VC')*1.3;
GenVC(c,'w4')=GENDATA(c,'VC')*0.7;
GenVC(c,'w5')=GENDATA(c,'VC')*1;
GenVC(c,'w6')=GENDATA(c,'VC')*1.3;
GenVC(c,'w7')=GENDATA(c,'VC')*0.7;
GenVC(c,'w8')=GENDATA(c,'VC')*1;
GenVC(c,'w9')=GENDATA(c,'VC')*1.3;
TABLE PRA(r,t)           'forecast renewable generation level'
         t1       t2       t3       t4
r1       60       20       40       10
r2       0        80       60       0
r3       90       70       80       50;
PARAMETER ICR(r)         'investment cost of stochastic renewable
     units'
/r2      7e6
r3       1e7/;
TABLE PDmin(d,t)         'lower demand bound'
         t1       t2       t3       t4
d1       0        0        10       15;
TABLE PDmax(d,t)         'upper demand bound'
         t1       t2       t3       t4
d1       20       20       25       30;
PARAMETER EDmin(d)      'minimum energy consumption'
/d1      60/;
```

```
PARAMETERS PDmin2(d,t,w),PDmax2(d,t,w),EDmin2(d,w);
PDmin2(d,t,'w1')=PDmin(d,t)*0.5;
PDmin2(d,t,'w2')=PDmin(d,t)*0.5;
PDmin2(d,t,'w3')=PDmin(d,t)*0.5;
PDmin2(d,t,'w4')=PDmin(d,t)*1;
PDmin2(d,t,'w5')=PDmin(d,t)*1;
PDmin2(d,t,'w6')=PDmin(d,t)*1;
PDmin2(d,t,'w7')=PDmin(d,t)*1.5;
PDmin2(d,t,'w8')=PDmin(d,t)*1.5;
PDmin2(d,t,'w9')=PDmin(d,t)*1.5;
PDmax2(d,t,'w1')=PDmax(d,t)*0.5;
PDmax2(d,t,'w2')=PDmax(d,t)*0.5;
PDmax2(d,t,'w3')=PDmax(d,t)*0.5;
PDmax2(d,t,'w4')=PDmax(d,t)*1;
PDmax2(d,t,'w5')=PDmax(d,t)*1;
PDmax2(d,t,'w6')=PDmax(d,t)*1;
PDmax2(d,t,'w7')=PDmax(d,t)*1.5;
PDmax2(d,t,'w8')=PDmax(d,t)*1.5;
PDmax2(d,t,'w9')=PDmax(d,t)*1.5;
EDmin2(d,'w1')=EDmin(d)*0.5;
EDmin2(d,'w2')=EDmin(d)*0.5;
EDmin2(d,'w3')=EDmin(d)*0.5;
EDmin2(d,'w4')=EDmin(d)*1;
EDmin2(d,'w5')=EDmin(d)*1;
EDmin2(d,'w6')=EDmin(d)*1;
EDmin2(d,'w7')=EDmin(d)*1.5;
EDmin2(d,'w8')=EDmin(d)*1.5;
EDmin2(d,'w9')=EDmin(d)*1.5;
TABLE STORAGEDATA(s,*)          'storage data'
      PCmax    PDmax    Emin     Emax     EFFC     EFFD     E0       IC
s1    20       20       0        40       0.9      0.9      0        5e5;
PARAMETER pi(w) 'weight of scenarios'
/w1       0.06
w2        0.18
w3        0.06
w4        0.08
w5        0.24
w6        0.08
w7        0.06
w8        0.18
w9        0.06/;
VARIABLES
Z             'objective function'
PE(t,w)       'power traded in the market in time period t';
POSITIVE VARIABLES
ES(s,t,w)     'energy stored in storage c in time period t'
PCPP(c,t,w) 'power production of conventional power plant c in
    time period t'
PD(d,t,w)     'power consumption of demand d in time period t'
PR(r,t,w)     'power production of renewable unit r in time period
    t'
PSC(s,t,w)    'charging power of storage s in time period t'
PSD(s,t,w)    'discharging power of storage s in time period t';
```

```
BINARY VARIABLES
UCI(c)       'investment in conventional power plant c'
URI(r)       'investment in stochastic renewable unit r'
USI(s)       'investment in storage unit unit s';
EQUATIONS
OF         'objective function'
Budget     'investment budget'
EMLim1     'market limits I'
EMLim2     'market limits II'
Bal        'power balance in the VPP'
DLim1      'power demand limits I'
DLim2      'power demand limits II'
EDlim      'minimum energy consumption in the day'
Clim1      'existing conventional power plant production limits'
Clim2      'candidate conventional power plant production limits'
Rlim1      'existing stochastic renewable generating unit limits'
Rlim2      'candidate stochastic renewable generating unit limits'
ESbal1     'energy balance in the storage units I'
ESbal2     'energy balance in the storage units II'
ESlim1     'energy limits in the storage unit I'
ESlim2     'energy limits in the storage unit II'
CPlim1     'charging power limits of existing storage units'
CPlim2     'charging power limits of candidate storage units'
DPlim1     'discharging power limits of existing storage units'
DPlim2     'discharging power limits of candidate storage units';
OF..                 z=e=sum(w,pi(w)*sigma*(sum(t,priceE(t)*PE(t,w)
    )-sum((c,t),GenVC(c,w)*PCPP(c,t,w))))-sum(c$candC(c),GENDATA(
    c,'IC')*UCI(c))-sum(r$candR(r),ICR(r)*URI(r))-sum(s$candS(s),
    STORAGEDATA(s,'IC')*USI(s));
Budget..             sum(c$candC(c),GENDATA(c,'IC')*UCI(c))+sum(
    r$candR(r),ICR(r)*URI(r))+sum(s$candS(s),STORAGEDATA(s,'IC')*
    USI(s))=l=IB;
EMLim1(t,w)..        -PEMAX=l=PE(t,w);
EMLim2(t,w)..        PE(t,w)=l=PEMAX;
Bal(t,w)..           PE(t,w)=E=sum(c,PC(c,t,w))+sum(r,PR(r,t,w))+
    sum(s,PSD(s,t,w)-PSC(s,t,w))-sum(d,PD(d,t,w));
DLim1(d,t,w)..       PDmin2(d,t,w)=l=PD(d,t,w);
DLim2(d,t,w)..       PD(d,t,w)=l=PDmax2(d,t,w);
EDlim(d,w)..         sum(t,PD(d,t,w))=g=EDmin2(d,w);
Clim1(c,t,w)$(not candC(c))..       PCPP(c,t,w)=l=GENDATA(c,'Pmax'
    );
Clim2(c,t,w)$(candC(c))..           PCPP(c,t,w)=l=GENDATA(c,'Pmax'
    )*UCI(c);
Rlim1(r,t,w)$(not candR(r))..       PR(r,t,w)=l=PRA(r,t);
Rlim2(r,t,w)$(candR(r))..           PR(r,t,w)=l=PRA(r,t)*URI(r);
ESbal1(s,w)..                       ES(s,'t1',w)=e=STORAGEDATA(s,'
    E0')+STORAGEDATA(s,'EFFC')*PSC(s,'t1',w)-PSD(s,'t1',w)/
    STORAGEDATA(s,'EFFD');
ESbal2(s,t,w)$(ord(t) GT 1)..       ES(s,t,w)=e=ES(s,t-1,w)+
    STORAGEDATA(s,'EFFC')*PSC(s,t,w)-PSD(s,t,w)/STORAGEDATA(s,'
    EFFD');
ESlim1(s,t,w)..                     ES(s,t,w)=l=STORAGEDATA(s,'
    Emax');


ESlim2(s,t,w)..                     ES(s,t,w)=g=STORAGEDATA(s,'
    Emin');
CPlim1(s,t,w)$(not candS(s))..      PSC(s,t,w)=l=STORAGEDATA(s,'
    PCmax');
CPlim2(s,t,w)$(candS(s))..          PSC(s,t,w)=l=STORAGEDATA(s,'
    PCmax')*USI(s);
DPlim1(s,t,w)$(not candS(s))..      PSD(s,t,w)=l=STORAGEDATA(s,'
    PDmax');
DPlim2(s,t,w)$(candS(s))..          PSD(s,t,w)=l=STORAGEDATA(s,'
    PDmax')*USI(s);
MODEL Deterministic /all/;
OPTION optcr=0;
SOLVE Deterministic using mip maximizing z;
DISPLAY z.l,UCI.l,URI.l,USI.l,PE.l,PCPP.l,PR.l,PD.l;
```

7.7.3 Stochastic Risk-Constrained Expansion Planning

An input GAMS [4] code for solving Illustrative Example 7.8 is provided below:

```
SETS
c          'conventional power plants'                /c1*c3/
candC(c)   'candidate conventional power plants'      /c2*c3/
d          'demands'                                  /d1*d1/
r          'stochastic renewable units'               /r1*r3/
candR(r)   'candidate stochastic renewable units'     /r2*r3/
s          'storage units'                            /s1*s1/
candS(s)   'candidate storage units'                  /s1/
t          'time periods'                             /t1*t4/
w          'scenarios'                                /w1*w9/;
SCALAR PEMAX      'maximum power traded in the market'        /200/;
SCALAR sigma      'factor to make costs comparable'
    /2190/;
SCALAR IB         'investment budget'                         /100e6
    /;
PARAMETER priceE(t)      'market price'
/t1       20
t2        50
t3        35
t4        31/;
TABLE GENDATA(c,*)              'conventional generating unit data'
          Pmax     VC       IC
c1        60       33       0
c2        40       22       2e6
c3        80       12       7e6;
PARAMETER GenVC(c,w)     'variable costs conventional power
    plants';
GenVC(c,'w1')=GENDATA(c,'VC')*0.7;
GenVC(c,'w2')=GENDATA(c,'VC')*1;
GenVC(c,'w3')=GENDATA(c,'VC')*1.3;
GenVC(c,'w4')=GENDATA(c,'VC')*0.7;
GenVC(c,'w5')=GENDATA(c,'VC')*1;
GenVC(c,'w6')=GENDATA(c,'VC')*1.3;
GenVC(c,'w7')=GENDATA(c,'VC')*0.7;
GenVC(c,'w8')=GENDATA(c,'VC')*1;
GenVC(c,'w9')=GENDATA(c,'VC')*1.3;
```

```
TABLE PRA(r,t)          'forecast renewable generation level'
          t1      t2      t3      t4
r1        60      20      40      10
r2        0       80      60      0
r3        90      70      80      50;
PARAMETER ICR(r)          'investment cost of stochastic renewable units'
/r2       7e6
r3        1e7/;
TABLE PDmin(d,t)          'lower demand bound'
          t1      t2      t3      t4
d1        0       0       10      15;
TABLE PDmax(d,t)          'upper demand bound'
          t1      t2      t3      t4
d1        20      20      25      30;
PARAMETER EDmin(d)        'minimum energy consumption'
/d1       60/;
PARAMETERS PDmin2(d,t,w),PDmax2(d,t,w),EDmin2(d,w);
PDmin2(d,t,'w1')=PDmin(d,t)*0.5;
PDmin2(d,t,'w2')=PDmin(d,t)*0.5;
PDmin2(d,t,'w3')=PDmin(d,t)*0.5;
PDmin2(d,t,'w4')=PDmin(d,t)*1;
PDmin2(d,t,'w5')=PDmin(d,t)*1;
PDmin2(d,t,'w6')=PDmin(d,t)*1;
PDmin2(d,t,'w7')=PDmin(d,t)*1.5;
PDmin2(d,t,'w8')=PDmin(d,t)*1.5;
PDmin2(d,t,'w9')=PDmin(d,t)*1.5;
PDmax2(d,t,'w1')=PDmax(d,t)*0.5;
PDmax2(d,t,'w2')=PDmax(d,t)*0.5;
PDmax2(d,t,'w3')=PDmax(d,t)*0.5;
PDmax2(d,t,'w4')=PDmax(d,t)*1;
PDmax2(d,t,'w5')=PDmax(d,t)*1;
PDmax2(d,t,'w6')=PDmax(d,t)*1;
PDmax2(d,t,'w7')=PDmax(d,t)*1.5;
PDmax2(d,t,'w8')=PDmax(d,t)*1.5;
PDmax2(d,t,'w9')=PDmax(d,t)*1.5;
EDmin2(d,'w1')=EDmin(d)*0.5;
EDmin2(d,'w2')=EDmin(d)*0.5;
EDmin2(d,'w3')=EDmin(d)*0.5;
EDmin2(d,'w4')=EDmin(d)*1;
EDmin2(d,'w5')=EDmin(d)*1;
EDmin2(d,'w6')=EDmin(d)*1;
EDmin2(d,'w7')=EDmin(d)*1.5;
EDmin2(d,'w8')=EDmin(d)*1.5;
EDmin2(d,'w9')=EDmin(d)*1.5;
TABLE STORAGEDATA(s,*)          'storage data'
      PCmax   PDmax   Emin    Emax    EFFC    EFFD    E0      IC
s1    20      20      0       40      0.9     0.9     0       5e5;
PARAMETER pi(w) 'weight of scenarios'
/w1       0.06
w2        0.18
w3        0.06
```

```
w4       0.08
w5       0.24
w6       0.08
w7       0.06
w8       0.18
w9       0.06/;
PARAMETER beta   'weighting parameter'     /1/;
PARAMETER alpha  'confidence level'        /0.95/;
VARIABLES
Z           'objective function'
PE(t,w)     'power traded in the market in time period t'
var         'value-at-risk';
POSITIVE VARIABLES
ES(s,t,w)    'energy stored in storage c in time period t'
PCPP(c,t,w) 'power production of conventional power plant c in time period t'
PD(d,t,w)    'power consumption of demand d in time period t'
PR(r,t,w)    'power production of renewable unit r in time period t'
PSC(s,t,w)   'charging power of storage s in time period t'
PSD(s,t,w)   'discharging power of storage s in time period t'
tau(w)       'auxiliary CVaR variable';
BINARY VARIABLES
UCI(c)       'investment in conventional power plant c'
URI(r)       'investment in stochastic renewable unit r'
USI(s)       'investment in storage unit unit s';
EQUATIONS
OF         'objective function'
Budget     'investment budget'
CVaREQ     'CVaR equation'
EMLim1     'market limits I'
EMLim2     'market limits II'
Bal        'power balance in the VPP'
DLim1      'power demand limits I'
DLim2      'power demand limits II'
EDlim      'minimum energy consumption in the day'
Clim1      'existing conventional power plant production limits'
Clim2      'candidate conventional power plant production limits'
Rlim1      'existing stochastic renewable generating unit limits'
Rlim2      'candidate stochastic renewable generating unit limits'
ESbal1     'energy balance in the storage units I'
ESbal2     'energy balance in the storage units II'
ESlim1     'energy limits in the storage unit I'
ESlim2     'energy limits in the storage unit II'
CPlim1     'charging power limits of existing storage units'
CPlim2     'charging power limits of candidate storage units'
DPlim1     'discharging power limits of existing storage units'
DPlim2     'discharging power limits of candidate storage units';
OF..          z=e=sum(w,pi(w)*sigma*(sum(t,priceE(t)*PE(t,w))-sum((c,t),GenVC(c,w)*PCPP(c,t,w))))-sum(c$candC(c),GENDATA(c,'IC')*UCI(c))-sum(r$candR(r),ICR(r)*URI(r))-sum(s$candS(s),STORAGEDATA(s,'IC')*USI(s))+beta*(var-(1/(1-alpha))*sum(w,pi(w)*tau(w)));
```

```
Budget..          sum(c$candC(c),GENDATA(c,'IC')*UCI(c))+sum(
    r$candR(r),ICR(r)*URI(r))+sum(s$candS(s),STORAGEDATA(s,'IC')*
    USI(s))=l=IB;
CVaREQ(w)..       var-(sigma*(sum(t,priceE(t)*PE(t,w))-sum((c,t)
    ,GenVC(c,w)*PCPP(c,t,w)))-sum(c$candC(c),GENDATA(c,'IC')*UCI(
    c))-sum(r$candR(r),ICR(r)*URI(r))-sum(s$candS(s),STORAGEDATA(
    s,'IC')*USI(s)))=l=tau(w);
EMLim1(t,w)..     -PEMAX=l=PE(t,w);
EMLim2(t,w)..    PE(t,w)=l=PEMAX;
Bal(t,w)..       PE(t,w)=E=sum(c,PC(c,t,w))+sum(r,PR(r,t,w))+
    sum(s,PSD(s,t,w)-PSC(s,t,w))-sum(d,PD(d,t,w));
DLim1(d,t,w)..   PDmin2(d,t,w)=l=PD(d,t,w);
DLim2(d,t,w)..   PD(d,t,w)=l=PDmax2(d,t,w);
EDlim(d,w)..     sum(t,PD(d,t,w))=g=EDmin2(d,w);
Clim1(c,t,w)$(not candC(c))..     PCPP(c,t,w)=l=GENDATA(c,'Pmax');
Clim2(c,t,w)$(candC(c))..         PCPP(c,t,w)=l=GENDATA(c,'Pmax'
    )*UCI(c);
Rlim1(r,t,w)$(not candR(r))..     PR(r,t,w)=l=PRA(r,t);
Rlim2(r,t,w)$(candR(r))..         PR(r,t,w)=l=PRA(r,t)*URI(r);
ESbal1(s,w)..                     ES(s,'t1',w)=e=STORAGEDATA(s,'
    E0')+STORAGEDATA(s,'EFFC')*PSC(s,'t1',w)-PSD(s,'t1',w)/
    STORAGEDATA(s,'EFFD');
ESbal2(s,t,w)$(ord(t) GT 1)..     ES(s,t,w)=e=ES(s,t-1,w)+
    STORAGEDATA(s,'EFFC')*PSC(s,t,w)-PSD(s,t,w)/STORAGEDATA(s,'
    EFFD');
ESlim1(s,t,w)..                   ES(s,t,w)=l=STORAGEDATA(s,'
    Emax');
ESlim2(s,t,w)..                   ES(s,t,w)=g=STORAGEDATA(s,'
    Emin');
CPlim1(s,t,w)$(not candS(s))..    PSC(s,t,w)=l=STORAGEDATA(s,'
    PCmax');
CPlim2(s,t,w)$(candS(s))..        PSC(s,t,w)=l=STORAGEDATA(s,'
    PCmax')*USI(s);
DPlim1(s,t,w)$(not candS(s))..    PSD(s,t,w)=l=STORAGEDATA(s,'
    PDmax');
DPlim2(s,t,w)$(candS(s))..        PSD(s,t,w)=l=STORAGEDATA(s,'
    PDmax')*USI(s);
MODEL Deterministic /all/;
OPTION optcr=0;
SOLVE Deterministic using mip maximizing z;
PARAMETERS ExpProfit, CVaR;
ExpProfit=sum(w,pi(w)*sigma*(sum(t,priceE(t)*PE.l(t,w))-sum((c,t)
    ,GenVC(c,w)*PCPP.l(c,t,w))))-sum(c$candC(c),GENDATA(c,'IC')*
    UCI.l(c))-sum(r$candR(r),ICR(r)*URI.l(r))-sum(s$candS(s),
    STORAGEDATA(s,'IC')*USI.l(s));
CVaR=var.l-(1/(1-alpha))*sum(w,pi(w)*tau.l(w));
DISPLAY z.l,UCI.l,URI.l,USI.l,PE.l,PCPP.l,PR.l,PD.l,ExpProfit,
    CVaR;
```

References

1. Birge, J.R., Louveaux, F: Introduction to Stochastic Programming, 2nd edn. Springer, New York, US (2011)
2. Conejo, A.J., Baringo, L., Kazempour, S.J., Siddiqui, A.S.: Investment in Electricity Generation and Transmission: Decision Making Under Uncertainty. Springer, Switzerland (2016)
3. Conejo, A.J., Carrión, M., Morales, J.M.: Decision Making Under Uncertainty in Electricity Markets. Springer, New York, US (2010)
4. GAMS.: Available at www.gams.com/ (2018)
5. Garver, L.L.: Transmission network estimation using linear programming. IEEE Trans. Apparat. Syst. **89**(7), 1688–1697 (1970)
6. Rockafella, R.T., Uryasev, S.: Optimization of conditional value-at-risk. J. Risk **2**(3), 21–41 (2000)
7. Sioshansi, R., Conejo, A.J.: Optimization in Engineering. Models and Algorithms. Springer, New York, US (2017)
8. Zhang, X., Conejo, A.J.: Coordinated investment in transmission and storage systems representing long- and short-term uncertainty. IEEE Trans. Power Syst. **33**(6), 7143–7151 (2018)

Correction to: Virtual Power Plant Model

Correction to:
Chapter 2 in: L. Baringo, M. Rahimiyan, *Virtual Power Plants and Electricity Markets*, https://doi.org/10.1007/978-3-030-47602-1_2

The original version of the chapter was inadvertently published with wrong placement of section headings and an incomplete GAMS code. The chapter has now been corrected and approved by the author.

The updated online version of the chapter can be found at
https://doi.org/10.1007/978-3-030-47602-1_2

L. Baringo, M. Rahimiyan, *Virtual Power Plants and Electricity Markets*, https://doi.org/10.1007/978-3-030-47602-1_8

Appendix A
Optimization Problems

This appendix explains the basic theory of mathematical optimization problems described throughout this book to model the decision-making problems for a virtual power plant (VPP) participating in electricity markets. Specifically, two optimization problems are described, namely linear programming (LP) and mixed-integer linear programming (MILP).

A.1 Introduction

The decision-making problems for VPPs participating in electricity markets described in this book can be represented in the form of optimization problems. The VPP seeks to achieve the maximum possible profit (or minimum cost), which is represented through an objective function. Additionally, these decision-making problems are constrained by the feasible operating region of the electricity generation and demand assets in the VPP, which is represented in the constraints of the optimization problem.

In general, a mathematical optimization problem can be formulated as:

$$\min_{\Phi} \quad f(\Phi) \tag{A.1a}$$

subject to

$$h_k(\Phi) = 0, \quad \forall k \in \Omega^{\mathrm{E}}, \tag{A.1b}$$

$$g_i(\Phi) \leq 0, \quad \forall i \in \Omega^{\mathrm{I}}, \tag{A.1c}$$

$$\Phi \in \mathbb{R}^N, \tag{A.1d}$$

where:

- $\mathbb{R}$ is the set of real numbers,
- $\Phi=\{x_j, \forall j\}$ is the set of optimization variables,
- $f(\Phi)$: $\mathbb{R}^N \rightarrow \mathbb{R}$ is the objective function (A.1a),
- $h_k(\Phi)$: $\mathbb{R}^N \rightarrow \mathbb{R}$, $\forall k \in \Omega^{\mathrm{E}}$, are the functions that define equality constraints (A.1b),
- $g_i(\Phi)$: $\mathbb{R}^N \rightarrow \mathbb{R}$, $\forall i \in \Omega^{\mathrm{I}}$, are the functions that define inequality constraints (A.1c),
- N is the number of optimization variables, and
- Ω^{E} and Ω^{I} are the sets of equality and inequality constraints, respectively.

The intersection of domains of the objective function f and constraining functions h_k, $\forall k \in \Omega^{\mathrm{E}}$, and g_i, $\forall i \in \Omega^{\mathrm{I}}$, is the domain of the optimization problem (A.1), denoted by D. Each value of optimization variables $\Phi \in D$ is a feasible solution of the optimization problem if it can satisfy all equality and inequality constraints. Accordingly, the feasible set Φ^{F} is defined as $\Phi^{\mathrm{F}}=\{\Phi \in D |\ h_k(\Phi) = 0, \forall k \in \Omega^{\mathrm{E}},\ g_i(\Phi) \leq 0, \forall i \in \Omega^{\mathrm{I}}\}$ that comprise all feasible solutions.

A feasible solution Φ is a *local* optimum if the value of the objective function at Φ is lower than or equal to the value of the objective function at any other feasible solutions belonging to its neighborhood. A feasible solution Φ is *global* optimum

L. Baringo, M. Rahimiyan, *Virtual Power Plants and Electricity Markets*, https://doi.org/10.1007/978-3-030-47602-1

if the objective function takes the minimum value throughout the feasible set. In other words, there is no any other feasible solution that provides a value of the objective function lower than that minimum value.

The mathematical structure of the optimization problem (A.1) depends on the properties of the optimization variables and the functions indicated in the objective and constraints. On one hand, the optimization variables can take continuous and discrete values. On the other hand, the functions can be linear or nonlinear with respect to the optimization variables. Through these different properties, different classes of the optimization problem such as LP, MILP, quadratic programming (QP), nonlinear programming (NLP), and mixed-integer nonlinear programming (MINLP) are formed. In this appendix, only LP and MILP problems are considered as they are the two types of problems used in this book.

The reminder of the appendix is organized as follows. Section A.2 describes LP problems. Specifically, it is explained how to obtain the dual optimization problem corresponding to an LP problem. Section A.3 provides the formulation of MILP problems and clarifies how it differs from LP problems. Section A.4 summarizes this chapter and suggests further reading for the interested readers.

A.2 Linear Programming

Let us consider the general formulation of an optimization problem (A.1). It is called an LP problem if the two following properties are identified:

1. The optimization variables in set Φ are continuous variables.
2. All the functions in the objective function and the constraints are linear with respect to the optimization variables.

Under these two properties, an LP problem can be formulated as the following optimization problem:

$$\min_{\Phi} \quad \sum_j C_j x_j \tag{A.2a}$$

subject to

$$\sum_j A^{\mathrm{E}}_{kj} x_j = B^{\mathrm{E}}_k \quad : \lambda_k, \quad \forall k \in \Omega^{\mathrm{E}}, \tag{A.2b}$$

$$\sum_j A^{\mathrm{I}}_{ij} x_j \leq B^{\mathrm{I}}_i \quad : \mu_i, \quad \forall i \in \Omega^{\mathrm{I}}, \tag{A.2c}$$

where:

- $C_j, \forall j$, are the cost coefficients of variables $x_j, \forall j$, in the objective function (A.2a),
- $A^{\mathrm{E}}_{kj}, \forall j$, and B^{E}_k are the coefficients that define equality constraints (A.2b), $\forall k \in \Omega^{\mathrm{E}}$,
- $A^{\mathrm{I}}_{ij}, \forall j$, and B^{I}_i are the coefficients that define inequality constraints (A.2c), $\forall i \in \Omega^{\mathrm{I}}$, and
- $\lambda_k, \forall k \in \Omega^{\mathrm{E}}$, and $\mu_i, \forall i \in \Omega^{\mathrm{I}}$, are the Lagrange multipliers corresponding to equality and inequality constraints (A.2b) and (A.2c), respectively. Note that the Lagrange multipliers of equality constraints are not restricted in sign, while those of inequality constraints are nonnegative.

LP problems can be solved using the simplex method [4] for which a detailed description is provided in [7]. In addition, LP problems can be solved using commercially available software tools, e.g., using the CPLEX solver [5] under GAMS [6]. In general, CPLEX solves efficiently LP problems in a limited time; however, it may have issues related to insufficient physical memory in large LP problems.

Illustrative Example A.1 LP Problem

In this example, an LP problem is presented. First, the objective function to be minimized is given below:

$$z = 10x_1 + 20x_2 - 30x_3 - 15x_4.$$

Second, the constraints are as follows:

- Equality constraints:

$$-x_1 - x_2 + x_3 + x_4 = 0 \quad : \lambda.$$

- Inequality constraints:

$$-x_1 \le 0 \quad : \mu_1,$$

$$x_1 \le 20 \quad : \mu_2,$$

$$-x_2 \le 0 \quad : \mu_3,$$

$$x_2 \le 20 \quad : \mu_4,$$

$$-x_3 \le 0 \quad : \mu_5,$$

$$x_3 \le 25 \quad : \mu_6,$$

$$-x_4 \le 0 \quad : \mu_7,$$

$$x_4 \le 25 \quad : \mu_8.$$

where set Φ=$\{x_1, x_2, x_3, x_4\}$ includes the optimization variables.

The above LP problem is solved using the CPLEX solver [5] under GAMS [6]. The optimal solution is provided in Table A.1. For the obtained optimal solution, the value of the objective function is equal to −450.

□

LP problem (A.2) is a convex optimization problem due to the following reasons:

1. The objective function is linear and continuous, and thus is a convex function.
2. All the functions in the objective function and constraints are linear and continuous, and thus convex functions, which results in a convex feasible set.

An important feature of the convex optimization problems is that a *local* optimal solution is also a *global* optimal solution. In other words, a solution that is optimal in its neighborhood feasible set is also optimal in the feasible set of the optimization problem. Therefore, in a convex optimization problem, the Karush-Kuhn-Tucker (KKT) optimality conditions are both necessary and sufficient conditions to achieve the *global* optimal solution.

To formulate the KKT conditions, first the Lagrange function is formulated below:

$$L(x_j, \lambda_k, \mu_i) = \sum_j C_j x_j + \sum_k \lambda_k \left(\sum_j A^{\mathrm{E}}_{kj} x_j - B^{\mathrm{E}}_k \right) + \sum_i \mu_i \left(\sum_j A^{\mathrm{I}}_{ij} x_j - B^{\mathrm{I}}_i \right). \tag{A.3}$$

The above Lagrange function is a function of the optimization variables $x_j, \forall j$, and the Lagrange multipliers $\lambda_k, \forall k \in \Omega^{\mathrm{E}}$, and $\mu_i, \forall i \in \Omega^{\mathrm{I}}$. Then, the KKT conditions are formulated below:

$$C_j + \sum_k \lambda_k A^{\mathrm{E}}_{kj} + \sum_i \mu_i A^{\mathrm{I}}_{ij} = 0, \quad \forall j, \tag{A.4a}$$

$$\sum_j A^{\mathrm{E}}_{kj} x_j = B^{\mathrm{E}}_k, \quad \forall k \in \Omega^{\mathrm{E}}, \tag{A.4b}$$

$$0 \le \mu_i \perp \left(\sum_j A^{\mathrm{I}}_{ij} x_j - B^{\mathrm{I}}_i \right) \le 0, \quad \forall i \in \Omega^{\mathrm{I}}. \tag{A.4c}$$

Table A.1 Illustrative Example A.1: optimal solution

x_1	x_2	x_3	x_4
20	5	25	0

Equations (A.4a) and (A.4b) are obtained by taking the derivative of the Lagrange function (A.3) with respect to the optimization variables x_j, $\forall j$, and the Lagrange multipliers λ_k, $\forall k \in \Omega^{\mathrm{E}}$, respectively. Equations (A.4c) are the complementarity constraints corresponding to the inequality constraints, which comprise:

1. the nonnegativity of Lagrange multipliers associated with inequality constraints, i.e., $0 \leq \mu_i$, $\forall i \in \Omega^{\mathrm{I}}$,
2. the inequalities $\sum_j A^{\mathrm{I}}_{ij} x_j - B^{\mathrm{I}}_i \leq 0$, $\forall i \in \Omega^{\mathrm{I}}$, and
3. the slackness conditions $\mu_i(\sum_j A^{\mathrm{I}}_{ij} x_j - B^{\mathrm{I}}_i) = 0$, $\forall i \in \Omega^{\mathrm{I}}$.

Solving Eqs. (A.4) provides the *global* optimal solutions of the LP problem (A.2). This means that the LP problem (A.2) can be equivalently replaced by the optimality KKT conditions (A.4).

Illustrative Example A.2 Solving LP Problem Using the KKT Conditions

Now, the LP problem introduced in the Illustrative Example A.1 is solved using the KKT conditions. To accomplish this task, first the Lagrange function is provided below:

$$\begin{aligned}
&L(x_1, x_2, x_3, x_4, \lambda, \mu_1, \mu_2, \mu_3, \mu_4, \mu_5, \mu_6, \mu_7, \mu_8) = 10x_1 + 20x_2 - 30x_3 - 15x_4 \\
&+ \lambda(-x_1 - x_2 + x_3 + x_4) - \mu_1 x_1 + \mu_2(x_1 - 20) - \mu_3 x_2 + \mu_4(x_2 - 20) \\
&- \mu_5 x_3 + \mu_6(x_3 - 25) - \mu_7 x_4 + \mu_8(x_4 - 25).
\end{aligned}$$

Then, the KKT conditions can be obtained as follows:

- The derivative of the Lagrange function with respect to the optimization variables x_j, $\forall j$:

$$\begin{aligned}
&10 - \lambda - \mu_1 + \mu_2 = 0, \\
&20 - \lambda - \mu_3 + \mu_4 = 0, \\
&- 30 + \lambda - \mu_5 + \mu_6 = 0, \\
&- 15 + \lambda - \mu_7 + \mu_8 = 0.
\end{aligned}$$

- The derivative of the Lagrange function with respect to the Lagrange multiplier λ:

$$- x_1 - x_2 + x_3 + x_4 = 0.$$

- The complementarity constraints:

$$\begin{aligned}
&0 \leq x_1 \perp \mu_1 \geq 0, \\
&0 \leq 20 - x_1 \perp \mu_2 \geq 0, \\
&0 \leq x_2 \perp \mu_3 \geq 0, \\
&0 \leq 20 - x_2 \perp \mu_4 \geq 0, \\
&0 \leq x_3 \perp \mu_5 \geq 0, \\
&0 \leq 25 - x_3 \perp \mu_6 \geq 0, \\
&0 \leq x_4 \perp \mu_7 \geq 0, \\
&0 \leq 25 - x_4 \perp \mu_8 \geq 0,
\end{aligned}$$

The above complementarity constraints are nonlinear and can be transformed into the linear constraints below:

$$\begin{aligned}
&0 \leq x_1 \leq u_1 M, \\
&0 \leq \mu_1 \leq (1 - u_1) N,
\end{aligned}$$

$$0 \le 20 - x_1 \le u_2 M,$$
$$0 \le \mu_2 \le (1 - u_2)N,$$
$$0 \le x_2 \le u_3 M,$$
$$0 \le \mu_3 \le (1 - u_3)N,$$
$$0 \le 20 - x_2 \le u_4 M,$$
$$0 \le \mu_4 \le (1 - u_4)N,$$
$$0 \le x_3 \le u_5 M,$$
$$0 \le \mu_5 \le (1 - u_5)N,$$
$$0 \le 25 - x_3 \le u_6 M,$$
$$0 \le \mu_6 \le (1 - u_6)N,$$
$$0 \le x_4 \le u_7 M,$$
$$0 \le \mu_7 \le (1 - u_7)N,$$
$$0 \le 25 - x_4 \le u_8 M,$$
$$0 \le \mu_8 \le (1 - u_8)N,$$
$$u_1, u_2, u_3, u_4, u_5, u_6, u_7, u_8 \in \{0, 1\},$$

where the optimization variables are the primal variables $\Phi=\{x_1, x_2, x_3, x_4\}$, as well as the Lagrange multipliers λ, μ_1, μ_2, μ_3, μ_4, μ_5, μ_6, μ_7, μ_8, and auxiliary binary variables u_1, u_2, u_3, u_4, u_5, u_6, u_7, u_8. The parameters M and N are large enough positive constants, which are considered to be equal to 50 and 100, respectively.

The values of the primal variables and the Lagrange multipliers are provided in Tables A.2 and A.3, respectively. As expected, the values of the optimization variables x_1, x_2, x_3, x_4 are equal to the optimal solution obtained in Illustrative Example A.1. The nonzero value of the Lagrange multipliers $\mu_2 = 10$, $\mu_6 = 10$, and $\mu_7 = 5$ means that the corresponding inequality constraints are active based on the equality of the slackness condition.

□

The LP optimization problem (A.1) is referred to as a primal optimization problem. The dual function corresponding to this primal optimization problem with Lagrange function $L(x_j, \lambda_k, \mu_i)$ can be defined as:

$$d(\lambda_k, \mu_i) = \operatorname*{Inf}_{\Phi \in D} \quad L(x_j, \lambda_k, \mu_i). \tag{A.5}$$

Considering the above definition, the dual optimization problem corresponding to problem (A.1) can be formulated as:

$$\max_{\Phi \mathrm{D}} \quad d(\lambda_k, \mu_i) \tag{A.6a}$$

subject to

$$\mu_i \ge 0, \quad \forall i \in \Omega^{\mathrm{I}}, \tag{A.6b}$$

Table A.2 Illustrative Example A.2: optimal primal variables

x_1	x_2	x_3	x_4
20	5	25	0

Table A.3 Illustrative Example A.2: optimal Lagrange multipliers

λ	μ_1	μ_2	μ_3	μ_4	μ_5	μ_6	μ_7	μ_8
20	0	10	0	0	0	10	5	0

where set $\Phi^{\mathrm{D}}=\{\lambda_k, \forall k \in \Omega^{\mathrm{E}}; \mu_i, \forall i \in \Omega^{\mathrm{I}}\}$ includes the optimization variables of the dual optimization problem (A.6), which are known as dual variables.

Applying the formulations (A.5) and (A.6) to the LP problem (A.2), its equivalent dual optimization problem can be formulated as follows:

$$\max_{\Phi^{\mathrm{D}}} \quad -\sum_k B_k^{\mathrm{E}} \lambda_k - \sum_i B_i^{\mathrm{I}} \mu_i \tag{A.7a}$$

subject to

$$C_j + \sum_k A_{kj}^{\mathrm{E}} \lambda_k + \sum_i A_{ij}^{\mathrm{I}} \mu_i = 0, \quad \forall j, \tag{A.7b}$$

$$\mu_i \geq 0, \quad \forall i \in \Omega^{\mathrm{I}}. \tag{A.7c}$$

The dual optimization problem (A.7) is also another LP problem, which is a convex optimization problem. Thus, solving the dual optimization problem (A.7) provides the optimal dual variables.

Illustrative Example A.3 Dual Optimization Problem

In this example, the dual optimization problem corresponding to the primal optimization problem introduced in Illustrative Example A.1 is formulated.

First, the objective function to be maximized is provided below:

$$z = -20\mu_2 - 20\mu_4 - 25\mu_6 - 25\mu_8.$$

Note that the dual variables λ, μ_1, μ_3, μ_5, and μ_7 are not included in the above dual function as they are multiplied by zero.

Next, the constraints are formulated as follows:

- Equality constraints:

$$10 - \lambda - \mu_1 + \mu_2 = 0,$$

$$20 - \lambda - \mu_3 + \mu_4 = 0,$$

$$-30 + \lambda - \mu_5 + \mu_6 = 0,$$

$$-15 + \lambda - \mu_7 + \mu_8 = 0.$$

- Inequality constraints:

$$\mu_1, \mu_2, \mu_3, \mu_4, \mu_5, \mu_6, \mu_7, \mu_8 \geq 0.$$

The dual optimization variables are included in set $\Phi^{\mathrm{D}}=\{\lambda, \mu_1, \mu_2, \mu_3, \mu_4, \mu_5, \mu_6, \mu_7, \mu_8\}$.

The dual optimization problem is an LP problem and can be solved using the CPLEX solver [5] under GAMS [6]. The optimal solution is presented in Table A.4. Note that the optimal value of the dual function is equal to −450.

□

As a criterion linking the primal and dual optimization problems, the duality gap is defined as the difference between the values of their objective functions, i.e., $f(\Phi)$-$d(\Phi^{\mathrm{D}})$. It can be easily proved that the duality gap for the optimal primal and dual solutions is nonnegative, which is known as the *weak duality* condition. Thus, the *weak duality* condition holds for LP problems. Particularly, a useful property of LP problems is that *strong duality* condition also holds. Under this condition, the duality gap is equal to zero for the optimal primal and dual solutions. In other words, the optimal value of the objective

Table A.4 Illustrative Example A.3: optimal solution

λ	μ_1	μ_2	μ_3	μ_4	μ_5	μ_6	μ_7	μ_8
20	0	10	0	0	0	10	5	0

function (A.2a) is equal to the optimal value of the objective function (A.7a). This means that the duality gap can be used as a convergence criterion to find optimal solutions. Taking the advantage of the *strong duality* condition, the LP problem (A.2) can be equivalently replaced by the following set of equations:

1. constraints of the LP problem (A.2),
2. constraints of the dual optimization problem (A.7), and
3. the *strong duality* condition.

Illustrative Example A.4 Strong duality Condition

In this example, the *strong duality* condition is used to find the optimal solution of the LP problem introduced in Illustrative Example A.1.

The following set of constraints needs to be solved to find the optimal solution:

- Constraints of the primal optimization problem:

$$\begin{aligned}
&-x_1 - x_2 + x_3 + x_4 = 0,\\
&-x_1 \leq 0,\\
&x_1 \leq 20,\\
&-x_2 \leq 0,\\
&x_2 \leq 20,\\
&-x_3 \leq 0,\\
&x_3 \leq 25,\\
&-x_4 \leq 0,\\
&x_4 \leq 25.
\end{aligned}$$

- Constraints of the dual optimization problem:

$$\begin{aligned}
&10 - \lambda - \mu_1 + \mu_2 = 0,\\
&20 - \lambda - \mu_3 + \mu_4 = 0,\\
&-30 + \lambda - \mu_5 + \mu_6 = 0,\\
&-15 + \lambda - \mu_7 + \mu_8 = 0,\\
&\mu_1, \mu_2, \mu_3, \mu_4, \mu_5, \mu_6, \mu_7, \mu_8 \geq 0.
\end{aligned}$$

- *Strong duality* condition:

$$10x_1 + 20x_2 - 30x_3 - 15x_4 = -20\mu_2 - 20\mu_4 - 25\mu_6 - 25\mu_8.$$

In this case, the optimization variables are the primal optimization variables included in set Φ={x_1, x_2, x_3, x_4} and the dual optimization variables included in set Φ^D={λ, μ_1, μ_2, μ_3, μ_4, μ_5, μ_6, μ_7, μ_8}. The left- and right-hand sides of the *strong duality* condition are the objective function of the primal and dual optimization problems, respectively.

The above system of equations is solved, which provides the optimal solution of the primal and dual optimization variables in Tables A.5 and A.6, respectively. For the optimal solution, both left- and right-hand sides of strong duality condition are equal to -450.

□

Table A.5 Illustrative Example A.4: optimal primal optimization variables

x_1	x_2	x_3	x_4
20	5	25	0

Table A.6 Illustrative Example A.4: optimal dual optimization variables

λ	μ_1	μ_2	μ_3	μ_4	μ_5	μ_6	μ_7	μ_8
20	0	10	0	0	0	10	5	0

Considering the *strong duality* condition, a *local* sensitivity analysis with respect to the parameters can be done. Consider a perturbed form of the LP problem with perturbation parameters E_k, $\forall k \in \Omega^{\mathrm{E}}$, and I_i, $\forall i \in \Omega^{\mathrm{I}}$ as below:

$$\min_{\Phi} \quad \sum_j C_j x_j \tag{A.8a}$$

subject to

$$\sum_j A^{\mathrm{E}}_{kj} x_j - B^{\mathrm{E}}_k = E_k \quad \forall k \in \Omega^{\mathrm{E}}, \tag{A.8b}$$

$$\sum_j A^{\mathrm{I}}_{ij} x_j - B^{\mathrm{I}}_i \leq I_i \quad \forall i \in \Omega^{\mathrm{I}}. \tag{A.8c}$$

In fact, the above LP problem (A.8) is similar to the original LP problem (A.2). The main differences are the additional perturbation parameters E_k, $\forall k \in \Omega^{\mathrm{E}}$, and I_i, $\forall i \in \Omega^{\mathrm{I}}$, in the right-hand side of equality and inequality constraints (A.8b) and (A.8c), respectively. If perturbation parameters are zero, i.e., $E_k = 0$, $\forall k \in \Omega^{\mathrm{E}}$, and $I_i = 0$, $\forall i \in \Omega^{\mathrm{I}}$, the perturbed LP problem (A.8) is equivalent to the original LP problem (A.2). For given values of E_k, $\forall k \in \Omega^{\mathrm{E}}$, and I_i, $\forall i \in \Omega^{\mathrm{I}}$, $F^*(E, I)$ is defined as the optimal value of the objective function (A.8a). Thus, $F^*(0, 0)$ is the optimal value of the objective function (A.8a) for $E_k = 0$, $\forall k \in \Omega^{\mathrm{E}}$, and $I_i = 0$, $\forall i \in \Omega^{\mathrm{I}}$. In other words, $F^*(0, 0)$ is the optimal value of the objective function in the original LP problem (A.2).

Given the above definition, the following conditions hold based on the derivative of $F^*(0, 0)$ with respect to the perturbation parameters [2]:

$$\lambda^*_k = -\frac{\partial F^*(0,0)}{\partial E_k}, \quad \forall k \in \Omega^{\mathrm{E}}, \tag{A.9a}$$

$$\mu^*_i = -\frac{\partial F^*(0,0)}{\partial I_i}, \quad \forall i \in \Omega^{\mathrm{I}}, \tag{A.9b}$$

where λ^*_k, $\forall k \in \Omega^{\mathrm{E}}$, and μ^*_i, $\forall i \in \Omega^{\mathrm{I}}$, are the optimal values of the dual variables corresponding to the LP problem (A.2).

Using the above theory, it is possible to find the change in the optimal value of the objective function due to the perturbation parameters as:

$$F^*(E, I) - F^*(0, 0) = -\sum_k \lambda^*_k E_k - \sum_i \mu^*_i I_i. \tag{A.10}$$

Illustrative Example A.5 Sensitivity Analysis

This example analyzes how changes in some parameters of Illustrative Example A.1 influence the optimal value of the objective function. These changes may occur due to errors in the estimation of parameters.

For example, consider the following perturbed form of the LP problem introduced in Illustrative Example A.1:

$$\min_{\Phi} = 10x_1 + 20x_2 - 30x_3 - 15x_4$$

subject to:

$$-x_1 - x_2 + x_3 + x_4 = 1,$$

$$-x_1 \leq 0,$$

$$x_1 \leq 19,$$

$$
\begin{aligned}
&-x_2 \leq 0,\\
&x_2 \leq 20,\\
&-x_3 \leq 0,\\
&x_3 \leq 25,\\
&-x_4 \leq 0,\\
&x_4 \leq 25.
\end{aligned}
$$

In the original LP problem introduced in Illustrative Example A.1, the right-hand side of equality constraint is 0, and the upper bound of variable x_1 is equal to 20. In the above LP problem, the right-hand side of the equality constraint is 1, while the upper bound of variable x_1 is 19. This is equivalent to a perturbed form with $E = 1$ and $I_2 = -1$. Note that in this perturbed form, $I_1=I_3=I_4=I_5=I_6=I_7=I_8=0$.

Using the theory of the sensitivity analysis, it is possible to find the optimal objective function of the above LP problem without solving the problem. The optimal objective function of the original LP problem is −450. Additionally, the dual variables of the original LP problem are obtained in Illustrative Example A.3. Then, the optimal objective function of the above LP problem can be calculated as:

$$F^*(1, 0, -1, 0, 0, 0, 0, 0, 0) = -450 - 20 \times 1 - 10 \times (-1) = -460.$$

□

A.3 Mixed-Integer Linear Programming

As in the case of LP problems, the objective functions and constraints of MILP problems are linear with respect to the optimization variables. However, MILP problems are characterized by including both continuous and integer optimization variables. Note that integer variables, including binary variables, are used in different decision-making problems throughout this book.

An MILP problem can be generally formulated as:

$$\min_{\Phi} \quad \sum_j C_j^{\mathrm{C}} x_j + \sum_\ell C_\ell^{\mathrm{I}} z_\ell \tag{A.11a}$$

subject to

$$\sum_j A_{kj}^{\mathrm{E,C}} x_j + \sum_\ell A_{k\ell}^{\mathrm{E,I}} z_\ell = B_k^{\mathrm{E}}, \quad \forall k \in \Omega^{\mathrm{E}}, \tag{A.11b}$$

$$\sum_j A_{ij}^{\mathrm{I,C}} x_j + \sum_\ell A_{i\ell}^{\mathrm{I,I}} z_\ell \leq B_i^{\mathrm{I}}, \quad \forall i \in \Omega^{\mathrm{I}}, \tag{A.11c}$$

$$x_j \in \mathbb{R}, \quad \forall j, \tag{A.11d}$$

$$z_\ell \in \mathbb{I}, \quad \forall \ell, \tag{A.11e}$$

where:

- $\mathbb{I}$ is the set of integer numbers,
- $\Phi=\{x_j, \forall j; z_\ell, \forall \ell\}$ is the set of optimization variables,
- $C_j^{\mathrm{C}}, \forall j$, and $C_\ell^{\mathrm{I}}, \forall \ell$, are respectively the cost coefficients of variables $x_j, \forall j$, and $z_\ell, \forall \ell$, in the objective function (A.11a),
- $A_{kj}^{\mathrm{E,C}}, \forall j$, $A_{k\ell}^{\mathrm{E,I}}, \forall \ell$, and B_k^{E} are the coefficients that define equality constraints (A.11b), $\forall k \in \Omega^{\mathrm{E}}$, and
- $A_{ij}^{\mathrm{I,C}}, \forall j$, $A_{i\ell}^{\mathrm{I,I}}, \forall \ell$,and B_i^{I} are the coefficients that define inequality constraints (A.11c), $\forall i \in \Omega^{\mathrm{I}}$.

Table A.7 Illustrative Example A.6: optimal solution

x_1	x_2	x_3	x_4	z_1	z_2
20	10	25	5	1	1

The above MILP problem is a non-convex optimization problem due to the presence of integer optimization variables. MILP problems can be solved using efficient branch-and-cut methods, for which a detailed description can be found in [7]. MILP problems can also be solved using commercial software tools such as the CPLEX solver [5] under GAMS [6].

Solving MILP models in large problems generally needs a high computation time and memory. A possible solution is to use a Benders' decomposition algorithm that decomposes the original problem into subproblems with only continuous variables, and an MILP master problem including the integer variables [3].

Illustrative Example A.6 MILP Problem

Consider an MILP problem that is formulated as follows. First, the objective function to be minimized is:

$$z = 10x_1 + 20x_2 - 30x_3 - 15x_4.$$

Second, the constraints are provided below:

- Equality constraints:

$$-x_1 - x_2 + x_3 + x_4 = 0.$$

- Inequality constraints:

$$\begin{aligned}
& -x_1 + 10z_1 \le 0, \\
& x_1 - 20z_1 \le 0, \\
& -x_2 + 10z_2 \le 0, \\
& x_2 - 20z_2 \le 0, \\
& -x_3 \le 0, \\
& x_3 \le 25, \\
& -x_4 \le 0, \\
& x_4 \le 25.
\end{aligned}$$

- Continuous and integer variables:

$$\begin{aligned}
& x_1, x_2, x_3, x_4 \in \mathbb{R}, \\
& z_1, z_2 \in \{0, 1\}.
\end{aligned}$$

Finally, set $\Phi=\{x_1, x_2, x_3, x_4, z_1, z_2\}$ includes the optimization variables.

The above MILP problem is solved using the CPLEX solver [5] under GAMS [6]. The optimal solution is provided in Table A.7. For the optimal solution, the value of the objective function is equal to -425.

A.4 Summary and Further Reading

This appendix provides a theoretical background on the optimization problems used throughout this book to formulate the different decision-making problems faced by VPPs. These optimization problems include both LP and MILP problems.

Additional information including more detailed theories about convex optimization and its application in engineering can be found in the monographs by Boyd and Vandengerghe [2], Arora [1], and Sioshamsi and Conejo [7]. Additionally, reference [3] provides applicable decomposition techniques for solving optimization problems with complicating variables or constraints.

References

1. Arora, J.S.: Introduction to Optimum Design, 2th edn. Elsevier Academic Press, Oxford (2004)
2. Boyd, S., Vandenberghe, L.: Convex Optimization, 7th edn. Cambridge University Press, Cambridge (2009)
3. Conejo, A.J., Castillo, E., Mínguez, R., García-Bertrand, R.: Decomposition Techniques in Mathematical Programming. Springer, New York (2006)
4. Dantzig, G.B.: Linear Programming and Extensions. Princeton Univ. Press, Princeton, NJ (1963)
5. ILOG CPLEX.: Available at www.ilog.com/products/cplex/ (2020)
6. GAMS.: Available at www.gams.com/ (2020)
7. Sioshansi, R., Conejo, A.J.: Optimization in Engineering. Models and Algorithms. Springer, New York, NY (2017)

Appendix B
Optimization Under Uncertainty

This appendix describes the main characteristics of the optimization models used in this book to handle uncertainties involved in decision-making problems of virtual power plants (VPPs), namely stochastic programming and robust optimization.

B.1 Introduction

Operation and expansion planning decisions for VPPs are generally made under an uncertain environment. In order to handle the uncertainties in the decision-making problems, two alternative optimization models are usually considered, namely stochastic programming and robust optimization.

Scenario-based stochastic programming uses a finite number of scenarios representing different realizations of the uncertain parameters. In addition, each scenario has associated a probability of occurrence. Then, the stochastic programming problem determines the decision variables that minimize (maximize) the expected cost (profit) over all scenarios, taking into account the constraints of the problem for each scenario realization.

In order to generate accurate scenarios, two important issues arise. On one hand, it is needed information about the probability distribution functions of uncertain parameters. On the other hand, a large enough number of scenarios are generally needed since the feasibility of the problem is only guaranteed for the considered scenarios.

An alternative to stochastic programming is robust optimization that uses uncertainty sets, e.g., based on confidence bounds, instead of considering some predefined scenarios. Robust optimization optimizes the worst case of the objective function considering that uncertain parameters can take any value within the considered uncertainty sets. Moreover, the feasibility of the problem is guaranteed for all realizations of the uncertain parameters within these uncertainty sets. Robust optimization is mainly useful when the probability distribution functions of uncertain parameters are not available.

Two different robust optimization models can be formulated depending on the decisions made in the decision-making process, namely robust optimization without recourse, also known as static robust optimization, and robust optimization with recourse, also known as two-stage robust optimization or adaptive robust optimization.

The reminder of this appendix is organized as follows. Section B.2 describes the main characteristics of stochastic programming. Section B.3 is devoted to robust optimization, including a description of both static robust optimization and robust optimization with recourse. Finally, Sect. B.4 summarizes the appendix and suggests some references for further reading.

B.2 Stochastic Programming

This section provides a brief tutorial introduction to stochastic programming. For the sake of clarity, a two-stage stochastic programming problem is considered, although the explanations below can be easily extended to multi-stage stochastic programming.

L. Baringo, M. Rahimiyan, *Virtual Power Plants and Electricity Markets*, https://doi.org/10.1007/978-3-030-47602-1

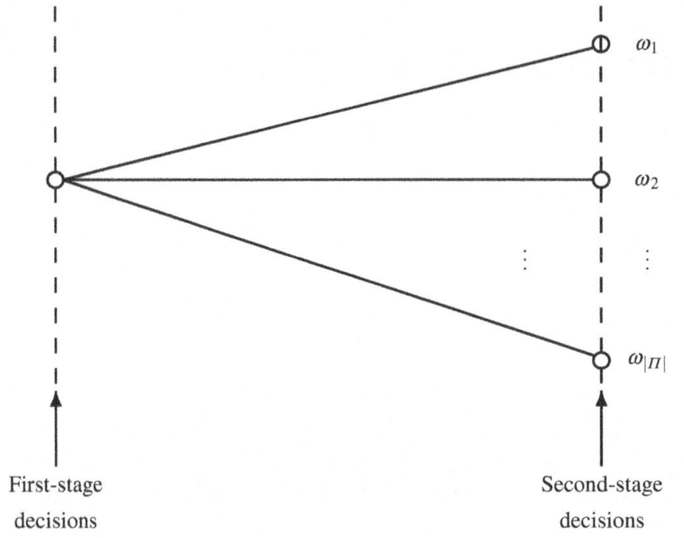

Fig. B.1 Scenario tree used in a two-stage decision-making problem

B.2.1 Uncertainty Model

Uncertain parameters are modeled using a number of scenarios indexed by ω with their probability of occurrence. Set Π includes the considered scenario realizations of uncertainties, while P_ω is the probability of occurrence of scenario ω.

Optimization variables can be classified into Φ^{I} and Φ^{II}, corresponding to the decision variables made at the first and second stages, respectively. Accordingly, the decision-making sequence carried out by a decision-maker is explained as below:

1. In the first stage, the decision-maker seeks to minimize (maximize) the first-stage cost (profit) plus the expected second-stage cost (profit) over the considered scenarios. The decision-maker implements the decision variables Φ^{I}. These decisions are made before the actual realization of uncertain data, and thus, do not depend on the future scenario realizations. In other words, these decisions are *here-and-now* decisions with respect to the uncertainties.
2. Once a scenario ω is realized, the decision-maker gets information about the actual realization of the uncertain data. Then, in the second stage, the decision-maker minimizes (maximizes) the cost (profit) and implements the decision variables Φ^{II}. As these decisions are made after knowing the scenario realization ω, they depend on the scenarios, and thus are *wait-and-see* decisions with respect to the scenario realization ω.

Figure B.1 shows the scenario tree used in the above two-stage decision-making sequence.

B.2.2 Formulation

A two-stage stochastic programming problem can be formulated as:

$$\max_{\Phi^{\mathrm{I}}, \Phi^{\mathrm{II}}_{\omega}, \forall \omega} \quad f^{\mathrm{I}}(\Phi^{\mathrm{I}}) + E_{\Pi}[f^{\mathrm{II}}_{\omega}(\Phi^{\mathrm{II}}_{\omega})] \tag{B.1a}$$

subject to

$$h^{\mathrm{I}}_{k}(\Phi^{\mathrm{I}}) = 0, \quad \forall k \in \Omega^{\mathrm{E,I}}, \tag{B.1b}$$

$$g^{\mathrm{I}}_{i}(\Phi^{\mathrm{I}}) \leq 0, \quad \forall i \in \Omega^{\mathrm{I,I}}, \tag{B.1c}$$

$$\Phi^{\mathrm{I}} \in \mathbb{R}^{N}, \tag{B.1d}$$

$$h^{\mathrm{II}}_{k'\omega}(\Phi^{\mathrm{I}}, \Phi^{\mathrm{II}}_{\omega}) = 0, \quad \forall k' \in \Omega^{\mathrm{E,II}}_{\omega}, \forall \omega \in \Pi, \tag{B.1e}$$

$$g^{\mathrm{II}}_{i'\omega}(\Phi^{\mathrm{I}}, \Phi^{\mathrm{II}}_{\omega}) \leq 0, \quad \forall i' \in \Omega^{\mathrm{I,II}}_{\omega}, \forall \omega \in \Pi, \tag{B.1f}$$

$$\Phi^{\mathrm{II}}_{\omega} \in \mathbb{R}^{M}, \tag{B.1g}$$

where:

- $\mathbb{R}$ is the set of real numbers,
- $\Phi^{\mathrm{I}}=\{x^{\mathrm{I}}_{j}, \forall j\}$ and $\Phi^{\mathrm{II}}_{\omega} = \{x^{\mathrm{II}}_{j'\omega}, \forall j'\}, \forall \omega \in \Pi$, are the set of optimization variables corresponding to the first- and second-stage decisions, respectively,
- $f^{\mathrm{I}}(\Phi^{\mathrm{I}})$: $\mathbb{R}^N \rightarrow \mathbb{R}$ and $f^{\mathrm{II}}_{\omega}(\Phi^{\mathrm{II}}_{\omega})$: $\mathbb{R}^M \rightarrow \mathbb{R}, \forall \omega \in \Pi$, are the terms of the objective function (B.1a) corresponding to first- and second-stage decisions, respectively,
- $E_{\Pi}(\cdot)$ is the expectation operator over scenario set Π,
- $h^{\mathrm{I}}_{k}(\Phi^{\mathrm{I}})$: $\mathbb{R}^N \rightarrow \mathbb{R}, \forall k \in \Omega^{\mathrm{E,I}}$, are the functions that define equality constraints (B.1b) at the first stage,
- $g^{\mathrm{I}}_{i}(\Phi^{\mathrm{I}})$: $\mathbb{R}^N \rightarrow \mathbb{R}, \forall i \in \Omega^{\mathrm{I,I}}$, are the functions that define inequality constraints (B.1c) at the first stage,
- $h^{\mathrm{II}}_{k'\omega}(\Phi^{\mathrm{I}}, \Phi^{\mathrm{II}}_{\omega})$: $\mathbb{R}^N \times \mathbb{R}^M \rightarrow \mathbb{R}, \forall k' \in \Omega^{\mathrm{E,II}}$, are the functions that define equality constraints (B.1e) under scenario ω at the second stage,
- $g^{\mathrm{II}}_{i'\omega}(\Phi^{\mathrm{I}}, \Phi^{\mathrm{II}}_{\omega})$: $\mathbb{R}^N \times \mathbb{R}^M \rightarrow \mathbb{R}, \forall i' \in \Omega^{\mathrm{I,II}}$, are the functions that define inequality constraints (B.1f) under scenario ω at the second stage,
- N is the number of optimization variables at the first stage,
- M is the number of optimization variables per scenario at the second stage,
- $\Omega^{\mathrm{E,I}}$ and $\Omega^{\mathrm{I,I}}$ are the sets of equality and inequality constraints at the first stage, respectively, and
- $\Omega^{\mathrm{E,II}}_{\omega}$ and $\Omega^{\mathrm{I,II}}_{\omega}$ are the sets of equality and inequality constraints under scenario realization ω at the second stage, respectively.

Illustrative Example B.1 Stochastic Programming

Consider a power producer that owns a conventional power plant with a capacity of 50 MW and a variable cost of \$18/MWh. The scheduling problem of the power producer can be formulated as a two-stage decision-making problem. In first stage, the power producer decides the power quantity x^{I} that can be traded through a selling contract characterized with a maximum power of 30 MW at a fixed price of \$18/MWh. Additionally, in the second stage, the power producer can buy or sell the power quantity x^{II}_{1} in the energy market with an uncertain price, and produce the power quantity x^{II}_{2} using the conventional power plant.

The uncertainty in the market price is represented through a number of scenarios in set $\Pi = \{15, 20, 8, 12, 21\}$\$/MWh. The probability of occurrence of scenarios is: $P = \{0.2, 0.4, 0.1, 0.1, 0.2\}$.

Considering the above data, the two-stage stochastic model is formulated. The objective function to be maximized includes two parts, namely the revenue obtained in the first stage and the expected profit that can be obtained in the second stage:

$$z = 18x^{\mathrm{I}} + 0.2(15x^{\mathrm{II}}_{11} - 18x^{\mathrm{II}}_{21}) + 0.4(20x^{\mathrm{II}}_{12} - 18x^{\mathrm{II}}_{22}) + 0.1(8x^{\mathrm{II}}_{13} - 18x^{\mathrm{II}}_{23})$$
$$+ 0.1(12x^{\mathrm{II}}_{14} - 18x^{\mathrm{II}}_{24}) + 0.2(21x^{\mathrm{II}}_{15} - 18x^{\mathrm{II}}_{25}).$$

The constraints of the problem are formulated as follows:

- Inequality constraints in the first stage:

$$0 \leq x^{\mathrm{I}} \leq 30.$$

- Equality constraints in the second stage:

$$x^{\mathrm{I}} + x^{\mathrm{II}}_{11} - x^{\mathrm{II}}_{21} = 0,$$
$$x^{\mathrm{I}} + x^{\mathrm{II}}_{12} - x^{\mathrm{II}}_{22} = 0,$$
$$x^{\mathrm{I}} + x^{\mathrm{II}}_{13} - x^{\mathrm{II}}_{23} = 0,$$
$$x^{\mathrm{I}} + x^{\mathrm{II}}_{14} - x^{\mathrm{II}}_{24} = 0,$$
$$x^{\mathrm{I}} + x^{\mathrm{II}}_{15} - x^{\mathrm{II}}_{25} = 0.$$

Table B.1 Illustrative Example B.1: optimal solutions [MW]

x^{I}	x_{11}^{II}	x_{12}^{II}	x_{13}^{II}	x_{14}^{II}	x_{15}^{II}	x_{21}^{II}	x_{22}^{II}	x_{23}^{II}	x_{24}^{II}	x_{25}^{II}
30	−30	20	−30	−30	20	0	50	0	0	50

- Inequality constraints in the second stage:

$$0 \le x_{21}^{\mathrm{II}} \le 50,$$

$$0 \le x_{22}^{\mathrm{II}} \le 50,$$

$$0 \le x_{23}^{\mathrm{II}} \le 50,$$

$$0 \le x_{24}^{\mathrm{II}} \le 50,$$

$$0 \le x_{25}^{\mathrm{II}} \le 50.$$

The optimization variables in the first- and second-stage problems are those included in sets $\Phi^{\mathrm{I}} = \{x^{\mathrm{I}}\}$ and $\Phi^{\mathrm{II}} = \{x_{11}^{\mathrm{II}}, x_{12}^{\mathrm{II}}, x_{13}^{\mathrm{II}}, x_{14}^{\mathrm{II}}, x_{15}^{\mathrm{II}}, x_{21}^{\mathrm{II}}, x_{22}^{\mathrm{II}}, x_{23}^{\mathrm{II}}, x_{24}^{\mathrm{II}}, x_{25}^{\mathrm{II}}\}$, respectively. The second subscript in the variables refers to the scenario realization.

The above optimization problem is solved in GAMS [7] using the CPLEX solver [8]. The optimal value of the objective function is \$94, which is obtained as the summation of \$540 and −\$446 as the profit related to the first and second stages, respectively. The optimal solutions are provided in Table B.1.

□

The stochastic model (B.1) seeks to optimize the expected value, while ignoring the risk associated with the uncertain data in the second stage. In fact, the objective function under uncertainty is a random variable, and thus can take values different from the expected value. For example, the variability of the objective function may result in values lower than the expected value with high probability. A risk-neutral decision-maker is indifferent to such risks and only optimizes the expected value considering model (B.1). However, a risk-averse decision-maker may be interested in using a risk metric to control the risk associated with its decisions. The risk-averse decision-maker optimizes the expected value, but is also willing to minimize undesired outcomes.

There are different risk metrics, and among them, the value-at-risk (VaR) and the conditional value-at-risk (CVaR) are the well-known ones in finance [10]. For a given confidence level $\alpha \in [0, 1)$, the VaR can be defined as the $(1 - \alpha)$-quantile of distribution of objective function, and the CVaR is the expectation of outcomes lower than the VaR. In fact, the VaR is an indicator of undesired outcomes that may be observed from the distribution of the objective function. For example, if a one-year 90% VaR of \$1000 is measured for a profit of financial choice, it is expected that the investor will not obtain a profit lower than \$1000 in 90% of the years.

Based on the definition of the CVaR at the confidence level α, the risk-constrained stochastic programming can be formulated below:

$$\max_{\Phi, \Phi^{\mathrm{C}}} \quad \sum_{\omega \in \Pi} P_{\omega} f_{\omega}(\Phi) + \beta \mathrm{CVaR} \tag{B.2a}$$

subject to

$$\Phi \in \Lambda, \tag{B.2b}$$

$$\mathrm{CVaR} = \rho - \frac{1}{1-\alpha} \sum_{\omega \in \Pi} P_{\omega} \mu_{\omega}, \tag{B.2c}$$

$$- f_{\omega}(\Phi) + \rho - \mu_{\omega} \le 0, \quad \forall \omega \in \Pi, \tag{B.2d}$$

$$\mu_{\omega} \ge 0, \quad \forall \omega \in \Pi, \tag{B.2e}$$

where:

- Φ includes the optimization variables in the stochastic problem (B.1),
- Λ is the feasible set of solutions in the stochastic problem (B.1),
- $\Phi^{\mathrm{C}} = \{\mathrm{CVaR}; \rho; \mu_{\omega}, \forall \omega \in \Pi\}$ includes additional optimization variables required to incorporate the CVaR metric.

- ρ is the VaR at the confidence level α, and
- β is a nonnegative weighting factor, which allows to manage risk.

The objective function comprises two terms, namely the expected profit and the weighted CVaR. Note that both terms of the objective function are to be maximized. In fact, a risk-averse decision-maker is willing to find the highest value of the CVaR. For $\beta = 0$, the above stochastic model is the same as the stochastic model (B.1). By increasing parameter β, the decision-maker chooses a more risk-averse strategy. Generally, the expected value versus CVaR is presented as the *Pareto frontier* for different values of β. Through the *Pareto frontier*, a higher expected value is obtained at the expense of a lower CVaR (i.e., higher risk).

Illustrative Example B.2 Risk-Constrained Stochastic Programming

Illustrative Example B.1 is extended below to consider a risk-averse power producer. The confidence level α is set equal to 0.9.

The objective function to be maximized includes two parts, namely the expected value and the weighted CVaR:

$$z = \{0.2f_1 + 0.4f_2 + 0.1f_3 + 0.1f_4 + 0.2f_5\} + \beta\text{CVaR}.$$

The constraints are presented as follows:

- Terms needed to calculate the expected profit:

$$f_1 = 18x^{\text{I}} + (15x_{11}^{\text{II}} - 18x_{21}^{\text{II}}),$$
$$f_2 = 18x^{\text{I}} + (20x_{12}^{\text{II}} - 18x_{22}^{\text{II}}),$$
$$f_3 = 18x^{\text{I}} + (8x_{13}^{\text{II}} - 18x_{23}^{\text{II}}),$$
$$f_4 = 18x^{\text{I}} + (12x_{14}^{\text{II}} - 18x_{24}^{\text{II}}),$$
$$f_5 = 18x^{\text{I}} + (21x_{15}^{\text{II}} - 18x_{25}^{\text{II}}).$$

- Inequality constraints in the first stage:

$$0 \leq x^{\text{I}} \leq 30.$$

- Equality constraints in the second stage:

$$x^{\text{I}} + x_{11}^{\text{II}} - x_{21}^{\text{II}} = 0,$$
$$x^{\text{I}} + x_{12}^{\text{II}} - x_{22}^{\text{II}} = 0,$$
$$x^{\text{I}} + x_{13}^{\text{II}} - x_{23}^{\text{II}} = 0,$$
$$x^{\text{I}} + x_{14}^{\text{II}} - x_{24}^{\text{II}} = 0,$$
$$x^{\text{I}} + x_{15}^{\text{II}} - x_{25}^{\text{II}} = 0.$$

- Inequality constraints in the second stage:

$$0 \leq x_{21}^{\text{II}} \leq 50,$$
$$0 \leq x_{22}^{\text{II}} \leq 50,$$
$$0 \leq x_{23}^{\text{II}} \leq 50,$$
$$0 \leq x_{24}^{\text{II}} \leq 50,$$
$$0 \leq x_{25}^{\text{II}} \leq 50.$$

Table B.2 Illustrative Example B.2: expected value and CVaR

β	Expected profit [\$]	CVaR [\$]
0	94	40
0.5	86	60
1	86	60
1.5	86	60
2	86	60

- Risk constraints:

$$
\begin{aligned}
&\text{CVaR} = \rho - \frac{1}{1-0.9}(0.2\mu_1 + 0.4\mu_2 + 0.1\mu_3 + 0.1\mu_4 + 0.2\mu_5),\\
&-f_1 + \rho - \mu_1 \leq 0,\\
&-f_2 + \rho - \mu_2 \leq 0,\\
&-f_3 + \rho - \mu_3 \leq 0,\\
&-f_4 + \rho - \mu_4 \leq 0,\\
&-f_5 + \rho - \mu_5 \leq 0,\\
&\mu_1 \geq 0,\\
&\mu_2 \geq 0,\\
&\mu_3 \geq 0,\\
&\mu_4 \geq 0,\\
&\mu_5 \geq 0.
\end{aligned}
$$

The optimization variables are those included in sets $\Phi = \{x^{\mathrm{I}}, x^{\mathrm{II}}_{11}, x^{\mathrm{II}}_{12}, x^{\mathrm{II}}_{13}, x^{\mathrm{II}}_{14}, x^{\mathrm{II}}_{15}, x^{\mathrm{II}}_{21}, x^{\mathrm{II}}_{22}, x^{\mathrm{II}}_{23}, x^{\mathrm{II}}_{24}, x^{\mathrm{II}}_{25}\}$ and $\Phi^{\mathrm{C}} = \{\text{CVaR}; \rho; \mu_1, \mu_2, \mu_3, \mu_4, \mu_5, f_1, f_2, f_3, f_4, f_5\}$.

The above optimization problem is solved in GAMS [7] using the CPLEX solver [8] for different values of parameter β. The expected value and CVaR in each case are provided in Table B.2. Note that as parameter β increases, the CVaR is also increased, at the expense of reducing the expected profit.

An input GAMS code [7] for solving Illustrative Example B.2 is provided below:

```
* Risk-constrained stochastic programming
SET j  'First-stage decisions' /j1*j1/;
SET jp 'Second-stage decisions' /jp1*jp2/;
SET i  'Inequality constraints in first stage' /i1*i1/;
SET kp 'equality constraints in second stage' /kp1*kp1/;
SET ip 'inequality constraints in second stage' /ip1*ip2/;
SET ome 'scenarios' /ome1*ome5/;
SCALAR beta 'weighting parameter' /1/;
SCALAR alpha 'confidence level to calculate CVaR' /0.9/;
PARAMETER C1(j) 'coeffients of first-stage decisions in the
    objective function'
/j1      18/;
PARAMETER C2(jp,ome) 'coeffients of second-stage decisions in the
     objective function'
/jp1.ome1      15
jp1.ome2       20
jp1.ome3       8
jp1.ome4       12
jp1.ome5       21
jp2.ome1       -18
jp2.ome2       -18
jp2.ome3       -18
jp2.ome4       -18
jp2.ome5       -18/;
PARAMETER AI1(i,j) 'coeffients that define inequality constraints
     in first stage'
/i1.j1    1/;
```

```
PARAMETER BI1(i) 'coeffients that define inequality constraints
    in first stage'
/i1       30/;
PARAMETER AE2(kp,j) 'coeffients that define equality constraints
    in second stage'
/kp1.j1      1/;
PARAMETER AEP2(kp,jp) 'coeffients that define equality
    constraints in second stage'
/kp1.jp1     1
kp1.jp2     -1/;
PARAMETER BE2(kp) 'coeffients that define equality constraints in
     second stage'
/kp1      0/;
PARAMETER AIP2(ip,jp) 'coeffients that define inequality
    constraints in second stage'
/ip1.jp1     0
ip1.jp2     1
ip2.jp1     0
ip2.jp2     -1/;
PARAMETER BI2(ip) 'coeffients that define inequality constraints
    in second stage'
/ip1      50
ip2       0/;
PARAMETER P(ome) 'probability of occurrence of scenarios'
/ome1      0.2
ome2       0.4
ome3       0.1
ome4       0.1
ome5       0.2/;
VARIABLES
z 'objective function',
z1 'objective function in first stage',
z2 'objective function in second stage',
x2(jp,ome) 'second-stage decision variables',
CVaR  'CVaR',
f(ome) 'objective in each scenario',
rho 'value at risk';
POSITIVE VARIABLES x1(j) 'first-stage decision variables',
mu(ome) 'auxiliary variable';
EQUATIONS obj 'objective function',
obj1 'objective function in first stage',
obj2 'objective function in second stage',
icon1 'inequality constraints in first stage',
econ2 'equality constraints in second stage',
icon2 'inequality constraints in second stage',
fcons  'objective in each scenario',
CVaR_cal 'CVaR',
CVaR_con 'CVaR constraint';
*
obj..     z=e=(z1+z2)+beta*CVaR;
obj1..    z1=e=SUM(j,C1(j)*x1(j));
obj2..    z2=e=SUM(ome,P(ome)*f(ome));
icon1(i)..  SUM(j,AI1(i,j)*x1(j))=L=BI1(i);
```

```
econ2(kp,ome).. SUM(j,AE2(kp,j)*x1(j))+SUM(jp,AEP2(kp,jp)*x2(jp,
    ome))=E=BE2(kp);
icon2(ip,ome).. SUM(jp,AIP2(ip,jp)*x2(jp,ome))=L=BI2(ip);
fcons(ome).. f(ome)=E=SUM(jp,C2(jp,ome)*x2(jp,ome));
CVaR_cal..    CVaR=E=rho-(1/(1-alpha))*SUM(ome,P(ome)*mu(ome));
CVaR_con(ome).. -(z1+f(ome))+rho-mu(ome) =L= 0;
*******************************************
******** Model definition *******************
MODEL SP       /all/;
OPTION OPTCR=0;
OPTION OPTCA=0;
OPTION iterlim=1e8;
********************************************
**** Parameter definition *******************
PARAMETER state 'Convergence status';
PARAMETER Exp 'Expected value';
SOLVE SP USING LP MAXIMIZING z;
state = SP.modelstat;
Exp=z1.l+z2.l;
display state,Exp;
```

Stochastic approach is used to solve operation and planning problems of VPPs in Chaps. 3, 4, 5, 6, and 7.

B.3 Robust Optimization

This section explains the basic theory of robust optimization to handle uncertainties in decision-making problems. For the sake of clarity, a two-stage decision-making problem is considered in which the decision-maker seeks to implement the first-stage decisions, while facing uncertainty in the second stage.

B.3.1 Static Robust Optimization

This section provides the theory of the robust optimization approach described in [4] and explains how it can be used to handle uncertainties in two-stage decision-making problems. In this type of problems, the decision-maker seeks to minimize (maximize) the first-stage cost (profit) plus the worst-case cost (profit) that may occur under uncertainty in the second stage.

Accordingly, the two-stage decision-making problem can be represented as a min-max optimization problem that can be formulated as the following bi-level optimization problem:

$$\min_{\Phi^{\mathrm{I}},\Phi^{\mathrm{II}}} \sum_{j} C^{\mathrm{I}}_{j} x^{\mathrm{I}}_{j} + \max_{\Pi^{\mathrm{O}}} \sum_{j'} C^{\mathrm{II}}_{j'} x^{\mathrm{II}}_{j'} \tag{B.3a}$$

subject to

$$\sum_{j} A^{\mathrm{E,I}}_{kj} x^{\mathrm{I}}_{j} = B^{\mathrm{E,I}}_{k}, \quad \forall k \in \Omega^{\mathrm{E,I}}, \tag{B.3b}$$

$$\sum_{j} A^{\mathrm{I,I}}_{ij} x^{\mathrm{I}}_{j} \leq B^{\mathrm{I,I}}_{i}, \quad \forall i \in \Omega^{\mathrm{I,I}}, \tag{B.3c}$$

$$\sum_{j} A^{\mathrm{E,II}}_{k'j} x^{\mathrm{I}}_{j} + \sum_{j'} A'^{\mathrm{E,II}}_{k'j'} x^{\mathrm{II}}_{j'} = B^{\mathrm{E,II}}_{k'}, \quad \forall k' \in \Omega^{\mathrm{E,II}}, \tag{B.3d}$$

$$\max_{\Pi^{\mathrm{C}}_{i'}} \left[\sum_{j} A^{\mathrm{I,II}}_{i'j} x^{\mathrm{I}}_{j} + \sum_{j'} A'^{\mathrm{I,II}}_{i'j'} x^{\mathrm{II}}_{j'} \right] \leq B^{\mathrm{I,II}}_{i'}, \quad \forall i' \in \Omega^{\mathrm{I,II}}, \tag{B.3e}$$

where:

- sets $\Phi^{\mathrm{I}}=\{x_j^{\mathrm{I}}, \forall j\}$ and $\Phi^{\mathrm{II}}=\{x_{j'}^{\mathrm{II}}, \forall j'\}$ include the optimization variables of the upper-level problem, corresponding to the first- and second-stage problems, respectively,
- sets Π^{O} and $\Pi_{i'}^{\mathrm{C}}, \forall i' \in \Omega^{\mathrm{I,II}}$, include the optimization variables of the lower-level problems; these variables are defined through the uncertainty sets including uncertain coefficients in the objective function (B.3a) and constraints (B.3e), respectively,
- coefficients $C_j^{\mathrm{I}}, \forall j$, are known, but coefficients $C_{j'}^{\mathrm{II}}, \forall j'$, are uncertain in the objective function (B.3a),
- all coefficients in equality and inequality constraints (B.3b) and (B.3c) in the first stage are known,
- equality constraints (B.3d) in the second stage do not include any uncertainty, and
- coefficients $A_{i'j}^{\mathrm{I,II}}, \forall j$, and $A_{i'j'}^{\mathrm{I,II}}, \forall j'$, are uncertain in inequality constraints (B.3e), $\forall i' \in \Omega^{\mathrm{I,II}}$, in the second stage.

Without loss of generality, if coefficients $B_{i'}^{\mathrm{I,II}}, \forall i' \in \Omega^{\mathrm{I,II}}$, in inequality constraints (B.3e) are uncertain, it is possible to add this uncertain coefficient in the uncertainty set $\Pi_{i'}^{\mathrm{C}}$, and define an auxiliary variable $y_{i'}$ in such a way that (B.3e) are rewritten as $\max_{\Pi_{i'}^{\mathrm{C}}} \left[\sum_j A_{i'j}^{\mathrm{I,II}} x_j^{\mathrm{I}} + \sum_{j'} A_{i'j'}^{\prime\mathrm{I,II}} x_{j'}^{\mathrm{II}} - B_{i'}^{\mathrm{I,II}} y_{i'}\right] \leq 0$ and $1 \leq y_{i'} \leq 1$.

The min-max problem (B.3) is a static robust problem that can be equivalently reformulated as the following problem:

$$\min_{\Phi} \quad \max_{\Pi^{\mathrm{O}}} \quad \sum_j C_j x_j \tag{B.4a}$$

subject to

$$\sum_j A_{kj}^{\mathrm{E}} x_j = B_k^{\mathrm{E}}, \quad \forall k \in \Omega^{\mathrm{E}}, \tag{B.4b}$$

$$\max_{\Pi_i^{\mathrm{C}}} \quad \sum_j A_{ij}^{\mathrm{I}} x_j \leq B_i^{\mathrm{I}}, \quad \forall i \in \Omega^{\mathrm{I}}, \tag{B.4c}$$

where:

- set $\Phi=\{x_j, \forall j\}$ includes the optimization variables of the upper-level problem,
- sets Π^{O} and $\Pi_i^{\mathrm{C}}, \forall i \in \Omega^{\mathrm{I}}$, include the optimization variables of the lower-level problems; these variables are defined through the uncertainty sets including uncertain coefficients in the objective function (B.4a) and constraints (B.4c), respectively,
- a number of coefficients $C_j, \forall j$, in the objective function (B.4a) are uncertain,
- all coefficients in the equality constraints (B.4b) are known, and
- a number of coefficients $A_{ij}^{\mathrm{I}}, \forall i \in \Omega^{\mathrm{I}}, \forall j$, in some inequality constraints (B.4c) are uncertain.

Thus, the two parts of the objective function (B.3a) are unified as the objective function (B.4a). Equality constraints (B.3b) and (B.3d) are merged into equality constraints (B.4b). Similarly, inequality constraints (B.3c) and (B.3e) are merged into inequality constraints (B.4c). This way, a number of coefficients in (B.4c) are uncertain.

B.3.1.1 Uncertainty Model

Uncertainty sets and some control parameters are defined to address uncertainties in the static robust optimization problem (B.4) described in the previous section.

Uncertainty set Π^{O} is defined as $\Pi^{\mathrm{O}}=\{C_j | C_j \in [\tilde{C}_j - \hat{C}_j, \tilde{C}_j + \hat{C}_j], \hat{C}_j > 0\}$ and comprises the unknown coefficients in the objective function. Note that parameter $\hat{C}_j$ is the deviation value from the nominal value $\tilde{C}_j$ that is anticipated for a given confidence level. Additionally, parameter Γ^{O} is defined and used to control the robustness of the solution with respect to the uncertain data in the objective function. This parameter can take nonnegative values within the interval $[0, |\Pi^{\mathrm{O}}|]$, in which the operator $|.|$ denotes the cardinality of the set. Accordingly, different values of Γ^{O} allow considering different strategies as explained below:

- A conservative strategy: if parameter Γ^{O} takes the maximum value of $|\Pi^{\mathrm{O}}|$, the solution is robust against the deviation of all coefficients in the objective function from their nominal value within the predefined confidence interval.
- A less conservative strategy: if parameter Γ^{O} is chosen lower than the maximum value, the impact of the deviation of Γ^{O} unknown coefficients from their nominal value within the confidence interval is considered.
- A risky strategy: the minimum value of Γ^{O}=0 ignores the impact of all uncertain coefficients in the objective function.

To address the uncertainty in the inequality constraints, uncertainty sets Π_i^{C}, $\forall i \in \Omega^{\mathrm{I}}$, are defined as $\Pi_i^{\mathrm{C}}=\{A_{ij}^{\mathrm{I}}|A_{ij}^{\mathrm{I}} \in [\tilde{A}_{ij}^{\mathrm{I}} - \hat{A}_{ij}^{\mathrm{I}}, \tilde{A}_{ij}^{\mathrm{I}} + \hat{A}_{ij}^{\mathrm{I}}], \hat{A}_{ij}^{\mathrm{I}} > 0\}, \forall i \in \Omega^{\mathrm{I}}$. Parameter $\hat{A}_{ij}^{\mathrm{I}}$ is the deviation from the nominal value $\tilde{A}_{ij}^{\mathrm{I}}$, which is obtained for a given confidence level. Each uncertainty set Π_i^{C} includes the unknown coefficients in each inequality constraint. To protect the constraints against a subset of uncertain coefficients, parameters Γ_i^{C}, $\forall i \in \Omega^{\mathrm{I}}$, are used. These parameters can take values within the interval $[0, |\Pi_i^{\mathrm{C}}|]$. According to the value of each parameter Γ_i^{C}, the following strategies are considered:

- A conservative strategy corresponds to the maximum value of $\Gamma_i^{\mathrm{C}}=|\Pi_i^{\mathrm{C}}|$, $\forall i \in \Omega^{\mathrm{I}}$. In this case, inequality constraints are protected against the deviation of all unknown coefficients from their nominal value within the predefined confidence intervals. Through this strategy, the obtained solution is guaranteed to be feasible if the actual realization of unknown coefficients takes value within the predefined confidence intervals.
- A less conservative strategy is taken if the value of Γ_i^{C} is selected lower than the maximum one, i.e., $\Gamma_i^{\mathrm{C}} < |\Pi_i^{\mathrm{C}}|$. Under such a condition, the feasibility of solution in constraint i is satisfied if up to $\lfloor \Gamma_i^{\mathrm{C}} \rfloor$ unknown coefficients vary within the considered confidence intervals, and one unknown coefficient A_{ij}^{I} change by at most $(\Gamma_i^{\mathrm{C}} - \lfloor \Gamma_i^{\mathrm{C}} \rfloor)\hat{A}_{ij}^{\mathrm{I}}$. Otherwise, the feasibility of solution is probabilistically guaranteed.
- A risky strategy is considered if parameter Γ_i^{C}=0. This way, the solution is guaranteed to be feasible in constraint i, if the actual realizations of its unknown coefficients are equal to their nominal value. Otherwise, the solution may not be feasible.

B.3.1.2 Formulation

The robust optimization approach proposed by Bertsimas [4] constrains the optimization problem with a subset of the uncertainty sets, which allows the decision-maker to control the level of conservatism. Accordingly, the standard robust counterpart of LP problem (B.4) is formulated as:

$$\min_{\Phi} \sum_j \tilde{C}_j x_j + \max_{\{S^{\mathrm{O}} \cup \{t^{\mathrm{O}}\} | S^{\mathrm{O}} \subseteq \Pi^{\mathrm{O}}, \lfloor \Gamma^{\mathrm{O}} \rfloor = |S^{\mathrm{O}}|, t^{\mathrm{O}} \in \Pi^{\mathrm{O}} \setminus S^{\mathrm{O}}\}} \left\{ \sum_{j \in S^{\mathrm{O}}} \hat{C}_j |x_j| + (\Gamma^{\mathrm{O}} - \lfloor \Gamma^{\mathrm{O}} \rfloor) \hat{C}_{t^{\mathrm{O}}} |x_{t^{\mathrm{O}}}| \right\} \tag{B.5a}$$

subject to

$$\sum_j A_{kj}^{\mathrm{E}} x_j = B_k^{\mathrm{E}}, \quad \forall k \in \Omega^{\mathrm{E}}, \tag{B.5b}$$

$$\sum_j \tilde{A}_{ij}^{\mathrm{I}} x_j + \max_{\{S_i^{\mathrm{C}} \cup \{t_i^{\mathrm{C}}\} | S_i^{\mathrm{C}} \subseteq \Pi_i^{\mathrm{C}}, \lfloor \Gamma_i^{\mathrm{C}} \rfloor = |S_i^{\mathrm{C}}|, t_i^{\mathrm{C}} \in \Pi_i^{\mathrm{C}} \setminus S_i^{\mathrm{C}}\}} \left\{ \sum_{j \in S_i^{\mathrm{C}}} \hat{A}_{ij}^{\mathrm{I}} |x_j| + (\Gamma_i^{\mathrm{C}} - \lfloor \Gamma_i^{\mathrm{C}} \rfloor) \hat{A}_{it_i^{\mathrm{C}}}^{\mathrm{I}} |x_{t_i^{\mathrm{C}}}| \right\} \le B_i^{\mathrm{I}}, \quad \forall i \in \Omega^{\mathrm{I}}. \tag{B.5c}$$

where:

- set S^{O} includes the optimization variables in the lower-level problem and is defined through a subset of the uncertainty set Π^{O} affecting the objective function (B.5a),
- set t^{O} includes one optimization variable in the lower-level problem and belongs to the uncertainty set Π^{O}, while it is not included in set S^{O},
- each set S_i^{C} contains the optimization variables in the lower-level problem related to constraint $i \in \Omega^{\mathrm{I}}$ and is defined over a subset of the uncertainty set Π_i^{C}, and

- each set t_i^{C} contains one optimization variable in the lower-level problem associated with constraint $i \in \Omega^{\mathrm{I}}$; the variable belongs to the uncertainty set Π_i^{C}, while it is not included in set S_i^{C}.

The robust optimization problem (B.5) is a bi-level problem:

1. In the upper-level problem, the decision-maker seeks to minimize the cost, anticipating the worst case of uncertainties.
2. For fixed upper-level decisions, a lower-level problem seeks to maximize the robustness term under a subset of uncertain data affecting the objective function (B.5a). The other lower-level problems aim at maximizing the protection levels against a subset of uncertain data in constraints (B.5c).

The above optimization problem is nonlinear due to the nonlinear function $|.|$. To linearize it, $|x_j|$ is replaced with auxiliary variable y_j, and new constraints $y_j \geq 0$ and $-y_j \leq x_j \leq y_j$ are added.

Then, by applying the strong duality theory, the robust optimization model (B.5) is transformed into the equivalent LP model below:

$$\min_{\Phi,\Phi^{\mathrm{R}}} \quad \sum_j \tilde{C}_j x_j + \Gamma^{\mathrm{O}} \upsilon^{\mathrm{O}} + \sum_{j \in \Pi^{\mathrm{O}}} \eta_j^{\mathrm{O}} \tag{B.6a}$$

subject to

$$\sum_j A_{kj}^{\mathrm{E}} x_j = B_k^{\mathrm{E}}, \quad \forall k \in \Omega^{\mathrm{E}}, \tag{B.6b}$$

$$\upsilon^{\mathrm{O}} + \eta_j^{\mathrm{O}} \geq \hat{C}_j y_j, \quad \forall j \in \Pi^{\mathrm{O}}, \tag{B.6c}$$

$$\sum_j \tilde{A}_{ij}^{\mathrm{I}} x_j + \Gamma_i^{\mathrm{C}} \upsilon_i^{\mathrm{C}} + \sum_{j \in \Pi_i^{C}} \eta_{ij}^{\mathrm{C}} \leq B_i^{\mathrm{I}}, \quad \forall i \in \Omega^{\mathrm{I}}, \tag{B.6d}$$

$$\upsilon_i^{\mathrm{C}} + \eta_{ij}^{\mathrm{C}} \geq \hat{A}_{ij}^{\mathrm{I}} y_j, \quad \forall i \in \Omega^{\mathrm{I}}, j \in \Pi_i^{\mathrm{C}}, \tag{B.6e}$$

$$-y_j \leq x_j \leq y_j, \quad \forall j, \tag{B.6f}$$

$$\upsilon^{\mathrm{O}} \geq 0, \tag{B.6g}$$

$$\upsilon_i^{\mathrm{C}} \geq 0, \quad \forall i \in \Omega^{\mathrm{I}}, \tag{B.6h}$$

$$\eta_j^{\mathrm{O}} \geq 0, \quad \forall j \in \Pi^{\mathrm{O}}, \tag{B.6i}$$

$$\eta_{ij}^{\mathrm{C}} \geq 0, \quad \forall i \in \Omega^{\mathrm{I}}, j \in \Pi_i^{\mathrm{C}}, \tag{B.6j}$$

$$y_j \geq 0, \quad \forall j, \tag{B.6k}$$

where set $\Phi^{\mathrm{R}}=\{y_j, \forall j; \upsilon^{\mathrm{O}}, \upsilon_i^{\mathrm{C}}, \forall i \in \Omega^{\mathrm{I}}; \eta_j^{\mathrm{O}}, \forall j \in \Pi^{\mathrm{O}}; \eta_{ij}^{\mathrm{C}}, \forall i \in \Omega^{\mathrm{I}}, \forall j \in \Pi_i^{\mathrm{C}}\}$ includes the optimization variables needed to convert the standard robust optimization model (B.5) into the equivalent LP model (B.6).

Finally, note that in order to implement decisions in a two-stage decision-making framework considering a robust optimization approach, a sequential procedure known as receding horizon can be carried out as explained below:

1. In the first stage, the decision-maker determines the optimal solutions of the two stages for the worst-case realization of uncertain data.
2. The decision-maker only implements the optimal solution at the first stage.
3. The uncertainty sets are updated considering new information available between the first and second stages.
4. In the second stage, the decision-maker finds the optimal solutions for the worst-case realization of uncertain data, and the decision-maker implements the optimal solution at the second stage.

Illustrative Example B.3 Static Robust Optimization

The power producer described in Illustrative Example B.1 is considered again. In this case, the aim is to decide the power traded x^{I} through a selling contract in the first stage, while scheduling the amount of power x_1^{II} bought from (negative)/sold to (positive) the market and dispatching the power production x_2^{II} of the conventional power plant under the worst-case realization of uncertainty in the market price and the variable production cost in the second stage.

First, uncertainty is disregarded, i.e., only the nominal value of the market price ($15/MWh) and the variable cost ($10/MWh) are considered. Accordingly, the two-stage decision-making problem is formulated below:

$$\min_{\Phi} \quad -18x^{\mathrm{I}} + (-15x_1^{\mathrm{II}} + 10x_2^{\mathrm{II}})$$

subject to:

$$0 \leq x^{\mathrm{I}} \leq 30,$$
$$x^{\mathrm{I}} + x_1^{\mathrm{II}} - x_2^{\mathrm{II}} = 0,$$
$$0 \leq x_2^{\mathrm{II}} \leq 50.$$

Note that the nominal values of the two coefficients C_1^{II} and C_2^{II} are $\tilde{C}_1^{\mathrm{II}} = -15$ and $\tilde{C}_2^{\mathrm{II}} = 10$. However, these two coefficients are subject to uncertainty, and thus, the uncertainty set is defined as Π^{O}={C_1^{II}, C_2^{II}}. These two coefficients can vary within the confidence bounds $C_1^{\mathrm{II}} \in [-20, -10]$ and $C_2^{\mathrm{II}} \in [8, 12]$, respectively.

Second, to handle the uncertainties in the objective function, the above problem is formulated using the static min-max optimization model below:

$$\min_{\Phi} \quad \max_{\Pi^{\mathrm{O}}} \quad -18x^{\mathrm{I}} + (C_1^{\mathrm{II}}x_1^{\mathrm{II}} + C_2^{\mathrm{II}}x_2^{\mathrm{II}})$$

subject to:

$$0 \leq x^{\mathrm{I}} \leq 30,$$
$$x^{\mathrm{I}} + x_1^{\mathrm{II}} - x_2^{\mathrm{II}} = 0,$$
$$0 \leq x_2^{\mathrm{II}} \leq 50.$$

The cost function is minimized over the optimization variables Φ for the worst case under the uncertainty set Π^{O}. Parameter Γ^{O} is defined within the interval [0,2] to control the robustness level of solution against the uncertainty in both market price and variable production cost. Then, the above min-max optimization model can be converted into the equivalent LP model.

The objective function is formulated below:

$$\min_{\Phi \cup \Phi^{\mathrm{R}}} \left\{ -18x^{\mathrm{I}} + (-15x_1^{\mathrm{II}} + 10x_2^{\mathrm{II}}) \right\} + \left\{ \Gamma^{\mathrm{O}}\upsilon^{\mathrm{O}} + \eta_1^{\mathrm{O}} + \eta_2^{\mathrm{O}} \right\}.$$

The constraints are provided below:

- Constraints of the original problem:

$$0 \leq x^{\mathrm{I}} \leq 30,$$
$$x^{\mathrm{I}} + x_1^{\mathrm{II}} - x_2^{\mathrm{II}} = 0,$$
$$0 \leq x_2^{\mathrm{II}} \leq 50.$$

- Constraints to handle the uncertainty in the objective function:

$$\upsilon^{\mathrm{O}} + \eta_1^{\mathrm{O}} \geq 5y_1^{\mathrm{O}},$$
$$\upsilon^{\mathrm{O}} + \eta_2^{\mathrm{O}} \geq 2y_2^{\mathrm{O}},$$
$$-y_1^{\mathrm{O}} \leq x_1^{\mathrm{II}} \leq y_1^{\mathrm{O}},$$
$$-y_2^{\mathrm{O}} \leq x_2^{\mathrm{II}} \leq y_2^{\mathrm{O}},$$

Table B.3 Illustrative Example B.3: optimal solution

Γ^{O}	x^{I}	x_1^{II}	x_2^{II}	υ^{O}	η_1^{O}	η_2^{O}	y_1^{O}	y_2^{O}
2	30	0	30	0	0	60	0	30

$$\upsilon^{O} \geq 0,$$

$$\eta_1^{O}, \eta_2^{O} \geq 0,$$

$$y_1^{O}, y_2^{O} \geq 0.$$

where Φ=$\{x^{I}, x_1^{II}, x_2^{II}\}$ and Φ^{R}=$\{\upsilon^{O},\eta_1^{O},\eta_2^{O},y_1^{O}, y_2^{O}\}$ are the optimization variables.

The above LP problem is solved using the CPLEX solver [8] under GAMS [7]. The optimal solution for the maximum value of parameter $\Gamma^{O} = 2$, i.e., considering a conservative strategy, is provided in Table B.3. In this case, the solution is optimal under the worst-case realization of the market price C_1^{II} and the variable cost C_2^{II} within the intervals $[-20, -10]$ and $[8, 12]$, respectively. Note that the decision-maker only implements the first-stage decision, i.e., x^{I}, at this stage. The optimal value of the objective function is equal to $-\$180$ as the summation of the nominal value $-\$240$ plus the robust term \$60.

An input GAMS code [7] used to solve the robust optimization problem is below:

```
* Static robust model
SET j  'First-stage decisions' /j1*j1/;
SET jp 'Second-stage decisions' /jp1*jp2/;
SET i  'Inequality constraints in first stage' /i1*i1/;
SET kp 'equality constraints in second stage' /kp1*kp1/;
SET ip 'inequality constraints in second stage' /ip1*ip2/;
SCALAR Gamma 'control parameter' /2/;
PARAMETER C1(j) 'coeffients of first-stage decisions in the
    objective function'
/j1        -18/;
PARAMETER C2(jp) 'coeffients of second-stage decisions in the
    objective function'
/jp1       -15
jp2        10/;
PARAMETER CH2(jp) 'deviation of coeffients of second-stage
    decisions in the objective function'
/jp1       5
jp2        2/;
PARAMETER AI1(i,j) 'coeffients that define inequality constraints
     in first stage'
/i1.j1     1/;
PARAMETER BI1(i) 'coeffients that define inequality constraints
    in first stage'
/i1        30/;
PARAMETER AE2(kp,j) 'coeffients that define equality constraints
    in second stage'
/kp1.j1       1/;
PARAMETER AEP2(kp,jp) 'coeffients that define equality
    constraints in second stage'
/kp1.jp1      1
kp1.jp2      -1/;
PARAMETER BE2(kp) 'coeffients that define equality constraints in
     second stage'
/kp1       0/;
PARAMETER AIP2(ip,jp) 'coeffients that define inequality
    constraints in second stage'
/ip1.jp1      0
ip1.jp2      1
ip2.jp1      0
ip2.jp2      -1/;
PARAMETER BI2(ip) 'coeffients that define inequality constraints
    in second stage'
/ip1       50
ip2        0/;
```

```
VARIABLES
z 'objective function',
z1 'objective function in first stage',
x2 'second-stage decision variables';
POSITIVE VARIABLES
x1 'first-stage decision variables',
upsilon 'dual variable',
eta     'dual variable'
y       'auxiliary variable';
EQUATIONS
obj 'objective function',
obj1 'objective function in first stage',
icon1 'inequality constraints in first stage',
econ2 'equality constraints in second stage',
icon2 'inequality constraints in second stage',
rcon1 'robust constraints',
rcon2 'robust constraints',
rcon3 'robust constraints';
obj..     z=e=z1+Gamma*upsilon+SUM(jp,eta(jp));
obj1..    z1=e=SUM(j,C1(j)*x1(j))+SUM(jp,C2(jp)*x2(jp));
icon1(i)..  SUM(j,AI1(i,j)*x1(j))=L=BI1(i);
econ2(kp).. SUM(j,AE2(kp,j)*x1(j))+SUM(jp,AEP2(kp,jp)*x2(jp))=E=
    BE2(kp);
icon2(ip).. SUM(jp,AIP2(ip,jp)*x2(jp))=L=BI2(ip);
rcon1(jp).. upsilon+eta(jp) =G= Ch2(jp)*y(jp);
rcon2(jp).. x2(jp) =L= y(jp);
rcon3(jp).. - y(jp) =L= x2(jp);
*************************************
*****Model definition ********************
MODEL RO        /all/;
OPTION OPTCR=0;
OPTION OPTCA=0;
OPTION iterlim=1e8;
****************************************
******** Parameter definition***************
PARAMETER state 'Convergence status';
SOLVE RO USING LP MINIMIZING z;
state = RO.modelstat;
display state;
```

Next, the impact of changing parameter Γ^{O} from 0 to 2 on the objective function is analyzed. The results in Table B.4 indicate that by the increase of parameter Γ^{O}, the objective function increases. This can be seen as the price of the robustness with respect to the uncertainty affecting the objective function.

□

Static robust optimization is used to solve scheduling problems of VPPs in Chap. 4.

B.3.2 Adaptive Robust Optimization

The main difference between the static robust optimization model described in the previous section and an adaptive robust optimization problem is that in the latter it is possible to take some corrective actions after the uncertainty realization. Therefore, solutions are not as conservative as in the case of static robust optimization. In particular, the decision-making process is as follows:

Table B.4 Illustrative Example B.3: optimal value of the objective function

Γ^{O}	Objective function [$]
0	−340
0.5	−290
1	−240
1.5	−190
2	−180

1. First-stage decisions are made with the aim of minimizing (maximizing) the worst-case cost (profit).
2. For fixed first-stage decisions, the worst-case uncertainty realization is identified, i.e., that maximizing (minimizing) the cost (profit).
3. For fixed first-stage decisions and fixed uncertainty realizations, second-stage decisions are made with the aim of minimizing (maximizing) the cost (profit), i.e., it is possible to take corrective actions.

Considering the above decision sequence, it is possible to formulate adaptive robust optimization problems as three-level programming models.

B.3.2.1 Uncertainty Set

Uncertain parameters are modeled by decision variables that take values within known confidence bounds:

$$\mathbf{v} \in [\tilde{\mathbf{v}} - \hat{\mathbf{v}}, \tilde{\mathbf{v}} + \hat{\mathbf{v}}], \tag{B.7}$$

where

- $\mathbf{v}$ is a vector representing the uncertain variables,
- $\tilde{\mathbf{v}}$ is a vector of the forecast levels of uncertain variables, and
- $\hat{\mathbf{v}}$ is a vector of the fluctuation levels of uncertain variables.

The adaptive robust optimization problem identifies the worst-case uncertainty realization. For the considered polyhedral uncertainty set (B.7), this worst-case realization corresponds to a vertex of the polyhedron [9]. Therefore, (B.7) can be replaced by the following equivalent binary-variable-based uncertainty set:

$$\Omega = \Big\{ \mathbf{v} = \tilde{\mathbf{v}} + \mathrm{diag}(\mathbf{u}^+)\hat{\mathbf{v}} - \mathrm{diag}(\mathbf{u}^-)\hat{\mathbf{v}}, \tag{B.8a}$$

$$\mathbf{u}^+, \mathbf{u}^- \in \{0, 1\}^m, \tag{B.8b}$$

$$\sum_{k=1}^{m} \left(u_k^+ + u_k^-\right) \leq \Gamma, \tag{B.8c}$$

$$u_k^+ + u_k^- \leq 1, \quad \forall k \Big\}, \tag{B.8d}$$

where:

- Ω is the uncertainty set,
- diag is the diagonal operator,
- m is the size of the vector of variables $\mathbf{v}$,
- Γ is the uncertainty budget, and
- u_k^+ and u_k^-, $\forall k$, are binary variables that indicate if the worst-case uncertainty realization of variable v_k corresponds to the lower, forecast, or upper level.

Constraints (B.8a) express the uncertain variables in vector $\mathbf{v}$ in terms of the forecast and fluctuation levels. Constraints (B.8b) define the binary character of vectors $\mathbf{u}^+$ and $\mathbf{u}^-$. Constraints (B.8c) allow controlling the robustness in the solution through the so-called uncertainty budget Γ. If $\Gamma = 0$, then all uncertain variables in vector $\mathbf{v}$ are equal to their forecast values, i.e., uncertainty is disregarded; however, as the value of this uncertainty budget increases, the uncertain variables in vector $\mathbf{v}$ are allowed to deviate from their forecast values, i.e., a comparatively more robust solution is considered. Finally, constraints (B.8d) impose that uncertain variables v_k, $\forall k$, cannot be simultaneously equal to its lower and upper bounds.

B.3.2.2 Problem Formulation

The adaptive robust optimization model is formulated as the following three-level programming problem:

$$\min_{\mathbf{x}} \mathbf{C}^T\mathbf{x} + \max_{\mathbf{v}\in\Omega} \quad \min_{\mathbf{y}\in\Xi(\cdot)} \mathbf{D}(\mathbf{v})^T\mathbf{y} \tag{B.9a}$$

subject to

$$\mathbf{A}^T\mathbf{x} \leq \mathbf{B}, \tag{B.9b}$$

where:

- $\mathbf{x}$ and $\mathbf{y}$ are the vectors of first-stage and second-stage decision variables, respectively,
- Ξ is the feasibility set,
- $\mathbf{C}$ and $\mathbf{D}$ are the vectors of cost coefficients related to first- and second-stage decisions, respectively,
- $\mathbf{A}$ and $\mathbf{B}$ are the matrices with coefficients of inequality constraints (B.9b).

Problem (B.9) involves three nested optimization problems:

1. The first level associated with the first-stage decisions, i.e., variables in vector $\mathbf{x}$.
2. The second level related to the worst-case uncertainty realizations, i.e., variables in vector $\mathbf{v}$.
3. The third level modeling the corrective actions against first- and second-level decisions, i.e., variables in vector $\mathbf{y}$.

Sets Ω and Ξ used in problem (B.9) are the uncertainty and the feasibility sets, respectively. Uncertainty set Ω is described in Sect. B.3.2.1, while set Ξ identifies the feasible space of the third-level optimization variables as explained in Sect. B.3.2.3.

B.3.2.3 Feasibility Set

Given first- and second-level decision variables, set Ξ models the feasible space of third-level optimization variables:
$\Xi\,(\mathbf{x}, \mathbf{v}) = \Big\{\mathbf{y}:$

$$\mathbf{Ex} + \mathbf{Fy} = \mathbf{G}(\mathbf{v}) \ : \ \boldsymbol{\lambda}, \tag{B.10a}$$

$$\mathbf{Hx} + \mathbf{Iy} \leq \mathbf{J}(\mathbf{v}) \ : \ \boldsymbol{\mu}, \tag{B.10b}$$

$\Big\}$, where:

- $\mathbf{E}$ and $\mathbf{F}$ are the matrices with coefficients of equality constraints (B.10a),
- $\mathbf{H}$ and $\mathbf{I}$ are the matrices with coefficients of inequality constraints (B.10b),
- $\mathbf{G}$ and $\mathbf{J}$ are the vectors of coefficients of equality constraints (B.10a) and inequality constraints (B.10b), respectively, and
- $\boldsymbol{\lambda}$ and $\boldsymbol{\mu}$ are the dual variables associated with equality constraints (B.10a) and inequality constraints (B.10b), respectively.

B.3.2.4 Solution Procedure

The three-level optimization problem (B.9) can be solved using a column-and-constraint generation algorithm [11] based on the iterative solution of a master problem and a subproblem.

Master Problem

The master problem at iteration υ is provided below:

$$\min_{\Phi^{\mathrm{M}}} \mathbf{C}^T\mathbf{x} + \theta \tag{B.11a}$$

subject to

$$\text{Constraints (B.9b)}, \tag{B.11b}$$

$$\theta \geq \mathbf{D}(\mathbf{v}^{(\upsilon')})^T\mathbf{y}_{\upsilon'}, \quad \forall \upsilon' \leq \upsilon, \tag{B.11c}$$

$$\mathbf{Ex} + \mathbf{Fy}_{\upsilon'} = \mathbf{G}(\mathbf{v}^{(\upsilon')}), \quad \forall \upsilon' \leq \upsilon, \tag{B.11d}$$

$$\mathbf{Hx} + \mathbf{Iy}_{\upsilon'} \leq \mathbf{J}(\mathbf{v}^{(\upsilon')}), \quad \forall \upsilon' \leq \upsilon, \tag{B.11e}$$

where:

- υ is the iteration counter,
- variables in set $\Phi^{\mathrm{M}} = \{\mathbf{x}, \theta, \mathbf{y}_{\upsilon'}, \forall \upsilon' \leq \upsilon\}$ are the optimization variables of master problem (B.11), and
- θ is an auxiliary variable that iteratively approximates the worst-case value of the second-level objective function.

Master problem (B.11) is a relaxed version of the three-level problem (B.9), whose size increases with the number of iterations. Parameters with superscript (υ') refer to the optimal values of variables yielded by the subproblem at iteration υ'.

Subproblem

At each iteration υ, the master problem (B.11) provides first-stage decision variables, i.e., variables in vector $\mathbf{x}$. Then, for fixed values of these variables, the subproblem determines the worst-case uncertainty realization:

$$\max_{\mathbf{v} \in \Omega} \quad \min_{\mathbf{y} \in \Xi(\cdot)} \mathbf{D}(\mathbf{v})^T \mathbf{y}. \tag{B.12}$$

Subproblem (B.12) is a bi-level model whose lower-level problem is convex on its decision variables. Therefore, it can be replaced by its dual constraints. In addition, using the strong duality equality, the objective function of the subproblem can be replaced by the dual lower-level objective function [2], and the subproblem can be formulated as a single-level mixed-integer linear programming problem.

Algorithm

Given the master problem (B.11) and subproblem (B.12), the column-and-constraint generation algorithm works as follows:

1. The iteration counter is initialized ($\upsilon \leftarrow 0$), the convergence tolerance (ϵ) is selected, and the lower bound (LB) and upper bound (UB) are set to $-\infty$ and $+\infty$, respectively.
2. The master problem (B.11) is solved.
3. The lower bound is updated:

$$\mathrm{LB} = z^{\mathrm{M}*}, \tag{B.13}$$

   where $z^{\mathrm{M}*}$ is the optimal value of the objective function (B.11a).
4. Vector $\mathbf{x}^{(\upsilon)}$ is set to $\mathbf{x}^{(\upsilon)} = \mathbf{x}^*$, where $\mathbf{x}^*$ denotes the optimal values of variables in vector $\mathbf{x}$ yielded by the solution of the master problem (B.11) in step 2.
5. The subproblem (B.12) is solved.
6. The upper bound is updated:

$$\mathrm{UB} = \min\{\mathrm{UB}, \mathbf{C}^T \mathbf{x}^* + z^{\mathrm{S}*}\}, \tag{B.14}$$

   where $z^{\mathrm{S}*}$ is the optimal value of the objective function (B.12).
7. If $\mathrm{UB} - \mathrm{LB} < \epsilon$, the algorithm stops. The optimal first-stage decisions are those in vector $\mathbf{x}^*$. Otherwise, go to step 8.
8. The iteration counter is updated $\upsilon \leftarrow \upsilon + 1$.
9. Vector $\mathbf{v}^{(\upsilon)}$ is set to $\mathbf{v}^{(\upsilon)} = \mathbf{v}^*$, where $\mathbf{v}^*$ denotes the optimal values of variables in vector $\mathbf{v}$ yielded by the solution of the subproblem (B.12) in Step 5.
10. Go to Step 2.

Illustrative Example B.4 Adaptive Robust Optimization Problem

A power producer owns a conventional power plant and a wind-power unit. The conventional power plant has a capacity of 100 MW and a variable cost of $25/MWh, while the expected available wind-power production levels for the next 2 hours are 50 MW and 60 MW. In order to consider the uncertainty in the available wind-power production level, it is assumed that in the next hour this level will be between 45 MW and 55 MW, while in the following one it will be between 50 MW and 70 MW.

The power producer decides the self-scheduling of these two units in an energy market. The prices in this market for the following 2 hours are \$20/MWh and \$24/MWh, being 200 MW the maximum power traded in the market.

Considering these data, the problem is formulated as the following adaptive robust optimization model:

$$\max_{\Phi^{\mathrm{UL}}} \min_{\Phi^{\mathrm{ML}}} \max_{\Phi^{\mathrm{LL}}} \; 20p_1^{\mathrm{E}} + 24p_2^{\mathrm{E}} - 25p_1^{\mathrm{C}} - 25p_2^{\mathrm{C}}$$

subject to

$$-200 \leq p_1^{\mathrm{E}} \leq 200,$$

$$-200 \leq p_2^{\mathrm{E}} \leq 200,$$

subject to

$$u_1^-, u_1^+, u_2^-, u_2^+ \in \{0, 1\},$$

$$p_1^{\mathrm{W,A}} = 50 - 5u_1^- + 5u_1^+,$$

$$p_2^{\mathrm{W,A}} = 60 - 10u_2^- + 10u_2^+,$$

$$u_1^- + u_1^+ \leq 1,$$

$$u_2^- + u_2^+ \leq 1,$$

$$u_1^- + u_1^+ + u_2^- + u_2^+ \leq \Gamma,$$

subject to

$$p_1^{\mathrm{E}} = p_1^{\mathrm{W}} + p_1^{\mathrm{C}},$$

$$p_2^{\mathrm{E}} = p_2^{\mathrm{W}} + p_2^{\mathrm{C}},$$

$$0 \leq p_1^{\mathrm{C}} \leq 100,$$

$$0 \leq p_2^{\mathrm{C}} \leq 100,$$

$$0 \leq p_1^{\mathrm{W}} \leq p_1^{\mathrm{W,A}},$$

$$0 \leq p_2^{\mathrm{W}} \leq p_2^{\mathrm{W,A}},$$

where the optimization variables of the upper-, middle-, and lower-level problems are those included in sets $\Phi^{\mathrm{UL}} = \left\{p_1^{\mathrm{E}}, p_2^{\mathrm{E}}\right\}$, $\Phi^{\mathrm{ML}} = \left\{p_1^{\mathrm{W,A}}, p_2^{\mathrm{W,A}}\right\}$, and $\Phi^{\mathrm{LL}} = \left\{p_1^{\mathrm{W}}, p_2^{\mathrm{W}}, p_1^{\mathrm{C}}, p_2^{\mathrm{C}}\right\}$, respectively.

The above three-level optimization problem can be solved using the column-and-constraint generation algorithm described in Sect. B.3.2.4. In this problem, parameter Γ is used to control the level of robustness in the solution. This parameter can take values between 0 and 2, i.e., the number of uncertain parameters. If it is equal to 0, it means that the available wind-power production level is considered equal to its expected value, therefore disregarding the uncertainty. If this parameter is equal to 1, then the available wind-power production level can deviate from its expected value in just one period. Finally, if this parameter is equal to 2, it is considered that the available wind-power production level can deviate from its expected value in both time periods. Note that this case corresponds with the most conservative solution. □

Adaptive robust optimization is used to solve operation problems for VPPs in Chaps. 4 and 5.

B.4 Summary and Further Reading

This chapter explains the basic theory of the optimization problems under uncertainty introduced throughout the book. First, the detailed formulation of the stochastic programming for a two-stage decision-making problem is described. Second, the main details of the theory of robust optimization are provided.

A detailed information on stochastic programming is described in the books by Birge et al. [5] and Conejo et al. [6]. Further information on the cardinality constrained robust optimization for LP problems under uncertainty can be found in [4]. Additionally, interested readers can find the extension of the robust optimization for mixed-integer linear programming problems proposed by Bertsimas and Sim [3]. The mathematical framework of the adaptive robust optimization approach is also developed and described in detail by Ben-Tal et al. [1].

References

1. Ben-Tal, A., Goryashko, A., Guslitzer, E., Nemirovski, A.: Adjustable robust solutions of uncertain linear programs. Math. Program. **99**(2), 351–376 (2004)
2. Bertsimas, D., Litvinov, E., Sun, X.A., Zhao, J., Zheng, T.: Adaptive robust optimization for the security constrained unit commitment problem. IEEE Trans. Power Syst. **28**1, 52–63, (2013)
3. Bertsimas, D., Sim, M.: Robust discrete optimization and network flows. Math. Program. Springer **98**(1-3), 49–71 (2003)
4. Bertsimas, D., Sim, M.: The price of robustness. Oper. Res. **52**(1), 35–53 (2004)
5. Birge, J.R., Louveaux, F: Introduction to Stochastic Programming, 2nd edn. Springer, New York, US (2011)
6. Conejo, A.J., Carrión, M., Morales, J.M.: Decision Making Under Uncertainty in Electricity Markets. Springer, New York, US (2010)
7. GAMS.; Available at www.gams.com/ (2020)
8. ILOG CPLEX.: Available at www.ilog.com/products/cplex/ (2020)
9. Jiang, R., Zhang, M., Li, G., Guan, Y.: Two-stage network constrained robust unit commitment problem. Eur. J. Oper. Res. **234**(3), 751–762 (2014)
10. Rockafellar, R.T., Uryasev, S.: Conditional value-at-risk for general loss distributions. J. Bank. Finance **26**7, 1443–1471, (2002)
11. Zeng, B., Zhao, L.: Solving two-stage robust optimization problems using a column-and-constraint generation method. Oper. Res. Lett. **41**5, 457–461 (2013)

Index

L. Baringo, M. Rahimiyan, *Virtual Power Plants and Electricity Markets*, https://doi.org/10.1007/978-3-030-47602-1

The manufacturer's authorised representative in the EU is Springer Nature Customer Service Centre GmbH, Europaplatz 3, 69115 Heidelberg, Germany. If you have any concerns regarding our products, please contact ProductSafety@springernature.com

Printed and bound by CPI Group (UK) Ltd, Croydon, CR0 4YY
30/09/2026
02225516-0009